ADVANCES IN
HEAT TRANSFER

Volume 39

Advances in
HEAT TRANSFER

Coordinating Technical Editor
George A. Greene
Energy Sciences and Technology
Brookhaven National Laboratory
Upton, New York

Serial Editors

James P. Hartnett[†]
Energy Resources Center
University of Illinois at Chicago
Chicago, Illinois

Young I. Cho
Department of Mechanical Engineering
Drexel University
Philadelphia, Pennsylvania

Avram Bar-Cohen
Department of Mechanical Engineering
University of Maryland
College Park, Maryland

Volume 39

Founding Editors
Thomas F. Irvine, Jr.[†]   State University of New York at Stony Brook, Stony Brook, NY
James P. Hartnett [†]   University of Illinois at Chicago, Chicago, IL

ELSEVIER

Amsterdam Boston London New York Oxford Paris
San Diego San Francisco Singapore Sydney Tokyo
Academic Press is an Imprint of Elsevier

ACADEMIC PRESS

Academic Press is an imprint of Elsevier
84 Theobald's Road, London WC1X 8RR, UK
Radarweg 29, PO Box 211, 1000 AE Amsterdam, The Netherlands
The Boulevard, Langford Lane, Kidlington, Oxford OX5 1GB, UK
30 Corporate Drive, Suite 400, Burlington, MA 01803, USA
525 B Street, Suite 1900, San Diego, CA 92101-4495, USA

First edition 2006

Copyright © 2006 Elsevier Inc. All rights reserved

No part of this publication may be reproduced, stored in a retrieval system
or transmitted in any form or by any means electronic, mechanical, photocopying,
recording or otherwise without the prior written permission of the publisher

Permissions may be sought directly from Elsevier's Science & Technology Rights
Department in Oxford, UK: phone (+44) (0) 1865 843830; fax (+44) (0) 1865 853333;
email: permissions@elsevier.com. Alternatively you can submit your request online by
visiting the Elsevier web site at http://elsevier.com/locate/permissions, and selecting
Obtaining permission to use Elsevier material

Notice
No responsibility is assumed by the publisher for any injury and/or damage to persons
or property as a matter of products liability, negligence or otherwise, or from any use
or operation of any methods, products, instructions or ideas contained in the material
herein. Because of rapid advances in the medical sciences, in particular, independent
verification of diagnoses and drug dosages should be made

ISBN-13: 978-0-12-020039-9
ISBN-10: 0-12-020039-2

ISSN: 0065-2717

For information on all Academic Press publications
visit our website at books.elsevier.com

Printed and bound in Great Britain

06 07 08 09 10 10 9 8 7 6 5 4 3 2 1

Working together to grow
libraries in developing countries

www.elsevier.com | www.bookaid.org | www.sabre.org

ELSEVIER   BOOK AID International   Sabre Foundation

CONTENTS

Contributors xi
Preface xiii

Sonoluminescence and the Search for Sonofusion

R.T. LAHEY, Jr., R.P. TALEYARKHAN,
R.I. NIGMATULIN, I.S. AKHATOV

I. Introduction 1
II. Discussion 5
   A. The Analysis of SBSL 5
   B. Experimental Techniques for SBSL 7
III. Experimental Results and Considerations 17
   A. Bubble Stability Considerations (SBSL) 21
   B. Lessons Learned from SBSL Experiments and Analysis .. 31
   C. Choice of Test Liquid (Sonofusion) 35
   D. Experimental System 36
   E. Description of the Bubble Dynamics Process 39
   F. Timing of Key Parameters 40
   G. Evidence of Nuclear Fusion 43
   H. Production of Tritium 45
   I. Production of D/D Neutrons 47
   J. Coincidence between Neutron and SL Signals 48
IV. Analytical Modeling 53
   A. Analysis of Bubble Dynamics 53
   B. Equations of State 54
   C. Low Mach Number Bubble Dynamics 60
   D. High Mach Number Bubble Dynamics 65
   E. Kinetics of Nuclear Reactions 79
   F. HYDRO Code Analysis of Sonofusion 82
   G. Nucleation Phenomena in Sonofusion 89
V. Potential Applications of Sonofusion Technology 97
   Appendix A: Linear and Non-linear Bubble Cluster Dynamics 101
      The Mathematical Model for a Bubbly Liquid 103
      Impact of Bubble Concentration on the Intensity of Bubble Cluster Collapse 125

Appendix B: Transient Phenomena during Bubble Implosions in an Acoustically Driven, Liquid-Filled Flask............ 130
    The Dynamics of a Spherical Liquid-Filled Flask without Dissipation.... 131
    The Dynamics of a Spherical Liquid-Filled Flask with Dissipation...... 134
    The Dynamics of a Spherical Liquid-Filled Flask with Dissipation...... 138

Appendix C: Nucleation of a Bubble in Tensioned Liquids Using High-Energy Neutrons......................... 151
    Mechanical Equilibrium of a Vapor Bubble in a Liquid.............. 152
    Thermodynamical Equilibrium of a Vapor Bubble in a Liquid......... 152
    Activation Threshold for Nucleation: The Critical Bubble Size......... 154
    The Role of Homogeneous Nucleation......................... 156
    Neutron-Induced Bubble Nucleation.......................... 156

References.. 161

Phonon Transport in Molecular Dynamics Simulations: Formulation and Thermal Conductivity Prediction

A.J.H. McGaughey, M. Kaviany

I. Introduction.................................... 169
    A. Challenges at Micro- and Nano-scales........................ 169
    B. Motivation for using Molecular Dynamics Simulations............. 170
    C. Scope...................................... 171
II. Conduction Heat Transfer and Thermal Conductivity of Solids.. 172
III. Real and Phonon Space Analyses..................... 175
    A. Molecular Dynamics Simulation.......................... 175
    B. Lennard-Jones System................................ 182
    C. Real Space Analysis................................. 184
    D. Phonon Space Analysis............................... 192
IV. Nature of Phonon Transport in Molecular Dynamics Simulations.. 208
    A. Quantum Formulation and Selection Rules................... 208
    B. Phonon Gas and Normal Modes......................... 210
V. Thermal Conductivity Prediction: Green-Kubo Method.... 211
    A. Formulation..................................... 211
    B. Case Study: Lennard-Jones Argon........................ 213
    C. Quantum Corrections................................ 228
VI. Thermal Conductivity Prediction: Direct Method......... 232
    A. Introduction..................................... 232
    B. Implementation................................... 235
    C. Size Effects..................................... 239
VII. Discussion...................................... 241
    A. Notable work.................................... 241
    B. Comparison of Green-Kubo and Direct Methods............... 244
    C. Expectations from Molecular Dynamics..................... 245

VIII. Concluding Remarks	246
Acknowledgements	247
Nomeclature	247
References	248

Heat and Mass Transfer in Fluids with Nanoparticle Suspensions

G.P. Peterson, C.H. Li

I. Introduction	257
II. Experimental Studies and Results	264
A. Preparation of the Nanoparticle Suspensions	264
B. Experimental Methods and Test Facilities	270
C. Measurements of the Effective Thermal Conductivity	276
D. Discussion and Conclusions	298
III. Theoretical Investigations	310
A. Development of Effective Thermal Conductivity Equations	310
B. The Effects of Brownian Motion Coupled with Thermal Phoresis	338
C. Other Transport Phenomena in Nanoparticle Suspensions	359
D. Discussion and Conclusions	367
IV. Summary	368
References	370

The Effective Thermal Conductivity of Saturated Porous Media

H.T. Aichlmayr, F.A. Kulacki

I. Introduction	377
II. The Effective Thermal Conductivity	377
A. Experimental Determination	379
B. Results And General Trends	392
III. Small Solid-Fluid Conductivity Ratios	394
A. Maxwell's Formulae	394
B. Mixture Rules	397
C. Physical Bounds for the Effective Conductivity	399
D. Closing Comments on Small Conductivity Ratio Media	400
IV. Intermediate Solid-Fluid Conductivity Ratios	400
A. Conduction Models	401
B. Lumped Parameter Models	407
C. Resistor Network Models	411
D. Closing Comments on Moderate Conductivity Ratios	414

V. Large Solid-Fluid Conductivity Ratios 415
   A. The Method of Volume Averaging 416
   B. Empirical Contact Parameter Models......................... 428
   C. Contact Models 440
   D. Closing Comments on Large Conductivity Ratio Media. 450
VI. Conclusion 452
   Acknowledgment 453
   Nomenclature 453
   References... 456

Mesoscale and Microscale Phase-Change Heat Transfer

P. Cheng, H.Y. Wu

I. Introduction 461
II. Meso-/Microdevices Involving Phase-Change Heat Transfer
    Processes... 462
    A. Classification of Meso- and Microsystems: The Bond Number 462
    B. Meso-/Microchannel Heat Exchangers........................... 465
    C. Meso-(Mini-)/Microheat Pipes 467
    D. Thermal Bubble Actuators 477
III. Meso- and Microscale Phase-Change Heat Transfer Phenomena   487
    A. Homogeneous and Heterogeneous Nucleation in Pool Boiling 487
    B. Subcooled Pool Boiling Under Constant Heat Flux................. 491
    C. Subcooled Pool Boiling under Pulse Heating or Transient Heating 497
    D. Flow Boiling in Meso-/Microchannels 509
    E. Flow Condensation in Meso-/Microchannels 529
IV. Concluding Remarks 556
    Acknowledgments 556
    References... 556

Jet Impingement Heat Transfer: Physics, Correlations, and Numerical Modeling

N. Zuckerman, N. Lior

I. Summary.. 565
II. Introduction 565
    A. Impinging Jet Regions...................................... 567
    B. Nondimensional Heat and Mass Transfer Coefficients 570
    C. Turbulence Generation and Effects 574
    D. Jet Geometry 580
    E. The Effect of Jet Pitch: The Jet–Jet Interaction.................... 580
    F. Alternate Geometries and Designs.............................. 584

III. Research Methods	588
A. Experimental Techniques	588
B. Modeling	591
C. Conclusions Regarding Model Performance	607
Conclusions	609
Appendix: Correlation Reference	610
A. Correlation List	610
B. Trends Within the Correlations	618
Nomenclature	625
References	626
Author Index	633
Subject Index	647

CONTRIBUTORS

Numbers in parentheses indicate the pages on which the author's contributions begin.

HANS T. AICHLMAYR (377), Sandia National Laboratories, 7011 East Ave., Livermore, CA 94550, USA

ISKANDER S. AKHATOV (1), North Dakota State University, Fargo, ND 58105, USA

PING CHENG (461), School of Mechanical & Power Engineering, Shanghai Jiaotong University, Shanghai 200030, P. R. China

MASSOUD KAVIANY (169), Department of Mechanical Engineering, University of Michigan, Ann Arbor, MI 48109-2125, USA

FRANCIS A. KULACKI (377), Thermodynamics and Heat Transfer Laboratory, Department of Mechanical Engineering, The University of Minnesota, 111 Church St. SE, Minneapolis, MN 55455, USA

RICHARD T. LAHEY, Jr. (1), Rensselaer Polytechnic Institute, Troy, NY 12180-3590, USA

CALVIN HONG LI (257), Department of Mechanical, Aerospace and Nuclear Engineering, Rensselaer Polytechnic Institute, Troy, NY 12180, USA

NOAM LIOR (565), Department of Mechanical Engineering and Applied Mechanics, The University of Pennsylvania, Philadelphia, PA 19104, USA

A. J. H. MCGAUGHEY (169), Department of Mechanical Engineering, Carnegie Mellon University, Pittsburgh, PA 15213-3890, USA

ROBERT I. NIGMATULIN (1), Baskortostan Branch, Russian Academy of Sciences, Ufa, Baskortostan 450077, Russia

G. P. PETERSON (257), Department of Mechanical, Aerospace and Nuclear Engineering, Rensselaer Polytechnic Institute, Troy, NY 12180-3590, USA

RUSI P. TALEYARKHAN (1), School of Nuclear Engineering, Purdue University, West Lafayette, IN 47907, USA

H. Y. WU (461), School of Mechanical & Power Engineering, Shanghai Jiaotong University, Shanghai 200030, P. R. China

NEIL ZUCKERMAN (565), Department of Mechanical Engineering and Applied Mechanics, The University of Pennsylvania, Philadelphia, PA 19104, USA

PREFACE

For more than 40 years, the serial publication *Advances in Heat Transfer* has filled the information gap between regularly published journals and university-level textbooks. The series presents review articles on topics of current interest. Each contribution starts from widely understood principles and brings the reader up to the forefront of the topic being addressed. The favorable response by the international scientific and engineering community to the 39 volumes published to date is an indication of the success of our authors in fulfilling this purpose.

In recent years, the editors have published topical volumes dedicated to specific fields of endeavor. Examples of such volumes are Volume 22 (Bioengineering Heat Transfer), Volume 28 (Transport Phenomena in Materials Processing) and Volume 29 (Heat Transfer in Nuclear Reactor Safety). The editors intend to continue publishing topical volumes as well as the traditional general volumes in the future. Volume 32, a cumulative author and subject index for the first 32 volumes, has become a valuable tool to search the series for contributions relevant to their current research interests.

The editorial board expresses its appreciation to the contributing authors of Volume 39 who have maintained the high standards associated with *Advances in Heat Transfer*. Lastly, the editors would like to acknowledge the efforts of the staff at Academic Press and Elsevier who have maintained the attractive presentation of the volumes over the years.

Sonoluminescence and the Search for Sonofusion

R.T. LAHEY Jr.[1], R.P. TALEYARKHAN[2], R.I. NIGMATULIN[3] and I.S. AKHATOV[4]

[1]Rensselaer Polytechnic Institute, Troy, NY 12180-3590, USA
[2]Purdue University, West Lafayette, IN 47907, USA
[3]Russian Academy of Science, Ufa, Baskortostan 450077, Russia
[4]North Dakota State University, Fargo, ND 58105, USA

I. Introduction

The field of multiphase flow and heat transfer has many important practical applications. An interesting and important subset of this field has to do with the bubble dynamics associated with sonofusion and/or sonoluminescence (SL) phenomena. The former will be the primary focus of this paper. It should be stressed that we will be concerned with conditions suitable for thermonuclear fusion (i.e., very high temperatures and densities) rather than the conditions normally associated with the so-called "cold fusion."

Many years ago it was observed that photographic plates became fogged when they were immersed in a liquid that was excited ultrasonically [1]. It soon became clear [2] that this was due to light pulses which were emitted during cavitation bubble collapse. Harvey [3] called this phenomenon sonoluminescence (i.e., the light emission associated with ultrasonically-induced bubble implosions). In particular, this phenomena was what we now call multibubble sonoluminescence (MBSL), in which multiple bubbles formed by cavitation (typically by using an acoustic horn), grow and collapse in the induced pressure field, giving off light pulses during bubble implosions.

Some interesting hypotheses and models concerning the origin of these light pulses have included mechanisms associated with

- *Triboluminescence* [4]. It was postulated that the sudden disruption of the quasi-crystalline structure of the liquid during bubble expansion (i.e., the creation of interface) gave rise to light emission. More recently, Prosperetti [5] postulated a related phenomenon, Fractoluminescence, in which light is emitted when a high-speed liquid jet from an unstable interface impacts and ruptures the adjacent interface during bubble implosions.

- *Electrical microdischarge* [3,6–9]. These hypotheses assume that an electric field builds up, creating an electric discharge during bubble implosion. These theories include the Balloelectric effect [3] and the Flexoelectrical effect [9]. More details on these interesting models have been summarized by Young [10].
- *Mechanochemical model* [11]. This model assumes that the light emissions are due to the discharge associated with the recombination of ions formed during the bubble expansion (not compression) process.
- *Chemiluminescence model* [12,13]. In this model, high-temperature chemical reactions give rise to oxidizing agents that dissolve in the liquid and create chemical reactions that give off light flashes.
- *Hot Spot models* [14,15]. In these models, the gas in an imploding bubble is assumed to be compression heated to incandescent temperatures. The Hot Spot model was subsequently extended by Jarman [16], who discussed the effect of shock waves on the heating process.

While research continues to improve our understanding of the interesting phenomena associated with SL, most researchers now believe the validity of the extended Hot Spot model. That is, that shock-induced compression heating results in an incandescent plasma which emits the observed light flashes [17–27]. Moss et al. [28–30] performed hydrodynamic shock code (e.g., KYDNA code) simulations, which indicated that peak temperatures as high as 10^6 K may occur during the implosion of non-condensable gas bubbles, and such temperatures have apparently been confirmed experimentally [31]. Moreover, assuming that due to chemical reactions of nitrogen and hydroxyl ions [32] within and around oscillating air bubbles, the entrapped gas quickly becomes essentially argon [25]. Moss et al. [30] showed that LASNEX code predictions agreed with the measured [23,33] light pulse spectra and width (∼50–250 ps), thus supporting the validity of the extended Hot Spot model.

There are numerous excellent review papers and publications on SL phenomenon [10,34–38,174,175]. In addition, much has been published about high-temperature chemistry [39–41,176] for the conditions associated with SL. Thus, only an overview of some of the key findings and important features of SL and sonochemistry will be presented herein. Nevertheless, these interesting findings will, in turn, allow us to better understand sonofusion phenomenon.

It is important to note that a revolution in the research associated with SL was precipitated by the seminal discovery[1] of Gaitan and Crum [21] that one

---

[1] Earlier indications of this phenomenon were found by Yosioka and Omura [42], but because their results were published in Japanese, they were largely ignored by the scientific community.

could levitate a single gas bubble in the pressure antinode of an acoustic standing wave, causing it to periodically grow and collapse (i.e., implode), emitting sonoluminescent light pulses. Indeed, for the first time, it became possible to study what is now known as single bubble sonoluminescence (SBSL). The basic phenomenon associated with SBSL is shown schematically in Figs. 1a–f. The bubbles typically oscillate in an impressed acoustic pressure field (i.e., the pressure antinode of a standing acoustic wave). During the rarefaction phase they grow and during the compression phase they rapidly collapse (i.e., they implode). Figure 1a shows the start of bubble implosion when the interfacial Mach number $\left(M_g \equiv |\dot{R}|/C_g\right)$ is much less than unity. As the interfacial Mach number approaches unity (i.e., when the interfacial speed, $|\dot{R}|$, approaches the local speed of sound in the gas, C_g), a compression shock wave is formed in the gas/vapor mixture and, as shown schematically in Fig. 1b, this shock wave (dashed line) moves toward the center of the bubble and, in doing so, greatly intensifies (e.g., $M_g \sim 1/r^{0.4}$ for air [43]).

Figure 1c shows the situation just after the shock wave has bounced off itself at the center of the bubble. This highly compresses and heats a small core region near the center of the gas/vapor bubble. At this point, we normally have an SL light pulse, and if the gas/vapor is of suitable composition, and the temperature, density and their duration are sufficiently high, we may also have conditions suitable for nuclear emissions (i.e., thermonuclear fusion). Interestingly, the pressurization process within the bubble continues until a short time later when the interface comes to rest (see Fig. 1d).

Figure 1e shows the onset of bubble expansion during the refraction phase of the impressed acoustic pressure field, and Fig. 1f shows that a relatively weak shock wave is formed in the liquid surrounding the bubble during bubble expansion. This shock wave is easily heard by the experimenter when it reaches the wall of the test section in which the experiment is being performed.

Analyses of the bubble dynamics associated with SBSL have been performed using the Rayleigh equation [44], the Rayleigh–Plesset equation [45] or the more detailed extended Rayleigh models of Herring [46], Flynn [47], Gilmore [48], Knapp et al. [49], Prosperetti and Lezzi [50] and Nigmatulin et al. [51], which account for acoustic scattering.

Various equations of state and thermodynamic processes have been assumed in the analysis of SBSL, including, among others, the perfect gas law [51], the Van der Waals equation of state [52] and the assumption of an isentropic compression process [53].

It is now known that the expansion process of a gas bubble and the early stages of bubble collapse are nearly isothermal (i.e., the gas is essentially at the liquid pool temperature), but as the implosion process proceeds it is

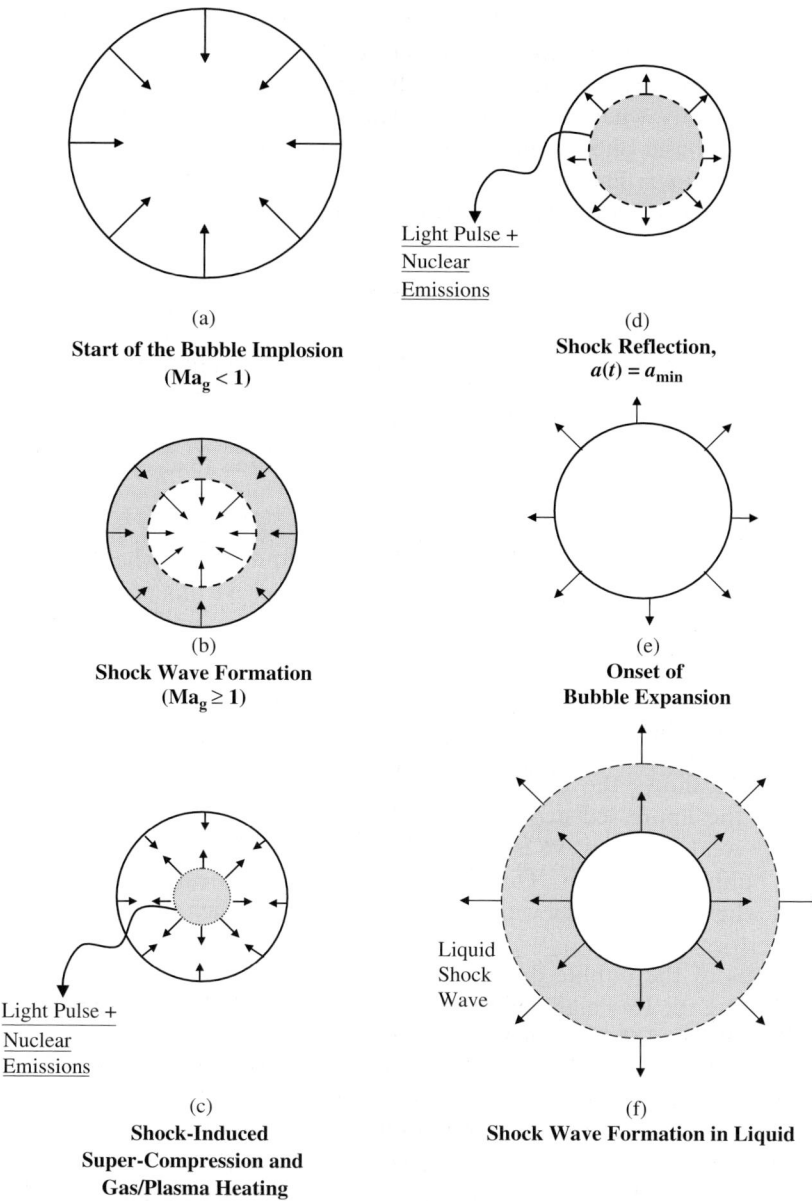

FIG. 1. Schematic of sonoluminescence and sonofusion phenomena.

better approximated by an adiabatic process. Moreover, Rayleigh equation-type models can predict most of the period of bubble dynamics, when the interface velocity ($\dot{R}$) is relatively small; however, they are insufficient for the final stages of the implosion process where shock waves may form and dominate the gas/plasma compression process. Indeed, full hydrodynamic shock code simulations are required during the final stages of the bubble implosion process, shown schematically in Figs. 1b–e.

II. Discussion

Let us now take a more in-depth look at some of the important phenomena and findings discussed above. While most of the research, which has been done to date, has focused on SL and sonochemistry, the "holy grail" of this type of research has always been to investigate ways to try and achieve acoustically-forced inertial confinement fusion (i.e., sonofusion). Indeed, Margulis [39], among others, had speculated about this possibility, and Flynn filed a U.S. Patent application in 1978 for one manifestation of this process (the Cavitation Fusion Reactor – CFR), and a patent was awarded to him in 1982. Nigmatulin and Lahey [177] analyzed bubble fusion (i.e., sonofusion) and proposed the initiation of a comprehensive bubble fusion research program, more recently, compelling experimental evidence for the achievement of sonofusion (i.e., bubble fusion) was announced by Taleyarkhan et al. [54] and detailed confirmatory data were subsequently published [56,179,180].

Sonofusion will be the main focus of this paper. Nevertheless, as noted previously, before discussing this phenomenon, we need to discuss SBSL in more detail to understand the similarities and differences between these two very interesting and related phenomena.

A. The Analysis of SBSL

Let us begin with a brief discussion of bubble dynamics. As noted previously, during bubble growth, and for most of the time associated with bubble collapse, the interface (at $r = R$ for a spherical bubble) between the liquid and gas/vapor mixture is moving at subsonic speeds (i.e., $\dot{R} \equiv dR/dt$ is relatively small). That is, the Mach number of phase-k, $M_k \equiv |\dot{R}|/C_k$ is such that $M_\ell < M_g << 1.0$. Thus, assuming spherical symmetry, to a good first approximation the motion of the bubble is given by the extended Rayleigh equation

$$R\ddot{R} + \frac{3}{2}\dot{R}^2 = \frac{(p_{\ell i} - p_{\text{I}})}{\rho_\ell} + \frac{R}{\rho_\ell C_\ell} \frac{d}{dt}[p_{\ell i} - p_{\text{I}}] \qquad (1)$$

where $R(t)$ is the instantaneous radius of the bubble, $p_{\ell i}$, the liquid pressure at the interface, ρ_ℓ is the density of liquid, C_ℓ is the speed of sound in liquid and p_I, is the incident acoustic pressure on a liquid boundary layer which surrounds the bubble [51]. We note that Eq. (1) is a non-linear ordinary differential equation, which must be evaluated numerically.

Once an appropriate equation of state has been specified for the liquid (normally considered to be incompressible) and the gas/vapor mixture, and the initial conditions are given, Eq. (1) can be evaluated (numerically) to give the bubble radius versus time, $R(t)$. Interestingly, when this equation is combined with a linear acoustics model for the propagation of pressure perturbations between the oscillating flask wall and the bubble, both forward and inverse solutions are possible [51]. That is, one can determine the bubble's response to a specified flask wall transient, or alternatively, one can determine what the flask wall transient must be to give the desired bubble response.

Figure 2 shows a typical comparison between data for an air bubble in a pool of water [18] and the predictions of the transient bubble radius, $R(t)$, using Eq. (1). It can be seen that, as expected, the gas bubble grows when the incident pressure (p_I) in the liquid is low and implodes, in a very non-linear fashion, when the incident liquid pressure increases.

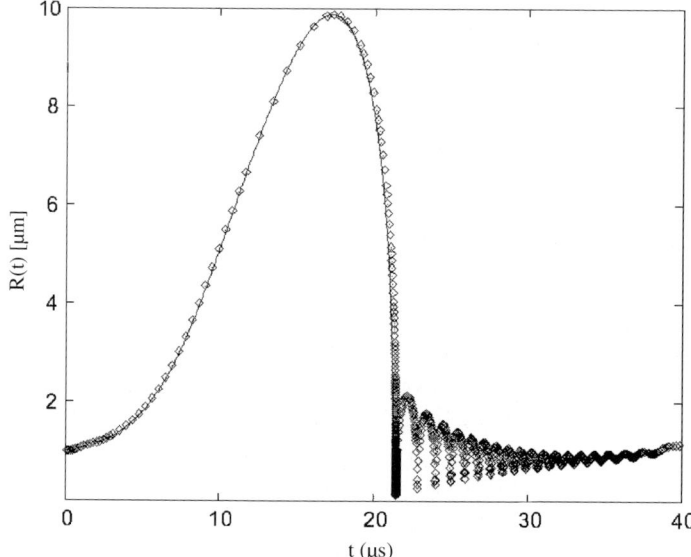

FIG. 2. Comparison of predictions (line) of the extended Rayleigh equation, Eq. (1), with the air/water data (diamonds) of Barber et al. [18].

When the proper conditions exist, this process is very repeatable and the bubble can oscillate in this manner for a very long time (i.e., for many millions of cycles). It is also interesting to note in Fig. 2 that the bubble may "bounce" after the initial implosion, since the natural frequency of the bubble (i.e., the so-called Minnaert frequency, $f_n = (3\gamma p_\ell/\rho_\ell)^{1/2}/2\pi R_0$, where $\gamma = c_p/c_v$, and R_0 is the bubble's equilibrium radius) is normally much higher than the acoustic forcing frequency. Thus, even though the bubble starts to expand, if the incident acoustic pressure is still high, it causes the bubble to collapse again and, as shown, damped high-frequency oscillations may occur. Finally, when the incident acoustic pressure is reduced (at the imposed acoustic forcing frequency) the gas bubble begins to grow again to a value about 10 times the equilibrium radius, R_0, and the cycle repeats.

The observed light flashes during SBSL occur during the final few nanoseconds of the implosion process. The primary implosion, associated with the so-called "giant response" of the bubble [57–59], is the strongest and this gives the most intense picosecond duration light pulse, however if the secondary implosions during the bubble "bounces" are sufficiently strong, light pulses may also be produced.

It is important to stress that even though Eq. (1) yields good agreement with the data for most of the bubble dynamics process, it is not able to give good predictions during the final stages of the bubble implosion (i.e., $M_g \geqslant 0.1$), since it is not able to accommodate shock wave phenomenon. In order to do this a hydrodynamic shock analysis is required, and such models will be considered later when sonofusion phenomenon are discussed.

There are many interesting things that can be said about the response of Eq. (1). For example, at a given acoustic forcing pressure, as the equilibrium bubble radius, R_0, is increased, higher order harmonics may be excited, and if the forcing pressure is large enough, a chaotic (i.e., oscillatory but non-periodic) response may be predicted above some critical R_0 [34,60]. A typical bifurcation diagram for a Poincaré section for 100 cycles is shown in Fig. 3. Note the classic cascade of period doubling bifurcations into chaos.

B. EXPERIMENTAL TECHNIQUES FOR SBSL

A typical SBSL experiment involves a glass flask (often of cylindrical or spherical shape) to which is attached (normally with epoxy) a piezoelectric transducer (e.g., a lead–zirconate–titanate, PZT, ceramic transducer) which has been polarized and has the property that it contracts or expands as the applied voltage is increased or decreased. The PZT is normally excited with a relatively high frequency (~10–20 kHz) AC electrical signal such that a standing acoustic wave is established in the liquid (typically water for SBSL experiments) in the acoustic chamber (i.e., the glass flask).

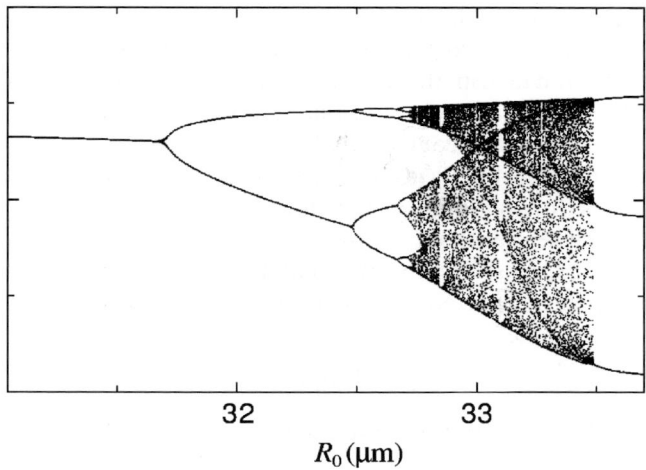

FIG. 3. A bifurcation diagram for bubble dynamics – Poincaré map for 100 cycles [60].

An effective acoustic chamber is typically designed such that the resonant frequency produces a pressure antinode at the center of the chamber and a pressure node (at which there is no significant pressure change) at the boundaries of the chamber. One important design consideration is that resonant frequency be low enough such that the gas bubble has a chance to grow sufficiently before it implodes and yet high enough so that it is above the audible range (to minimize the need for noise isolation).

It is instructive to consider the design of a typical acoustic chamber using a transient, multidimensional, finite element fluid/structure interaction (FSI) digital computer code, ATILA^{TM} [61]. In order to study the effect of cavitation bubble-induced ultrahigh pressures and temperatures during bubble implosions, we need a high-Q acoustic chamber with a highly peaked pressure–frequency distribution. In any resonant system, the Q-factor relates the energy in the system to the rate at which energy is lost by dissipative processes. The Q-factor is defined as

$$Q = \frac{f_0}{f_2 - f_1} \qquad (2)$$

where f_0 is the resonant frequency and f_2-f_1 the full-width of the peak at half-maximum (FWHM).

Cavitation bubbles may be produced when the liquid is under sufficient tension. After nucleation and growth, these bubbles can implode when the liquid is put into compression. A typical ultrasonic chamber consists of a liquid-filled glass flask, which is driven acoustically with a piezoelectric transducer creating a standing acoustic wave in the liquid. The shape of the standing wave depends on the geometry, flask material, the driving frequency and the surroundings and resonant frequency of an acoustic chamber depends mainly on its geometry and the sound speed of the liquid in the chamber. As noted previously, an ideal acoustic chamber would be one with an acoustic pressure antinode at the center and nodes at the sidewalls and top and bottom of the chamber, such that the minimum and maximum pressure values only occur near the center of the chamber. This avoids undesirable bubble nucleation (i.e., cavitation) at the walls of the chambers.

1. *Acoustic Chamber Design Considerations*

In Fig. 4, we see a cutaway section of a typical high-Q cylindrical acoustic chamber. It consists of a Pyrex^{TM} cylinder with two end fittings (hereafter called the top and bottom piston). The shape of these end fittings is designed to decouple the deformation dynamics of the vertical wall from the ends of

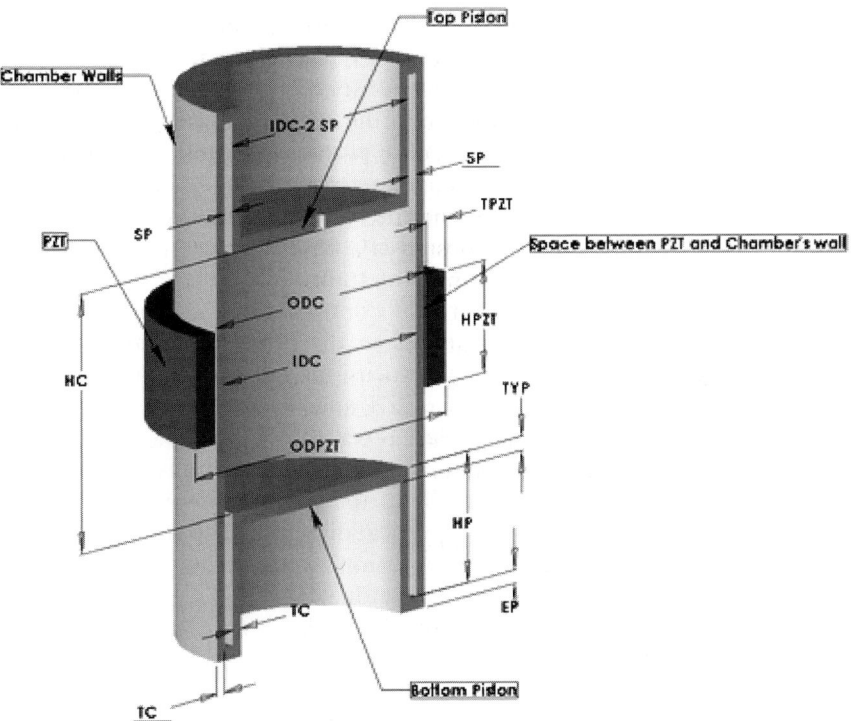

FIG. 4. Sketch of a typical acoustic chamber design.

the cylinder. A similar design strategy was previously successfully used by West [62] in his attempts to design a neutron detector.

When the two end pistons have the same thickness, we have a symmetric boundary condition that promotes a pressure antinode at the center and nodes at the ends of the chamber. The choice of the dimensions of the acoustic chamber depends on the particular application for which the chamber is intended. To minimize any perturbations introduced by the specific application requirements, the dimensions of the chamber should produce a stationary wave which has a large wavelength compared to the dimensions of any perturbing objects (e.g., holes for filling, evacuating or inserting a hydrophone), therefore the larger the chamber the better. However, the larger the size, the lower the resonant frequency and we normally want to avoid frequencies that are within the audible range to avoid having to enclose the chamber in an acoustic isolation box. Taking into account these considerations a typical outside diameter of the Pyrex glass cylinder (ODC) is 64 mm.

The height of the chamber (HC) is normally chosen to be equal to the inside diameter (IDC), because this minimizes the wall surface for a particular fluid volume. Therefore, the losses, which are strongly related to the surface area, are minimized. There are two types of losses, namely radiation and dissipative losses. The former are a consequence that the enclosing walls of the chamber radiate sound energy to the surrounding environment; the latter are due to dissipative effects in the piezoelectric and structure. Since dissipative losses in the fluid are negligible, the ratio of stored energy to lost energy in a thin wall acoustic chamber is roughly proportional to the volume-to-area ratio. Thus, the selection of HC = IDC maximizes the Q-factor of the chamber. A cylindrical piezoelectric transducer is attached to the wall of the Pyrex glass acoustic chamber. For example, a lead–zirconate–titanate ring (C5400, PZT1), which is polarized in the radial direction, is a good piezoelectric transducer (PZT). The dimensions of the PZT are chosen in order to have a small gap between the PZT and the glass walls ($\sim$1 mm). To attach the PZT to the Pyrex cylinder, Stycast^{TM} 1264 epoxy [63], an adhesive, which is particularly good for ultrasonic applications, may be used. The PZT is often placed in the middle of the chamber to give a symmetric acoustic pressure distribution in the vertical direction. The test liquid, which was used in successful previous sonofusion experiments [54,56,180], was acetone (both normal and deuterated acetone).

2. Parameter Optimization

The geometrical parameters other than the ODC and HC, were optimized using the finite element FSI code, ATILA. The goal was to obtain a high-pressure amplitude at the center of the chamber, a high Q-factor and the largest electromechanical coupling possible within the constrains of the specific geometry and materials used.

The coupling factor (k) measures the relationship between the mechanical energy in the acoustic chamber and the input electrical energy to the PZT, and is given by [64]:

$$k = \frac{U_{ME}}{\sqrt{U_{MM} U_{EE}}} \qquad (3)$$

where U_{ME} is the energy converted from electrical into mechanical form, U_{MM} the resultant mechanical energy and U_{EE} the input electrical energy. In the case of an electromechanical system, such as an acoustic chamber, we can obtain the electromechanical coupling factor by calculating an effective value of k as [65]

$$k_{eff} = \sqrt{1 - \left(\frac{f_r}{f_a}\right)^2} \qquad (4)$$

where f_r is the resonant frequency and f_a the anti-resonance frequency.

The ATILA^{TM} code evaluates the mass and momentum equations for the liquid and structures for linear acoustic approximations, the elasticity equations for the solid materials and the electric field within the piezoelectric. ATILA^{TM} also evaluates the quasi-static piezoceramic equations [66].

The ATILA^{TM} code yields the normal modes of the system and the forced harmonic response. For the forced harmonic analysis, the PZT was driven by a sinusoidal voltage at the resonance frequency of the system, which was obtained by modal analysis using ATILA^{TM}. The amplitude of the harmonic driving potential was typically 200 V.

For the acoustic chamber, optimization was done on the following geometrical (see Fig. 4) parameters: SP (piston gap), HP (piston height) and TYP (piston thickness). The smaller the thickness of the chamber walls the better; however, the thickness of the chamber wall (TC) is normally fixed at the thickness of commercially available glass cylinders (e.g., 2.4 mm). The thickness of the PZT (TPZT) in a typical design was 3 mm and the internal diameter (IDPZT) was 65 mm.

A non-uniform mesh with second-order rectangular finite elements was used to discretize the computational domain and nodal convergence was achieved. Most material properties, except those explicitly stated below, were taken from ATILA's database.

Several simulations were performed changing the above-mentioned parameters, and the final selection of design parameters is given in Table I.

TABLE I
Design Parameters for Acoustic Chamber

Parameters	Acoustic chamber (mm)
EP (wall thickness at base of gap)	3
HC (chamber height)	59.2
ODC (outer diameter of the chamber)	64
IDC (chamber internal diameter)	59.2
TC (wall thickness)	2.4
ODPZT (outer diameter of the PZT)	71
TPZT (thickness of the PZT)	3
HPZT (PZT height)	25.4
SP (gap between the piston and chamber walls)	2.6
HP (piston length)	30
TYP (thickness of the piston)	3.175

3. Results of Simulations of the Acoustic Chamber

Using the geometry specified with these optimized parameters, a modal analysis gives the value of the natural frequencies of the system. A harmonic analysis can then be performed using a forcing frequency at or near the natural frequency.

In Figs. 5 and 6, we see the predicted pressure profiles and displacements of the chamber.

In Fig. 7, we see the pressure amplitude and electric impedance of the PZT for our coupled system. The minimum value of the impedance (actually the minimum power supplied to the PZT) corresponds to the resonance frequency (f_r), where we get the maximum conversion of electrical into mechanical energy. The maximum value of the impedance is near where we have a so-called anti-resonance (f_a).

In order to better understand the dynamic response of this system, let us consider the equivalent electrical circuit of a PZT [65] shown in Fig. 8. Here, L_1 represent the structural inertia, C_1 the structural stiffness, R_1 the dissipative mechanical losses and C_2 the electrical capacitance of the PZT. The behavior of the system at the resonance frequency is dominated by the upper branch of the equivalent circuit of the PZT. However, if the resistance caused by mechanical losses is negligible (i.e., $R_1 \cong 0$), the impedance of the equivalent circuit of the PZT is very large, and the resonant frequency will be

$$f_r = \frac{1}{2\pi} \frac{1}{\sqrt{L_1 C_1}} \tag{5}$$

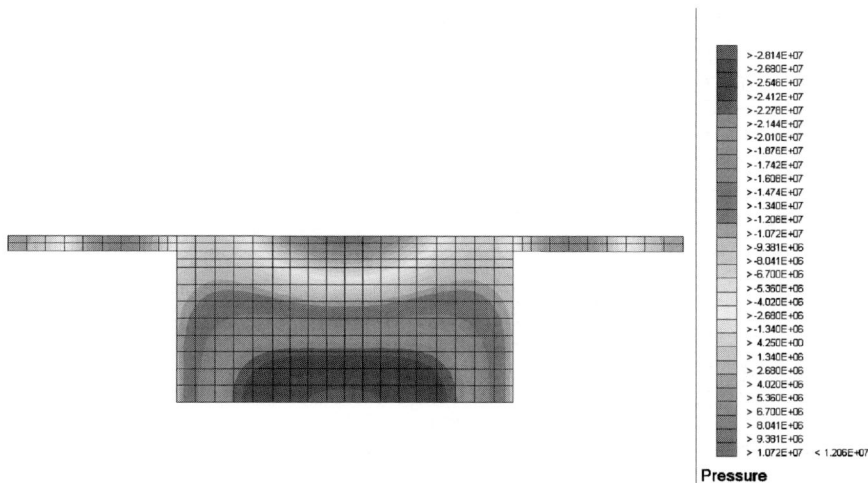

FIG. 5. Pressure profile in the acoustic chamber when the forcing frequency is 18.00 kHz. (Top of acoustic chamber on right – pressures in Pascal.)

FIG. 6. Displacement of the acoustic chamber when the forcing frequency is f = 18.00 kHz. (The displacement was magnified to make it visible.)

The behavior at the anti-resonance is dominated by the lower branch of the equivalent circuit of the PZT and the impedance value will be infinite in the absence of mechanical losses (R_1). The expression for the anti-resonance frequency in the case of $R_1 \cong 0$ is given by

$$f_a = \frac{1}{2\pi} \sqrt{\frac{C_1 + C_2}{L_1 C_1 C_2}} \qquad (6)$$

We may include losses in the material properties for the PZT, epoxy and the glass. The losses are normally included in the simulations as a loss angle, δ

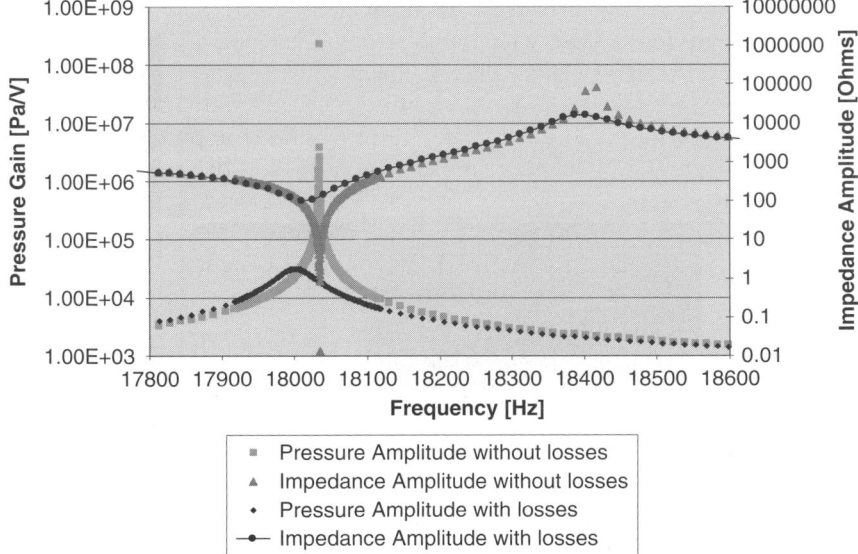

FIG. 7. Acoustic chamber simulations with and without losses.

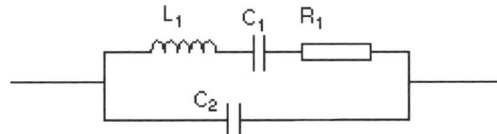

FIG. 8. Equivalent circuit of a PZT.

[64,65]. The loss angle is defined as

$$\tan \delta = \frac{E^{\text{loss}}}{E} \qquad (7)$$

where E^{loss} is the value of the mechanical or electrical property associated with the losses and E the value of the particular electrical or mechanical properties under consideration.

In Fig. 7, we can see the frequency response for cases with and without losses. The pressure is normalized by the amplitude of the sinusoidal voltage used to excite the PZT.

Losses strongly affect the pressure amplitude at the resonant frequency and the Q-factor. The Q-factor value for the case in which losses were considered was about 212. For the case in which losses were not considered, the value of the Q-factor is very large and strongly depends on the frequency resolution (Δf), since, in theory, for a linear system with no losses, the Q-factor should be infinite. A systematic discretization study must be done to determine which mesh and frequency resolutions are appropriate.

For the case of the acoustic chamber simulations without losses, three different frequency resolutions, 15, 1 and 0.3 Hz, were used. The results are shown in Fig. 9. The Q-factor for a frequency resolution of 15 Hz is 400, for the case of 1 Hz resolution it is 1127, and for a frequency resolution of 0.3 Hz the Q-factor was larger than 36,000. Thus, as expected, we see a growing trend in the Q-factor value.

For the simulations with losses, a similar study was done for frequency resolution and the results are shown in Fig. 10. We see that in the case in which losses are considered there is practically no difference between the values for the Q-factor at the different frequency resolutions. The difference in the Q-factor value between the case of a 15 Hz frequency resolution and 1 Hz frequency resolution was 3%; and the difference between 1 Hz frequency resolution and 0.3 Hz frequency resolution was within 0.002%

The results of the acoustic chamber simulations with losses for two different mesh discretizations, one using 1871 nodes and the other one using 5284 nodes, are shown in Fig. 11. They indicate that for a finer nodalization

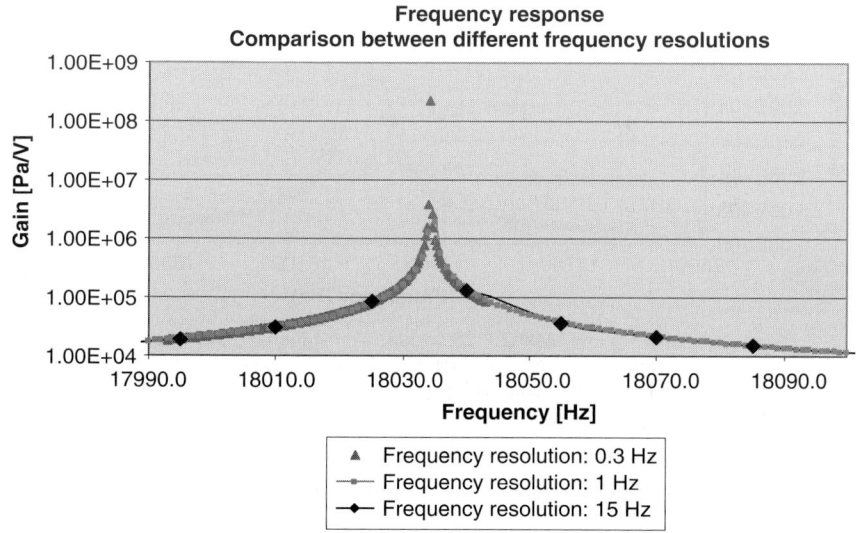

FIG. 9. Comparison of the frequency response for the case of no losses for three different frequency resolutions. The mesh used in this case contained 1871 nodes.

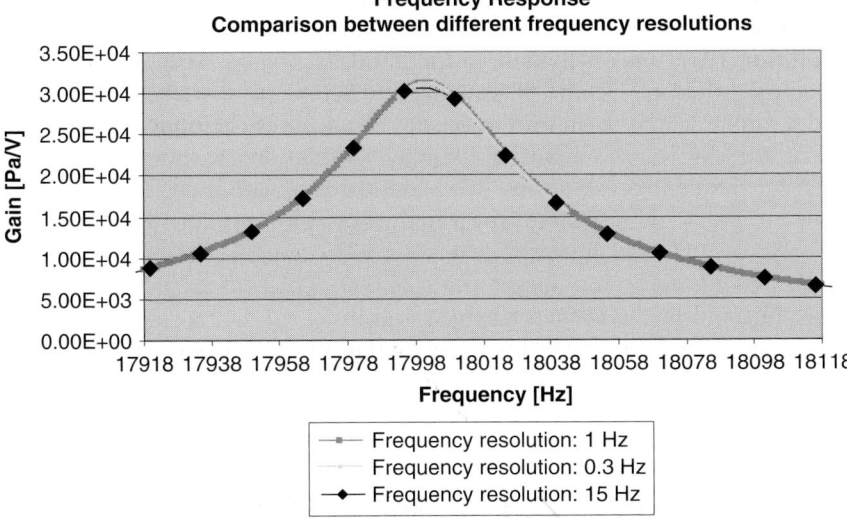

FIG. 10. Comparison of the frequency response for the case in which losses were considered for three different frequency resolutions. The mesh used in this case contained 5284 nodes.

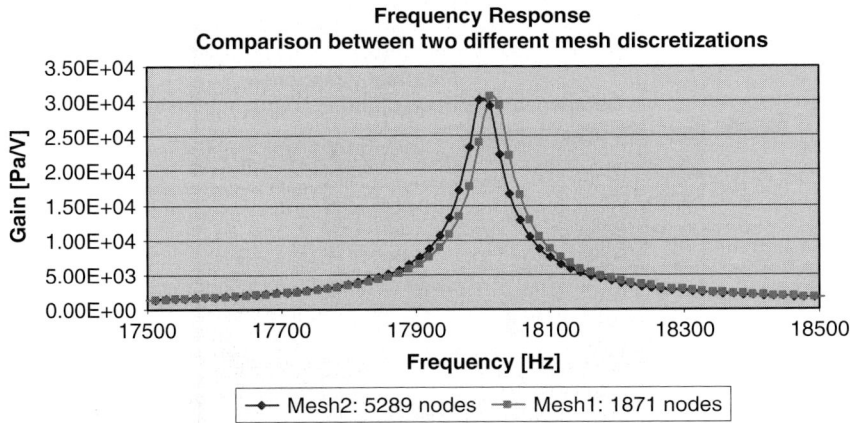

FIG. 11. Comparison of the frequency response for the case in which losses were introduced using two different mesh discretizations.

the Q-factor was 0.01%smaller, the maximum value of the pressure was 8%smaller, the frequency was 0.01%smaller and the coupling factor was reduced by 2%where all the percentage values are with respect to the results using a mesh with 1871 nodes. The results using different mesh discretizations and frequency resolutions show that appropriate discretization values for simulations with losses were chosen.

The resonant frequency with and without losses for a mesh discretization of 5284 nodes were $f_r = 18.001$ and $18.034\,\text{kHz}$, respectively. It can be seen that the losses do not have a significant effect on the resonant frequency of the chamber.

III. Experimental Results and Considerations

The acoustic chamber design discussed in the previous section was built and tested. The as-built chamber is shown in Fig. 12. The acoustic chamber was mounted by hanging it from three wires to avoid interactions with the supporting structure. The PZT transducer was driven by a wave generator

FIG. 12. As-built acoustic chamber.

connected to an audio amplifier. The liquid was well degassed with a vacuum system attached to the chamber and the pressure field inside the chambers was measured with a traversing PCB Piezotronics^{TM}, model S113A26, 6 mm OD pressure transducer. We also measured the frequency response keeping the hydrophone at the position of the pressure antinode by changing the excitation frequency. Unfortunately, this transducer introduced a perturbation to the standing wave, as the ATILA^{TM} simulations and calibrations with other smaller hydrophones indicated. Nevertheless, the basic measured mode shape and frequency response were still valid.

The results of the frequency response data and simulations are shown in Fig. 13. As noted previously, the losses were included as a loss angle. For the case of Pyrex, the loss angle (δ) is around 0.005 rad and the loss angle for the dielectric constants is 0.02 rad [65].

The resonant frequency obtained with ATILA^{TM} simulations was $f_r = 18.001$ kHz for the case in which losses were included, and the value of the measured resonant frequency was quite close, $f_r = 18.23$ kHz. The Q-factors for the data and the simulations were 152 and 212, respectively, and we see that the predicted resonant frequency is somewhat smaller (by about 1.2%) than the experimental value, presumably due to geometric modeling errors and/or possible errors in the values for the losses of the materials used in the simulations. Nevertheless, this is a quite satisfactory agreement since it allows the experimentalist to easily identify the proper resonant frequency at which to operate the acoustic chamber.

The measurement of the transient bubble radius (e.g., see data in Fig. 2) is normally done using laser scattering and is based on Mie scattering theory

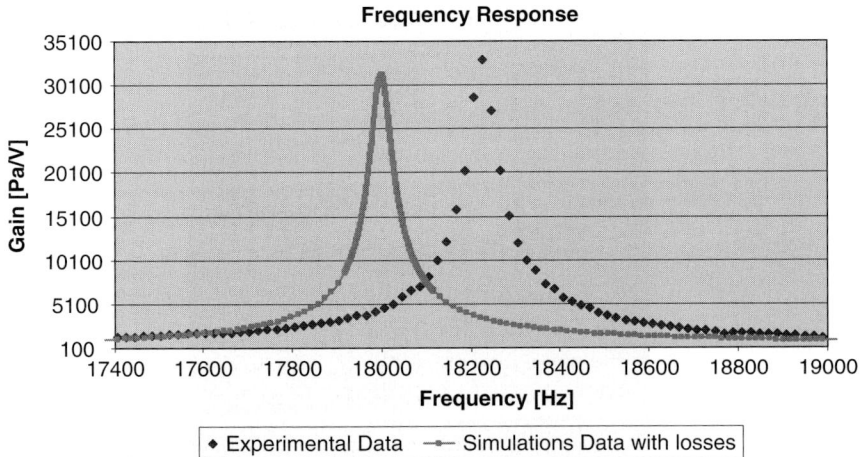

FIG. 13. Frequency response comparison between the experimental data and simulations.

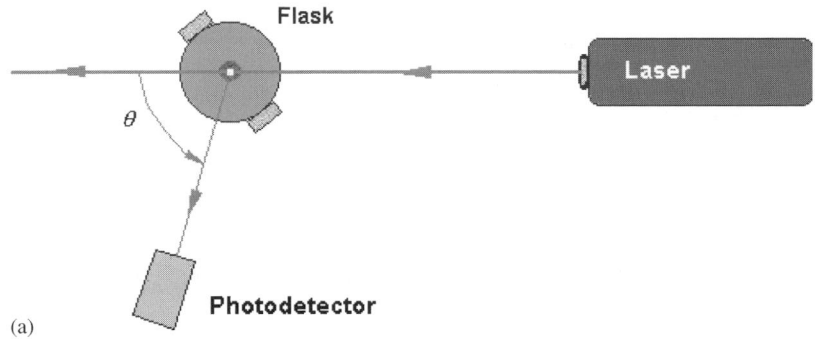

FIG. 14a. A laser beam polarized parallel to the paper illuminating a spherical bubble.

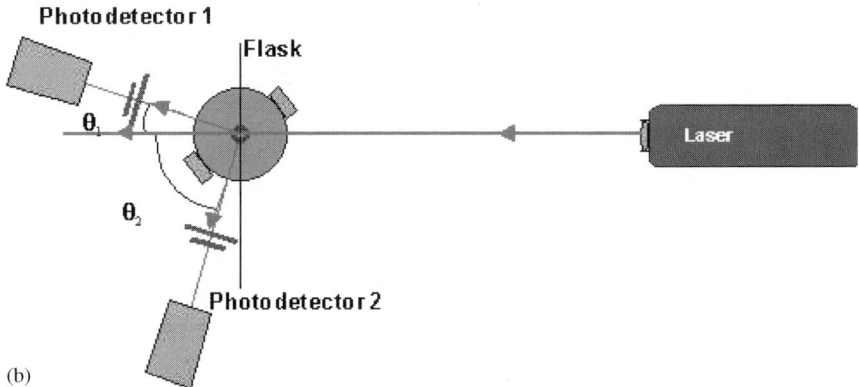

FIG. 14b. Typical set-up for a double detector technique. A polarized laser beam illuminates a bubble acoustically trapped in the acoustic antinode generated in the liquid-filled acoustic chamber. The two photodetectors collect scattered light at locations where intensity modulations are expected.

[106]. Figure 14a gives a typical measurement system configuration, where for an air/water system the critical angle (i.e., where only the reflected laser beam light from the interface of a spherical bubble will reach the photodetector) is $\theta = 83°$. While this type of measurement system is easy to set up and use, it requires an external calibration of the bubble size. If, on the other hand, as shown in Fig. 14b, two photodetectors are used [67,68], then there is no need for an external calibration; however, neither of these Mie scattering-based methods work very well during the final phase of the implosion, when the bubble size is very small and changes very rapidly. Another optical method, which has been used, is based on the Doppler effect. This method

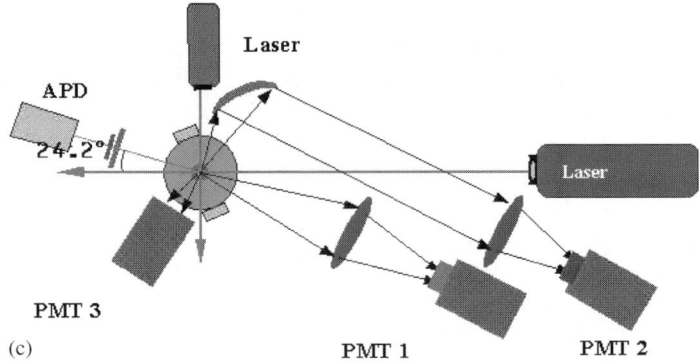

FIG. 14c. Bubble radius and photon detection systems. The spectral response of the PMTs is obtained using band-pass optical filters. In this way, the ratio of intensities yields a relative temperature measurement [Delgadino et al, 2000].

does not require an independent calibration and is very fast. One implementation of this method has been used to directly measure supersonic interfacial velocities (i.e., $M_g > 1.0$) during bubble implosions [69].

It is also interesting to note that the optical emission spectrum has been measured using photomultiplier tubes (PMTs) and spectrum analyzers for both MBSL and SBSL, and have been found to be distinctly different [37]. In particular, Flint and Suslick [70] found that the emission spectrum for acoustic-horn-induced MBSL was characterized by spectral peaks that correspond to electronic transitions of the rotational and vibrational bands within particular elements of the liquid surrounding the bubble (rather than the gas/vapor within the bubbles), and implied temperatures on the order of 5000 K (i.e., high enough for sonochemistry).

In contrast, Hiller et al. [33] found that the optical emission spectrum of SBSL was smooth and could be fitted to a Bremsstrahlung spectrum for small bubbles driven at high frequencies (~10 MHz) or a Blackbody spectrum for larger bubbles driven at lower frequencies (~20–30 kHz). Moreover, the inferred mean gas/vapor temperatures for SBSL was much higher than for MBSL [31,71].

A physical mechanism for the substantial difference between MBSL and SBSL spectra and temperatures has been proposed by Crum [37]. Briefly, Crum proposed that in MBSL, non-symmetries may cause the formation of energetic liquid jets within the bubbles, which might lead to elevated local liquid temperatures. In contrast, in SBSL, it is the highly compressed gas/vapor that produces the light flashes. Since we are more concerned in this paper with SBSL and sonofusion phenomena, the fundamental differences between MBSL and SBSL spectra will not be discussed at length herein.

Suffice it to say that the temperatures, which can be achieved in typical SBSL experiments, are much higher than in acoustic horn-induced MBSL experiments. Indeed, as previously noted, peak temperatures as high as 10^6 K were predicted by Moss et al. [28–30] and measured by Camara et al. [31] for SBSL conditions.

Many attempts have been made to directly measure the peak gas/vapor temperature during bubble implosions in SBSL experiments. Figure 14c shows a typical optical device that was used [72] to simultaneously measure the bubble size (using a dual detector[2], Mie scattering, technique) and the "mean" vapor/gas temperature using two photomultipliers (PMT-1 and PMT-2) having 450 and 300 nm optical filters, respectively [72]. This device was basically an optical pyrometer and showed that while the SL intensity increased with the acoustic driving pressure, the "mean" vapor/gas temperature did not [72]. This is presumably because the highly compressed core of the gas/vapor bubble is optically thick (i.e., opaque) and thus only the surface of the core region, and the optically thin gas surrounding it, were measured by the pyrometer during SL events. It should be recalled that the sun's surface temperature is $\sim$7000 K and it can be measured with a pyrometer, but the temperatures in the opaque interior of the sun cannot be measured in this way. A similar situation exists within an imploding bubble, unless the bubble being studied is very small [31].

A. Bubble Stability Considerations (SBSL)

The analysis of the stability of the bubbles undergoing bubble dynamics is difficult, since the phenomena leading to bubble distortion and breakup are inherently non-linear and likely are strongly influenced by the implosion process (i.e., the high Mach number phase of bubble collapse), where Rayleigh-type equations are inadequate. Nevertheless, it is instructive to consider the linear stability of an oscillating bubble by applying spherical harmonic perturbations to the Rayleigh–Plesset equation.

Let us assume that the time-dependent solution of a spherical oscillatory bubble, $R_s(t)$, is related to that of the corresponding distorted bubble by

$$R(t, \theta, \phi) = R_s(t) + \sum_{n,m} a_{nm}(t) Y_n^{(m)}(\theta, \phi) \qquad (8)$$

where $a_{nm}(t)$ is the amplitude of the shape perturbation of each node and $Y^{(m)}(\theta, \phi) = C_{n,m} P_n^{(m)}(\cos \theta) \exp(im\phi)$ are the so-called spherical harmonics, where $P_n^{(m)}$ are Legendre polynomials. We are normally only interested

---
[2] An avalanche photodiode (APD) and a photomultiplier (PMT-3).

in zonal harmonics (i.e., θ dependence only), thus we may omit the m-index summation.

For small amplitude perturbations (i.e., $|a_n(t)| << R_s(t)$), Eq. (8) and the extended Rayleigh–Plesset equation yields an equation of the form

$$\ddot{a}_n + A_n(R(t), \dot{R}(t))\dot{a}_n + B_n(R(t), \ddot{R}(t), \sigma, v_\ell; n)a_n = 0 \qquad (9)$$

where σ is the interfacial surface tension and v_ℓ the kinematic viscosity of the liquid. Equation (9) indicates that the instability of an oscillating bubble depends not only on interfacial acceleration $(\ddot{R})$, as in classical Rayleigh–Taylor instabilities (which occurs when the vapor/gas phase accelerates into the liquid phase), but also on the interfacial velocity $(\dot{R})$. This is a natural consequence of the initially spherical geometry under consideration.

Equation (9) has been evaluated for various levels of complexity by numerous authors [34,35,73–76,82], and thus only some of the salient results and conclusions will be given here.

Since significantly different time scales are involved, Brenner et al. [82] have proposed that the instabilities of interest can be considered as being either Rayleigh–Taylor-type instabilities (i.e., interfacial instabilities which have a relatively short time scale, $\sim 10^{-9}$ s), or parametric/Faraday instabilities (i.e., shape instabilities, which have a longer time scale, $\sim 10^{-3}$ s). In any event, linear stability analyses of oscillating bubbles indicate that the bubbles will be unstable to shape instabilities during implosions, and unstable to interfacial instabilities during the rapid interfacial deceleration associated with the latter stages of bubble implosion and the early rebound (growth) phase of the implosion process [10]. These phenomena will be discussed in more detail later when the results of non-linear, three-dimensional (3D) direct numerical simulations (DNSs) of bubble dynamics are presented. Significantly, these non-linear simulations imply that the conditions which can be achieved for energetic bubble implosions during sonofusion experiments should not be limited by these instabilities.

It should be noted that in typical SBSL experiments, unlike in sonofusion experiments, a gas bubble might oscillate for very many (i.e., millions of) cycles. This implies that interfacial instabilities may build-up over repeated cycles and can have a significant effect on the bubble dynamics. This, in turn, can limit our ability to reach significant implosion-induced temperatures and pressures in SBSL experiments.

In any event, in a typical SBSL experiment (e.g., an air/water system), for a given ambient pressure, acoustic forcing pressure, liquid pool temperature and dissolved gas content, a gas bubble is trapped in the antinode of the acoustic standing wave. As the acoustic pressure amplitude in the antinode is increased, the maximum amplitude of the oscillating bubble increases as does its equilibrium radius (R_0). As the bubble grows, it normally becomes unstable (parametric/shape instability) and small pockets of gas are detached from the bubble. This causes the bubble to erratically "dance" [74] near the center of the acoustic antinode. If the conditions are proper, when the acoustic forcing pressure is further increased, the bubble's equilibrium radius and maximum radius will suddenly decrease and the bubble ceases erratic motion. This is the point at which SL typically begins.

As the acoustic forcing pressure is progressively increased, the brightness of the light pulses increases until erratic motion of the bubble reappears, followed by bubble disappearance. Three thresholds can be identified in terms of the acoustic forcing pressure's amplitude.

- The first is where R_0 shrinks and SL begins. This is called the non-static Blake threshold [57,58].
- The second is where the bubble may grow in size due to rectified diffusion (to be discussed subsequently), but is stabilized due to the pinch off of small pockets of gas (the parametric instability boundary).
- The third threshold limits the magnitude of the acoustic forcing pressure that can be applied before the bubble ceases to exist (i.e., the polarity of the Bjerknes force).

The latter threshold may involve an interaction between rectified diffusion and parametric instabilities. In any event, there is a breakdown in the ability to entrap the bubble in the acoustic pressure antinode [57,58]. To investigate this important threshold further, let us next consider the forces, which entrap a bubble in acoustic antinodes.

1. *The Bjerknes Force*

The force on a bubble of volume V that experiences a pressure gradient in the liquid, ∇p, is

$$\underline{F} = -\iint_{S_B} p\underline{n} \, da = -\iiint_{V_B} \nabla p \, dv \cong -V \nabla p \qquad (10a)$$

If the bubble is oscillating, the net force, called the primary Bjerknes force (F_B), is given by the time average of F over the period of the oscillation

$$\underline{F}_B = -\langle V(t) \nabla p(t) \rangle \qquad (10b)$$

When a bubble of equilibrium size R_0 is undergoing radial oscillations at a frequency which is below its resonance frequency (i.e., the Minnaert frequency), then the acoustic pressure and bubble volume oscillations are out-of-phase [78], in that bubble size decreases when the liquid pressure increases. In contrast, for a forcing frequency that is higher than the bubble's resonant frequency, these oscillations will be in-phase.

For the normal case, the bubble is out-of-phase with the acoustic pressure field and is located in the pressure antinode. When the acoustic pressure is large, the bubble will be small, and a relatively small force acts in the same direction as the buoyancy force. On the other hand, when the acoustic pressure is negative, the bubble will be large and a force acts on the bubble, which is opposite to the buoyancy force. Thus, the time-average of the net force (i.e., the buoyancy and the primary Bjerknes force) can entrap the bubble in the pressure antinode [79].

It should also be noted that for a spherical acoustic antinode, if F_B is negative, then the oscillating bubble is trapped in the pressure antinode, while if F_B is positive, it is repelled. It has been shown [57,58] that, for a typical air/water SBSL experiment at an acoustic forcing frequency of 20 kHz, the limiting value of the acoustic forcing pressure (p_a) before there is a breakdown in the ability to entrap the bubble, decreases as the bubble's equilibrium radius (R_0) increases (or, for a given R_0, as the acoustic forcing frequency increases). For most practical purposes, this implies that it is not possible to repetitively force the bubbles with an acoustic pressure amplitude any higher than about 1.7 bar. That is, for $p_a > 1.7$ bar, there will be a change in the sign of the primary Bjerknes force, resulting in an expulsion of the bubble from the antinode of the standing acoustic wave. This appears to be a fundamental limitation, which controls the maximum bubble compression and induced temperatures in SBSL experiments.

Let us next investigate another fundamental limitation on the ability of SBSL experiments to achieve conditions suitable for ultrahigh bubble compressions; namely, rectified diffusion.

2. Rectified Diffusion

Mass transfer will take place across an oscillating bubble's interface. Even in the absence of an acoustic pressure field, a free gas bubble in a liquid will dissolve due to the mass transfer driving potential associated with the excess internal gas pressure required to balance surface tension, $2\sigma/R_0$, unless the liquid is sufficiently supersaturated with dissolved gas.

When a spherical bubble is forced to oscillate due to an impressed acoustic pressure field, so-called rectified diffusion takes place. This phenomenon is due to the bubble volume and gas pressure oscillations that take place as the bubble expands (the internal gas pressure decreases) and contracts (the internal gas pressure increases), and the associated changes in the interfacial area, $4\pi R^2(t)$. Since the gas diffusion rate across the interface is directly proportional to the interfacial area, more gas will tend to enter the expanding bubble (when the bubble is large and its pressure is low) than leave during bubble contraction (when the bubble is small, and the gas pressure is high). If so, the bubble will grow and may become unstable. However, what actually happens will depend on the dissolved gas concentration in the liquid, since this will establish the mass transfer driving potential (i.e., the difference in the instantaneous gas concentration at the bubble's interface and in the liquid).

In any event, the mass transfer boundary layer in the liquid, which surrounds the bubble, grows as the bubble contracts and shrinks as the bubble expands. The spherically symmetric conservation equation that quantifies the diffusion of the dissolved gas in the liquid, and thus defines the instantaneous mass transfer boundary layer, is given by

$$\frac{DC}{Dt} \equiv \frac{\partial C}{\partial t} + \underline{u}_\ell \cdot \nabla C = \nabla \cdot D_g \nabla C \qquad (11)$$

where $C(\underline{x}, t)$ is the local, instantaneous concentration of dissolved gas in the liquid, $\underline{u}_\ell$ the velocity of the liquid which is induced by the bubble's motion and D_g the binary mass diffusion coefficient of the gas dissolved in the liquid.

In accordance with Fick's law, the rate of transport of gas mass across the bubble's interface into the bubble is given for a spherical bubble by

$$\frac{dm_g}{dt} = 4\pi R^2 D_g \frac{\partial C}{\partial r} \qquad (12)$$

and the instantaneous gas concentration at the bubble's interface is given by Henry's law, $C_i(R, t) = p_g/H$, where H is Henry's constant.

Finally, at the outer edge of the mass transfer boundary layer (δ_m) we have

$$C(r = R + \delta_m, t) = C_0 \qquad (13)$$

Normally, $C_0 \leq C_{0_s}(T)$, where $C_{0_s}(T)$ is the saturation gas concentration in the liquid at temperature T.

There have been many studies, which have dealt with rectified diffusion, including the pioneering work of Eller and Flynn [80] and enhancements and use of this theory by Fyrillas and Szeri [81], Brenner et al. [82] and Akhatov et al. [57,58]. It has been shown that, to a good first approximation, diffusional equilibrium (i.e., when the average bubble size does not increase or decrease) is given by [80]

$$\frac{\langle p(t) \rangle_\tau}{p_0} = \frac{C_0}{C_{0_s}} \qquad (14)$$

here p_0 is the static liquid pressure (i.e., $p_1(t) = p_0 + p_a \sin(\omega t)$, where $\omega = 2\pi f$) and the non-linear weighted-average pressure in the gas bubble is given by

$$\langle p(t) \rangle_\tau = \frac{\int_0^{1/f} p(t) R^4(t) \, dt}{\int_0^{1/f} R^4(t) \, dt} \tag{15}$$

When $\langle p(t) \rangle_\tau > p_0 C_0/C_{0_s}$, the bubble will dissolve and if $\langle p(t) \rangle_\tau < p_0 C_0/C_{0_s}$ it will grow. Moreover, the criterion for the diffusional equilibrium point to be stable is $d\langle p(t) \rangle_\tau / dR_0 > 0$.

The Eller and Flynn model is particularly easy to use since the extended Rayleigh equation, Eq. (1), can be used to calculate $R(t)$ and thus Eq. (15) can be numerically evaluated. Indeed, many experimenters have used Eqs. [14] and [15] to verify their concentration ratio, C_0/C_{0_s}.

Equations (11)–(13) can also be evaluated numerically in conjunction with a suitable hydrodynamic shock (i.e., HYDRO) code. Alternately, we may use the extended Rayleigh equation, Eq. (1), and an appropriate thermodynamic process during bubble dynamics (i.e., an isothermal process during bubble expansion and bubble contraction until $Mg \geq 0.1$, after which an adiabatic process may be assumed). Since the vast majority of the mass transfer takes place during low interfacial Mach numbers [$Mg < 0.1$], both a HYDRO code and the Rayleigh equation yield essentially the same results.

Typical SBSL predictions are shown in Fig. 15 for $R_0 = 5 \, \mu m$ air bubble in room temperature water for $C_0/C_{0_s} = 0.34\%$ at a forcing acoustic pressure of amplitude $p_a = 1.25$ bar and frequency of 20.6 kHz. The various times shown in Fig. 15 are for the expansion phase (times t_1–t_5) and the implosion phase (t_6–t_{10}) of a typical SBSL experiment. The corresponding air concentration ratios in the water, $C(r,t)/C_0$, are given in Figs. 16a and b for the bubble expansion process, and in Figs. 17a and b for the implosion.

As expected, it can be seen that the instantaneous concentration gradient implies gas transfer into the bubble for times t_1–t_5, and out of the bubble for times t_6–t_{10}. If the mass transfer is averaged over the period of bubble oscillation ($1/f = 48.5 \, \mu s$), we find that there is no net mass transfer into the bubble and thus it will continue to oscillate as shown in Fig. 15 (i.e., the time-average bubble size will neither grow or shrink with time).

It should be noted that this dynamic diffusional equilibrium is a characteristic of SBSL experiments. However, it only occurs for very special conditions. In particular, if the rectified diffusion analysis discussed above is repeated for $C_0/C_{0_s} < 0.34\%$ the bubble will dissolve, while for $C_0/C_{0_s} > 0.34\%$ it will grow and become unstable. Moreover, the dynamic diffusional equilibrium conditions will be a strong function of R_0 and p_a.

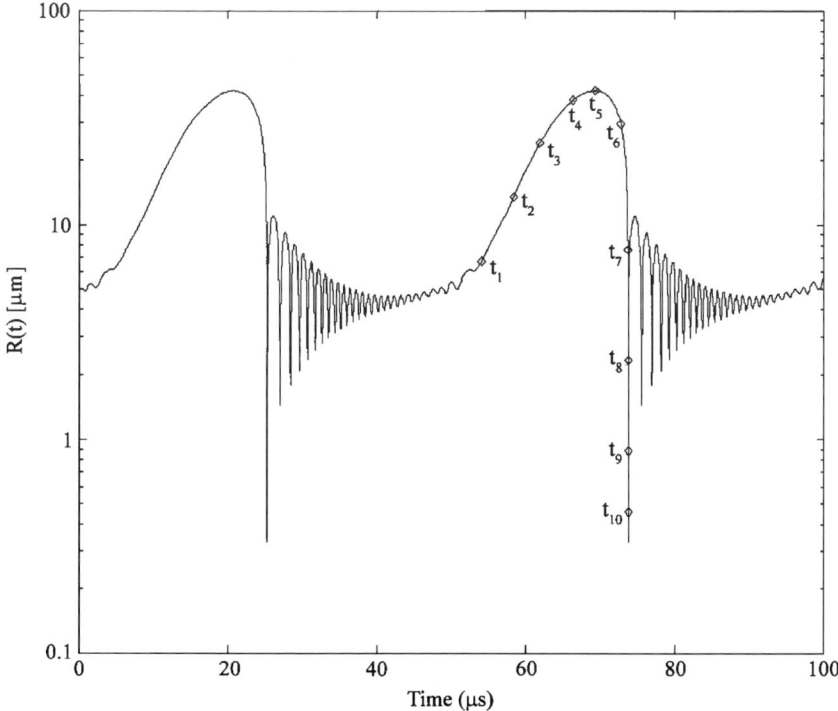

FIG. 15. Bubble dynamics during rectified diffusion, R(t) (C_0/C_{0_s} = 0.34% air in water, p_a = 1.25 bar, T = 20 °C, f = 20.6 kHz, R_0 = 5 μm).

If parametric computations are made using the coupled thermal-hydraulic and mass transfer models just discussed, we can determine the region in parameter space where SBSL would be expected. Figure 18 shows that in C_0/C_{0s} versus R_0 space, the rectified diffusion stability boundary is given by the locus of the minimum of lines of constant acoustic forcing pressure (p_a). That is, for a given C_0/C_{0s}, if we have an equilibrium bubble size (R_0) which is to the right of the minimum, then as R_0 increases at a given p_a, we have less gas concentration in the liquid than is required for equilibrium and thus the bubble will shrink back to the solid line (for that p_a). In contrast, if for the same conditions, R_0 decreases to the left of the solid line, then we have excess concentration and the bubble will grow until it reaches the solid line.

If we now look at a point on the left of the rectified diffusion stability boundary (where $\frac{\partial(C_0/C_{0s})}{\partial R_0} < 0$), then if the R_0 decreases from a point on the particular dashed line in question, then the bubble will dissolve since the gas concentration in the liquid is insufficient to support a stable bubble of that

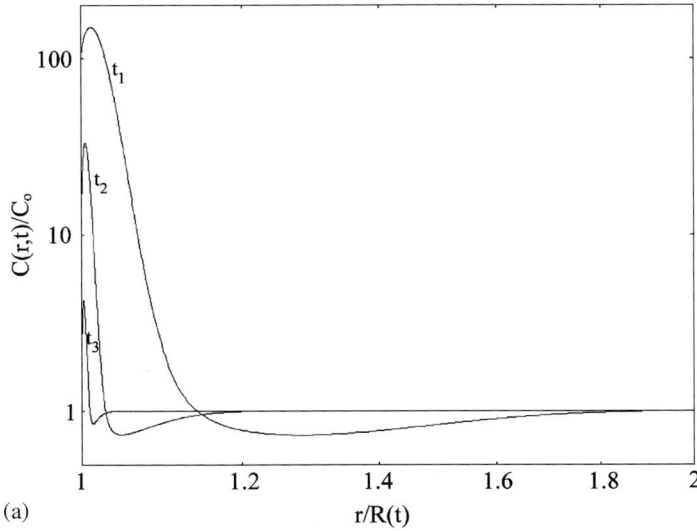

FIG. 16a. Air concentration profiles in water during the early stage of bubble growth (times t_1–t_3 in Fig. 15).

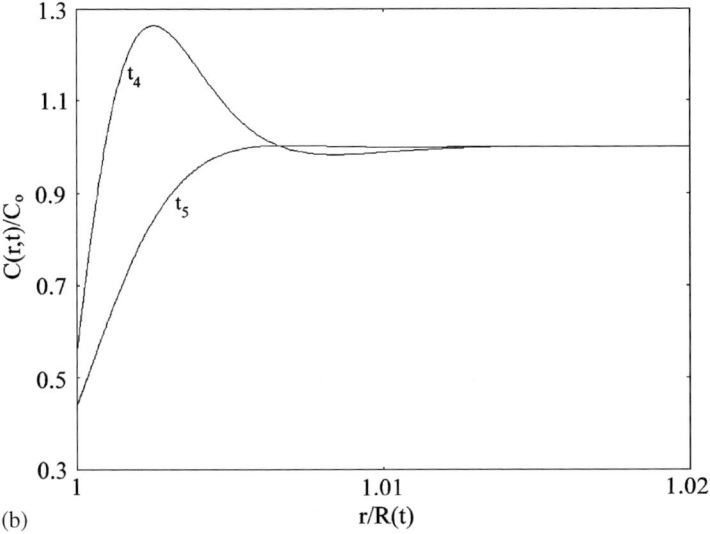

FIG. 16b. Air concentration profiles in water during the final stage of bubble growth (times t_4 and t_5 in Fig. 15).

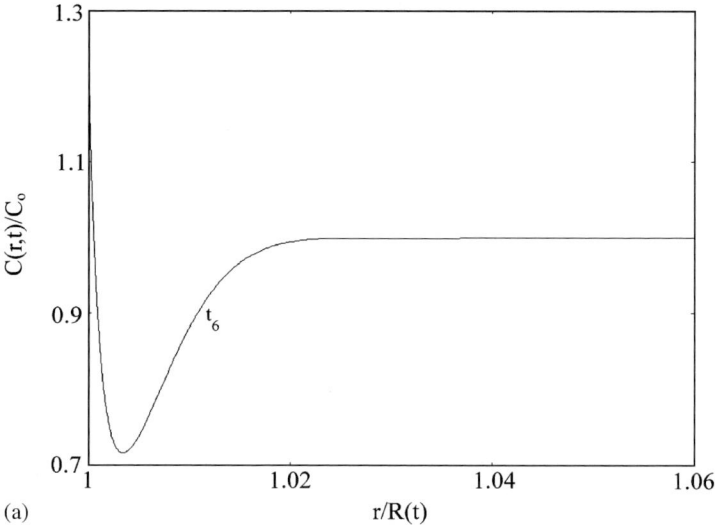

FIG. 17a. Air concentration profiles in water during the start of bubble implosion (time t_6 in Fig. 15).

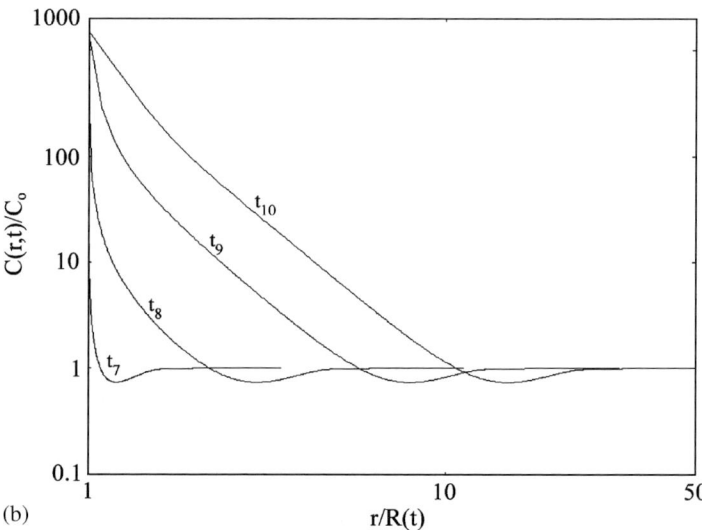

FIG. 17b. Air concentration profiles in water during the bubble implosion process (times t_7–t_{10} in Fig. 15).

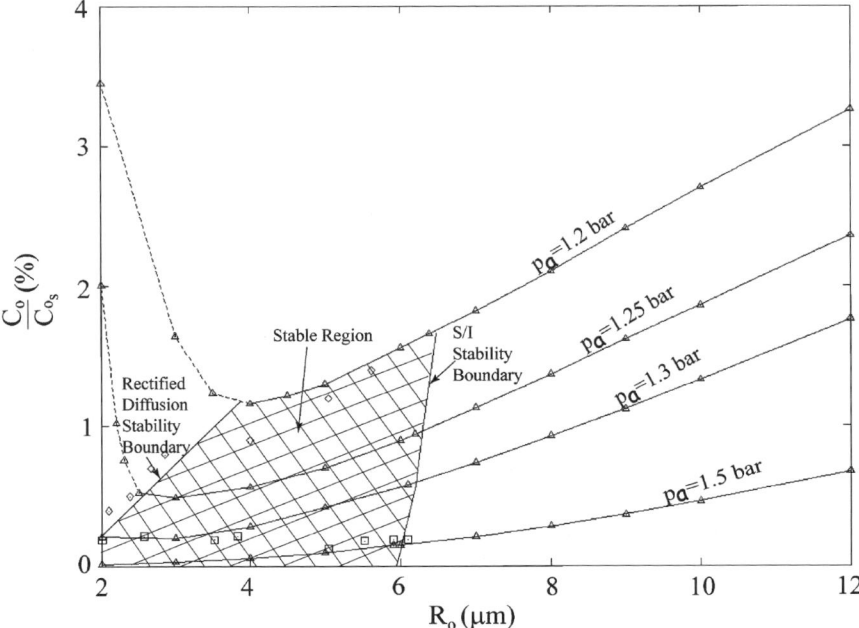

FIG. 18. Stable region for SBSL (air/water at STP, f = 20.6 kHz) compared with typical SBSL data (diamonds and squares).

size at the imposed forcing (p_a). Similarly, if R_0 increases from the dashed line, the bubble will grow due to the excess gas dissolved in liquid. Depending on where the bubble ends up compared to the shape/interfacial (S/I) stability boundary, it may become stable at a larger size if it is in the shaded stable region of the figure. Otherwise, the bubble will become unstable and break up. It should also be recalled that the ability to trap a bubble in the pressure antinode breaks down at $p_a \sim 1.7$ bar [57,58], thus there is a relatively limited region in parameter space for which SBSL can occur.

We see in Fig. 18 that the data (squares and diamonds) of Holt and Gaitan [83] are generally consistent with this analysis. This implies that researchers interested in SBSL can use this approach to determine, for a particular liquid and operating conditions, where SBSL would be expected.

It should also be noted that some progress has been made to enhance the bubble implosion process using mixed frequency excitation [72,84,85], in which the maximum bubble radius and speed of collapse can be increased; however, to date, only incremental enhancements have been possible.

Given these discussions on the conditions required to achieve SBSL, it is quite amazing that Gaitan was able to empirically find the conditions for which a gas bubble would stably oscillate for days and emit picosecond duration light pulses with precise timing. Nevertheless, the above discussion clearly indicates that this particular experimental technique has inherent limitations on the test parameters, such as how hard the bubble can be driven (e.g., $p_a \leqslant 1.7$ bar), and thus the strength of the implosion process. It should be abundantly clear that in order to achieve the conditions suitable for sonofusion, an alternate experimental technique must be sought. This will be considered in the next section.

B. Lessons Learned from SBSL Experiments and Analysis

Let us now assess what we can learn from SBSL experiments that may be useful for enhancing the bubble implosion process. As noted previously, except for very small bubbles [31], it is impossible to make direct experimental measurements of the highest temperatures achieved during SL, since for high temperatures (i.e., above $\sim 10^4$ K) the electromagnetic emissions corresponding to these temperatures (or more precisely to the temperature of the electrons radiating photons due to electron dynamics) are strongly absorbed by the optically thick (opaque) dense gas core in the bubble and in the first few millimeters of the liquid. For example, as can be seen in Fig. 19, the macroscopic absorbtion coefficient (i.e., cross section) for water is very large outside the range of visible radiation. Thus, the peak temperatures achievable during SL, which have relatively small (ultraviolet) wavelengths, can normally only be inferred by calculations. Moss et al. [28–30][3] have performed hydrodynamic shock code calculations for the spherically symmetric implosion of a gas bubble in acoustic fields with frequencies of 27 and 45 kHz, and liquid pressure amplitudes near the bubble of 0.25 bar (or 1.5 bar). These conditions are typical of SBSL experiments, when the bubble grows from an initial radius of about $R_0 = 4.5$ μm (or 10 μm), in equilibrium with the average pressure in the fluid, to about 40 μm (or 80 μm) during the acoustic rarefaction phase, and then implodes to a radius of about 0.3 μm

---

[3] Some points of the investigations reported by Moss et al., are invalid. In particular, as shown in the paper by Nigmatulin et al. [86], the adiabatic state of the gas assumed in their calculations was in error. Actually, the gas remains isothermal for almost the whole period of bubble expansion and compression and its temperature is about the same as in the liquid. It is not difficult to show that during the 30 to 50 μs long period of bubble oscillations, thermal waves have a chance to equalize the gas temperature in a bubble of $\sim$50 μm radius to that of the surrounding fluid. Only during the implosion (which is about 10^{-9} s = 10^{-3} μs long), when the radius decreases from about 2 to 3 R_0, to its minimum, R_{min}, with an interface velocity of $\sim 10^3$ m/s, will there be adiabatic gas heating. As a result, the bubble's thermal cycle corresponds to a thermal pumping cycle. In addition, Moss et al., only calculated the very first oscillation after rest and not the asymptotic oscillations, when the amplitude, fluid kinetic energy and temperature achieved in the hot nucleus are several times higher than those for the first implosion. Nevertheless, their basic conclusions were correct.

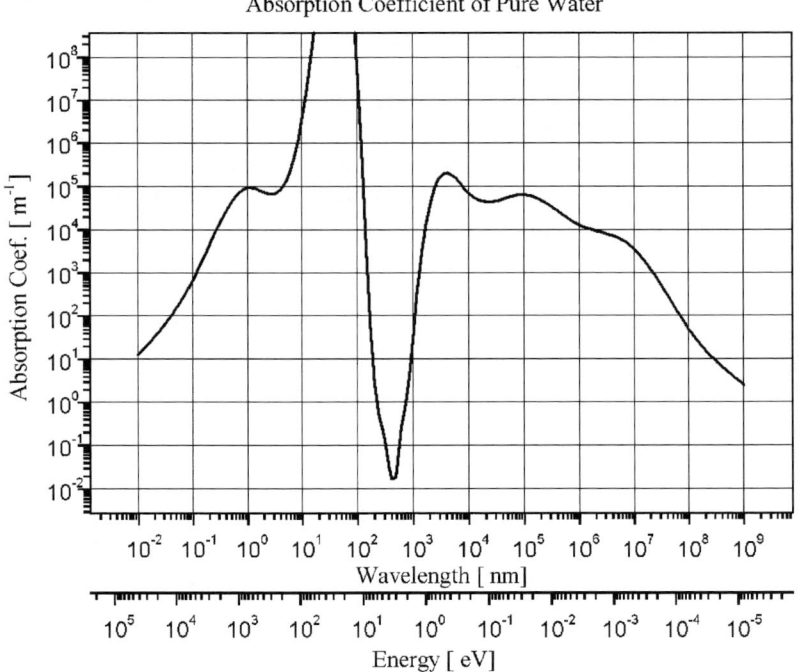

FIG. 19. Macroscopic absorption coefficient for water.

(or 3 μm) during the acoustic compression phase. Significantly, the final stage of collapse was only about $\sim 10^{-3}$ μs = 1 ns long. During bubble implosion, a shock wave is initiated because of the fluid acceleration toward the center of the bubble. While focusing (i.e., cumulating) toward the center of the bubble and then bouncing back, the shock wave compresses the gas (due to the inertia of the fluid accelerated toward the center) and heats it up to high temperatures. When the shock wave is focused at the center, a high-density ionized gas/vapor core is produced near the bubble's center. According to Moss' calculations, this plasma core has radius, r_*; lifetime of the highly compressed state (before dissipation), Δt_*; gas density, ρ_*; pressure, p_*; and peak ion temperature, T_{i*}, of about

$$r_* \sim 2 \text{ nm}; \Delta t_* \sim 10^{-11} \text{s}; \rho_* \sim 10^4 \text{ kg/m}^3; p_* \sim 10^9 \text{ bar}; T_{i*} \sim 10^6 \text{ K} \tag{16}$$

Significantly, during Δt_* the electron temperature, T_{e*}, that governs SL was found to be much less than the ion temperature (T_{i*}).

These calculations gave an important confirmation: the duration of a (SL) light flash, which is determined by the lifetime of the highly compressed, hot central core ($\sim 10^{-11}$ s$\equiv$10 ps), was in good agreement with experimental measurements of the duration of SBSL light flashes. However, an assessment [29] of the possibility for thermonuclear fusion to occur under the stated conditions for a bubble containing a deuterium–nitrogen mixture (nitrogen was added to slow down the sound speed in the gas) gave a very small rate of neutron production of about 0.1/h, which corresponds to only 10^{-5} neutrons per implosion[4].

It should be kept in mind that such an abnormally high heating of a tiny central zone in the bubble occurs by transforming into internal energy part of the kinetic energy of the fluid accelerated toward the bubble's center. The greater the kinetic energy of the liquid, the higher will be the compression and heating of the central gas/vapor core.

The aim of sonofusion experiments is to reach intense heating, which is at least two orders of magnitude higher (i.e., to 10^8 K) than in typical SBSL experiments, creating conditions suitable for thermonuclear fusion. For this purpose, a novel experimental procedure must be employed that enables a large increase in the kinetic energy of the fluid accelerating toward the bubble's center, thus enhancing the effect of the cumulative shock wave compression of the central core region of the bubble. Moreover, much greater amplitudes of the acoustic field must be used than the conventional 1.0–1.5 bar amplitudes used in SBSL experiments (which are limited due to the sign of the Bjerknes force, rectified diffusion and/or inherent shape and interfacial instability considerations).

In the seminal sonofusion experiments run at ORNL [54,56] and later confirmed at Purdue [180], the harmonic acoustic fields had an amplitude of 15–40 bar (and within an imploding bubble cluster, the local liquid pressure will be significantly intensified). To achieve this, two difficulties had to be surmounted. The first task was to choose a liquid that could accommodate the repetitive high-frequency states of liquid tension without cavitation. The second was to develop and focus intense acoustic fields in a rather small region within the liquid. This required the use of a different experimental technique than that used in SBSL experiments. In particular, a technique that was originally developed for neutron detection [62,88,89] was modified and used.

Moreover, instead of a non-condensable gas bubble (as in SBSL experiments) vapor cavitation bubbles (i.e., the bubbles were filled with vapor of the surrounding liquid) were used. This is important since the fluid being

---
[4] The invalid points in the calculations stated previously lead to a further reduction of the predicted D/D fusion neutron production rate by a factor of 10 to 100.

accelerated toward the gas bubble's center is cushioned by the increasing gas pressure caused by fluid-induced gas compression, thus reducing the ultimate pressure and temperature achievable. In contrast, in a vapor bubble the effect of cushioning is mitigated due to vapor condensation at the bubble's interface (i.e., during bubble compression much of the vapor is condensed into liquid). In addition, when compressing a non-condensible gas bubble of constant mass, its pressure grows monotonically while an imploding vapor bubble's pressure remains almost constant in time due to vapor condensation until the final phase of collapse. This also mitigates the effect of gas/vapor cushioning during bubble implosion.

In the ORNL bubble fusion experiments [54,56], vapor bubbles were nucleated at the point of maximum liquid tension by using energetic external neutrons, which collided with the liquid's atomic nuclei. The bubble's nucleation centers had an initial radius, R_0, of ~ 0.01–0.1 μm, and these bubbles grew rapidly during the phase of liquid rarefaction induced by the external acoustic field, and reached a maximum radius of $R_m \sim 500$–800 μm, which was much greater than in typical SBSL experiments (where, $R_m \sim 50$–80 μm). Thus, the maximum bubble volume was larger by a factor of three orders of magnitude in comparison to typical SBSL experiments. Because of the above-mentioned reduced cushioning due to vapor condensation, the much higher compression pressure of the acoustic field, and larger the maximum bubble volume, the liquid near the vapor bubble's interface was accelerated toward the center to relatively high radial velocities, w (i.e., up to 7–8 km/s, which is at least three times higher than in SBSL experiments). As a result, the fluid's kinetic energy, determined by the product $R^3 \dot{R}^2$, was at least 10^4 times higher than the kinetic energy in typical SBSL experiments. Moreover, this energy was imparted not to the total amount of vapor that filled the bubble at the instant of its maximum size, but only to the vapor left after condensation took place. Analysis indicates [90,114] that 50–90% of the vapor is condensed (i.e., by the time of maximum compression there is only 50–10% of the vapor left in the bubble relative to the vapor mass at the time of maximum bubble size). Thus, in our experiments, the amount of vapor left by the time of bubble collapse is only about 100 times (not 1000 times) more than in typical SBSL experiments. As a result, there was about a $10^4/(10^2) = 100$ times more fluid kinetic energy imparted to a unit mass of vapor in the imploding bubbles in the ORNL experiments [54,56] than in typical SBSL experiments. Thus, if the gas ion temperature during SBSL experiments is $T_{i*} \sim 10^6$ K, in sonofusion experiments using cavitation-induced vapor bubbles, the ion temperature could be expected to reach $T_{i*} \sim 10^8$ K, which implies conditions suitable for D/D nuclear fusion.

The influence of the maximum bubble size, R_m, determined by the amplitude of acoustic forcing, and the radial implosion velocity, w, on higher gas compression, were also supported by independent experiments [72,85]. In particular, it was noted in SBSL experiments that increasing R_m modestly (by ~50%) or increasing the rate of collapse, w, could result in very large increases in light emission during bubble implosion. Indeed, these two SBSL experimental observations guided the design of the sonofusion experiments at ORNL.

C. Choice of Test Liquid (Sonofusion)

Acetone was chosen as the test liquid (density, $\rho_\ell = 0.79$ g/cm^3, sound velocity, C_ℓ ~1190 m/s) in the ORNL [54,56] and confirmatory [180] sonofusion experiments. In contrast to SBSL experiments, it was well degassed so that it was essentially free from non-condensable gases in the liquid. There were two isotopic compositions tested, normal acetone (C_3H_6O, hereafter called H-acetone) was the control fluid, which was without nuclei capable of easily undergoing nuclear fusion, and deuterated acetone (C_3D_6O, hereafter called D-acetone) was the primary test fluid. Significantly, D-acetone contains deuterium (D) nuclei, which are capable of undergoing D/D nuclear fusion reactions at sufficiently high temperatures and densities.

There were a number of reasons to choose acetone. First, it is relatively safe, readily available and not too expensive. Also, like most hydrocarbons, acetone has a high cavitation strength, (i.e., it permits the attainment of large tensile states without premature cavitation). This, in turn, permits large values of liquid superheat to be attained, and thus rapid evaporation rates and maximum bubble sizes subsequent to nucleation. Second, in the vapor phase of acetone (D-acetone's molecular weight is 64) the sound velocity is relatively low. Thus, a fixed interface velocity compressing the bubble results in the formation of strong shock wave compressions. Third, organic liquids have relatively large accommodation coefficients for condensation. As will be described later, this is very important since it allows for intense condensation during implosive bubble collapses. Certainly, the presence of foreign atoms or nuclei (i.e., 4 nuclei and 52 g of carbon and oxygen out of the 10 nuclei and 64 g of acetone, respectively) leads to losses in energy due to carbon and oxygen ion heating and poses impediments to D-nuclei approach and collision. Nevertheless, the advantage of a large accommodation coefficient and relatively low speed of sound far outweighs these parasitic losses.

Unless otherwise noted, the liquid in the acoustic chamber was maintained at ~0 C, which was the lowest value obtainable with the equipment which was used in the ORNL experiments [54,56]. The test liquid was degassed and subjected to an acoustic pressure field that oscillated at the resonance frequency of the liquid sample in the test section (i.e., a acoustic standing wave having a single pressure antinode was formed).

D. Experimental System

As shown schematically in Fig. 20, the acoustic chamber used in the ORNL experiments [54,56] was a cylindrical test section made of Pyrex (i.e., borosilicate glass), which was 65 mm in diameter and 20 cm high. The test section (i.e., acoustic chamber) was filled with a test liquid (either degassed D- or H-acetone), and a PZT piezoelectric transducer ring was attached to the outer surface of the test section with epoxy.

The flask walls and the liquid within the acoustic chamber were harmonically driven with the PZT. For D-acetone, the acoustic frequency was $f = 19.3\,\text{kHz}$ (i.e., an acoustic period of about 52 μs). The acoustic chamber under consideration was a high-Q acoustic system, and as such required careful tuning for optimal performance (i.e., to achieve about 30–50 energetic bubble cluster implosions per second).

To detect the shock waves caused by the bubble cloud implosions, two pill microphones were installed on diametrically opposite outer sides of the test section.

All experiments were conducted in the test section shown schematically in Fig. 20 using either normal acetone (100 at.% pure C_3H_6O) or D-acetone (99.92 at.% C_3D_6O), which were initially filtered through 1 μm filters. Degassing was performed at vacuum conditions (~10 kPa) and acoustic

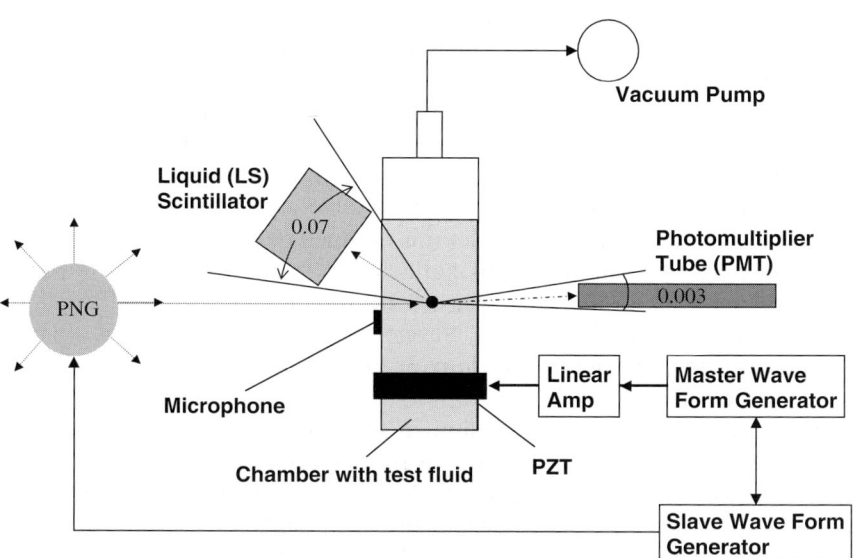

FIG. 20. ORNL experimental configuration [54,56].

cavitation of the liquid was performed for about 2 h to promote gas release from the liquid.

Nucleation of the vapor bubbles was initiated with fast neutrons from either an isotopic source (Pu–Be) or a pulsed neutron generator (PNG), which was fired on demand at a predefined phase of the acoustic pressure field. The Pu–Be source produced neutrons at a rate of $\sim 2 \times 10^6$ n/s, where most of the neutron energy was below 5 MeV (i.e., 75–80%) and about 10 MeV was the maximum energy emitted.

The PNG generated 14.1 MeV neutrons at a rate of $\sim 5 \times 10^5$ n/s and it was operated at a frequency of 200 Hz (i.e., $\sim$2500 neutrons per pulse were emitted isotropically). Each PNG neutron pulse was about 12 μs long and was phase matched with the acoustic pressure oscillations in the test section. In particular, the PNG was fired during the liquid expansion phase when the pressure achieved its minimum (i.e., maximum negative) value. As follows from the given acoustic ($\sim$20 kHz) and PNG (200 Hz) frequencies, the PNG neutron pulse was produced not in each acoustic cycle but once every $\sim$100 acoustic oscillations. The reason for using a pulse neutron source with a relatively low frequency for initiating cavitation was to minimize the background due to the 14.1 MeV PNG neutrons when 2.45 MeV neutron emissions were produced during bubble implosions. It should be noted that successful sonofusion experiments have also been run in which a dissolved alpha particle emitter (i.e., a uranium salt) was used instead of an external PNG neutron source [179]. However, since the timing of the alpha-emission-induced bubble cluster formation process could not be controlled, this method was less efficient (i.e., it produced less D/D neutrons then when a PNG was used [54,56]).

Either a plastic (PS) or liquid (LS) scintillation detector (i.e., scintillator) was used for detection of the neutron and gamma signals. In particular, an Elscint LS, which had dimensions 5 cm (diameter) × 5 cm (thickness), was used for most of the neutron data [54,56]. Pulse-shape discrimination (PSD) was also used [91–94]. This discriminator made it possible to count neutrons separately from γ-ray-induced scintillations on the basis of differences in the signal decay time of each type of scintillation. The discrimination of neutrons was important since it allows one to determine the energy of each neutron-induced scintillation, or more precisely, the energy of the resultant recoil proton responsible for the given scintillation. When striking a proton, a neutron of energy E_n, gives up some of its energy to a proton in the LS. The energy of this proton (called a recoil proton), E_p, will thus be in the range $0 \leq E_p \leq E_n$. Hence, the 2.45 MeV D/D neutrons that were measured by the LS registered an energy of less than or equal to 2.45 MeV [54,56].

The LS detector was calibrated with cobalt-60 and cesium-137 γ-ray sources by relating the Compton edge of each spectrum with the energy of the proton, which would produce the same amount of light in the scintillator [91,93]. Using a multichannel pulse height analyzer (MCA) with 256 channels, the energy of a 2.45 MeV proton was determined [54] to lie near channel 40; in particular, in our measuring system a 2.45 MeV neutron had a proton recoil 'edge' at channel 32. Similarly, measurements with the 14.1 MeV PNG neutrons showed a proton recoil edge at channel 115. There was a 21-channel offset in the MCA such that zero energy corresponded to channel 21 [54]. Thus, 21 channels must be subtracted from the pulse height data before comparing energies of Compton and proton recoil edges. Doing this, the ratio of 14.1 to 2.45 MeV neutrons was $(115–21)/(32–21) \sim 9$ which is consistent with previously reported values [91,93,95]. In subsequent sonofusion experiments [56], this offset was eliminated through the use of a more advanced detection system.

The efficiency of the Elscint LS detector (designated as η_{ET}) was determined using a Pu–Be neutron source with a known intensity of 2×10^6 n/s, by positioning the source near the face of the LS detector (i.e., at a distance of ~ 1 cm) and the efficiency of the detector was found [54] to be $\sim 5 \times 10^{-3}$. This was corrected for the true distance (5–7 cm) between the LS and the 2.45 MeV neutron source (i.e., the zone of cavitation) in the ORNL experiments [54] and also for the corresponding solid angles (see Fig. 20), giving an estimated efficiency which was less by a factor of $(5/1)^2 = 25$ to $(7/1)^2 = 49$. When estimating the neutron production in the zone of cavitation, it should be noted that about 50% of the 2.45 MeV neutrons were attenuated in the acetone-filled flask, so the efficiency of neutron detection[5] from the zone of cavitation in the actual configuration (Fig. 20) was [90]

$$\eta_{ET} \sim 5 \times 10^{-3} \times (0.02 - 0.04) \times 0.5 = (0.5 - 1) \times 10^{-4} \quad (17)$$

In subsequent experiments [56], a more efficient detector is used, and thus the efficiency was higher. Nevertheless, the final results (i.e., 4×10^5 n/s) were essentially the same.

The total "light pulses" (i.e., the actual SL light pulses from the cavitation zone and the flashes occurring due to the interaction of neutrons and γ-rays with the PMT) were detected [54] with a Hammamatsu R212 PMT, which had a 2 ns rise time. The timing of the SL light flashes relative to the PS or LS nuclear-induced scintillations were made using an MCA and a high-speed digital storage scope.

---

[5] The correction for neutron losses in the liquid filled flask was made after the original publication in *Science* by Taleyarkhan *et al.* [54].

E. Description of the Bubble Dynamics Process

To ensure robust nucleation and significant bubble growth and implosions, the drive voltage to the PZT was set to be about double that needed for the onset of neutron-induced cavitation in the acetone in the test section. The negative pressure threshold for the onset of bubble nucleation by neutrons and α particles in acetone is -7 to -8 bar [96,97]. A pressure map of the liquid was obtained using a calibrated hydrophone. Using a linear scale factor for the induced pressures in the acoustic chamber versus drive voltage to the PZT, and gradually increasing the drive amplitude of the acoustic field pressure, it was verified that neutron-induced acoustic cavitation began at a negative pressure in the test section of about -7 bar. In the ORNL sonofusion experiments [54,56], the pressure amplitude within the zone of bubble cluster nucleation was $\geqslant 15$ bar.

A 14.1 MeV neutron pulse from the PNG induces cavitation in the liquid at the position in the acoustic chamber where the acoustic pressure antinode is located and at that the instant when the liquid at that position is under the greatest tension. At this time, the neutron scintillator detects, as expected, the neutron pulse from the PNG. The interaction of 14.1 MeV neutrons with tensioned liquid gave rise to the nucleation of visible bubble clusters (occurring at a rate of ~ 30–$50/s$) and photographic evidence and analysis suggest that the bubble clusters that were formed consisted of about 1000 microbubbles (see the section entitled "Nucleation Phenomena in Sonofusion" and Appendix-C for more details). During rarefaction conditions these microbubbles grow rapidly within the bubble cluster, until the increasing acoustic pressure in the liquid during the second half of the acoustic cycle arrests their growth and causes them to begin to collapse, resulting in an implosion. If the implosion is robust enough, the bubbles emit SL light flashes, which can be detected by the PMT. The Rayleigh equation for an inviscid incompressible liquid shows that if the pressure in a bubble does not increase during compression (e.g., due to complete vapor condensation), the speed of the liquid at the bubble's interface tends to infinity. At some point in time, however, the rate of condensation will not be able to compensate for the bubble's volume reduction, and from this time on, the remaining vapor will begin to be compressed and grow in pressure. When the liquid speed at the interface approaches the speed of sound in the vapor, the liquid will generate compressive shock waves in the vapor directed toward the bubble's center. These waves, focusing and concentrating at the bubble's center, and reflecting from it, induce in the remaining uncondensed vapor/gas core, ultrahigh temperature, pressure and density for a very short period of time ($\sim$ps). If the vapor contains deuterium (D) and/or tritium (T) atoms, and the temperature and vapor density in the highly compressed core near the bubble's center are high enough for a sufficiently long time interval, then D/D (and D/T) fusion may occur in addition to the SL flash. As a result, nuclear particles (i.e., neutrons, protons, He-3 and T nuclei and γ-rays) may be produced. If so, the emitted neutrons and γ-rays will be detected owing to their collisions with the nuclei (protons) in the LS detector.

The region of high pressure, density and temperature produced as a result of the imploding bubble causes the generation of a reflected shock wave radiating outward from the bubble's center. This rarefying shock wave travels in the liquid at about the speed of sound and can be easily detected at the test section walls with pill microphones that are attached there.

F. TIMING OF KEY PARAMETERS

The slaved electronic timing system shown schematically in Fig. 20 ensures an accurate determination of the moment of the liquid's greatest tension (i.e., the maximum negative acoustic pressure) at the acoustic antinode on the flask axis. This time instant (see Fig. 21) may be taken as the starting point for the time reference ($t = 0$). Just before this time instant (i.e., at $t \sim -2\,\mu s$), a neutron "burst" was initiated in the PNG. Some of the neutrons from this short isotropic burst make it to the acoustic antinode. When analyzing the time spectrum of the neutrons from the PNG, it was found that most of the PNG neutrons were emitted over the interval $t \approx -2$ to $10\,\mu s$ (which corresponded to a pulse FWHM of $\sim 6\,\mu s$). The pressure amplitude at the antinode was more negative than that required for nucleation; hence, because of the PNG pulse bubble nucleation could occur at $t = -2\,\mu s$ and persist for $t > 0$, (i.e., somewhat after the minimum liquid pressure was reached). Nevertheless, the effect of the PNG neutrons was minimal by the time ($\sim 27 \pm 3\,\mu s$) that D/D neutrons were produced during implosion of the bubbles. In any event, as shown in Fig. 21, a cluster of cavitation microbubbles was nucleated by the PNG neutrons in the tensioned liquids at the acoustic antinode. As discussed previously, Fig. 21 shows that these microbubbles grow rapidly due to the negative acoustic pressure,[6] and their maximum radius reached $R_m \sim 500$–$800\,\mu m$. When the acoustically forced liquid pressure around the bubble cluster became positive, the bubbles imploded eventually causing adiabatic compression heating of the uncondensed vapor remaining in the bubbles. During the implosion of the various bubbles, SL light flashes are emitted from the cluster of bubbles. The initial SL flash due

---

[6] As shown schematically in Fig. 21 (upper figure; dashed line), right after the bubbles are nucleated the liquid pressure surrounding them rapidly approaches the saturation pressure, after which the pressure surrounding the interior bubbles increases significantly due to bubble cluster dynamics Ref [98–100,114]. This important phenomena is discussed in more depth in Appendix-A.

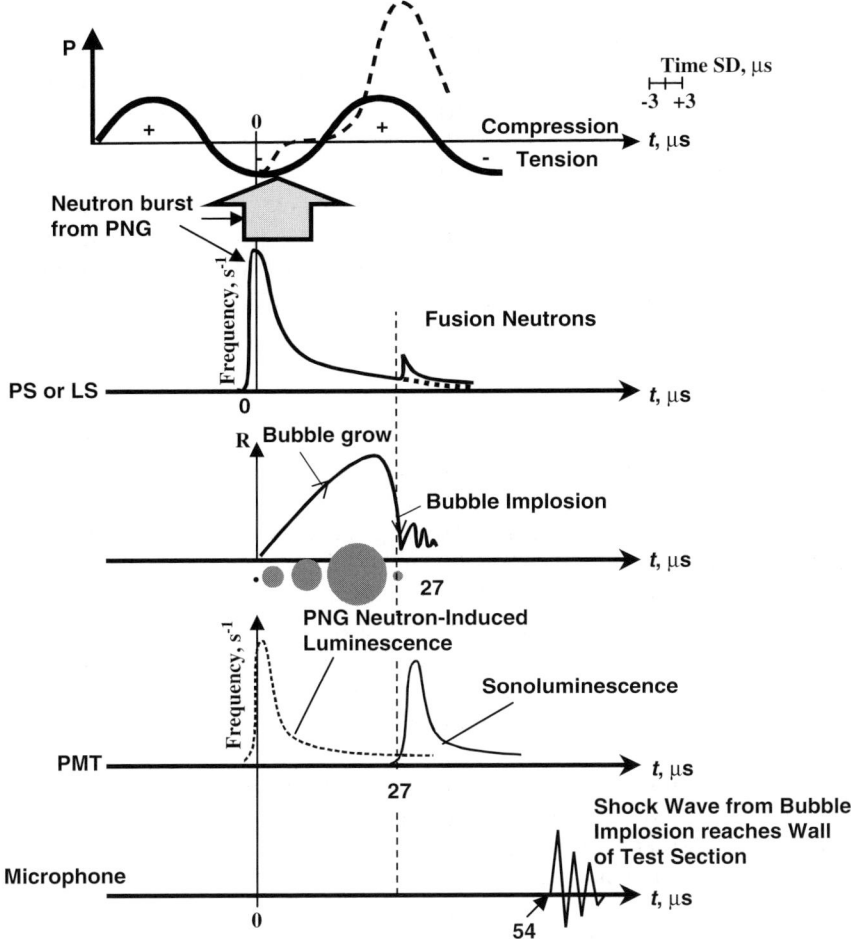

FIG. 21. Schematic sequence of events during the initial implosion in the ORNL experiments.

to the first bubble to collapse took place at $t_{1C} = 27 \pm 3\,\mu s$ (see Fig. 21). Other SL flashes were observed during the subsequent 15 μs interval, with the highest intensity within the first 5 μs.

Separate measurements using a different detection system [1 0 1] gave inconclusive results for nuclear emissions due to the detector used [55], nevertheless they showed secondary SL signals being emitted much later and over a much longer time span (i.e., 500–2500 μs). This implies that there were subsequent acoustically driven bubble cluster implosions. Subsequent measurements done at ORNL

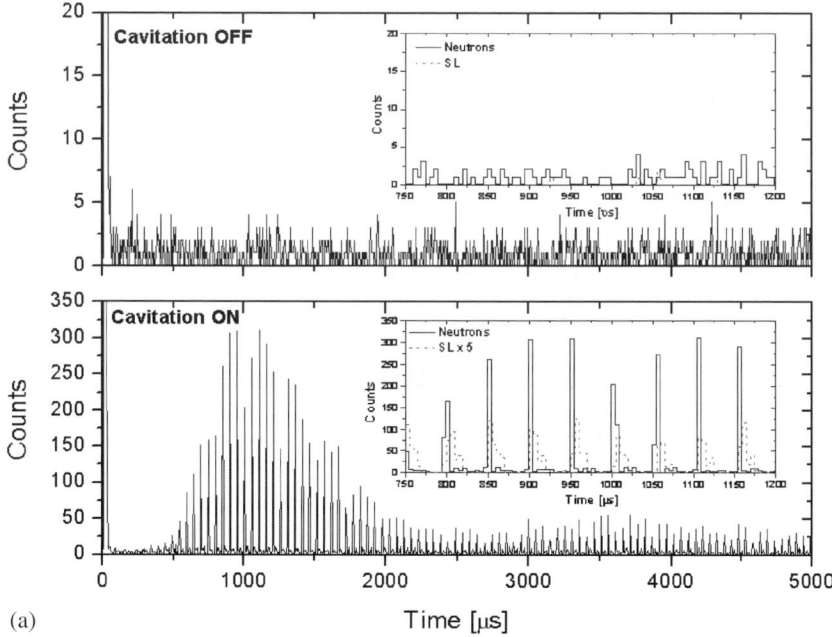

FIG. 22a. Data for chilled (0°C) irradiated D-acetone (C_3D_6O) without and with cavitation [56].

[56] confirmed these data trends and showed that 2.45 MeV D/D neutrons were also emitted well after the initial bubble cluster implosion. These cavitation ON data are shown in Fig. 22a.

As shown schematically in Figs. 1 and 21, bubble cluster implosions result in a shock pulse being reflected from the bubble's center; the shock wave in the liquid reaching the flask wall at $t_{1W} = 54 \pm 3\,\mu s$ (see Fig. 21). This pressure pulse was recorded by two pill microphones installed on diametrically opposite sides of the test section wall. The time for the shock wave to travel from the center of the chamber to its walls (~32 mm away) is $t_{1W} - t_{1C} \approx 27\,\mu s$, which is in agreement with the fact that a shock wave travels in the liquid (acetone) at about the speed of sound, $C_L \approx 1190\,m/s$.

Because the high energy PNG neutrons did not always reach the pressure antinode in the test section, the production rate of bubble clusters was several times less than the frequency of the neutron pulses from the PNG, and bubble cluster implosions varied from ~30 to 50 implosions per second, depending on PZT excitation tuning. Moreover, as discussed in Appendix B, subsequent to each bubble cluster implosion, shock wave interaction with the flask walls, and the resultant defocusing of the acoustic standing wave, lasted for about

10 acoustic cycles, ~520 μs. During this time the coherent standing wave, and thus the bubble dynamics, were disturbed. This response is seen in Fig. 22a (lower figure).

The time spectrum of the above-mentioned events showed that the initial LS scintillations corresponding to the PNG activation (lasting ~12 μs) was also followed by SL flashes (lasting ~15 μs because of the dynamics of the various bubbles involved), which started ~27 μs later. As noted previously, the PNG neutrons were emitted isotropically (i.e., over a 4π solid angle). Those that went in the direction of the pressure antinode may collide with the atomic nuclei of the apparatus and structures around it, thus losing some of their energy. Some of them reached the scintillator directly and produced successively less frequent scintillations for ~20 μs. Anyway, the primary neutrons from the PNG created a relatively high neutron background for several tens of microseconds. Because of this, we needed to separate the scintillations from the primary (PNG) and secondary neutrons (generated due to D/D fusion). To achieve this, the former were fired for only a short interval of time (~12 μs), and as discussed above, with a frequency about 100 times less than that of the pressure acoustic field (i.e., at 200 Hz). Since the PNG neutron signal was essentially background ~20 μs after the PNG burst, the D/D fusion neutrons were well separated in time from the PNG neutrons, and this continued over the entire 5000 μs interval before the PNG was again fired. This is shown schematically in Fig. 21 and in the data presented in Fig. 22a.

Not all the SL emissions are accompanied with neutron emissions. Since only the highly compressed bubbles inside the bubble cluster achieved thermonuclear fusion conditions. Fig. 22b shows a similar cross plot of γ-ray emissions compared with the 2.45 MeV neutron emissions [102]. We see that, as expected, the γ-ray emissions during bubble implosions are time-delayed, since the D/D fusion process produces 2.45 MeV neutrons (but no γ-rays), which interact with the materials present and in doing so slow down (a process which takes ~10–20 μs) allowing some of the neutrons to be absorbed in the surrounding hydrogen atoms (including those in the (LS) detector) which then become excited, emitting 2.2 MeV prompt γ-rays. In addition, in another sonofusion experiment [179] there were also ~1 MeV γ-rays associated with neutron capture by the chlorine atoms in the C_2Cl_4 which was present.

G. Evidence of Nuclear Fusion

Each D/D fusion event can lead to one of two almost equally probable nuclear reactions. These are

- The production of a 1.01 MeV tritium (T) nucleus and a 3.02 MeV proton.
- The production of a 0.82 MeV helium-3 (^{3}He) nucleus and a 2.45 MeV neutron.

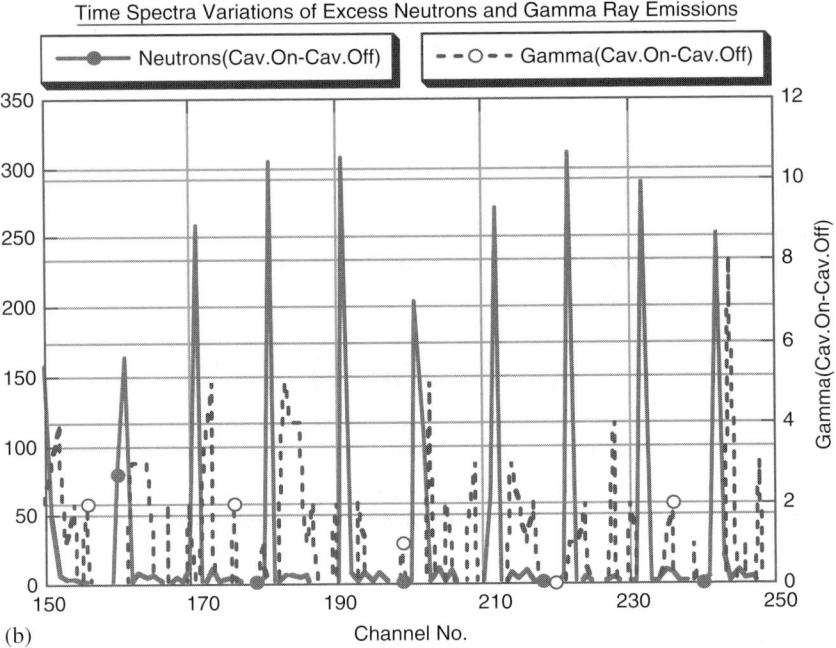

FIG. 22b. Time variation of neutron and gamma emissions during sonofusion [102].

The evidence that D/D fusion occurred during the implosion of cavitation vapor bubbles in chilled D-acetone includes

1. A statistically significant increase in tritium nuclei content.
2. A statistically significant number of scintillations from 2.45 MeV D/D neutrons.
3. An approximately equal number of D/D neutrons and T nuclei produced.
4. The generation of D/D neutrons coincident with SL flashes during bubble cluster implosions and the subsequent emission of 2.2 MeV γ-rays due to D/D neutron capture by the hydrogen (D) atoms in the LS neutron detector and the surrounding structures (e.g., the ice packs and the paraffin shielding blocks).

Significantly, these effects only occurred in chilled, cavitated D-acetone. The use of D-acetone in the absence of any one of the parameters, for example, PNG or Pu–Be neutron irradiation, without cavitation (e.g., no PZT excitation), or in the absence of sufficient liquid pool cooling, did not result in statistically significant nuclear emissions. Moreover, the use of H-acetone never resulted in any nuclear emissions.

H. Production of Tritium

Tritium measurements are an independent way to confirm D/D fusion. To detect tritium (T) production in the liquid used in the ORNL sonofusion experiments, one can measure the tritium decay in a liquid sample taken from the test chamber. This was performed in the ORNL experiments using a Beckman LS 6500 scintillation counter, calibrated to detect 5–18 keV β-ray decay from tritium [54,56].

Before and after the experiments a 1.0 ml liquid sample was withdrawn from the test section and mixed with a 15 ml sample of Ecolite scintillation cocktail. The total T content in the sample was found by measuring the tritium decay rate. Subtracting the T content before the experiment (if any) from that after the experiment we determined the T production in H- and D-acetone at different liquid pool temperatures, both with and without cavitation and irradiation with PNG or Pu–Be neutrons. All other experimental conditions were identical, including placing the acoustic chamber under standard vacuum conditions. In this way, a series of well-controlled experiments were conducted, changing only one parameter at a time. The chamber was initially filled with H-acetone and irradiated with PNG neutrons for 7 h without cavitation, and then the change in T activity was measured. Thereafter, H-acetone experiments with neutron irradiation and cavitation were performed for 7 h and again the change in T activity was measured. The same process was repeated for 12 h. After verifying the absence of T production in the control tests with H-acetone, the experiments were repeated with D-acetone in the same manner without cavitation, and again no measurable increase in T activity was found. Indeed, all the experiments with H-acetone and the experiments with D-acetone, which were irradiated with the PNG fast neutrons, but had no cavitation, showed that irradiation of H and D nuclei with fast neutrons did not result in statistically significant T production.

The PNG irradiation and cavitation experiments of 7 and 12 h duration with H- and D-acetone were repeated several times at a liquid pool temperature of $T_0 \sim 0\,^\circ\text{C}$. Tests were also conducted to assess the impact of liquid pool temperature on T activity build-up by testing with D-acetone at room temperature ($T_0 \sim 22\,^\circ\text{C}$). A separate test was also conducted over 5 h, using a Pu–Be source producing primary neutrons with constant intensity, constant energy spectrum and in an uninterrupted fashion (i.e. not pulsed as in the case with the PNG). Tritium production was only found for the case of chilled (0 °C), irradiated and cavitated D-acetone.

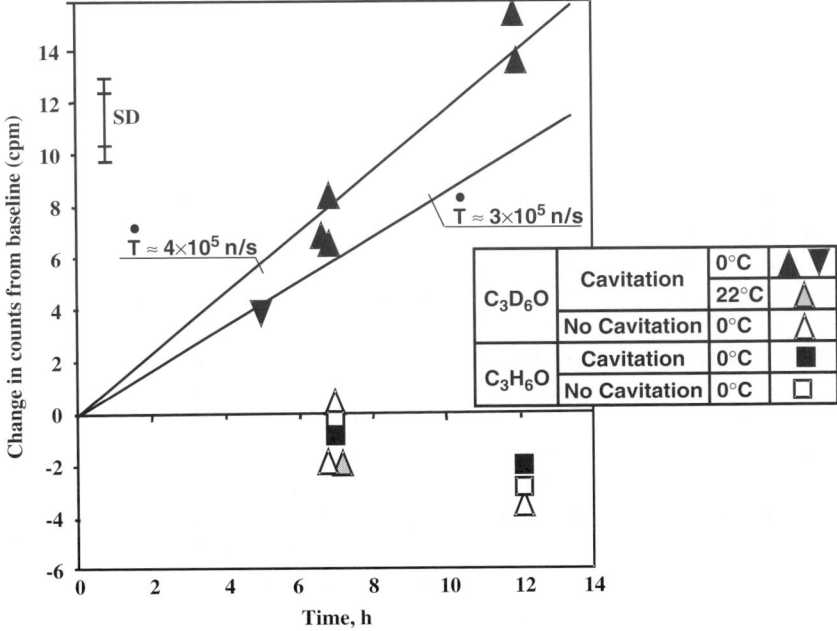

FIG. 23. Tritium measurements [90].

The results of these experiments are summarized in Fig. 23, which also includes the values of standard deviation (SD). The background (i.e., initial) values of T radioactivity because of detector resolution and some T constituent in the acetone were characterized by a scintillation frequency equal to 17 ± 1.3 counts per minute (cpm), and by 53.3 ± 2.3 cpm in D-acetone (where because of the deuteration process, the T concentration is higher than in H-acetone). These measurements revealed no significant change in T activity for H-acetone under PNG neutron irradiation, with or without cavitation. Similarly, for the same experimental conditions, irradiation of D-acetone samples with 14.1 MeV neutrons from the PNG, or with neutrons from a Pu–Be source, without cavitation, did not result in any statistically significant change in T content. It should be stressed that no appreciable T was produced when uncavitated D-acetone were irradiated with fast neutrons. In contrast, in the experiments with cavitated D-acetone at $T_0 \sim 0°C$ and pulsed (from the PNG) or uninterrupted (Pu–Be) neutron irradiation, a measurable increase in T activity was detected. In particular, as can be seen in Fig. 23, the 5, 7 and 12 h experiments revealed T production, which was directly proportional to the duration of the test. The accuracy of the T content measurements at these times was ~ 1, 2.5 and 4.5 SD, respectively.

Finally, the same experiment with D-acetone with cavitation and irradiation from the PNG, over 7 h, but at ~22 C liquid pool temperature, did not reveal any significant change in T activity. As will be discussed subsequently, this apparent paradox agrees with the results of hydrodynamic shock code simulations of the process. Indeed, these calculations showed a lower speed of the surrounding fluid toward the center of the imploding bubble, and thus less compression, because of a lower condensation rate at higher pool temperature.

The experiment with a Pu–Be source of neutrons showed (see the experimental point in Fig. 23 corresponding to 5 h) that the continuous production of neutrons was not as effective (i.e., 3×10^5 vs 4×10^5 n/s) as for pulsed external neutron production. This is presumably due to the fact that the bubble clusters were not always nucleated at the minimum acoustic pressure and thus the D/D neutron production rate varied.

If we assume that none of the resultant T atoms fused with D atoms, an inverse calculation based on the observed T activity showed that the intensity of T nuclei generation due to D/D fusion implies a 2.45 MeV D/D neutron production rate of $\sim 7 \times 10^5$ n/s [54]. Subsequently, more thorough and precise estimates and measurements reduced the value of T activity to [90] the value plotted in Fig. 23

$$Q_T \sim 4 \times 10^5 n/s \qquad (18)$$

I. Production of D/D Neutrons

The neutron intensity was detected by counting the scintillations on the LS detector [54,56]. The counts for each mode were recorded over an interval of 100–300 s. The intensity of scintillations was about 500/s. In this case neutron output varied within $\pm 0.2\%$ for each given condition (D- or H-acetone, with or without cavitation, and at liquid pool temperature, T_0).

The scintillations were counted in various energy ranges. The first range covered the energy range of scintillations between the lowest detectable value and 2.5 MeV. The second was from 2.5 and upward to 14 MeV. Subtracting the number of the scintillations without cavitation from the number of scintillations with cavitation yielded the increase in the neutron counts, ΔN_1 and ΔN_2, in the two energy ranges.

As shown in Fig. 24, for chilled D-acetone ($T_0 \sim 0°C$), but not for H-acetone, or room temperature D-acetone, cavitation results in an increase of ≤ 2.5 MeV neutron scintillations by $\Delta N_1 = 0.037 \times N_1$, (i.e., by 3.7% of the total number of scintillations in this energy range). In the second energy range, there was no statistically significant increase of scintillations.

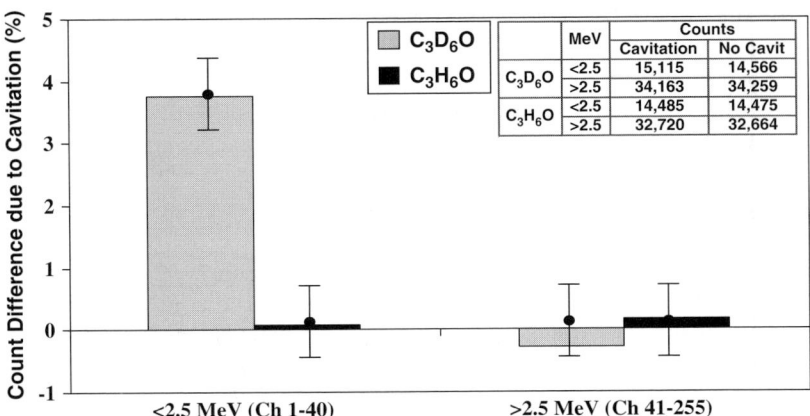

FIG. 24. Emitted neutron energy distribution [54].

These experiments were repeated many times; hence, the change in the number of scintillations by $(3.7 \pm 0.4)\%$ is a statistically significant increase of about 10 SD in comparison with the background. Subsequent experiments [56] increased the statistical accuracy to about 60 SD. This increase of scintillations corresponds to the detection of 5–6 n/s. Considering that the estimation given in Eq. (17) for the net efficiency of the detector for the neutrons emitted from the acetone-filled flask was $(0.5–1) \times 10^{-4}$, the measured intensity of neutron production was [54]

$$Q_n = (5 \text{ to } 6)/(0.5 \text{ to } 1 \times 10^{-4}) = (0.5 \text{ to } 1.2) \times 10^5 \, n/s \qquad (19)$$

These results were later corrected [90] for neutron attenuation to $\sim(3–5) \times 10^5$ n/s.

Subsequent measurements using a more efficient detection system [56] also gave a neutron production rate of $\sim 4 \times 10^5$ n/s. These values are fully consistent with previous data [54] and the neutron generation inferred from the measurements of T production (i.e., Eq. (18), which also gives $\sim 4 \times 10^5$ n/s).

J. Coincidence between Neutron and SL Signals

Let us now consider the coincidence between the SL (i.e., the PMT) and LS (neutron/gamma) pulses.

Three modes were tested for data acquisition and the number of coincidences between the PMT and PS/LS signals [54]. In mode-1, records of such coincident signals were obtained by direct analysis of the oscillograms

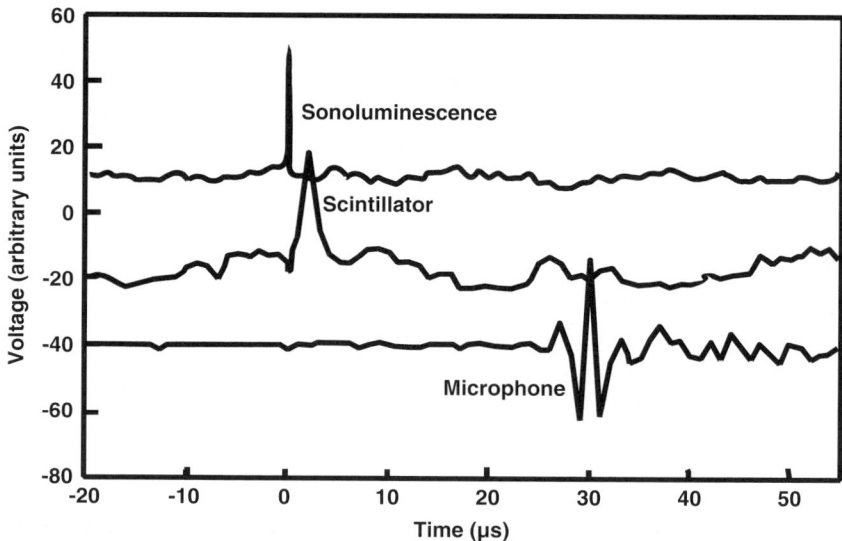

FIG. 25. Conicident light/neutron emissions and subsequent shock wave signals for D-acetone at 0° C [54].

on a digital storage oscilloscope triggered by the SL signal, at a low bias voltage for the PMT (−300 V), which resulted in the elimination of false signals but had reduced PMT sensitivity.[7] In this mode, no SL signals were detected by the PMT during PNG operation (i.e., neutron-induced false SL activity was excluded); however, many of the genuine SL signals due to bubble implosions were also rejected. In this mode of operation, coincidence between SL signals in the PMT and scintillations on the LS repeatedly took place for cavitated D-acetone at $T_0 \sim 0°C$. Figure 25 gives a typical data trace for tests with chilled, cavitated D-acetone, showing coincidence between the SL flash and the neutron pulse, and the subsequent acoustic (shock) signal on the flask wall, as was noted in the discussion of Fig. 21. No such coincidences were seen for tests with H-acetone. However, data acquisition in mode-1 operation (using a 100 MHz four-channel digital storage oscilloscope) was slow, and, as noted previously, many genuine SL signals were not recorded at such a reduced PMT sensitivity.

---

[7] The frequency of luminescent signals in the PMT was ~0.1/s for a bias voltage of −300 V, and these were SL flashes during bubble implosion. For a bias voltage of −450 V the frequency of luminescent signals was 1–5/s, among which about 30% were false signals, being induced by so-called "dark current" [92] and/or the PNG neutrons.

In the second mode of operation, we considered only double coincidences (SL signal in the PMT, and PS or LS pulse) with $\pm 2\,\mu s$ accuracy. The bias voltage to the PMT was -450 V, which resulted in higher sensitivity. As a consequence, the PMT recorded weaker light signals as well, among them not only the SL flashes from bubble implosions, but also light flashes induced by neutrons or γ-ray emissions. Each signal in the PMT triggered scintillation counting in the PS or LS and was binned in $\Delta t = 2\,\mu s$ bins from $t_L - 3\Delta t$ to $t_L + 3\Delta t$, where t_L is the moment of the noted luminescence (in particular, SL) signal in the PMT. The number of scintillations in each of those six time intervals was counted twice; with and without cavitation. The luminescence signals induced by neutrons or γ-rays in the PMT could be random, almost simultaneous, or close in time ($\pm 2\,\mu s$) with the neutron scintillations in the LS and gave a number of false coincidences. The number and intensity of such false, or background, coincidences were determined by the number of scintillations in the above-mentioned intervals relative to the detected luminescence signals in the PMT under PNG operation without cavitation (i.e., without PZT operation), leaving all other parameters the same. The genuine number of coincidences (GNC) with cavitation was found by subtracting the number of false, or background, scintillations taken in each interval, without cavitation, from the corresponding total number of scintillations taken with cavitation.

Figure 26 gives the distribution of the number of such scintillations within time intervals in relation to luminescence signals in the PMT for D- and

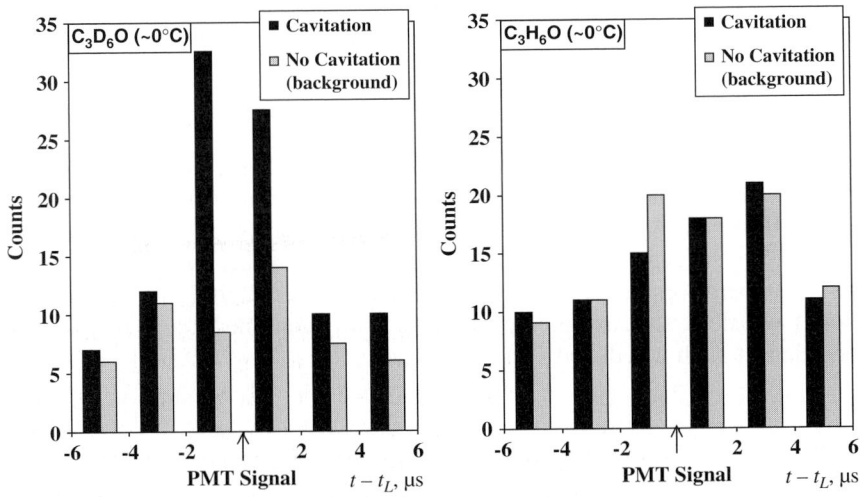

FIG. 26. Coincidence light/neutron time spectrum [54].

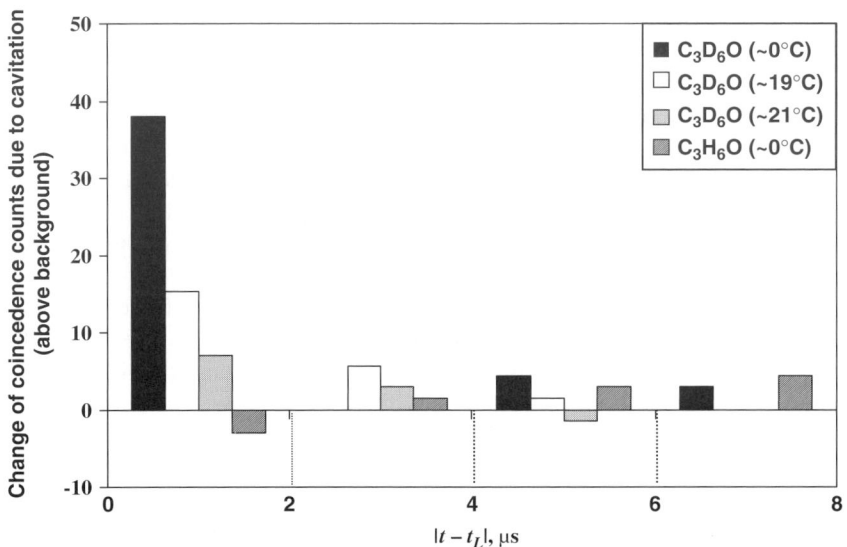

FIG. 27. The effect of pool temperature on light/neutron coincidence distribution [54].

H-acetone, with and without cavitation. Figure 27 gives the GNC above background for D- and H-acetone at different temperatures (including the conditions shown in Fig. 26). It can be seen that only for tests with chilled (i.e., $T_0 \sim 0°C$), cavitated D-acetone and, only within the time interval $t_L \pm 2\,\mu s$, is there a sharp GNC peak. Outside this interval the GNC was within 1 SD, and thus was not statistically significant. No such GNC peaking was seen for H-acetone. Similar data for D-acetone were obtained at $T_0 \sim 19$ and $\sim 21°C$, and at these liquid pool temperatures no significant GNC were seen. That is, as in the case of the measurements for T production, these data show that no measurable neutron production took place in room temperature-cavitated D-acetone during an SL flash emission due to bubble implosion.

The time required to obtain 100 coincidences was ~ 30 min on the average. The SD was calculated by extracting the square root of the sum of the counts in each time interval (2 μs) during bubble implosion and SL light emission. Using MCA time spectrum data, the instantaneous scintillation frequency from neutrons and γ-rays during bubble implosion and SL light emission was calculated, and it was about 1–50/s [54]. For the time window of 20 μs, the frequency of SL signals was about one flash per second, and the counting period of coincidences equal to ~ 1800 s, thus the number of random coincidences was estimated to be negligibly small: [(20 × 10^{-6}) × (1–50) × 1 × 1800] ~ 0.036–1.8.

In mode-2 operation, data were obtained in a two-channel 500 MHz oscilloscope. Simultaneous time spectra data were not possible for the SL and scintillator signals; however, these were measured separately under identical experimental conditions. These series of experiments revealed insignificant deviations in the nuclear and SL signal counts. The data were then used to estimate random coincidences. It was found that coincidences occurring during the time of PNG operation were all random. However, as discussed above, the random coincidences during bubble implosions appear to be insignificant.

In mode-3 operation, using an MCA the time history of the number of nuclear scintillations (due to neutrons and γ-rays) during the first 100 µs after initiation of a PNG pulse (at $t = 0.0$ µs) was measured [54]. Figure 28 shows that for the case of chilled, cavitated D-acetone, an increase in the number of these scintillations falls in the time interval from $t_{1C} = 27 \pm 2$ to $t \approx 42$ µs (i.e., channels 96–130), where the SL signals appear to be intense during bubble implosion. No increase was seen with H-acetone and room temperature D-acetone. This increase is statistically significant (about 5 SD). In this case, the background (without cavitation) values of the number of scintillations that are found in channels 70–255, and correspond to the time after the completion of a PNG neutron pulse, were 198 ± 3.

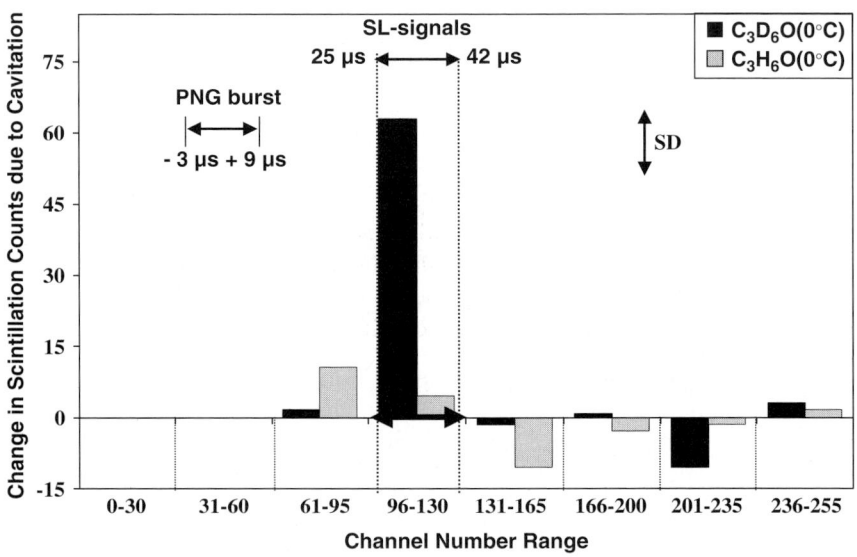

FIG. 28. Expanded time distribution for nuclear and SL signals [54].

IV. Analytical Modeling

A. ANALYSIS OF BUBBLE DYNAMICS

As can be seen in Fig. 21, since they are in a highly superheat liquid and exposed to a negative pressure field (e.g., -15 bar), the cavitation bubbles start to grow rapidly. The bubbles are filled with vapor whose content changes with time due to evaporation/condensation kinetics. Bubble growth is slow relative to the speed of sound in vapor. Thus a low Mach number model, the extended Rayleigh equation, Eq. (1), may be used to describe the liquid and vapor motion during bubble growth [86,87,103]. As noted previously, the pressure inside the bubble is essentially uniform, although the temperature is not, and an incompressible model for the liquid adjacent to the interface may be used. The bubbles grow until increasing pressure in the liquid during the compression phase of the acoustic cycle arrests bubble growth and causes them to contract. At the beginning of bubble collapse, the velocity of the vapor/liquid interface is still small relative to the local speed of sound in vapor and the low Mach number model remains valid.

To model the latter stages of bubble implosion (and bubble rebound), a high Mach number, hydrodynamic shock code (i.e., HYDRO code) model, based on the full set of fluid dynamics conservation equations must be employed. This model can be derived using the "bubble-in-cell" model, which has been successfully used in many prior studies of bubble dynamics [86,87,104]. According to this scheme the entire region is divided into three zones, namely: (1) $0 \le r \le R(t)$ – the gas/vapor zone, where r is the radial coordinate and $R(t)$ the bubble radius; (2) $R(t) \le r \le R_\ell(t)$ – the liquid boundary layer zone, filled with a compressible liquid, whose external radius $R_\ell(t)$ is larger than the bubble radii, but much smaller than the characteristic size of the acoustic chamber; (3) $R_\ell(t) \le r \le R_w$ – the liquid zone between the outer part of the cell (i.e., boundary layer zone) and the acoustic chamber wall (R_w). In this model, the radial velocity of the cell, $\dot{R}_\ell(t)$, is always small with respect to the speed of sound in the liquid. Therefore, the dynamics of $R_\ell(t)$ can be described by the generalized Rayleigh–Plesset equation, Eq. (1). Hence, it is necessary to conduct detailed calculations only in zones (1) and (2).

During bubble expansion, evaporation takes place keeping the vapor in the bubble almost at its saturation state. This leads to an almost constant vapor pressure. During the contraction stage, the vapor starts to condense. It is important to note that during bubble collapse non-equilibrium condensation takes place due to the thermal inertia. The faster the bubble implosion is, the farther the vapor is from thermodynamic equilibrium with the liquid, and higher the pressure and temperature of the vapor will be.

When the temperature on the vapor side of the interface reaches the thermodynamic critical temperature, condensation ceases. From this moment on the bubble contracts as if it were filled with a "non-condensable vapor." This phenomenon has been previously modeled and discussed by Akhatov et al. [104]. Hence, careful modeling of the evaporation/condensation process is important because it controls the amount of vapor, which is left in a collapsing bubble and will be involved in the subsequent compression process.

During implosion of the bubble, along with an almost adiabatic compression in the bubble's interior, a shock wave, converging to the bubble's center, may be generated. This shock wave ultimately compresses and heats a very small central part of the bubble. Vapor molecules in this region begin to dissociate, ionize and create a two-component, two-temperature fluid of ions and electrons, giving rise to plasma interactions.

The compression rate of the plasma is very high and the ion temperature increases rapidly. In contrast, as will be discussed in more detail subsequently, the electrons, which when ionization occurs, have essentially the same initial velocity as the ions, but much less mass, do not have enough time to exchange energy with the ions, and therefore stay relatively "cold" and thus have negligible impact on the energy and momentum of the plasma. In particular, the well-known energy loss mechanisms [105] associated with the electrons (i.e., Bremsstrahlung, line losses, recombination losses, etc.) are relatively small compared to those in laser-induced inertia confinement experiments [138]. Finally, the ion temperature and density reach conditions that are suitable for D/D thermonuclear fusion in a very small zone near the center of the collapsed bubble during a very short time interval. The neutron yield due to these thermonuclear fusion reactions may be calculated using the predicted, local, instantaneous thermal-hydraulic conditions and a suitable neutron kinetics model [105], Eq. (89).

It should be noted here that a realistic equation of state for the fluid, which accounts for the impact of vapor dissociation and ionization and liquid dissociation on the thermodynamic properties, is crucial in the theoretical predictions of experimentally observed sonofusion phenomena. Thus, we begin a discussion of the analytical model with a detailed description of the equation of state.

B. Equations of State

1. *Equations of State for the Low Mach Number Period of Bubble Dynamics*

During the low Mach number period of bubble dynamics, when the velocity of the interface is much less than the local speed of sound in vapor, the vapor density and pressure are not very high, and the vapor parameters satisfy the ideal gas equation of state

$$\varepsilon_v = c_v T_v, \quad p_v = \rho_v R_v T_v, \quad \gamma_v = \frac{(c_v + R_v)}{c_v} \qquad (20)$$

where ρ_v, p_v, T_v, ε_v, c_v, R_v and γ_v are the density, pressure, temperature, internal energy, heat capacity, gas constant and adiabatic exponent of the vapor, respectively.

The liquid in the region around the bubble during the low Mach number period can be treated as nearly incompressible (i.e., linearly compressible), because its density does not deviate significantly from its initial value. In this case the liquid pressure, p_ℓ, is calculated from the equations of liquid motion and the liquid internal energy is

$$\varepsilon_\ell = c_\ell T_\ell \qquad (21)$$

where T_ℓ, c_ℓ are temperature and heat capacity of the liquid, respectively.

According to thermodynamic matching conditions for the internal energies of the vapor and liquid on the saturation line, the following correction for vapor internal energy can be used [103]

$$\varepsilon_v(\rho_{vS}(T), T) = h_{\ell v}(T) + p_S(T)\left(\frac{1}{\rho_{vS}(T)} - \frac{1}{\rho_{\ell S}(T)}\right) + \varepsilon_\ell(\rho_{\ell S}(T), T) \qquad (22)$$

where $\varepsilon_v(\rho_{vS}(T), T)$, $\varepsilon_\ell(\rho_{\ell S}(T), T)$, $\rho_{vS}(T)$ and $\rho_{\ell S}(T)$ are the values of the internal energy and density of vapor and liquid on the saturation line, respectively; $p_S(T)$ is the saturation pressure and $h_{\ell v}(T)$ the latent heat of vaporization.

2. *Equations of State for the High Mach Number Period*

During the high Mach number period, when the velocity of the interface is comparable to, or higher than, the local speed of sound in vapor, the vapor density and pressure during a bubble implosion may be very high, and the ideal gas equation of state, Eq. (20), is no longer valid. The liquid around a bubble is also highly compressed and heated, and the incompressible (or linearly compressible) approximation is no longer valid.

To describe the thermodynamic properties of the liquid, vapor and the condensed matter (i.e., supercritical) phases, the Mie–Grüneisen equation of state can be used [106,107]. In this model, the pressure and the internal energy of a fluid are treated as the sum of potential, or cold, (p_p, ε_p), and thermal, or hot, (p_T, ε_T), components.

$$p = p_p + p_T, \quad \varepsilon = \frac{e}{\rho} - \frac{v^2}{2} = \varepsilon_p + \varepsilon_T + \varepsilon_c \qquad (23)$$

The cold, or potential components (p) quantity the short range atomic forces and are responsible for the elastic properties of the fluid and the hot, thermal components (T) quantify the effect of the oscillations of the atoms in a fluid [103,108–112]. The so-called chemical component (c) quantifies the chemical binding energy.

The thermal components of pressure and internal energy can be written as

$$p_T = \rho \Gamma(\rho) \varepsilon_T, \quad \varepsilon_T = \bar{c}_v T \qquad (24)$$

where it may be assumed that the Grüneisen coefficient, Γ, depends only on fluid density ρ, and the mean heat capacity, $\bar{c}_v$, is a constant.

In general, the internal energy of a fluid is the sum of the energy of the translational motion of the particles (i.e., molecules, atoms, ions and electrons), the rotational and vibrational energies of the molecules, the chemical energy (i.e., the electronic excitation energy and energy of dissociation of the molecules), and the energy of ionization of the various atoms. Excitation of any of the degrees of freedom and the establishment of thermodynamic equilibrium requires a finite time, the so-called relaxation time. Each degree of freedom has its own relaxation time. The above-mentioned degrees of freedom, listed in order from fast to slow, are: translational motion of the particles, rotational motion of the molecules, vibrational motion of the molecules, dissociation of the molecules and ionization of the atoms. When the temperature and density increase during bubble implosion, the relaxation times get shorter for all degrees of freedom.

Let us consider a fluid, which contains molecules of a single type having N atoms. We denote β_d as the dissociation coefficient, which is that portion of the dissociated molecules assuming (for simplicity) that each molecule breaks up into N identical atoms. This dissociation coefficient is a variable according to the appropriate dissociation kinetics. For the density of the fluid the following formulae are useful:

$$\rho = \rho_m + \rho_a = n_m m_m + n_a m_a = n(1 - \beta_d) m_m + n \beta_d N m_a$$

Therefore,

$$\rho = n[m_m + \beta_d(N m_a - m_m)] = n m_m \qquad (25)$$

Here ρ_m, ρ_a are the partial densities of the molecules and atoms; m_m, m_a the masses of the molecules and atoms; n_m, n_a the number densities of the molecules and atoms; and n is the number density of molecules in the non-dissociated state.

The hot (i.e., thermal) components of pressure and internal energy may be estimated using an ideal gas law approximation for the non-dissociated (nd) and dissociated (d) components of the hot gas.

$$p_T = p_T^{(nd)} + p_T^{(d)} = (n_m + n_a)kT = [n(1 - \beta_d) + n\beta_d N]kT \\ = n[1 + (N-1)\beta_d]kT \quad (26)$$

$$\varepsilon_T = \varepsilon_T^{(nd)} + \varepsilon_T^{(d)} = n(1 - \beta_d)\left[\frac{3}{2}kT + (\Delta E)_{rot} + (\Delta E)_{vib}\right] + n\beta_d N\frac{3}{2}kT + n\beta_d E_d \quad (27)$$

Here the expression in brackets in Eq. (27) represents the sum of the energies of the translational, rotational and vibrational motions of a single molecule, and E_d is the energy of dissociation. Shock tube data for liquid acetone [113,139] is given in Fig. 29. Non-dissociated liquid acetone is much stiffer than when it is dissociated, thus less strain energy will be built up in the liquid due to shock wave compression if it is not fully dissociated. This, in turn, implies stronger shocks and more compression of the vapor/plasma in the bubble. We shall return to this point later.

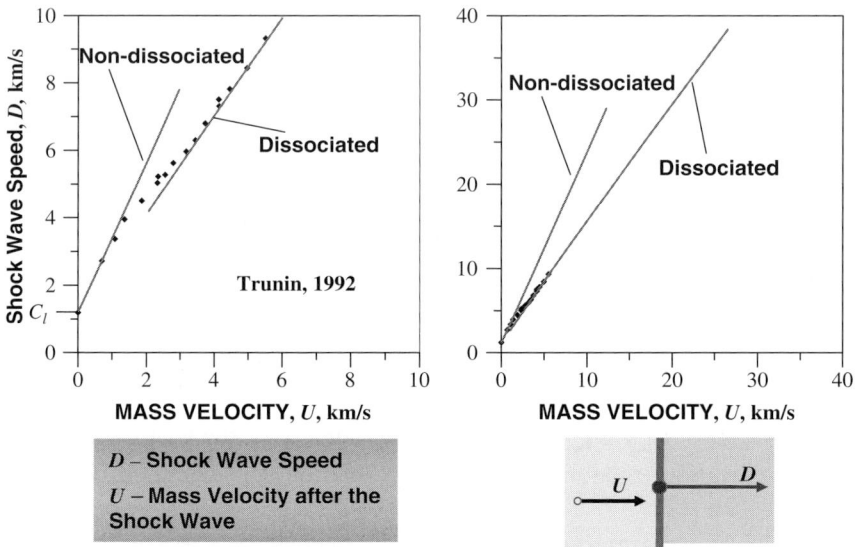

FIG. 29. Shock adiabat (D–U plot) for liquid acetone [139].

Let us next consider the dynamics of the dissociation process. The equilibrium dissociation state is characterized by a coefficient, $\beta_d^{(e)}$, which is a function of density and temperature and may be derived from minimum of the Gibbs free energy function. The exact formula depends on the detailed set of chemical reactions for the dissociation of particular molecule, but it always satisfies the following [112]:

$$\beta_d^{(e)} \sim \frac{1}{\sqrt{\rho}} \exp\left(-\frac{E_d}{2k_b T}\right) \ll 1 \quad (28)$$

Thus, the following simple approximation may be used:

$$\beta_d^{(e)} = \exp\left(-\frac{2T_d}{T}\right) \quad (29)$$

where T_d is the characteristic temperature of dissociation (e.g., $T_d \approx 46{,}500$ K for C_3D_6O).

The dissociation dynamics can be modeled as a first-order relaxation process.

$$\frac{d\beta_d}{dt} = \frac{(\beta_d^{(e)} - \beta_d)}{\tau_d} \quad (30)$$

where the relaxation time of dissociation, τ_d, is determined by the reaction constants of the corresponding dissociation reactions. Alternative models for the dissociation dynamics have also been used [114].

The chemical component of the internal energy of a dissociated (d) and ionized (i) gas is given by:

$$\varepsilon_c = \varepsilon_c^{(d)} + \varepsilon_c^{(i)} \quad (31a)$$

where, for D-acetone (C_3D_6O),

$$\varepsilon_c^{(i)} = \frac{12}{64} R_D T_{D1} f_{D1} + \frac{16}{64} R_0 \sum_{j=1}^{n} T_{0j} f_{0j} + \frac{36}{64} R_c \sum_{j=1}^{6} T_{c_j} f_{c_j} \quad (31b)$$

and $R_D = 4157 \, \text{J/kg} - \text{K}$; $R_0 = 519.6 \, \text{J/kg} - \text{K}$; $R_c = 692.8 \, \text{J/kg} - \text{K}$

The ionization fraction of atom-A, f_{Ak}, can be approximated by

$$f_{Ak} = 0.5\left\{\tanh\left[\frac{5(T - T_{Ak})}{T_{Ak}}\right] + \tanh(5)\right\} \quad (31c)$$

where, for C_3D_6O: Ak = D1, O1, O2, O3, O4, O5, O6, O7, C1, C2, C3, C4, C5, C6 and, recalling that 1 eV = 11,600 K, the ionization energies are

$T_{D1} = 13.60$ eV	$T_{O1} = 13.69$ eV
$T_{C1} = 11.26$ eV	$T_{O2} = 35.19$ eV
$T_{C2} = 23.38$ eV	$T_{O3} = 54.95$ eV
$T_{C3} = 47.89$ eV	$T_{O4} = 77.41$ eV
$T_{C4} = 64.49$ eV	$T_{O5} = 113.9$ eV
$T_{C5} = 392.0$ eV	$T_{O6} = 138.1$ eV
$T_{C6} = 490.0$ eV	$T_{O7} = 739.3$ eV

More details on the kinetics of the dissociation and ionization processes have been given by Nigmatulin et al. [114], where it is shown that for D-acetone (C_3D_6O), the energy required for complete dissociation is 28.2×10^6 J/kg and for complete ionization, 7.663×10^9 J/kg. The implications of these large endothermic energy requirements will be considered later when Fig. 36 is discussed.

A complete account of the phase and structure transitions, taking place in the fluids, must be based on the principles of the mechanics of a multiphase media [103]. Namely, the existence of three atomic states of the fluid should be assumed: the molecular state of liquid and vapor; the dissociated state, consisting of a mixture of dissociated components of molecules, which appear when the temperature is higher than some characteristic dissociation temperature $(T > T_d)$; and, the ionized state, consisting of atomic nuclei, which appear when the temperature exceeds some ionization temperature $(T > T_i)$. Transformation from one atomic state to another happens due to the kinetics of the transition process which, in turn, depends on the temperature of the substance and on the extent of the transition process.

To model the cold components of pressure and internal energy, which describe the elastic interactions between the atoms at zero temperature $(T = 0)$, one may use a Born–Mayer potential, which accounts for the intermolecular attraction–repulsion forces in a condensed medium [103,109,112,114,115–117]:

$$p_p = A\left(\frac{\rho}{\rho_0}\right)^{2/3} exp\left[b\left(1 - \left(\frac{\rho_0}{\rho}\right)^{1/3}\right)\right] + E\left(\frac{\rho}{\rho_0}\right)^{m+1} - K\left(\frac{\rho}{\rho_0}\right)^{n+1} + \Delta p_p \tag{32a}$$

$$\varepsilon_p = \frac{3A}{\rho_0 b} exp\left[b\left(1 - \left(\frac{\rho_0}{\rho}\right)^{1/3}\right)\right] + \frac{E}{m\rho_0}\left(\frac{\rho}{\rho_0}\right)^m - \frac{K}{n\rho_0}\left(\frac{\rho}{\rho_0}\right)^n + \Delta\varepsilon_p + \varepsilon^0 \tag{32b}$$

Here A, K, E, b, n and m are fluid-specific constants, and $\Delta\varepsilon_p$ is an integration constant of potential energy, chosen to satisfy the Gibbs equation at constant entropy. That is

$$\varepsilon_p(\rho) = \int_{\rho_*}^{\rho} \frac{p_p(\rho)}{\rho^2}\, d\rho \qquad (33)$$

where ρ_* is the fluid density at which the cold pressure is zero ($p_p(\rho_*) = 0$), and a minimum value of potential energy (i.e., ε^0 is chosen so that $\varepsilon_p(\rho_*) = 0$) are achieved. These constants must be evaluated using shock adiabat (i.e., Hugoniot) data for the fluid of interest [114]. In particular, Fig. 29 gives shock tube data for acetone [113]. It is interesting to note that if $A = 0$, we obtain the well known Lennard–Jones potential.

The Born–Mayer potential was first proposed for the description of the elastic behavior of metals, ion crystals and condensed media in shock waves [109]. For strong compressions the behavior of liquids differs only slightly from solids, and therefore, a Born–Mayer potential is also applicable [116].

The necessity for a unified modeling of a fluid in both the vapor and liquid phases appears in a wide range of problems, related to vapor/liquid systems. That is why one must extend the applicability of the Born–Mayer potential into the range of high specific volumes. For example, Cowperthwaite and Zwisler [117] matched the liquid equation of state with experimental isotherms of the vapor phase.

At present there are no theoretical models for the elastic potentials of complex organic liquids, like acetone. Indeed, in order to predict adequately the behavior of a fluid under strong compression (liquid or vapor phases) as well as at large specific volumes (vapor phase), one must use appropriate experimental data to evaluate the parameters in Eqs. (32a) and (32b).

Interpolation of the Grüneisen coefficient as a function of density can be made to better fit the experimental data on the shock adiabat and isothermal compressibility of liquid phase, and to fit the isotherms of vapor phase in the range of large specific volumes.

Figure 30 presents the resultant equation of state (EOS) for D-acetone. In particular, the isotherms and Grüneisen function (Γ) for both non-dissociated and dissociated D-acetone are shown.

Let us next turn our attention to the analysis of the bubble dynamics associated with sonofusion. In the high Mach number calculations the equation of state discussed above plays a crucial role.

C. Low Mach Number Bubble Dynamics

To decrease computational requirements a homobaric approximation for the vapor inside the bubble and an incompressibility assumption for the

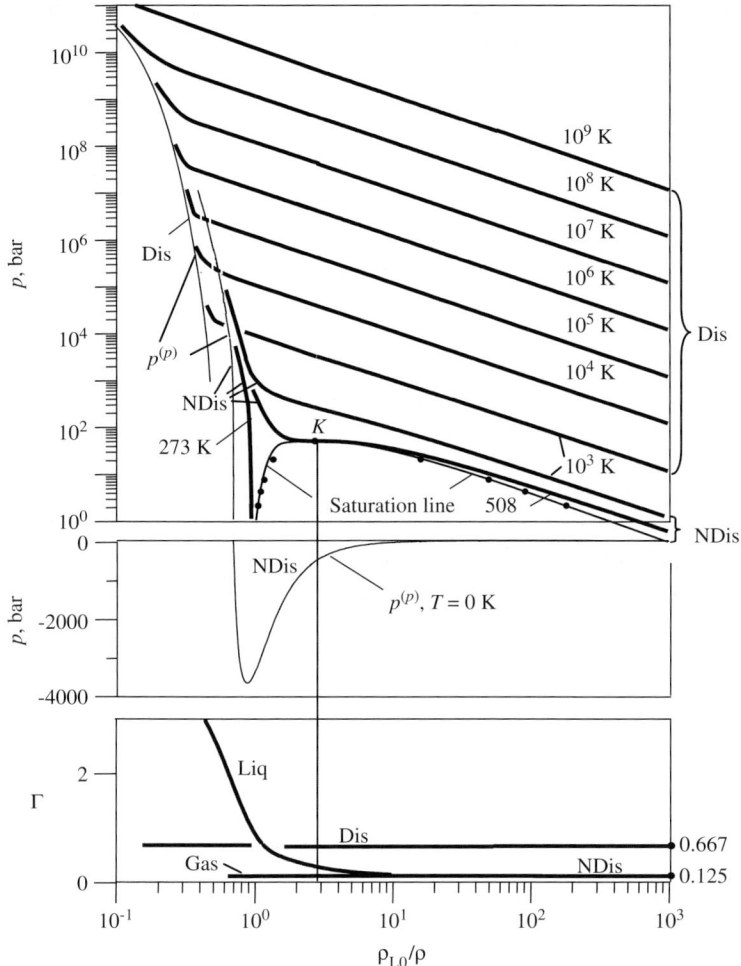

FIG. 30. EOS – Isotherms, Grüneisen Function (Γ) for dissociated (Dis) and non-dissociated (NDis) D-acetone.

surrounding liquid may be used when the velocity in the vapor, v_v, is much smaller than the speed of sound in the vapor, C_v. These assumptions and approximations have been verified using HYDRO code simulations [114]. We explicitly use this scheme when $|v_v|/C_v \leq M_{v,\,crit} \approx 0.1$. Significantly, it has been shown numerically that the results are not sensitive to slight changes in the threshold, $M_{v,\,crit}$.

In a homobaric model, the pressure inside the bubble is spatially uniform, but time-dependent, while the temperature and velocity of the vapor are functions of time and space. A mathematical model describing the low Mach number period of bubble dynamics was proposed by Nigmatulin and Khabeev [118] and Prosperetti et al. [119]:

$$\frac{\gamma p_v}{(\gamma-1)T_v}\left(\frac{\partial T_v}{\partial t}+v_v\frac{\partial T_v}{\partial r}\right)=\frac{1}{r^2}\frac{\partial}{\partial r}\left(k_v r^2\frac{\partial T_v}{\partial r}\right)+\frac{dp_v}{dt} \qquad (34)$$

$$\frac{dp_v}{dt}=-\frac{3\gamma p_v}{a}v_v+\frac{3(\gamma-1)}{a}k_v\frac{\partial T_v}{\partial r}\bigg|_{r=a} \qquad (35)$$

$$v_v=\frac{(\gamma-1)}{\gamma p_v}k_v\frac{\partial T_v}{\partial r}-\frac{r}{3\gamma p_v}\frac{dp_v}{dt} \qquad (36)$$

Here p_v, T_v and v_v are pressure, temperature and radial velocity of the vapor, respectively; k_v and γ are the thermal conductivity and the adiabatic exponent of the vapor, respectively. The formulae in Eqs. (34)–(36) have been derived, assuming that the vapor parameters satisfy an ideal gas equation of state.

The liquid in the region around the bubble can be treated as nearly incompressible because its density does not deviate significantly from its initial value. In this case, the Bernoulli equation [120], derived from the liquid equation of motion in spherical geometry, can be employed to calculate the pressure distribution in liquid:

$$p_\ell = p_\ell\big|_{r=R} - \rho_{\ell 0}\left(R\frac{dv_{\ell i}}{dt}+2\frac{dR}{dt}v_{\ell i}-\frac{v_{\ell i}^2}{2}\right)+\frac{\rho_{\ell 0}}{r}\frac{d}{dt}\left(R^2 v_{\ell i}\right)-\frac{\rho_{\ell 0}R^4}{2r^4}v_{\ell i}^2 \qquad (37)$$

where $\rho_{\ell 0}$ is the initial liquid density, and p_ℓ and v_ℓ are liquid pressure and radial velocity, respectively, and $R(t)$ is the bubble's radius. The liquid velocity distribution in the cell region surrounding the bubble is taken as the exact solution of the liquid continuity equation for an incompressible liquid.

$$v_\ell = \frac{v_{\ell i}R^2}{r^2} \quad (r \geq R) \qquad (38)$$

The temperature distribution around the bubble, $T_\ell(r,t)$, is calculated using the energy equation

$$c_\ell \rho_{\ell 0}\left(\frac{\partial T_\ell}{\partial t}+v_\ell\frac{\partial T_\ell}{\partial r}\right)=\frac{1}{r^2}\frac{\partial}{\partial r}\left(k_\ell r^2\frac{\partial T_\ell}{\partial r}\right) \qquad (39)$$

where c_ℓ is the heat capacity of the liquid.

The bubble radius, $R(t)$, is determined by evaluating the generalized Rayleigh–Plesset equation, Eq. (1), which takes into account acoustic radiation by the bubble due to liquid compressibility and the effect of mass transfer through the bubble wall due to evaporation/condensation [51,103,104,124]. This equation can be rewritten as

$$R\frac{dv_{\ell i}}{dt} + \frac{3}{2}v_{\ell i}^2 + 2\frac{v_{\ell i}\dot{m}_v''}{\rho_{\ell 0}} = \frac{(p_{\ell i} - p_I)}{\rho_{\ell 0}} + \frac{R}{\rho_{\ell 0} C_\ell}\frac{d}{dt}(p_{\ell i} - p_I) \quad (40)$$

The liquid and vapor velocities at the bubble's interface and the radial velocity of the interface itself differ due to phase transition (i.e., evaporation/condensation). This fact is taken into account by the following mass jump conditions at the interface:

$$v_{vi} = \dot{R} - \frac{\dot{m}_v''}{\rho_v}, \quad v_{\ell i} = \dot{R} - \frac{\dot{m}_v''}{\rho_\ell} \quad (41)$$

where $\dot{m}_v''$ is the evaporation flux, which is negative if condensation takes place. Similarly, matching of the vapor and liquid pressure at the interface is done by using the following momentum jump condition:

$$p_{vi} = p_{\ell i} + \frac{2\sigma}{R} + \frac{4\mu_\ell v_{\ell i}}{R} \quad (42)$$

where σ is the surface tension, and μ_ℓ the dynamic viscosity of the liquid.

1. *Evaporation and Condensation Kinetics*

To describe evaporation/condensation processes at the interface (i.e., at $r = R$) the following energy jump condition is used:

$$k_\ell \frac{\partial T_\ell}{\partial r}\bigg|_{r=R} - k_v \frac{\partial T_v}{\partial r}\bigg|_{r=R} = \dot{m}_v'' h_{\ell v}(p_{vi}) \quad (43)$$

The thermal conductivity coefficients for the vapor and liquid, k_v and k_ℓ, respectively, and the latent heat, $h_{\ell v}(p_{vi})$, are determined from appropriate experimental data.

Analysis of the transport process of the vapor molecules at a liquid/vapor interface is based on the Hertz–Knudsen–Langmuir formula for phase transition dynamics [103,121,122]:

$$\dot{m}_v'' = \frac{\alpha}{\sqrt{2\pi R_v}}\left(\frac{p_S(T_{\ell i})}{\sqrt{T_{\ell i}}} - \frac{\chi_v p_{vi}}{\sqrt{T_{vi}}}\right) \quad (44)$$

where α is the so-called accommodation coefficient (i.e., the phase change coefficient), which for condensation defines the portion of the vapor molecules hitting the liquid/vapor interface that are absorbed; the remaining portion, $(1 - \alpha)$, being reflected. For instance, if $\alpha = 1$, then all the molecules hitting the interface are absorbed (i.e., condensed). Also, p_S is the vapor saturation pressure which is a known function of liquid temperature at the liquid/vapor interface, $T_{\ell i}$, and p_{vi} and T_{vi} are the actual vapor pressure and temperature at the liquid/vapor interface.

The correction factor, χ_v, results from the net motion of the vapor toward the surface, which is superimposed on an assumed Maxwellian velocity distribution of the vapor molecules. It is given by

$$\chi_v = \exp(-\Omega^2) - \Omega\sqrt{\pi}\left(1 - \frac{2}{\sqrt{\pi}}\int_0^{\Omega} \exp(-x^2)\,dx\right) \quad (45a)$$

where

$$\Omega = \frac{\dot{m}_v''}{\sqrt{2}p_{vi}}\sqrt{R_v T_{vi}} = \frac{\dot{R} - v_{vi}}{\sqrt{2R_v T_{vi}}} \quad (45b)$$

In general, the phasic temperatures on the interface boundary may undergo a jump

$$T_{\ell i} - T_{vi} \equiv [T] \quad (46)$$

where $[T]$ denotes the temperature jump. Analysis of the transfer processes in a thin Knudsen layer of vapor (having a thickness of a few mean free paths) leads to [103]

$$[T] = \frac{0.45\,\dot{m}_v'' T_S(p_{vi})}{\sqrt{2R_v T_S(p_{vi})}\rho_{vi}} \quad (47)$$

where ρ_{vi} is the vapor density at the interface.

In general, Eqs. (44)–(47) do not provide an explicit formula for the evaporation flux, $\dot{m}_v''$, and one must solve this coupled non-linear system of equations. This would be a computationally time-consuming procedure. Fortunately, the value of the temperature jump does not strongly influence the solution of the problem, and thus one may assume:

$$T_{\ell i} = T_{vi} = T_i \quad (48)$$

where T_i denotes the temperature of the interface.

Also, the parameter Ω, which represents the ratio of the relative speed of the vapor, $v_{vi} - \dot{R}$, to a characteristic molecular velocity, $\sqrt{2R_v T_v|_{r=R}}$, is normally quite small. So, instead of Eqs. (45a) and (45b) an approximation for $\Omega \ll 1$ can be used.

$$\chi_v \cong 1 - \sqrt{\pi\Omega} \qquad (49)$$

Substituting Eq. (49) into Eqs. (45a) and (44) yields

$$\dot{m}''_v = \frac{\alpha}{\sqrt{2\pi R_v T_i}} \left[p_s(T_i) - p_v|_{r=R} \right] + \frac{\alpha}{2} \dot{m}''_v \qquad (50)$$

or equivalently

$$\dot{m}''_v = f \frac{(p_s(T_i) - p_v|_{r=R})}{\sqrt{2\pi R_v T_i}}, \quad f = \frac{2\alpha}{(2-\alpha)} \qquad (51)$$

Equation (51) represents an explicit formula for the phase change rate, where f is an effective "accommodation coefficient." It is easy to see, that for small α there is no difference between α and f, but if $\alpha = 1$ (which corresponds to a maximum value for the accommodation coefficient), then $f = 2$.

A more detailed derivation for the effective accommodation coefficient accounting for the temperature jump at the interface is given by Kucherov and Rikenglaz [123]. This analysis supports the approximations discussed above.

D. High Mach Number Bubble Dynamics

Many investigators have assumed the validity of spherically symmetric 1D bubble dynamics to study SL, sonochemistry and sonofusion [28–30, 51,114,125]. However, as discussed previously, linear stability analyses show that shape and interfacial instabilities are expected during bubble implosions and rebounds. Fortunately, the 3D DNS of an imploding air bubble has shown [126] that this may not cause a fundamental problem. Indeed, these results imply that the use of 1D hydrodynamic shock code models, which implicitly assume spherical symmetry during bubble dynamics, may be quite sufficient to analyze sonofusion phenomenon. Naturally, this greatly reduces the computational requirements for HYDRO code simulations.

In order to assess the validity of a 1D assumption, let us first consider the DNS of the 3D bubble dynamics of an imploding air bubble. In particular, a Galerkin least-squares finite element method (FEM) was used for the differencing of the various conservation equations, and the associated interfacial jump conditions. The interface was resolved using a level set algorithm [127].

This model was based on extensive experience with Rensselaer's DNS code, PHASTA [128–130]. PHASTA is a 3D FEM code, which is based on a fully unstructured, adaptive grid. The time step is dynamically readjusted to yield a specified accuracy and the computational grid is refined in regions where steep gradients occur, while being less fine in other regions. This enhances accuracy and resolution while significantly reducing computational time. In addition, hierarchical basis functions [131,132] may be used to achieve higher order accuracy (i.e., > third-order resolution) in space.

PHASTA-2C is a digital computer code intended for the analysis of compressible fluids. It can be spatially nodalized down to the local Kolmogorov scale and used to evaluate the Navier–Stokes equation and the corresponding phasic continuity and energy equations, which written in matrix form, are

$$U_{,t} + F^{adv}_{i,i} - F^{diff}_{i,i} = S \qquad (52)$$

where, for each phase, we have

$$U = \begin{Bmatrix} U_1 \\ U_2 \\ U_3 \\ U_4 \\ U_5 \end{Bmatrix} = \rho \begin{Bmatrix} 1 \\ u_1 \\ u_2 \\ u_3 \\ e_{tot} \end{Bmatrix}, \quad F^{adv}_i = u_i U + p \begin{Bmatrix} 0 \\ \delta_{1i} \\ \delta_{2i} \\ \delta_{3i} \\ u_i \end{Bmatrix}, \quad F^{diff}_i = \begin{Bmatrix} 0 \\ \tau_{1i} \\ \tau_{2i} \\ \tau_{3i} \\ \tau_{ij}u_j - q''_i \end{Bmatrix} \qquad (53)$$

and S is a body force (or source) vector, such as gravity and surface tension, and

$$\tau_{ij} = 2\mu \left(S_{ij}(u) - \frac{1}{3} S_{kk}(u)\delta_{ij} \right), \quad S_{ij}(u) = \frac{u_{i,j} + u_{j,i}}{2}$$

$$q''_i = -kT_{,i}, \quad e_{tot} = e + \frac{u_i u_i}{2}, \quad e = c_v T \qquad (54)$$

The state variables are the phasic velocity components, u_i; pressures, p; densities, ρ; temperatures, T; and the total specific energy, e_{tot}. Constitutive laws relate the stress, τ_{ij}, to the deviatoric part of the strain rate tensor, $S^d_{ij} = S_{ij} - \frac{1}{3}S_{kk}\delta_{ij}$, through a molecular viscosity, μ. Similarly, the heat flux, q''_i, is proportional to the gradient of temperature with the proportionality constant given by the thermal conductivity, k. It should be noted that because the spatial grid resolves the local Kolmogorov scales, turbulence modeling is not required. This yields true DNS predictions for each interacting phase and, using a level set model, accounts for the various (continuous/dispersed) deformable interfaces.

Closure of the DNS model is achieved by the use of appropriate equations of state for the liquid and vapor phases, and the interfacial jump conditions, which include surface tension, σ, (e.g., Marangoni effects), which are modeled as an interfacial force density vector

$$\underline{M}_i = \kappa\sigma\delta(\underline{x}-\underline{x}_i)\underline{n}_i + \left[\frac{d\sigma}{dT}|\nabla_s T| + \sum_{j=1}^{N}\frac{d\sigma}{dC_j}|\nabla_s C_j| + \dot{m}''_v(\underline{u}_{vi} - \underline{u}_{\ell i})\right]\delta(\underline{x}-\underline{x}_i)\underline{t}_i \tag{55}$$

where κ is the local instantaneous interfacial curvature, $\nabla_s = (\underline{I} - \underline{n}\,\underline{n}).\nabla$ the gradient operator along the interface, C_j the concentration of component-j (i.e., if we have a multicomponent mixture), $\delta(\underline{x}-\underline{x}_i)$ a Dirac delta function (which is only non-zero at the position of the interface, x_i), and $\underline{n}_i$ and $\underline{t}_i$ are the unit interfacial normal and tangential vectors, respectively.

It is convenient to define the quasi-linear operator associated with Eq. (52) as

$$\boldsymbol{L} \equiv \boldsymbol{A}_0\frac{\partial}{\partial t} + \boldsymbol{A}_i\frac{\partial}{\partial x_i} - \frac{\partial}{\partial x_i}\left(\boldsymbol{K}_{ij}\frac{\partial}{\partial x_j}\right) \tag{56a}$$

which can be decomposed into time, advective and diffusive portions.

$$\boldsymbol{L} = \boldsymbol{L}_t + \boldsymbol{L}_{adv} + \boldsymbol{L}_{diff} \tag{56b}$$

Using this notation, we can rewrite Eq. (52) as

$$\boldsymbol{L}\boldsymbol{Y} = \boldsymbol{S} \tag{57}$$

To derive the finite element form of Eq. (57), the entire equation is dotted with a vector of weight functions, $\boldsymbol{W}$, and integrated over the spatial domain. Integration by parts is then performed to move the spatial derivatives onto the weight functions. This process leads to the following integral equation:

$$\int_W \left(\boldsymbol{W}\cdot\boldsymbol{A}_0\boldsymbol{Y}_{,t} - \boldsymbol{W}_{,i}\cdot\boldsymbol{F}_i^{adv} + \boldsymbol{W}_{,i}\cdot\boldsymbol{F}_i^{diff} + \boldsymbol{W}\cdot\boldsymbol{S}\right)dW - \int_\Gamma \boldsymbol{W}\cdot\left(-\boldsymbol{F}_i^{adv} + \boldsymbol{F}_i^{diff}\right)n_i\,d\Gamma$$
$$+ \sum_{e=1}^{n_{el}}\int_{W^e}\boldsymbol{L}^T\boldsymbol{W}\cdot\tau(\boldsymbol{L}\boldsymbol{Y} - \boldsymbol{S})dW = 0$$

$$\tag{58}$$

where

$$\boldsymbol{Y} = (p, u_1, u_2, u_3, T)^T \tag{59}$$

The first two terms in Eq. (58) contain the Galerkin nodalization (interior and boundary) and the third term contains a least-squares stabilization. It is interesting to note that the well-known SUPG (Streamline Upwind Petrov Galerkin) stabilization can be obtained by replacing L^T with L_{adv}^T. The stabilization matrix, τ, is an important part of this method and it has been well documented by Franca and Frey [133]. The integrals in Eq. (58) are evaluated using Gauss quadratures resulting in a system of non-linear ordinary differential equations, which can be written as

$$M_1 \underline{\dot{Y}} = N_1(\underline{Y}) \tag{60}$$

where the under bar is added to make clear that $\underline{Y}$ is the vector of solution values at discrete points (spatially interpolated with the finite element basis functions). The vapor/liquid interfaces are resolved using a level set method. The level set approach of Sussman et al. [134] and Sethian [127], represents the interface as the zero level set of a smooth function, ϕ, which is the signed distance from the interface. Hence instead of explicitly tracking the interface, we implicitly capture the interface within a field, which is interpolated between the nodes like any other state variable. This enables one to predict the shape of the interface between the two phases accurately and to track its position (including any breakup and coalescence) on a fixed spatial grid.

Since the interface moves with the fluid, the evolution of ϕ is governed by the following transport equation:

$$\frac{D\phi}{Dt} \equiv \frac{\partial \phi}{\partial t} + \underline{u} \cdot \nabla \phi = 0 \tag{61}$$

The fluid properties, for example density, are given by

$$\rho(\phi) = \rho_1 H(\phi) + \rho_2(1 - H(\phi)) \tag{62}$$

where $H(\phi)$ is a smoothed heaviside function [135].

The additional transport equation for the level set scalar is solved in a manner similar to the flow equations; however, appropriate redistancing must be done at each step [134] to account for the non-uniform convection of ϕ and to assure mass conservation. In particular, in order to correct the distance, d, between the level sets, at each time step, Δt, we solve to steady state (i.e., $|\nabla d| = 1$) the following partial differential equation:

$$\frac{\partial d}{\partial \tau} = s(\phi)(1 - |\nabla d|) \tag{63a}$$

where τ is a pseudo-time step and

$$s(\phi) = \begin{Bmatrix} -1, & \text{if } \phi < 0 \\ 0, & \text{if } \phi = 0 \\ +1, & \text{if } \phi > 0 \end{Bmatrix} \qquad (63b)$$

Mass conservation is obtained by another iteration to assure that the interface (i.e., $\phi = 0$) was not moved during the redistancing procedure.

The FEM formulation of Eq. (61) is

$$\int_W (w\phi_{,t} + wu_i\phi_{,i} + wS)\,dW + \sum_{e=1}^{n_{e\ell}} \int_{W^e} L^T w\tau(\phi_{,t} + u_i\phi_{,i} - S)\,dW = 0 \qquad (64)$$

These integrals are also evaluated using Gauss quadratures, resulting in a system of non-linear ordinary differential equations, which can be written as

$$\boldsymbol{M}_2\dot{\phi} = \boldsymbol{N}_2(\phi) \qquad (65)$$

Finally, the system of coupled non-linear ordinary differential equations given by Eqs. (60) and (65) are discretized in time via a generalized α time integrator, resulting in a non-linear system of algebraic equations. This system of equations is then linearized, which yields a set of linear algebraic system of equations, which are solved using Newton's method. Newton's iterations continue until the local residuals are sufficiently small at each time step, after which the method proceeds to the next time step, where the process is started all over again. This DNS code (i.e., PHASTA-2C) has been applied to the 3D analysis of bubbly flows [126] and a ghost fluid numerical technique [136] was employed to improve the algorithm.

Figures 31a–e show a time sequence of the implosion process for an air/water bubble suddenly exposed to 100 atm over-pressure, in which the perfect gas law was used as the equation of state [126]. As seen in Figs. 31c and d, the rapidly imploding air bubble experiences significant shape and interfacial instabilities, but these instabilities do not lead to bubble breakup and tend to wash-out during the re-expansion phase (see Fig. 31e).

Significantly, the pressure field deep within the imploding bubble remains essentially spherical and this leads to high compression near the center of the bubble. Figure 32a shows the transient bubble response in a vertical (y-axis) cut of the 3D bubble simulations. It can be seen that the results for bubble radius agree well with the extended Rayleigh equation, Eq. (1), except during the final stages of the implosion process where the bubble becomes

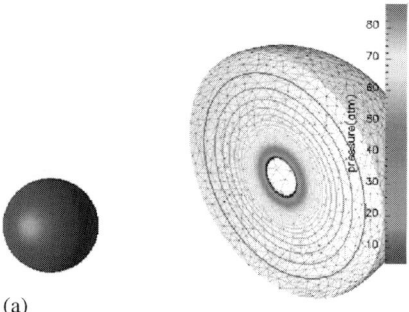

FIG. 31a. An imploding air bubble in room temperature water and subjected to 100 atm overpressure in the early stage of implosion process (t = 0.06 μs) [126].

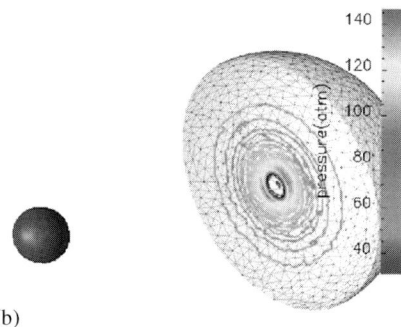

FIG. 31b. Subsonic; ($\dot{R} < C_g$) stage of bubble implosion (t = 0.085 μs) [126].

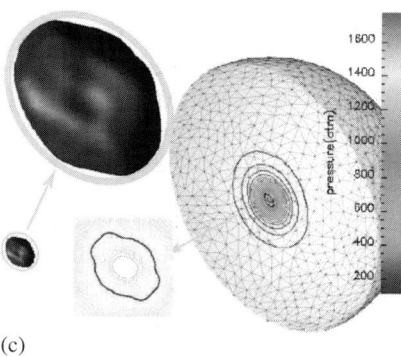

FIG. 31c. High Mach number ($M_\alpha = |\dot{R}|/C_g > 1.0$) stage of bubble implosion (t = 0.093 μs) [126].

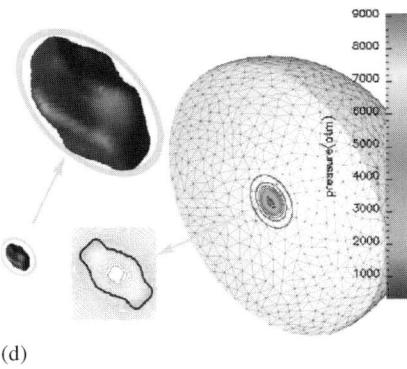

(d)

FIG. 31d. Final stage of bubble implosion (t = 0.095 μs) [126].

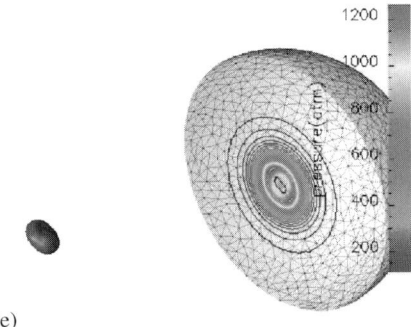

(e)

FIG. 31e. Dynamic conditions during bubble rebound (t = 0.1 μs) [126].

distorted and compressibility in the liquid phase becomes important. The corresponding predicted pressure profiles are also shown in Fig. 32b. A shock wave can be seen to be developing within the gas bubble and this steepening and intensifying pressure front bounces off itself at the center of the bubble resulting in, as can be seen in Figs. 32b and c, relatively high local pressures, and temperatures near the center of the collapsed air bubble.

While these simulations were done for an air/water bubble with a very simple equation of state and no heat losses, they appear to indicate conditions suitable for SL. Moreover, it appears that sonofusion conditions may occur for sufficiently energetic implosions, since, even though the bubble distorts during the implosion process, it may remain intact. Finally, these results show why a 1D HYDRO code can yield realistic sonofusion results (i.e., the initial shock waves formed remains spherical within the bubble even though the bubble severely distorts during an implosion).

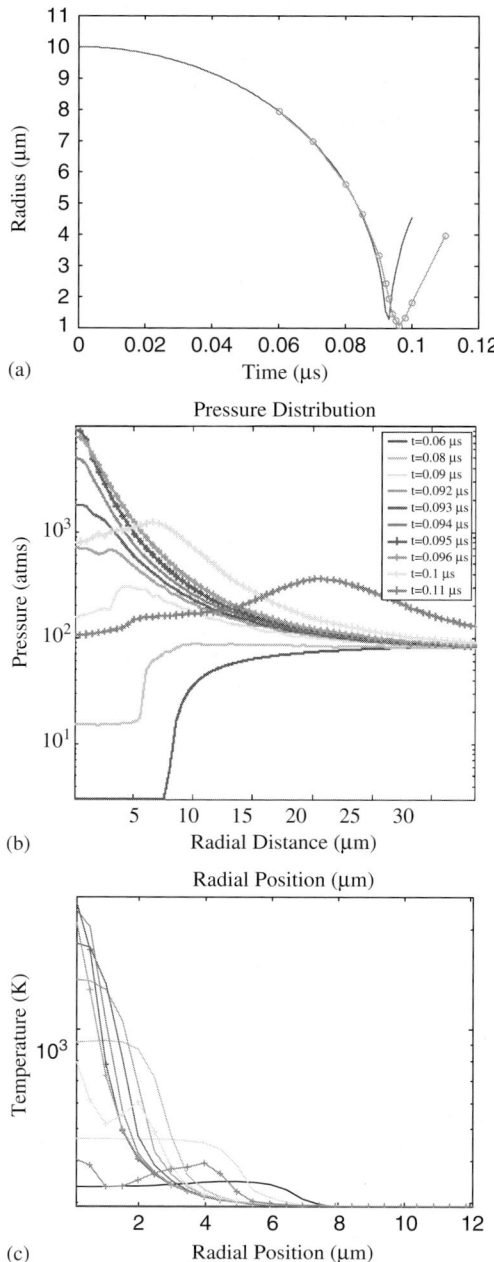

FIG. 32. Air bubble implosion results along the vertical (y-axis) [126].

1. Conservation Equations for the Molecular and Dissociated States of a Fluid

Let us now consider the 1D equations of motion that comprise a state-of-the-art hydrodynamic shock (HYDRO) code [114]. We may assume that all particles, molecules and atoms, have the same mean velocity and temperature. In this case, the conservation equations for a dissociated or non-dissociated fluid look identical, the only difference is in their equations of state. In the absence of viscosity and diffusion processes the azimuthally symmetric 1D phasic ($k = v, \ell$) conservation equations for the conservation of mass, momentum and energy in spherical coordinates are

$$\frac{\partial \rho_k}{\partial t} + \frac{1}{r^2}\frac{\partial}{\partial r}\left(r^2 \rho_k v_k\right) = 0 \tag{66}$$

$$\frac{\partial \rho_k v_k}{\partial t} + \frac{1}{r^2}\frac{\partial}{\partial r}\left(r^2 \rho_k v_k^2\right) + \frac{\partial p_k}{\partial r} = 0 \tag{67}$$

$$\frac{\partial e_k}{\partial t} + \frac{1}{r^2}\frac{\partial}{\partial r}\left[r^2 v_k (e_k + p_k)\right] = \frac{1}{r^2}\frac{\partial}{\partial r}\left(k_k r^2 \frac{\partial T_k}{\partial r}\right) \tag{68}$$

Here ρ_k, v_k, p_k, T_k, and $e_k \equiv \rho_k(\varepsilon_k + v_k^2/2)$, are the density, velocity, pressure, temperature and the total convected energy per unit volume, respectively, ε_k the internal energy per unit mass and k_k the thermal conductivity. The indices $k = \ell$ and v denote the liquid and vapor phases, respectively. Equations (66)–(68) describe the vapor dynamics inside the bubble ($k = v$, $0 \leq r \leq R(t)$) and the compressible fluid dynamics in the liquid surrounding the bubble ($k = \ell$, $R(t) \leq r \leq R_\ell(t)$).

The motion of the surface of the spherical cell, $R_\ell(t)$, can be described by the generalized Rayleigh equation, Eq. (40), with R replaced by $R_\ell, v_{\ell i}$ by $\dot{R}_\ell$, and $p_{\ell i}$ by $p_\ell|_{r=R_\ell}$. Also, the interfacial evaporation term, $\dot{m}_v''$, is omitted. This approximation is appropriate for the cell because the radial velocity of the outer radius of the cell, $\dot{R}_\ell$, is always small with respect to the speed of sound in liquid.

2. Conservation Equations for the Plasma State of a Fluid

Let us consider a fluid, which, after dissociation is complete, contains (for simplicity) atoms of one type. Let Z_m be a maximum number of electrons in one atom (a constant), and Z be the number of electrons ripped off the atom due to ionization process (a variable according to the ionization kinetics). The plasma is considered to be electrically neutral and the density of the plasma is

$$\begin{aligned}\rho &= \rho_i + \rho_e = n_i m_i + n_e m_e = n_i[m_{i0} + (Z_m - Z)m_e] + Z n_i m_e \\ &= n_i(m_{i0} + Z_m m_e) = n_i m_a\end{aligned} \tag{69}$$

Here ρ_i, ρ_e are the partial densities of the ions and electrons; n_i, n_e the number densities of electrons and ions; and m_i, m_e the masses of ions and electrons, respectively; m_{i0} is the mass of a fully ionized atom, and m_a the mass of a non-ionized atom.

Let us assume that all the particles in the plasma (i.e., the atoms, ions and electrons) have the same mean velocity, v; the temperature of the atoms and ions are equal (i.e., $T = T_i$), but, in our case, the temperatures of ions and electrons differ from each other ($T_i \neq T_e$).

The 1D equations for mass and momentum conservation for such a plasma in spherical coordinates are

$$\frac{\partial \rho}{\partial t} + \frac{1}{r^2}\frac{\partial}{\partial r}(r^2 \rho v) = 0 \qquad (70)$$

$$\frac{\partial \rho v}{\partial t} + \frac{1}{r^2}\frac{\partial}{\partial r}(r^2 \rho v) + \frac{\partial}{\partial r}(p_i + p_e + p_r) = 0 \qquad (71)$$

where p_i, p_e, p_r are the pressures due to the ions, electrons and the thermal radiation (i.e., photons), respectively.

The corresponding energy conservation equation is

$$\frac{\partial}{\partial t}\left[\rho(\varepsilon_i + \varepsilon_e) + e_r + \frac{\rho v^2}{2}\right] + \frac{1}{r^2}\frac{\partial}{\partial r}\left[r^2 \rho v\left(\varepsilon_i + \varepsilon_e + \frac{e_r}{\rho} + \frac{(p_i + p_e + p_r)}{\rho} + \frac{v^2}{2}\right)\right]$$
$$= \frac{1}{r^2}\frac{\partial}{\partial r}\left[r^2\left(q^{(i)} + q^{(e)} + S^{(r)}\right)\right] \qquad (72)$$

where ε_i, ε_e, e_r are the specific internal energies of the ions and electrons, and the photons (i.e., radiation), respectively; $q^{(i)}$, $q^{(e)}$, $S^{(r)}$ are the ion and electron heat fluxes, and the flux of radiation energy density, respectively.

Since $T_i \neq T_e$, it is necessary to derive an additional equation for temperature of electrons. In order to do that let us formulate the energy conservation law for the electron gas coupled with radiation.

$$\frac{\partial}{\partial t}\left[\rho\left(\varepsilon_e + \frac{e_r}{\rho} + \frac{\rho_e}{\rho}\frac{v^2}{2}\right)\right]$$
$$+ \frac{1}{r^2}\frac{\partial}{\partial r}\left\{r^2\left[\rho v\left(\varepsilon_e + \frac{e_r}{\rho} + \frac{p_e + p_r}{\rho} + \frac{\rho_e}{\rho}\frac{v^2}{2}\right) + q^{(e)} + S^{(r)}\right]\right\}$$
$$= q_{ie}(T_i, T_e) + f_{ie}v + n_i \frac{dZ}{dt}\left(E_{e(ie)} + \frac{1}{2}m_e v^2\right) \qquad (73)$$

Here, $q_{ie}(T_i, T_e)$ is the heat flux from the ions to electrons and $E_{e(ie)}$ the internal energy of the electron just ripped off an atom.

It is logical to assume that an electron just ripped off the ionized atom has the same velocity as the ion. Accounting for the small electron mass compared to the mass of the ion, these electrons possess a thermally induced kinetic energy corresponding to a temperature, which is much lower than that of the ions. That is (k_b is the Boltzmann constant),

$$E_{e(ie)} = \frac{3}{2} k_b T_{e(ie)}, \quad T_{e(ie)} = \frac{m_e}{m_i} T_i \ll T_i \qquad (74)$$

The "external force," f_{ie}, acting on the electrons from the ion gas in the presence of radiation tries to make the velocities of the ions and electrons equal. This force can be calculated from the momentum balance equation for an electron gas

$$\rho_e \frac{Dv}{Dt} = -\frac{\partial}{\partial r}(p_e + p_r) + f_{ie} \qquad (75)$$

where, $D/Dt = \partial/\partial t + v \partial/\partial r$ is the material derivative.

Substituting f_{ie} from Eq. (75) into Eq. (73), and taking into account the mass conservation equation, Eq. (70), one obtains the following equation for the internal energy of the electrons coupled with the energy density of radiation:

$$\frac{\partial}{\partial t}(\rho \varepsilon_e + e_r) = -\frac{1}{r^2} \frac{\partial}{\partial r} \left\{ r^2 \left[\rho v \left(\varepsilon_e + \frac{e_r}{\rho} \right) + q^{(e)} + S^{(r)} \right] \right\}$$
$$- (p_e + p_r) \frac{1}{r^2} \frac{\partial}{\partial r}(r^2 v) + q_{ie}(T_i, T_e) + n_i \frac{dZ}{dt} E_{e(ie)} \qquad (76)$$

Thus, to model two-component, two-temperature plasma dynamics, one must solve the set of Eqs. (70)–(72) and (76) together with the respective closure equations for $p_i, p_e, p_r, \varepsilon_i, \varepsilon_e, e_r, q^{(i)}, q^{(e)}, S^{(r)}, q^{(ie)}$ and Z. Fortunately, for the case of a violent bubble implosion (e.g., during sonofusion experiments) the complete problem may be substantially simplified, and reduced to Eqs. (70)–(72).

Let us now discuss some details of the plasma model, which was used in our HYRDO code.

3. Conductive Thermal Exchange in a Plasma

In general, an electron gas obeys Fermi–Dirac quantum statistics, which only in case of sufficiently high temperatures turns into classical Boltzmann statistics. This happens when the electron temperature is much higher than the temperature of quantum degeneration. That is, when

$$T_e \gg T_{\text{crit1}} = \frac{h^2}{4\pi^2 m_e k_b} n_e^{2/3} \tag{77}$$

Here m_e is mass of an electron, h and k_b are Planck and Boltzmann constants, respectively, n_e the number density of electrons, which is related to the fluid density, ρ, as

$$n_e = \frac{\rho N_A}{M} N_e \times 10^{-6} \tag{78}$$

where M is the molecular weight of the fluid (for D-acetone, 64 kg/kmol), N_e the mean number of electrons per molecule of the fluid (for D-acetone, 32), N_A the Avogadro number (6.023×10^{26}/kmol), and the factor 10^{-6} is used to give the number density of electrons in cm^{-3}, since fluid density is normally measured in kg/m^3.

In addition to the criterion that the electron gas is classical there is also an important criterion that it is ideal. This implies that the average kinetic energy of an electron must be large in comparison to the average energy of interaction of the electron with its neighbor.

$$T_e \gg T_{\text{crit2}} = \frac{e^2 n_e^{1/3}}{k_b} \tag{79}$$

A similar criterion for the ions is

$$T_i \gg T_{\text{crit3}} = \frac{(Ze)^2 n_i^{1/3}}{k_b} \tag{80}$$

where Z is an average charge of the ions (for D-acetone, 3.2), n_i the number density of ions, which are related to fluid density ρ as

$$n_i = \frac{\rho N_A}{M} N_i \times 10^{-6} \tag{81}$$

where N_i is the mean number of ions per molecule of the fluid (for D-acetone, 10).

Depending on the combination of the compression rate of matter and the energy exchange between the electrons and ions during bubble implosion, the electron temperature and density may be above or below the critical values.

It turns out that if an electron gas is classical and ideal, $T_e \gg (T_{\text{crit1}}, T_{\text{crit2}})$, the heat transfer rate due to electron–electron collisions is determined by the following relaxation time [112]:

$$\tau_{ee} \sim \frac{m_e^{1/2}(k_b T_e)^{3/2}}{n_e e^4 \ln \Lambda_e}, \quad \Lambda_e = \frac{\ell_{De}}{\delta_e}, \quad \ell_{De} = \sqrt{\frac{k_b T_e}{n_e e^2}}, \quad \delta_e = \frac{e^2}{k_b T_e} \quad (82)$$

where $\ln \Lambda_e$ is the so-called Coulomb logarithm, Λ_e characterizes the ratio of the Debye length ℓ_{De} (a distance, where screening of the Coulomb field takes place in an electron gas) to the characteristic impact parameter of electron–electron collisions, δ_e.

Similarly, the time of relaxation for the ionic component of the plasma toward the equilibrium state is [112]

$$\tau_{ii} \sim \sqrt{\frac{m_i}{m_e}} \tau_{ee} \gg \tau_{ee} \quad (83)$$

Finally, the relaxation time required to establish equilibrium between the electrons and the ions is of order

$$\tau_{ei} \sim \frac{m_i}{m_e} \tau_{ee} \gg \tau_{ii} \gg \tau_{ee} \quad (84)$$

Hence, the transfer of energy between the electrons and ions and the corresponding equalization of the electron and ion temperatures is a much slower process than the approach of either the electrons or ions to equilibrium separately.

Thus, it is reasonable to model the plasma in an imploding bubble as a two-temperature, two-component fluid, where the ionic and electronic components have different temperatures. In the interior core region behind a converging shock wave the ion temperature goes up very rapidly; however, due to energy exchange delays between the ions and electrons, the electrons stay relatively cold. Further into the implosion there is a very rapid rise in the density as well. Thus, the relatively cold electron gas may lose its ideal and classical properties. Moreover, as the densities become higher the Debye length decreases and the relaxation time in Eq. (84) goes up before the classical approximations fail, and the electron gas is likely trapped in its quantum degenerate state. This allows us to make an important approximation, which is not normally valid in laser-induced inertial confinement fusion analysis. In particular, during the final stage of implosion the electron temperature can be neglected in the model, because both the photons and cold electrons carry almost no mass, momentum and energy at that time.

Therefore, there is no need to consider the energy exchange between ions and electrons and the thermal conductivity of, and energy losses from, the electron gas are small. Moreover, the effective thermal conductivity of such a plasma is only due to the ions.

In contrast to the electrons, the ions are very hot and may be easily modeled as a classic ideal gas of positively charged particles. The cross section for ion collisions is calculated as [112]

$$\sigma_{ii} \approx \pi \frac{Z^4 e^4}{(k_b T_i)^2} \ell n \, \Lambda_i, \quad \Lambda_i = \frac{\ell_{Di}}{\delta_i}, \quad \ell_{Di} = \sqrt{\frac{k_b T_i}{n_i Z^2 e^2}}, \quad \delta_i = \frac{Z^2 e^2}{k_b T_i} \quad (85)$$

and the mean free path of the ions is calculated as

$$L_{ii} = \frac{1}{n_i \sigma_{ii}} \quad (86)$$

According to kinetic theory, the ion–ion thermal conductivity is given by [112]

$$k_{ii} = \frac{k_b}{2\sigma_{ii}} \left(\frac{8 k_b T_i}{\pi m_i} \right)^{1/2} \quad (87)$$

4. Radiative Thermal Exchange in the Plasma

The hot matter (plasma) emits light by an energy cascade directed from the ions, to the electrons, and further to the photons. Although pdV-type compression work is done on both the ions and electrons as the bubble collapses, most of the mechanical energy from the shock goes into the ions [112], which results in high ion temperatures, but relatively low electron temperatures. The physical mechanisms responsible for electron–photon coupling are free–free transitions (i.e., Bremsstralung emission and absorption), bound-free transitions (i.e., photoelectric ionization and recombination), and bound–bound (i.e., discrete) transitions. The first two types of electron transitions are responsible for a broad band spectrum of radiation and the third one for a line spectrum. In order to model the electron–photon coupling only one parameter must be specified; that is a photon absorption coefficient, which, in general, is a function of plasma density, radiation frequency and, what is of most importance, the electron temperature, T_e.

If the radiation mean free path is much smaller than the bubble size, the approximation of local thermodynamic equilibrium between the radiation and matter is applicable. This is the so-called optically thick (i.e., opaque) approximation. In this case, the radiation energy per unit volume, e_r, and pressure of thermal radiation, p_r, are [112]

$$e_r = e_r^{(e)} = \frac{4\sigma T_e^4}{c}, \quad p_r = \frac{e_r}{3} = p_r^{(e)} = \frac{4\sigma T_e^4}{3c} \tag{88}$$

where σ is the Stefan–Boltzmann constant and c the speed of light.

Estimations show that the energy of radiation and the radiation pressure become comparable with the internal energy and pressure of the plasma, when electron temperature is around 10^6 K and higher. Since the maximum electron temperatures during bubble implosion are around 10^5, we can neglect the associated energy of radiation, the radiation pressure and, consequently electron thermal conductivity. Thus, the energy losses during sonofusion experiments are much less than those during laser-induced inertial confinement fusion experiments and those in magnetically confined plasma experiments.

Such a simplification is also appropriate in the case of the so-called optically thin approximation, when the radiation mean free path is comparable or much larger than the bubble size or the region in the bubble where light is emitted. This is easy to understand because in this case the energy of radiation and radiation pressure are nearly equal to zero, and the thermal energy emitted by a hot plasma (i.e., the integrated emission coefficient) is proportional to σT_e^4.

It has been shown [30] that electron thermal conduction and the opacity (i.e., light absorption) of the plasma in a collapsing gas bubble are the mechanisms responsible for the picosecond duration of SBSL emissions. Nevertheless, the above model of a partially ionized plasma with distinct but coupled ion and electron temperature fields, associated losses by plasma thermal conduction, an emission model for coupling the ionization energy to the radiation field, and a model for the opacity of the radiating matter, is generally suitable for the analysis of sonofusion experiments, such as the sonofusion experiments performed at ORNL [54,56].

E. Kinetics of Nuclear Reactions

In order to estimate the production of fusion neutrons, the neutron kinetics model given by Gross [105] can be used in conjunction with the local, instantaneous HYDRO code evaluations of the thermal–hydraulic parameters during the bubble implosion process. For deuterium/deuterium (D/D) fusion reactions the kinetics model is given by [105]

$$J_n \equiv \frac{dn_n'''}{dt} = \frac{1}{2} \langle \sigma v \rangle (n_D''')^2 \tag{89}$$

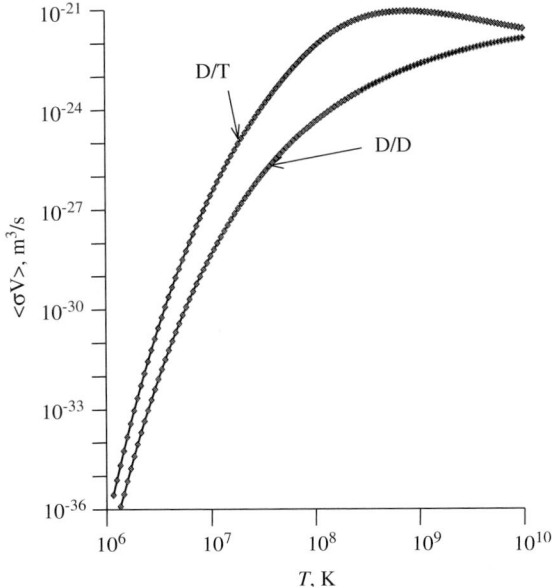

FIG. 33. Variation of weighted fusion cross sections with plasma temperature [137].

where, n_n''' is the concentration of fusion neutrons produced (i.e., neutrons/m^3), n_D''' the concentration of the deuterium ions in question and $\langle \sigma v \rangle$ the weighted cross section for D/D neutron fusion reactions [137]. The weighted cross sections for D/D and D/T fusion reactions are given in Fig. 33.

It can be seen in this figure that there will be a very large change in D/D neutron yield with temperature (e.g., about a nine order of magnitude increase when the temperature goes from 10^6 to 10^7 K, and about a twelve order of magnitude increase in going from 10^6 to 10^8 K). This, in turn, implies that any D/D neutron produced at temperatures below $\sim 10^8$ K would be difficult, if not impossible, to detect. Moreover, we see that, at the same temperature, the D/T reaction will yield more fusion neutrons than the D/D reaction.

Before discussing the results of HYDRO code simulations, it is interesting to estimate the expected number of neutrons produced during a single bubble collapse. The neutron production per bubble per implosion may be determined from

$$n_n = \int_{\Delta t_{\text{implosion}}} dt \int_{V_{\text{bubble}}} \frac{dn_n'''}{dt} \, dV \quad (90)$$

This integral may be estimated using Eq. (89) and the Mean Value Theorem

$$n_n \sim (n_D''')_\bullet^2 \langle \sigma v \rangle_\bullet R_\bullet^3 \Delta t_\bullet \tag{91}$$

where, $R_\bullet$ is the radius of the highly compressed, hot central core region of the bubble, $\Delta t_\bullet$ the time interval over which fusion conditions are maintained, $(n_D''')_\bullet$ the characteristic concentration of deuterium ions in the supercompressed/superheated central core region of the bubble and $\langle \sigma v \rangle_\bullet$ the weighted nuclear cross section.

As will be discussed in more detail subsequently, depending on the modeling assumptions made, the range of the characteristic values for sonofusion are the following:

$$(n_D''')_\bullet \approx 10^{30} \text{ D/m}^3 \text{(i.e., } \rho_\bullet \approx 10^4 \text{ kg/m}^3\text{)}$$
$$\langle \sigma v \rangle_\bullet \approx 10^{-25} \text{m}^3/\text{s(i.e., } T_\bullet \approx 10^8 \text{K)}$$
$$R_\bullet \approx 60 \text{ nm}; \ \Delta t_\bullet \approx 1.0 \text{ ps} \tag{92}$$

Thus, a reasonable estimate for D/D neutron production is

$$n_n \approx 10 \text{ neutrons/implosion/bubble} \tag{93}$$

It is expected [114] that only a few percent of the $\sim$1000 bubbles (e.g., $\sim$15 bubbles) in the bubble clusters in the ORNL experiments [54,56] experienced conditions suitable for fusion (i.e., the interior bubbles). Also, each bubble cluster experienced up to 50 implosions/s, and as seen in Fig. 22a, had $\sim$50 subsequent implosive "bounces" at the acoustic frequency. Thus, the estimated average neutron production rate is about

$$\dot{n}_n \approx (10)(15)(50)(50) \sim 4 \times 10^5 \text{ n/s} \tag{94}$$

which is in good agreement with what was measured in the ORNL experiments [54,56].

It is interesting to note that the so-called Lawson criterion for D/D fusion ignition at $\sim 10^8$ K is [105] $(n_D''')_\bullet \tau \cong 10^{22}$ s/m^3. Noting that $\tau \equiv \Delta t_\bullet$, the data in Eq. (92) yield, $n_D''' \Delta t_\bullet = 10^{18}$ s/m^3. Thus, the sonofusion experiments run at ORNL produced conditions which are about four orders of magnitude below those required for the type of explosive fusion burn experienced in thermonuclear weapons and desired in laser-induced inertial confinement fusion experiments [138]. Rather, the conditions achieved to-date in our sonofusion experiments imply D/D fusion "sparks." However, it appears that there are significant opportunities to scale-up the neutron yield in future sonofusion experiments. For example, the corresponding Lawson criterion for D/T fusion ignition is $\sim 10^{20}\,\text{s/m}^3$ at $10^8\,\text{K}$, which is only two orders of magnitude above what has been achieved to-date in the ORNL sonofusion experiments.

F. HYDRO Code Analysis of Sonofusion

To quantify the implosive bubble collapse conditions and the D/D neutron yield in the experiments described previously, a transient 1D hydrodynamic, shock (HYDRO) code was developed which accounted for the heat and mass exchange processes in and around a vapor bubble. As discussed previously, this model includes the appropriate partial differential conservation equations for mass, momentum and energy conservation and the interface conditions, which account for non-equilibrium phase change (i.e., evaporation and condensation). As noted previously, the phase change process was evaluated using Hertz–Knudsen–Langmuir kinetics with an accommodation coefficient α. We have also used the Mie–Gruniesen equations of state [112] which account for intermolecular and/or interatomic interactions in the form of a Born–Mayer potential function and its extensions [114]. As noted previously, these equations of state are known to be valid for highly compressed fluids. Parameters for the equation of state for acetone were calculated based on available thermodynamic data (in particular, in the two-phase region[8]) and the experimental shock wave adiabat for liquid acetone measured by Trunin et al. [139]. Account was taken of the effects of dissociation and ionization, and for the strong temperature dependence on thermal conductivity [Zeldovich & Raizer, 1963] during plasma formation within the imploding bubbles. Account was also taken of the state of temperature non-equilibrium among the ions and electrons formed during a 10^{-13}–$10^{-12}\,\text{s}$ time interval associated with the final stage of the bubble implosion process. Figure 30 shows the equation of state (EOS) for D-acetone, including the isotherms (note the isotherm corresponding to $T = 0\,\text{K}$, which reflects the cold components (p) in Eq. (23)). The Grüniesen coefficient (Γ) for the degree of acetone dissociation (Dis) and non-dissociation (NDis) is also shown. The temperature dependence of vapor thermal conductivity, including the plasma state, is consistent with Eq. (87) and is of the form

$$k \equiv k_{ii} = k_0\left[1 - k'_0 + k'_0(T/T_0)^\vartheta\right], \quad k_0 = 8.23 \times 10^{-3}\,\text{kg} \times \text{m}/(\text{s}^3 \times \text{K}), \quad k'_0 = 7.5, \quad \vartheta = 0.5$$

(95a)

---

[8] Moss et al. [29] recognized that they did not know how to account for phase change with vapor (in their case it was water vapor added to deuterium) and the equations of state related to the two-phase region. Thus they considered the vapor as air. In our investigation phase transformations of the vapor/liquid type were taken into account explicitly and this phase transformation was crucial during the implosion process.

Moreover, for the high Mach number phase of the bubble implosion process, Fourier's law becomes a relaxation equation, which is given by [140]

$$\tau_{ii}\frac{\partial q}{\partial t} + q = -k\nabla T_i \qquad (95b)$$

where, q is the heat flux, T_i the ion temperature and $\tau_{ii} \sim 10^{-13}$ s [140].

Using these equations a HYDRO code was developed based on Godunov's numerical integration technique, and the bubble dynamics were evaluated in acetone for the conditions in the ORNL experiments [114].

It is very important to remember that bubble growth and almost all of the compression process (during $\sim 25\,\mu s$ at 19.3 kHz) occurs with values of radial interface velocity, $|\dot{R}|$, less than $\sim 50\,m/s$. Under these conditions of small Mach number ($M_g \equiv |\dot{R}|/C_g$), the gas pressure in the bubble is uniform and does not exceed 10 bar. The process can then be described by Eqs. (34)–(36), (40)–(43) and (51). In this case, the low-density gas behaves in accordance with the perfect gas law. Typical computational results for this relatively slow (low-Mach) stage are shown in Fig. 34. During the bubble expansion stage of bubble dynamics an accumulation of liquid kinetic energy takes place. Part of this will later be transformed into internal energy in the liquid around, and the vapor inside, a bubble. It is important to note that a significant amount of the vapor created by evaporation at the time of maximum bubble size is condensed during the subcritical phase of the implosion process. Moreover, as can be seen in Fig. 34, the incident pressure, p_I, of the liquid surrounding the interior bubbles within the imploding bubble clusters increases due to the pressure intensification mechanisms discussed in Appendix A.

Close to the final stage of bubble collapse, during a nanosecond time span, compressibility effects are important even in the liquid (i.e., the liquid density increases by a factor of two or three near the interface), and there are shock waves, high gas densities, pressures and vapor temperatures near the bubble's center. The vapor transforms into a state of high-density dissociated and ionized plasma. During this phase of bubble implosion, the imploding vapor bubbles reach a minimum radius of about $R_{\min} \approx 10\,\mu m$, and the radial velocity of the fluid moving toward the bubble's center is $\sim 7\,km/s$.

A picosecond time range follows next, where the liquid at the interface is accelerated to enormous speeds, initiating shock and/or quasi-shock compression waves in the vapor, which converge and focus toward the bubble's center. As the shock wave is focused, it strengthens and the plasma temperature and pressure reach very high levels as the wave arrives at the center. Thereafter the shock wave, which is reflected from the center, diverges and becomes weaker with radial distance. Nevertheless, the bubble continues to implode for a brief interval and this further compression leads to an additional

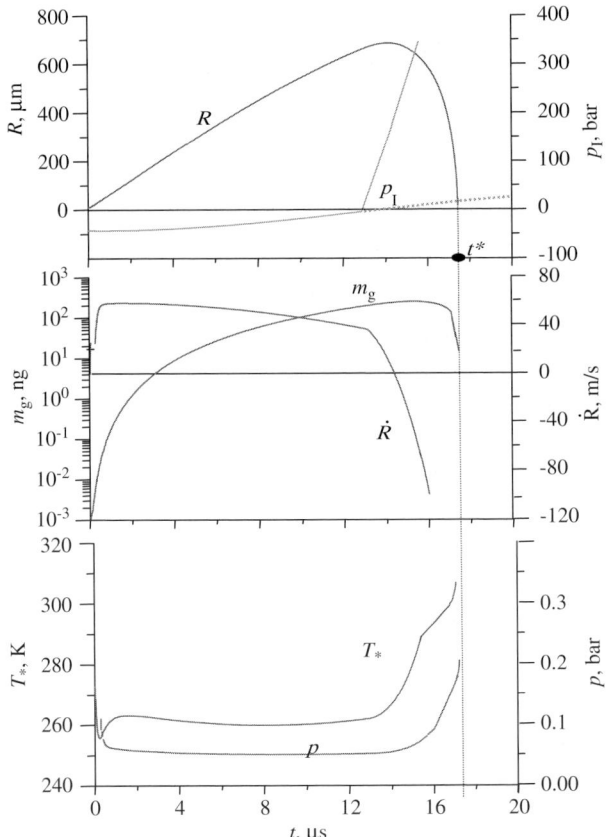

FIG. 34. Low Mach number phase of bubble dynamics [90].

increase in temperature and vapor/plasma density. The closer to the center, the higher is the temperature, vapor/plasma density and potential for thermonuclear reactions, but the shorter is the duration of this state, and thus the reaction time, and the amount of material to react is much less. A typical high Mach number sequence of events is shown in Fig. 35 for two different assumptions for thermal conductivity [90]. A strong shock is seen during bubble collapse (t_1, t_2 and t_3) and shock rebound can be seen at t_5. It can also be seen that, due to the speed of the process, the peak density, temperature and pressure are relatively insensitive to the modeling of thermal conductivity (i.e., using Eq. (95a) or assuming a constant thermal conductivity).

To estimate the neutron and tritium (T) production rate per unit volume due to D/D nuclear fusion, we can employ a well-known neutron kinetics

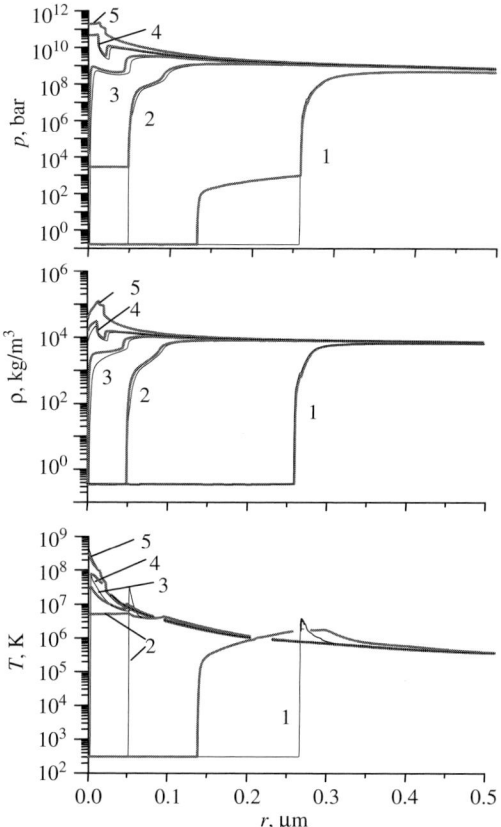

FIG. 35. Shock wave propagation and cumulation within a bubble during the high Mach number phase of bubble implosion (thick red lines use Eqs. (95a) and (95b) and thin lines use a constant thermal conductivity). The numbers indicate the spatial distributions at times $t_1 = 0.0$ (Arbitrary), $t_2 = 0.61$, $t_3 = 0.68$, $t_4 = 0.72$ and $t_5 = 0.76$ ps [90].

equation for the fusion rate due to D/D nuclei collisions [105]

$$J_n = 1/2 \langle \sigma v \rangle (n_D''') \approx J_T \qquad (96)$$

As in Eq. (89), n_D''' is the D-ion concentration, $\langle \sigma v \rangle$ the mean reactivity, which is equal to the averaged product of the cross section, σ and the D-ion speed, v [137]. As seen in figure 33, this reactivity depends strongly on the ion temperature, T_i. For example, at $T_i = 10^8$ K, the reactivity, $\langle \sigma v \rangle$, for D/D fusion is about 3×10^{-25} m^3/s. For D-acetone (C$_3$D$_6$O) n_D''' is determined by

$$n_D''' = \frac{6\rho_g N_A}{M} = \rho_g N_m (N_m = 6N_A/M = 0.565 \times 10^{26}/\text{kg}) \qquad (97)$$

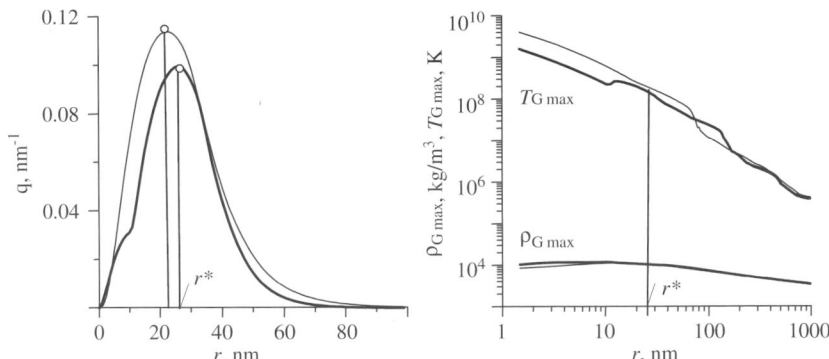

FIG. 36. The maximum neutron production distribution (q), and maximum temperatures ($T_{G\ max}$) and densities ($\rho_{G\ max}$) with (heavy lines) and without (thin lines) endothermic "chemical energy" losses from the dissociation and ionization of C_3D_6O molecules.

where N_A is the Avogadro number, M the molecular weight ($M = 64$ kg/kmol for D-acetone with six deuterium nuclei in each molecule).

The total neutron and tritium production per implosion is defined by the integral taken over by the bubble's volume, V_b, and the phasic period for the expression for fusion rate density, J_n, given by Eq. (96),

$$Q = \int_{V_b} \int_0^{1/f} J_n \, dt \, dV = \int_0^R q(r) \, dr, \quad q(r) = 4\pi r^2 \int_0^{1/f} J(r,t) \, dt \tag{98}$$

Here $q(r)$ characterizes the D/D fusion rate at each radial position. Detailed HYDRO code calculations [114] show that the neutron and tritium production have a peak at $r_* \approx 20$–30 nm, and the total production takes place within a radius of, $r_c \approx 80$ nm (see Fig. 36). Despite the fact that the maximum temperatures and densities are higher at $r < r_*$ than at r_*, these extreme conditions at smaller radii exist only for a very short time and occupy a very small volume. This is why there is a maximum value of the kernel function q(r) at a finite radius. In our calculations, characteristic values for parameters at radius r_*, namely: the time interval Δt_*, density ρ_*, pressure p_*, ion temperature, T_{i*} and interfacial velocity, w_*, were[9]

$$r_* \sim 20 \text{ nm}, \quad \Delta t_* \sim 10^{-12} - 10^{-13}, \quad \rho_* \sim 10^4 kg/m^3, \quad p_* \sim 10^{11} \text{bar},$$
$$T_{i*} \sim 10^8 K, \quad w_* \sim 900 km/s \tag{99}$$

---

[9] Note that these pressures, temperatures and interface velocity are several orders of magnitude larger than for typical SBSL experiments (i.e., see Eq. (16)).

With a decrease in liquid pool temperature from room temperature ($T_0 \approx 20\,°C$) to $T_0 = 0\,°C$ and below, the neutron and tritium (T) production grows manyfold. This is because, during bubble growth less vapor is evaporated in a low-temperature liquid, and during bubble implosion more vapor is condensed. This results in less vapor cushioning and higher kinetic energy of liquid by the time of initiating a shock wave that moves toward the bubble's center through the remaining vapor. As a consequence, for a low-temperature liquid pool, an increase is found in the intensity of the focusing shock wave and the extent to which the substance in the plasma sphere at radius r_* is compressed and heated. As already discussed, this paradoxical effect of liquid temperature has been experimentally verified [54]. This clearly shows the synergy of the experiments and analysis, and how they aid each other in studying and interpreting sonofusion phenomenon.

Similarly, larger values of the condensation (i.e., accommodation) coefficient, α, yields more condensation and less increase in pressure during the early phase of bubble collapse which is favorable to higher interfacial acceleration and ultimately to more intense vapor compression. Significantly, organic fluids, with their relatively large and bulky molecules, have greater coefficients of condensation, approaching $\alpha \sim 1.0$ [141]. Therefore, deuterated organic fluids offer advantages over, for example, heavy water (D_2O) that has a relatively small coefficient, $\alpha \sim 0.075$ [104,142,143]. Moreover, it is difficult to expose water (as opposed to organic fluids such as acetone) to large levels of tension without having premature cavitation. Thus, D-acetone (C_3D_6O), which was used in the ORNL experiments, appears to be a much better test fluid for sonofusion than heavy water (D_2O). Indeed, Geisler et al. [144] were unable to obtain D/D fusion conditions in experiments on laser-induced cavitation in heavy water. This is not only due to their use of heavy water, but, a laser does not cavitate a suitable cluster of spherical bubbles.

HYDRO code simulations have also shown that for the case of a single bubble subjected to acoustic forcing with an amplitude of 15–40 bar, neutron and T production is insignificant. It is important to stress that there are bubble clusters formed by the PNG neutrons, and a many-fold increase in the liquid pressure occurs in the central region of an imploding bubble cluster [77,100,114]. This was taken into account by applying an increase in the incident pressure acting on the interior bubbles during acoustic compression (e.g., see Fig. 34). Appendix A gives a more thorough discussion of bubble cluster dynamic phenomena, including a prediction of the non-linear pressure intensification process within an imploding bubble cluster.

Another important point is the absence of "cold" fluid dissociation at the interface for pressures of order 10^5 bar. This makes the fluid more rigid (i.e., less compressible because the bulky structure of the molecules is preserved) as compared to the equilibrium dissociated atomic structure. For the liquid, which is relatively "cold," dissociation of the molecules requires a dissociation time of, $t_\text{D} \sim 10^{-7}$ s. However, the HYDRO code simulations indicate that highly compressed conditions on the liquid side of the interface are sustained for only $\sim 10^{-9}$ s. Thus significant dissociation of the liquid does not have sufficient time to take place, and the resultant stiffer non-dissociated liquid creates a stronger shock wave in the vapor, leading to significantly higher peak pressures and temperatures in the interior of the imploding vapor bubbles.

Also in typical MBSL and SBSL experiments with non-condensable gas bubbles endothermic chemical reactions (associated with the dissociation and ionization processes within the gas/vapor bubble) significantly decrease the temperature rise [145,146].[10] In sonofusion experiments, there are also endothermic chemical reactions within the imploding bubble; however, as can be seen in Fig. 36, they do not significantly reduce the peak temperatures, densities and D/D neutron production rates, since the reduction in peak temperature is less than about 5%, that is, $\sim 10^6$–10^7 K or so out of at least 10^8 K [114].

It is also important to recall that the ionized plasma remains essentially in non-equilibrium, with the electron temperatures being significantly below the ion temperatures (i.e., $T_\text{e} << T_\text{i}$), over the period of the fusion reactions ($\sim 10^{-12}$–10^{-13} s), since the electrons have insufficient time to be heated by the ions. This reduces the energy losses significantly (e.g., Bremsstrahlung, line and recombination losses, etc.), and one only needs to evaluate the ion temperature, T_i (i.e., a two temperature plasma dynamics model is not needed).

For the experimental conditions reported by Taleyarkhan et al. [54,56], detailed HYDRO code simulations [114] indicate a neutron production of about 10 D/D neutrons per implosion per bubble. Photographic evidence of the bubble clusters suggests that there were about 1000 bubbles in each bubble cluster in our experiments, and up to 50 clusters/sec were nucleated during the ORNL experiments [54,56]. As noted previously, an analysis of the bubble cluster dynamics [114] indicates that only about 15 bubbles undergo energetic implosion, and Fig. 22a shows that there are about 50 subsequent implosions (i.e., "bounces") of the bubble cluster after each initial implosion. Thus, HYDRO code predictions [114] imply a D/D neutron production rate of about 4×10^5 neutrons (and tritium nuclei) per second (i.e., $10 \times 15 \times 50 \times 50$), which is consistent with measurements for the T and D/D fusion neutron production rates [54,56] and the estimate is given in

---

[10] The maximum energy consumed by these endothermic "chemical reactions" is equivalent to $\sim 10^6$ K–10^7 K.

Eq. (94). In any event, these predictions clearly indicate conditions within the bubbles which are suitable for D/D fusion.

In subsequent investigations more detailed models will likely be developed to improve our understanding of the plasma physics (e.g., the kinetics of the dissociation and ionization processes), the nuclear physics (e.g., the n/T branching ratio), and the pressure intensification process within a bubble cluster. Nevertheless, the basic conclusions discussed above are expected to remain valid.

G. Nucleation Phenomena in Sonofusion

Most of the sonofusion analysis and experiments that have been conducted to date have focused on the use of either deuterated acetone (D-acetone) or heavy water as the primary test fluid. Successes have been reported when D-acetone was used [54,56,180], and when a mixture of deuterated and nondeuterated liquids were used, in which alpha particle emissions from a dissolved uranium salt (UN) nucleated the bubble clusters rather than an external neutron source [179]. However it is far from clear that the optimum liquids have been chosen. Indeed, there are many different deuterated hydrocarbons that are commercially available and high-temperature liquids (containing, for example, silicon or boron and hydrogen) that have more attractive properties from the point of view of the thermodynamics of energy conversion systems.

In order to screen these various candidate test liquids one needs to consider not only the inherent thermodynamic and transport properties of the liquid (e.g., vapor pressure, compressibility, etc.) but also the bubble nucleation process, since, as discussed in Appendix A, there appears to be an optimum number of bubbles in a bubble cluster, which will produce the maximum pressure intensification within an imploding bubble cluster. Thus, let us next consider how various test fluid candidates can be effectively screened using bubble nucleation analyses.

In general if a high-energy neutron interacts with the liquid within an acoustic chamber, which is in tension, it will normally have only a single interaction. Since neutrons are neutral particles, they do not participate in Coulomb scattering, which is the primary mechanism for energy deposition to a liquid. Thus, the bulk of the energy deposition to the liquid comes from the "knock-on" ions that the neutron knocks out of their molecular structure. Hence our problems is one of both neutron and ion transport.

It is clear that the knock-on ions deposit most of the energy to the liquid. The most important parameters in the bubble nucleation model given in Appendix C are critical radius (R_*) and critical energy (E_*). The significance of these parameters is relatively simple; the model says, if the critical energy

(E_*) is deposited within the critical diameter ($2R_*$) then there will be bubbles formed at the size of the critical radius (which will grow in the negative pressure field of the standing acoustic wave). The values of E_* and R_* are properties of the liquid (in this case D-acetone), which can be seen in the following equations (see Appendix C):

$$R_* \approx \frac{2\sigma(T_\ell)}{p_s(T_\ell) - p_\ell} \tag{100a}$$

$$E_* \approx \frac{16\pi\sigma^3(T_\ell)}{3\left(p_s(T_\ell) - p_\ell\right)^2} + \frac{4}{3}\pi R_*^3 \rho_g(T_\ell) h_{fg}(T_\ell) \tag{100b}$$

For the analysis of neutron collisions it is important to catalogue each neutron collision of interest in D-acetone. These collisions were simulated with a well-known Monte Carlo code, MCNP [147].

A control volume was selected about the acoustic antinode (the point of highest/lowest pressure). The nominal value of radius of the control volume was 1 cm. The instantaneous pressure profile should be fairly flat within the control volume, and there may be a number of neutron-induced interactions within this sphere. This control volume is shown in Fig. 37a.

The neutron transport code MCNP was run, and information about each individual collision and energy deposition was obtained. These results were then utilized to begin the knock-on ion transport analysis using the SRIM code [148].

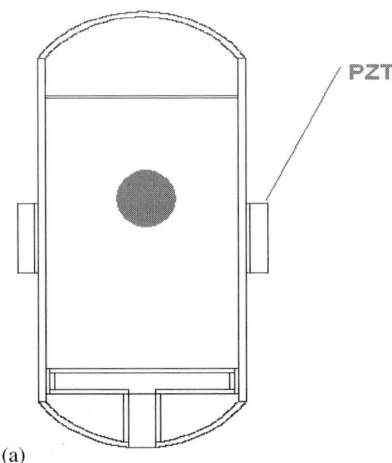

FIG. 37a. Control volume (red sphere) centered in the acoustic chamber's antinode.

Before neutron transport calculations were made it is necessary to define the pressure-related parameters at the point of collision because they have some effect on how bubbles are formed. We are interested in the pressure at a given point in the chamber when bubbles are being formed (normally the minimum pressure). Thus, one can use the amplitude of the pressure wave without any time dependence. This simplifies the actual numerical calculations of this pressure by a significant amount, and one can be sure that the static pressure distribution will vary from the time-dependent distribution by at most a phase factor of $e^{i\omega t}$, where ω is the angular frequency of the imposed acoustic oscillation.

The ATILA code (discussed previously) can be used to determine the pressure field, however, a reasonable approximation to the pressure profile of an acoustic chamber having a single antinode is given by the solution to the wave equation in cylindrical coordinates. The spatial part of the wave equation yields a profile, which includes a zero-order Bessel function, J_0, in the radial direction and a cosine function in the axial direction. That is

$$p(r,z) = AJ_0(B_0 r)\cos(\pi z/H) \tag{101}$$

The above equation may be used to estimate the pressure at a specific point in the chamber (during bubble nucleation). The variable 'r' in Eq. (101) represents the distance from the acoustic antinode's center to the cylindrical wall, while 'z' represents the distance above or below the antinode. Figure 37b is a contour plot of this pressure distribution.

The MCNP code allowed for source definitions as continuous spectra. If a PNG was used, this D/T source will produce 14.1 MeV neutrons. In contrast, if an LINAC (such as the one at RPI) was used to produce the source neutrons (by the interaction of high energy electrons with a suitable target), there will be a distribution of neutron energies. A Watt fission spectrum is

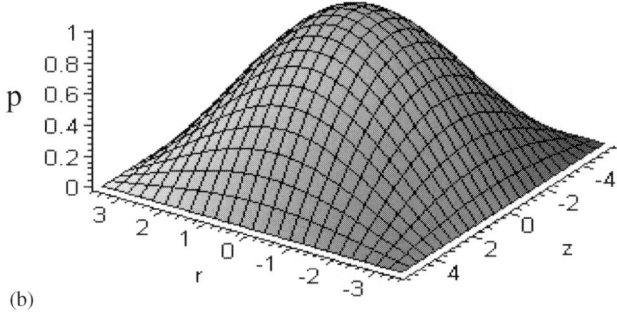

(b)

FIG. 37b. Typical acoustic pressure profile.

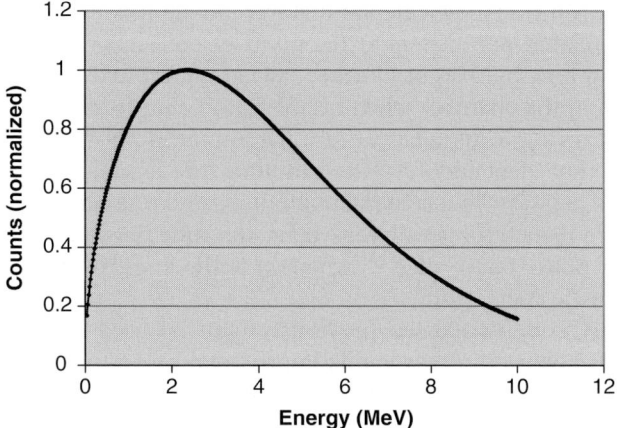

Fig. 38. Normalized LINAC neutron spectrum.

close to the neutron energy spectrum produced by an LINAC. A Watt fission spectrum has the following form:

$$N(E) = Ce^{(-E/a)} \sinh\left(\sqrt{bE}\right) \qquad (102)$$

where, N is the relative, or normalized, counts as a function of energy and 'a', 'b', and 'C' are adjustable parameters. In general 'a' and 'b' govern the shape of the spectrum, while 'C' is determined by normalization. The curve in Fig. 38 shows a typical LINAC's neutron spectrum approximation. Here the normalization was chosen so that the largest value of the spectrum takes a value of unity and other locations are reported relative to that.

1. Ion Transport

Once neutron collision information is accumulated in MCNP (especially the directed energy transferred to specific ions) the essential problem is then a problem of ion transport in D-acetone rather than neutron transport. The energy that the neutron deposits to "knock-on" ions during each collision in the control volume (centered in the acoustic antinode) will become the starting energy for that ion as it moves through the material. It should be noted that since the range of these "knock-on" ions are in the order of microns,[11] this justifies using the instantaneous pressure at the position of

---

[11] The range of these "knock-on" ions can be as much as 1.8 mm for 15 MeV energy transfer. However, a typical energy transfer in our system is ~2 MeV and the range would be about 11 microns. Therefore it is valid to assume a constant pressure exists within the region of the antinode of interest.

neutron collision. Also, the liquid temperature may be considered to be constant in this analysis.

SRIM code simulations generated energy loss (and stopping power) curves for an ion (either deuterium, carbon, or oxygen in D-acetone; C_3D_6O). These curves do not change, therefore the entire energy loss curve can be generated once and for all. D-acetone curves are given in Fig. C.5 (Appendix C). It should be noted that if the test liquid is changed new energy loss curves for ion transport will need to be generated using SRIM.

One can numerically integrate the curves generated by SRIM to find the energy deposited as a function of path length, x. This has been done for the knock-on ions of deuterium, oxygen and carbon (i.e., for D-acetone), and the results are given in Figs. 39 and 40. In addition, Fig. C.6 (Appendix C) shows that the ions produced by a single impact of a high-energy neutron have a range which is much larger than the critical diameter, $2R^*$, for bubble cavitation, which indicates that the knock-on ions may nucleate bubbles at various locations.

By generating these curves for all of the ions the number of bubbles can, in principle, be calculated. The only fundamental question that remains is: if more than the critical energy (E_*) gets deposited in the critical diameter ($2R_*$), what happens? We know what happens when less than the critical energy is deposited (i.e., nothing), but when more is deposited it is less clear. At least two assumptions are possible. First, there is a linear model (the upper bound), which says that the extra energy deposited (above E_*) makes more than one bubble by some integer factor. The threshold model (the lower bound) is just the opposite. The threshold model assumes that all the deposited energy goes into making just one energetic bubble. To decide which model is closer to reality, neutron-induced bubble cavitation data must be examined and compared with predictions.

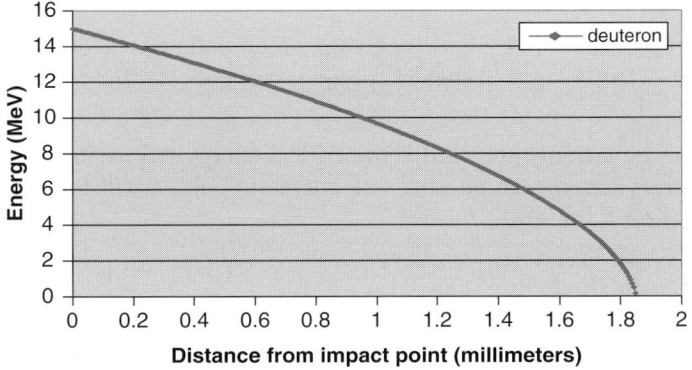

FIG. 39. Energy deposition by deuterium ions as a function of path length.

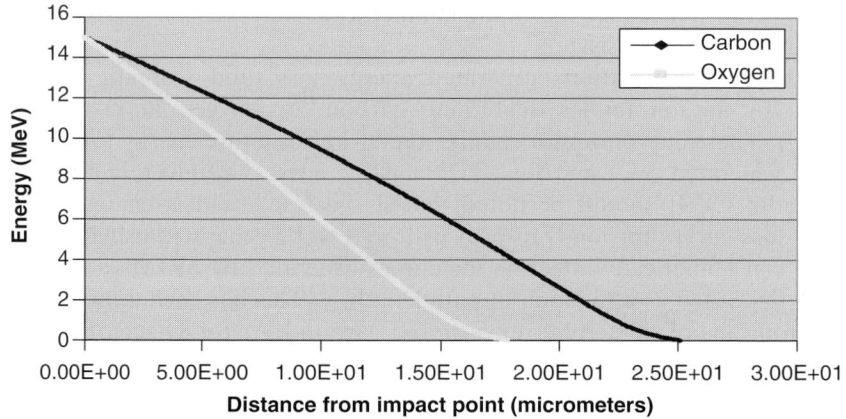

Fig. 40. Energy deposition for carbon and oxygen ions as a function of distance.

The two models for bubble nucleation can be described simply. First, for the linear model (upper bound) the following equation for the number (N) of vapor bubbles formed will hold true:

$$N = \begin{cases} \text{Integer}(E_{\text{dep}}/E_*), & E_{\text{dep}} \geq E_* \\ 0.0, & \text{otherwise} \end{cases} \quad (103)$$

In Eq. (103), E* refers to the critical energy that needs to be deposited in the critical diameter such that a bubble will form. The parameter $E_{\text{dep}} = dE/dx \, 2R_*$ refers to the calculated energy deposited (i.e., stopping power) in one critical diameter. One can easily see that Eq. (103) can yield more than one bubble for a given critical diameter. Similarly, the equation that holds for the threshold model (lower bound) of bubble nucleation is

$$N = \begin{cases} 1.0, & E_{\text{dep}} \geq E_* \\ 0.0, & E_{\text{dep}} < E_* \end{cases} \quad (104)$$

The above equation states that if the critical energy requirement for bubble nucleation is met or exceed only one energetic bubble will form, otherwise no cavitation bubble will form.

2. Neutron Interactions

Another natural question to ask about neutron transport is, how many neutrons will interact in the control volume? This is perhaps the most straightforward question to answer and some results can be used to compare

TABLE II
Neutron Source Comparison Data

	LINAC watt spectrum	14.1 MeV (D/T)	2.45 MeV (D/D)
Average energy deposition (MeV)	1.02235215	1.884643406	0.541635131
Maximum energy deposition (MeV)	7.32686	12.4902	2.12248
Fraction of emitted particles interacting in the acoustic antinode	0.0356	0.00041	0.00136

various neutron sources. Table II shows some statistics for three external neutron sources where the type and energies vary.

The 14.1 MeV isotropic D/T neutron source does not necessarily appear to be the best candidate (at least compared to the other two), the reason is because of the low interaction rate. This neutron energy is such that many neutrons stream directly through the control volume without interacting (on the average). A better neutron source appears to be the 2.45 MeV isotropic source (i.e., from a D/D source), though it has a lower maximum energy deposition and average energy deposition, about 0.14% of the emitted neutrons will interact in the antinode. Note that all these sources were placed at a position of 4 cm away from the antinode just outside the acoustic chamber. Finally, an LINAC neutron source looks pretty good and indicates that ~3.6% of the emitted neutrons will interact with the D-acetone in the acoustic antinode.

3. Bubble Nucleation

As a result of neutron interactions, bubble cluster nucleation may occur, which is essential for sonofusion. The ion transport results were used to find how many bubbles would be nucleated using either the linear or threshold model. These results are shown in Figs. 41–43.

Figure 41 shows the SRIM predicted distribution of the number of bubble clusters (or clouds) created for each source neutron emitted, separated by the ion that creates the bubble clusters. Note that if each source is summed over all the ions one recovers exactly the same results as Table II, because each neutron which interacts in the control volume forms an associated bubble cluster. Using this information, one can examine the average number of bubbles formed per bubble cluster using each of the two bounding energy partition models.

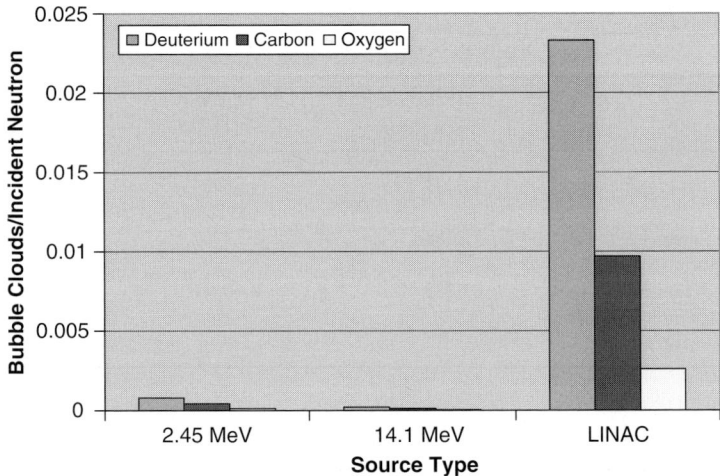

Fig. 41. Histogram of number of bubble clusters (i.e., clouds) created per source neutron.

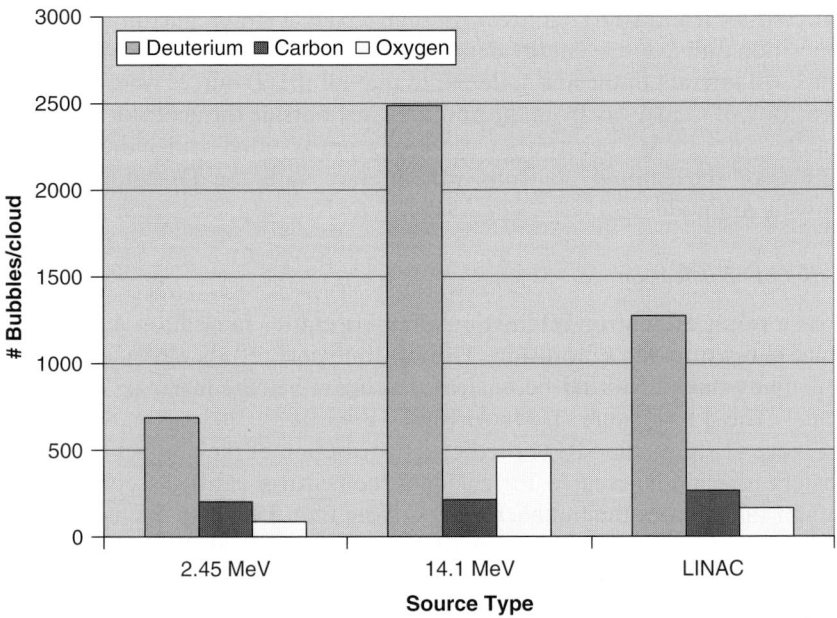

Fig. 42. Histogram of the average number of bubbles within each bubble cluster formed using the linear model, separated by ion and neutron source type.

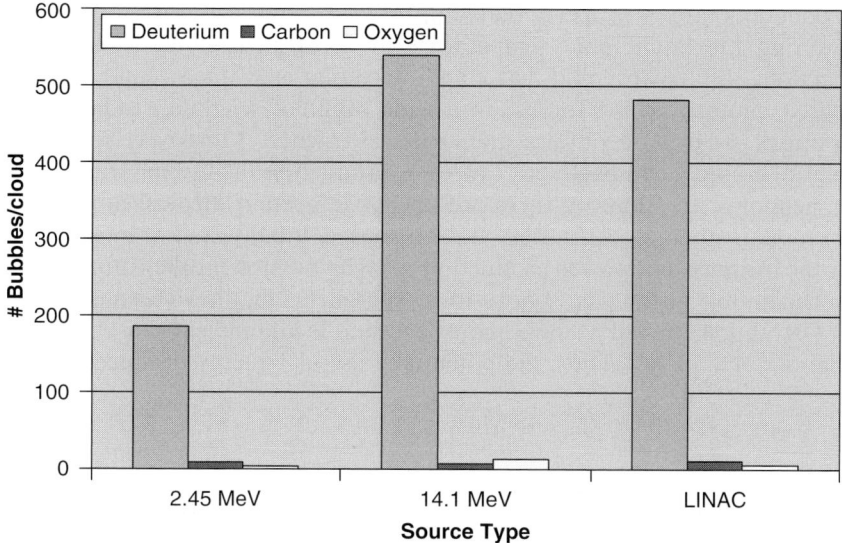

FIG. 43. Histogram of the average number of bubbles within each bubble cluster formed using the threshold model, separated by ion and neutron source type.

Figures 42 and 43 give the average number of bubbles per bubble cluster using various neutron sources for each of these models; Eqs. (103) and (104), respectively.

As expected, the linear model forms many more bubbles than the threshold model. As discussed in Appendix A, the number of bubbles in a cluster has important implications on the sonofusion process since it affects the pressure intensification within the bubble clusters during an implosion. Finally, it should be noted that, based on high speed video, Taleyarkhan et al. [54,56] estimated that there were about 1000 bubbles in each bubble cluster produced in the ORNL sonofusion experiments. This is generally consistent with Figs. 42 and 43 for the 14.1 MeV PNG neutron source used at ORNL, and these results indicate that the number of bubbles in a typical bubble cluster is bounded by the predictions of the linear and threshold models.

V. Potential Applications of Sonofusion Technology

It is too early to understand the full implications of sonofusion technology, however, it appears to be very promising. For example, it may offer important new ways to parametrically study the plasma physics and neutron cross sections associated with nuclear fusion processes. Also, if the neutron production rate is properly scaled-up (e.g., by optimizing the test liquid, lowering the liquid pool temperature or using deuterium–tritium (D/T) reactions, etc.), we could have an interesting new picosecond duration pulsed neutron source for use in a wide range of scientific studies in, for example, solid state physics and material science. Moreover, sonofusion might be used as a novel new, low cost, production facility for tritium and/or helium-3. In addition, there are always important medical applications for a relatively low cost pulsed neutron source. Finally, and of most interest to the authors, net power production may be possible in the future.

The bubble fusion (i.e., sonofusion) experiments that have been performed at ORNL [54,56] had a single acoustic antinode and indicated a D/D fusion rate of $\sim 4 \times 10^5$ n/s. Thus, the potentially useful[12] energy produced was

$$\dot{E}_{\text{out}} = (4 \times 10^5 \text{ n/s})(2.45 \text{ MeV/n})(1.6 \times 10^{-13} \text{ J/MeV}) \approx 2 \times 10^{-7} \text{ J/s} \tag{105}$$

The energy required to perform those experiments was about 5 W, thus

$$\dot{E}_{\text{in}} \sim 5 \text{ W} = 5 \text{ J/s} \tag{106}$$

Hence, using deuterium (D) as the fuel, we are currently about seven orders of magnitude from break-even. If a significant amount of tritium were also present in the liquid such that D/T fusion reactions would take place, then, because the mean reactivity cross section ($\langle \sigma v \rangle$) for D/T fusion is several orders of magnitude larger than that for D/D fusion (and 14.1 MeV D/T neutrons rather than 2.45 MeV D/D neutrons would be emitted), we could expect about a three order of magnitude increase in $\dot{E}_{\text{out}}$. Nevertheless, we would still be about four orders of magnitude from break-even. Thus, it is clear that in order to achieve break-even (an important goal in fusion power technology), the neutron yield will need to be significantly scaled-up; fortunately there appears to be opportunities to do so. For example, the number of possible deuterium ions per unit mass of D-acetone (C_3D_6O) is

$$\hat{n}_D = 6 \frac{N_A}{M} = \frac{6(6.02 \times 10^{26} \text{ kmol}^{-1})}{64 \text{ kg/kmol}} = 0.56 \times 10^{26}/\text{kg} \tag{107}$$

For the conditions achieved to-date in ORNL type sonofusion experiments, Eq. (92), the number of deuterium ions in the super-compressed core region

---

[12] The total energy release is about three (3) times larger but the recoil energy would be deposited in the liquid pool, which would, in the current generation of sonofusion experiments, need to be cooled to promote the fusion reaction.

of an imploded bubble is

$$n_D = V_{CORE} \rho_{CORE} \hat{n}_D = (4\pi/3)(R_\bullet^3)\rho_\bullet \hat{n}_D \quad (108a)$$

Thus,

$$n_D = (0.905 \times 10^{-21} \text{ m}^3)(10^4 \text{ kg/m}^3)(0.56 \times 10^{26}/\text{kg}) \quad (108b)$$

and

$$n_D \cong 5 \times 10^8 \text{ deuterons} \quad (108c)$$

If all of the deuterium ions fused, we would have 0.25×10^9 D/D fusion events (i.e., it takes two deuterium ions to fuse), and about half of these events would yield 2.45 MeV neutrons. Thus, the D/D fusion neutron yield would be increased about four orders of magnitude (i.e., to $\sim 10^9$ n/s) rather than $\sim 10^5$ n/s, which, when combined with the D/T reactions, would be near break-even.

Let us next consider what it would take to develop the neutron yield required for a commercial nuclear reactor. In order to have a 1000 MW$_e$ fusion power reactor ($\sim$3000 MW$_T$) we would need a D/D fusion neutron production rate ($\dot{n}$) of at least

$$\dot{n} = \frac{3 \times 10^9 \text{ J/s}}{(2.45 \text{ MeV/n})(1.6 \times 10^{-13} \text{ J/MeV})} = 0.77 \times 10^{22} \text{ n/s} \quad (109)$$

Thus, to use sonofusion technology for power production, we would need many active acoustic antinodes, in which fusion occurs.

It is also interesting to evaluate the energy density that has been achieved to date.

Assuming that the maximum packing of the bubbles within a bubble cluster is a void fraction of, $\varepsilon_{max} = 0.74$, and that the maximum diameter (D_b) of each of the $\sim$1000 bubbles within the clusters is $\sim$1 mm, we find that the maximum size of the bubble clusters (D_c) is about

$$\varepsilon = \frac{n_b V_b}{V_c} = \frac{n_b \frac{1}{6}\pi D_b^3}{\frac{1}{6}\pi D_c^3} \quad (110)$$

Thus,

$$0.74 = \frac{(1000)(1 \times 10^{-3} \text{ m})^3}{D_c^3}$$

which gives

$$D_{c_{max}} = 11 \text{ mm} \tag{111}$$

Hence

$$V_c = \frac{\pi}{6} D_{c_{max}}^3 = 0.7 \text{ cm}^3$$

Now the local power density $\left(\dot{E}'''\right)$ achieved to date during the ORNL bubble fusion experiments was

$$\dot{E}''' = \left(\frac{2 \times 10^{-7} \text{ J/s}}{0.7 \text{ cm}^3}\right) \cong 3 \times 10^{-7} \text{ W/cm}^3 = 0.3 \text{ W/m}^3 \tag{112}$$

which is obviously quite small.

However, if break-even could be achieved the local power density would be 3 MW/m^3, which is quite interesting even compared to the power densities in existing commercial nuclear fission reactors.

In any event, what is needed is, using improved experimental techniques, advanced working fluids and other fusion reactions (e.g., D/T), to increase the neutron production rate in each antinode to at least break-even and to establish a chain reaction between various adjacent (out-of-phase) antinodes [178]. Indeed, scale-up of the neutron yield is the current thrust of sonofusion research.

As noted at the start of this chapter, the "holy grail" of all fusion research is to find a new safe, environmentally friendly, way to produce energy for practical applications (e.g., electric energy production). At present, we are in a situation, which is somewhat analogous to being between the discovery of nuclear fission by Otto Hahn and Lisa Meitner and the demonstration of a nuclear chain reaction by Enrico Fermi. Much more research will be needed to determine whether sonofusion can be scaled-up sufficiently to be able to produce commercially interesting amounts of energy. Nevertheless, this exciting new technology appears to be inherently safe (i.e., no significant decay heat will be present after reactor shut-down) and environmentally friendly (i.e., the tritium (T) produced is a fuel which can be consumed in D/T reactions, and thus it should be possible to greatly minimize radioactive waste products).

Moreover, about 1.5%(by mass) of sea water contains deuterium and the earth's oceans contain about 20×10^{10} TW-years (i.e., 20×10^{16} MW-years) of energy if all the deuterium in sea water could be extracted and burned in a thermonuclear D/D fusion reactor. It is also interesting to note that the projected energy needs of the world by the year 2025 are about 27 TW-years, thus there is potentially about 10 orders of magnitude more energy available from D/D fusion than would be required at that time. Thus, if nuclear fusion reactors can be successfully developed, they will provide an essentially infinite supply of clean energy to meet the needs of mankind, and will greatly mitigate the expensive, politically dangerous, and environmentally damaging, dependence we have on fossil fuels. Clearly, it is worth the effort to pursue further research on fusion technology.

Only time will tell what the practical significance of sonofusion technology may be, however, it appears to be an interesting and challenging application of multiphase thermal-hydraulics and nuclear engineering & physics technologies, and as noted above, it offers great promise for mankind.

Appendix A: Linear and Non-linear Bubble Cluster Dynamics

The achievement of sonofusion implies that the local liquid pressure surrounding the imploding bubbles is not just the impressed external ($\sim \pm 15$) bar acoustic pressure field (which, by itself, is not large enough to create conditions suitable for D/D fusion [114]), rather it is coupled with the complicated dynamics of the entire bubble cluster, which may greatly intensify the liquid pressure within the imploding bubble cluster. The purpose of this appendix is to analyze the physical mechanisms, which can cause liquid-phase pressure intensification within an imploding bubble cluster.

During the last few decades there has seen substantial progress in the modeling of the dynamics of bubbly liquids [75,103,149,150]. It was shown that only for extremely low void fractions could the acoustic properties of bubbly liquids be analyzed on the basis of single-bubble dynamics. Otherwise, the bubble volume changes are strongly coupled with the liquid pressure distribution and velocity field of the interacting bubbles within bubble arrays or clusters. This also leads to a drastic modification of the sonic speed in the mixture, which is frequency-dependent.

The theoretical work on bubble cluster dynamics presented in the literature has been mainly focused on a continuum description of wave propagation in bubble clusters, assuming that these clusters contain a large number of bubbles. The acoustic properties of the bubble clusters for small amplitude bubble oscillations has been analyzed by Omta [151] and D'Agostino and Brennen [152]. They investigated a spherical bubble cloud subject to harmonic far-field pressure excitation and showed that the natural frequency of the bubble cluster is always lower than the natural frequency of the individual bubbles.

In this appendix, we consider various aspects of the dynamics of a cavitation bubble cluster in an acoustic pressure field [144]. In particular, we seek to understand the effect of the coupling between the bubble cluster dynamics and the flask dynamics on the resonance properties of the entire system. Moreover, we seek to quantify the mechanisms for the pressure intensification process during bubble cluster implosions.

One of the most appealing physical mechanisms for pressure intensification during cavitation bubble cluster implosions was proposed by Morch [98,153,154] and Hanson et al. [155]. They suggested that the implosion of a cluster of bubbles involves the formation and inward propagation of a shock wave and that the geometric focusing of this shock (i.e., cumulation) at the center of the bubble cluster creates significant liquid pressure intensification on some of the imploding bubbles within the collapsing bubble cluster.

This hypothesis was confirmed numerically by Wang and Brennen [99] and Reisman et al. [156]. In these papers, a collapsing bubble cluster was treated within the framework of a classical continuum bubbly fluid model [77,103], which is appropriate only for bubble clusters containing a large number of bubbles, but a low void fraction. Very complicated non-linear phenomena were obtained when the shock wave propagates to the center of the bubble cluster and produces a rebounding shock wave within the cluster. It was found that sometimes the central bubbles collapse first and form outgoing shocks, which start at the center and weaken as they propagate outward, while in other cases the outer bubbles collapse first, yielding intensifying shock waves which propagate to the center of the bubble cluster. It should be noted that the predicted intensities of the incoming and outgoing shocks in these papers were relatively low because the theoretical model used was only valid for moderate shock waves since the liquid was considered to be incompressible.

A more advanced continuum model, in which the compressibility of the liquid was taken into account, has been applied by Shimada et al. [157,158]. Numerical results show that high impulsive pressures are emitted from each bubble in the center region of the bubble cluster and pressure oscillations of high frequency was observed at the cluster's center.

In this appendix, a mathematical model of the bubble cluster dynamics and an effective numerical algorithm that allows liquid compression was used. This model indicates strong pressure intensification within the imploding bubble cluster.

The continuum models discussed above assume that there are many bubbles in the cluster and all the bubbles are spherical. Actually, the bubbles within a cluster may experience various disturbances, which can lead to severe departures from sphericity. Indeed, the pressure pulses emitted by neighboring bubbles in the cluster can induce an asymmetric collapse [159].

Theoretical investigations were also undertaken by Oguz [160], who reported extreme variability in the behavior of two bubbles close together and oscillating in an acoustic field of 20 kHz. This variability can be caused by slight changes in the inter-bubble distance, the bubble radii and the acoustic pressure. Chahine and Duraiswami [161] provided theoretical calculations of the behavior of bubbles belonging to small bubble clusters undergoing pressure variations. They conducted numerical simulations using five discrete bubbles. Initially the inter-bubble distance, the internal pressure and the size were the same for each bubble. These calculations indicated that the internal bubbles grow similarly to the peripheral bubbles, but this terminates with bubble distortion. Significantly, it was found that the bubbles on the periphery of the bubble cluster developed inwardly directed re-entrant jets, which can induce high pressures in the interior of the collapsing bubble cluster. These numerical calculations were successfully compared with analytical results obtained by asymptotic expansions.

The systematic mathematical modeling of vapor bubble cluster dynamics, which is applicable for high-intensity bubble implosions, various void fractions, bubble distortions and various numbers of bubbles, requires detailed 3-D hydrodynamic shock (i.e., HYDRO) code computations and thus requires significant computer resources. These multiscale calculations represent a major numerical challenge; indeed it is a "grand challenge."

In this appendix, we have used some simplified mathematical models to illuminate the essential pressure intensification mechanisms during the implosion of a bubble cluster, and to demonstrate that there may be an optimal number of bubbles in a cluster, which leads to significant pressure intensification during bubble cluster implosion.

Let us now consider the model and the results of linear and non-linear analyses of this model.

THE MATHEMATICAL MODEL FOR A BUBBLY LIQUID

Let α_ℓ and α_g be the volume fractions and ρ_ℓ and ρ_g the densities, of the liquid and the gas/vapor, respectively. Then the density of the two-phase mixture, in the bubbly liquid mixture is given by

$$\bar{\rho} = \alpha_\ell \rho_\ell + \alpha_g \rho_g \qquad (A.1)$$

where, $\alpha_\ell + \alpha_g = 1$. Note that the gas/vapor volume fraction depends on the instantaneous bubble radius (R)

$$\alpha_g = \frac{4}{3}\pi R^3 N''' \qquad (A.2)$$

where N''' is the number density of bubbles in the two-phase mixture.

Bubble-in-Cell Dynamics

The bubbly fluid model is based on the well-known "bubble-in-cell" technique [103], which divides the bubbly liquid mixture into cells, with each cell consisting of a liquid sphere of radius R_{LS}, with a bubble of radius R at its center. A similar approach has also been used to successfully model wet foam drop dynamics in an acoustic field [162].

In the bubble-in-cell technique, the radius, R_{LS}, of the liquid sphere comprising each cell and the embedded bubble, are related to the local void fraction at any instant by

$$\frac{R}{R_{LS}} = \alpha_g^{1/3} \tag{A.3}$$

We note that R_{LS}, R and α_g are variable in time, and $R < R_{LS} << R_c$, where R_c is the instantaneous radius of the bubble cluster.

Assuming spherical symmetry, the liquid conservation of mass equation and the assumed liquid incompressibility in the vicinity of the bubble, imply that the radial velocity around a single bubble is:

$$v' = \frac{R^2 \dot{R}}{r'^2}, \quad R \leq r' \leq R_{LS} \tag{A.4}$$

where $R(t)$ is the instantaneous radius of the bubble, R_{LS} the radius of the liquid cell of interest, the primed quantities denote local (bubble-in-cell) variables, namely, r' is the radial coordinate with origin at the bubble's center, v' the radial velocity of the liquid at radial location r' and the dot denotes a time derivative (i.e., $\dot{R} = dR/dt$).

The dynamics of the surrounding liquid is analyzed by assuming spherical symmetry and by writing the momentum equation for an incompressible, inviscid liquid as

$$\frac{\partial v'}{\partial t} + v' \frac{\partial v'}{\partial r'} + \frac{1}{\rho_\ell} \frac{\partial p'}{\partial r'} = 0 \tag{A.5}$$

Integrating this equation over $r' \leq R_{LS}$, from R to some arbitrary r', and using Eq. (A.4) yields

$$p' = p_R - \rho_\ell \left[R\ddot{R} + \frac{3}{2}\dot{R}^2 - \left(R\ddot{R} + 2\dot{R}^2\right)\frac{R}{r'} + \frac{1}{2}\dot{R}^2\left(\frac{R}{r'}\right)^4 \right] \tag{A.6}$$

where p' is the liquid pressure at location r' around a bubble, and p_R is the liquid pressure at the bubble's interface (i.e., $p_R \equiv p_\ell(r = R)$).

Evaluating Eq. (A.6) at the outer radius of the liquid cell, $r' = R_{LS}$, and using Eq. (A.3), results in

$$\frac{(p_R - p_{LS})}{\rho_\ell} = \left(1 - \alpha_g^{1/3}\right) R\ddot{R} + \frac{3}{2}\left(1 - \frac{4}{3}\alpha_g^{1/3} + \frac{1}{3}\alpha_g^{4/3}\right)\dot{R}^2 \quad \text{(A.7)}$$

This equation bears a strong resemblance to the well-known Rayleigh–Plesset equation [75] that governs the motion of a single bubble in an infinite liquid. Indeed, in the limit of vanishing void fraction, $\alpha_g \to 0$, the pressure at the outer radius of the cell (p_{LS}) becomes the liquid pressure far from the bubble ($p_{LS} \to p_\infty$), and Eq. (A.7) reduces to the Rayleigh–Plesset equation.

Let us assume that the gas pressure inside the bubble, p_g, is spatially uniform. The momentum jump condition requires that the gas pressure differ from the liquid pressure at the bubble wall due to surface tension and viscous terms, according to

$$p_R = p_g - \frac{2\sigma}{R} - \frac{4\mu_\ell \dot{R}}{R} \quad \text{(A.8)}$$

where σ is the surface tension and μ_ℓ the liquid viscosity. Furthermore, the gas pressure may be assumed to be governed by a polytropic gas law of the form

$$p_g = \left(p_0 + \frac{2\sigma}{R_0}\right)\left(\frac{R}{R_0}\right)^{-3\gamma} \quad \text{(A.9)}$$

where R_0 is the equilibrium radius of the bubble (i.e., the bubble radius at ambient pressure) and γ a polytropic exponent. Combining Eqs. (A.7)–(A.9) gives the following relationship between the liquid pressure at the edge of the cell (p_{LS}) and the bubble's instantaneous radius (R):

$$p_{LS} = \left(p_0 + \frac{2\sigma}{R_0}\right)\left(\frac{R}{R_0}\right)^{-3\gamma} - \frac{2\sigma}{R} - \frac{4\mu_\ell \dot{R}}{R}$$
$$- \rho_\ell\left[\left(1 - \alpha_g^{1/3}\right)R\ddot{R} + \frac{3}{2}\left(1 - \frac{4}{3}\alpha_g^{1/3} + \frac{1}{3}\alpha_g^{4/3}\right)\dot{R}^2\right] \quad \text{(A.10)}$$

Equation (A.10) can be integrated to find the bubble radius, $R(t)$, given the time-dependent pressure p_{LS} at $r' = R_{LS}$ and the initial conditions.

Liquid Compressibility

When the pressure waves are very intense, and the void fraction is small (i.e., $\alpha_g \to 0$), one should take into account liquid compressibility, which may lead to acoustic radiation by the oscillating bubbles. In order to account for these effects, the Rayleigh–Plesset equation may be extended [50,51]

$$R\ddot{R} + \frac{3}{2}\dot{R}^2 = \frac{(p_R - p_I)}{\rho_{\ell 0}} + \frac{R}{\rho_{\ell 0} C_\ell} \frac{d}{dt}(p_R - p_I) \qquad (A.11)$$

where C_ℓ is the speed of sound in the liquid, and p_I is the incident pressure on the bubbles. We note that since $p_R \equiv p_{\ell i}$, this is just Eq. (1).

Finally, for moderate compression levels, the equation of state of the liquid can be approximated as

$$p_\ell = p_0 + C_\ell^2(\rho_\ell - \rho_{\ell 0}) \qquad (A.12)$$

Dynamics of Bubbly Liquids

Next, we connect the various cells at points on their outer radii and replace the assemblage with an equivalent fluid whose dynamics approximate those of the bubbly liquid. We assume that the velocity of the translational motion of the bubbles in such a fluid is equal to the velocity of the bubbly fluid, $\underline{v}$. The fluid is required to conserve mass, thus

$$\frac{\partial \bar{\rho}}{\partial t} + \nabla \cdot (\bar{\rho}\,\underline{v}) = 0 \qquad (A.13)$$

where $\bar{\rho}$ is given by Eq. (A.1). The fluid also satisfies Euler's equation

$$\bar{\rho}\frac{D\underline{v}}{Dt} + \nabla p = 0 \qquad (A.14)$$

where $D/Dt = \partial/\partial t + \underline{v}\nabla$, the so-called material derivative, and p the mean fluid pressure, which is approximately the pressure at the outer radius of a cell, p_{LS}.

The mass and bubble densities are related by requiring that the total mass of each cell does not change in time. The initial volume of one cell is $V_0 = 1/N_0'''$, so the initial mass of the cell is $M_0 = \bar{\rho}_0/N_0'''$, where $\bar{\rho}_0 = \alpha_{\ell 0}\rho_{\ell 0} + \alpha_{g0}\rho_{g0}$. Requiring that the mass of each cell remain constant gives

$$\frac{\bar{\rho}}{N'''} = \frac{\bar{\rho}_0}{N_0'''} \qquad (A.15)$$

This set of equations describes the dynamics of a bubbly liquid mixture over a wide range of pressures and temperatures.

The Linear Dynamics of a Bubble Cluster. We can now linearize these equations by assuming that the time-dependent quantities vary slightly from their equilibrium values. Specifically, we write, $\bar{\rho} = \bar{\rho}_0 + \tilde{\rho}$, $N''' = N_0''' + \tilde{N}'''$, $p = p_0 + \tilde{p}$, $\underline{v} = \underline{\tilde{v}}$, $\alpha_g = \alpha_{g0} + \tilde{\alpha}_g$ and $R = R_0 + \tilde{R}$. The perturbed ($\sim$) quantities are assumed to be small, so that any product of the perturbations may be neglected. In a linear analysis we may assume that the liquid is incompressible, thus $\rho_\ell =$ constant. When these assumptions are introduced into Eq. (A.10), we arrive at the following linearized equation:

$$\tilde{p} = -\left[\left(p_0 + \frac{2\sigma}{R_0}\right)\frac{3\gamma}{R_0} - \frac{2\sigma}{R_0^2}\right]\tilde{R} - \frac{4\mu_\ell}{R_0}\frac{\partial \tilde{R}}{\partial t} - \rho_\ell\left(1 - \alpha_{g0}^{1/3}\right)R_0\frac{\partial^2 \tilde{R}}{\partial t^2} \quad (A.16)$$

Now, linearizing Eq. (A.15) gives

$$\tilde{N}''' = \frac{N_0'''}{\bar{\rho}_0}\tilde{\rho} \quad (A.17)$$

Similarly linearizing the density in Eq. (A.1), taking into account that $\rho_g/\rho_\ell \ll 1$, and using Eq. (A.17) yields

$$\tilde{\rho} = -4\pi R_0^2 \bar{\rho}_0 N_0''' \tilde{R} \quad (A.18)$$

where, $\bar{\rho}_0 \cong \rho_\ell(1 - \alpha_{g0})$. By combining Eqs. (A.16) and (A.18) we obtain a pressure–density relationship for the bubbly fluid

$$\tilde{p} = \frac{C_b^2}{\omega_b^2}\left(\omega_b^2 \tilde{\rho} + 2\eta\frac{\partial \tilde{\rho}}{\partial t} + \frac{\partial^2 \tilde{\rho}}{\partial t^2}\right) \quad (A.19)$$

where C_b represents the sound speed in the bubbly liquid

$$C_b^2 = \frac{3\kappa p_0 + (3\kappa - 1)2\sigma/R_0}{3\alpha_{g0}(1 - \alpha_{g0})\rho_\ell} \quad (A.20)$$

and ω_b is the natural frequency of a bubble in the cell

$$\omega_b^2 = \frac{\omega_M^2}{(1 - \alpha_{g0}^{1/3})} \quad (A.21)$$

We note that ω_M is the natural frequency of a single bubble in an infinite liquid (i.e., the so-called Minnaert frequency) is

$$\omega_M^2 = \frac{3\kappa p_0 + (3\kappa - 1)2\sigma/R_0}{\rho_\ell R_0^2} \quad (A.22)$$

In Eq. (A.19), the parameter η represents the dissipation coefficient due to the bulk liquid viscosity

$$\eta = \frac{2\mu_\ell}{\rho_\ell R_0^2 (1 - \alpha_{g0}^{1/3})} \quad \text{(A.23)}$$

Actually, acoustic wave damping in bubbly liquids occurs due to several different physical mechanisms: bulk liquid viscosity, gas/vapor thermal diffusivity, acoustic radiation, etc. The contribution of each of these dissipation mechanisms to the total dissipation during bubble oscillations depends on the frequency, bubble size, type of gas in a bubble and the liquid compressibility [103]. For convenience, we may thus use an effective dissipation coefficient in the form of a viscous dissipation.

$$\eta = \frac{2\mu_c}{\rho_\ell R_0^2 (1 - \alpha_{g0}^{1/3})} \quad \text{(A.24)}$$

where μ_c denotes the effective viscosity (instead of just the bulk liquid viscosity), which implicitly includes all the above-mentioned dissipation mechanisms. The value for μ_c must be chosen to fit experimental observations and the theoretical data of amplitude vs. frequency for a single bubble.

We can also linearize Eqs. (A.13) and (A.14) to obtain

$$\frac{\partial \tilde{\rho}}{\partial t} + \bar{\rho}_0 \nabla \cdot \underline{\tilde{v}} = 0 \quad \text{(A.25)}$$

$$\bar{\rho}_0 \frac{\partial \underline{\tilde{v}}}{\partial t} + \nabla \tilde{p} = 0 \quad \text{(A.26)}$$

Combining the time derivative in Eq. (A.25) with the divergence of Eq. (A.26) results in

$$\frac{\partial^2 \tilde{\rho}}{\partial t^2} = \nabla^2 \tilde{p} \quad \text{(A.27)}$$

Combining this result with the Laplacian of Eq. (A.19), and introducing a velocity potential ($\underline{v} = \nabla \varphi$), gives a wave equation for the bubbly liquid of the form

$$\frac{\partial^2 \varphi}{\partial t^2} - C_b^2 \nabla^2 \left(\varphi + \frac{2\eta}{\omega_b^2} \frac{\partial \varphi}{\partial t} + \frac{1}{\omega_b^2} \frac{\partial^2 \varphi}{\partial t^2} \right) = 0 \quad \text{(A.28)}$$

Dynamics of the Bubbly Liquid in an Acoustic Chamber. The wave equation for the velocity potential in the liquid between a bubble cluster and the flask's wall is [163]

$$\frac{\partial^2 \varphi}{\partial t^2} - C_\ell^2 \nabla^2 \left(\varphi + \frac{\beta}{C_\ell^2} \frac{\partial \varphi}{\partial t} \right) = 0 \qquad (A.29)$$

where β is the dissipation coefficient for the acoustic waves in the flask.

In general, there are several mechanisms causing acoustic wave damping in an acoustic chamber (i.e., flask) filled with a liquid. All these dissipative mechanisms may be divided into two groups: (1) dissipation in the liquid and (2) dissipation in the structure (i.e., flask, the acoustic transducer, etc.) and the associated electric circuit. The first group includes liquid viscosity, liquid thermal conductivity, liquid relaxations due to possible chemical reactions and excitation of the different molecular degrees of freedom, and any boundary layer effects. It is known that viscosity is the principal dissipation mechanism for liquids [164]. Dissipation mechanisms associated with the second group are difficult to estimate theoretically, although these mechanisms are often more important than liquid dissipation. That is why it is convenient to write down the dissipation coefficient, β, in the form of an effective viscous dissipation

$$\beta = \frac{4\mu_f}{3\rho_\ell} \qquad (A.30)$$

where μ_f denotes an effective viscosity (instead of bulk liquid viscosity), which, as before, implicitly includes all the above-mentioned dissipation mechanisms. The value for μ_f should be chosen to fit experimental observations of amplitude vs. frequency based on the acoustic chamber's characteristics.

Standing Acoustic Wave Field in the Bubble Cluster and Liquid. First, let us consider a spherical bubble cluster of radius R_c at the center of a spherical acoustic chamber (i.e., flask) of radius R_f. We denote all variables related to the bubble cluster and to the liquid in the flask (around the cluster) by subscripts c and f, respectively. Equations (A.28) and (A.29) must be solved together with the boundary conditions at the flask wall ($r = R_f$) and at the bubble cluster/liquid interface ($r = R_c$). To derive formulae for a standing acoustic wave field, let the velocity potentials inside the bubble cluster and in the liquid between the cluster and flask wall be harmonic in time

$$\varphi_j = \Phi_j \exp(i\omega t), \quad (j = c, f) \qquad (A.31)$$

where ω is the driving frequency. The substitution of Eq. (A.31) into Eqs. (A.28) and (A.29) results in Helmholtz equations for the complex amplitudes of the velocity potential

$$\nabla^2 \Phi_j + k_j^2 \Phi_j = 0, \quad (j = c, f) \tag{A.32}$$

where

$$k_c^2 = \frac{\omega^2}{C_b^2 \left(1 + i\frac{2\eta\omega}{\omega_b^2} - \frac{\omega^2}{\omega_b^2}\right)}, \quad k_f^2 = \frac{\omega^2}{C_\ell^2 \left(1 + \frac{i\beta\omega}{C_\ell^2}\right)} \tag{A.33}$$

The general solutions for the Helmholtz equation, Eq. (A.32), for spherical symmetry are

$$\varphi_c = \left[A \frac{\mathrm{Sin}(k_c r)}{k_c r} + B \frac{\mathrm{Cos}(k_c r)}{k_c r}\right] \exp(i\omega t) \tag{A.34a}$$

$$\varphi_f = \left[C \frac{\mathrm{Sin}(k_f r)}{k_f r} + D \frac{\mathrm{Cos}(k_f r)}{k_f r}\right] \exp(i\omega t) \tag{A.34b}$$

and the fluid velocities and pressures are calculated as

$$v_c = \frac{\partial \varphi_c}{\partial r} \tag{A.35a}$$

$$p_c = p_0 - \rho_\ell \alpha_{\ell 0} \frac{\partial \varphi_c}{\partial t} \tag{A.35b}$$

$$v_f = \frac{\partial \varphi_f}{\partial r} \tag{A.36a}$$

$$p_f = p_0 - \rho_\ell \frac{\partial \varphi_f}{\partial t} \tag{A.36b}$$

The constants A, B, C and D must be determined so that we satisfy the following boundary conditions:

$$\begin{aligned} v_c &= 0, & r &= 0 \\ p_f &= p_0 + \Delta p \exp(i\omega t), & r &= R_f \\ v_c &= v_f, & r &= R_c \\ p_c &= p_f, & r &= R_c \end{aligned} \tag{A.37}$$

The resulting pressure distribution in the bubble cluster is

$$p_c(r,t) = p_0 + \frac{\Delta p}{F(\omega)} \frac{\mathrm{Sin}(k_c r)}{k_c r} \exp(i\omega t) \qquad (A.38)$$

where

$$F(\omega) = \frac{k_f R_c}{\alpha_{\ell 0}} \left\{ \alpha_{\ell 0} \frac{\mathrm{Cos}(k_f R_f - k_f R_c)}{k_f R_f} \frac{\mathrm{Sin}(k_c R_c)}{k_c R_c} \right. \\
\left. + \frac{\mathrm{Sin}(k_f R_f - k_f R_c)}{k_f R_f k_f R_c} \left[\mathrm{Cos}(k_c R_c) - \alpha_{g 0} \frac{\mathrm{Sin}(k_c R_c)}{k_c R_c} \right] \right\} \qquad (A.39)$$

In the case of a liquid-filled spherical acoustic chamber (i.e., one having no bubble cluster) the pressure distribution in the flask has a similar functional form to that of a bubble cluster:

$$p_\ell(r,t) = p_0 + \Delta p \frac{k_f R_f}{\mathrm{Sin}(k_f R_f)} \frac{\mathrm{Sin}(k_f r)}{k_f r} \exp(i\omega t) \qquad (A.40)$$

Linear Results

Figure A.1 gives the non-dimensional liquid pressure amplitude at the center of a liquid-filled spherical flask (having no bubble cluster) vs. frequency, for an effective viscosity of $\mu_f = 10^3 \, \mathrm{kg/(m\,s)}$. Note that the higher harmonics have a lower amplitude than the fundamental. It is expected that $\mu_f = 10^3 \, \mathrm{kg/(m\,s)}$ is an appropriate value of effective viscosity, because it provides a good match with experimental observations that for a water-filled flask, the acoustic pressure amplitude at the center of a spherical flask (at resonance) is about 70 times higher than the pressure at the flask wall [165].

Let us now consider a liquid-filled spherical flask with a bubble cluster at the center. The dimensionless pressure amplitude at the center of a bubble cluster is plotted as a function of dimensionless frequency in Figs. A.2–A.4. The following set of parameters has been used for these calculations: $R_f = 10^{-1}\,\mathrm{m}$, $R_c = 10^{-1}\,\mathrm{m}$, $R_0 = 10^{-4}\,\mathrm{m}$, $p_0 = 10^5\,\mathrm{Pa}$, $\rho_\ell = 10^3\,\mathrm{kg/m^3}$, $C_\ell = 1.5 \times 10^3\,\mathrm{m/s}$, $\gamma = 1.4$, $\mu_c = 10^{-1}\,\mathrm{kg/(m\,s)}$, $\mu_f = 10^3\,\mathrm{kg/(m\,s)}$, $\sigma = 0.0725\,\mathrm{N/m}$. The plots were done for different numbers of bubbles in the cluster. The resonances related to flask and bubble cluster resonances are marked as f_k and c_k, respectively, where k denotes the order of the harmonic.

When the number of bubbles in the cluster is large (Fig. A.2), the resonance frequencies related to flask and cluster resonances are well separated; all significant bubble cluster resonance frequencies are below the first

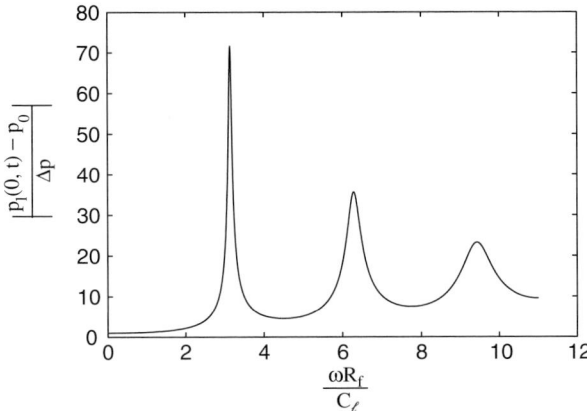

FIG. A.1. Dimensionless liquid pressure amplitude at the center of a liquid-filled spherical flask vs. dimensionless frequency: $\mu_f = 103 \text{ kg/(m s)}$, $R_f = 10^{-1}$ m, $\rho_\ell = 10^3 \text{ kg/m}^3$, $C_\ell = 1.5 \times 10^3$ m/s.

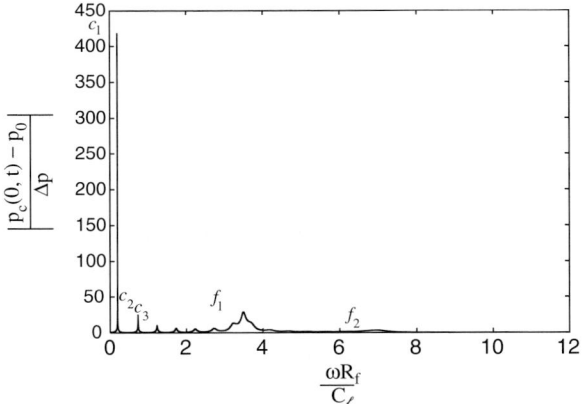

FIG. A.2. The dimensionless pressure amplitude at the center of the bubble cluster vs. dimensionless frequency: $R_f = 10^{-1}$ m, $R_c = 10^{-2}$ m, $R_0 = 10^{-4}$ m, $p_0 = 10^5$ Pa, $\rho_\ell = 10^3 \text{ kg/m}^3$, $C_\ell = 1.5 \times 10^3 \text{m/s}$, $\kappa = 1.4$, $\mu_c = 10^{-1} \text{ kg/(m s)}$, $\mu_f = 10^3 \text{ kg/(m s)}$, $\sigma = 0.0725$ N/m. Number of bubbles in the cluster $N = 5 \times 10^5$. The resonances related to flask and cluster resonances are marked as f_k and c_k, respectively, where k denotes the harmonic.

flask resonance (f_1), and those that are higher than the first flask resonance are strongly damped.

For a smaller number of bubbles the bubble cluster resonance frequencies become larger so that some of them start to overlap with the first flask

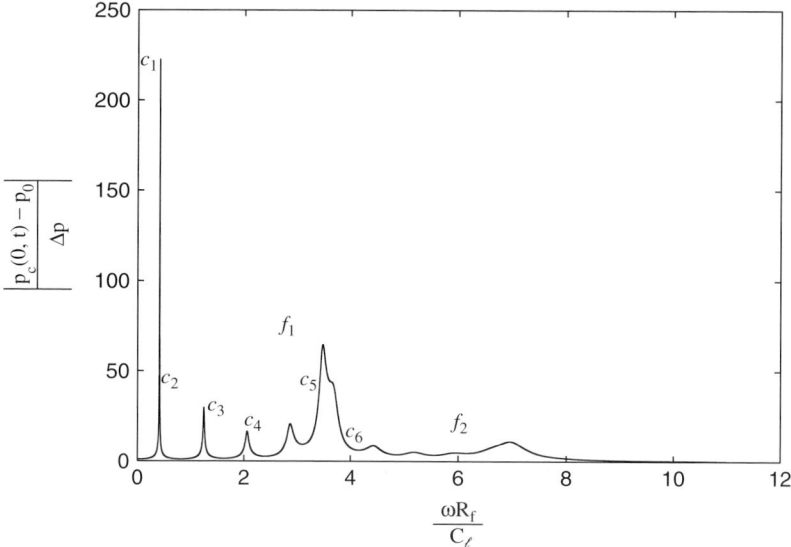

FIG. A.3. The dimensionless pressure amplitude at the center of the cluster vs. dimensionless frequency: $R_f = 10^{-1}$ m, $R_c = 10^{-2}$ m, $R_0 = 10^{-4}$ m, $p_0 = 10^5$ Pa, $\rho_\ell = 10^3$ kg/m^3, $C_\ell = 1.5 \times 10^3$ m/s, $\kappa = 1.4$, $\mu_c = 10^{-1}$ kg/(m s), $\mu_f = 10^3$ kg/(m s), $\sigma = 0.0725$ N/m. Number of bubbles in the cluster $N = 10^5$. The resonances related to flask and cluster resonances are marked as f_k and c_k, respectively, where k denotes the harmonic.

resonance frequency and pass through it (e.g., see resonances f_1 and c_5 in Fig. A.3).

As can be seen in Fig. A.4, with a further decrease in the number of bubbles the first flask resonance interacts with cluster resonances whose intensity is comparable with it. As a result, the first flask resonance moves to higher frequencies, being replaced by the dominant higher cluster resonances.

Gradually, when the number of bubbles is further reduced, the flask resonances are replaced by the cluster resonances. This process is clearly seen in Fig. A.5.

The pressure amplitude for the first cluster resonance is presented in Fig. A.6 as a function of the number of bubbles in the cluster. There is a local maximum of the pressure amplitude when the number of bubbles is around 1200. This maximum happens when the first flask resonance is replaced by the first cluster resonance. The shaded area in Fig. A.6 is due to rapid changes in pressure amplitude with number of bubbles (for relatively a large number of bubbles).

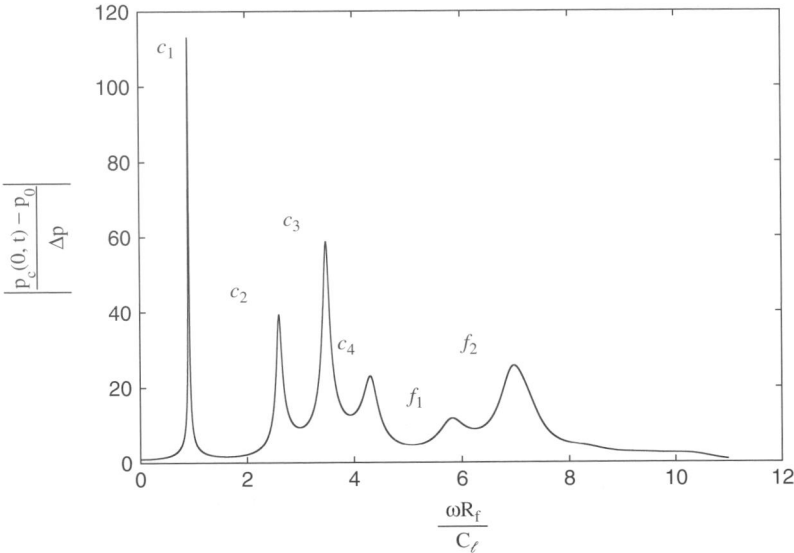

FIG. A.4. The pressure dimensionless amplitude at the center of the cluster vs. dimensionless frequency: $R_f = 10^{-1}$ m, $R_c = 10^{-2}$ m, $R_0 = 10^{-4}$ m, $p_0 = 10^5$ Pa, $\rho_\ell = 10^3$ kg/m^3, $C_\ell = 1.5 \times 10^3$ m/s $\kappa = 1.4$, $\mu_c = 10^{-1}$ kg/(m s), $\mu_f = 10^3$ kg/(m s), $\sigma = 0.0725$ N/m. Number of bubbles in the cluster $N = 2 \times 10^4$. The resonances related to flask and cluster resonances are marked as f$_k$ and c$_k$, respectively, where k denotes the harmonic.

These are interesting and potentially important observations; indeed, since the number of bubbles in the bubble clusters created in the ORNL bubble fusion experiments [54,56] was ~1000, these clusters would be expected to experience strong implosions (due to interaction of the flask and bubble cluster resonance modes). Moreover, if alternative liquids are to be tested, calculations should be made of the resonance frequencies and the number of bubbles expected in the cluster (e.g., formed by neutron-induced nucleation) to determine if these bubble cluster implosions would give significant pressure intensification.

Not only can we get global pressure intensification (due to the resonant interaction phenomena discussed above) we also expect significant local pressure intensification within the imploding bubble cluster. However, a non-linear analysis is required to predict this phenomena and it will be considered in the next section.

The Non-linear Dynamics of a Bubble Cluster: The Initial Stage. Here, we simulate a bubble cluster implosion when the liquid pressure around the cluster is taken to be constant, p_c, and equal to the initial liquid pressure in

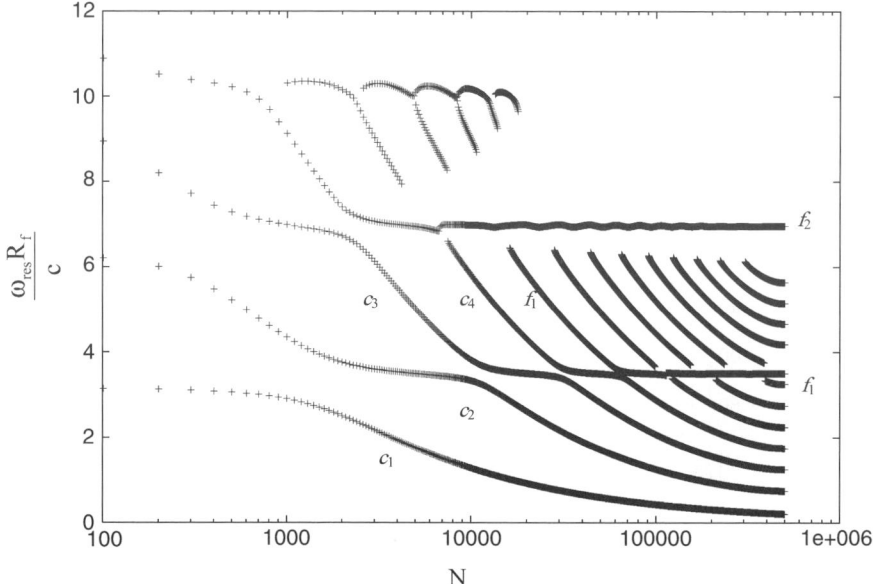

FIG. A.5. Dimensionless resonance frequencies vs. number of bubbles in the cluster: $R_f = 10^{-1}$ m, $R_c = 10^{-2}$ m, $R_0 = 10^{-4}$ m, $p_0 = 10^5$ Pa, $\rho_\ell = 10^3$ kg/m^3, $C_\ell = 1.5 \times 10^3$ m/s, $\kappa = 1.4$, $\mu_c = 10^{-1}$ kg/(m s), $\mu_f = 10^3$ kg/(m s), $\sigma = 0.0725$ N/m. The resonances related to flask and cluster resonances are marked as f_k and c_k, respectively, where k denotes the harmonic.

the cluster. The vapor cavities at the beginning of cluster contraction may be modeled as being empty because the vapor pressure is negligibly small relative to p_c almost all the time during the implosion process except during the final stage of implosion.

The liquid may be treated as incompressible, accounting for the characteristic time of single bubble collapse, t_c, which is much larger than the time duration it takes for sound to travel between the neighboring bubbles.

$$t_c \approx \sqrt{\frac{3\rho_\ell}{2p_c}} R_0 \gg \frac{R_c}{C_\ell N^{1/3}} \qquad (A.41)$$

Thus, the approximation of an incompressible liquid is valid when

$$\frac{R_0}{R_c} \gg \sqrt{\frac{2p_c}{3\rho_\ell}} \frac{1}{C_\ell N^{1/3}} \qquad (A.42)$$

For example, according to Eq. (A.42), for a 1 cm radius bubble cluster having N = 1000 bubbles in water under atmospheric pressure conditions,

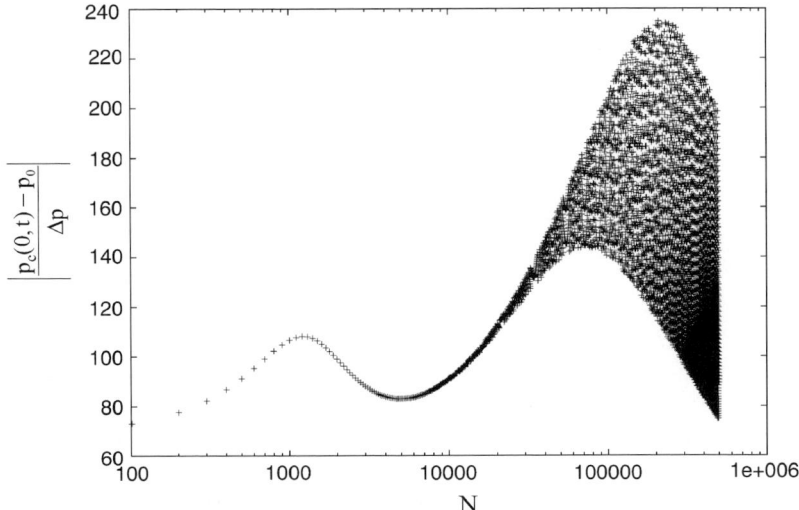

FIG. A.6. The dimensionless pressure amplitude at first cluster resonance (c_1) vs. number of bubbles: $R_f = 10^{-1}$ m, $R_c = 10^{-2}$ m, $R_0 = 10^{-4}$ m, $p_0 = 10^5$ Pa, $\rho_\ell = 10^3$ kg/m^3, $C_\ell = 1.5 \times 10^3$ m/s, $\kappa = 1.4$, $\mu_c = 10^{-1}$ kg/m s, $\mu_f = 10^3$ kg/m s, $\sigma = 0.0725$ N/m.

the initial bubble size (R_0) must be much more than 5 μm (which corresponds to void fractions $\gg 10^{-7}$).

Such a mechanical system with zero pressure in the bubbles has no elasticity and therefore has zero speed of sound. This means that the global fluid motion in the cluster is controlled only by liquid inertia and the volume of the bubbles, which is coupled with the local liquid pressure and local liquid inertia.

The density of the bubble cluster is calculated using

$$\bar{\rho} = \rho_\ell(1 - \alpha_g), \quad \alpha_g = \frac{4}{3}\pi R^3 N''', \quad \rho_\ell = \text{constant} \quad \text{(A.43)}$$

Conservation of mass and the number of bubbles in the case of spherically symmetric motion are described by the following equations (compare with Eq. (A.13)):

$$\frac{D\bar{\rho}}{Dt} + \frac{\rho}{r^2}\frac{\partial}{\partial r}(r^2 v) = 0 \quad \text{(A.44)}$$

or using Eq. (A.15):

$$\frac{DN'''}{Dt} + \frac{N'''}{r^2}\frac{\partial}{\partial r}(r^2 v) = 0 \quad \text{(A.45)}$$

where v is the radial velocity of liquid, and D/Dt the material derivative $\left(D/Dt = \partial/\partial t + v\partial/\partial r\right)$.

Equations (A.43)–(A.45) may be transformed into

$$\frac{1}{r^2}\frac{\partial}{\partial r}\left(r^2 v\right) = 4\pi R^2 N'''(R)\frac{dR}{dt} \qquad \text{(A.46a)}$$

$$N'''(R) = \frac{N_0'''}{1 - \alpha_{g0}\left[1 - (R/R_0)^3\right]} \qquad \text{(A.46b)}$$

where it was assumed that initial bubble number density, N_0''', and bubble radius, R_0, are uniform in the cluster and α_{g0} is the initial void fraction.

Substitution of Eq. (A.46b) into Eq. (A.46a) leads to the following equation coupling the volume of the bubbles with the global liquid motion:

$$\frac{1}{r^2}\frac{\partial}{\partial r}\left(r^2 v\right) = \frac{d}{dt}\left\{\ell n\left[1 - \alpha_{g0}\left(1 - \frac{R^3}{R_0^3}\right)\right]\right\} \qquad \text{(A.47)}$$

To analyze qualitatively the initial stage of bubble cluster contraction, let us consider a zero-order approximation (which may be called the "homodisperse" approximation) and assume that, in the beginning, all the bubbles in the cluster contract synchronously. Mathematically, this means that bubble radius depends on time but not on the spatial coordinate (i.e., $R = R(t)$). In this case, the right-hand side of Eq. (A.46a) depends only on time. Let us denote it as

$$A(t) \equiv 4\pi R^2 N'''(R)\frac{dR}{dt} \qquad \text{(A.48)}$$

Then Eq. (A.46a) can be easily integrated leading to the following velocity distribution within a cluster induced by contraction of the bubbles:

$$v = \frac{1}{3}A(t)r \qquad \text{(A.49)}$$

When the bubbles are contracting A is negative and Eq. (A.49) describes a motion converging to the center of the cluster. As example, Eq. (A.49) leads to the following formula for the evolution of the radius of a bubble cluster:

$$R_c = R_{c0}\exp\left(\frac{1}{3}\int_0^t A(t')\,dt'\right) = R_{c0}\left[1 - \alpha_{g0}\left(1 - \frac{R^3}{R_0^3}\right)\right]^{1/3} \qquad \text{(A.50)}$$

To calculate the pressure field coupled with such an induced motion one should use the momentum conservation equation (Eq. (A.14))

$$\bar{\rho} \frac{Dv}{Dt} = -\frac{\partial p}{\partial r} \tag{A.51}$$

which together with Eqs. (A.49), (A.48) and (A.43) leads to the following equation for the pressure gradient:

$$-\frac{\partial p}{\partial r} = B(t)r \tag{A.52a}$$

where

$$B(t) = \frac{\rho_\ell}{9}\left[1 - \frac{4}{3}\pi R^3 N'''(R)\right]\left(3\frac{dA}{dt} + A^2\right) \tag{A.52b}$$

Integration of Eq. (A.52a) gives a parabolic spatial pressure distribution

$$p = p_c + \frac{1}{2}B(t)(R_c^2 - r^2) \tag{A.53}$$

and the following formula for the pressure at the center of the bubble cluster:

$$p|_{r=0}(t) = p_c + \Delta p(t), \quad \Delta p(t) = \frac{1}{2}B(t)R_c^2 \tag{A.54}$$

Thus if one knows how the bubble radii changes in time during the collapse, it is possible to calculate the pressure everywhere in the cluster. In order to simulate a synchronous implosion for all the bubbles in a cluster one should evaluate the Rayleigh equation with the driving liquid pressure equal to some mean cluster pressure, $\bar{p}_c$, which is experienced by all the bubbles in the cluster. As far as the pressure in the cluster is coupled with the bubble dynamics, one can consider the mean cluster pressure to be a function of bubble radius, $\bar{p}_c(R)$.

$$R\ddot{R} + \frac{3}{2}\dot{R}^2 = -\frac{\bar{p}_c(R)}{\rho_\ell} \tag{A.55}$$

The Rayleigh equation, Eq. (A.55), has the following energy integral:

$$\left(\frac{dR}{dt}\right)^2 = \frac{2}{R^3}\int_R^{R_0} \frac{R^2 \bar{p}_c(R)}{\rho_\ell} dR \tag{A.56}$$

To consider the initial stage of bubble cluster collapse, one can make a linear approximation for $\bar{p}_c(R)$.

$$\bar{p}_c(R) \cong p_c^{(0)} + p_c^{(1)}\tilde{R}, \quad R = R_0(1+\tilde{R}) \tag{A.57}$$

where $p_c^{(0)}$, $p_c^{(1)}$ are unknown coefficients, which will be evaluated later. Then the solution of Eq. (A.56) may be simplified as follows:

$$\dot{R} \cong -\sqrt{\frac{2p_c^{(0)}}{\rho_\ell}}\left[\sqrt{-\tilde{R}} + \left(1 - \frac{p_c^{(1)}}{4p_c^{(0)}}\right)(-\tilde{R})^{3/2}\right] \tag{A.58}$$

Note that for a collapsing bubble cluster, $\tilde{R} < 0$. Substituting Eq. (A.58) into Eq. (A.48) one obtains an approximate formula for A(t).

$$A \cong -4\pi R_0^2 N_0''' \sqrt{\frac{2p_c^{(0)}}{\rho_\ell}}\left[\sqrt{-\tilde{R}} + \left(3\alpha_{g0} - 1 - \frac{p_c^{(1)}}{4p_c^{(0)}}\right)(-\tilde{R})^{3/2}\right] \tag{A.59}$$

Using Eq. (A.59) it is easy to show that

$$\dot{A} \cong -\frac{4\pi R_0 N_0''' p_c^{(0)}}{\rho_\ell}\left[1 + \left(9\alpha_{g0} - 2 - \frac{p_c^{(1)}}{p_c^{(0)}}\right)(-\tilde{R})\right] \tag{A.60}$$

$$A^2 \cong \frac{32\pi^2 R_0^4 N_0'''^2 p_c^{(0)}}{\rho_\ell}(-\tilde{R}) \tag{A.61}$$

Substituting Eqs. (A.60), (A.61) and (A.52b) into Eq. (A.53) yields

$$p(\eta, t) \cong p_c - \tfrac{1}{2}\alpha_{g0}\alpha_{\ell 0}p_c^{(0)}\left(\frac{R_{c0}}{R_0}\right)^2(1-\eta^2)\left[1 + \left(8\alpha_{g0} - 2 - \frac{p_c^{(1)}}{p_c^{(0)}}\right)(-\tilde{R})\right],$$
where, $\eta = \frac{r}{R_c}$

$$\tag{A.62}$$

This pressure distribution may be used in the following integral for pressure:

$$\bar{p}_c = \frac{3}{4\pi}\int_0^1 p(\eta, t)4\pi\eta^2 \, d\eta = p_c^{(0)} + p_c^{(1)}\tilde{R} \tag{A.63}$$

Eqation (A.63) is a natural equation to evaluate the parameters $p_c^{(0)}$ and $p_c^{(1)}$. That is

$$p_c^{(0)} = \frac{p_c}{1 + \frac{1}{5}\alpha_{g0}\alpha_{\ell 0}(R_{c0}/R_0)^2}, \quad p_c^{(1)} = \frac{\frac{8}{5}\alpha_{g0}\alpha_{\ell 0}(R_{c0}/R_0)^2(\alpha_{g0} - \frac{1}{4})}{\left[1 + \frac{1}{5}\alpha_{g0}\alpha_{\ell 0}(R_{c0}/R_0)^2\right]^2} p_c$$

(A.64)

So, at the initial moment the mean cluster pressure, $\bar{p}_c$, drops down to a value equal to $p_c^{(0)}$, $(0 < p_c^{(0)} < p_c)$, due to the pressure distribution which has a slope directed to the center of the cluster. This pressure gradient initiates implosive fluid motion toward the cluster's center.

It is easy to see from Eq. (A.64) that there is a critical void fraction

$$\alpha_* = \frac{1}{4}$$

(A.65)

For relatively small void fractions ($\alpha_{g0} < \alpha_*$), the mean bubble cluster's liquid pressure, $\bar{p}_c$, increases during bubble collapse. This indicates the tendency of the pressure inside the cluster to become uniform, leading to synchronous bubble collapse throughout the cluster. This agrees with D'Agostino and Brennen [152] who also did a linear analysis of gas bubble cluster dynamics. It was shown there that when the void fraction in the bubble cloud was small enough, the natural frequency of the bubble cluster was close to that of the individual gas bubbles in the cluster. This implies that the bubbles in the cluster tend to behave as individual gas bubbles in an infinite fluid, because the pressure is almost uniform inside the cluster and bubble/bubble interaction effects are minor.

In contrast, when the void fraction is large (i.e., $\alpha_{g0} > \alpha_*$), the mean cluster pressure decreases during bubble cluster collapse. This indicates the tendency of pressure to form a boundary layer near the cluster's interface. Moreover, this implies that the bubbles near the cluster's interface experience higher liquid pressure than the ones at the cluster's center and therefore collapse first, eventually emitting shock waves, which propagate inwardly.

The Non-linear Dynamics of a Bubble Cluster: The Final Stage. Here, we consider the bubble cluster as a bubbly/liquid mixture, with the non-linear equations, which describe bubble cluster dynamics written in Lagrangian coordinates

Mass conservation

$$\frac{\rho}{\rho_0}\left(\frac{r}{\xi}\right)^2 \frac{\partial r}{\partial \xi} = 1, \quad \frac{\partial r}{\partial t} = v$$

or equivalently [103]

$$\frac{\partial \rho}{\partial t} + \frac{2\rho v}{r} + \frac{\rho^2}{\rho_o}\left(\frac{r}{\xi}\right)^2 \frac{\partial v}{\partial \xi} = 0 \quad (A.66a)$$

Momentum conservation

$$\frac{\partial v}{\partial t} = -\frac{1}{\rho_0}\left(\frac{r}{\xi}\right)^2 \frac{\partial p}{\partial \xi} \quad (A.66b)$$

where r and $\xi = r - \int_0^t v(t',\xi)\,dt'$ are the Eulerian and Lagrangian coordinates, respectively, and $(0 \leq \xi \leq R_{c0})$.

The pressure within the vapor bubbles, p_g, is equal to the saturated pressure of the vapor, p_v, at the liquid pool temperature, with the exception of during the final stage of the implosion, when the vapor compression inside the bubble is almost adiabatic

$$p_g = \begin{cases} p_v, & R > R_* \\ p_v\left(\frac{R_*}{R}\right)^{3\gamma}, & R \leq R_* \end{cases} \quad (A.67)$$

where the equilibrium bubble radius is given by $R_* = R_{\max}/\sqrt[3]{10}$, and $R_{\max}$ is the maximum radius that the bubble achieves in the process of expansion. We note that, based on detailed hydrodynamic shock (i.e., HYDRO) code simulations of single bubble collapse, the final mass of the non-condensed vapor is about 10 times less than its maximum mass [104,114,166].

The appropriate initial and boundary condition are $v = 0$ for $\xi = 0$, and $p = p_c$ for $\xi = R_{c0}$, where p_c is the pressure on the surface of the bubble cluster, which is assumed to be

$$p_c(t) = p_s + \Delta p_c \sin(\omega t) \quad (A.68)$$

Let us rewrite the equations of the bubbly liquid dynamics, Eqs. (A.66a) and (A.66b), in a form which is more convenient for numerical integration [103]. We derive the second derivative of velocity with respect to time and the Lagrangian coordinate from the equation of motion (Eq. (A.66b))

$$\frac{\partial^2 v}{\partial t \partial \xi} = \frac{\partial}{\partial \xi}\left[-\frac{1}{\rho_0}\left(\frac{r}{\xi}\right)^2 \frac{\partial p}{\partial \xi}\right] = -\frac{1}{\rho_0}\left(\frac{r}{\xi}\right)^2 \left[\frac{\partial^2 p}{\partial \xi^2} - \frac{2}{\xi}\left(1 - \frac{\rho_0}{\rho}\left(\frac{r}{\xi}\right)^3\right)\frac{\partial p}{\partial \xi}\right]$$

(A.69)

The same derivative may also be calculated from the mass conservation equation, Eq. (A.66a)

$$\frac{\partial^2 v}{\partial t \partial \xi} = \frac{\rho_0}{\rho}\left(\frac{\xi}{r}\right)^2 \left[\frac{2}{\rho^2}\left(\frac{\partial \rho}{\partial t}\right)^2 + \frac{2}{\rho}\frac{\partial \rho}{\partial t}\frac{2v}{r} - \frac{1}{\rho}\frac{\partial^2 \rho}{\partial t^2} + \frac{2}{r}\frac{1}{\rho_0}\left(\frac{r}{\xi}\right)^2 \frac{\partial p}{\partial \xi} + \frac{6v^2}{r^2}\right] \tag{A.70}$$

Thus, equating Eqs. (A.69) and (A.70), the spherically symmetric equation for the liquid pressure within the bubble cluster is

$$\frac{\partial^2 p}{\partial \chi^2} - \frac{2}{\chi}\left[1 - \frac{2\rho_0}{\rho}\left(\frac{\chi}{r}\right)^3\right]\frac{\partial p}{\partial \chi} = \frac{\rho_0^2}{\rho}\left(\frac{\chi}{r}\right)^4 \left[\frac{1}{\rho}\frac{\partial^2 \rho}{\partial t^2} - \frac{2}{\rho^2}\left(\frac{\partial \rho}{\partial t}\right)^2 - \frac{2}{\rho}\frac{\partial \rho}{\partial t}\frac{2v}{r} - \frac{6v^2}{r^2}\right] \tag{A.71}$$

Equation (A.71) is valid for both the case of a pure liquid and for a bubbly mixture. Note for the latter case that

$$\frac{1}{\rho}\frac{\partial \rho}{\partial t} = \frac{1}{\bar{\rho}_\ell}\frac{\partial \bar{\rho}_\ell}{\partial t}, \quad \frac{1}{\rho}\frac{\partial^2 \rho}{\partial t^2} = \frac{1}{\bar{\rho}_\ell}\frac{\partial^2 \bar{\rho}_\ell}{\partial t^2}, \quad \frac{\partial \bar{\rho}_\ell}{\partial t} = \alpha_\ell \frac{\partial \rho_\ell}{\partial t} + \rho_\ell \frac{\partial \alpha_\ell}{\partial t}$$

$$\frac{\partial \alpha_\ell}{\partial t} = -\frac{\partial \alpha_g}{\partial t}, \quad \frac{\partial \alpha_g}{\partial t} = \alpha_\ell \alpha_g \left(\frac{3\dot{R}}{R} + \frac{1}{\rho_\ell}\frac{\partial \rho_\ell}{\partial t}\right)$$

$$\frac{\partial^2 \alpha_g}{\partial t^2} = \frac{\alpha_\ell - \alpha_g}{\alpha_\ell \alpha_g}\left(\frac{\partial \alpha_g}{\partial t}\right)^2 + \alpha_\ell \alpha_g \left[R\ddot{R} - \frac{3\dot{R}^2}{R^2} - \frac{1}{\rho_\ell^2}\left(\frac{\partial \rho_\ell}{\partial t}\right)^2 + \frac{1}{\rho_l}\frac{\partial^2 \rho_l}{\partial t^2}\right]$$

Finally, we may obtain the equation for the liquid pressure within the bubble cluster in the following form:

$$\frac{\partial^2 p}{\partial \xi^2} - \frac{2}{\xi}\left[1 - \frac{2\rho_0}{\rho}\left(\frac{\xi}{r}\right)^3\right]\frac{\partial p}{\partial \xi}$$
$$= \frac{\rho_0^2}{\rho}\left(\frac{\xi}{r}\right)^4 \left\{\alpha_\ell\left[\frac{1}{\rho_\ell}\frac{\partial^2 \rho_\ell}{\partial t^2} - \frac{2}{\rho_\ell^2}\left(\frac{\partial \rho_\ell}{\partial t}\right)^2 - \frac{4v}{r}\frac{1}{\rho_\ell}\frac{\partial \rho_\ell}{\partial t}\right] - \frac{6v^2}{r^2} \right. \tag{A.72}$$
$$\left. -\alpha_g\left[\frac{3}{R^2}\left(R\ddot{R} + 2\dot{R}^2\right) - \frac{12v\dot{R}}{rR}\right]\right\}$$

where

$$\frac{\partial \rho_\ell}{\partial t} = \frac{d\rho_\ell}{dp}\frac{\partial p}{\partial t} = \frac{1}{C_\ell^2}\frac{\partial p}{\partial t} \tag{A.73}$$

We note that if we assume that the liquid is incompressible (i.e., C_ℓ is infinite), then Eq. (A.72) becomes an ordinary differential equation which is readily integrated numerically. In general, however, Eqs. (A.72) and (A.73) give more accurate results.

Let $t = 0$ be the moment of nucleation of the microbubbles in a bubble cluster due to, for example, PNG neutron-induced cavitation [54,56]. If the external pressure applied to the cluster is such that $\Delta p_c = 15\,\text{bar}$, $f = 20\,\text{kHz}$, cavitation bubbles appear and grow in the acoustic field and reach a maximum radius of $\sim 400\,\mu\text{m}$ at about $t \approx 12\,\mu\text{s}$. This instant may be considered to be the initial condition for our calculations.

The pressure equations, Eqs. (A.72) and (A.73), were solved using an implicit finite-difference scheme and the other equations were solved using the Runge–Kutta method. The calculations were conducted for the following set of parameters: $R_0 = 400\,\mu\text{m}$, $\alpha_{g0} = 0.05$, $N = 50$, $R_{c0} = 4\,\text{mm}$, and the properties of deuterated acetone. The numerical results are shown in Figs. A.7 and A.8, where, at $t = 12\,\mu\text{s}$, the cluster radius is 400 μm, the void fraction is 0.05 and the number of bubbles in the cluster is 50. The initial pressure in the bubble is equal to the saturated pressure of vapor, $p(\text{r},0) \approx p_v(T_0)$, where T_0 is the temperature of the liquid pool, 0 °C [54,56].

Figure A.7 shows that for the conditions evaluated, bubbles at the bubble cluster's periphery implode earlier than the ones at the bubble's cluster center. These collapsing bubbles emit a converging shock wave propagating through the bubble cluster. This shock wave intensifies due to cumulation and reaches the center of the bubble cluster at about $t \sim 28.5\,\mu\text{s}$ (see Fig. A.8). Comparison of the time of the first bubble implosion, when this shock wave is emitted, with the time when the shock reaches the cluster's center shows that this shock wave propagates with the speed of sound in a pure liquid (D-acetone).

The collapsing peripheral bubbles also initiate a second converging shock wave, which propagates through the liquid/bubble mixture, inducing motion

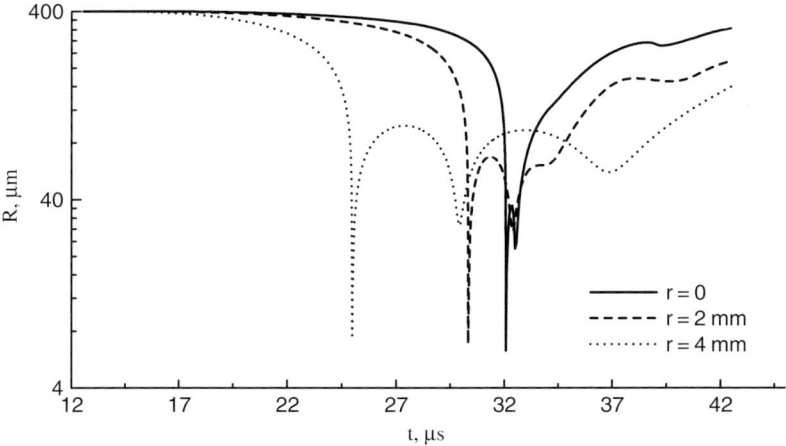

FIG. A.7. Bubble radii vs. time at different distances from the center of collapsing cluster.

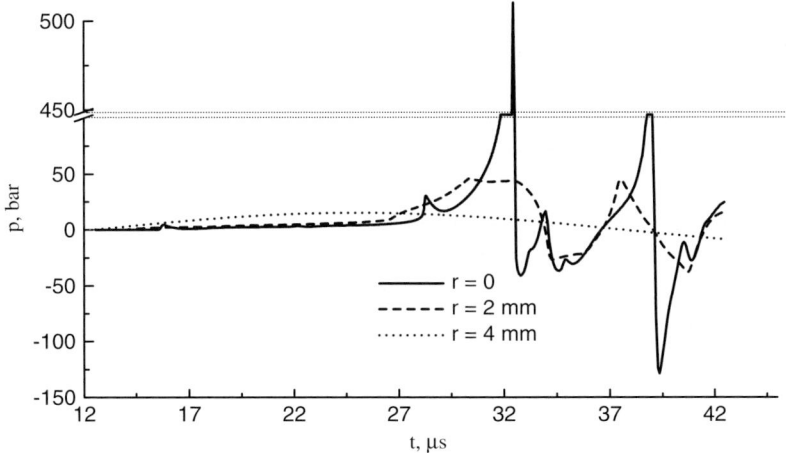

Fig. A.8. Liquid pressure vs. time at different distances from the center of the collapsing cluster.

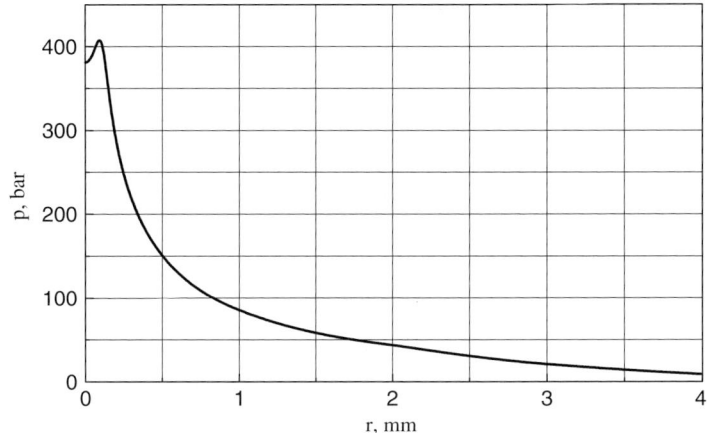

Fig. A.9. Liquid pressure distribution in the collapsing bubble cluster at $t = 32\,\mu s$.

in both the liquid and bubbles. This shock wave is slower than the initial one, and propagates with the two-phase speed of sound. It carries almost all the energy and drives the interior bubbles into a violent implosion. At the time of bubble implosion the liquid pressure exceeds 500 bar. After that, a complicated wave interaction takes place throughout the cluster. However, the liquid pressure is never as high as before. Moreover, as can be seen in Fig. A.9, the maximum instantaneous liquid pressure is very localized near the center of the bubble cluster. In any event, this relatively large incident liquid pressure (≥ 500 bar) causes the bubbles near the center of the bubble cluster to violently implode. Other, similar, results have been given by Nigmatulin et al. [114].

IMPACT OF BUBBLE CONCENTRATION ON THE INTENSITY OF BUBBLE CLUSTER COLLAPSE

As discussed previously, during bubble implosion, when conditions within the bubble are suitable for D/D fusion, the mean cluster liquid pressure at time $t = 0$ may suddenly decrease (from p_c to $p_c^{(0)} < p_c$). This agrees with La Chatelier's principle, according to which a system reacts to some external influence so as to reduce the result of this action [167]. In the case of a bubble cluster implosion this leads to a non-uniform pressure distribution, where the pressure at the bubble cluster's center may be momentarily lower than the one at the cluster's boundary. Such a pressure distribution has a slope directed to the center of the cluster and this pressure gradient induces fluid motion toward the cluster's center.

The analysis shows (see Eq. (A.64)) that the non-uniformity of the pressure distribution within the bubble cluster is controlled by the parameter

$$\beta = \alpha_{g0}\alpha_{\ell 0}\left(\frac{R_{c0}}{R_0}\right)^2 \tag{A.74}$$

The importance of this parameter was apparently first noted by D'Agostino and Brennen [152], who did a linear analysis of gas bubble cluster dynamics. It was shown that the bubble cluster's natural frequency was strongly dependent on this parameter. If β is small, the natural frequency of the bubble cluster is close to that of the individual gas bubbles in the cluster. This implies that the bubbles in the cluster tend to behave as individual gas bubbles in an infinite fluid, where bubble/bubble interaction is minimal. In contrast, bubble/bubble interaction dominates when the value of β is greater than order one. For this case, the collective oscillations of bubbles in the cluster result in a bubble cluster natural frequency which is lower than the natural frequency of individual bubbles.

Later Wang and Brennen [99] and Reisman et al. [156] found that the "cluster interaction" parameter, β, is also crucially important for the non-linear dynamics of the bubble cluster. They presented numerical calculations of the non-linear growth and subsequent collapse of a spherical cluster of gas bubbles in an acoustic field. Three modes of collapse were identified.

(1) At large values of β, the implosion involves the formation of an inward propagating shock wave, which initially forms at the periphery of the cluster. This shock dominates the first mode of

collapse and strengthens rapidly due to geometric focusing (i.e., cumulation) and the coupling of the bubble dynamics with the flow. In this mode, a large pressure pulse is produced by the arrival of the shock wave at the center of the bubble cluster. Moreover, there are other weaker shocks, which arrive at the cluster's center and thus produce a train of acoustic impulses, which eventually, leads to a regular oscillation of the cluster at the first natural frequency (c_l).

(2) At low values of β, the self-shielding effect of the outer bubbles causes the bubbles in the core of the bubble cluster to grow to a smaller maximum size and collapse first. This creates a shock wave, which reduces in intensity as it propagates outward.

(3) At intermediate values of β, bubble collapse first occurs at an intermediate radius and shock waves then propagate both outward and inward from this location.

Thus, in order to obtain an inwardly propagating and intensifying shock in a bubble cluster, and thus to obtain significant local liquid pressure intensification, we have to fulfill the following criterion:

$$\beta = \frac{R_0}{R_{c0}} N \left[1 - \left(\frac{R_0}{R_{c0}} \right)^3 N \right] \geq \beta_{cr} \tag{A.75}$$

The critical value for β may be determined only by detailed numerical analysis. Reisman et al. [156] have calculated that, for the assumptions implicit in their model, at $\beta = 3$ there was no converging shock and a relatively non-energetic (mode-2) bubble cluster collapse was observed, while at $\beta = 300$ there was a converging shock wave, and, an energetic (mode-1) bubble cluster implosion was observed, thus, $3 < \beta_{cr} < 300$. Function $\beta(N)$ was calculated and is presented in Fig. A.10. The two critical bubble numbers are

$$N_{\pm} = \frac{R_{c0}^3}{2R_0^3} \left\{ 1 \pm \sqrt{1 - \frac{4R_0^2}{R_{c0}^2} \beta_{cr}} \right\} \tag{A.76}$$

Mode-1 bubble cluster collapse is realized when $N_- < N < N_+$. For easy estimation one can take into account that normally $R_0/R_{c0} \ll 1$, thus

$$N_- \cong \frac{R_{c0}}{R_0} \beta_{cr}, \quad N_+ \cong \left(\frac{R_{c0}}{R_0} \right)^3 - \frac{R_{c0}}{R_0} \beta_{cr} \tag{A.77}$$

For example, if $R_{c0} = 1$ cm, $R_0 = 100\,\mu$m and $\beta_{cr} = 10$, then $N_- = 1000$ (shown in Fig. A.10 by the shaded circle). Thus, for this case, in order to get a converging shock wave one needs at least 1000 bubbles in a cluster. The

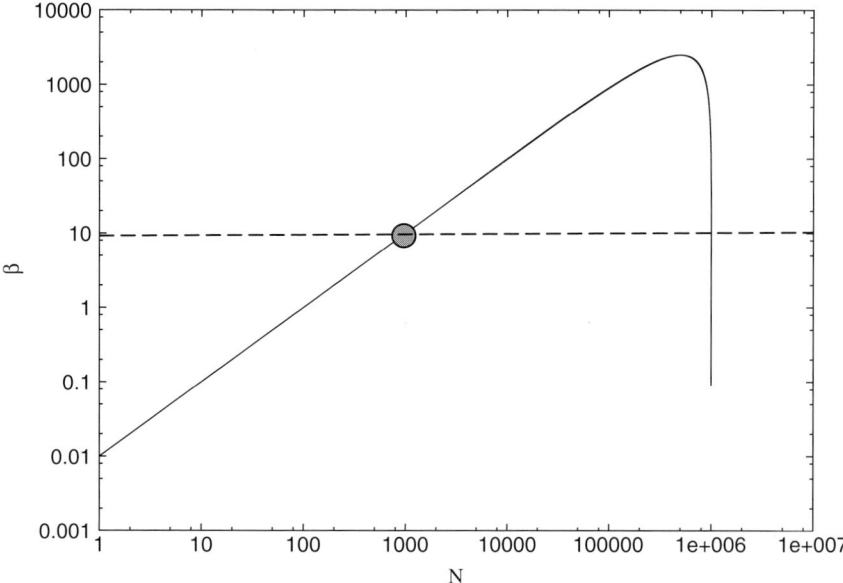

FIG. A.10. "Cluster interaction" parameter β vs. N (number of bubbles in the cluster) for parameter values: $R_{\max} = 100\,\mu\text{m}$, $R_c = 1\,\text{cm}$. Dashed line shows the level of $\beta_{cr} = 10$.

upper critical number of bubbles, N_+, is normally so large that it is never reached in practice.

It is natural to assume that the intensity of bubble cluster collapse is proportional to the number of bubbles generating the converging shock wave. The thickness of the bubble shell at the bubble cluster's periphery, which generates the converging shock wave, can be estimated as

$$\delta \cong \frac{R_0}{\sqrt{3\alpha_{g0}(1-\alpha_{g0})}} \qquad (A.78)$$

and the number of bubbles collapsing in this shell is

$$N_c \cong 4\pi R_c^2 \delta \frac{N}{\frac{4}{3}\pi R_c^3} = \frac{3\delta}{R_c} N \approx \sqrt{\frac{(R_c/R_0)N}{1-(R_0/R_c)^3 N}} \approx \sqrt{\frac{R_c}{R_0}N} \qquad (A.79)$$

Assuming that all bubbles in this shell make an equal contribution to the energy of bubble cluster's collapse, one can conclude that the efficiency of bubble cluster collapse grows as $N^{1/2}$, however, this growth is necessarily bounded.

Actually, the bubbles in the cluster may undergo various disturbances, which can lead to severe departures from sphericity. Indeed, for larger N, bubbles in the bubble cluster are necessarily close to each other and their collapse may not be spherical [159]. Asymmetrically collapsing bubbles were experimentally investigated by Ohl et al. [168,169], in which laser-generated bubbles collapsed asymmetrically in the vicinity of a solid boundary. The dependency of the intensity of the sonoluminescence light flash emitted at bubble collapse on the distance from the solid wall was measured and is given in Fig. A.11 and the following fit:

$$\frac{E_c(\chi)}{E_c(\infty)} = \frac{1}{2}\left[1 + \tan\left(\frac{\chi - \chi_{cr}}{\Delta\chi}\right)\right], \quad \chi_{cr} = 9, \quad \Delta\chi = 2 \quad (A.80)$$

where $\chi = s/R_{\max}$ is the normalized distance to the wall, and s the distance of the bubble cluster's nucleation site from the solid boundary.

These empirical results can be applied to a bubble cluster by using the following simple correlation between χ and α_g:

$$\chi = \alpha_{g0}^{-1/3} = \left[\left(\frac{R_0}{R_c}\right)^3 N\right]^{-1/3} = \frac{R_c}{R_0 N^{1/3}} \quad (A.81)$$

Using Eq. (A.81) the normalized light intensity is plotted as a function of the number of bubbles in the cluster (N) in Fig. A.12.

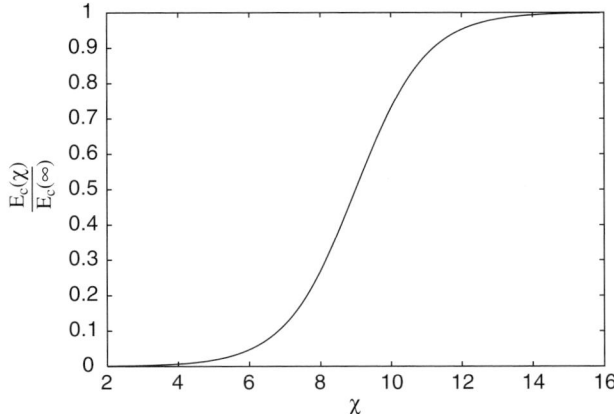

FIG. A.11. Normalized energy of light flash emitted during bubble collapse as a function of the distance from a solid wall (Eq. (A.80)) for $\chi_{cr} = 9$, $\Delta\chi = 2$ [169].

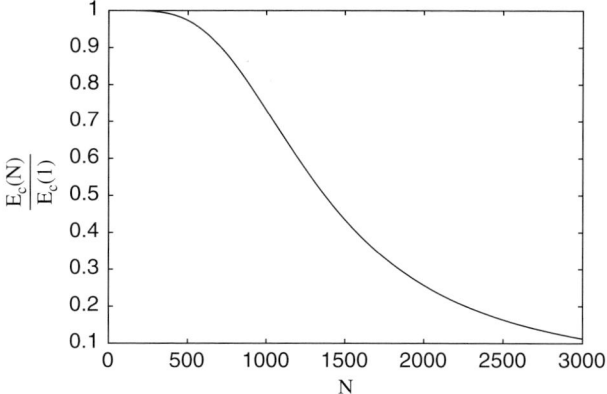

FIG. A.12. Approximation of normalized energy of the light flash emitted during bubble collapse on the number of bubbles in the cluster ($\chi_{cr} = 9$, $\Delta\chi = 2$) for parameter values: $R_{max} = 100\,\mu m$, $R_c = 1\,cm$.

Thus, it appears that one can characterize the normalized efficiency of bubble cluster collapse by using Eqs. (A.79) and (A.80), where χ is replaced by Eq. (A.81).

$$F(N) \approx \sqrt{\frac{\frac{R_c}{R_0}N}{1 - \left(\frac{R_0}{R_c}\right)^3 N}} \tan\left[\frac{\left(\frac{R_c}{R_0 N^{1/3}} - \chi_{cr}\right)}{\Delta\chi}\right] \quad (A.82)$$

This function is presented in Fig. A.13. The maximum of this function characterizes the optimal number of bubbles in a cluster that one needs to obtain the most violent bubble cluster collapse. We note that this value is around 1000 bubbles.

While both the linear and non-linear models used in these analyses of bubble cluster implosion are fairly simple, we see that they both imply that there is an optimal number of bubbles within a bubble cluster which can lead to maximum pressure intensification, and that the number observed in the experiments of Taleyarkhan et al. [54,56] using D-acetone (i.e., $N \approx 1000$) is near the predicted optimum. This type of analysis should be taken into account when selecting alternate test liquids and external neutron energies. For example, when performing a nucleation analysis, which leads to the number of bubbles, N, expected within each imploding bubble cluster. That is, if the number of nucleation bubbles produced in a bubble cluster are too small, the implosion process may not yield conditions which are suitable for thermonuclear bubble fusion (i.e., sonofusion).

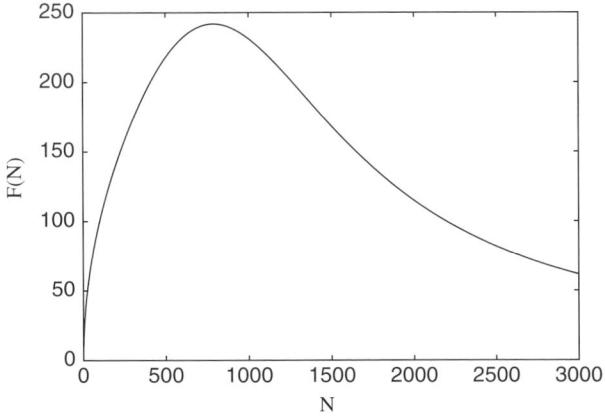

FIG. A.13. Normalized efficiency of bubble cluster collapse as a function of number of bubbles ($\chi_{cr} = 9$, $\Delta\chi = 2$) for parameter values: $R_{max} = 100\,\mu m$, $R_c = 1\,cm$.

Appendix B: Transient Phenomena during Bubble Implosions in an Acoustically Driven, Liquid-Filled Flask

Deuterium/deuterium (D/D) neutrons, and associated tritium production, were observed when cavitation bubbles imploded in well-degassed, chilled deuterated acetone [54]. Subsequently, more detailed sonofusion experiments were performed [56] in which data were taken well after the initial implosion of the bubble cluster. After the initial energetic implosion, the SL pulses continued at the acoustic forcing frequency (i.e., at the natural frequency of the D-acetone liquid-filled flask; $f = 19.3\,\text{kHz}$, $T \equiv 1/f = 52\,\mu s$) and, as can be seen in Fig. 22a, the neutron counts ceased for about 10 acoustic cycles (i.e., $\sim 520\,\mu s$), after which they built up to a fairly large level and then decayed off to a lower level until the PNG was again activated at $\sim 5\,ms$.

This interesting phenomenon was also seen in the sonoluminescence data of Shapira and Saltmarsh [55,101] and it implies that the strong compression shock wave associated with the initial implosion process interacts with the vibrating test section and PZT transducer, disrupting the acoustic standing wave until the coupled electromechanical system recovers. The purpose of the study presented herein is to assess this hypothesis.

For simplicity, a spherically symmetric problem was considered in which a single gas bubble at the center of a spherical flask filled with a compressible liquid was excited by periodic pressure variations due to motion of the flask wall. The purpose of this study was to provide a theoretical explanation for the experimental observations in the sonofusion (i.e., bubble fusion) experiments shown in Fig. 22a [56]. In particular, to explain why after the first energetic bubble cluster implosion, the acoustically driven D/D neutron emissions cease for a while, then build up and later decayed to a lower level before new external PNG pulse neutrons reinitiate the process (i.e., at 5 ms).

THE DYNAMICS OF A SPHERICAL LIQUID-FILLED FLASK WITHOUT DISSIPATION

First, let us consider an analysis of the forced bubble oscillations in a spherical flask for the case when there is no energy dissipation in the flask [51], which is a reasonable approximation to the actual case of interest. The analytical approach is based on the assumption that liquid viscosity is important only close to the bubble, and the rest of the liquid in the flask can be treated as inviscid, but weakly compressible.

Two asymptotic solutions have been found [51] for the low Mach number bubble dynamics ($|\dot{R}|/C_v \ll 1$, where $\dot{R}$ is the radial velocity of the bubble's interface and C_v the speed of sound in the vapor). The first one is an asymptotic solution far from the bubble (i.e., the far liquid field), and it corresponds to the linear wave equation with no damping. The second one is an asymptotic solution near the bubble (i.e., the near liquid field), which corresponds to the solution of the Rayleigh–Plesset equation for an incompressible, viscous liquid. For the analytical solution of the coupled problem, matching of these asymptotic solutions has been done, yielding a generalization of the Rayleigh–Plesset equation [51].

This generalization takes into account liquid compressibility and includes a difference equation for both the lagging and leading times and a generalized Rayleigh–Plesset equation. These equations are

$$p_{\mathrm{I}}\left(t + \frac{2R_{\mathrm{f}}}{C_\ell}\right) = p_{\mathrm{I}}(t) + \frac{2R_{\mathrm{f}}}{C_\ell}\frac{dp_{\mathrm{w}}}{dt}\left(t + \frac{R_{\mathrm{f}}}{C_\ell}\right) - \frac{2\rho_\ell}{C_\ell}\frac{d^2 Q(t)}{dt^2} \quad \text{(B.1a)}$$

where

$$Q = R^2 \frac{dR}{dt} \equiv R^2 \dot{R} \quad \text{(B.1b)}$$

and, as discussed in the text of the chapter, Eq. (1) gives the generalized Rayleigh–Plesset equation as

$$R\frac{d^2 R}{dt^2} + \frac{3}{2}\left(\frac{dR}{dt}\right)^2 = \frac{p_{\mathrm{R}} - p_{\mathrm{I}}}{\rho_\ell} + \frac{R}{\rho_\ell C_\ell}\frac{d}{dt}(p_{\mathrm{R}} - p_{\mathrm{I}}) \quad \text{(B.2)}$$

In the above equations R_f is the instantaneous radius of the flask wall, C_ℓ and ρ_ℓ are the speed of sound and density of the liquid, respectively, p_I the incident pressure (i.e., the liquid pressure at $r \sim 50R(t)$, which is still fairly close to the bubble's interface), p_w the liquid pressure at the flask wall, R the instantaneous bubble radius and $p_R \equiv p_{\ell_i}$ is the liquid pressure at the bubble's interface.

In order to close the system of Eqs. (B.1) and (B.2), one may use the following jump and state equations:

$$p_R = p_g(R) - \frac{2\sigma}{R} - \frac{4\mu}{R}\frac{dR}{dt}, \quad p_g(R) = \left(p_0 + \frac{2\sigma}{R_0}\right)\left(\frac{R}{R_0}\right)^{3\gamma} \quad \text{(B.3)}$$

Here, p_g is the gas pressure in the bubble, σ the surface tension, μ the effective liquid viscosity (including the effect of thermal dissipation), γ the polytropic exponent, p_0 the initial pressure of the system and R_0 the initial bubble radius.

Normally, in the low Mach number regime the last term in Eq. (B.1a) is small and may be neglected. Thus, to a good first order approximation, we have

$$p_I\left(t + \frac{2R_f}{C_\ell}\right) = p_I(t) + \frac{2R}{C_\ell}\frac{dp_w}{dt}\left(t + \frac{R_f}{C_\ell}\right) \quad \text{(B.4)}$$

This indicates that the influence of flask wall oscillations on the incident pressure, p_I, which is approximately the liquid pressure at the center of the flask without a bubble.

Transients in the flask cannot be adequately described by Eq. (B.4), because it does not account for dissipation mechanisms. For instance, let us consider a liquid-filled flask which is initially at rest at time $t = 0$ and then is suddenly excited by a periodic pressure at the flask wall which is given by

$$p_w(t) = \begin{cases} p_0, & t \leq 0 \\ p_0 - \Delta p_w \sin(\omega t), & t > 0 \end{cases} \quad \text{(B.5)}$$

In the following analyses, we have used the parameters: $R = 5\,\text{cm}$, $C_\ell = 1500\,\text{m/s}$ and $\rho_\ell = 1000l\,\text{kg/m}^3$. The dynamics of the incident liquid pressure (which is essentially the liquid pressure at the center of the flask in the absence of a bubble) are calculated from Eq. (B.4) and are presented in Figs. B.1. Here the driving frequency is taken very close to flask's primary resonance frequency, $\omega R/C_\ell = \pi$. One can see that the amplitude of the incident pressure linearly increases with time. This is unrealistic and clearly

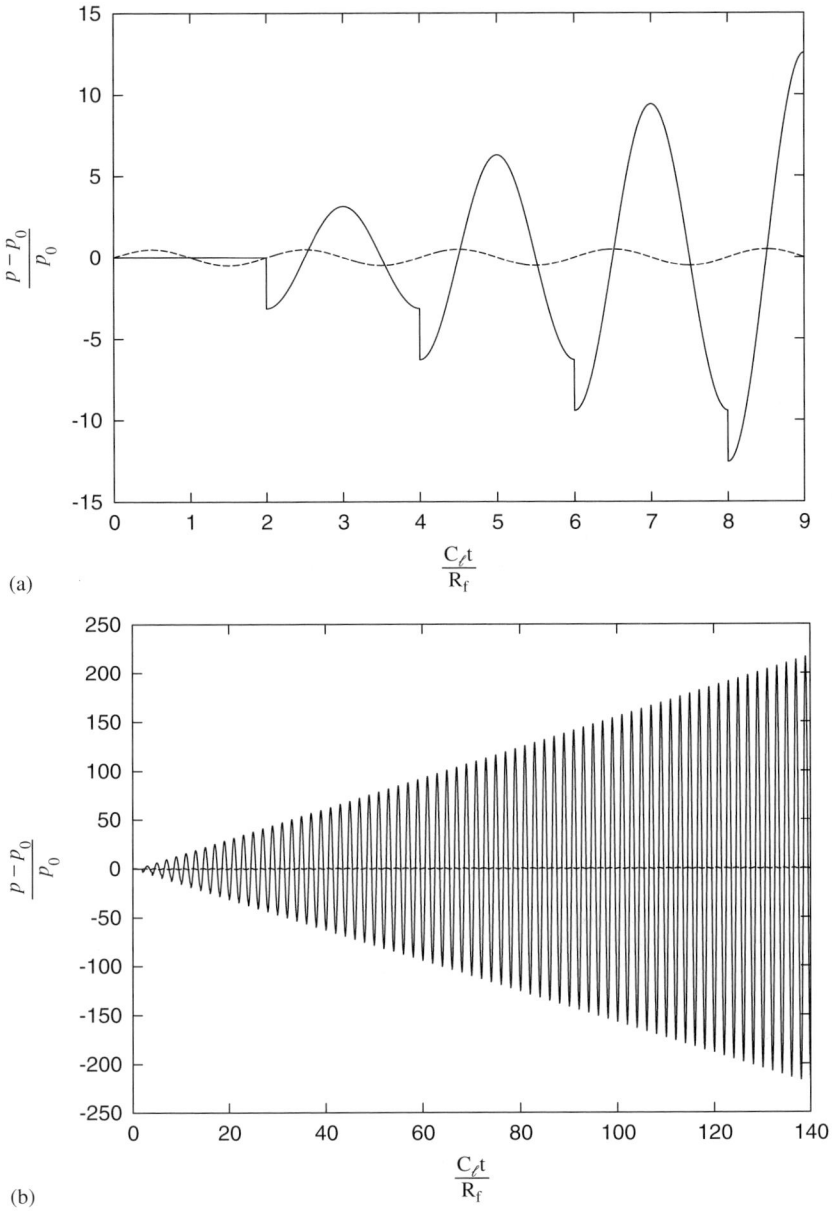

FIG. B.1 (a, b). Normalized liquid pressure at the center of a spherical flask (p_I: solid line) and at the flask wall (p_w: dashed line) vs. normalized time $C_\ell t/R_f$ for the case of no dissipation ($R_f = 5\,\text{cm}$, $C_\ell = 1500\,\text{m/s}$, $\rho_\ell = 1000\,\text{kg/m}^3$, $\omega R_f/C_\ell = 3.1409$, $\Delta p_R/p_0 = 0.5$).

shows that Eq. (B.4) cannot be used to describe realistic transients in liquid-filled flasks.

The Dynamics of a Spherical Liquid-Filled Flask with Dissipation

We must generalize Eq. (B.4) to account for dissipation in a liquid-filled flask. To this end, let us consider the wave equation with damping [163]

$$\frac{\partial^2 \tilde{p}}{\partial t^2} - C_\ell^2 \nabla^2 \left(\tilde{p} + \frac{\beta}{C_\ell^2} \frac{\partial \tilde{p}}{\partial t} \right) = 0 \qquad (B.6)$$

where $\tilde{p}$ is a pressure perturbation and β the dissipation coefficient for the acoustic waves in the flask.

In general, there are several mechanisms causing acoustic wave damping in the liquid-filled flask. These mechanisms may be divided into two groups: (1) dissipation in the bulk liquid, and, (2) dissipation in the structure and the transducing electric circuit (i.e., the flask, PZT transducer, etc.). The first group includes liquid viscosity, liquid thermal dissipation, liquid relaxations due to possible chemical reactions and excitation of the different molecular degrees of freedom, and boundary layer effects. It is well-known that viscosity is the principal dissipation mechanism for liquids [164]. Dissipation mechanisms in the structures and circuitry are very difficult to estimate theoretically, although these mechanisms may be more important than the liquid bulk dissipation. For this reason, as in Eg. (A.30) it is convenient to define the dissipation coefficient, β, in the form of an effective viscous dissipation

$$\beta = \frac{4\mu_f}{3\rho_\ell} \qquad (B.7)$$

where μ_f denotes an effective viscosity (instead of just the bulk liquid viscosity), which implicitly includes all the significant dissipation mechanisms. The value for μ_f must be chosen to fit experimental observations of amplitude vs. frequency based on the acoustic chamber's characteristics.

Unfortunately, it is impossible to directly apply the method of matching the asymptotic solutions used by Nigmatulin et al. [51] to generalize Eq. (B.4) to account for damping, because Eq. (B.6) has no exact solution (when $\beta \neq 0$). Here, we propose a novel indirect method for the generalization of Eq. (B.4).

To derive a solution of Eq. (B.6) for a standing acoustic wave field, let us assume that the liquid pressure perturbation in the flask is harmonic in time

$$\tilde{p} = P \exp(i\omega t) \qquad (B.8)$$

where ω is the angular driving frequency. The substitution of Eq. (B.8) into Eq. (B.6) results in a Helmholtz equation for the complex pressure amplitude, P

$$\nabla^2 P + k^2 P = 0 \tag{B.9}$$

where

$$k^2 = \frac{\omega^2}{C_\ell^2 \left(1 + \frac{i\beta\omega}{C_\ell^2}\right)} \tag{B.10}$$

The solution of the Helmholtz equation, Eq. (B.9), leads to the following relationship between the complex amplitudes of the liquid pressure at the center of the flask, Δp_I, and the liquid pressure on the flask wall, Δp_w:

$$\sin(kR_\mathrm{f})\Delta p_\mathrm{I} = kR_\mathrm{f}\Delta p_\mathrm{w} \tag{B.11}$$

For convenience, let us replace t by $t - R_\mathrm{f}/C_\ell$ in Eq. (B.4), yielding

$$p_\mathrm{I}\left(t + \frac{R_\mathrm{f}}{C_\ell}\right) - p_\mathrm{I}\left(t - \frac{R_\mathrm{f}}{C_\ell}\right) = \frac{2R_\mathrm{f}}{C_\ell} \frac{dp_\mathrm{w}}{dt}(t) \tag{B.12}$$

After substituting p_w from Eq. (B.5), Eq. (B.12) becomes

$$p_\mathrm{I}\left(t + \frac{R_\mathrm{f}}{C_\ell}\right) - p_\mathrm{I}\left(t - \frac{R_\mathrm{f}}{C_\ell}\right) = -\frac{2R_\mathrm{f}}{C_\ell} \omega \Delta p_\mathrm{w} \cos(\omega t) \tag{B.13}$$

Eqs (B.13) has the following solution:

$$p_\mathrm{I}(t) = -\frac{\omega R_\mathrm{f}/C_\ell}{\sin(\omega R_\mathrm{f}/C_\ell)} \Delta p_\mathrm{w} \sin(\omega t) \tag{B.14}$$

which, in terms of amplitudes, implies

$$\sin\left(\frac{\omega R_\mathrm{f}}{C_\ell}\right) \Delta p_\mathrm{I} = \frac{-\omega R_\mathrm{f}}{C_\ell} \Delta p_\mathrm{w} \tag{B.15}$$

Comparing Eq. (B.15) with Eqs. (B.10) and (B.11), we note that Eq. (B.15) is the exact solution for case of no damping ($\beta = 0$).

Let us now consider how to modify Eq. (B.12) so that it properly (i.e., with accuracy $O(\beta\omega/C_\ell^2)$), approximates the damping of the standing acoustic waves. This equation will then be used to calculate the transient of interest.

To this end we make a Taylor series expansion of Eq. (B.10)

$$k \cong \frac{\omega}{C_\ell} - \frac{i\beta\omega^2}{2C_\ell^3} \tag{B.16}$$

Next, the Taylor series expansion of Eq. (B.11) is

$$\left[\sin\left(\frac{\omega R_f}{C_\ell}\right) - \cos\left(\frac{\omega R_f}{C_\ell}\right) \frac{i\beta\omega^2 R_f}{2C_\ell^3}\right] \Delta p_1 \cong \left[\frac{\omega R_f}{C_\ell} - \frac{i\beta\omega^2 R_f}{2C_\ell^3}\right] \Delta p_w \tag{B.17}$$

We now generalize Eq. (B.12) to the following:

$$\left(1 + \alpha_1 \tfrac{d^2}{dt^2}\right) p_1\left(t + \tfrac{R_f}{C_\ell}\right) - \left(1 + \alpha_2 \tfrac{d^2}{dt^2}\right) p_1\left(t - \tfrac{R_f}{C_\ell}\right)$$
$$= \tfrac{2R_f}{C_\ell} \left[\tfrac{dp_w}{dt}(t) + \alpha_3 \tfrac{d^2 p_w}{dt^2}(t)\right] \tag{B.18}$$

and select the parameters α_1, α_2 and α_3 so that its solution for forced oscillations coincides with Eq. (B.17). This requires

$$\alpha_1 = -\alpha_2 = -\frac{\beta R_f}{2C_\ell^3}, \quad \alpha_3 = -\frac{\beta}{2C_\ell^2} \tag{B.19}$$

Thus, for the case of finite damping, Eq. (B.12) may be replaced by

$$p_1\left(t + \tfrac{R_f}{C_\ell}\right) - p_1\left(t - \tfrac{R_f}{C_\ell}\right) = \tfrac{\beta R_f}{2C_\ell^3} \tfrac{d^2}{dt^2}\left[p_1\left(t + \tfrac{R_f}{C_\ell}\right) + p_1\left(t - \tfrac{R_f}{C_\ell}\right)\right]$$
$$+ \tfrac{2R_f}{C_\ell}\left(1 - \tfrac{\beta}{2C_\ell^2}\tfrac{d}{dt}\right)\tfrac{dp_w}{dt}(t) \tag{B.20}$$

This equation approximates the solution of the wave equation in Eq. (B.6) to accuracy $O(\beta\omega/C_\ell^2)$.

However, the proper approximation is not the only requirement for the generalized equation we are looking for. It must also be well posed (i.e., avoid non-physical exponential growth of the solution in time). The first term on the right-hand side of Eq. (B.20)

$$\frac{\beta R_f}{2C_\ell^3} \frac{d^2}{dt^2} p_1\left(t + \frac{R_f}{C_\ell}\right) \tag{B.21}$$

causes exponential growth of the solution. This is easier to see if one rewrites Eq. (B.20) as

$$-\frac{\beta R_{\rm f}}{2C_\ell^3}\frac{d^2}{dt^2}\left[p_{\rm I}\left(t+\frac{R_{\rm f}}{C_\ell}\right)\right]+p_{\rm I}\left(t+\frac{R_{\rm f}}{C_\ell}\right)$$
$$=p_{\rm I}\left(t-\frac{R_{\rm f}}{C_\ell}\right)+\frac{\beta R_{\rm f}}{2C_\ell^3}\frac{d^2}{dt^2}\left[p_{\rm I}\left(t-\frac{R_{\rm f}}{C_\ell}\right)\right]+\frac{2R_{\rm f}}{C_\ell}\left(1-\frac{\beta}{2C_\ell^2}\frac{d}{dt}\right)\frac{dp_{\rm w}}{dt}(t) \quad (\text{B.22})$$

Equation (B.22) is a second-order ordinary differential equation for $p_{\rm I}(t+R_{\rm f}/C_\ell)$, which has an exponentially diverging solution. This fact may be easily verified numerically. Thus we need to eliminate this term to avoid exponential growth, however, we must preserve the order of the approximation of the equation. From Eq. (B.20) it follows that

$$p_{\rm I}\left(t+\frac{R_{\rm f}}{C_\ell}\right)=p_{\rm I}\left(t-\frac{R_{\rm f}}{C_\ell}\right)+\frac{2R_{\rm f}}{C_\ell}\frac{dp_{\rm w}}{dt}(t)+O\left(\frac{\beta\omega}{C_\ell^2}\right) \quad (\text{B.23})$$

Substituting this representation of $p_{\rm I}(t+R_{\rm f}/C_\ell)$ into the right-hand side of Eq. (B.20), we obtain

$$p_{\rm I}\left(t+\frac{R_{\rm f}}{C_\ell}\right)-p_{\rm I}\left(t-\frac{R_{\rm f}}{C_\ell}\right)=\frac{\beta R_{\rm f}}{C_\ell^3}\frac{d^2}{dt^2}\left[p_{\rm I}\left(t-\frac{R_{\rm f}}{C_\ell}\right)\right]$$
$$+\frac{2R_{\rm f}}{C_\ell}\left[1-\frac{\beta}{2C_\ell^2}\frac{d}{dt}+\frac{\beta R_{\rm f}}{2C_\ell^3}\frac{d^2}{dt^2}\right]\frac{dp_{\rm w}}{dt}(t) \quad (\text{B.24})$$

This equation replaces Eq. (B.12) when calculating standing waves and transients in the flask when there is damping.

Let us now transform all the variables to dimensionless form

$$\tilde{t}=\frac{C_\ell t}{R_{\rm f}}, \quad \tilde{p}=\frac{p}{p_0}, \quad \tilde{\omega}=\frac{\omega R_{\rm f}}{C_\ell} \quad (\text{B.25})$$

Then Eq. (B.24) may be rewritten as follows (the "tilde" has been omitted for simplicity):

$$p_{\rm I}(t+1)-p_{\rm I}(t-1)=\xi\frac{d^2}{dt^2}[p_{\rm I}(t-1)]+\left(2-\xi\frac{d}{dt}+\xi\frac{d^2}{dt^2}\right)\frac{dp_{\rm w}}{dt}(t) \quad (\text{B.26})$$

where

$$\alpha=\frac{4\mu_{\rm f}}{3\rho_\ell C_\ell R_{\rm f}}=\frac{\tau_{\rm ac}}{\tau_{\rm dis}} \quad (\text{B.27})$$

We note that α is the ratio of the characteristic time of propagation of the acoustic waves through the liquid-filled flask, τ_{ac}, to the characteristic time of energy dissipation in the coupled fluid/structural system, τ_{dis}. Normally, ξ is small.

The standing wave solution of Eq. (B.26) is

$$\Delta p_{\rm I} = A(\omega)\Delta p_{\rm w}, \quad A(\omega) = \frac{i\omega(2 - i\alpha\omega - \alpha\omega^2)}{\exp(i\omega) - (1 - \alpha\omega^2)\exp(-i\omega)} \quad (\text{B.28})$$

The amplitude and phase angle of standing wave oscillations according to the exact model, Eq. (B.11), and the approximate model, Eq. (B.28), for different effective damping coefficients of the coupled fluid/structural system are shown in Figs. B.2a and b. It can be seen that both curves essentially coincide for the first few resonances. Hence Eq. (B.26) may be used to calculate the transient solutions for the range of frequencies of interest.

THE DYNAMICS OF A SPHERICAL LIQUID-FILLED FLASK WITH DISSIPATION

Here, we model a typical situation seen in the sonofusion experiments of Taleyarkhan et al. [54,56]. The initial violent bubble implosion produces a pulse of D/D fusion neutrons and the resultant bubble dynamics induces a compression shock wave which, after an acoustic time delay (i.e., $R_{\rm f}/C_\ell$), interacts with the flask wall and the PZT attached to it, disrupting its motion and thus detuning the standing acoustic wave field. For simplicity, the state of this electromechanical system just after this event may be approximated as being at rest (i.e., having no appreciable motion). The PZT transducer on the flask wall that is driven by a function generator goes through some electrical circuit dynamics but continues to try and cause the PZT to contract and expand harmonically, which leads to complicated fluid/structure dynamics until the standing wave pressure field in the liquid-filled flask is restored.

Ignoring the transient structural dynamics, the pressure at the center of the flask ($p_{\rm I}$) may be calculated using Eq. (B.26). The numerical algorithm involves the calculation of the second time derivative of the delayed incident pressure. Mathematically, this is an inverse problem. Thus, to evaluate Eq. (B.26) a special algorithm for numerical differentiation was developed and used.

We have plotted the liquid pressure amplitude at the center of a liquid-filled spherical flask vs. frequency for different effective viscosities, $\mu_{\rm f}$, to obtain an estimation of the values, which provide realistic amplitude/frequency diagrams (see Figs. B.3a and b). One notes that for our assumed parameters $\mu_{\rm f} = 500\,{\rm kg}/({\rm m\,s})$ would be the appropriate value, because it

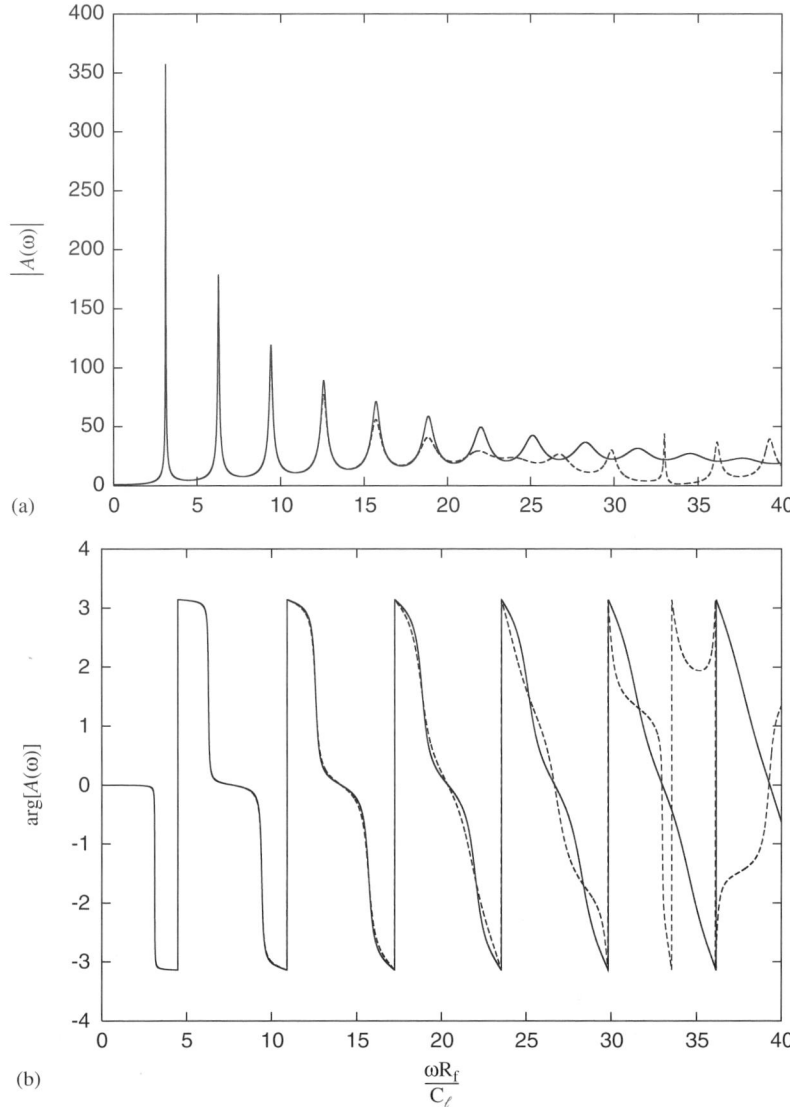

FIG. B.2 (a, b). Normalized amplitude and phase angle of standing wave pressure oscillations at the center of a spherical flask vs. normalized driving frequency $\omega R_f/C_\ell$ ($R_f = 5$ cm, $C_\ell = 1500$ m/s, $\mu_f = 100$ kg/(m s), $\rho_\ell = 1000$ kg/m^3). Exact model – solid line; approximate model – dashed line. The first resonance data (exact model: $\omega_{res} R_f/C_\ell = 3.14161$, $|A_{res}| = 358.095$; approximate model: $\omega_{res} R_f/C_\ell = 3.14157$, $|A_{res}| = 354.96$).

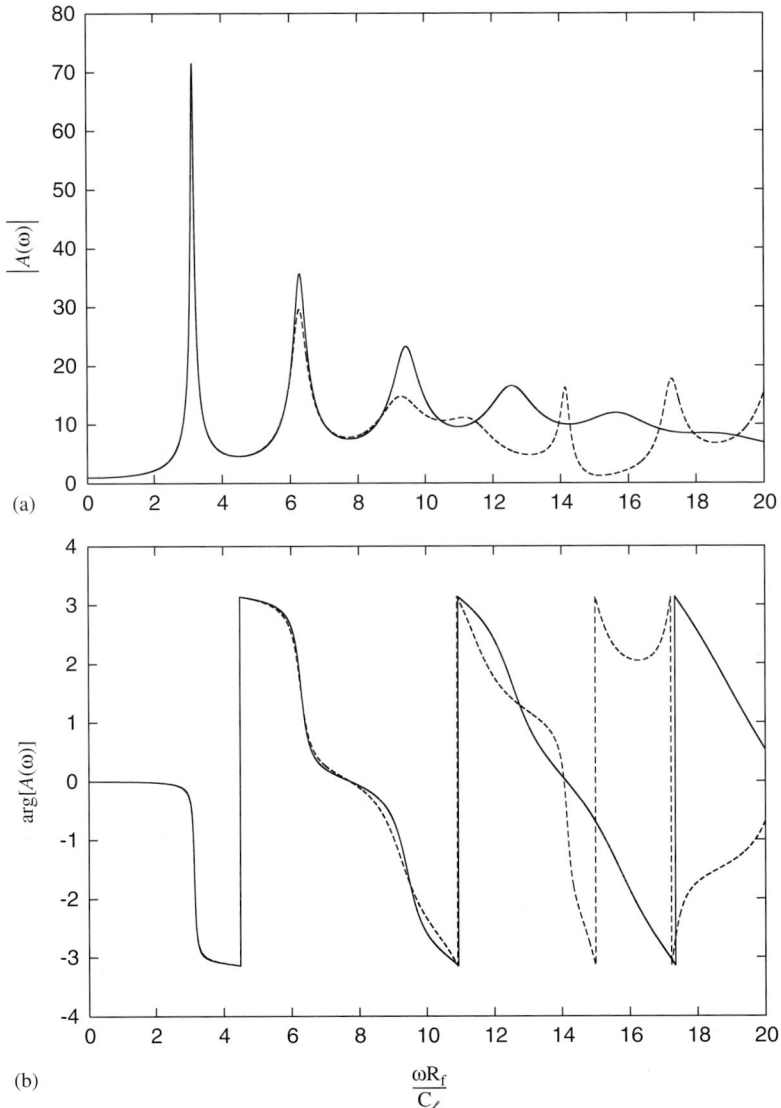

FIG. B.3 (a, b). Normalized amplitude and phase angle of standing wave pressure oscillations at the center of a spherical flask vs normalized driving frequency $\omega R_f/C_\ell$ ($R = 5\,\text{cm}$, $C_\ell = 1500\,\text{m/s}$, $\mu_f = 500\,\text{kg/(m s)}$, $\rho_\ell = 1000\,\text{kg/m}^3$). Exact model – solid line; approximate model–dashed line. The first resonance data (exact model: $\omega R_f/C_\ell = 3.1419$; $|A_{res}| = 71.604$; approximate model: $\omega_{res} R_f/C_\ell = 3.1409$, $|A_{res}| = 68.494$).

provides a good match with experimental observations that the acoustic pressure amplitude at the center of a liquid-filled flask (at resonance) is about 70 times higher than the pressure at the flask wall [165].

Figures B.4a–d show the results of calculations for the acoustic startup transient of a standing wave for an effective viscosity of the flask-liquid system of $\mu_f = 500 \, \text{kg}/(\text{m s})$. The initial stage of pressure oscillations at the center of the flask is very similar to the one we obtained for case of no damping (i.e., compare Figs. B.1a and B.4a). However, later in time Figs. B.4c and d show that the growth of pressure amplitude is arrested by damping, which leads to a limit cycle (i.e., a standing wave). It is important to note that during the transient the shape of the pressure oscillations are close to sinusoidal, with some abrupt periodic jumps. These pressure jumps are relatively large during the initial stage of the process, but later on, with the increase of pressure amplitude, these jumps become negligibly small, although the amplitude of these jumps does not change with time. This response is similar to that seen in Fig. 22a after a 10 acoustic cycle "dead time."

Figures B.5 and B.6 show the results of transient calculations for a standing wave having an effective viscosity of the coupled system of $\mu_f = 100 \, \text{kg}/(\text{m s})$. The results are very similar to Figs. B.4a–d, however because the effective viscosity is smaller than that used in the previous plots, the duration of the transient is longer.

To demonstrate how this model also simulates the decay of standing wave oscillations we have performed calculations of the dynamics of the liquid pressure in the flask after the driving pressure at the flask wall is detuned (e.g., due to shock wave interaction with the wall). The results are presented in Fig. B.7, and, as expected, the amplitude of the standing wave decays with time. This response is less abrupt than in Fig. 22a just after the initial bubble cluster implosion, presumably because it neglects the effect on the PZT circuitry of the energetic shock wave associated with the initial bubble cluster implosion.

In order to calculate the bubble dynamics in a flask one needs to use the pressure at the center of the flask, $p_I(t)$, obtained from Eq. (B.26), as the driving pressure for the bubble dynamics implied by Eq. (B.2). This is, of course, possible to do, however, for simplicity, we may approximate $p_I(t)$ by a sinusoid modulated with an exponential transient function.

$$p_I(t) = 1 - \Delta p_I \left[1 - \exp\left(-\frac{t}{\tau}\right) \right] \sin(\omega t) \tag{B.29}$$

The evaluation of Eq. (B.29) is shown in Fig. B.8, and we see results which are very similar to the results in Figs. B.4c. This driving pressure has been

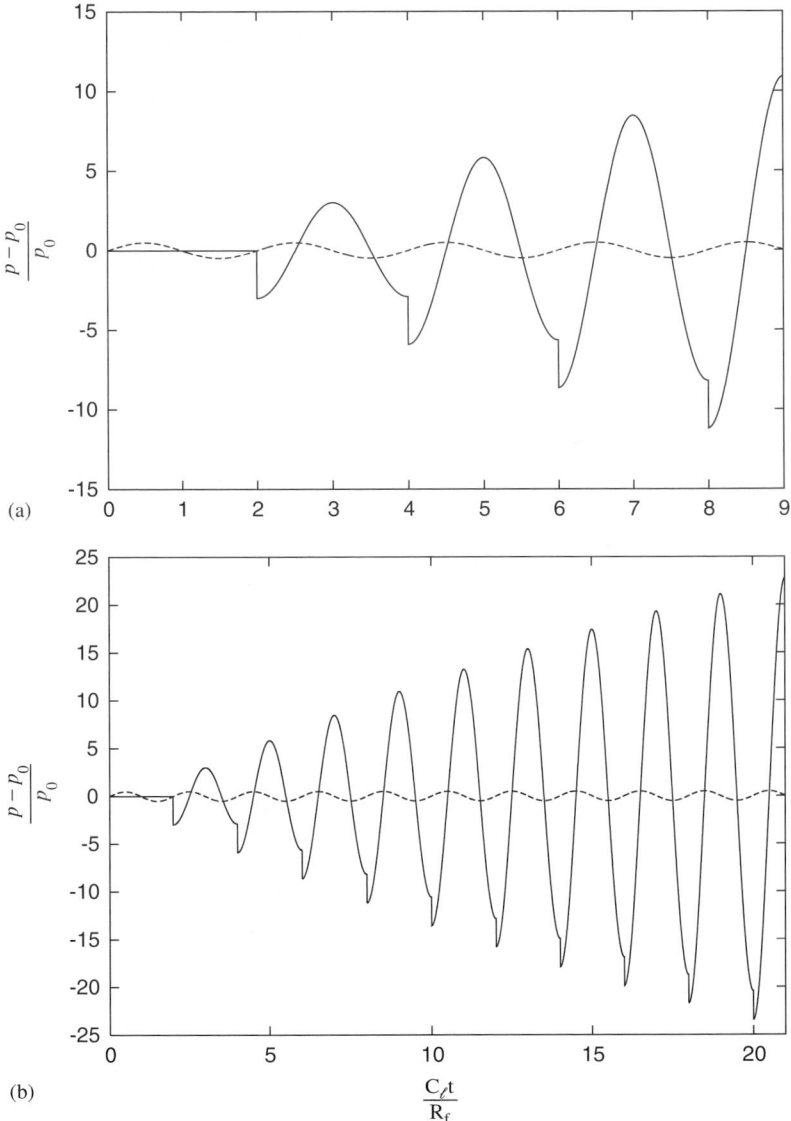

FIG. B.4 (a, b). A transient to a standing acoustic wave state. Normalized pressure at the center of a spherical flask (p_I: solid line) and at the flask wall (p_w: dashed line) vs. normalized time, $C_\ell t/R_f$ ($R_f = 5$ cm, $C_\ell = 1500$ m/s, $\mu_f = 500$ kg/(m s), $\rho_\ell = 1000$ kg/m^3, $\omega R_f/C_\ell = 3.1409$, $\Delta p_R/p_0 = 0.5$, $N = 1000$, $n_d = 250$).

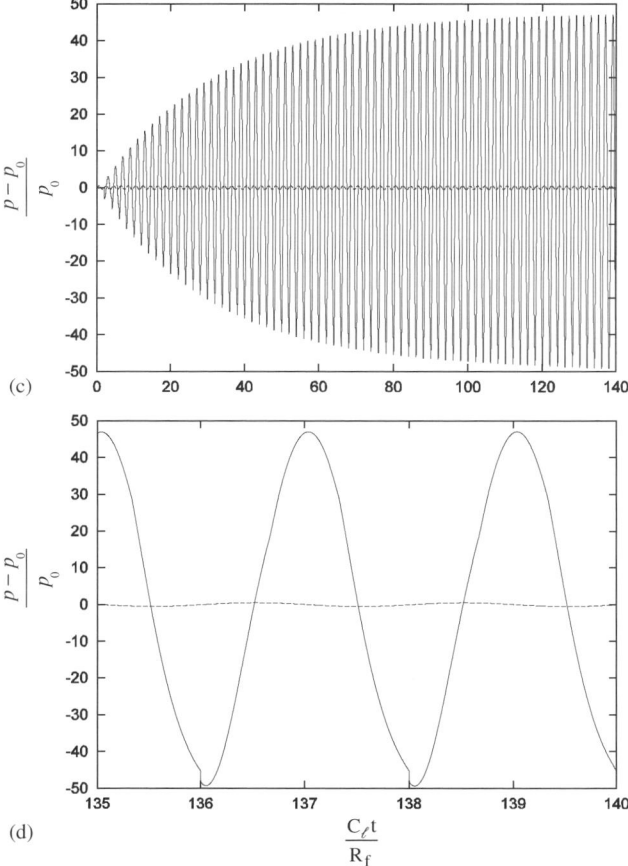

FIG. B.4 (c, d). A transient to a standing acoustic wave state. Normalized pressure at the center of a spherical flask (p_1: solid line) and at the flask wall (p_w: dashed line) vs. normalized time $C_\ell t/R_f$ ($R_f = 5$ cm, $C_\ell = 1500$ m/s, $\mu_f = 500$ kg/(m s), $\rho_\ell = 1000$ kg/m^3, $\omega R_f/C_\ell = 3.1409$, $\Delta p_w/p_0 = 0.5$, $N = 1000$, $n_d = 250$).

applied to bubbles of different equilibrium radii. At initial time ($t = 0$), the bubble is assumed to be at rest and has an equilibrium radius (R_0).

Figure B.9a shows the beginning of bubble oscillations for a bubble having an equilibrium radius (R_0) of 10 μm. During the first period of driving, the bubble does not respond very much because the amplitude of the forcing incident pressure is low. During the next period of driving, the bubble expands its radius about 35 times and induces an implosion followed by several after-bounces, which is quite similar to the bubble dynamics often

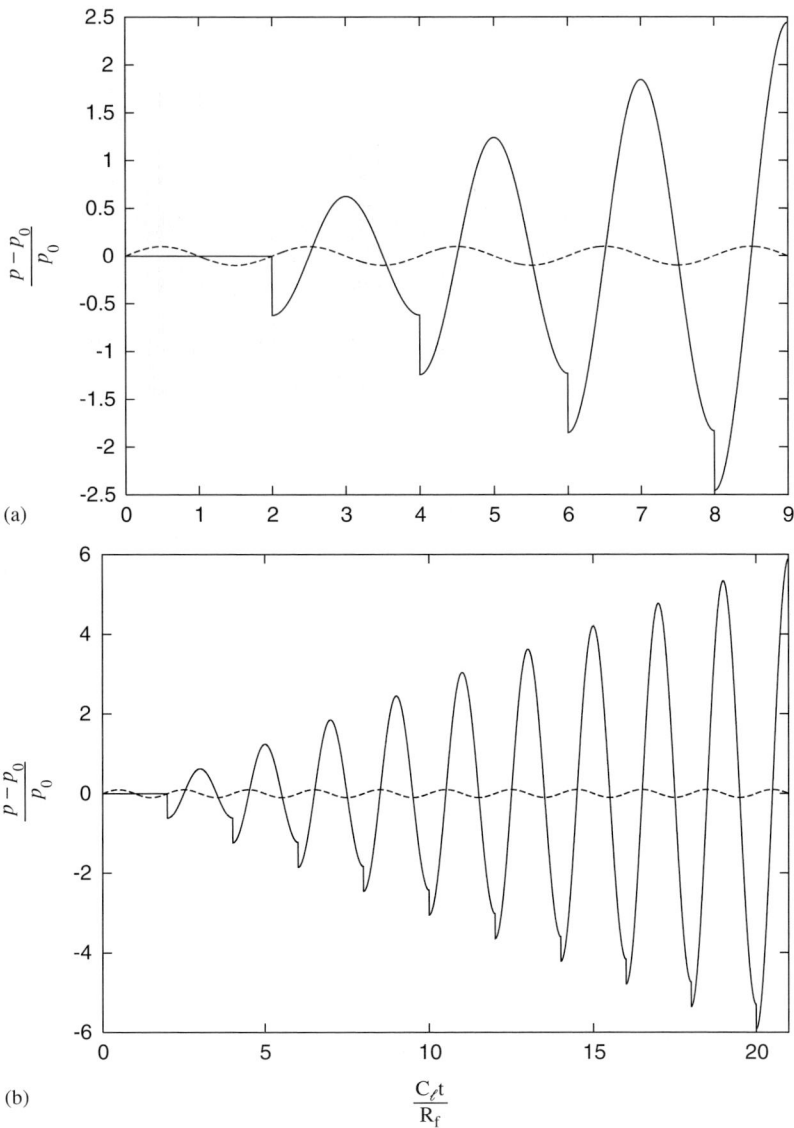

Fig. B.5 (a, b). A transient for standing acoustic wave. Normalized pressure at the center of a spherical flask (p_l: solid line) and at the flask wall (p_w: dashed line) vs. normalized time $C_\ell t/R_f$ ($R_f = 5$ cm, $C_\ell = 1500$ m/s, $\mu_f = 100$ kg/(m s), $\rho_\ell = 1000$ kg/m^3, $\omega R_f/C_\ell = 3.14157$, $\Delta p_w/p_0 = 0.1$, N = 1000, $n_d = 250$).

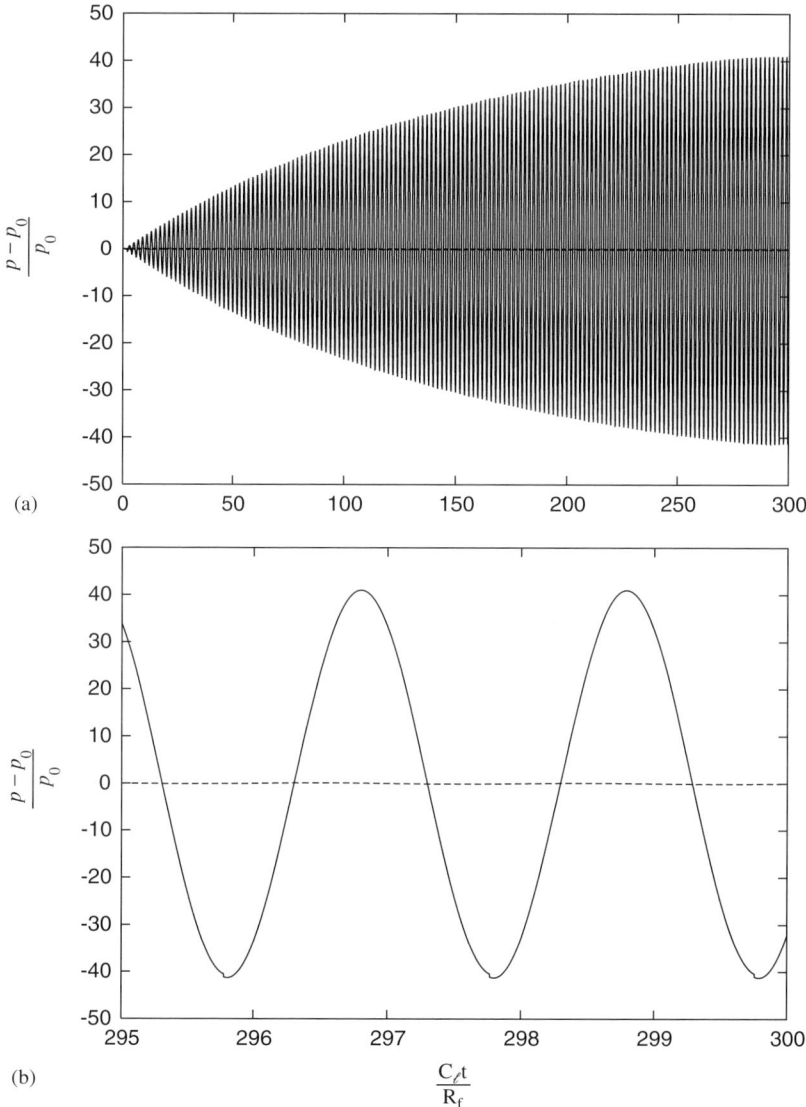

FIG. B.6 (a, b). A transient for standing acoustic wave. Normalized pressure at the center of a spherical flask (p_I: solid line) and at the flask wall (p_w: dashed line) vs. normalized time $C_\ell t/R_f$ ($R_f = 5$ cm, $C_\ell = 1500$ m/s, $\mu_f = 100$ kg/(m s), $\rho_\ell = 1000$ kg/m^3, $\omega R_f C_\ell = 3.14157$, $\Delta p_w/p_0 = 0.1$, $N = 1000$, $n_d = 250$).

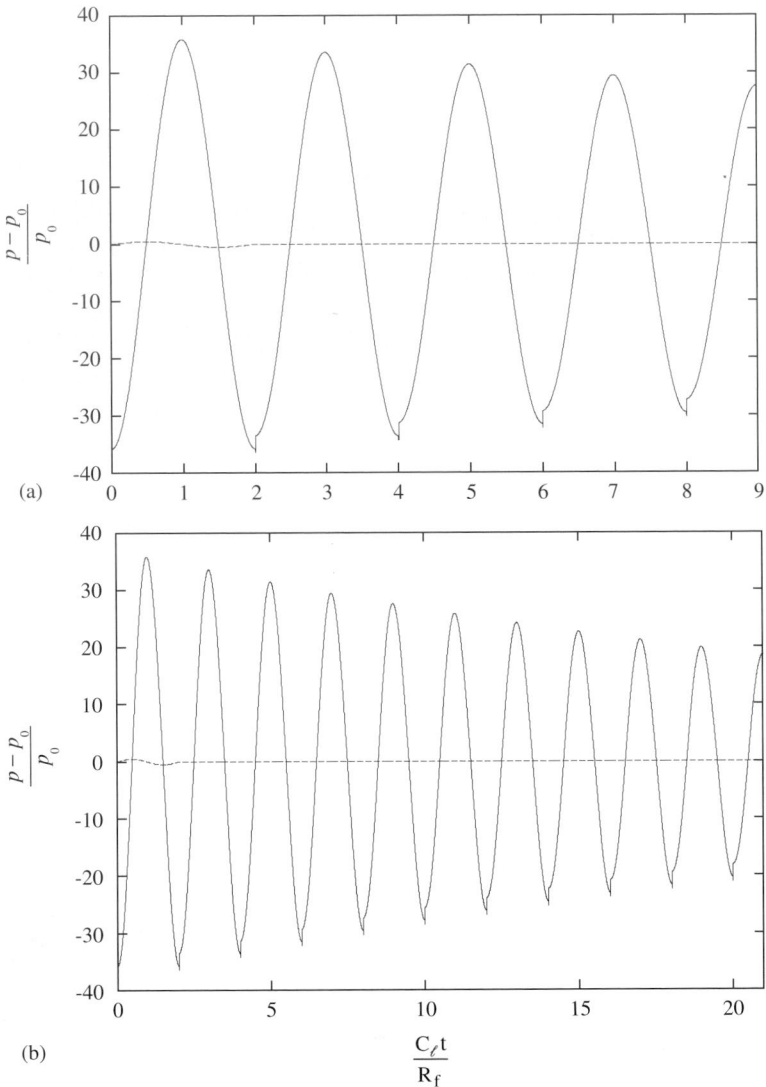

FIG. B.7 (a, b). Standing wave decay after the driving pressure at the flask wall is turned off. Normalized pressure at the center of a spherical flask (p_I: solid line) and at the flask wall (p_w: dashed line) vs. normalized time $C_\ell t/R_f$ ($R_f = 5$ cm, $C_\ell = 1500$ m/s, $\mu_f = 500$ kg/(m s), $\rho_\ell = 1000$ kg/m^3, $\omega R_f/C_\ell = 3.1409$, $\Delta p_w/p_0 = 0.5$, $N = 1000$, $n_d = 250$).

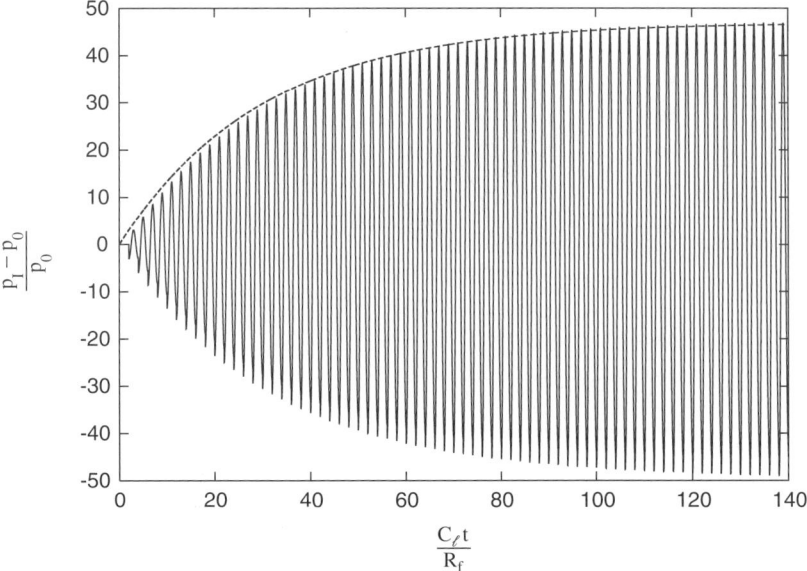

FIG. B.8. A transient to a standing acoustic wave state. Normalized pressure at the center of a spherical flask ($R_f = 5$ cm, $C_\ell = 1500$ m/s, $\mu_f = 500$ kg/(m s), $\rho_\ell = 1000$ kg/m^3, $\omega R_f/C_\ell = 3.1409$, $N = 1000$, $n_d = 250$) vs. normalized time, $C_\ell t/R_f$, compared with the exponential approximation of amplitude, $(p_1 - p_0)/p_0 = -(\Delta p/p_0)[1 - \exp(-C_\ell t/R_f \tau)]$ (dashed line, $\Delta p/p_0 = 47$, $\tau = 30$).

seen in SBSL experiments. With a further increase in incident pressure amplitude the bubble expansion period also increases, which reduces the time for post-implosion bounces. After three driving periods the bubble starts to oscillate with one collapse per driving period without any afterbounces.

The bubble expansion period and intensity of the bubble collapse increases from period to period. Fig. B.9b shows that after a while the bubble response loses its periodicity and a period doubling bifurcation takes place. Although, the bubble expands and collapses every period of driving, the intensity of the expansion and collapse varies from period to period. Later in time, the bubble starts to collapse only one time per two periods of acoustic driving. After a complicated transient, the bubble oscillations achieve a limit cycle. A typical transient is shown in Fig. B.10a.

In order to help demonstrate the dynamics and the intensity of energetic bubble collapses (i.e., implosions), the gas temperature in the bubble was calculated using an adiabatic compression approximation. It can be seen in Fig. B.10b that, as expected, temperature spikes only occur during energetic bubble implosions.

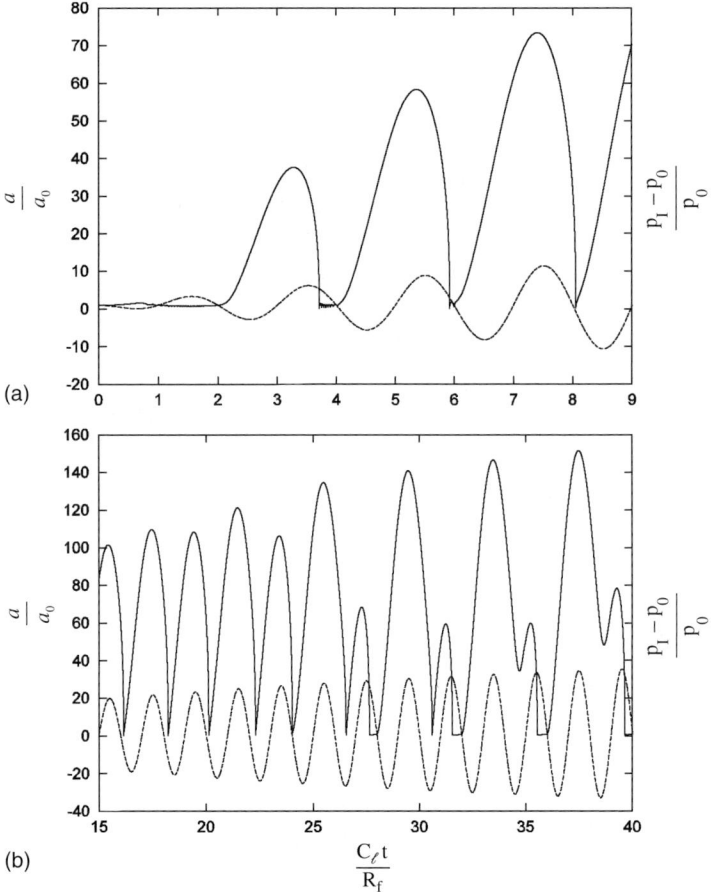

FIG. B.9 (a, b). Radius of a bubble (a(t): solid line) of initial radius $R_0 = 10\,\mu m$ driven by incident pressure (p_I: dashed line), $(p_I - p_0)/p_0 = -(\Delta p/p_0)\left[1 - \exp(-C_\ell t/R_f \tau)\right]\sin(\omega C_\ell t/R_f)$ ($\Delta p/p_0 = 47$, $\tau = 30$, $\omega R_f/C_\ell = 3.1409$) vs. normalized time $C_\ell t/R_f$.

These calculations have been conducted for the bubbles of different equilibrium radii. However, all the calculations demonstrated similar bubble behavior. In particular, after a start-up transient the bubbles, instead of collapsing every period of acoustic driving, start to collapse less frequently. As shown in Figs. B.10a and b, a 20 µm diameter bubble eventually implodes one time every two periods of acoustic driving, while Fig. B.11 shows that a 400 µm diameter bubble oscillates more chaotically and collapses about one time per three periods of acoustic driving.

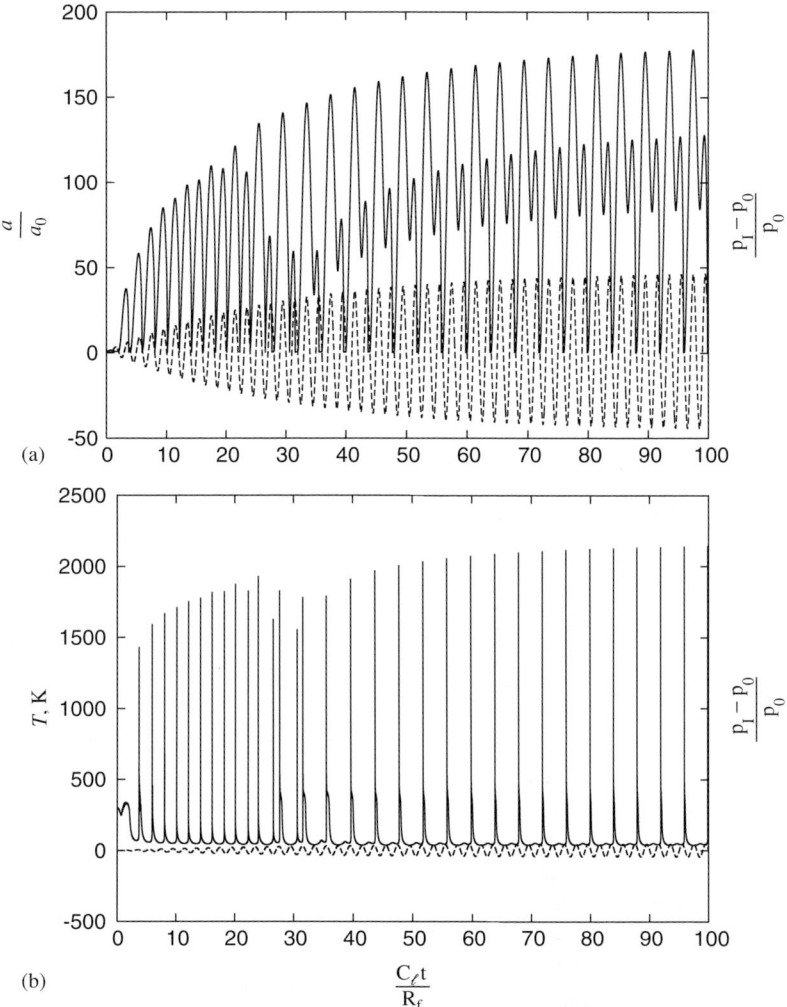

FIG. B.10 (a, b). Radius of the bubble and temperature in the bubble (a(t) and T(t): solid lines) of initial radius $R_0 = 10\,\mu$m driven by incident pressure (p_I: dashed line) $(p_I - p_0)/p_0 = -(\Delta p/p_0)[1 - \exp(-C_\ell t/R_f \tau)] \sin(\omega C_\ell t/R_f)$, ($\Delta p/p_0 = 47$, $\tau = 30$, $\omega R_f/C_\ell = 3.1409$) vs. normalized time, $C_\ell t/R_f$.

In sonofusion experiments, we may have many bubbles of different equilibrium radii in a bubble cluster [54,56] and the implosion of this bubble cluster produces an energetic shock wave, which can detune the acoustic chamber. During the first part of the recovery process of the standing acoustic wave, all the bubbles collapse at the acoustic frequency and the intensity of

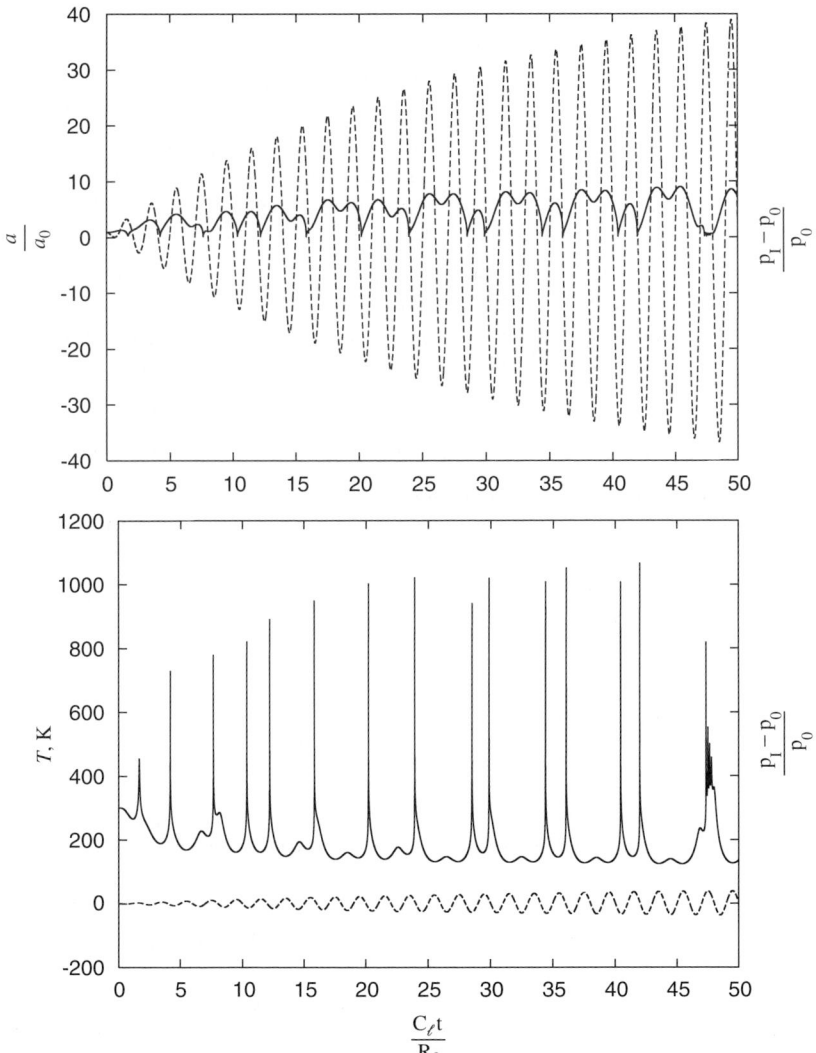

FIG. B.11. Radius of the bubble and temperature in the bubble (solid lines) of initial radius $R_0 = 200\,\mu\text{m}$ driven by incident pressure (dashed line), $(p_\text{I} - p_0)/p_0 = -(\Delta p/p_0)\left[1 - \exp(-C_\ell t/R\tau)\right]\sin(\omega C_\ell t/R_\text{f})$, ($\Delta p/p_0 = 47$, $\tau = 30$, $\omega R_\text{f}/C_\ell = 3.1409$) vs. normalized time, $C_\ell t/R_\text{f}$.

resultant D/D neutron emissions grow. Later on, the bubbles may undergo a bifurcation and start to implode less frequently. Some bubbles implode one time per two acoustic cycles, others one time per three cycles and some others may be even less. The phases of their collapse are not correlated and they collapse for different acoustic cycles. Thus, the amount of bubbles collapsing during one acoustic cycle may not be equal to the total amount of bubbles in the bubble cluster. This, coupled with condensation of the vapor bubbles in the highly subcooled liquid pool, and, more importantly, a strong repulsion of the bubble cluster from the acoustic antinode due to the primary Bjerknes force, can lead to the reduction of D/D neutron yield per acoustic cycle seen in Fig. 22a for times greater than about 1.1 ms.

In summary, a simple mathematical model and numerical algorithm for the transient dynamics in a spherical flask/liquid/bubble system has been developed. This model takes into account acoustic wave propagation in a liquid, damping of the acoustic energy in the coupled flask/liquid system and non-linear bubble dynamics phenomena.

The model appears to explain many of the phenomena seen in sonofusion experiments [56]. In particular, it appears that the first violent bubble implosion produces a compression shock, which interacts with the flask and momentarily disrupts the standing acoustic wave field in the flask. However, the PZT transducer at the flask wall continues to be driven to oscillate sinusoidally, and this leads to some complicated fluid/structure dynamics, which eventually restores the standing wave acoustic pressure field in the flask (after about 10 acoustic cycles). Later, due to bifurcation phenomena, possible bubble condensation and Bjerknes-force-induced ejection of the bubble clusters from the acoustic antinode, the resultant D/D neutron yield decays (see Fig. 22a).

This transient was evaluated numerically, and the calculated pressure at the center of the flask was applied to an equivalent single bubble. It was shown that during this transient the intensity of the bubble implosions initially increases in time. This corresponds with the observed increase in D/D neutron yield [56]. After that, due to bifurcations, condensation phenomena and the primary Bjerknes force, there is a reduction of D/D neutron yield per acoustic cycle. This analysis appears to explain the experimental observations in sonofusion experiments [56] and opens the possibility for future optimization of sonofusion phenomenon.

Appendix C: Nucleation of a Bubble in Tensioned Liquids Using High-Energy Neutrons

Let us consider a superheated liquid, when its pressure is less than the saturation vapor pressure at a particular temperature, $p_\ell < p_s(T)$. Such a meta-stable situation may occur in a liquid subject to tension in a high-intensity acoustic field (i.e., during the rarefaction period the local liquid pressure may drop way below the saturation pressure). For example, water at room temperature has a saturation vapor pressure of $p_s \cong 2300$ Pa, which is negligibly small compared to atmospheric pressure ($p_0 = 10^5$ Pa = 1 bar). Therefore, in an acoustic field with amplitude $\Delta p \geqslant 1$ bar, the liquid will experience a superheated state during some part of rarefaction period. Liquid in this state is unstable and the presence of some perturbation may cause cavitation (i.e., the appearance of small vapor bubbles). These perturbations in the liquid may be initiated by various means: thermo-fluctuations, which leads to homogeneous nuclei formation; particles of suspended impurities and/or roughness of the walls of a vessel, which may lead to heterogeneous nuclei formation, or, alternatively, a high-energy particle (e.g., a neutron) interacting with the fluid may cause nucleation.

MECHANICAL EQUILIBRIUM OF A VAPOR BUBBLE IN A LIQUID

Let us consider a single spherical bubble of radius R in a liquid in which there exists mechanical equilibrium. The pressures in the liquid and vapor are p_ℓ and p_v, respectively. The conditions for such equilibrium may be derived in a following way. If the bubble experiences an infinitesimal displacement, δR, of its interface (assuming the bubble remains spherical) then neglecting viscous and liquid inertia-induced losses, the work done on the vapor is

$$\delta W = (p_\ell - p_v) 4\pi R^2 \delta R + \overbrace{8\pi R \sigma \delta R}^{\delta(4\pi R^2 \sigma)} \quad \text{(C.1)}$$

where σ is the surface tension. The first term is the work associated with the displacement of the bubble's interface, and second term is the work needed for the creation of new interface. For mechanical equilibrium this perturbation in work must be equal to zero. This leads to the classical Laplace's formula

$$p_v = p_\ell + \frac{2\sigma}{R} \quad \text{(C.2)}$$

THERMODYNAMICAL EQUILIBRIUM OF A VAPOR BUBBLE IN A LIQUID

At fixed values of liquid pressure, p_ℓ, and temperature, T, of the liquid/vapor mixture, for a fixed mass mixture the system will approach the state characterized by the minimum value of the specific Gibbs potential, $\varphi(p,T)$.

We may take saturation thermodynamic conditions T, $p_s(T)$ as a starting point, and account for the equality of the phasic Gibbs potentials in a state of saturation $\varphi_\ell(p_s(T), T) = \varphi_v(p_s(T), T)$. If a vapor bubble is created, and $\varphi_\ell(p_s(T), T) > \varphi_v(p_v, T)$, then the state of the liquid in the mixture is unstable, and it will transform into vapor, (i.e., the vapor bubble will grow). If, in contrast, $\varphi_\ell(p_\ell(T), T) < \varphi_v(p_v, T)$, then the thermodynamic state of the vapor in the mixture is also unstable, and it will transform into liquid phase (i.e., the vapor bubble will condense). Thus, for thermodynamic equilibrium, it is necessary that

$$\varphi_\ell(p_\ell, T) - \varphi_\ell(p_s(T), T) = \varphi_v(p_v, T) - \varphi_v(p_s(T), T) \tag{C.3}$$

Recalling that

$$\left(\frac{\partial \varphi}{\partial p}\right)_T = \frac{1}{\rho} \tag{C.4}$$

the equilibrium condition, Eq. (C.3), can be expressed in terms of integrals along the isotherms

$$\int_{p_s}^{p_\ell} \frac{dp}{\rho_\ell(p, T)} = \int_{p_s}^{p_v} \frac{dp}{\rho_v(p, T)} \tag{C.5}$$

Since the vapor density along an isotherm is proportional to pressure and the liquid is almost incompressible

$$\rho_v = \rho_{vs} \frac{p}{p_s}, \quad \rho_\ell \cong \text{constant} \tag{C.6}$$

we obtain the equation of thermodynamic equilibrium that interrelates the deviation of the phasic pressures, p_ℓ, p_v, for the liquid and vapor, respectively, and the saturation pressure, p_s

$$\frac{p_v}{p_s} = \exp\left[\frac{\rho_{vs}}{\rho_\ell}\left(\frac{p_\ell}{p_s} - 1\right)\right] \tag{C.7}$$

Accounting for mechanical equilibrium condition at the spherical interface, Eq. (C.2), one obtains

$$\frac{p_\ell}{p_s} + \frac{2\sigma}{p_s R} = \exp\left[\frac{\rho_{vs}}{\rho_\ell}\left(\frac{p_\ell}{p_s} - 1\right)\right] \tag{C.8}$$

If there is a vapor bubble of radius R in a liquid of temperature T (where p_s, ρ_{vs}, p_ℓ and σ are known), then liquid and vapor pressures at thermodynamic equilibrium can be calculated from Eqs. (C.8) and (C.7). The solution of Eq. (C.8) shows that the liquid pressure may be positive or negative depending on bubble radius; however, the vapor pressure is always positive. If one knows the liquid pressure, p_ℓ, then Eq. (C.8) may be rewritten as the equation for the equilibrium (mechanical and thermodynamic) bubble radius

$$R_* = \frac{2\sigma}{p_s \exp[\rho_{vs}/\rho_\ell(p_\ell/p_s - 1)] - p_\ell} \tag{C.9}$$

When the liquid superheating is not very high (i.e., p_v and p_ℓ are not very far from p_s), it is useful to use an approximation for Eq. (C.7)

$$\frac{p_v}{p_s} - 1 \cong \frac{\rho_{vs}}{\rho_\ell}\left(\frac{p_\ell}{p_s} - 1\right) \tag{C.10}$$

which leads to the following well-known approximate formula for equilibrium bubble radius:

$$R_* = \frac{2\sigma}{(p_s - p_\ell)(1 - \rho_{vs}/\rho_\ell)} \tag{C.11}$$

The approximation in Eq. (C.10) obviously breaks down when it leads to a negative vapor pressure. Therefore it is valid only when liquid pressures satisfy

$$\frac{p_\ell}{p_s} > -\frac{\rho_\ell}{\rho_{vs}} + 1 \tag{C.12}$$

ACTIVATION THRESHOLD FOR NUCLEATION: THE CRITICAL BUBBLE SIZE

In order to calculate the energy necessary to nucleate a vapor bubble, let us consider a single bubble in a liquid that is generally not in mechanical and thermodynamical equilibrium. If a bubble experiences an infinitesimal displacement, δR, of its interface then, as before, the work done on the vapor is

$$\delta W = (p_\ell - p_v)4\pi R^2 \delta r + 8\pi R\sigma \delta R \tag{C.13}$$

Here δr is the liquid displacement at the bubble's interface. The first term is the work done to displace the liquid, and the second term is the work required for the creation of new interface. The liquid displacement, δr, may differ from the interfacial displacement, δR, because of evaporation (δR) or condensation ($\delta R < 0$)

$$\delta r = \delta R - \delta \xi = \delta R - \frac{\rho_v}{\rho_\ell}\delta R = \left(1 - \frac{\rho_v}{\rho_\ell}\right)\delta R \qquad (C.14)$$

where $\delta \xi$ is the bubble radius perturbation due to phase change. Thus, Eq. (C.13) takes the form

$$\delta W = (p_\ell - p_v)\left(1 - \frac{\rho_v}{\rho_\ell}\right)4\pi R^2 \delta R + 8\pi R \sigma \delta R \qquad (C.15)$$

In order to calculate the work that must be done by the liquid to create a vapor bubble of a certain size, one should integrate Eq. (C.15) over bubble's radius. This is impossible to do accurately, because, in general, vapor pressure and density depend on the kinetics of the evaporation/condensation processes and rate of bubble expansion. Obviously, such a problem must be coupled with the bubble dynamics involving the heat and mass exchange. Nevertheless, one can do an approximate solution assuming the vapor pressure and density are equal to their saturation values

$$\delta W = -(p_s - p_\ell)\left(1 - \frac{\rho_{vs}}{\rho_\ell}\right)4\pi R^2 \delta R + 8\pi R \sigma \delta R \qquad (C.16)$$

Such an approximation is valid for relatively low superheat. The integration of Eq. (C.16) yields

$$W(R) = 4\pi \sigma R^2 - (p_s - p_\ell)\left(1 - \frac{\rho_{vs}}{\rho_\ell}\right)\frac{4\pi}{3}R^3 \qquad (C.17)$$

In the case of superheated liquid ($p_\ell < p_s$), $W(R)$ is not a monotonous function of R, rather it has a maximum at

$$R_* = \frac{2\sigma}{(p_s - p_\ell)(1 - \rho_{vs}/\rho_\ell)} \sim \frac{2\sigma(T_\ell)}{(p_s(T_\ell) - p_\ell)} \qquad (C.18a)$$

Thus,

$$W_* \equiv A_* = \frac{4}{3}\pi \sigma R_*^2 = \frac{16\pi\sigma^3(T_\ell)}{3(p_s(T_\ell) - p_\ell)^2 (1 - \rho_{vs}(T_\ell)/\rho_\ell)} \sim \frac{16\pi\sigma^3(T_\ell)}{3(p_s(T_\ell) - p_\ell)^2} \qquad (C.18b)$$

where A_* is often called the activation energy.

The Role of Homogeneous Nucleation

The rate the of homogeneous nucleation of bubbles of critical size (i.e., J, bubbles/m^{3}s) is calculated as follows [170]:

$$J = J_0 \exp\left[-\frac{16\pi\sigma^3(T_\ell)}{3(p_s(T_\ell) - p_\ell)^2 k_b T_\ell}\right], \quad J_0 = N'''\left(\frac{2\sigma(T_\ell)}{\pi m}\right)^{1/2} \quad (C.19)$$

Here k_b, m and N''' are the Boltzman's constant, the mass of the molecules in the fluid, and the bubble number density, respectively.

Measurements on a large number of organic compounds have been made [171] which appear to be consistent with the theoretical nucleation rate, Eq. (C.19). Measured limits of superheat have been compared with limits calculated from Eq. (C.19) for $J = 10^6$ bubbles/cm^{3}s. In particular, for acetone it was found that at $p_{\ell*} = 1$ bar, the limiting superheat was $T_{\ell*} = 447$ K. These data allow one to calculate the tensile strength, p_ℓ, for acetone at any liquid temperature, T_ℓ.

One can equate the exponential arguments for the respective states of acetone, namely

$$\frac{\sigma^3(T_\ell)}{[p_s(T_\ell) - p_\ell]^2 T_\ell} = \frac{\sigma^3(T_{\ell*})}{[p_s(T_{\ell*}) - p_{\ell*}]^2 T_{\ell*}} \quad (C.20)$$

We may then solve this equation for p_ℓ yielding

$$p_\ell = p_s(T_\ell) - \left[\frac{\sigma(T_\ell)}{\sigma(T_{\ell*})}\right]^{3/2} \left(\frac{T_{\ell*}}{T_\ell}\right)^{1/2} [p_s(T_{\ell*}) - p_{\ell*}] \quad (C.21)$$

Of course, one must also know the dependence of saturation pressure and surface tension on temperature (e.g., $p_s(T_\ell), \sigma(T_\ell)$). For acetone these data are presented in Table C.I. For instance, if $T_\ell = 273$ K, then Eq. (C.21) gives $p_\ell \cong -180$ bar for acetone.

It is clear that if acetone is put in tension at a negative pressure of about -15 bar, which is much less than its ultimate tensile strength (-180 bar), homogeneous nucleation does not play any significant role in bubble nucleation by neutrons.

Neutron-Induced Bubble Nucleation

The minimum energy, E_*, which is necessary to nucleate a "critical size" bubble is the sum of the activation energy, A_*, and the energy associated with the latent heat, Q_*. That is

$$E_* = A_* + Q_* \quad (C.22)$$

TABLE C.I
Properties for Acetone

T_ℓ (K)	p_s (Pa)	σ (N/m)	h_{fg} (J/kg)
225	380.21	0.03700	330,000.0
250	2,378.81	0.03100	290,000.0
273	9,203.60	0.02616	264,804.2
300	33,050.50	0.02281	249,768.2
325	86,991.31	0.01976	234,888.1
350	195,746.74	0.01675	218,839.3
375	390,222.00	0.01381	201,466.2
400	707,594.58	0.01093	182,301.2
450	1,887,470.83	0.00544	129,558.5
508	4,648,365.96	0.00001	0.0

where

$$Q_* = \frac{4}{3}\pi R_*^3 \rho_{vs}(T_\ell) h_{fg}(T_\ell) \qquad (C.23)$$

and h_{fg} is specific latent heat of the liquid.

We have calculated the activation energy, A_*, the energy associated with phase change, Q_* and total energy, E_*, which is needed to nucleate acetone at different temperatures and pressures. The properties of liquid acetone (i.e., $p_s(T_\ell)$, $\sigma(T_\ell)$, $h_{fg}(T_\ell)$) are presented in Table C.I. For various levels of liquid tension (i.e., under pressures) linear interpolation of these data has been used, and the density of saturated acetone vapor, $\rho_{vs}(T_\ell)$, was calculated using the perfect gas law (the gas constant for acetone vapor is $R = 130$ J/kg K). The results are presented in Figs. C.1–C.4. We see that Q_* is only significant at low levels of underpressure. In particular, at -15 bar, and the liquid temperatures of interest in sonofusion (~ 273 K), $E_* \sim A_*$.

The critical radius, R_*, is the minimal radius a bubble must have in order to start to nucleate and grow and is given by Eq. (C.18a) as

$$R_* \approx \frac{2\sigma(T_\ell)}{p_s(T_\ell) - p_\ell} \qquad (C.24)$$

where σ is the liquid's surface tension, $p_s(T_\ell)$ liquid's saturation pressure at temperature T_ℓ and p_ℓ the liquid's static pressure. As discussed previously, the activation energy required for the nucleation of a bubble with radius R_*

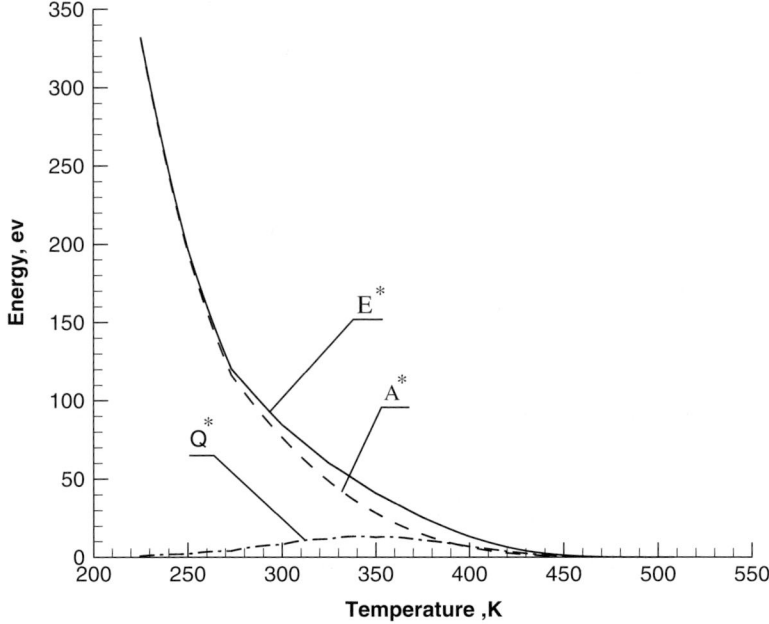

Fig. C.1. Acetone at −40 bar underpressure.

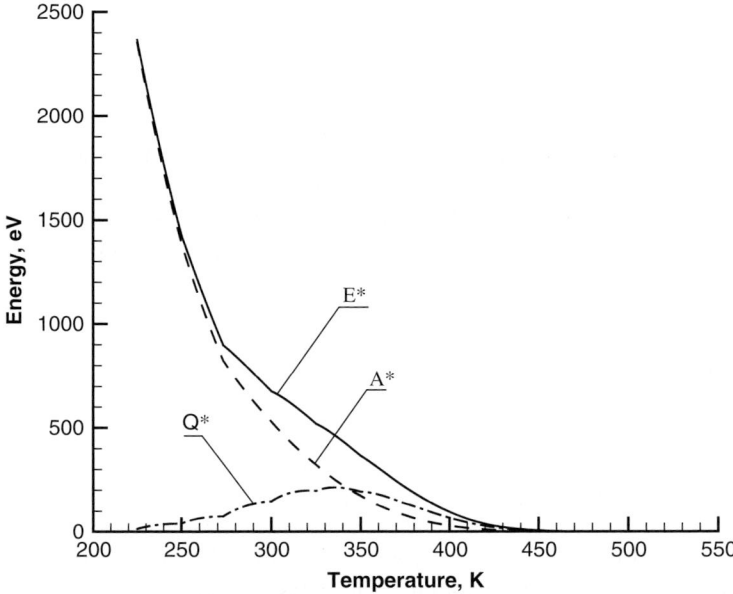

Fig. C.2. Acetone at −15 bar underpressure.

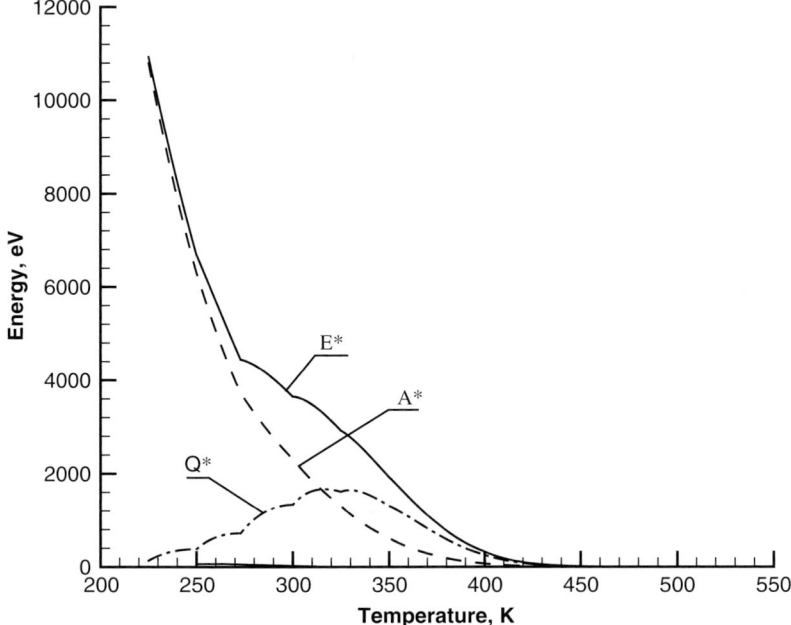

Fig. C.3. Acetone at −7 bar underpressure.

is given by Eqs. (C.18b), (C.22) and (C.23) as

$$E_* \approx \frac{16\pi\sigma^3(T_\ell)}{3(p_s(T_\ell) - p_\ell)^2} + \frac{4}{3}\pi R_*^3 \rho_{vs}(T_\ell) h_{fg}(T_\ell) \quad \text{(C.25)}$$

where ρ_{vs} is the vapor's saturation density and $h_{fg}(T_\ell)$ the specific latent heat.

It has been shown [172,173] that a bubble will nucleate if the energy deposited along a specific length (i.e., the diameter of a critical size bubble, $2R_*$), is larger than or equal to E_*. Thus, the criteria for bubble formation is

$$2R_* \frac{dE}{dx} \geq E_* \quad \text{(C.26)}$$

where dE/dx is the recoiled (i.e., knock-on) ion's energy loss per unit path length (i.e., the stopping power), which, in turn, is the energy gained by the liquid molecules.

It can be noted that dE/dx has two components; the dominant one is from coulombic interactions and the other from particle interactions. As noted previously, calculations of dE/dx and the range curves were done with the SRIM code [148].

It can be seen in Fig. C.5 that the stopping power, dE/dx, is smaller for the ions of the lighter isotope, deuterium (D), and that it drops off as the

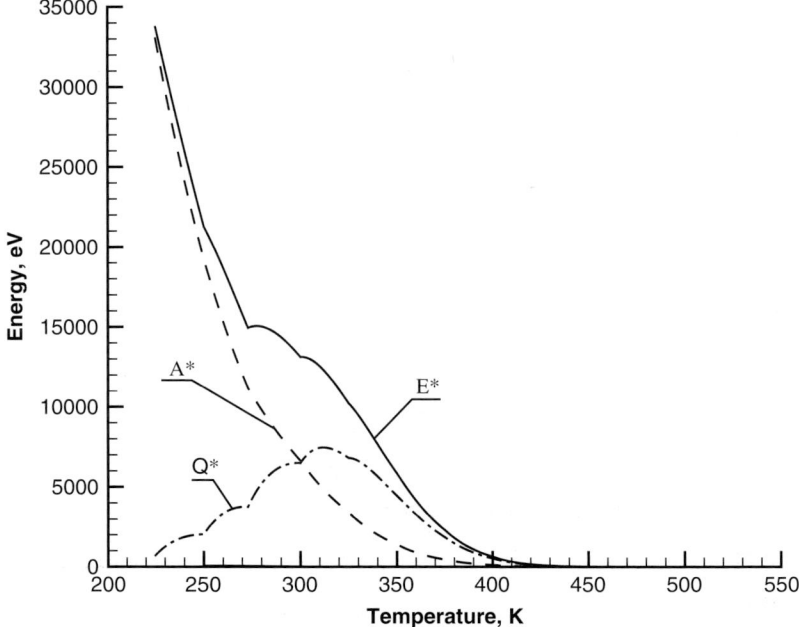

FIG. C.4. Acetone at −4 bar underpressure.

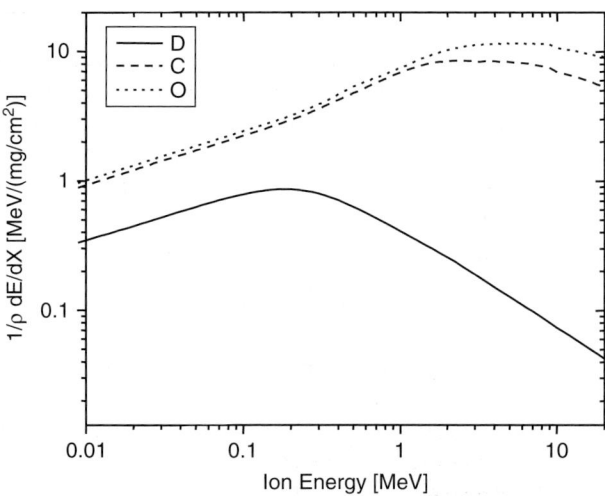

FIG. C.5. Energy loss of D, C and O ions in C_3D_6O.

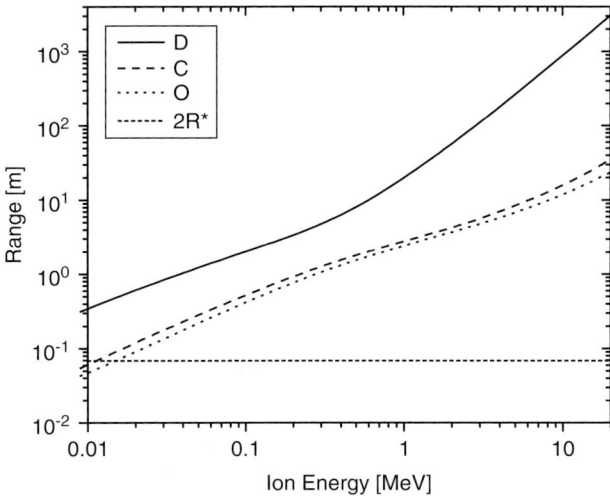

FIG. C.6. Ranges of D, C and O ions in C_3D_6O.

energy of the knock-on ions increases. This implies that the production of bubbles by deuterons occurs mainly at the lower ion energies.

Finally, Fig. C.6 shows that the range of all the ions in deuterated acetone is much greater than the critical bubble diameter ($2R^*$), except for carbon and oxygen ions at low energies. Thus the knock-on ions produced by a single high-energy neutron can produce many bubbles (i.e., a bubble cluster) in the liquid pool.

References

1. Marinesco, N. and Trillat., J. J. (1933). Chimie physique – action des ultrasons sur les plaques photographiques. *Comptes Rendues Acad. Sci. Paris* **196**, 858.
2. Frenzel, J. and Schultes., H. (1934). Luminescenz im ultraschallbeschickten Wasser. *Kurze Mitteilung. Zeit fur Phys. Chem. B* **27**, 421.
3. Harvey., E. N. (1939). Sonoluminescence and Sonic Chemiluminescences. *J. Am. Chem. Soc.* **61**, 2392.
4. Chalmers, L. A. (1936). The emission of visible light from pure liquids during acoustic excitation. *Phys. Rev.* **49**, 881.
5. Prosperetti, A. (1997). A new mechanism for sonoluminescence. *J. Acoust. Soc. Am.* **101**, 2003.
6. Frenkel, J. (1940). On electrical phenomena associated with cavitation due to ultrasonic vibrations in liquids. *Acta Phys – Chim. USSR* **12**, 317, (in Russian).
7. Degrois, L. A. and Baldo, P. (1974). A new electrical hypothesis explaining sonoluminescence, chemical actions and other effects produced in gaseous cavitation. *Ultrasonics* **12**, 25.

8. Margulis, M. A. (1985). Study of electrical phenomena associated with cavitation: II. Theory of the development of sonoluminescence in acoustochemical reactions. *Russ. J. Phys. Chem.* **59**, 882.
9. Garcia, N. and Levanyuk, A. P. (1996). Sonoluminescence: A new electrical breakdown hypothesis. *JETP Lett* **64**, 907.
10. Young, F. R. (2004). "Sonoluminescence". CRC Press, Boca Raton, FL.
11. Weyl, W. A. and Marboe, E. C. (1949). Some electro-chemical properties of water. *Research* **2**, 19.
12. Griffing, V. (1950). Theoretical exploration of chemical effects of ultrasonics. *J. Chem. Phys.* **18**, 997.
13. Griffing, V. and Sette, D. (1952). The chemical effects of ultrasonics. *J. Chem. Phys.* **20**, 939.
14. Noltinigk, B. E. and Neppiras, E. A. (1950). Cavitation produced by ultrasonic. *Proc. Phys. Soc. B* **63**, 674.
15. Hickling, R. (1963). Effects of thermal conduction in sonoluminescence. *J. Acoust. Soc. Am.* **35**, 967.
16. Jarman, P. (1960). Sonoluminescence: A discussion. *J. Acoust. Soc. Am.* **32**, 1459.
17. Barber, B. P., Hiller, R., Arisaka, K., Fetterman, H., and Putterman, S. J. (1992). *J. Acoust. Soc. Am.* **91**, 3061.
18. Barber, B. P., Wu, C. C., Lofstedt, R., Roberts, P. H., and Putterman, S. J. (1994). Sensitivity of sonoluminescence to experimental parameters. *Phys. Rev. Lett.* **72**, 380.
19. Crum, L. and Gaitan, D. F. (1990). "Frontiers of Nonlinear Acoustics, 12th International Symposium on Nonlinear Acoustics". Elsevier Applied Science, New York.
20. Crum, L. and Matula, T. (1997). Shocking revelations. *Science* **276**, 1348.
21. Gaitan, D. F. and Crum, L. A. (1990). *J. Acoust. Soc. Am.* **87**(Suppl 1), 141.
22. Gaitan, D. F., Crum, L. A., Church, C. C., and Roy, R. A. (1992). *J. Acoust. Soc. Am.* **91**, 3166.
23. Hiller, R., Putterman, S., and Weninger, K. (1998). Time-resolved spectra of sonoluminescence. *Phys. Rev. Lett.* **80**, 1090.
24. Kaiser, J. (1995). Inferno in a bubble: Turning sound into light poses a tantalizing puzzle. *Sci. News.* **147**(17), 262.
25. Lohse, D. (1998). Lasers blow a bigger bubble. *Nature* **392**, 21.
26. McNamara, W. B., Didenko, Y. T., and Suslick, K. S. (1999). Sonoluminescence temperatures during multi-bubble cavitation. *Nature* **401**, 772.
27. Metcalf, H. (1998). That flashing sound. *Science* **279**, 1322.
28. Moss, W. C., Clarke, D. B., White, J. W., and Young, D. A. (1994). Hydrodynamic simulations of bubble collapse and picosecond sonoluminescence. *Phys. Fluids* **6**(9), 2979.
29. Moss, W. C., Clarke, D. B., White, J. W., and Young, D. A. (1996). Sonoluminescence and the prospects for table-top micro-thermonuclear fusion. *Phys. Lett. A* **211**, 69.
30. Moss, W. C., Clarke, D. B., and Young, D. A. (1997). Calculated pulse widths and spectra of a single sonoluminescing bubble. *Science* **276**, 1398.
31. Camara, C., Putterman, S., and Kirilov, E. (2004). Sonoluminescence from a single bubble driven at 1 MHz. *Phys. Rev. Lett.* **92**(12), 124301–1.
32. Beckett, M. A. and Hua, I. (2001). Impact of ultrasonic frequency on aqueous sonoluminescence and sonochemistry. *J. Phys. Chem-A* **105**, 3796.
33. Hiller, R., Putterman, S., and Barber, B. (1992). Spectrum of synchronous picosecond sonoluminescence. *Phys. Rev. Lett.* **69**(8), 1182.
34. Lauterborn, W., Kurz, T., Mettin, R., and Ohl, C. D. (1999). Experimental and theoretical bubble dynamics. *Adv. Chem. Phys.* **110**, 295.

35. Putterman, S. J. and Weninger, K. R. (2000). Sonoluminescence: How bubbles turn sound into light. *Annu. Rev. Fluid Mech.* **32**, 445.
36. Walton, A. J. and Reynolds, G. T. (1984). Sonoluminescence. *Adv. Phys.* **33**, 595.
37. Crum, L. (1994). Sonoluminescence, sonochemistry, and sonophysics. *J. Acoust. Soc. Am.* **95**(1), 559.
38. Margulis, M. A. (2000). *Phys. Sci. Adv.* **170**(3), 263, (in Russian).
39. Margulis, M. A. (1976). Modern views of acoustic-chemical reactions. *Russ. J. Phys. Chem.* **50**(1), 1.
40. Suslick, K.S., Ed. (1988). "Ultrasound: Its Chemical, Physical and Biological Effects," p. 230. VCH Publishers, New York.
41. Suslick, K. S. and Crum, L. A. (1997). Sonochemistry and Sonoluminescence. *In* "Encyclopedia of Acoustics" (M.J. Crocker, ed.), p. 271. Wiley, New York.
42. Yosioka, K. and Omura, A. (1962). The light emission from a single bubble driven by ultrasound and the spectra of acoustic oscillations. *Proc. Annual Meeting–Acoust. Soc. Jpn.* **126**, (in Japanese).
43. Whitham., G. B. (1974). "Linear and Nonlinear Waves". Wiley Interscience, New York.
44. Rayleigh, L. (1917). On the pressure developed in a liquid during the collapse of a spherical cavity. *Phil. Mag.* **34**, 94.
45. Plesset, M. S. and Hseih, D. Y. (1960). Theory of gas bubble dynamics in oscillating pressure fields. *Phys. Fluids* **3**, 882.
46. Herring, C. (1941). Theory of the pulsations of the gas bubble produced by an underwater explosion, OSRD Rep. 236.
47. Flynn, H. G. (1964). Physics of Acoustic Cavitation. *In* "Physical Acoustics" (W.P. Mason, ed.), p. 57. Academic Press, New York.
48. Gilmore, F. R. (1952). The collapse of a spherical bubble in a viscous compressible liquid. *Cal. Tech. Hydrodynamics Lab. Rep.* **26-24**.
49. Knapp, R. T., Daily, J. W., and Hammit, F. G. (1970). "Cavitation". McGraw-Hill, New York.
50. Prosperetti, A. and Lezzi, A. (1986). Bubble dynamics in a compressible liquid. Part 1. First order theory. *J. Fluid Mech.* **168**, 457.
51. Nigmatulin, R. I., Akhatov, I. S., Vakhitova, N. K., and Lahey, R. T., Jr. (2000). On the forced oscillations of a small gas bubble in a spherical liquid-filled flask. *J. Fluid Mech.* **414**, 47.
52. Wu, C. C. and Roberts, P. H. (1993). Shock-wave propagation in a sonoluminescing gas bubble. *Phys. Rev. Lett.* **70**, 3424.
53. Barber, B. P. and Putterman, S. J. (1991). Observations of synchronous picosecond sonoluminescence. *Nature* **352**, 318.
54. Taleyarkhan, R. P., Cho, J. S., West, C. D., Lahey, R. T., Jr., Nigmatulin, R. I., and Block, R. C. (2002). Evidence for nuclear emissions during acoustic cavitation. *Science* **295**, 1868.
55. Taleyarkhan R. P., Block, R. C., West, C. and Lahey, R. T., Jr. (2002). Comments on the Shapira/Saltmarsh Paper. [101] htpp://www.rpi.edu/∼laheyr/SciencePaper.pdf.
56. Taleyarkhan, R. P., West, C., Cho, J. S., Lahey, R. T., Jr., Nigmatulin, R. I., and Block, R. C. (2004). Additional evidence of nuclear emissions during acoustic cavitation. *Phys. Rev. E.* **69**, 036109.
57. Akhatov, I., Gumerov, N., Ohl, C. D., Parlitz, V., and Lauterborn, W. (1997). The role of surface tension in stable single bubble sonoluminescence. *Phys. Rev. Lett.* **78**, 227.
58. Akhatov, T., Mettin, R., Ohl, C. D., Parlitz, V., and Lauterborn, W. (1997). Bjerkens force threshold for stable single bubble sonoluminescence. *Phys. Rev. E* **55**(3), 3747.

59. Akhatov, I., Ohl, C. D., Mettin, R., Parlitz, V. and Lauterborn, W. (1998). Giant response in dynamics of small bubbles. Proceedings of the 16th International Congress on Acoustics & 135th Meeting of the ASA, Seattle, WA, 2285.
60. Konovalova, S. and Akhatov, I. (2005). Structure Formation in Acoustic Cavitation. *J. Multiphase Sci. Technol.* **17**(4), 343.
61. Magsoft Corporation. ATILA version 5.1X, User's Manual, (2000).
62. West, C. (1967). "Cavitation Nucleation by Energetic Particles, Topical Report–Electronics and Applied Physics Division". AERE, Harwell, Berkshire, UK.
63. Emerson & Cuming: Stycast 1264 A/B technical data sheet. www.emersoncuming.com (2003).
64. Uchino, K. (2000). "Ferroelectric Devices". Marcel Dekker, New York.
65. Wilson, O. B. (1989). "Introduction to the Theory and Design of Sonar Transducers," Reprint edition. Peninsula Publishers.
66. Leighton, T. G. (1997). "Acoustic Bubble". Academic Press, New York.
67. Stroud, J. S. and Marston, P. L. (1993). Optical detection of transient bubble oscillations associated with underwater noise of rain. *J. Acoust. Soc. Am.* **94**, 20.
68. Lentz, W. J., Atchley, A. A., and Gaitan, D. F. (1995). Mie scattering from a sonoluminescing air bubble in water. *Appl. Optics* **34**(15), 2648.
69. Delgadino, G. A. and Bonetto, F. J. (1997). Velocity interferometry technique used to measure the expansion and compression phases of a sonoluminescent bubble. *Phys. Rev. E* **56**, 6248.
70. Flint, E. B. and Suslick, K. S. (1991). The temperature of cavitation. *Science* **253**, 1397.
71. Vazquez, G., Camara, C., Putterman, S., and Weininger, K. (2001). Sonoluminescence: Nature's smallest blackbody. *Optics Lett* **26**(9), 575.
72. Delgadino, G., Bonetto, F., and Lahey, R. T., Jr. (2002). The relationship between the method of acoustic excitation and the stability of single bubble sonoluminescence for various noble gases. *Chem. Eng. Commun.* **189**, 786.
73. Plesset, M. S. and Mitchell, T. P. (1956). On the stability of the spherical shape of a vapor cavity in a liquid. *Quarterly Appl. Math.* **13**(4), 419.
74. Eller, A. I. and Crum, L. A. (1969). Instability of the motion of a pulsating bubble in a sound field. *J. Acoust. Soc. Am.* **47**(3–2), 762.
75. Plesset, M. S. and Prosperetti, A. (1977). Bubble dynamics and cavitation. *Ann. Rev. Fluid Mech.* **9**, 145.
76. Neppiras, E. A. (1980). Acoustic cavitation. *Phys. Rep.* **61**, 159.
77. Brennen, C. E. (1995). "Cavitation and Bubble Dynamics". Oxford University Press, New York.
78. Crum, L. A. and Prosperetti, A. (1983). Nonlinear Oscillations of Gas Bubbles in Liquids: An Interpretation of Some Experimental Results. *J. Acoust. Soc. Am.* **73**, 121.
79. Matula, T. J., Cordry, S. M., Roy, R. A., and Crum, L. A. (1997). Bjerknes force and bubble levitation under single-bubble sonoluminescence conditions. *J. Acoust. Soc. Am.* **102**, 1522.
80. Eller, A. I. and Flynn, H. G. (1965). Rectified diffusion during nonlinear pulsations of cavitation bubbles. *J. Acoust. Soc. Am.* **37**, 493.
81. Fyrillas, M. M. and Szeri, A. J. (1994). Dissolution or growth of oscillating bubbles. *J. Fluid Mech.* **277**, 381.
82. Brenner, M. P., Lohse, P., Oxtoby, D., and Dupont, T. F. (1995). Mechanisms for stable single bubble sonoluminescence. *Phys. Rev. Lett.* **76**, 1158.
83. Holt, R. G. and Gaitan, D. F. (1996). Observation of stability boundaries in the parameter space of single bubble sonoluminescence. *Phys. Rev. Lett.* **77**, 3791.

84. Holzfuss, J., Ruggeberg, M., and Mettin, R. (1998). Boosting sonoluminescence. *Phys. Rev. Lett.* **81**, 1961.
85. Moraga, F., Taleyarkhan, R. P., Bonetto, F. J., and Lahey, R. T., Jr. (2000). Role of very-high-frequency excitation in single-bubble sonoluminescence. *Phys. Rev.* **62**, 2233.
86. Nigmatulin, R., Akhatov, I., Vakhitova, N., and Lahey, R. T., Jr. (1999). Hydrodynamics, acoustics and transport in sonoluminescence phenomena. *In* "Sonochemistry and Sonoluminescence" (L.A. Crum, *et al.*, eds.). Kluwer Academic Publishers, Dordrecht.
87. Nigmatulin, R., Akhatov, I., Vakhitova, N., and Topolnikov, A. (1999). Bubble collapse and shock wave formation in sonoluminescence. *In* "Nonlinear Acoustics at the Turn of the Millennium" (W. Lauterborn and T. Kurz, eds.). ISNA-15, Göttingen, Germany.
88. West, C. D. and Howlet, R. (1967). Timing of sonoluminescence flash. *Nature* **215**, 727.
89. West, C. D. and Howlet, R. (1968). Some experiments on ultrasonic cavitation using a pulsed neutron source. *Brit. J. Appl. Phys.* **1**, 247.
90. Nigmatulin, R. I., Taleyarkhan, R. P., and Lahey, R. T., Jr. (2004). The evidence for nuclear emissions during acoustic cavitation revisited. *J. Power Energy* **128**(A), 345.
91. Abramov, A. I., Kazansky, A. Yu, and Matusevich E. S. (1985). Principles of experimental methods in nuclear physics, *Energoatomoizdat* (in Russian).
92. Knoll, G. F. (1989). "Radiation Detection and Measurement". J. Wiley & Sons, New York.
93. Harvey, J. and Hall, N. (1979). Scintillation detectors for neutron physics research. *Nucl. Instrum. Methods Phys. Res.* **162**, 507.
94. Lee, J. H. and Lee, C. S. (1998). Response function of NE213 scintillator for 0.5–6 Mev neutrons measured by an improved pulse shape discrimination. *Nucl. Instrum. Methods Phys. Res. A* **402**, 147.
95. Hawkes, N. P., Adams, J. M., Dond, D. S., Jarvis, O. N., and Watkins, N. (2002). Measurements of the proton light output function of the organic liquid scintillator NE213 in several detectors. *Nuc. Instrum. Methods Phys. Res. A* **476**, 190.
96. Greenspan, M. and Tschiegg, C. E. (1967). Radiation-induced acoustic cavitation; apparatus and some results. *J. Res. Nat. Bureau Standards 71C* **4**, 299.
97. Hahn, B. (1961). The fracture of liquids under stress due to ionizing particles. *Nuevo Cimento* **22**, 650–653.
98. Morch K. A., (1980). On the Collapse of Cavity Clusters in Flow Cavitation, Cavitation and Inhomogeneities in Underwater Acoustics, W. Lauterborn (Ed.), Springer, Berlin, Series-Electrophysics, **4**, 95.
99. Wang, Y. C. and Brennen, C. E. (1994). Shock Wave Development in the Collapse of a Cloud of Bubbles, ASME FED. *Cavitation and Multiphase Flow* **194**, 15.
100. Akhatov, I., Nigmatulin, R. I. and Lahey, R. T., Jr. (2005). An analysis of linear and nonlinear bubble cluster dynamics. *Multiphase Sci. Technol*, **17**(3), 225–256.
101. Shapira, D. and Saltmarsh, M. (2002). Nuclear fusion in collapsing bubbles – is it there? An attempt to repeat the observation of nuclear emissions from sonoluminescence. *Phys. Rev. Lett.* **89**, 104302.
102. Taleyarkhan, R. P., Lahey, R. T., Jr., and Nigmatulin, R. I. (2005). Bubble nuclear fusion – status and challenges. *Multiphase Sci. Technol.* **17**(3), 191.
103. Nigmatulin, R. I. (1991). "Dynamics of Multiphase Media," Vols. 1&2. Hemisphere, New York.
104. Akhatov, I., Lindau, O., Topolnikov, A., Mettin, R., Vakhitova, N., and Lauterborn, W. (2001). Collapse and rebound of a laser-induced cavitation bubble. *Phys. Fluids* **13**(10), 2805.
105. Gross, R. A. (1984). "Fusion Energy". Wiley, New York.

106. Mie, G. (1903). Zur kinetischen Theorie der einatomigen Könper. *Ann. Phys. Leipzig* **11**, 657, (in German).
107. Grüneisen, E. (1908). Zusammenhang zwischen Kompressibilität, thermischer Ausdehnung, Atomvolumen und Atomwärme der Metalle. *Ann. Phys.* **26**, 293, (in German).
108. Kuznetsov, N. M. (1961). Equation of state and heat capacity of water in wide range of thermodynamic parameters. *Prikladnaya mekhanika I tekhnicheskaya fizika* **No. 1**, 112, (in Russian).
109. Zharkov, V. N. and Kalinin, V. A. (1968). Equations of State of Solids at High Pressures and Temperatures, *Nauka* (in Russian).
110. Zamishlyaev, B. V. and Menzhulin, M. G. (1971). Interpolation equation of state of water and water vapor. *Prikladnaya Mekhanika I Tekhnicheskaya Fizika* **113**(3), (in Russian).
111. Gurtman, G. A., Kirsch, J. W., and Hasting, C. R. (1971). Analytical equation of state for water compressed to 300 Kbar. *J. Appl. Phys.* **42**, 851.
112. Zeldovich, Y. B. and Raizer, Y. P. (1966). "Physics of Shock Waves and High-Temperature Hydrodynamic Phenomena". Academic Press, New York.
113. Trunin, R. F. (1994). Shock compressibility of condensed substances in strong shock waves of underground nuclear explosions. *Uspekhi Fizicheskikh Nauk* **164**, 1215, (in Russian).
114. Nigmatulin, R. I., Akhatov, I. S., Topolnikov, A. S., Bolotnova, R. K., Vakhitova, N. K., Lahey, R. T., Jr., and Taleyarkhan, R. P. (2005). Theory of supercompression of vapor bubbles and nanoscale thermonuclear fusion. *Phys. Fluids* **17**, 107106.
115. Born, M. and Mayer, J. E. (1932). Zur Gittertheorie der Ionenkristalle. *Z. Phys.* **75**, 1, (in German).
116. Jacobs, S. J. (1968). "On the Equation of State of Compressed Liquids and Solids". NOLTR 68-214, United States Naval Ordnance Laboratory, White Oak, MD.
117. Cowperthwaite, M. and Zwisler, W. K. (1976). The JCZ Equations of state for Detonation Products and Their Incorporation into the TIGER Code, Sixth International Symposium on Detonation, ACR-221/ ONR, Naval Surface Weapons Center.
118. Nigmatulin, R. I. and Khabeev, N. S. (1974). Heat exchange between a gas bubble and a liquid. *Fluid Dyn* **9**, 759.
119. Prosperetti, A., Crum, L. A., and Commander, K. W. (1988). Nonlinear bubble dynamics. *J. Acoust. Soc. Am.* **83**, 502.
120. Lamb, H. (1932). "Hydrodynamics," 6th edn. Dover Publications, New York.
121. Volmer, M. (1939). "Kinetik der Phasenbildung". Steinkopff, Dresden-Leipzig.
122. Schrage, R. W. (1953). "A Theoretical Study of Interphase Mass Transfer". Columbia University Press.
123. Kucherov, Ya. R. and Rikenglaz, L. E. (1960). The problem of measuring the condensation coefficient. *Doklady Akad. Nauk. SSSR* **133**, 1130–1131, (in Russian).
124. Prosperetti, A. (1987). The Equation of bubble dynamics in a compressible liquid. *Phys. Fluids* **30**, 3626.
125. Bae, S-H. and Lahey, R. T., Jr. (1999). On the use of nonlinear filtering, artificial viscosity and artificial heat transfer for strong shock computations. *J. Comp. Phys.* **153**, 575.
126. Nagrath, S., Jansen, K. E., Lahey, R. T., Jr., and Akhatov, I. S. (2006). Hydrodynamic simulation of air bubble implosion using a FEM-based level set approach. *J. Comput. Phys.* **215**(1), 98–132.
127. Sethian, J.A. (1999). Level Set Methods and Fast Marching Methods, Cambridge University Press.
128. Jansen, K. E. (1999). A stabilized finite element method for computing turbulence. *Comp. Meth. Appl. Mech. Eng.* **174**, 299.

129. Jansen, K. E., Whiting, C. H., and Hulbert, G. M. (2000). A generalized-α method for integrating the filtered Navier–Stokes equations with a stabilized finite element method. *Comp. Meth. Appl. Mech. Eng.* **190**(3–4), 305–319.
130. Whiting, C. H. and Jansen, K. E. (2000). A stabilized finite element method for the incompressible Navier–Stokes equations using a hierarchical basis. *Int. J. Numer. Methods Fluids* **35**, 1055.
131. Szabo, B. and Sahrmann, G. J. (1988). Hierarchic plate and shell models based on p-Extension. *Int. J. Numer. Methods Eng.* **26**, 1855.
132. Babuska, I. and Suri, M. (1990). The p and h–p versions of the finite element method – an overview. *Comp. Methods Appl. Mech. Eng.* **80**, 5.
133. Franca, L. P. and Frey, S. (1992). Stabilized finite element methods-II: The incompressible Navier–Stokes equations. *Comp. Meth. Appl. Mech. Eng.* **99**, 209.
134. Sussman, M., Smereka, P., and Osher, S. J. (1994). A level set approach for computing solutions to incompressible two-phase flows. *J. Comput. Phys.* **14**, 146.
135. Sussman, M., Almgren, A. S., Bell, J. B., Howell, L. H., Colella, Ph., and Welcome, W. L. (1999). A level set approach for computing solutions to incompressible two-phase flows. *J. Comp. Phys.* **114**, 146.
136. Osher, S. and Fedkiw, R. (2003). Level Set Methods and Dynamic Implicit Surfaces, Applied Math. Series-153, Springer, Berlin.
137. Bosch, H.-S. and Hale, G. M. (1992). Improved formulas for fusion cross-sections and thermal reactivities. *Nucl. Fusion* **32**, 611–631.
138. Duderstadt, J. J. and Moses, G. A. (1981). "Inertial Cofinement Fusion". Wiley, New York.
139. Trunin, R. F., Zhernokletov, M. V., Kuznetsov, N. F., Radchenko, O. A., Sichevskaya, N. V., and Shutov, V. V. (1992). Compression of liquid organic substances in shock waves. *Khimicheskaya Fizika* **11**, 424, (in Russian).
140. Tien, C.L., Majumdar, A. and Gerner, F.M., eds. (1998). "Microscale Energy Transport". Taylor & Francis, Wasington, DC.
141. Paul, B. (1962). Compilation of evaporation coefficients. *ARS J* **32**, 1321.
142. Chodes, N., Warner, J., and Gagin, A. (1974). A determination of the condensation coefficient of water from the growth rate of small cloud droplets. *J. Atmos. Sci.* **31**, 1351.
143. Hagen, D. E. (1989). *J. Atmos. Sci.* **46**, 803.
144. Geisler, R., Schmidt-Ott, W. D., Kurz, T., and Lauterborn, W. (2004). *Europhys. Lett.* **66**(3), 435.
145. Didenko, Y. T. and Suslick, K. S. (2002). The energy efficiency of the formation of photons, radicals and ions during single bubble cavitation. *Nature* **418**, 394.
146. Lohse, D. (2002). Sonoluminescence: Inside a micro-reactor. *Nature* **418**, 381.
147. LANL Staff; Briemeister, J.F., Ed. (2002). Monte Carlo N-particle code, Version 4C, LA–13709–M.
148. Ziegler, F. Ed. (2000). Stopping Range of Ions in Matter Code. www.srim.org.
149. Nakoryakov, V. E., Shreiber, I. R., and Pokusaev, B. G. (1993). "Wave Propagation in Gas-Liquid Media". CRC Press, Boca Raton, FL.
150. Wijngaarden, L. van. (1972). One-dimensional flow of liquids containing small gas bubbles. *Ann. Rev. Fluid Mech.* **4**, 369.
151. Omta, R. (1987). Oscillations of a Cloud of bubbles of Small and Not So Small Amplitude. *J. Acoust. Soc. Am.* **82**, 1018.
152. D'Agostino, L. and Brennen, C. E. (1989). Linear Dynamics of Spherical Bubble Clouds. *J. Fluid Mech.* **199**, 155.
153. Morch, K. A. (1981). Cavity Cluster Dynamics and Cavitation Erosion, Proceedings of ASME Cavitation and Polyphase Flow Forum, 1–10.

154. Morch, K. A. (1984). Energy Consideration on the Collapse of Cavity Clusters. *Appl. Sci. Res.* **38**, 313.
155. Hanson, I., KI. Kedrinskii, V., and Morch, K. A. (1981). On the Dynamics of Cavity Clusters. *J. Appl. Phys.* **15**, 1725.
156. Reisman, G. E., Wand, Y. C., and Brennen, C. E. (1998). Observation of Shock Waves in Cloud Cavitation. *J. Fluid Mech.* **355**, 255.
157. Shimada, M., Kobayashi, T., and Matsumoto, Y. (1999). Dynamics of Cloud Cavitation and Cavitation Erosion. *Trans. Jpn. Soc. Mech. Eng.* **65**(634B), 1934, (in Japanese).
158. Shimada, M., Matsumoto, Y., and Kobayashi, T. (2000). Influence of the Nuclei Size Distribution on the Collapsing Behavior of the Cloud Cavitation. *JSME Int. J. Series B* **43**(3), 380.
159. Tomita, Y. and Shima, A. (1990). High Speed Photographic Observations of Laser-Induced Cavitation Bubbles in Water. *Acustica* **71**.
160. J. R. Blake, J. M. Boulton-Stone, and N. H. Thomas, Eds. (1994). Bubble dynamics and interface phenomena, IUTAM Symposium Proceedings, Birmingham, U K. Kluwer Academic, Dordrecht.
161. Chahine, G. L. and Duraiswami, R. (1992). Dynamical interactions in a multibubble cloud. *J. Fluids Eng.* **114**, 680.
162. McDaniel, J. G., Akhatov, I., and Holt, R. G. (2002). Inviscid dynamics of a wet foam drop with monodisperse bubble size distribution. *Phys. Fluids* **14**, 1886.
163. Blackstock, D. T. (2000). "Fundamentals of Physical Acoustics". Wiley, New York.
164. Pinkerton, J. M. M. (1947). A pulse method for the measurement of ultrasonic absorption in liquids: Results for water. *Nature* **160**, 128.
165. F. Moraga–RPI, Private communication (2002).
166. Akhatov, I., Topolnikov, N., Vakhitova, A., Zakirov, K., Woplfrum, B., Kurz, T., Lindau, O., Mettin, R., and Lauterborn, W. (2002). Dynamics of laser-induced cavitation bubbles. *Exp. Thermal Fluid Sci.* **26**, 731.
167. Landau, L. D. and Lifshits, E. M. (1959). "Fluid Mechanics". Pergamon Press, London.
168. Ohl, C. D., Lindau, O., and Lauterborn, W. (1998). Luminescence from spherically and aspherically collapsing laser induced bubbles. *Phys. Rev. Lett.* **80**, 393.
169. Ohl, C. D., Kurz, T., Geisler, R., Lindau, O., and Lauterborn, W. (1999). Bubble dynamics, shock waves and sonoluminescence. *Phil. Trans. Roy. Soc. Lond. A* **357**, 269.
170. Van Stralen, S. and Cole, R. (1979). "Boiling Phenomena," Vol. 1. McGraw-Hill, New York.
171. Blander, M. and Katz, J. L. (1975). Bubble nucleation in liquids. *AIChE J* **21**(5), 833.
172. Apfel, R. E., Roy, S. C., and Lo, Y. C. (1985). *Phys. Rev. A* **31**, 3194.
173. Harper, M. J. and Nelson, M. E. (1990). *Rad. Prot. Dos.* **47**, 535.
174. Camara, C., Putterman, S., Vazquez, G., and Weininger, K. (2001). Sonoluminescence: Nature's Tiniest Blackbody. *Optics & Photonics* **45**.
175. Gaitan, D. F. (1999). *Phys. World* **12**(3), 20 March.
176. Margulis, M. A. (1990). The Nature of Sonochemical Reactions and Sonoluminescence. *Adv. Sonochemi.* **1**, 39.
177. Nigmatulin, R. and Lahey, R. T. Jr. (1995). Prospects for Bubble Fusion, NUREG/CP-0142 (U.S. Nuclear Regulatory Commission Symposium Report).
178. Lahey, R. T., Jr., Taleyarkhan, R.P., and Nigmatulin, R.I. (2005). "Sonofusion–Fact or Fiction," Proc. NURETH–11, Avignon, France.
179. Taleyarkhan, R. P., West, C. D., Lahey, R. T., Jr., Nigmatulin, R. I., Block, R. C., and Xu, Y. (2006). Numerical emissions during self-nucleated acoustic cavitation. *Phys. Rev. Lett.* **96**, 034301.
180. Xu, Y. and Butt, A. (2005). Confirmatory experiments for nuclear emissions during acoustic cavitation. *Nuc. Eng & Design* **235**, 1317.

Phonon Transport in Molecular Dynamics Simulations: Formulation and Thermal Conductivity Prediction

A.J.H. McGAUGHEY[1] and M. KAVIANY[2]

[1] Department of Mechanical Engineering, Carnegie Mellon University, Pittsburgh, PA 15213-3890, USA; E-mail: mcgaughey@cmu.edu
[2] Department of Mechanical Engineering, University of Michigan, Ann Arbor, MI 48109-2125, USA; E-mail: kaviany@umich.edu

I. Introduction

A. Challenges at Micro- and Nano-scales

The past decade has seen rapid progress in the design, manufacturing, and application of electromechanical devices at micron and nanometer length scales. While advances in fabrication techniques, material characterization, and system integration continue, a lag exists in theoretical approaches that can successfully predict how these devices will behave. As classical and continuum theories reach their limits, phenomena that may be insignificant at larger length and time scales (such as interfacial effects) can become dominant. Only a basic, qualitative understanding of the observed behavior exists in many cases. In part, one can attribute this lack of knowledge to the difficulty in solving the Schrödinger equation exactly for anything more than a hydrogen atom, and to the enormous computational resources needed to solve it numerically for a system with more than a few hundred atoms.

By ignoring the electrons and instead moving to an atomic-level description, the computational demands are greatly reduced. Though neglecting electrons removes the ability to model the associated electrical and thermal transport, one can still consider many of the relevant thermal issues in devices. These include the transport of phonons in superlattices and across material interfaces and grain boundaries, the dissipation of heat in integrated circuits (the dimensions of FETs are approaching tens of nanometers), and low-dimensional effects in structures such as quantum dots and nanotubes. Descriptions of the current challenges in experiments, theory, and computations can be found in Refs. [1] and [2].

Molecular dynamics (MD) simulations, Monte Carlo methods, the Boltzmann transport equation (BTE), Brownian dynamics, and dissipative particle dynamics are useful in investigations of systems beyond the scope of first principle calculations. The approach chosen depends on the length and time scales associated with the problem of interest, and what level of detail is required in the analysis. Not all of the methods listed allow for complete atomic-level resolution or an investigation of the system dynamics. One often selects a suitable methodology by considering the relative magnitudes of the carrier mean free path and system dimensions (i.e., the layer thickness in a superlattice). This comparison will indicate if the transport is primarily diffusive, ballistic, or a combination of the two, and if a continuum or discrete technique is required.

B. MOTIVATION FOR USING MOLECULAR DYNAMICS SIMULATIONS

We will focus here on dielectric solids, where the valence electrons are tightly bound to the atomic nuclei, at temperatures low enough that photonic contributions to thermal transport are negligible[1]. Phonons, energy waves associated with the lattice dynamics, will dominate the thermal transport. In such systems, one gains an advantage by moving from the continuum to the atomic scale in that the lattice dynamics can be completely and explicitly modeled [3–6]. Analysis of thermal transport in dielectrics is typically done in the phonon space (also referred to as momentum space or frequency space), which is a wave-based description of the lattice dynamics. For a harmonic solid, the phonon system corresponds to a set of independent harmonic oscillators. This system is far simpler to analyze than the coupled motions of the atoms in real space. The harmonic theory is only exact at zero temperature, however. As temperature increases, anharmonic (higher-order) effects, which are difficult to model theoretically, become important.

While phonon space is convenient for analysis, the design and synthesis of new materials is performed in real space. It can be difficult to move from criteria in phonon space to a crystal structure that will have the desired behavior. For example, the link between the positions of the atoms in a large unit cell and the relaxation times of the associated phonon modes is not intuitive. A research environment has thus developed where design is done in real space, while analysis is performed in phonon space. From a thermal transport standpoint, there are no easy ways to move between these paradigms without losing important information. To proceed, one must

---

[1] The analysis techniques presented can also be taken to isolate phonon effects in materials with mobile electrons. Caution must be taken in interpreting the results, however, due to the possible strong coupling between the phonon and electron systems.

move either the analysis to real space, or the design to phonon space, or develop new tools to bridge the existing approaches.

MD simulations are a suitable tool for the analysis of dielectrics at finite temperature and for the bridging of real and phonon space analysis techniques. Most importantly, MD simulations allow for the natural inclusion of anharmonic effects and for atomic-level observations that are not possible in experiments. As large-scale computing capabilities continue to grow and component sizes decrease, the dimensions of systems accessible with MD approach those of real devices. Simulations run in parallel (e.g., on a Beowolf cluster) can handle systems with tens of millions of atoms, and supercomputers have modeled systems with billions of atoms [7,8].

C. Scope

The objective of this review is to describe how MD simulations can be used to further the understanding of solid-phase conduction heat transfer at the atomic level. The simulations can both complement experimental and theoretical work, and give new insights.

In Section II continuum-level thermal transport analysis is reviewed, and its limitations are used to motivate atomic-level analysis. Through this discussion, thermal conductivity, the material property that is the focus of much of the review, is introduced.

In Section III the MD method is described, and general details on setting up and running simulations are presented. This is done in the context of the Lennard-Jones (LJ) potential, and the associated face-centered cubic (fcc) crystal, amorphous, and liquid phases of argon. Investigating a simple system allows one to elucidate results that might not be evident in a more complex structure. Equilibrium system parameters, such as the density, specific heat, and root mean square (RMS) atomic displacement are calculated, and compared to predictions made directly from the interatomic potential and from other theories. The harmonic description of the phonon system is introduced, and used in conjunction with simulation results to calculate phonon dispersion curves and relaxation times. The MD simulations provide data that allow for the incorporation of anharmonic effects.

Moving toward issues of thermal transport, Section IV contains a discussion of the nature of phonon transport in the MD system. The small simulation cells typically studied and the classical nature of the simulations lead to a different description of phonon transport than in the standard quantum-particle-based model. Methods to predict thermal conductivity using MD simulations are then reviewed in Sections V–VII. Attention is given particularly to the Greek–Kubo (GK) method and an approach based on a direct application of the Fourier law of conduction (the direct method).

Consideration is given to crystalline and amorphous phases, predictions for bulk phases and finite structures, and what can be gained from the simulations beyond the thermal conductivity value. Limitations of each approach, and of MD in general, are assessed by considering size effects, quantum corrections, and the nature of the interatomic potential.

II. Conduction Heat Transfer and Thermal Conductivity of Solids

Conduction heat transfer in a solid can be realized through the transport of phonons, electrons, and photons. The individual contributions of these carriers can vary widely depending on the material studied and its temperature. The thermal conductivity, k, of a substance indicates the ease with which thermal energy can be transferred through it by conduction. Unlike the specific heat, which has an equilibrium definition based on classical thermodynamics, the thermal conductivity is defined as the constant of proportionality relating the temperature gradient, ∇T, and heat flux, $\mathbf{q}$, in a material by

$$\mathbf{q} = -\mathbf{k}\nabla T \qquad (1)$$

This is the Fourier law of conduction, and was originally formulated from empirical results. The thermal conductivity is generally a second-order tensor, but in a material with cubic isotropy it reduces to a scalar. The thermal conductivity is an intensive property (i.e., it can vary from point to point in a continuum) and is a function of both pressure and temperature. In the context of engineering heat transfer, an object's thermal resistance is a function of both its thermal conductivity and its geometry.

By combining the Fourier law and the energy equation, and assuming no mass transfer and constant properties, one obtains the partial differential equation

$$k\nabla^2 T = \rho C_v \frac{\partial T}{\partial t} \qquad (2)$$

Here, ρ is the mass density, C_v the specific heat (J/kg K), and t the time. Given the appropriate boundary and initial conditions, this equation can be solved to give the temperature distribution in a system of interest. Equations (1) and (2) are the standard basis for describing conduction heat transfer. The validity and applicability of each is based on these assumptions:

- The system behaves classically and can be modeled as a continuum.
- The energy carrier transport, be it a result of phonons, electrons, or photons, is diffuse (i.e., the scattering of the carriers is primarily a

result of interactions with other carriers). Cases where interface and/or boundary effects dominate correspond to ballistic transport, and cannot be considered in this formulation.
- The material properties are known.

At small length scales the suitability of the first two criteria may be questionable, and experimental property data may not always be available. Alternative approaches (such as the BTE or MD) may be required.

Consider the required properties in Eq. (2). Given the chemical composition of a material and some atomic-level length scales (e.g., the lattice constant for a crystal), predicting its density is straightforward. For the specific heat, the Debye approach [3] (based on quantum mechanics) models most solids well, although the Debye temperature must first be fit from the experimental data. One can approximate the high-temperature specific heat from the Dulong–Petit theory [3], which is based on classical statistical thermodynamics, and requires only the molecular mass for its evaluation. The thermal conductivity is a more elusive quantity, however. It is a material property related to energy transport, unlike the density and specific heat, which are associated with structure and energy storage, respectively. As opposed to thinking of the thermal conductivity in terms of a static, equilibrium system, one typically envisions this property in the context of nonequilibrium, although it can be determined in an equilibrium system by looking at the decay of energy fluctuations (the GK method, described in Section V). Experimental techniques for determining the thermal conductivity directly based on the Fourier law have been developed, and analogous computational techniques exist (see Section VI). Experimentally, it is also possible to determine the thermal diffusivity, α, from which the thermal conductivity can be inferred ($\alpha = k/\rho C_p$, where C_p is the volumetric specific heat at constant pressure, equal to C_v for a solid).

In Fig. 1, experimental thermal conductivities of a number of solids are plotted over a wide temperature range [9,10]. The energy carriers represented are electrons and phonons. The range of thermal conductivities available at essentially any temperature covers five orders of magnitude. While some overall trends are evident (e.g., amorphous materials have a much lower thermal conductivity and a different temperature dependence of this property than crystals), one cannot easily explain much of the behavior. The finite thermal conductivity of a perfect crystal is a result of anharmonic effects. Modeling these exactly is difficult, making it a challenge to realize simple and accurate expressions for the thermal conductivity. In Ref. [11], methods for predicting the thermal conductivity are reviewed. The available models often require simplifying assumptions about the nature of the thermal transport and/or require that one fit the predictions to experimental

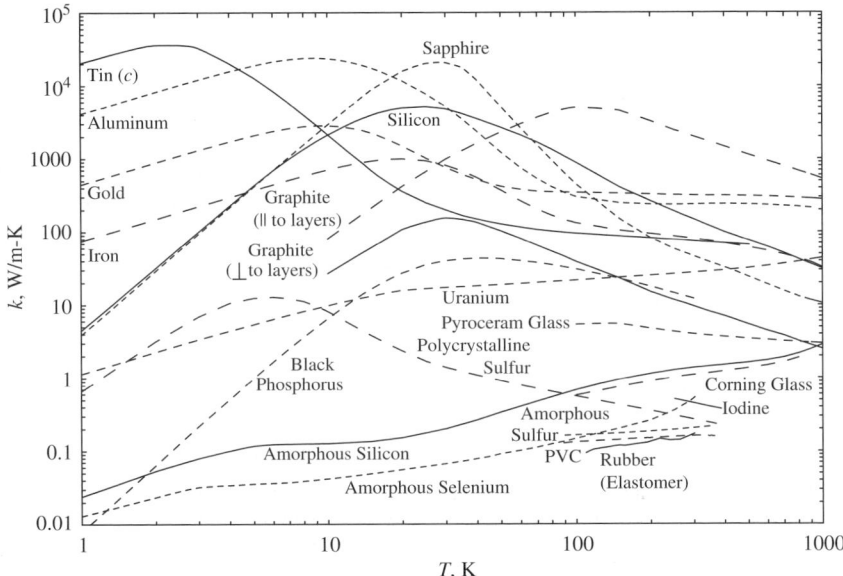

FIG. 1. Experimental thermal conductivities of crystalline and amorphous solids plotted as a function of temperature [9,10]. The materials presented are examples of dielectrics, semiconductors, and metals. They cover the spectrum of electron-dominated to phonon-dominated thermal transport. Note the wide range of thermal conductivity values available at almost any temperature.

data. Omini and Sparavigna [12] have developed a solution method based on an iterative solution of the BTE free of some of these assumptions. While the predictions are in reasonable agreement with experimental data, the required calculations are complex. As such, discussions of thermal conductivity typically revolve around the simple kinetic theory expression [6]

$$k = \frac{1}{3}\rho C_v v \Lambda \qquad (3)$$

where v is a representative carrier velocity and Λ the carrier mean free path, the average distance traveled between collisions. Reported values of the mean free path are often calculated using Eq. (3) and experimental values of the other parameters, including the thermal conductivity. This expression for the thermal conductivity assumes a similar behavior for all carriers, which may mask a significant amount of the underlying physics, especially in a solid.

In Fig. 2, the thermal conductivities of materials composed of carbon atoms are plotted as a function of temperature [9,13]. Clearly, thermal

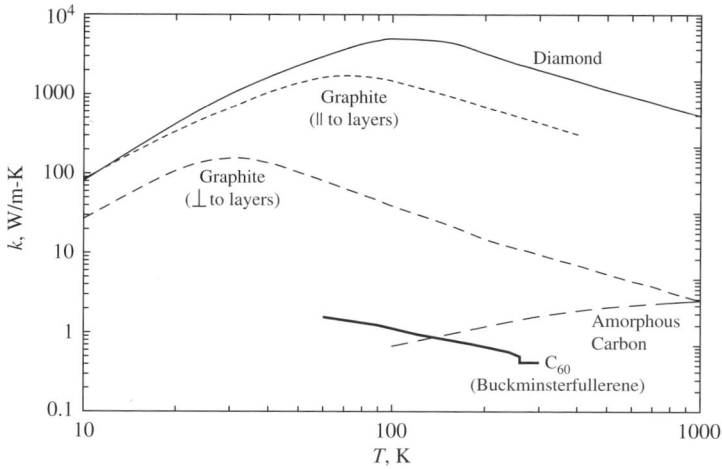

FIG. 2. Thermal conductivities of carbon-based solids plotted as a function of temperature [9,13]. The discontinuity in the C_{60} curve is a result of a phase transition.

transport relates to both what a material is made of, and how the constituent atoms are arranged. An explanation of the second effect is not as obvious as it would be for either the density or specific heat [i.e., the more atoms in a given volume, the higher the density, and higher the specific heat (on a volumetric basis)].

Without elaborate expressions for the mean free path (which can vary depending on wavelength, temperature, etc.), seeing how a relation such as Eq. (3) could be used to explain the trends in Figs. 1 and 2 is difficult. Many of the theoretical challenges can be bypassed in the MD formulation (e.g., anharmonic effects are naturally included), making this approach a good option for investigating atomic-level thermal transport in dielectrics.

III. Real and Phonon Space Analyses

A. Molecular Dynamics Simulation

In an MD simulation, one predicts the position and momentum space trajectories of a system of classical particles using the Newton laws of motion. The only required inputs are an atomic structure and an appropriate interatomic potential. Using the positions and momenta, it is possible to investigate thermal transport at the atomic level [14], and a variety of other problems, including, for example, molecular assembly, chemical reaction, and material fracture.

Early MD work focused on the fluid phase, and this is reflected in the contents of books written in the area (see, e.g. [15,16]). Only in the last 15 years has the solid state, and in particular, crystals (as opposed to amorphous materials), been studied extensively. This may be a result of the long correlation times that can exist in crystals. Extensive computational resources may be required to obtain sufficient data to observe trends in what might initially appear to be noisy data. Also, many solid state applications are concerned with electrons, which cannot be explicitly included in MD simulations. In the last five years, smaller system sizes and the associated high power densities have heightened interest in phonon thermal transport issues in semiconductor devices, and led to a significant increase in MD-related work. The analysis of new materials, such as carbon nanotubes, which are predicted to have very high thermal conductivities, has also motivated the development of the field.

1. *Simulation Setup*[2]

To perform an MD simulation, one first needs to specify the interatomic potential (referred to hereafter as simply the potential) that will be used to model the system of interest. The potential, ϕ, is an algebraic (or numerical) function used to calculate the potential energies and forces (through its derivative) associated with the particles in the system. It can include terms that account for two-body, three-body, etc. effects. Potentials can be obtained by fitting a chosen functional form to experimental data, or to the results of *ab initio* calculations. This process is not trivial.

The discussion here is left in terms of a general two-body (pair) potential, $\phi(r)$, where r is the distance between the two particles in question. In this case, the total potential energy in the system, Φ, is given by

$$\Phi = \sum_i \Phi_i = \frac{1}{2} \sum_i \sum_{i \neq j} \phi(r_{ij}) \tag{4}$$

where the summations are over the particles in the system,[3] Φ_i is the potential energy associated with particle i, and the factor of one half accounts for double counting. It is more computationally efficient to calculate the total potential energy as

$$\Phi = \sum_i \sum_{j>i} \phi(r_{ij}) \tag{5}$$

---

[2] Much of the material in Sections III.A.1–III.A.3 is taken from Refs. [15] and [16]. Additional references from the literature are included where appropriate.

[3] Where possible, i and j are used as indices for particle summations, and k is used for normal mode summations.

The total force on a particle i, $\mathbf{F}_i$, is given by

$$\mathbf{F}_i = \sum_{i \neq j} \mathbf{F}_{ij} = -\sum_{i \neq j} \frac{\partial \phi(r_{ij})}{\partial \mathbf{r}_{ij}} = -\sum_{i \neq j} \frac{\partial \phi(r_{ij})}{\partial r_{ij}} \frac{\partial r_{ij}}{\partial \mathbf{r}_{ij}} = -\sum_{i \neq j} \frac{\partial \phi(r_{ij})}{\partial r_{ij}} \hat{\mathbf{r}}_{ij} \quad (6)$$

where $\mathbf{F}_{ij}$ is the force exerted on particle i by particle j, $\mathbf{r}_{ij}$ the position vector between particles i and j ($= \mathbf{r}_i - \mathbf{r}_j$), and $\hat{\mathbf{r}}_{ij}$ the unit vector along $\mathbf{r}_{ij}$. Thus, for a pair potential, is $\mathbf{F}_{ij}$ parallel to $\mathbf{r}_{ij}$.

All results of an MD simulation are, at their most fundamental level, related to the suitability of the chosen potential. Even if the potential is initially well formed, this does not guarantee good-quality predictions. For example, if a potential has been constructed using the experimental lattice constant and elastic constants for a crystal phase, should one expect it to properly model thermal transport, or the associated liquid phase, or a variation of the crystal structure? The answer, unfortunately, is maybe, but as potentials are difficult and time intensive to construct, one often proceeds with what is readily available in the literature, and hopes it will be suitable.

The contribution to the potential energy of an atom from other atoms beyond a radius R is given by approximately

$$\int_R^\infty 4\pi r^2 \phi(r)\, dr \quad (7)$$

where it is assumed that the system is spherically symmetric with the particles evenly distributed. This is a good assumption for fluids, and reasonable for solids beyond the fourth or fifth nearest neighbor. One can check the validity of this assumption by constructing the radial distribution function (RDF, see Section III.C.2). Many pair potentials are of the form r^{-n}, so that the energy given by Eq. (7) will be proportional to

$$\int_R^\infty r^{2-n}\, dr \quad (8)$$

which is bounded for $n > 3$. If this condition is satisfied, contributions to the potential energy of a particle from beyond a certain radius can be safely neglected. For computational efficiency, one can then apply a cutoff to the potential at a radius R_c. One should choose the cutoff so that the magnitude of the energy at the cutoff is small compared to the largest magnitude the energy can take on (i.e., the energy of a pair of nearest-neighbor atoms). To ensure that energy is conserved in the system, the potential must go to zero at the cutoff. This is accomplished by forming a shifted potential, $\phi_c(r)$, defined as

$$\phi_c(r) = \phi(r) - \phi(R_c) \quad (9)$$

Note that the forces are not affected [see Eq. (6)]. Other techniques for implementing the cutoff exist [17]. For the LJ work to be presented here, Eq. (9) is applied. When dealing with potentials that are not bounded, such as electrostatic interactions (where $\phi \sim 1/r$), applying a cutoff is not a well-defined operation. Instead, such potentials require special computational techniques such as the Ewald summation, the cell multipole method [18,19], or the Wolf (direct summation) method [20].

To obtain sufficient statistics for thermal transport calculations, simulations on the order of 1 ns may be required. Typically, systems with hundreds or thousands of atoms are considered. The actual number of atoms considered should be taken to be the smallest number for which there are no size effects. One should choose the time step so that all timescales of interest in the system can be resolved. In simulations of LJ argon, for example, where the highest frequency is 2 THz, a time step of 5–10 ps is typically chosen.

Using periodic boundary conditions and the minimum image convention allow for modeling of a bulk phase in an MD simulation. The idea is to reproduce the simulation cell periodically in space in all directions, and have a pair of particles only possibly interact between their images that are the closest together (i.e., a particle only interacts with another particle once). Success of this technique requires that the potential cutoff be no larger than one half of the simulation cell side length. In normal mode analysis or when calculating RMS displacements, one must carefully calculate displacements from equilibrium for the particles near the boundaries. In very large systems, where some particles will never interact with certain others, binning techniques and neighbor lists can significantly reduce computation times, and allow for the implementation of parallel MD code.

2. Energy, Temperature, and Pressure

The total system energy, E, is given by $\Phi + KE$, where KE is the total kinetic energy. The potential energy is given by Eq. (5). The kinetic energy is

$$KE = \sum_i \frac{1}{2} \frac{|\mathbf{p}_i|^2}{m_i} \tag{10}$$

where $\mathbf{p}_i$ and m_i are the momentum vector and mass of particle i. From kinetic theory, the expectation value of the energy of one degree of freedom is $k_B T/2$, where k_B is the Boltzmann constant. The expectation value of the kinetic energy of one atom is $3k_B T/2$ (based on the three degrees of freedom associated with the momentum). The temperature of the MD system can be obtained by equating the average kinetic energy with that predicted from kinetic theory, such that

$$T = \frac{\langle \sum_i |\mathbf{p}_i|^2 / m_i \rangle}{3(N-1)k_B} \tag{11}$$

where N is the total number of atoms in the system. $3(N-1)$ degrees of freedom have been used in Eq. (11) as the MD simulation cell is assumed to be fixed in space (i.e., the total linear momentum is set to zero, removing three degrees of freedom). One should only use the expression for the temperature given by Eq. (11) with the expectation value of the kinetic energy (i.e., the temperature cannot really be defined at an instant in time). For the purpose of temperature control, however, it is assumed that the instantaneous temperature is a well-defined quantity.

When working in the canonical (NVT) ensemble, where the independent variables are the system mass, volume (V), and temperature, it is also possible to calculate the temperature based on the fluctuations of the total system energy as

$$T = \left[\frac{\langle (E - \langle E \rangle)^2 \rangle}{3(N-1)k_B c_v} \right]^{1/2} \tag{12}$$

where c_v is the specific heat per mode (J/K). The temperature defined as such can only be evaluated as an ensemble average. Similar expressions can be derived for different ensembles [21].

The pressure, P, of the system can be calculated from

$$P = \frac{Nk_B T}{V} + \frac{1}{3V} \langle \sum_i \sum_{j>i} r_{ij} \cdot \mathbf{F}_{ij} \rangle \tag{13}$$

which is based on the virial equation for the pressure. The temperature is calculated from Eq. (11). As with the temperature, this quantity should, in theory, only be calculated as an ensemble average. For the purposes of pressure control, however, it is assumed to be valid at an instant in time.

3. Equations of Motion

The equations of motion to be used in an MD simulation are dependent on which thermodynamic ensemble one wishes to model. The most natural ensemble is the NVE (microcanonical), where the independent variables are mass, volume, and energy. In this case, the equations of motion for a particle i are

$$\frac{d\mathbf{r}_i}{dt} = \frac{\mathbf{p}_i}{m_i} \tag{14}$$

$$\frac{d\mathbf{p}_i}{dt} = \mathbf{F}_i \tag{15}$$

To implement these equations in the MD simulations, they must be discretized. Different schemes are available for this procedure, which have varying levels of accuracy and computational requirements. While some higher-order approaches allow for the use of long-time steps (e.g., the Gear predictor–corrector method), if the dynamics of the system are of interest these may be computationally inefficient. When a small time step is needed, a scheme such as the Verlet leapfrog algorithm is suitable.

To reproduce the canonical ensemble, the temperature must be controlled with a thermostat. Rescaling the velocities at every time step to get the desired temperature is a straightforward method. This approach, however, does not reproduce the temperature fluctuations associated with the NVT ensemble [21]. Allowing the MD system to interact with a thermal reservoir at constant temperature is another possibility. To do so, the equations of motion are modified with a damping parameter η such that

$$\frac{d\mathbf{r}}{dt} = \frac{\mathbf{p}_i}{m_i} \qquad (16)$$

$$\frac{d\mathbf{p}_i}{dt} = \mathbf{F}_i - \eta \mathbf{p}_i \qquad (17)$$

The damping parameter changes in time according to

$$\frac{d\eta}{dt} = \frac{1}{\tau_T^2}\left(\frac{T}{T_{set}} - 1\right) \qquad (18)$$

where τ_T is the reservoir-system time constant and T_{set} the desired temperature. This is the Nose–Hoover thermostat [22–24], which produces the appropriate temperature fluctuations, and reduces to the NVE ensemble when η is set to zero. When the system temperature is above that desired, η is positive, and kinetic energy is removed from the system. When the system temperature is below the desired value, η will be negative, leading to an increase of the system kinetic energy. By setting η to be positive and constant, energy can be continually removed from the system. This concept can be used to simulate a quench from a liquid state to a solid state, possibly forming an amorphous phase, or to obtain the zero-temperature structure. The thermostat time constant determines the strength of the coupling between the MD system and the thermal reservoir. If τ_T is very large, the response of the MD system will be slow. The limit of $\tau_T \to \infty$ corresponds to the NVE ensemble. One should choose the time constant to reproduce the temperature as defined in the NVT ensemble [Eq. (12)]. From a dynamics standpoint, the equations of motion in the NVT ensemble have been modified in a nonphysical manner. Thus, one must use caution when extracting dynamical properties from a canonical MD system. Static properties (those not based on the time progression of quantities, but on their average values, such as the zero-pressure unit cell) are not affected.

To simulate the NPT ensemble, where the independent variables are the system mass, pressure, and temperature, the system volume must be allowed to change (i.e., a barostat is implemented). Simulations in this ensemble are useful for determining pressure-dependent cell sizes for use in either the NVE or NVT ensembles. Numerous techniques exist with modified equations of motion, notably those of Anderson [25], and Parrinello and Rahman [26].

4. Initialization

It is easiest to initialize the particle positions based on their equilibrium locations in a known solid state (be it an amorphous phase or a crystal). Such an initialization is also appropriate for a fluid, as the temperatures to be studied will quickly induce melting. If the particles are placed in their equilibrium positions each will experience no net force. For something to happen the initial momenta must be nonzero, so as to move the system away from equilibrium. A convenient method is to give each atom an extremely small, but nonzero, random momentum and then initially run the simulation in the NVT ensemble to obtain the desired temperature. If post-processing of the positions and momenta into normal mode coordinates is required, starting as such is beneficial as the equilibrium positions are exactly known. Many thermal properties require averaging over multiple simulations. One can initialize distinct simulations with the same independent variables by the initial momenta distribution. At equilibrium, the momenta of the particles will take on a Maxwell–Boltzmann distribution. Sufficient time must be given for this to occur. In the LJ case study reported here, at least 10^5 time steps are taken for all reported results. One could also initialize the momenta in the Maxwell–Boltzmann distribution.

To determine the equilibrium unit cell size at a specified temperature and pressure, simulations can be run in the NPT ensemble. A few hundred thousand time steps should be sufficient to generate enough data for a good average. The average potential energy of the system as a function of temperature can also be determined from the simulations. To set the temperature for runs in the NVE ensemble, we suggest the following scheme. The system is first run in the NVT ensemble until the momenta are properly initialized. The potential energy of the system is then monitored at every time step. When it reaches a value within 10^{-4}% of the desired potential energy as calculated from the NPT simulations, the ensemble is switched to NVE, and the velocities are scaled to the desired temperature. This procedure is less invasive than other temperature-setting techniques such as velocity scaling, as the particles are allowed to move within the equations of motion for a given ensemble at all times other than when the switch occurs. In the NVE ensemble, the total energy is a function of temperature. By simply setting the kinetic energy to a specific value without considering the potential energy, one will not achieve the desired temperature unless the potential energy at that time has its average value.

B. Lennard-Jones System

To begin the presentation of what MD simulations can reveal about the nature of atomic-level thermal transport in dielectrics, we consider materials described by the LJ potential. Choosing a simple system allows for the elucidation of results that may be difficult to resolve in more complex materials, where multi-atom unit cells (and thus, optical phonons) can generate additional effects. The LJ atomic interactions are described by the pair potential [3]

$$\phi_{LJ,ij}(r_{ij}) = 4\varepsilon_{LJ}\left[\left(\frac{\sigma_{LJ}}{r_{ij}}\right)^{12} - \left(\frac{\sigma_{LJ}}{r_{ij}}\right)^{6}\right] \quad (19)$$

The depth of the potential energy well is ε_{LJ}, and corresponds to an equilibrium particle separation of $2^{1/6}\sigma_{LJ}$. The LJ potential describes the noble elements well, and is plotted in dimensionless form in Fig. 3.

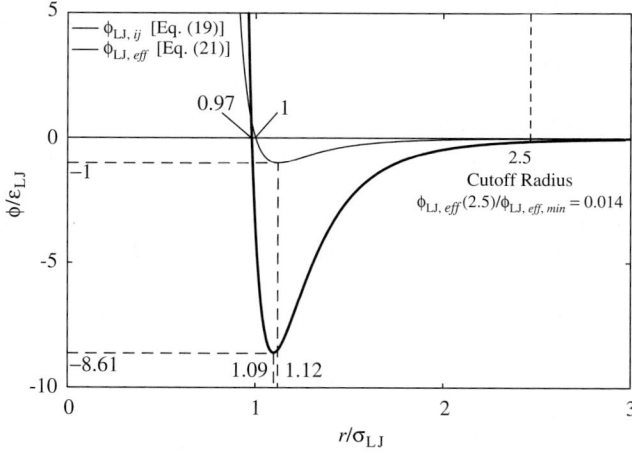

FIG. 3. A dimensionless plot of the LJ potential. Both the pair [Eq. (19)] and effective [Eq. (21)] curves are shown, along with the values of the energy and separation distance at important points.

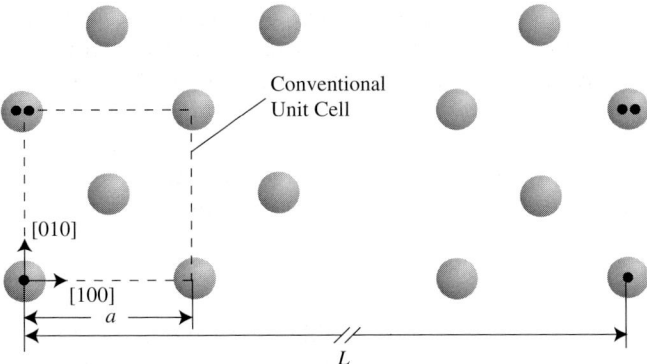

FIG. 4. A plane in the fcc crystal. The atoms with black dots are equivalent through the use of periodic boundary conditions.

Argon, for which σ_{LJ} and ε_{LJ} have values of 3.40×10^{-10} m and 1.67×10^{-21} J [3], is chosen for the current investigation. The fcc crystal, amorphous, and liquid phases are considered. The plane formed by the [1 0 0] and [0 1 0] axes in the crystal is shown in Fig. 4. In the figure, a is the side length of the conventional unit cell (which contains four atoms) and L the side length of the simulation cell (which is taken to be cubic). This leads to $\eta = L/a$ unit cells in each of the [1 0 0], [0 1 0], and [0 0 1] directions, and $N = 4\eta^3$ total atoms. Here, values of η of 4, 5, and 6 are considered, which correspond to 256, 500, and 864 total atoms. The true fcc unit cell, containing one atom, is rhombohedral, and not as suitable for analysis.

Unless noted, all reported data correspond to simulations of the 256 atom unit cell in the NVE ensemble at zero pressure with a time step of 4.285 fs (0.002 LJ units) [11,27,28]. This time step sufficiently resolves the phenomena of interest [e.g., the smallest timescale of interest in the heat current (see Section V) is about 20 time steps]. In similar simulations, Kaburaki et al. [29] found good agreement between the fcc crystal thermal conductivities predicted from cells containing 256 and 500 atoms. Tretiakov and Scandalo [30] found no size effects in simulation cells with as few as 128 atoms. Periodic boundary conditions are imposed in all directions. The equations of motion are integrated with a Verlet leapfrog algorithm. The atomic interactions are truncated and shifted at a cutoff radius equal to $2.5\sigma_{LJ}$. For the fcc crystal, temperatures between 10 and 80 K are considered in 10 K increments. Melting occurs at a temperature of 87 K. The amorphous phase is generated by starting with the desired number of atoms placed on an fcc lattice, running at a temperature of 300 K for 10^4 time steps to eliminate any memory of the initial configuration, and then quenching at 8.5×10^{11} K/s back to a temperature of 10 K. The amorphous phase is stable up to a temperature of 20 K. Above this point, the equilibrium thermal fluctuations in the system are large enough to return the atoms to the fcc crystal structure. This is consistent with the findings of Li [17]. Temperatures of 10, 15, and 20 K are considered. Three different amorphous phases (each with 250 atoms) were formed to check if the systems are truly disordered, and cells with 500 and 1000 atoms were created to investigate size effects. The liquid phase is obtained by first heating the 256 atom crystal phase to a temperature of 100 K to induce melting, then lowering the temperature. Using this approach, a stable liquid is found to exist at temperatures as low as 70 K. Owing to the small length and timescales used (necessary for reasonable computational times), the melting/solidifying temperature is not well defined, and it is possible to have stable fcc crystal and liquid phases at the same temperature and pressure, although the densities differ. Temperatures of 70, 80, 90, and 100 K are considered for the liquid simulations.

C. Real Space Analysis

1. *Prediction of System Parameters from Lennard-Jones Potential*

a. Unit Cell Size. When relaxed to zero temperature, the MD fcc crystal unit cell parameter, a (the lattice constant), is 5.2686 Å. The experimental value for argon is 5.3033 Å [3]. Li [17] has found values of 5.3050 Å for a cutoff radius of $2.5\sigma_{LJ}$, and 5.2562 Å for a cutoff radius of $5\sigma_{LJ}$. The discrepancy between the current result and that of Li can be attributed to differences in how the potential energy and force are cutoff. Whenever performing simulations, the question often arises of whether or not one should strive to match certain experimental parameters (such as the lattice constant). For the LJ argon system, this can be done by choosing a cutoff radius of about $3\sigma_{LJ}$. This by no means, however, guarantees that agreement with experimental data for the temperature dependence of the unit cell size, or other properties (e.g., elastic constant, thermal conductivity, etc.) will follow. The other option is to choose suitable simulation parameters and strive for consistency. These can be chosen based on previous work, to allow for comparison-making. The choice of a cutoff of $2.5\sigma_{LJ}$, used here, is standard.

The lattice constant can be predicted from the analytical form of the LJ potential. To do this, one must consider the total potential energy associated with one atom, Φ_i. If the energy in each pair interaction is assumed to be equally distributed between the two atoms, Φ_i will be given by

$$\Phi_i = \frac{1}{2} \sum_{i \neq j} \phi_{LJ,ij} \qquad (20)$$

which for the fcc crystal lattice can be expressed as [3]

$$\Phi_i = 2\varepsilon_{LJ}\left[A_{12}\left(\frac{\sigma_{LJ}}{r_{nn}}\right)^{12} - A_6\left(\frac{\sigma_{LJ}}{r_{nn}}\right)^{6}\right] \equiv \phi_{LJ,eff} \qquad (21)$$

where A_{12} and A_6 have values of 12.13 and 14.45, respectively, and r_{nn} is the nearest neighbor (*nn*) separation. This effective LJ potential is plotted in Fig. 3 alongside the pair potential, given by Eq. (19). By setting

$$\frac{\partial \Phi_i}{\partial r_{nn}} = 0 \qquad (22)$$

the equilibrium value of r_{nn} is found to be

$$r_{nn,equ} = \left(\frac{2A_{12}}{A_6}\right)^{1/6}\sigma_{LJ} = 1.09\sigma_{LJ} \qquad (23)$$

The location of the minimum is slightly shifted from that in the pair potential, and the energy well is deeper and steeper. For argon, the equilibrium separation of Eq. (23) corresponds to a unit cell parameter of 5.2411 Å, which agrees with the zero temperature MD result to within 0.6%.

b. Period of Atomic Oscillation and Energy Transfer. In a simplified, real space model of atomic-level behavior, the energy transfer between neighboring atoms can be assumed to occur over one half of the period of oscillation of an atom [31–33]. The associated time constant, τ_D, can be estimated from the Debye temperature, T_D, as

$$\tau_D = \frac{2\pi\hbar}{2k_B T_D} \qquad (24)$$

where $\hbar$ is the Planck constant divided by 2π. The factor of 2 in the denominator is included as one half of the period of oscillation is desired. By fitting the specific heat (as predicted by the MD zero-temperature phonon density of states, see Section III.D.3) to the Debye model using a least-square method, the Debye temperature is found to be 81.2 K. This compares well with the experimental value of 85 K [3]. The MD result is used in subsequent calculations, and gives a τ_D value of 0.296 ps (~69 time steps).

Using the LJ potential, an estimate of this time constant can also be made. The time constant is related to the curvature of the potential well that an atom experiences at its minimum energy. Assuming that the potential is harmonic at the minimum, the natural angular frequency, ω, of the atom will be given by

$$\omega = \left(\frac{1}{m}\frac{\partial^2 \Phi_i}{\partial r_{nn}^2}\bigg|_{r_{nn}=r_{nn,equ}}\right)^{1/2} = 22.88\left(\frac{\varepsilon_{LJ}}{\sigma_{LJ}^2 m}\right)^{1/2} \qquad (25)$$

where m is the mass of one atom, which for argon is 6.63×10^{-26} kg, and Φ_i and $r_{nn,equ}$ are taken from Eqs. (21) and (23). One half of the period of oscillation is then

$$\tau_{LJ} = \frac{1}{2}\frac{2\pi}{\omega} = 0.137\left(\frac{\sigma_{LJ}^2 m}{\varepsilon_{LJ}}\right)^{1/2} \tag{26}$$

which evaluates to 0.294 ps, within 1% of τ_D.

One can further investigate the physical significance of this time constant by considering the flow of energy between atoms in the MD simulation cell [27,34]. This is done by constructing energy correlation functions between an atom and its 12 nearest neighbors. The calculations are based on the deviations of the particle energies from their mean values. As all the atoms in the fcc crystal simulation cell are at equivalent positions, the results can be averaged over neighbors, space, and time. The resulting correlations for the fcc crystal are shown in Fig. 5 for all temperatures considered. In the figure, E denotes the particle energy, and the subscripts o and nn refer to a particle and one of its nearest neighbors. The curves are normalized against their zero time value to allow for comparison between the different temperatures. The first peak locations are denoted as τ_{nn}. The value of τ_{nn} increases with temperature, which is due to the decreasing density. As the

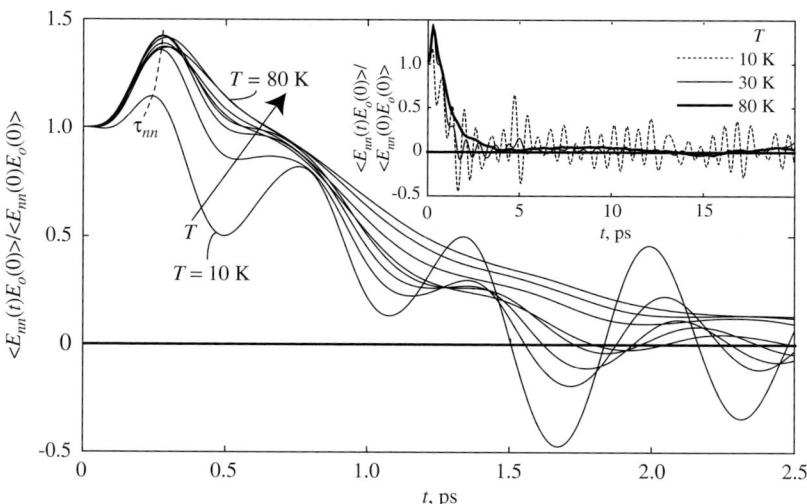

FIG. 5. Nearest-neighbor particle–particle energy correlation functions for the LJ fcc crystal. The energy data correspond to deviations from the mean values. A longer time scale is shown for $T = 10$, 30, and 80 K in the inset plot, where the decrease in the long time coherence at higher temperatures is evident.

atomic separation increases, it takes longer to transfer energy between two atoms. Except at a temperature of 10 K, the time constants τ_D, τ_{LJ}, and τ_{nn} agree to within 10%.

Ohara [35] has developed a technique in the same spirit for investigating the flow of energy between atoms in an MD simulation, and applied it to an LJ argon fluid. The technique does not involve correlation functions as described here, but instead looks at the time rate of energy change between pairs of atoms.

2. Lennard-Jones Phase Comparisons

The densities and potential energies per particle of the zero pressure LJ fcc crystal, liquid, and amorphous phases are plotted as a function of temperature in Figs. 6(a) and (b). As would be expected, the crystal phase has the lowest potential energy at a given temperature. Note the consistent trend between the amorphous and liquid phases in both density and potential energy. This is consistent with the idea of an amorphous phase being a fluid with a very high viscosity.

The RDF, $g(r)$, describes the distribution of the atoms in a system from the standpoint of a fixed, central atom. Its numerical value at a position r is the probability of finding an atom in the region $r - dr/2 < r < r + dr/2$ divided by the probability that the atom would be there in an ideal gas [i.e., $g(r) = 1$ implies that the atoms are evenly distributed]. The calculation includes no directional dependence. The Fourier transform of the RDF is the structure factor, which can be determined from scattering experiments. The RDFs for the fcc crystal at temperatures of 20, 40, 60, and 80 K, the amorphous phase at a temperature of 10 K, and the liquid phase at temperatures of 70 and 100 K are shown in Fig. 7. The results presented are based on 10^5 time steps of NVE simulation, with data extracted every five time steps. The RDF is calculated with a bin size of 0.034 Å for all atoms at each time step, then averaged over space and time. The RDF can only be determined up to one half of the simulation cell size, which here is about $3.25\sigma_{LJ}$ for the solid phases, and slightly larger for the liquid.

The RDF of the fcc crystal phase shows well-defined peaks that broaden as the temperature increases, causing the atomic displacements to grow. Each peak can be associated with a particular set of nearest-neighbor atoms. The locations of the peaks shift to higher values of r as the temperature increases and the crystal expands. In the amorphous phase, the first peak is well defined, but after that, the disordered nature of the system leads to a much flatter RDF. There is no order beyond a certain point. The splitting of the second peak is typical of amorphous phases, and consistent with the results of Li for LJ argon [17]. The presence of only short-range order is also

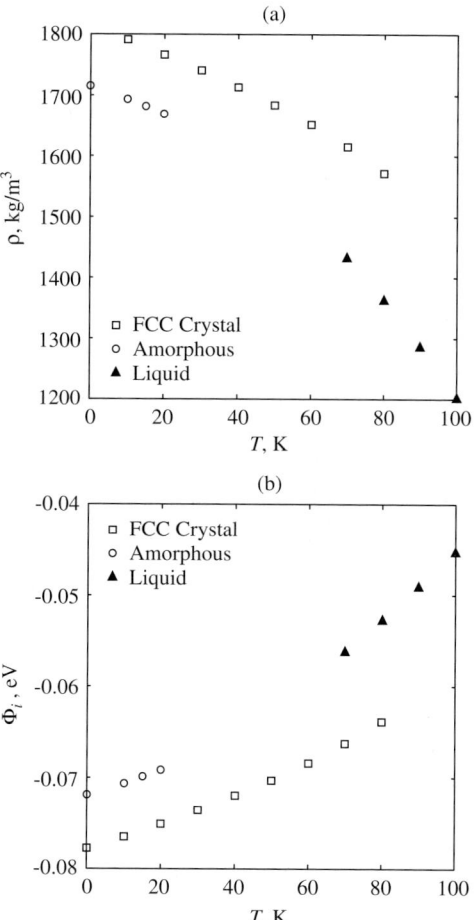

FIG. 6. Temperature dependencies of the LJ phase (a) densities and (b) per particle potential energies.

true for the liquid phase, where only the first neighbor peak is well defined. Since the physical size of the atoms defines the minimum distance over which they may be separated, this is expected.

The RMS displacement, $\langle |\mathbf{u}_i|^2 \rangle^{1/2}$, where $\mathbf{u}_i$ is the displacement of atom i from its equilibrium position, of the atoms in the fcc crystal is shown as a function of temperature in Fig. 8. The results presented are based on 10^5 time steps of *NVE* simulation, with data extracted every five time steps. The RMS displacement can be predicted from a quantum-mechanical

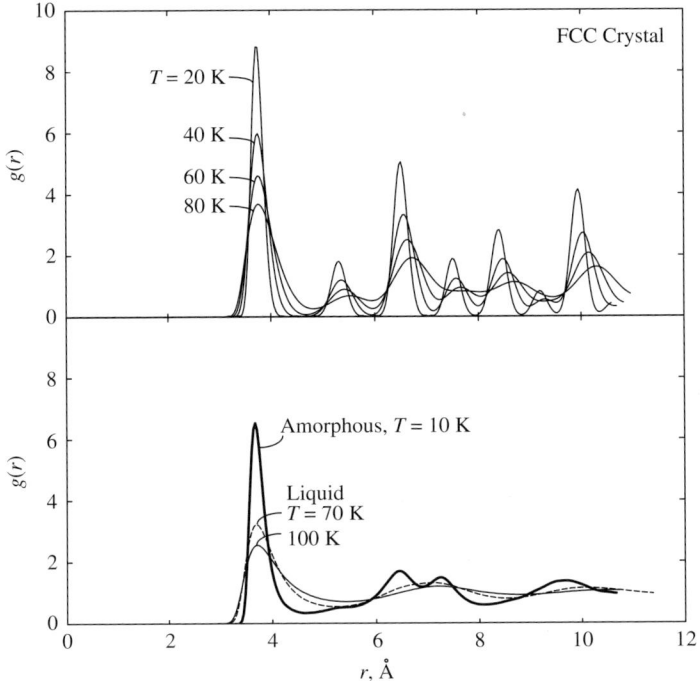

FIG. 7. Lennard-Jones RDFs for the fcc crystal at $T = 20, 40, 60$, and $80\,K$, the amorphous phase at $T = 10\,K$, and the liquid phase at $T = 70$ and $100\,K$.

description of the system under the Debye approximation as [36]

$$\langle |\mathbf{u}_i|^2 \rangle^{1/2} = \left\{ \frac{3\hbar}{m\omega_D} \left[\frac{1}{4} + \left(\frac{T}{T_D} \right)^2 \int_0^{T_D/T} \frac{x\,dx}{\exp(x) - 1} \right] \right\}^{1/2} \quad (27)$$

where $\omega_D = k_B T_D/\hbar$. This relation is also shown in Fig. 8. Considering the minimal input required in the theoretical model (only the atomic mass and the Debye temperature), the agreement between the two curves is fair. Note that while the quantum model predicts the finite zero-point motion $(3\hbar/4m\omega_D)^{1/2}$, the MD results show a trend toward no motion at zero temperature. One expects this in a classical system, as the phase space approaches a single point when motion ceases.

The specific heat is defined thermodynamically as the rate of change of the total system energy (kinetic and potential) as a function of temperature at

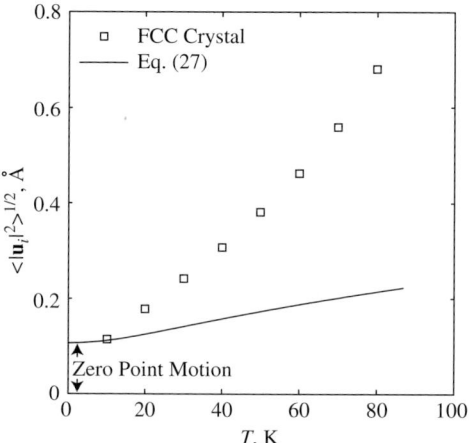

FIG. 8. Face-centered cubic crystal RMS data, and comparison to Eq. (27), a quantum-mechanical prediction. The zero-point RMS value is $(3\hbar/4m\omega_D)^{1/2}$.

constant volume [3],

$$c_v = \frac{\partial E}{\partial T}\bigg|_V \qquad (28)$$

Such a calculation can be explicitly performed using the results of MD simulations. The predicted specific heats for the fcc crystal, amorphous, and liquid phases are plotted in Fig. 9. The values given correspond to the specific heat per mode (i.e., per two degrees of freedom), of which there are $3(N-1)$. The calculation is performed by varying the temperature in 0.1 K increments over a ± 0.2 K range around the temperature of interest. Five simulations are performed at each of the five increments, with energy data averaged over 3×10^5 time steps. The resulting 25 data points are fit with a linear function, the slope of which is the specific heat. For the fcc crystal at temperatures of 70 and 80 K, and for all liquid phase calculations, 10 simulations were performed at all of the temperature increments due to larger scatter in the data. The spread of the energy data in these calculations increases with increasing temperature. The overall trends are good though, so any errors present are not likely more than 1% (based on linear fits to the data). The specific heat can also be predicted from system fluctuations (energy or temperature, depending on the ensemble), which are easily accessed in the simulations.

Not surprisingly, the fcc crystal and amorphous data are close. The crystal structure should not significantly affect the specific heat, especially at

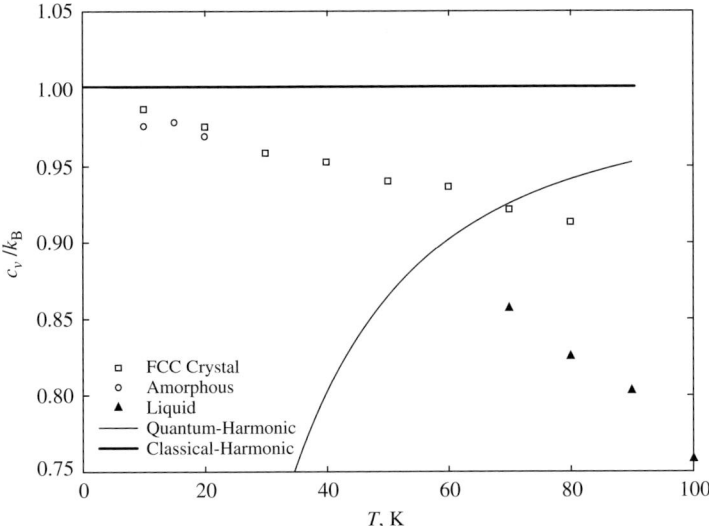

FIG. 9. The classical-anharmonic specific heat per degree of freedom predicted from the MD simulations, and the classical- and quantum-harmonic curves for the crystal phase (all scaled by k_B). The theoretical predictions are stopped at a temperature of 87 K, the melting point of the MD system.

low temperatures, where the harmonic approximation is still reasonable. There is a definite drop in the liquid values, and the specific heat would continue to decrease as the temperature is increased. The lower limit for the specific heat per mode is $0.5 k_B$, when potential energy effects have been completely eliminated (i.e., an ideal gas).

The specific heat predicted from the MD simulations is a classical-anharmonic value. Also shown in Fig. 9 are the classical- and the quantum-harmonic specific heats for the crystal phase. The classical-harmonic value, k_B, is based on an assumption of equipartition of kinetic and potential energy between the degrees of freedom. The equipartition assumption is always valid for the kinetic energy (i.e., it contributes $k_B/2$ to c_v, which has been verified). For the potential energy, however, it is only true under the harmonic approximation, which itself is only valid at zero temperature (this is discussed in Section III.D.1). The deviations of the classical anharmonic results from the classical harmonic model are significant. It is sometimes assumed that the mode specific heat of solids in MD is equal to k_B, which is not the case, and will lead to errors at high temperatures. The quantum-harmonic specific heat is based on the zero-temperature phonon density of states (calculated with normal mode analysis as discussed in Section III.D.3) and is given by [37]

$$c_{v,quant-harm} = k_B \sum_k \frac{x_k^2 \exp(x_k)}{[\exp(x_k) - 1]^2} \qquad (29)$$

where x_k is $\hbar\omega_k/k_B T$, and the summation is over the normal modes of the system. As expected, the classical and harmonic specific heats are significantly different at low temperatures, where quantum effects are important. Prediction of the quantum-anharmonic specific heat (that which would be measured experimentally) would require the temperature dependence and coupling of the normal modes into account. The results would be expected to converge with the classical-anharmonic value at high temperatures (i.e., on the order of the Debye temperature).

D. Phonon Space Analysis

1. *Harmonic Approximation*

One foundation of phonon analysis is the harmonic approximation (i.e., that the phonon modes are equivalent to independent harmonic oscillators). Even when anharmonicities are accounted for, it is usually as a perturbation to the harmonic solution of the lattice dynamics problem. In these cases, phonon–phonon interactions are modeled as instantaneous events, preceded and followed by the independent propagation of phonons through the system (i.e., the phonons behave harmonically except when they are interacting). In this section, we briefly review the harmonic description of the lattice dynamics, and present calculations on the zero-temperature unit cell. A discussion follows of how MD simulations can be used to incorporate anharmonic effects.

At zero temperature in a classical solid, all the atoms are at rest in their equilibrium positions. This is evident from the trend in the atomic RMS data shown in Fig. 8. The potential energy of the system, which is a function of the atomic positions, can only take on one value (i.e., the phase space consists of a single point). As the temperature of the system is raised, the atoms start moving, and the extent of the associated phase space increases.

In lattice dynamics calculations, a frequency space description is sought through which one can predict and analyze the motions of the atoms. Instead of discussing the localized motions of individual atoms, one describes the system by energy waves with given wave vector (κ), frequency (ω), and polarization vector (**e**). The formulation of lattice dynamics theory is described in detail in numerous books (see, e.g. [3–6]). Here, we will discuss a few specific points of interest.

Suppose that the equilibrium zero-temperature potential energy of a system with N atoms is given by Φ_0. If each atom i is moved by an amount $\mathbf{u}_i$, the resulting energy of the system, Φ, can be found by expanding around the equilibrium energy with a Taylor series as

$$\Phi = \Phi_0 + \sum_i \sum_\alpha \left.\frac{\partial \Phi}{\partial u_{i,\alpha}}\right|_0 u_{i,\alpha} + \frac{1}{2}\sum_{i,j}\sum_{\alpha,\beta}\left.\frac{\partial^2 \Phi}{\partial u_{i,\alpha}\partial u_{j,\beta}}\right|_0 u_{i,\alpha}u_{j,\beta}$$
$$+ \frac{1}{6}\sum_{i,j,k}\sum_{\alpha,\beta,\gamma}\left.\frac{\partial^3 \Phi}{\partial u_{i,\alpha}\partial u_{j,\beta}\partial u_{k,\gamma}}\right|_0 u_{i,\alpha}u_{j,\beta}u_{k,\gamma} + \cdots \quad (30)$$

Here, the i, j, and k sums are over the atoms in the system, and the α, β, and γ sums are over the x-, y-, and z-directions. Both Φ and Φ_0 are only functions of the atomic positions. The first derivative of the potential energy with respect to each of the atomic positions is the negative of the net force acting on that atom. Evaluated at equilibrium, this term is zero. Thus, the first non-negligible term in the expansion is the second-order term. The harmonic approximation is made by truncating the Taylor series at the second-order term. The $\partial^2 \Phi/\partial u_{i,\alpha}\partial u_{j,\beta}$ terms are the elements of the force constant matrix.

The harmonic approximation is valid for small displacements ($u_{i,\alpha} \ll r_{nn}$) about the zero-temperature minimum. Raising the temperature will cause deviations for two reasons: as the temperature increases, the displacements of the atoms will increase beyond what might be considered small (say, $0.05r_{nn}$), and the lattice constant will change, so that the equilibrium separation does not correspond to the well minimum. These two ideas can be illustrated using results of the MD simulations for the LJ fcc crystal.

The effective LJ potential (that which an atom experiences in the crystal) is given by Eq. (21). The area around the minimum is plotted in Fig. 10. Superimposed on the potential is the associated harmonic approximation, given by

$$\phi_{LJ,eff,harm} = -8.61\varepsilon_{LJ} + 260.67\frac{\varepsilon_{LJ}}{\sigma_{LJ}^2}(r_{nn} - 1.09\sigma_{LJ})^2 \quad (31)$$

The match between the harmonic curve and the effective potential, reasonable to the left of the minimum, is poor to its right. Also shown in Fig. 10 are the equilibrium atomic separations (equal to the lattice constant a divided by $2^{1/2}$, see Fig. 4) and the distribution of the nearest-neighbor atomic separations at temperatures 20, 40, 60, and 80 K. The distributions are based on 10^5 time steps of *NVE* simulation with data extracted every five time steps, and the probability density function $p(r)$ is defined such that $\int p(r)\,dr = 1$.

The crystal expands as the temperature increases. The sign of the thermal expansion coefficient for a solid is related to the asymmetry of the potential

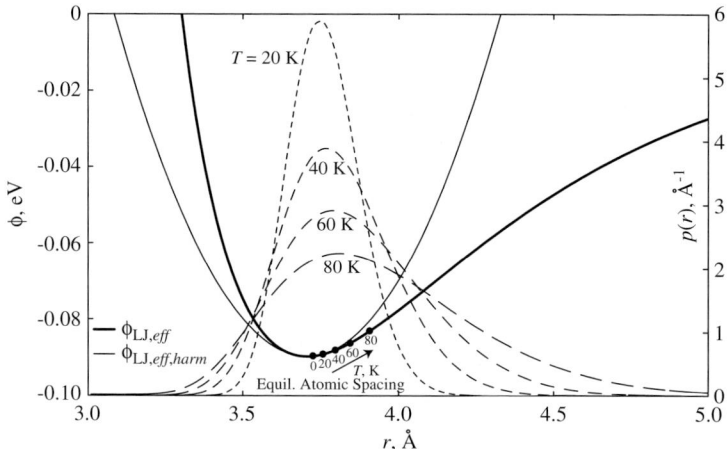

Fig. 10. The effective LJ potential, the associated harmonic approximation, and the resulting atomic separations at temperatures of 20, 40, 60, and 80 K for the fcc crystal.

well. At very low temperatures, this effect can be quantitatively related to the third-order term in the expansion of the potential energy [38]. For the LJ system, and most other solids, the potential is not as steep as the atomic separation is increased, leading to an expansion of the solid with increasing temperature. This is not always the case, however, as materials with a negative coefficient of thermal expansion exist (e.g., some zeolites [39], cuprous oxide [40], and ZrW_2O_8 [41]). As the temperature is raised, the spread of the nearest neighbor separation distance increases, and the distribution becomes significantly asymmetrical. This effect can also be interpreted based on the shape of the potential well.

For the case of the LJ system, it is clear that the harmonic approximation is not strictly valid, even at the lowest temperature considered (10 K). In order to work with phonons and normal modes, however, the harmonic approximation is necessary.

2. *Normal Modes*

A challenge in working with Eq. (30) under the harmonic approximation is the coupling of the atomic coordinates in the second-order derivatives. A transformation exists on the $3N$ real space coordinates (three for each of the N atoms) to a set of $3N$ new coordinates S_k (the normal modes) such that [5]

$$\Phi - \Phi_0 = \frac{1}{2}\sum_{i,j}\sum_{\alpha,\beta}\frac{\partial^2 \Phi}{\partial u_{i,\alpha}\partial u_{j,\beta}}\bigg|_0 u_{i,\alpha}u_{j,\beta} = \frac{1}{2}\sum_k \frac{\partial \Phi^2}{\partial S_k^2}\bigg|_0 S_k^* S_k \qquad (32)$$

The normal modes are equivalent to harmonic oscillators, each of which has an associated wave vector, frequency, and polarization. They are completely nonlocalized spatially. The specification of the normal modes (which is based on the crystal structure and system size) is known as the first quantization. It is not related to quantum mechanics, but indicates that the frequencies and wave numbers available to a crystal are discrete and limited. This idea can be understood by considering a one-dimensional arrangement of four atoms in a periodic system, as shown in Fig. 11. The atoms marked with black dots are equivalent as a result of the application of periodic boundary conditions, which any allowed vibrational mode must satisfy. Two such waves are shown, with wavelengths of $4a$ and $2a$. An important distinction between the lattice dynamics problem and the solution of a continuum system (e.g., elastic waves) is that the smallest allowed wavelength is restricted by the spacing of the atoms. A mode with a wavelength of a will be indistinguishable from the mode with wavelength $2a$. The minimum wavelength (i.e., the maximum wave number) defines the extent of the first Brillouin zone (BZ), the frequency space volume accessed by the system. There is no lower limit to the wavelength in a continuum. The longest allowed wavelength in any case is determined by the size of the system.

By introducing the idea of a quantum harmonic oscillator, the second quantization is made. In this case, in addition to quantizing the allowed normal modes, the energy of these modes is also quantized in units of $\hbar\omega$. The second quantization cannot be made in MD simulation due to their classical nature. The energy of a given mode is continuous.

Starting from Eq. (32), and noting that the second derivative terms can be considered as the spring constants, K_k, of the harmonic oscillators, the

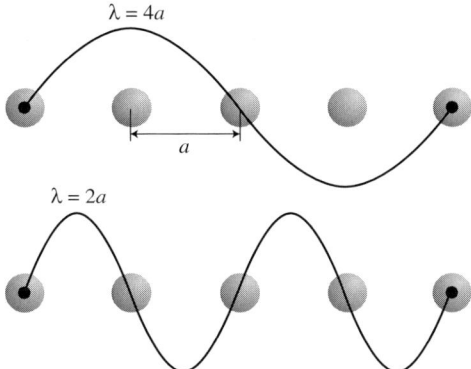

FIG. 11. One-dimensional example of how allowed wave vectors are determined and the first quantization is realized.

energy of one normal mode can be expressed as

$$\Phi_k = \frac{1}{2} \frac{\partial \Phi^2}{\partial S_k^2}\bigg|_0 S_k^* S_k = \frac{1}{2} \frac{K_k}{m_k} S_k^* S_k = \frac{1}{2} \omega_k^2 S_k^* S_k \qquad (33)$$

as the mass, frequency, and spring constant are related through $\omega_k = (K_k/m_k)^{1/2}$. The average potential energy will be

$$\langle \Phi_k \rangle = \frac{1}{2} \omega_k^2 \langle S_k^* S_k \rangle \qquad (34)$$

This is the expectation value of the potential energy of one degree of freedom. The expectation value for one degree of freedom in a classical-harmonic system is $k_B T/2$.

The total kinetic energy, KE, in the real and normal mode spaces is given by

$$KE = \sum_i \frac{1}{2} \frac{|\mathbf{p}_i|^2}{m_i} = \sum_k \frac{1}{2} \dot{S}_k^* \dot{S}_k \qquad (35)$$

As the kinetic energy of a particle in a classical system is proportional to the square of the magnitude of its velocity (and no higher order terms), this expression for the kinetic energy is valid in anharmonic systems. The classical-harmonic expectation value of the mode kinetic energy is $k_B T/2$ leading to a contribution of $k_B/2$ to the mode specific heat, as discussed in Section III.C.2.

In a classical-harmonic system there is an equipartition of energy between all degrees of freedom, so that the average kinetic energy of a mode will be equal to its average potential energy. Thus,

$$\langle E_k \rangle_{harm} = \omega_k^2 \langle S_k^* S_k \rangle = k_B T \qquad (36)$$

The instantaneous energy in a given mode is readily calculated in the MD simulations. It is crucial to note, however, that while these expressions are based on a harmonic theory, the MD simulations are anharmonic. We will discuss some of the consequences of this fact in Section III.D.5.

3. *Lattice Dynamics*

Given the crystal structure of a material, the determination of the allowed wave vectors (whose extent in the wave vector space make up the first BZ) is straightforward. One must note that points on the surface of the BZ that are separated by a reciprocal lattice vector, **G**, are degenerate. For the fcc crystal, the reciprocal lattice vectors are $2\pi/a(2,0,0)$, $2\pi/a(1,1,1)$, and appropriate rotations. For the 256 atom fcc crystal, application of the degeneracies reduces 341 points down to the expected 256, each of which has three polarizations [11]. Specifying the frequencies and polarizations of these modes is a more involved task. The polarizations are required to transform the atomic positions into the normal mode coordinates, and the frequencies are needed to calculate the normal mode potential energies [Eq. (33)]. We outline the derivation presented by Dove [5].

Consider a general crystal with an n-atom unit cell, such that the displacement of the jth atom in the lth unit cell is denoted by $\mathbf{u}(jl, t)$. The force constant matrix [made up of the second order derivatives in Eq. (30)] between the atom (jl) and the atom $(j'l')$ will be denoted by $\boldsymbol{\Phi}\binom{jj'}{ll'}$. Note that this matrix is defined for all atom pairs, including the case of $j = j'$ and $l = l'$. Imagining that the atoms in the crystal are all joined by harmonic springs, the equation of motion for the atom (jl) can be written as

$$m_j \ddot{\mathbf{u}}(jl, t) = -\sum_{j'l'} \boldsymbol{\Phi}\binom{jj'}{ll'} \cdot \mathbf{u}(j'l', t) \tag{37}$$

Now assume that the displacement of an atom can be written as a summation over the normal modes of the system, such that

$$\mathbf{u}(jl, t) = \sum_{\kappa,\nu} m_j^{-1/2} \mathbf{e}(j, \boldsymbol{\kappa}, \nu) \exp\{i[\boldsymbol{\kappa} \cdot \mathbf{r}(jl) - \omega(\boldsymbol{\kappa}, \nu)t]\} \tag{38}$$

At this point, the wave vector is known, but the frequency and polarization vector are not. Note that the index k introduced in Eq. (30) has been replaced by $(\boldsymbol{\kappa}, \nu)$. The polarization vector and frequency are both functions of the wave vector and the dispersion branch, denoted by ν. Substituting Eq. (38) and its second derivative into the equation of motion, Eq. (37), leads to the eigenvalue equation

$$\omega^2(\boldsymbol{\kappa}, \nu) \mathbf{e}(\boldsymbol{\kappa}, \nu) = \mathbf{D}(\boldsymbol{\kappa}) \cdot \mathbf{e}(\boldsymbol{\kappa}, \nu) \tag{39}$$

where the mode frequencies are the square roots of the eigenvalues and the polarization vectors are the eigenmodes. They are obtained by diagonalizing the matrix $\mathbf{D}(\boldsymbol{\kappa})$, which is known as the dynamical matrix, and has size $3n \times 3n$. It can be broken down into 3×3 blocks (each for a given jj' pair), which will have elements

$$D_{\alpha,\beta}(jj', \boldsymbol{\kappa}) = \frac{1}{(m_j m_{j'})^{1/2}} \sum_{l'} \Phi_{\alpha,\beta}\binom{jj'}{ll'} \exp\{i\boldsymbol{\kappa} \cdot [\mathbf{r}(j'l') - \mathbf{r}(jl)]\} \tag{40}$$

The LJ crystal phase considered is monatomic, so that the dynamical matrix has size 3×3. There are thus three modes associated with each wave

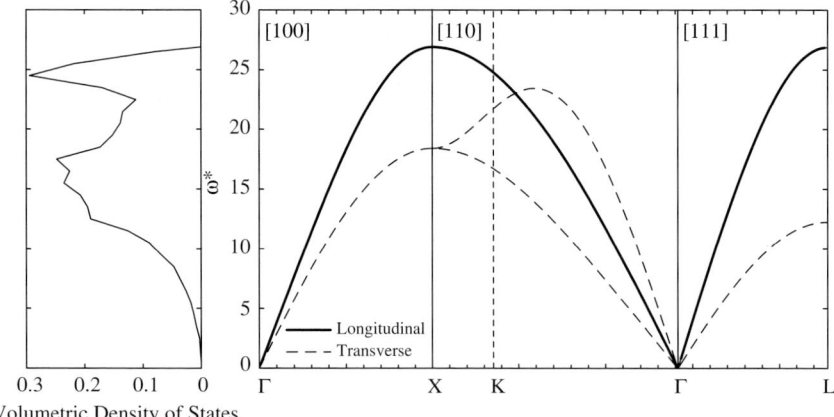

FIG. 12. Dimensionless dispersion curves and density of states for the LJ fcc crystal at zero temperature.

vector. Given the equilibrium atomic positions and the interatomic potential, one can find the frequencies and polarization vectors by substituting the wave vector into the dynamical matrix and diagonalizing. While this calculation can be performed for any wave vector, it is important to remember that only certain values are relevant to the analysis of a particular MD simulation cell.

The phonon dispersion curves are obtained by plotting the normal mode frequencies as a function of the wave number in different directions. These are shown in dimensionless form for the [1 0 0], [1 1 0], and [1 1 1] directions in Fig. 12 for the zero-temperature simulation cell. The first BZ for the fcc lattice is shown in Figs. 13(a) and (b). A dimensionless wave vector, κ^*, has been defined as

$$\kappa^* = \frac{\kappa}{2\pi/a} \qquad (41)$$

such that κ^* will vary between 0 and 1 in the [1 0 0] direction in the first BZ. In Fig. 12, the divisions on the horizontal axis (the wave number) are separated by $0.1 \times 2\pi/a$ (i.e., one-twentieth of the size of the first BZ in the [1 0 0] direction). Note the degeneracies of the transverse branches in the [1 0 0] and [1 1 1] directions, but not in the [1 1 0] direction. Also, as seen in the [1 1 0] direction, the longitudinal branch does not always have the highest frequency of the three branches at a given point. The frequencies of the longitudinal and transverse branches at the X point for argon are 2.00 and 1.37 THz, which compare very well to the experimental values of 2.01

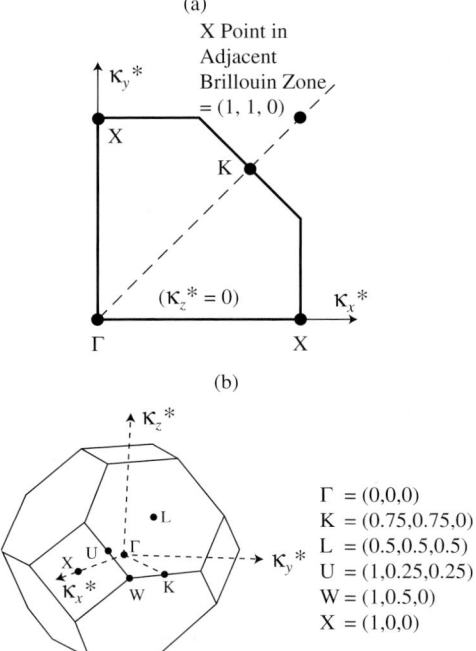

FIG. 13. (a) The [0 0 1] plane in the fcc crystal BZ showing how the Γ–X and Γ–K–X curves in the LJ dispersion (Fig. 12) are defined. (b) The first BZ for the fcc lattice, and important surface symmetry points. Each of the listed points has multiple equivalent locations.

and 1.36 THz, obtained at a temperature of 10 K [42]. This good agreement is somewhat surprising, as the LJ parameters ε_{LJ} and σ_{LJ} are obtained from the properties of low-density gases [43].

The phonon density of states describes the distribution of the normal modes as a function of frequency, with no distinction of the wave vector direction. It can be thought of as an integration of the dispersion curves over frequency. The volumetric density of states for the zero temperature LJ fcc crystal is plotted in Fig. 12 alongside the dispersion curves. The data are based on a BZ with a grid spacing of $1/21 \times 2\pi/a$. This leads to 37,044 distinct points (each with three polarizations) covering the entire first BZ. The frequencies are sorted using a histogram with a bin size of unity, and the resulting data are plotted at the middle of each bin. The density of states axis is defined such that an integration over frequency gives $3(N-1)/V \simeq 12/a^3$ (for large V, where N is the number of atoms in volume V).

As suggested by Eq. (38), the mode polarizations are required to transform the atomic coordinates to the normal modes (and vice versa). The relevant transformations are

$$S_k(\boldsymbol{\kappa}, v) = N^{-1/2} \sum_i m_i^{1/2} \exp(-i\boldsymbol{\kappa} \cdot \mathbf{r}_{i,o}) \mathbf{e}_k^*(\boldsymbol{\kappa}, v) \cdot \mathbf{u}_i \quad (42)$$

$$\mathbf{u}_i = (m_i N)^{-1/2} \sum_k \exp(i\boldsymbol{\kappa} \cdot \mathbf{r}_{i,o}) \mathbf{e}_k^*(\boldsymbol{\kappa}, v) S_k \quad (43)$$

As discussed, the normal modes are a superposition of the positions that completely delocalize the system energy, and are best thought of as waves.

4. *Phonon Relaxation Time*

While the harmonic analysis of the preceding section establishes some methodology, it is not directly applicable to a finite temperature crystal, where anharmonic effects lead to thermal expansion and mode interactions (which lead to finite thermal conductivities). In this section and the next, we show how results of MD simulations can be used to predict finite temperature phonon relaxation times and dispersion curves.

The phonon relaxation time, τ, as originally formulated in the BTE models of Callaway [44] and Holland [45], gives an indication of how long it will take a system to return to equilibrium when one phonon mode has been perturbed. One can also think of the relaxation time as a temporal representation of the phonon mean free path if the phonon-particle description is valid (i.e., $\tau = v\Lambda$), or as an indication of how long energy stays coherent in a given vibrational mode.

Ladd *et al.* [46] present a method that finds the relaxation time of the kth mode, $\tau_{k,r}$, using the time history of the mode potential energy, Φ_k. This method has been modified by considering the total energy (potential and kinetic) of each mode, E_k [28]. Under the harmonic approximation, the instantaneous, total energy of each mode of a classical system is given by

$$E_k = \frac{\omega_k^2 S_k^* S_k}{2} + \frac{\dot{S}_k^* \dot{S}_k}{2} \quad (44)$$

where the first term corresponds to the potential energy and the second to the kinetic energy. The temporal decay of the autocorrelation of E_k is related to the relaxation time of that mode [46]. The resulting curve for the transverse polarization at $\kappa^* = (0.5, 0, 0)$ for the $\eta = 4$ simulation cell at a temperature of 50 K is shown in Fig. 14. The required ensemble average is realized by averaging the autocorrelation functions (10^4 time steps long, based on 2×10^5 time steps of data) over the [1 0 0], [0 1 0], and [0 0 1] directions over five independent simulations. The relaxation time is obtained by fitting the data with an exponential decay. Based on this formulation, the calculated time constant must be multiplied by 2 to get the relaxation time

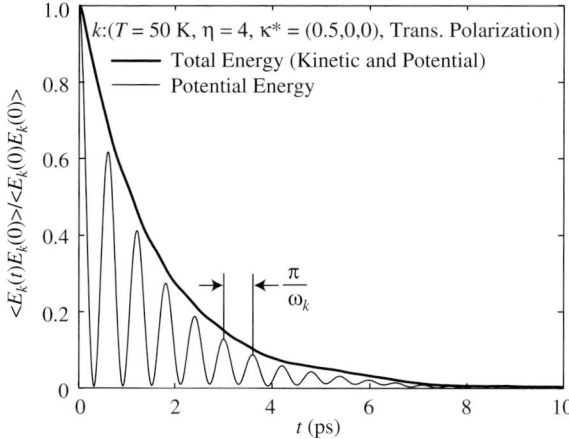

FIG. 14. Autocorrelation curves for the relaxation time and anharmonic phonon dispersion calculation methods. The data correspond to deviations from the mean energy values, and have been normalized against the zero time value of the autocorrelations. Shown are the total mode energy (used in the relaxation time calculation) and the potential energy (used to obtain the anharmonic phonon frequencies). The frequency of the oscillations in the potential energy curve is double that of the phonon mode in question because of the squaring operations in Eq. (44). From McGaughey and Kaviany [28]. Copyright (2004), with permission from America Physical Society.

used in the Callaway–Holland BTE formulation [28]. This is the value reported here. Most of the modes considered show a general behavior consistent with a single relaxation time. For some modes, a secondary decay is evident in the very early stages of the overall decay. In these cases, one neglects this portion of the autocorrelation when fitting the exponential. Alternatively, one could calculate the integral of the autocorrelation and from that deduce an effective relaxation time [46]. Owing to the short extent of the observed deviation from a single exponential decay, and the subsequent fitting of a continuous function to the discrete relaxation times, the difference between this approach and that used here is negligible.

To obtain a sufficient number of points within the first BZ to form continuous relaxation time and dispersion functions in a particular direction (required for upscaling to the BTE [28]), one must consider different-sized simulation cells. Having obtained a set of discrete $\tau_{k,r}$ values for a given temperature and polarization using the 256, 500, and 864 atom simulation cell, a continuous function, τ_r, can be constructed in the principle directions (where there is sufficient data). The discrete and continuous results for the LJ fcc crystal in the [1 0 0] direction at temperatures of 20, 35, 50, 65, and 80 K are plotted as $1/\tau_{k,r}$ (or $1/\tau_r$) vs. ω in Figs. 15(a) and (b). The data for each polarization can be broken down into three distinct regions. The first

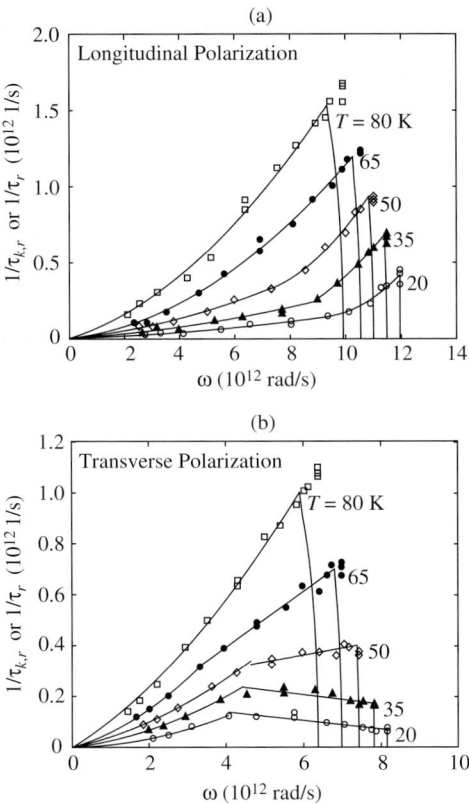

FIG. 15. Discrete relaxation times ($\tau_{k,r}$) and continuous curve fits (τ_r) for the LJ fcc crystal for (a) longitudinal and (b) transverse polarizations. From McGaughey and Kavinay [28]. Copyright (2004), with permission from America Physical Society.

two are fit with low-order polynomials. For the longitudinal polarization, the first region is fit with a second-order polynomial through the origin, and the second region with a second-order polynomial. For the transverse polarization, the first region is fit with a second-order polynomial through the origin, and the second region with a linear function. The resulting functions are also shown in Figs. 15(a) and (b) and are considered satisfactory fits to the MD data. The parts of the relaxation time curves are not forced to be continuous. For both the longitudinal and transverse polarizations, any resulting discontinuities are small, and are purely a numerical effect. The relaxation time functions do not contain the orders of magnitude discontinuities found in the Holland relaxation times for germanium, which result from the assumed functional forms, and how the fitting parameters are determined [47]. As the temperature increases, the behavior in the two regions becomes similar. For both polarizations at a temperature of 80 K, and for the longitudinal polarization at a temperature of 65 K, a single second-order polynomial through the origin is used to fit all data. In the third region, the continuous relaxation time functions are taken up to the maximum frequency ($\omega_{L,max}$ or $\omega_{T,max}$) using the Cahill–Pohl (CP) high scatter limit [32,33], which requires that the relaxation time correspond to at least one half of the mode's period of oscillation.

Dimensionless relaxation times for all distinct points in the first BZ of the 256 atom simulation cell (18 points describing 30 unique modes [11]) are plotted as (τ_r^* vs. κ^*) and as ($1/\tau_r^*$ vs. ω^*) in Figs. 16(a) and (b). The continuous

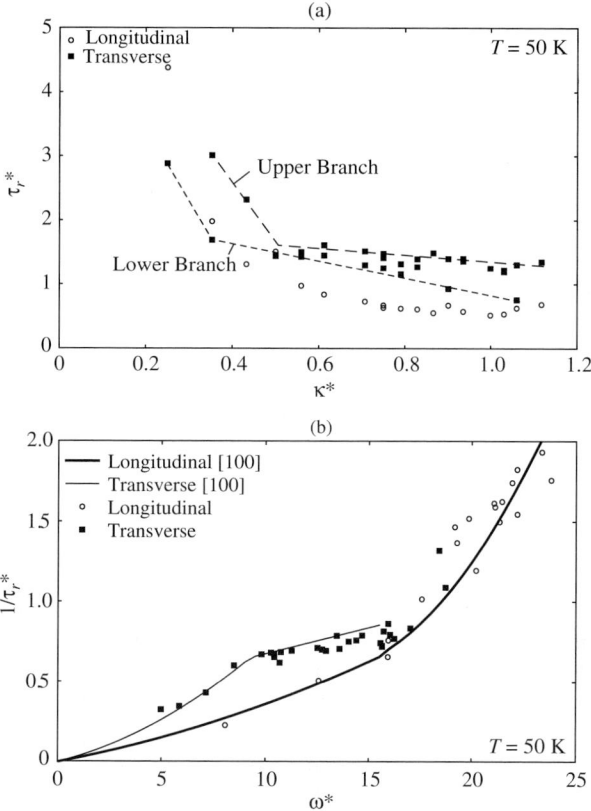

FIG. 16. Full BZ relaxation times at $T = 50$ K for the $\eta = 4$ simulation cell. (a) Relaxation time as a function of the wave number. Note the two distinct branches for the transverse polarization. (b) Inverse of the relaxation time as a function of frequency [compare to Fig. 15(a)]. The separation of the transverse data seen in (a) does not manifest. All results are dimensionless.

functions from Fig. 15 are also plotted in Fig. 16(b). For the longitudinal modes, the relaxation times show a common trend and agree well with the [1 0 0] curve. The discrepancy is larger for the transverse modes. In fact, two independent trends are clear in the wave number plot. All things being equal with the phonon dispersion, the fact that the secondary branch is lower than the main trend of the data is consistent with the isotropic assumption resulting in an over-prediction of the thermal conductivity, as found in Ref. [28]. The two transverse branches are not obvious in the frequency plot. The splitting is also most evident at the intermediate temperatures. At the low temperature (20 K), the larger uncertainties in the relaxation times make trends harder to discern. At the high temperature (80 K) all the relaxation times (longitudinal and transverse) appear to follow the same trend.

5. *Phonon Dispersion*

The zero-temperature phonon dispersion shown in Fig. 12 is harmonic, and can be determined exactly at any wave vector using the MD equilibrium atomic positions and the interatomic potential. Thus, one can obtain a continuous dispersion relation (see Section III.D.3). Deviations from this calculation at finite temperature result from two effects [5]. Based on the higher order terms in the expansion of the potential energy about its minimum, a solid will either expand (as seen here) or contract as the temperature increases. An expansion causes the phonon frequencies to decrease. Re-calculating the dispersion harmonically with the new lattice constant is known as the quasi-harmonic approximation.

The second effect is a result of anharmonicities in the atomic interactions, which become increasingly important as the temperature increases. It is difficult to model these effects exactly due to the complexity of the higher order terms in Eq. (30). The inclusion of anharmonic effects is important in any application that requires phonon velocities, such as using the BTE to predict the thermal conductivity under the single-mode relaxation time approximation [28]. To account for the anharmonic effects, the autocorrelation data for the mode potential energy are used to calculate the frequencies of the discrete modes present in the MD simulations. This is shown in Fig. 14. While the total energy autocorrelation shows a monotonic decay, that for the potential energy oscillates. This results from the total energy having both potential and kinetic energy components. One obtains an estimate of the anharmonic frequency by averaging over all non-negligible oscillations in the autocorrelation, generating a set of discrete anharmonic frequency data. By comparing these values to the associated quasi-harmonic frequencies, a second-order polynomial scaling function is constructed. This function is then applied to the continuous quasi-harmonic data to obtain the full anharmonic dispersion. The excellent quality of the mapping from the quasi-harmonic data to the anharmonic data (the R^2 values of the scaling functions are ~0.999) suggests that this procedure introduces minimal error.

The frequencies used for the horizontal axes of Figs. 15(a) and (b) are based on the anharmonic dispersion. Note, however, that the frequencies used in the phonon energy calculations [Eqs. (33) and (44)] must be those corresponding to the quasi-harmonic dispersion. This is a result of the phonon dynamics being based on a harmonic theory, while the BTE expression for the thermal conductivity is not. We can justify the need to use the quasi-harmonic frequencies in the phonon energy calculation by calculating the total, average phonon potential energy and comparing the result to that directly calculated with the LJ potential. This is shown for a 4 ps time interval in Figs. 17(a) and (b) for temperatures of 20 and 80 K. To allow

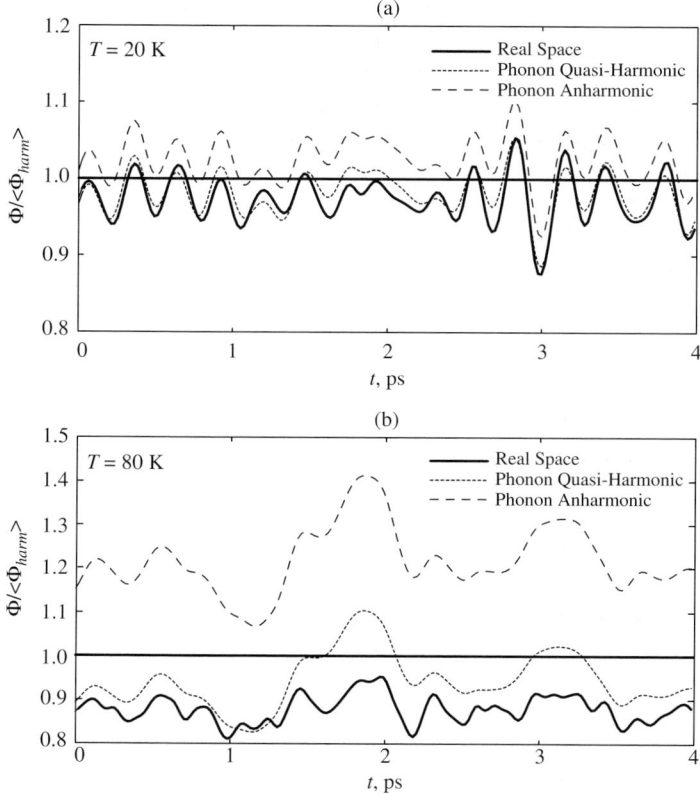

FIG. 17. Real and phonon space energy calculations at temperatures of (a) 20 K and (b) 80 K. The phonon space energy is calculated using both the quasi-harmonic and anharmonic frequencies. While the quasi-harmonic data match the real space calculation reasonably well in both cases, the anharmonic data diverge as the temperature increases.

for a direct comparison of the real space and phonon space energies, the real space value has been shifted by the potential energy that would exist if all the atoms were at their equilibrium positions. The data have been normalized against the classical-harmonic expectation value, $\langle \Phi_{harm} \rangle = 3(N-1)k_BT/2$.

In Fig. 18(a), the average energy for each case, along with data for temperatures of 35, 50, and 65 K (based on 10^5 time steps of *NVE* MD simulation) is plotted as a function of temperature. Using the quasi-harmonic frequencies results in an energy that matches the magnitude and temperature trend of the exact calculation to within 5% over the entire temperature range considered. The anharmonic results diverge from the

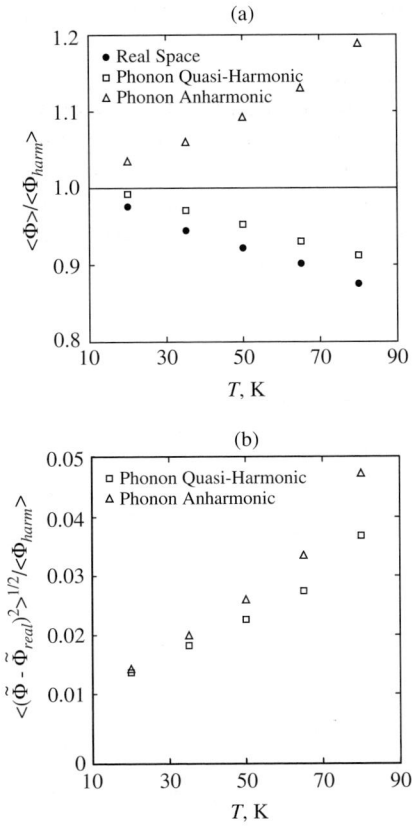

FIG. 18. (a) Average energies calculated using the quasi-harmonic and anharmonic frequencies compared to that predicted directly from the LJ potential in real space. (b) Root mean square deviations of the phonon space energies from that in the real space. In both plots, the energies have been scaled by the expectation value in the classical-harmonic system.

exact calculation, and are 36% larger at the highest temperature. This trend is evident from the plots in Fig. 17.

In Fig. 18(b), the RMS deviation of the phonon space energies (anharmonic and quasiharmonic) compared to the real space energy are plotted. To perform this calculation, all three sets of energy data were shifted to an average value of zero. From Fig. 17, the fluctuations in the quasiharmonic and anharmonic calculations appear to be about the same. Both curves seem to pick up an equivalent amount of the detail in the real space energy. But, as seen in Fig. 18(b), the deviations in the anharmonic data are larger. Therefore, based on both the magnitude and fluctuations in the energy calculations, it is the quasi-harmonic frequencies that need to be used when calculating the phonon space energy. This is not surprising, as the model in use is based in harmonic theory.

The phonon dispersion for the [1 0 0] direction is shown in Fig. 19(a) for the zero-temperature simulation cell, and for the quasi-harmonic and anharmonic predictions at a temperature of 50 K. The effect of the unit cell size is significant, and increases with increasing temperature. The anharmonic effects are significant for the longitudinal polarization at all temperatures, and increase with increasing temperature. For the transverse polarization, the deviations from the quasi-harmonic values are found to be significant only at a temperature of 80 K. The discrete anharmonic normal mode frequencies for the entire BZ for the $\eta = 4$ simulation cell at a temperature of 50 K are plotted in dimensionless form in Fig. 20 as a function of the magnitude of the associated wave vector. The anharmonic

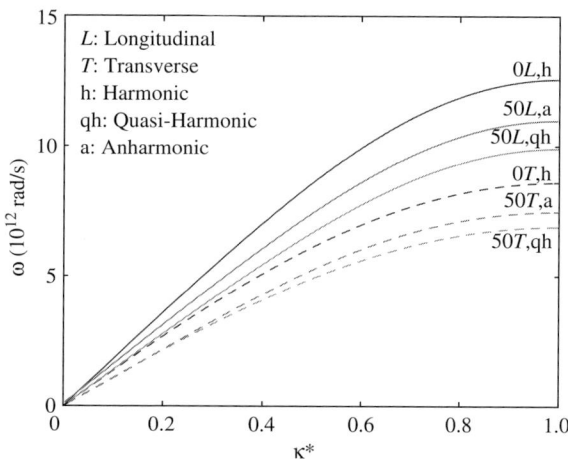

FIG. 19. (a) Phonon dispersion in the [1 0 0] direction. The curves are identified by the temperature in Kelvin, the polarization, and the nature of the calculation. From McGaughey and Kaviany [28]. Copyright (2004), with permission from American Physical Society.

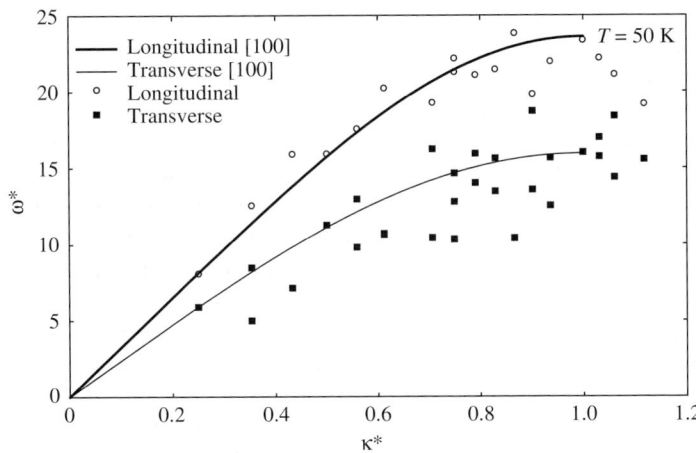

FIG. 20. Full BZ dispersion at $T = 50\,\text{K}$ for the $\eta = 4$ simulation cell. All data are dimensionless.

dispersion curves in the [1 0 0] direction are also shown. The [1 0 0] data represent the trend in the data for the entire first BZ reasonably well.

IV. Nature of Phonon Transport in Molecular Dynamics Simulations

A. Quantum Formulation and Selection Rules

Traditionally, phonon transport and interactions have been described in the context of a quantum-particle-based model. The existence of a particle implies localization. A phonon mode in a crystal is, in fact, completely nonlocalized, but one can imagine that in a large system (where the resolution of the BZ is fine), a wave packet can be created by superimposing phonon modes of similar wavelengths. Such a wave packet, or phonon particle, can then move around the system of interest. These phonon particles are assumed to propagate ballistically in between collisions with other phonons, impurities, and boundaries. Under these conditions, the phonon particles are often described as making up a phonon gas.

In MD simulations of a perfect crystal, the only scattering present is a result of inter-phonon interactions. To model such interactions theoretically, it is necessary to consider third order and higher (anharmonic) terms in the expansion of the potential energy about its minimum [see Eq. (30)]. Anharmonicities in the atomic interactions are required to obtain a finite bulk phase thermal conductivity for crystals. Interestingly, a disordered MD system modeled with a harmonic potential can produce a finite thermal conductivity [48,49]. The third order term in the expansion of the potential energy is related to three-phonon interactions, and is the basis for standard analysis techniques. The mathematics at this level are involved, and the level of complexity increases for fourth order and higher terms. These higher-order effects, however, are generally thought to be insignificant [4].

In the three-phonon interaction formulation there are two types of allowed events. In a type I interaction, one phonon decays into two others. In a type II interaction, two phonons combine to form a third. To satisfy conservation of energy, processes in which three phonons are either created or destroyed are not allowed. Two selection rules exist for the allowed phonon interactions.

First, from the translational invariance of the lattice potential energy, the wave vectors of the phonon modes in question must satisfy [4]

$$\kappa_1 = \kappa_2 + \kappa_3 + \mathbf{G} \quad \text{(type I)} \quad (45)$$

$$\kappa_1 + \kappa_2 = \kappa_3 + \mathbf{G} \quad \text{(type II)} \quad (46)$$

where $\mathbf{G}$ is either equal to zero [corresponding to a Normal (N) process] or a reciprocal lattice vector [corresponding to an Umklapp (U) process]. These criteria are valid in both the classical and quantum descriptions of the phonon system. These wave vector conservation equations are only dependent on the crystal structure, and are not affected by temperature. By multiplying through by $\hbar$, terms with the units of momentum are obtained, so that this criterion is often referred to as the conservation of crystal (or phonon) momentum. This is not real momentum, however, as no phonon modes (other than that at the center of the BZ) can carry physical momentum. With respect to thermal conductivity, the N and U processes play different roles [44,45]. Note that selection rules for the classical system cannot contain $\hbar$.

The second selection rule, which only applies to the quantum system, is based on conservation of energy. The second quantization in the formulation of the lattice dynamics theory results in the energy of the phonon modes being discretized into packets of size $\hbar\omega$. For the type I and II interactions, conservation of energy then leads to

$$\hbar\omega_1 = \hbar\omega_2 + \hbar\omega_3 \quad \text{(type I)} \quad (47)$$

$$\hbar\omega_1 + \hbar\omega_2 = \hbar\omega_3 \quad \text{(type II)} \quad (48)$$

With respect to the discussion of Section III.D.5, it is the anharmonic frequencies that should be used here. The second quantization cannot be made in the MD system. Energy is still conserved in the NVE ensemble, but not as described by Eqs. (47) and (48). In the quantum-particle description, the interactions are assumed to occur instantaneously (i.e., they are discrete events). In the MD system the energy in a given mode is a continuous function of time, and discrete energy exchange events will not occur. One must instead think of a continuous flow of energy between the modes in the frequency space. Considered from another standpoint, the MD system is a nonlinear, many-body dynamics problem being solved using the Newton laws of motion.

The wave vector and energy selection rules only indicate what three-phonon interactions are possible. The rate at which a given interaction takes place is related to the intrinsic scattering rate, and the degree of departure of the mode populations from the equilibrium distribution at an instant in time [4].

B. Phonon Gas and Normal Modes

To describe a phonon system as a phonon gas in which the particles are interacting weakly (i.e., kinetically), a number of criteria must be satisfied regarding the length scales in the system. These are shown schematically in Fig. 21.

To treat a phonon as a particle, a wave packet must be formed, the size of which, l, is much greater than the wavelength, λ, of the mode of interest. For the wave packets to have distinct interactions with each other, the distance they travel between collisions (the average value of this distance is the mean free path, Λ) must be much greater than the size of the wave packets. Finally, if inter-phonon interactions are to dominate the phonon scattering (i.e., diffuse transport), the size of the system, L, must be much greater than the mean free path.

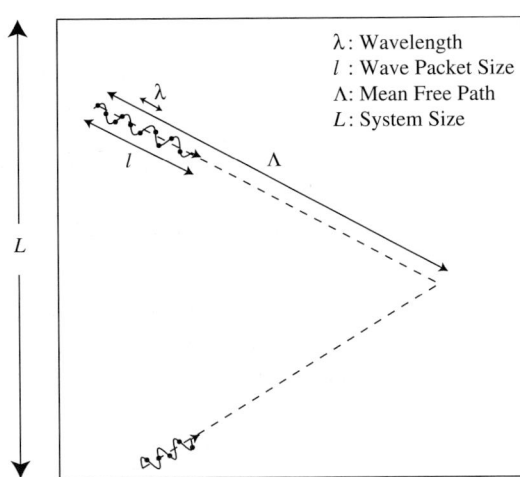

Fig. 21. Length scales in the phonon gas. For the phonon particle and phonon gas concepts to be valid, $\lambda \ll l \ll \Lambda \ll L$.

To form a wave packet, modes in the near vicinity of the mode of interest are superimposed. This is not possible in an MD system with a few hundred atoms. For example, in the 256 atom fcc crystal simulation cell, only four nonzero points can be resolved in the [1 0 0] direction. Considered from a different viewpoint, the system is too small to form a wave packet, as the side length is on the order of all the available wavelengths. Very large MD system sizes are required to form a wave packet. This has been done by Schelling et al. [50], Schelling and Phillpot [51], and Sinha et al. [52], who, for silicon, used simulation cells very long in one direction.

Having established that a wave packet cannot be formed in an MD simulation cell with a few hundred atoms, one can eliminate the possibility of treating the system as a collection of phonon particles. One must thus treat the phonon modes as they are defined in the lattice dynamics theory: as completely nonlocalized modes. The energy within a given phonon mode, while corresponding to a point in frequency space, cannot be spatially resolved in real space.

If a phonon mode cannot be spatially resolved, the concept of the mean free path becomes questionable. It is better to think of the energy in a given mode from a purely temporal perspective in the context of a relaxation time. From the phonon particle perspective, the relaxation time is the average time between collisions. In the nonlocalized description, the relaxation time is an indication of the timescale over which the energy in a mode stays correlated with itself. That is, the time over which a certain percentage of the energy in a mode is a part of the initial, and not the result of energy scattering into the mode as its original energy scatters to other modes. Within this framework, McGaughey [11] has directly observed phonon interactions in MD simulations of LJ argon. The results are interpreted in the context of internal resonance.

V. Thermal Conductivity Prediction: Green-Kubo Method

Two main techniques have been developed to predict the thermal conductivity of a dielectric material using MD simulations. These are the GK approach (an equilibrium method) and a direct application of the Fourier law (a steady state, nonequilibrium approach called the direct method). The implementation and interpretation of the results of these two approaches are discussed in this section and in Section VI.

A. FORMULATION

The development of a time-correlation function expression for the thermal conductivity (the GK approach) is based in classical statistical thermodynamics [53,54]. Multiple methods can be used to arrive at the final result [55]. Similar approaches can be used to develop expressions for the self- diffusion coefficient, the shear viscosity, and the bulk viscosity. These are all transport coefficients that cannot be obtained by applying a perturbation to the system Hamiltonian, as can be done for some properties (e.g. the electrical conductivity) where there is a real force that drives the transport. A step-by-step derivation using the method of Helfand [56] as outlined by McQuarrie [21], is presented in Ref. [11].

The net flow of heat in a solid, described by the heat current vector, **S**, fluctuates about zero at equilibrium. In the GK method, the thermal conductivity is related to how long it takes these fluctuations to dissipate, and for an isotropic material is given by [21]

$$k = \frac{1}{k_B V T^2} \int_0^\infty \frac{\langle \mathbf{S}(t) \cdot \mathbf{S}(0) \rangle}{3} dt \qquad (49)$$

where $\langle \mathbf{S}(t) \cdot \mathbf{S}(0) \rangle$ is the heat current autocorrelation function (HCACF). The GK approach is valid in the case of small disturbances from equilibrium and for long times (i.e., the hydrodynamic limit), and should be applied to simulations in the microcanonical ensemble. In ionic systems at high temperature, the charge flux may also contribute to the thermal conductivity. Lindan and Gillan [57] address this issue, which is generally irrelevant, by considering a more formal description of the transport theory. The heat current should be differentiated from the energy current, which is the expression that appears in a formal statement of the transport theory and contains terms related to the mass flux [58].

The heat current vector is given by [21]

$$\mathbf{S} = \frac{d}{dt} \sum_i \mathbf{r}_i E_i \qquad (50)$$

where the summation is over the particles in the system, and E_i is the energy (kinetic and potential) of a particle. Either the total or excess energy can be used. The units of the heat current are W m. Those of the heat flux, **q**, are W/m^2 such that $\mathbf{S} = \mathbf{q}V$. One should avoid using the symbol **q** for the heat current.

Li [17] has derived a general expression for the heat current for an n-body potential. For a pair potential ($n = 2$), Eq. (50) can be recast as

$$\mathbf{S} = \sum_i E_i \mathbf{v}_i + \frac{1}{2} \sum_{i,j} (\mathbf{F}_{ij} \cdot \mathbf{v}_i) \mathbf{r}_{ij} \qquad (51)$$

where **v** is the velocity vector of a particle. This form of the heat current is readily implemented in an MD simulation, and can be used for both rigid-ion and core-shell potentials [59]. The first term is associated with convection, and the second with conduction. For a three-body potential, an additional conduction term,

$$\frac{1}{6}\sum_{i,j,k} (\mathbf{F}_{ijk} \cdot \mathbf{v}_i)(\mathbf{r}_{ij} + \mathbf{r}_{ik}) \qquad (52)$$

where $\mathbf{F}_{ijk}$ is a three-body force, is added to Eq. (51). In many-body potentials ($n > 2$), the division of potential energy between particles is not unique. How the energy is divided has not been found to affect the predicted thermal conductivity [60,61].

Developing a physical interpretation of Eq. (49) is useful. The argument of the derivative in the heat current can be taken as the energy center of mass of the system. It is a vector that indicates the direction of energy transfer in the system at an instant in time. How long this quantity stays correlated with itself is related to the thermal conductivity. In a material with a high thermal conductivity, the correlation will be long lasting (i.e., fluctuations from equilibrium dissipate slowly). For a material with a low thermal conductivity, the correlation will be short-lived. One interesting aspect of the Green–Kubo approach is that with it, transport properties can be obtained from an equilibrium system. This is an important point for the thermal conductivity, since it is generally thought of in a nonequilibrium system with a temperature gradient.

The GK approach was first used to calculate the thermal conductivity of a three- dimensional-solid by Ladd *et al.* [46], who considered an fcc crystal modeled with an inverse 12th-order potential. Since then, researchers from a wide range of disciplines (e.g., mechanical engineering, physics, materials science, and nuclear engineering) have addressed methodological issues and used MD and the GK method as a predictive tool for many different materials. These include LJ argon [17,27–30,62–68], silicon [61,69–73], β-silicon carbide [60,74], silica structures [34,75], silicon nitride [76], diamond [77], carbon nanotubes [78–82], germanium-based materials [83,84], uranium dioxide [57,58,64,85], uranium nitride [86], calcium fluoride [57,58], potassium chloride [59], clathrate hydrates [87], nanofluids [88], and superlattices [89].

B. Case Study: Lennard-Jones Argon

All simulations used in the LJ thermal conductivity calculations presented here consist of 10^6 time steps over which the heat current vector is calculated every five time steps. A correlation length of 5×10^4 time steps with 2×10^5 time origins is used in the autocorrelation function. For all cases, five independent simulations (with random initial velocities) are performed and the HCACFs are averaged before finding the thermal conductivity. This ensures a proper sampling of phase space [60]. For the fcc crystal at a temperature of 10 K, where the correlation time is long, 10 independent simulations are performed. The main challenge at all conditions is the specification of the integral in Eq. (49).

1. *Heat Current Autocorrelation Function and Thermal Conductivity*

The HCACF and its integral [whose converged value is related to the thermal conductivity through Eq. (49)] are shown in Figs. 22(a) and (b) for the LJ argon fcc crystal, amorphous phase, and liquid described in Section III.B. The HCACF is normalized by its zero-time value to allow for comparison between the different temperatures. The integral is calculated using the trapezoidal rule. Note that as the temperature increases, the HCACFs of the three phases are approaching each other. The predicted thermal conductivities are shown in Fig. 23 along with experimental data [9], and the MD predictions of Li [17].

The zero-time value of the fcc crystal HCACF shows little temperature dependence (it increases by 10% over the entire temperature range). Li *et al.* [60] observe similar behavior for β-silicon carbide, and suggest that this quantity can be interpreted as a susceptibility. At finite time, the HCACF shows a two-stage decay. There is an initial drop, similar for all cases, followed by a longer decay, whose extent decreases as the temperature increases. The oscillations in the secondary decay are believed to result from the periodic boundary conditions [11]. For all cases considered, the integral of the HCACF converges well, and the thermal conductivity can be specified directly by averaging the integral over a suitable range. To remove the subjective judgment, Li *et al.* [60] proposed two methods by which one can specify the thermal conductivity. In the first dip (FD) method, the integral is evaluated at the first place where the HCACF goes negative. In the exponential fit (EF) method, an exponential function is fitted to the HCACF beyond a certain point (determined on a case-by-case basis), and this function is then used to calculate the contribution of the tail to the integral. Up to that point the integral is evaluated directly. In their investigation of β-silicon carbide, no significant differences were found between the predictions of these two methods.

Based on the observed shape of the fcc crystal HCACF, it can be fitted to a sum of two exponential functions as

$$\frac{\langle \mathbf{S}(t) \cdot \mathbf{S}(0) \rangle}{3} = A_{ac,sh}\exp(-t/\tau_{ac,sh}) + A_{ac,lg}\exp(-t/\tau_{ac,lg}) \qquad (53)$$

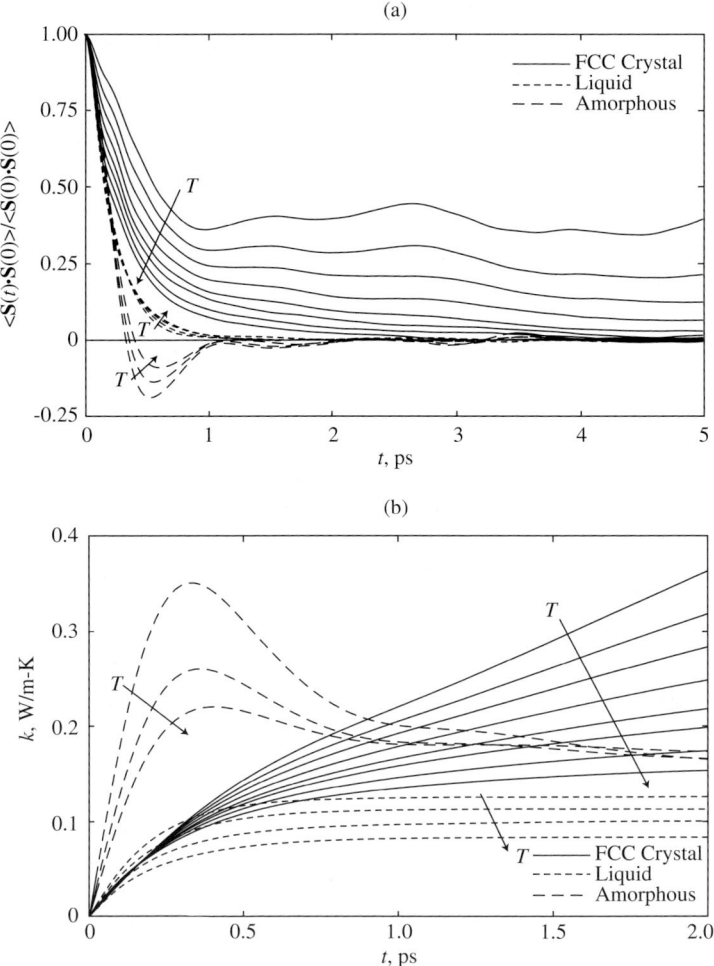

FIG. 22. Time dependence of (a) the raw HCACF and (b) its integral (the thermal conductivity) for LJ phases. Note the different timescales in the two plots.

as suggested by Che et al. [77]. With Eq. (49), the thermal conductivity is then

$$k = \frac{1}{k_B V T^2}(A_{ac,sh}\tau_{ac,sh} + A_{ac,lg}\tau_{ac,lg})$$
$$\equiv k_{ac,sh} + k_{ac,lg} \qquad (54)$$

In Eqs. (53) and (54) the subscripts ac, sh, and lg refer to acoustic, short-, and long range. Ladd et al. [46] first observed the two-stage decay in the HCACF.

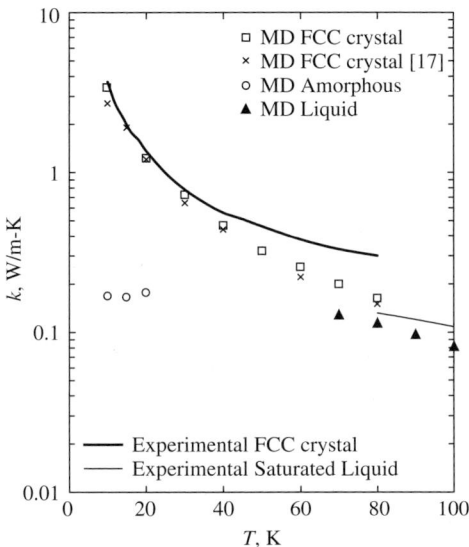

FIG. 23. Temperature dependence of the experimental and MD- predicted LJ argon thermal conductivities.

It is in contrast to the Peierls theory of thermal conductivity, which is consistent with a single- stage decay of the HCACF [17,46]. Kaburaki *et al.* [29] suggest that the two stages in the HCACF represent contributions from local dynamics and the dynamics of phonon transport, each having a time constant τ and strength A. The use of the term "local" is questionable, as in a crystal, there are no localized vibrational modes. Che *et al.* [77] associate the initial, fast decay of the HCACF with optical phonons, which cannot be the case here, as the unit cell is monatomic. Fitting the HCACF according to Eq. (53) captures the two-stage decay very well at all temperatures [27]. The only region where the exponential does not fit is at very short times, when Markovian processes are not yet present [77]. The fits of a single exponential function with time constant τ_1, according to

$$\frac{\langle \mathbf{S}(t) \cdot \mathbf{S}(0) \rangle}{3} = A_1 \exp(-t/\tau_1) \tag{55}$$

are reasonable at low and high temperatures, but poor at the intermediate temperatures [11].

For a series of five sets of five simulations at a temperature of 50 K for the fcc crystal, the thermal conductivities calculated with Eq. (54) fall within a range of 6.0% of their average value. This error is expected to increase as the temperature decreases, and longer correlation times are required for convergence. The thermal conductivity predicted by the fit of a single exponential function [Eq. (55)] to the fcc crystal HCACF agrees with the prediction of the fit of the sum of two exponentials [Eq. (53)] to within 2.2% and 12.2% at the temperature extremes of 10 and 80 K. At the intermediate temperatures, the difference between the two predictions is as much as 31% (at a temperature of 50 K). This confirms the importance of considering the two-stage decay. The success of the single exponential function at low temperatures is due to the dominance of $k_{ac,lg}$. At high temperatures, the single exponential succeeds because the two time constants have similar values [11].

For the fcc crystal, the direct specification of the integral and the FD method agree to within 3.6% at all temperatures. The thermal conductivities calculated by the fit of Eq. (53) to the HCACF for all temperatures agree with the direct specification and FD predictions to within 2.0%. The fcc crystal MD results are in reasonable agreement with the trend and magnitude of the experimental data (a decrease above the experimental peak value, which is near a temperature of 6 K), justifying neglecting quantum effects. The data are in good agreement with those of Li [17]. Tretiakov and Scandolo [30] have reported LJ argon fcc crystal thermal conductivity results that are in better agreement with the experimental data. This may be a result of their simulation setup, which resulted in higher densities than those found here [90].

Schelling *et al.* [61] have found a long tail in the HCACF of silicon at a temperature of 1000 K that is not well described by an exponential function. They find that a single exponential fit to the HCACF gives a thermal conductivity value 42% lower than that found from the direct specification of the integral. The thermal conductivities studied are around 50 W/m K, so that based on the predicted order of $k_{ac,sh}$, considering two exponentials would not affect the results. They indicate that such exponential fits are unsuitable. Evidence to support the two-stage exponential fit is found in simulations of LJ argon [27], a series of silica structures [34], silicon nitride [76], and carbon structures [77,78]. In the work of Che *et al.* [77], the thermal conductivity of diamond is predicted to be 1220 W/m K, a very large value. Thus, the failure of the exponential fit, as found for silicon, is not solely related to a large value of the thermal conductivity.

The two-stage behavior of the fcc crystal HCACF, and the resulting decomposition of the thermal conductivity into two distinct components, has been interpreted in the context of the mean phonon relaxation time [27]. While the relaxation time is generally taken to be an averaged quantity (over all phonons in a system or over those of a given phonon mode), it can be applied to an individual phonon. For a given phonon mode, there will thus be some continuous distribution of relaxation times. Physically, the lower bound on the relaxation time corresponds to a phonon with a mean free path equal to one half of its wavelength. This is the CP limit, a thermal conductivity model developed for amorphous materials [32,33]. The first part of the thermal conductivity decomposition ($k_{ac,sh}$) takes into account those phonons with this limiting value of the relaxation time. Phonons with longer relaxation times are accounted for by the second term ($k_{ac,lg}$), which has a longer decay time.

The liquid HCACF shows a single-stage decay, with a timescale comparable to that of the initial drop in the fcc crystal HCACF. Both the FD and EF methods are suitable for specifying the thermal conductivity. The fit of a single exponential to the liquid HCACF predicts a thermal conductivity that agrees with that predicted by the direct specification and FD methods to within 5.7% for all temperatures. The experimental liquid data correspond to saturation conditions, and agree reasonably well with the MD predictions.

The amorphous phase HCACF shows a very different behavior. It drops below zero in the initial decay, and oscillates between positive and negative as it converges to zero. The velocity autocorrelation function for amorphous LJ argon shows a similar form [91]. This behavior can be interpreted as follows. In the fcc crystal, each atom experiences the same local environment. By averaging over time, the same is true for the liquid. This is not the case for the amorphous solid, where each atom has a distinct local environment. At short timescales, atoms near their equilibrium positions experience the free trajectory of a liquid atom. When the atom eventually feels the effects of the other atoms, the trajectory changes. Because the intended trajectory cannot be completed, the correlation goes negative. The timescale for this behavior is comparable to that of the liquid HCACF. The FD and EF methods are not appropriate here, and the thermal conductivity must be found from a direct specification of the integral. The amorphous phase thermal conductivity results are independent of temperature. This is attributed to the small temperature range studied, the approximately constant specific heat in the classical MD simulations, and the attainment of the CP limit (i.e., the mean free path is a minimum, and equal to one half of the wavelength of a given mode). Two additional amorphous phases with 250 atoms gave thermal conductivities of 0.170 and 0.166 W/m K at a temperature of 10 K (the plotted value is 0.170 W/m K), indicating that the phases considered are truly disordered. Amorphous phases with 500 and 1000 atoms both gave thermal conductivities of 0.165 W/m K, indicating that the periodic boundary conditions do not introduce undesired effects.

2. *Thermal Conductivity Decomposition*

In Fig. 24, the decomposition of the fcc crystal thermal conductivity into $k_{ac,sh}$ and $k_{ac,lg}$ is shown along with the classical limit of the CP limit, k_{CP},

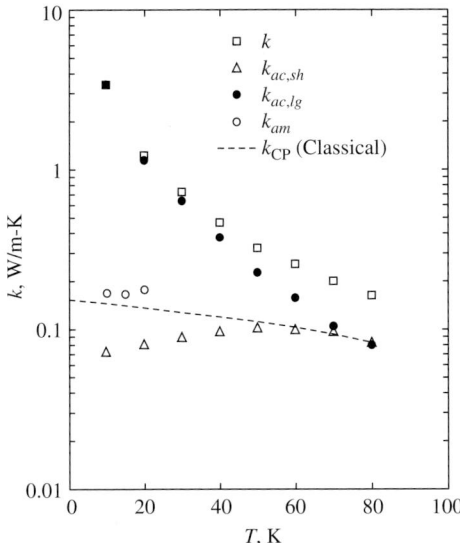

FIG. 24. Decomposition of the LJ fcc crystal thermal conductivity as described by Eq. (54).

and k_{am}. The idea of the crystal-phase thermal conductivity being made up of temperature-dependent and independent components has been explored previously in attempts to interpret experimental data [92,93]. In the decomposition given by Eq. (54), all of the temperature dependence of the thermal conductivity is contained in $k_{ac,lg}$, while the short-range component, $k_{ac,sh}$, shows little temperature dependence. We believe this behavior to be a result of the coordination of the atoms remaining constant as the density changes; it has also been found in silica structures [34] and carbon structures [77]. It has been proposed as a thermal conductivity limit [34].

Based on the association of $k_{ac,sh}$ with k_{CP} it might be expected that $k_{ac,sh}$, k_{am}, and k_{CP} would be the same. As shown in Fig. 24, while an exact equality is not observed, this statement is reasonable. The amorphous phase thermal conductivities are quite close to the CP limit. The values of $k_{ac,sh}$ at the higher temperatures agree will with k_{CP}. The disagreement at low temperatures may be a result of the fitting procedure for $k_{ac,sh}$. In the liquid phase (see Fig. 23), the thermal conductivity drops below the fcc crystal value to near $k_{ac,sh}$. Once the solid phase has been eliminated, only short-range interactions are important. The lack of fixed atomic positions in the liquid leads to an improved efficiency of these interactions in the transfer of heat, a shorter time constant, and a slightly higher thermal conductivity than $k_{ac,sh}$.

In Fig. 25, $\tau_{ac,sh}$ is plotted along with τ_{LJ}, τ_D, τ_{nn} (see Section III.C.1 for a description of these quantities), and τ_l for the liquid phase. Between

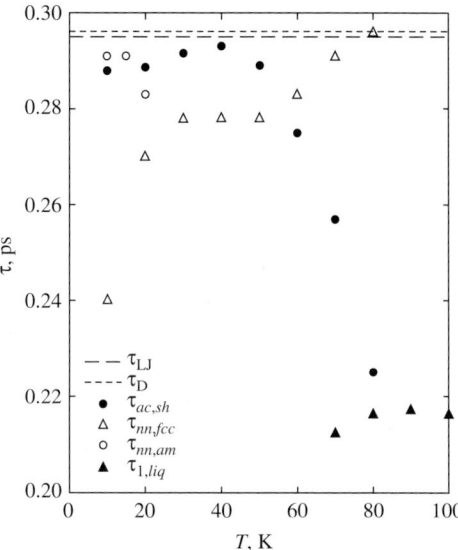

FIG. 25. Comparison of the short-time constants extracted from the MD simulations and from analytical calculations.

temperatures 20 and 60 K, there is an agreement to within 7% between $\tau_{ac,sh}$ and τ_{nn}. Thus, the first timescale in the HCACF decomposition is related to how long it takes for energy to move between nearest-neighbor atoms. No simulation cell size effects on the magnitude of $\tau_{ac,sh}$ or $k_{ac,sh}$ in the LJ argon system have been found, suggesting that the associated phonons are in the higher frequency range of the acoustic branches (i.e., those with wavelengths on the order of a few atomic spacings). From a real space perspective, one can imagine the movement of a phonon through a system as a series of energy transfers between neighboring atoms. For a phonon with a mean free path on the order of its wavelength (as assumed for $k_{ac,sh}$), this will correspond to a few $\tau_{ac,sh}$, which explains why this is the timescale found in the decomposition. Thus, this component of the thermal conductivity is strongly a function of the coordination of the atoms.

3. *Multi-atom Unit Cell Decomposition*

In the LJ fcc argon crystal the HCACF decays monotonically (see Fig. 22). Small oscillations can be attributed to the periodic boundary conditions. In other materials, such as β-silicon carbide [60] and diamond [77], larger oscillations are present, but their magnitudes are small compared to the total value of the HCACF. In such cases the thermal conductivity can be specified using different approaches, as discussed in Section V.B.1.

The HCACFs of some materials, such as silica structures [34], germanium structures [83], and carbon structures [77] do not decay monotonically. As shown in Figs. 26(a) and (b), for quartz(a) and quartz(c) at temperatures of 250 and 200 K, there are large oscillations in the HCACF. Similar oscillations have been attributed to the relative motion of bonded atoms with different masses [77]. Such behavior, however, has also been observed in an all germanium clathrate structure [83]. This suggests the more general explanation that the oscillations are a result of optical phonons. Lindan and Gillan [57] attribute the oscillations specifically to transverse optical phonons in CaF_2 and UO_2 by comparing the one observed oscillation frequency to phonon dispersion curves. In many cases more than one mode of oscillation is present [34], so their statement should not be generalized. The FD and EF methods are not suitable for determining the thermal

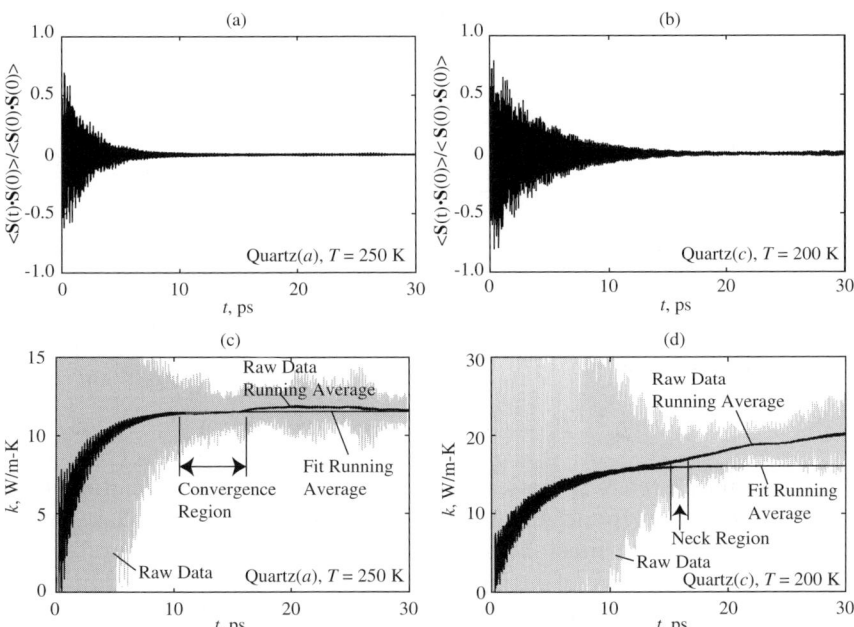

FIG. 26. Time dependence of the HCACF [(a) and (b)] and its integral [(c) and (d)] (whose converged value is proportional to the thermal conductivity) for quartz(a) at $T = 250$ K and quartz(c) at $T = 200$ K. From McGaughey and Kavinany [34]. Copyright (2004), with permission from Elsevier.

conductivity. Even the direct specification of the integral in Eq. (49) is not trivial. Noise in the HCACF can result in no obvious convergence region.

The following scheme has been proposed for the direct specification of the thermal conductivity [34] and is described here with respect to simulations of silica structures. First, the integral is averaged in overlapping blocks of 2500 time steps. The resulting curves related to Figs. 26(a) and (b) are shown in Figs. 26(c) and (d), along with the raw data. When the convergence is clear [Fig. 26(c)], a region of at least 5000 time steps is chosen over which the integral is deemed to have a constant value. The integral is averaged over this region, and this value is used to determine the thermal conductivity. When the convergence is not clear [Fig. 26(d)], it has been observed that the oscillations reach a minimum (i.e., a neck) before the divergence begins. Through comparison to the cases where the integral clearly converges, it is found that the HCACF function beyond this point does not make a significant contribution to the integral. An average of the integral is taken over 1000 time steps around the neck. This is the value used in the specification of the thermal conductivity.

The thermal conductivity decomposition can be applied in such a case by fitting the HCACF to a function of the form

$$\frac{\langle \mathbf{S}(t) \cdot \mathbf{S}(0) \rangle}{3} = A_{ac,sh} \exp(-t/\tau_{ac,sh}) + A_{ac,lg} \exp(-t/\tau_{ac,lg})$$
$$+ \sum_i B_{op,i} \exp(-t/\tau_{op,i}) \cos(\omega_{op,i} t) \quad (56)$$

so that, from Eq. (49),

$$k = \frac{1}{k_B V T^2} \left(A_{ac,sh} \tau_{ac,sh} + A_{ac,lg} \tau_{ac,lg} + \sum_i \frac{B_{op,i} \tau_{op,i}}{1 + \tau_{op,i}^2 \omega_{op,i}^2} \right)$$
$$\equiv k_{ac,sh} + k_{ac,lg} + k_{op} \quad (57)$$

The subscript op refers to optical phonon modes, which have a strength B. This procedure was found unsuitable for an amorphous silica phase, whose thermal conductivity must be specified directly from the integral of the HCACF [34]. The task of fitting the silica HCACFs to a function of the form of Eq. (56) is not trivial. A general procedure has been outlined [34], which is based on taking the Fourier transform of the HCACF, and fitting the optical modes in the frequency space, where the peaks are found to be well defined. Domingues et al. [94] used a similar approach to show that the thermal transport in β-cristobalite is a result of rigid unit modes.

4. Convective and Conductive Contributions

Defining, for a pair potential,

$$\mathbf{S}_k = \frac{1}{2} \sum_{i,j} (\mathbf{F}_{ij} \cdot \mathbf{v}_i) \mathbf{r}_{ij} \qquad (58)$$

$$\mathbf{S}_u = \sum_i e_i \mathbf{v}_i \qquad (59)$$

where $\mathbf{S}_k$ and $\mathbf{S}_u$ are the conductive and convective contributions to the heat current, the autocorrelation in Eq. (49) become

$$\langle \mathbf{S}(t) \cdot \mathbf{S}(0) \rangle = \langle \mathbf{S}_k(t) \cdot \mathbf{S}_k(0) \rangle + 2 \langle \mathbf{S}_k(t) \cdot \mathbf{S}_u(0) \rangle + \langle \mathbf{S}_u(t) \cdot \mathbf{S}_u(0) \rangle \qquad (60)$$

where there are individual terms for conduction and convection, and a cross term. Some investigations of solids have neglected the convection term in Eq. (51) [65,71]. A breakdown of the LJ fcc HCACF at a temperature of 50 K into the three components in Eq. (60) is shown in Fig. 27. While the convection contribution is indeed small, the cross term is not insignificant. It also displays a peculiar shape that warrants further study. The relative contributions of the conductive, convective, and cross terms to the thermal conductivity are 0.892,

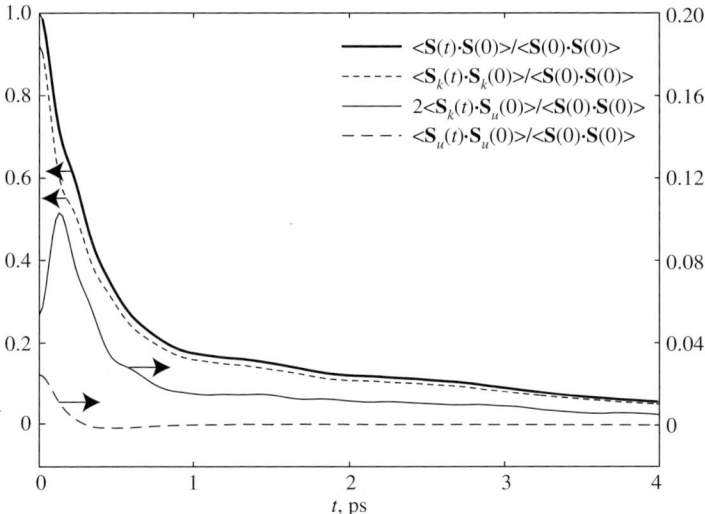

FIG. 27. Breakdown of the HCACF of the LJ fcc argon crystal at a temperature of 50 K into contributions from the conduction and convection parts of the heat current. Note that the total HCACF and the conduction component correspond to the left vertical axis, while the cross and convection curves correspond to the right axis.

0.005, and 0.103. The relative contributions of the convective and cross terms increase as the temperature goes up and anharmonic effects inhibit "pure" conduction. The absolute values of the conduction and cross term thermal conductivities both decrease with increasing temperature, while the convection contribution remains the same. While the convective term may not be as important for materials with higher thermal conductivity, in general, this contribution (and that of the cross term) should be checked before assuming they are negligible. This is particularly true at higher temperatures. Vogelsang et al. [62] investigated a similar breakdown for the LJ liquid. To identify heat transfer mechanisms in clathrate hydrates, Inoue et al. [87] have considered the contributions to the thermal conductivity from the conduction and convection terms, plus a torsional term that appears because of the rigid nature of their water molecule model. They did not consider any cross terms (three in this case), but the analysis allowed them to isolate the effect of the caged species on the thermal transport. A different breakdown of the HCACF has been performed by Keblinski et al. [88], who considered heat transfer in a single LJ nanoparticle suspended in an LJ fluid. By considering the liquid and solid HCACFs separately, they identified backscattering of phonons at the liquid–vapor interface.

One should not confuse neglecting the convection part of the heat current with an alternative definition of the heat current where the average atomic positions are used [46]:

$$\mathbf{S} = \frac{1}{2} \sum_{i,j} (\mathbf{F}_{ij} \cdot \mathbf{v}_i) \mathbf{r}_{o,ij} \qquad (61)$$

As discussed by Li [17], using Eq. (61) should lead to the same thermal conductivity as Eq. (51) for a solid, as there is not a unique way to define the heat current. Equation (61) will not be valid in a liquid where there can be net motion of particles. Ladd et al. [46] define two other heat currents based on the harmonic approximation: one in real space and the other in phonon space. Li [17] presents a thorough investigation and discussion of all four forms of the heat current. From a standpoint of extracting as much information as possible from the simulations, Eq. (51) is most useful.

5. Spectral Methods

Two different spectral methods exist for predicting the thermal conductivity using the GK method. As first described by Lee et al. [69], and later derived in detail by Li and Yip [74], the thermal conductivity can be related to the power spectrum of the Fourier transform of the heat current, $S(\omega)$, by

$$k = \frac{1}{2t_{sim}} \frac{1}{k_B T V^2} \lim_{\omega \to 0} |S(\omega)|^2 \qquad (62)$$

where t_{sim} is the length of the simulation. This method is useful because it reduces the computational load required to perform the autocorrelation, which may be substantial in materials with high thermal conductivities and long correlation times. The disadvantage of this approach is that the limit in Eq. (62) may not be trivial to evaluate. The smallest frequency that is physically meaningful is determined by the size of the simulation cell (i.e., the longest allowed wavelength, see Section III.D.2). Data at lower frequencies will come out in the Fourier transform due to long simulation times, but are meaningless. This is evident in the results of Lee et al. [69], shown in Fig. 28. In order to find the thermal conductivity, they perform an extrapolation of the Fourier transform power spectrum curve. The error associated with such an extrapolation is likely on the same order as the error introduced by different methods for evaluating the integral of the HCACF. Li and Yip [74] have used this approach with β-silicon carbide. They obtained reasonable values for the thermal conductivity (when compared to time integration). The computational time required was an order of magnitude less than that required to generate the HCACF.

The second approach to predicting the thermal conductivity with a spectral method is to take the Fourier transform of the HCACF,

$$\int_0^\infty \langle \mathbf{S}(t) \cdot \mathbf{S}(0) \rangle \exp(i\omega t) dt \qquad (63)$$

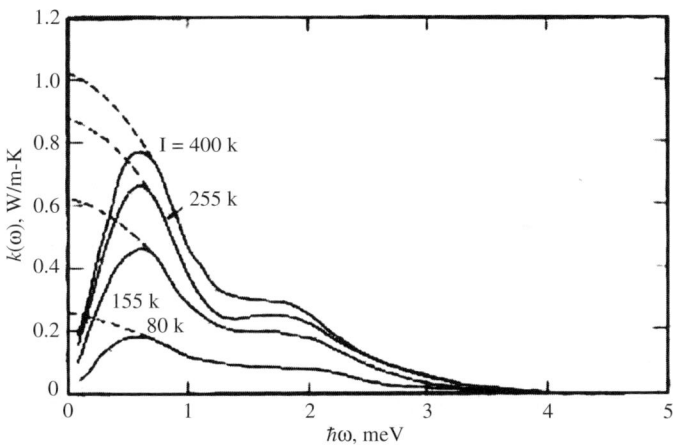

FIG. 28. Frequency-dependent thermal conductivity of amorphous silicon. The dotted lines show the extrapolation required to obtain the static thermal conductivity. From Lee et al. [69]. Copyright (1991), with permission from American Physical Society.

and then define the frequency-dependent thermal conductivity as

$$k(\omega) = \frac{1}{3k_B V T^2} \int_0^\infty \langle \mathbf{S}(t) \cdot \mathbf{S}(0) \rangle \exp(i\omega t) dt \qquad (64)$$

The static thermal conductivity, that normally predicted and reported, is the zero-frequency limit of Eq. (64). In taking this limit, similar issues as with the first approach exist, in that the data will be unreliable below the minimum frequency allowed in the simulation cell. Extrapolation techniques are required. By assuming that the HCACF follows a single exponential decay as given by Eq. (55), the frequency-dependent thermal conductivity is found to be

$$k(\omega) = \frac{A_1 \tau_1}{3k_B V T^2} \frac{\tau_1}{1 + i\tau_1 \omega} \qquad (65)$$

The imaginary part can be interpreted as the lag between the heat flux and the temperature gradient in a system excited at frequency ω.

For analysis purposes, the frequency-dependent thermal conductivity is often presented as the modulus of Eq. (65) so that

$$k(\omega) = \frac{A_1 \tau_1}{3k_B V T^2} \frac{\tau_1}{(1 + \tau_1^2 \omega^2)^{1/2}} \qquad (66)$$

This function can be fit to the power spectrum of the HCACF Fourier transform. Two exponentials can also be used [76]. Only the data beyond the minimum system frequency should be used in the fit. One could also fit the real part of Eq. (65) to the real part of the Fourier transform (the imaginary part goes to zero at zero frequency). Not surprisingly, this procedure has been found to give good agreement with time integration results when the HCACF is well modeled with the sum of two exponentials [76]. When a long tail is present in the HCACF, and an exponential fit is inappropriate, similar disagreement is found at the low frequencies in the frequency space fit [61]. This approach does not appear to present any significant advantages over time integration for HCACFs that decay smoothly.

6. *Nonequilibrium Green-Kubo*

To reduce the required computational time, a nonequilibrium technique that uses the GK expression for the thermal conductivity has been developed [95–98]. A fictitious field, $\mathbf{F}(t)$, that induces an energy flux while not creating a temperature gradient is applied to the MD system. This allows for the use of periodic boundary conditions in a reasonably sized system, and is accomplished by modifying the equations to motion to

$$\frac{d\mathbf{r}_i}{dt} = \frac{\mathbf{p}_i}{m_i} \qquad (67)$$

$$\frac{d\mathbf{p}_i}{dt} = \mathbf{F}_i + (E_i - \bar{E})\mathbf{F}(t) - \frac{1}{2}\sum_j \mathbf{F}_{ij}[\mathbf{r}_{ij} \cdot \mathbf{F}(t)] + \frac{1}{2N}\sum_{j,k} \mathbf{F}_{jk}[\mathbf{r}_{jk} \cdot \mathbf{F}(t)] \qquad (68)$$

where $\bar{E}$ is the instantaneous average particle energy. If desired, a thermostat can be included. The second term on the right-hand side of the momentum equation shows how the heat current is driven. A particle with an energy greater than the average will move with the applied field, while a particle with an energy less than the average value will move against the field. The last term on the right-hand side of the momentum equation ensures that momentum is conserved. These equations are valid for a system that contains a single mass species. More complicated expressions are required for multi-component systems [99].

By assuming that the applied field is constant and in a convenient direction [i.e., $\mathbf{F}(t) = (F_x, 0, 0)$] linear response theory and the GK expression for the thermal conductivity, Eq. (49), can be used to show that the thermal conductivity can be given by

$$k = \lim_{t \to \infty} \lim_{F_x \to 0} \frac{\langle S_x(t) \rangle}{F_x T V} \qquad (69)$$

This technique has been used in simulations of a one-dimensional anharmonic lattice [100], clathrate hydrates [87], carbon nanotubes [79,82], LJ argon [64,98], uranium dioxide [64], and zeolites [75].

In the evaluation of the time limit of Eq. (69), Berber et al. [79] find good convergence for carbon nanotubes, while the results of Motoyama et al. [64] show large fluctuations about the mean value. At the triple point of argon, Evans [98] finds a linear relation between the thermal conductivity and the applied field, making the evaluation of the F_x limit in Eq. (69) straightforward. Motoyama et al. also observe this behavior. Berber et al., on the other hand, find that the thermal conductivity of carbon nanotubes increases as the applied field is decreased, and have a more difficult extrapolation to find the thermal conductivity. This may explain why their predictions are significantly higher than those from either the GK or direct methods [77,101]. Inoue et al. [87] and Motayama et al. do not find good agreement between the nonequilibrium GK predictions are those from the equilibrium GK implementation. The reasons for the discrepancies are unclear. This result suggests that some caution should be exercised when using the nonequilibrium method.

C. QUANTUM CORRECTIONS

Before moving on to a discussion of the direct method for predicting the thermal conductivity, we first address the use of quantum corrections in the classical MD simulations. From the standpoint of lattice dynamics, there are two significant points to consider. First, as described in Section VI, the energy of the phonon modes is quantized in units of $\hbar\omega_k$. This is not true of the classical system, where the mode energies are continuous. The second point, and the focus of this section, is the temperature dependence of the mode excitations. As predicted by the Bose–Einstein distribution, there are significant temperature effects in the quantum system that are not present in a classical description. The MD approach is thus not suitable near and below the maximum in the crystal phase thermal conductivity (observed experimentally around one-tenth of the Debye temperature [102], and for argon at a temperature of 6 K [103]), where quantum effects on the phonon mode populations are important. The thermal conductivity in this region is also strongly affected by impurities and boundary effects, which are not considered here. As such, an MD simulation of a perfect crystal with periodic boundary conditions will lead to an infinite thermal conductivity at zero temperature, as opposed to the experimental value, which goes to zero.

The classical nature of the MD simulations is perhaps most evident when considering the predicted specific heats (see Fig. 9), and how they differ from the quantum-mechanical calculations. The reason for the discrepancy is that in a classical-anharmonic system at a given temperature, all modes are excited approximately equally. The expectation value of the mode energy is about $k_B T$. In a harmonic system, the excitation is exactly the same for all modes, and the expectation value of the energy is exactly $k_B T$. In the quantum system, there is a freezing out of high- frequency modes at low temperatures. Only above the Debye temperature are all modes excited approximately equally. The quantum system also has a zero-point energy not found in the MD system.

There is no simple way to explicitly include quantum effects in the MD simulations. In fact, the whole idea behind the simulations is to save significant computational resources by ignoring quantum effects. That being said, some effort has been made to address the classical-quantum issue by mapping the results of MD simulations onto an equivalent quantum system. Using the results for the LJ fcc crystal, one of these approaches [60,69,71,104] is presented and assessed here. The main idea is to scale the temperature and thermal conductivity (after the simulations have been completed) using simple quantum-mechanical calculations and/or arguments. For the remainder of this section, T_{MD} and k_{MD} are used to represent the temperature and thermal conductivity of the MD system, and T_{real} and k_{real} are used to represent the values for the "real", quantum system.

The temperature in the MD system is calculated from the relation

$$\langle \sum_i \frac{1}{2} m_i |\mathbf{v}_i|^2 \rangle = \frac{3}{2}(N-1) k_B T_{MD} \tag{70}$$

which equates the average kinetic energy of the particles (summed over the index i) to the expectation value of the kinetic energy of a classical system [see Eq. (11)]. For a harmonic system, where equipartition of energy exists between the kinetic and potential energies, and between the modes, the total system energy will be given by $3(N-1) k_B T_{MD}$. The temperature of the real system is found by equating this energy to that of a quantum phonon system, such that [60,69]

$$3(N-1) k_B T_{MD} = \sum_k \hbar \omega_k \left[\frac{1}{2} + \frac{1}{\exp(\hbar \omega_k / k_B T_{real}) - 1} \right] \tag{71}$$

where the summation is over the k normal modes of the system. A similar relation has been proposed without the zero-point energy included [i.e., the factor of $\hbar \omega_k / 2$ on the right-hand side of Eq. (71) is not considered] [71,104].

For the thermal conductivity, it has been proposed [60,69,71] that the heat flux, $\mathbf{q}$, in the classical and quantum systems should be the same. Written in one dimension,

$$q = -k_{MD} \frac{dT_{MD}}{dx} = -k_{real} \frac{dT_{real}}{dx} \tag{72}$$

such that

$$k_{real} = k_{MD} \frac{dT_{MD}}{dT_{real}} \tag{73}$$

The predicted T_{MD} and dT_{MD}/dT_{real} curves for the cases of both including and neglecting the zero-point energy are shown in Fig. 29(a). The data are plotted up to a temperature of 87 K, the melting point of the MD system.

When the zero-point energy is included, the MD simulations are only of interest at temperatures of 31.8 K and higher. This is an indication of the magnitude of the zero-point energy. The T_{MD} curve approaches T_{real} as the temperature is increased and more modes are excited in the quantum

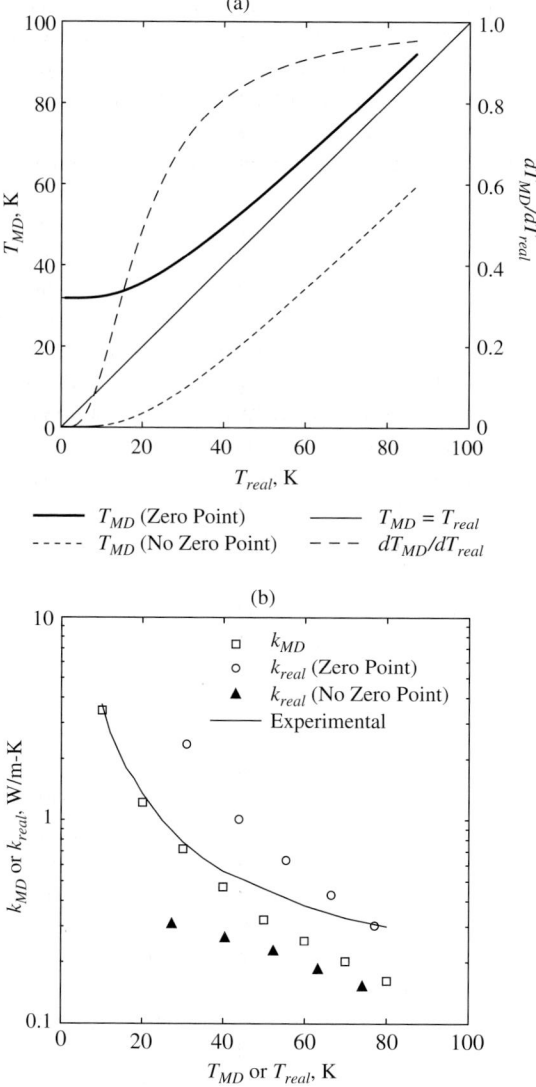

FIG. 29. (a) Temperature and thermal conductivity scaling relations. (b) Scaled thermal conductivities with raw MD predictions and experimental data.

system. The value of T_{MD} will always be higher than T_{real} because of the zero-point energy. The thermal conductivity scaling factor starts at zero. This ensures that the thermal conductivity will be zero at zero temperature. As the temperature increases, the scaling factor approaches unity.

When the zero-point energy is not included, the temperature scaling has the same shape as before, but has been shifted downward. In this case, T_{MD} will always be lower than T_{real} because of the manner in which the energy distributes in the modes of the quantum system. The T_{MD} value of zero is relevant in this case, because the associated quantum system can have zero energy. The thermal conductivity scaling factor is identical to when the zero-point energy is included, as the energies only differ by a constant.

The scaled thermal conductivities and unscaled MD predictions are shown in Fig. 29(b) along with the experimental data. To obtain these results, the T_{real} values corresponding to the available T_{MD} values are obtained. Owing to the nature of the scaling relation, not all the T_{MD} values have a corresponding T_{real}. The appropriate thermal conductivity scaling factor is then determined. Overall, the agreement with the experimental data worsens for either of the scaling possibilities compared to the raw MD data. Others have found an improved agreement (Li et al. [60] for β-silicon carbide including the zero-point energy, and Volz and Chen [71] for silicon, not including the zero-point energy). This lack of consistency raises a high level of doubt about the validity of this approach, and its possible widespread acceptance.

The main idea behind this somewhat ad hoc temperature scaling procedure is to map the classical MD results onto an equivalent quantum system. By not including the zero-point energy, a true quantum system is not being considered. For this reason, if such corrections are to be used, the zero-point energy should be included.

As it stands, there are a number of ways by which this method could be improved. These are related to the harmonic nature of the energy calculations on both sides of Eq. (71). The classical energy is based on an assumption of equipartition of energy. The average, total energy of the MD system is in fact less than $3(N-1)k_B T_{MD}$ due to anharmonic effects (see Fig. 17). As shown there, at a temperature of 20 K, the deviation is 2.6%, and increases to 12.6% at a temperature of 80 K. This correction is straightforward to implement. The phonon space energy is most easily calculated using the zero temperature, harmonic dispersion relation (see Section III.D.3). As shown in Section III.D.5, temperature has a significant effect on the phonon dispersion, and temperature-dependent normal modes would make the temperature scaling more rigorous.

That being said, it is unlikely that these modifications would lead to a much improved model. The main drawback of this temperature/thermal conductivity scaling approach, as discussed by Li [17], is that it is a post-processing step that maps the entire MD system onto a quantum description. The effects are in practice manifested on a mode by mode basis, and how the energy is distributed, therefore making corrections on an integrated level simply not suitable. Li goes on to suggest a way by which the classical MD system can be linked to a quantum description through the BTE.

Che et al. [77] have taken a more general approach to investigate the classical-quantum issue by comparing the general forms of the classical and quantum HCACFs. They do not find evidence to support the use of quantum corrections with MD thermal conductivity predictions. They argue that this is because long wavelength phonons are the dominant contributors to the thermal transport, which are active even at low temperatures. This is in contrast to the specific heat, where it is the high-frequency (short wavelength) modes that get excited as the temperature of the quantum system is increased, and lead to the significant temperature dependence up to the Debye temperature.

As discussed, the MD simulations are classical because it is within this framework that computational costs become reasonable enough to perform simulations of big systems, or for long times. When comparing the results of MD simulations to experiments, there are additional factors beyond their classical nature that need to be considered. These include the interatomic potential used, size effects, and the simulation procedures. It is difficult to isolate these effects. Efforts are needed on all fronts to increase the confidence in the results of MD simulations.

VI. Thermal Conductivity Prediction: Direct Method

A. INTRODUCTION

The direct method is a nonequilibrium, steady-state approach to predicting the thermal conductivity. By imposing a one-dimensional temperature gradient on a simulation cell and measuring the resulting heat flux (or, imposing a heat flux and measuring the resulting temperature gradient), one can predict the thermal conductivity using the Fourier law, Eq. (1). The direct method has a physical intuitiveness not present in the GK approach, and is analogous to an experimental technique. While experimentally it is easiest to apply a known heat flux and let the temperature distribution develop, in MD simulations different setups have been investigated, and are discussed in this section.

In all implementations of the direct method, energy flows from a hot region (the hot slab) to a cold region (the cold slab) of the simulation cell. Between the hot and cold slabs is the sample region, where a temperature gradient develops. This is realized by modifying the equations of motion in the hot and cold slabs. How this is done, and how the boundaries/interfaces in the system are treated has been extensively studied, and is reviewed here.

The modifications to the equations of motion must conserve momentum, or the system will drift. Energy conservation is not critical. The boundary modeling should be as physical as possible (i.e., there should be no effect in bulk phase predictions, but an effect should be present in a finite structure, such as a thin film). The goal is to obtain a result as close as possible to that which would be measured experimentally.

In the GK method, the MD system runs in the NVE ensemble and the thermal conductivity is predicted in a post-processing calculation. In the direct method, the thermal conductivity is predicted directly from the simulations, and the simulator has a greater impact on the value obtained. One must make decisions regarding simulation parameters. Consider the case of applying a temperature difference across the sample region by fixing the temperatures of the hot and cold slabs. Thermal conductivity is a temperature-dependent quantity; if a large temperature difference is present, there will be a gradient in the thermal properties across the sample region. It is not obvious if the predicted thermal conductivity will correspond to that for the average temperature in the system. A large temperature difference may also result in a nonlinear response, making the Fourier law, which requires a linear temperature profile, invalid. Given the large temperature gradients and heat fluxes manifested in such simulations (of order 10^9 K/m and 10^{12} W/m^2), it is somewhat surprising that the response does in fact turn out to be generally linear. Conversely, if too small a temperature difference is present, it may be difficult to get a good prediction of the thermal conductivity because of fluctuations inherent to the system.

Consider Fig. 30(a), where temperature profiles obtained by Muller-Plathe [105] for an LJ fluid are shown. For the smallest temperature difference considered, shown in Fig. 30(b), the uncertainty at each point is larger than the total temperature change. As the temperature difference increases, the magnitude of these errors should not be significantly affected, which will allow for a more accurate prediction. In Fig. 31, the thermal conductivity of t-ZrO$_2$ at a temperature of 1450 K as a function of the applied temperature difference is shown [106]. There is a clear effect. One should ideally create such a plot for all cases considered, and perform an extrapolation to zero-temperature difference. At the same time, this will significantly increase the required computational time. There are other corrections than can be made in the direct method to get the best possible value for the thermal conductivity. These include accounting for size effects, as will be discussed in Section VI.C. In the end, there is always a compromise to be made between accuracy and reasonable computation times. The works of Schelling *et al.* [61], Lukes *et al.* [107] and Chantrenne and Barret [108] address many of the important methodological issues associated with the direct method.

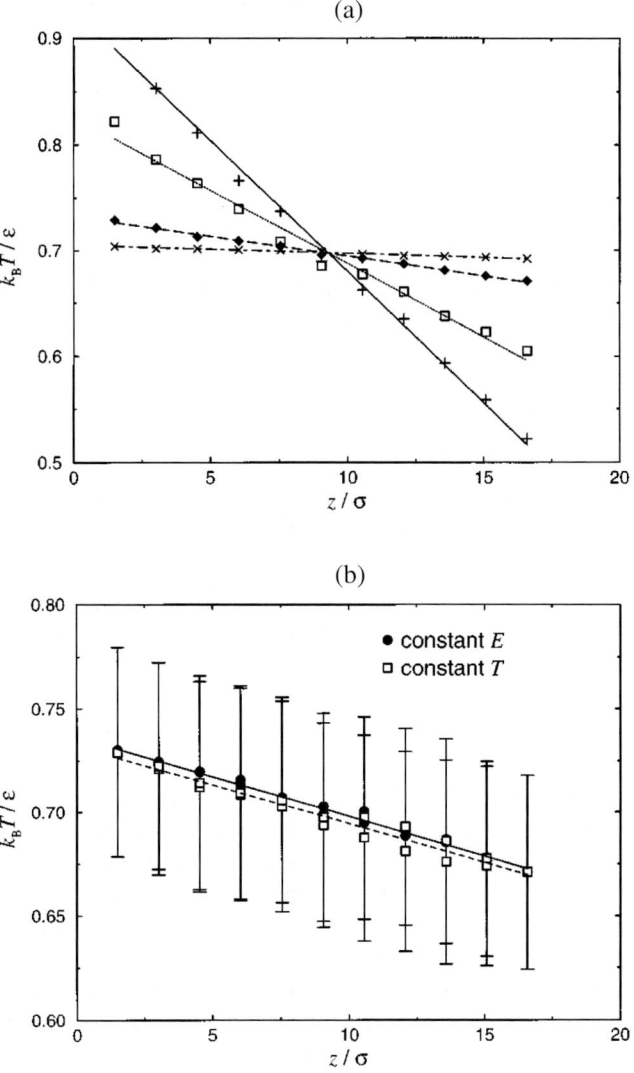

FIG. 30. (a) Temperature profiles in an LJ fluid. The profiles remain reasonably linear at large overall temperature differences. (b) Uncertainty in the temperatures for a small temperature difference. The uncertainty is larger than the overall temperature difference. The results show that applying either NVE or NVT statistics to the sample region does not affect the results. From Muller-Plather [105]. Copyright (1997), with permission from American Institute of Physics.

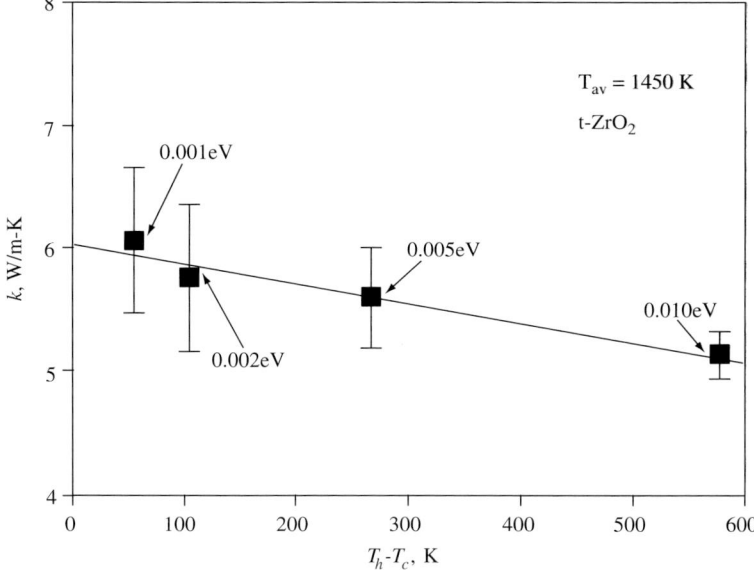

FIG. 31. Effect of the applied temperature difference on the thermal conductivity of t-ZrO_2. The average temperature is 1450 K. From Schelling and Phillpot [106]. Copyright (2001), with permission from Blackwell.

B. IMPLEMENTATION

One can classify implementations of the direct method into two types: applied temperature gradient or applied heat flux. Within each of these, two different simulation cell setups can be used. In one case, shown in Fig. 32(a), a system of finite length in the direction of the heat flow is used, with periodic boundary conditions in the transverse directions. Such an approach is useful for finite-sized structures, such as thin films. Energy flows in one direction. Alternatively, periodic boundary conditions can be applied in the direction of the heat flow, as shown in Fig. 32(b). This is advantageous for the simulation of a bulk phase, as one does not need to consider how the atoms at the ends will be modeled. Here, there is energy flow from both sides of the hot slab.

Because of boundary effects, the temperature gradient in the sample region may not be the same as that suggested by the hot and cold slab temperatures. A temperature discontinuity or nonlinearity may exist at the interfaces with the sample region. An example of this is shown for silicon at a temperature of 500 K in Fig. 33 [61]. As such, the temperature gradient should be calculated from the temperature profile away from the slabs. Measurement of the

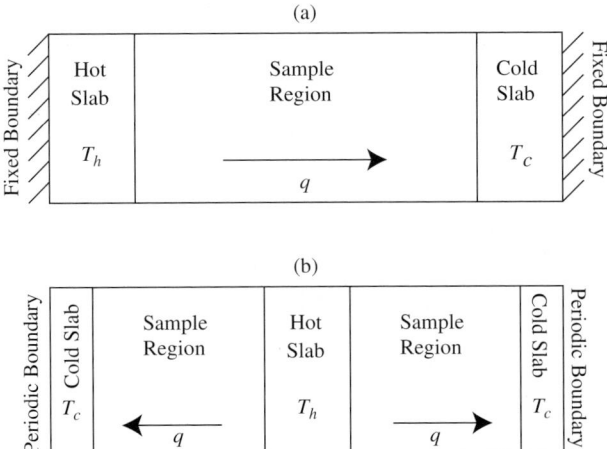

Fig. 32. Implementation of the direct method with (a) fixed end boundaries and (b) periodic boundary conditions.

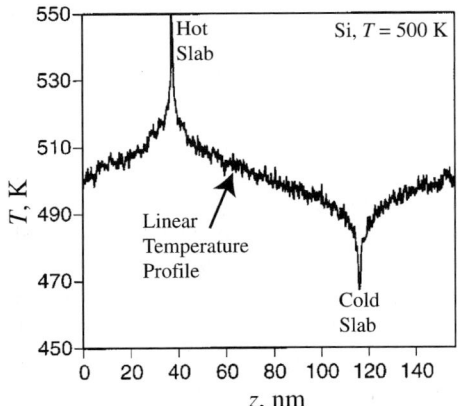

Fig. 33. Temperature profile in silicon at an average temperature of 500 K. The temperature profile around the hot and cold slabs is strongly nonlinear. The temperature gradient should be fit in the linear region. From Schelling et al. [61]. Copyright (2002), with permission from American Physical Society.

temperature along the sample region is thus required in both the temperature gradient and heat flux methods. The temperature gradient is found by applying a linear fit to the temperature data. The sample region is typically divided into thin slabs, each containing a few layers of atoms. Over such distances and in the imposed state of nonequilibrium, the validity of a local temperature may be questionable. It has been found, however, that provided a long enough time is considered (typically on the order of a few hundred thousand time steps) and the layer is wide enough (at least four unit cells) that reasonable results and convergence are found. Chantrenne and Barrat [108] find that the departure from equilibrium is in fact quite small. The phonon densities of states for an equilibrium system and one that is thermostated are found to be indistinguishable, and the velocity distributions throughout the simulation cell are found to follow Maxwell–Boltzmann statistics. In general, for a given length of simulation cell, one must make a choice between the width of the slabs (more atoms will give better statistics at a given point) and the number of slabs (more points in the temperature profile will give a better fit to the gradient).

1. *Applied Temperature Gradient*

Two general approaches have been used to apply a temperature gradient to an MD cell. In simulations of a two-dimensional LJ lattice, Mountain and MacDonald [109] scale the velocities in the hot and cold slabs to achieve the desired temperatures. This is equivalent to a thermostat that keeps the kinetic energy constant. Such an approach has also been used for simulations of a two-dimensional Penrose tiling [110], amorphous and crystalline selenium and silica [111], carbon nanotubes [101,112,113], and LJ materials [108]. Poetzsch and Bottger [114] impose a temperature difference by placing the hot and cold slabs in contact with hot and cold reservoirs using Nose–Hoover thermostats. They use this approach to study a two-dimensional elastic percolating system and the roles played by anharmonicity and disorder in the associated thermal transport. The results are in agreement with predictions from the Green–Kubo method. Imamura [115] uses this method to study GaAs/AlAs superlattices. This method is less intrusive than the direct temperature scaling approach, and will produce the temperature fluctuations associated with the NVT ensemble. These fluctuations may be detrimental to achieving convergence, however. A similar strategy has been developed by Tenenbaum *et al.* [116] in which fixed walls at constant temperature at the system boundaries interact with the nearby atoms. Maruyama's [117] phantom technique is of the same spirit, and was used to model carbon nanotubes [118]. In all of the above approaches, the heat flux is found by calculating the difference in kinetic energy before and after the adjustments of the velocities. The heat flux will not be constant at every time step, and needs to be averaged over time. Michalski [110] finds good agreement between the heat current calculated with the kinetic energy change and that from Eq. (49).

2. Applied Heat Flux

Studies have reported slow convergence times for the applied temperature difference method [107]. The alternative is to apply a known heat flux. In one approach, this is accomplished by adding/removing a constant amount of energy, $\Delta\varepsilon$ to/from the cold/hot slab at a regular interval. By taking this interval to be one time step, a constant heat flux will be achieved. Jund and Julien [119], in a study of amorphous silica, do this by computing the new velocities in the hot and cold regions, $\mathbf{v}_i'$, from

$$\mathbf{v}_i' = \mathbf{v}_G + \left(1 \pm \frac{\Delta\varepsilon}{E_G}\right)^{1/2} \qquad (74)$$

Here, $\mathbf{v}_G$ is the velocity of the center of mass of the hot/cold slab, and E_G the difference between the slab's actual kinetic energy and that which its center of mass would have if it were a single particle. This implementation ensures that momentum is conserved. It has been used in studies of zirconia and yttria-stabilized zirconia [106], silicon [61], quartz [120], and metals [121].

Alternatively, new velocities can be calculated from

$$\mathbf{v}_i' = R\mathbf{v}_i + \mathbf{v}_{sub} \qquad (75)$$

where R and v_{sub} are different for the hot and cold slabs, change at every time step, and are chosen to conserve momentum as described by Ikeshoji and Hafskjold [122]. Lukes et al. [107] used this approach to predict the thermal conductivity of thin LJ films and the corresponding bulk phase. A detailed analysis of the errors in the predicted temperature gradient, heat flux, and thermal conductivity are presented. They find that the applied heat flux does not always match that measured in the sample region due to leakage in the transverse directions. By running sufficiently long simulations, agreement between the two values is found to within 4%. Subsequent work using this approach in LJ systems has considered the effect of porosity [123], and two-component systems and nanostructures [124–126]. It has also been used to investigate the phonon contribution to the thermal conductivity of zirconium hydride [127]. Kotake and Wakuri [128] used a modified version of this approach to model a two-dimensional system where both the kinetic and potential energies are modified.

An alternative to the energy addition/removal has been used by Muller-Plathe [105]. Instead of specifying the amount of energy to be added or removed, the velocity vector of the coldest atom in the hot region and the hottest atom in the cold region are exchanged at a regular time interval. This approach conserves both energy and momentum, although it results in a variable heat flux, which must be averaged over time. It has been used to predict the thermal conductivity of fluids [105,129].

3. *Transient Thermal Diffusivity Approach*

Daly *et al.* [130,131] have developed a variation on the direct method where a sinusoidal temperature perturbation is applied to a long MD simulation cell. The decay of the perturbation is observed, and used with the energy equation to calculate the thermal diffusivity, from which the thermal conductivity can be calculated. Agreement between their predictions for the cross-plane thermal conductivity of a simplified model of a GaAs/AlAs superlattice and predictions from the temperature gradient method has been found by Imamura *et al.* [115].

C. SIZE EFFECTS

In the GK method, all phonon modes are nonlocalized, and size effects are most likely an effect of not having enough modes present to establish an accurate description of the scattering processes [11]. A few thousand atoms are typically enough to overcome such size effects. When the direct method is used to predict the bulk thermal conductivity, the predicted value is found to increase as the length of the sample in the direction of the heat flow increases. One finds size effects even when tens of thousands of atoms are considered, and in amorphous materials, where the mean free path is expected to be small. Size effects are a result of three possible factors. First, the finite size of the simulation cell may not allow phonon modes to naturally decay (i.e., the length is shorter than the mean free path). Second, long wavelength modes cannot exist. Third, there is a thermal boundary resistance at the interface between the hot and cold slabs and the sample region. Based on GK predictions, the first and third factors are likely responsible for observed size effects. To find the true bulk thermal conductivity, the results must be extrapolated to what would be present in a simulation cell of infinite length.

To motivate how this can be done, Schelling *et al.* [61] consider a simple description of the phonon scattering where the mean free path is calculated using the Matheson rule, taking into account inter-phonon interactions and boundary scattering such that

$$\frac{1}{\Lambda} = \frac{1}{\Lambda_\infty} + \frac{4}{L} \qquad (76)$$

Here, Λ is the mean free path in the finite system of length L, and Λ_∞ the mean free path in an infinite system. The factor of 4 in the numerator of the boundary scattering term is a result of their simulation cell setup. By taking the thermal conductivity to be given by Eq. (3), Eq. (76) can be recast as

$$\frac{1}{k} = \frac{3}{\rho C_v v}\left(\frac{1}{\Lambda_\infty} + \frac{4}{L}\right) \tag{77}$$

suggesting that one should plot $1/k$ vs. $1/L$ and extrapolate to $L = 0$. The slope of this curve will have units of W/K, and can be interpreted as the boundary thermal resistance [110,114], which should be constant for large unit cells. This procedure has been found to work quite well (the curve is often very close to linear), and has been used in numerous investigations [61,108,110,111,114,120,121]. Results for silicon and diamond from Schelling et al. [61] are shown in Fig. 34.

That being said, some investigations of bulk systems have ignored size effects, and the presented results should be viewed cautiously. In some cases the authors acknowledge the existence of size effects, and instead of using computational resources to get extrapolated thermal conductivity values for all conditions considered, instead focus on one size, and vary other parameters, such as the number of layers in a superlattice.

In simulations of amorphous silica, Jund and Jullien [119] do not take size effects into account as discussed above, but do correct their low-temperature results (between 8 and 20 K) to account for the low-frequency modes that are not present in the simulation cell, but are expected to play a significant role in the low-temperature transport. This results in a marked improvement in the agreement between their results and experimental data. At very low temperatures (<8 K) the quantum effects in the experimental data are not

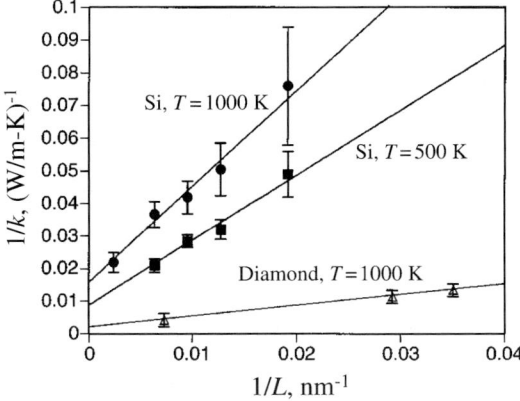

FIG. 34. System size dependence of thermal conductivity for silicon and diamond. From Schelling et al. [61]. Copyright (2002), with permission from American Physical Society.

reproducible, as would be expected. The temperature profiles they obtain at the low temperatures have strong nonlinearities, which seem to be a result of the dominance of boundary scattering. They use all of the temperature data in fitting the temperature gradient, however, which may have led to some of their results being overpredicted (see their Fig. 2). As discussed, one should neglect the temperatures near the boundaries when fitting the temperature profile. Oligschleger and Schon [111] find size effects for amorphous silica (although, interestingly, not for amorphous selenium), albeit with a different interatomic potential. The Jund and Jullien thermal conductivities are higher than those predicted with the GK method with the same potential by McGaughey and Kaviany [34] but in good agreement with experimental data. Considering size effects would increase their predicted thermal conductivities.

For an LJ system, increasing the lateral size beyond four conventional unit cells has not been found to have a significant effect on the predicted thermal conductivity [107]. Computational resources are better spent on longer simulation times to achieve better convergence. This result can be interpreted in the same way that the general lack of size effects in the GK method is explained; the system is bulk in the two transverse direction, and only a few modes are needed to establish representative scattering. In a finite structure, such as a nanowire, is to be simulated, this size will have a significant effect.

VII. Discussion

A. Notable work

Many GK studies have considered the effect of defects on thermal transport, including potassium chloride doped with rubidium [59], krypton impurities and vacancies in LJ argon [67], defects in diamond [77], and yttria-stabilized zirconia [106]. Li et al. [60] considered a range of point defects in β-SiC, and investigated the resulting density of states at both system and species levels. Ishii et al. [84] investigated isotope effects in germanium, and Murakawa et al. [73] performed a similar study with silicon. By choosing a sample with the typical experimental concentration of isotopes, Murakawa et al. find much better agreement with experimental thermal conductivity data than that found with pure Si^{28} [70].

In all cases, as is found experimentally, defects reduce the thermal conductivity and its temperature dependence. In the GK method, the change is observed in the form of the HCACF, which decays much faster than in the perfect crystal. A similar thermal conductivity reduction and temperature independence was found in zeolite crystals [34]. In such materials, the large unit cells localize energy, preventing long-range correlations from developing. This phenomenon leads to a small, near temperature-independent value of the thermal conductivity, which has been interpreted as a thermal conductivity limit. The temperature independence does not appear to be a result of the classical nature of the simulations as temperature-dependent amorphous phase thermal conductivities have been reported [34,69,111,119]. The increasing thermal conductivity of amorphous materials with increasing temperature can be attributed to the temperature dependence of the specific heat (a quantum effect). The mean free path is assumed constant. One cannot use this explanation to analyze results from MD simulations.

Simulations of amorphous materials have allowed for an investigation of the role of disorder and anharmonicity in phonon transport. Unlike in the crystal phase, where all modes are nonlocalized (diffusive), in an amorphous structure localized, diffusive, and propagating modes exist (locons, diffusions, and propagons). These can be identified in MD simulations and used to interpret the thermal transport [61,110,132]. In a crystal, increased anharmonicity at higher temperature increases phonon scattering, reducing the thermal conductivity. In an amorphous material, anharmonicity enhances the energy exchange between localized modes, resulting in an increase in the thermal conductivity.

Dong et al. [83] used MD and the GK method to address the connection between dispersion and thermal transport in the prediction of the thermal conductivity of germanium-based materials. In their Fig. 1, they show the dispersion curves for a diamond structure, a clathrate cage, and for the same cage structure filled with strontium atoms. While the range of frequencies accessed by the vibrational modes in these structures is comparable, the dispersion characteristics differ. The large unit cell of the clathrate cage significantly reduces the frequency range of the acoustic phonons, the carriers generally assumed to be most responsible for thermal transport. This results in an order of magnitude reduction in the thermal conductivity. In the filled cage, the guest atoms have a natural frequency that cuts directly through the middle of what would be the acoustic phonon branch. The thermal conductivity is reduced by a further factor of 10. McGaughey et al. [66], by analyzing the predicted thermal conductivities together with the frequency spectra of the HCACF and quasi-harmonic phonon dispersion curves, link the thermal transport in a two-atom LJ system to the lattice dynamics.

The direct method applies well to the study of thin films, where size effects are of great interest. Anderson and Tamma [133] have reviewed techniques for predicting the thermal conductivity of thin films using BTE and MD approaches. In such simulations, the treatment of the film boundaries and the hot and cold slabs is important, and different methods have been considered [107,108]. Generally, the conductivity of a thin film is less than the corresponding bulk value; it increases as the film thickness increases and the relative contribution of boundary scattering to the thermal resistance decreases.

A natural extension of the study of thin films is an investigation of superlattices. Interfacial scattering strongly affects the thermal transport, and the periodicity of the superlattice modifies the dispersion through zone folding. Qualitative differences exist between experimental results and theoretical predictions. Studies have been performed on LJ materials [124–126], GaAs/AlAs [115,130,131] and Ge/Si [89], investigating effects including interfacial strain, number of layers, layer thickness, and interface roughness. Temperatures discontinuities are found at the interfaces, indicative of thermal boundary resistance. In general, the number of layers has an insignificant effect. The layer thickness does have an effect though, as that parameter affects the relative contributions of the interfaces and the layers to the total thermal resistance.

Carbon nanotubes of differing size and chirality have been investigated using both the GK and direct methods [78–82,101,118]. The predicted thermal conductivities are very large (room temperatures values of thousands of W/m K) and there is significant scatter between different investigations. Different factors might be responsible for this scatter, many of which were discussed in previous sections. As the length of the nanotube increases, its thermal conductivity diverges. With the exception of the work of Maruyama [118], all studies reporting temperature dependence see a peak in the thermal conductivity. This behavior is strange, as the peak is generally interpreted as a quantum effect. In this case, it is most likely related to the two-dimensional nature of the nanotube. Osman and Srivastava [101] find that the location of the peak shifts to higher temperature as the diameter of the nanotube increases, a result they attribute to the effects of Umklapp processes and radial phonons. A comparison of a (10,0) tube and a (5,5) tube, which have the same diameter, shows no significant chirality dependence of the thermal conductivity. Che *et al.* [78] considered the effects of defects and vacancies in a (10,10) nanotube.

Beyond straight, single-walled nanotubes, Noya *et al.* [112] studied the effect of placing fullerene (C_{60}) molecules inside a (10,10) nanotube (a peapod). The thermal conductivity is larger than the bare nanotube value, consistent with experimental results. They attribute this result to energy transfer to the fullerene molecules by the radial vibrations of the nanotube, and subsequent transfer between the fullerenes. Cummings *et al.* [113] studied a Y-junction nanotube [(14,0) → (7,0), (7,0)] using the direct method. They find a temperature discontinuity at the junction, lowering the thermal conductivity. The addition of defects to the straight (14,0) nanotube also produces a temperature discontinuity. While the electrical conductivity of such junctions has been found to be current-direction-dependent, this effect is not found for the thermal conductivity.

B. Comparison of Green–Kubo and Direct Methods

Both the GK and direct methods have been used extensively to study thermal transport in dielectric materials and to predict their thermal conductivities. Surprisingly, few studies exist that directly compare the two approaches. Schelling *et al.* [61] predicted the thermal conductivity of silicon at a temperature of 1000 K using both methods, and found good agreement. Poetzsch and Bottger [114] also found good agreement between the two approaches in a two-dimensional elastic percolating system. Below, we qualitatively compare the two methods.

- The GK calculations are based on an equilibrium system. Implementation has no effect on the particle dynamics, and the system temperature will be uniform and constant. In the direct method, the system is perturbed to a steady-state, nonequilibrium configuration. The equations of motion of the atoms in the hot and cold slabs are modified. A temperature gradient exists, and one must check for linear system response.
- Implementation of the GK method is well defined. One must use subjective judgement in deciding how to specify the integral of the HCACF (see Section V.B.1). In the direct method, one must make more decisions. Because of this, significant time may be needed to "tune" the simulations, and comparing results of different investigations may be difficult.
- In the GK method, hundreds or thousands of atoms are required to eliminate size effects. Multiple simulations from different initial conditions (each on the order of one million time steps) are needed for good convergence of the HCACF and thermal conductivity. When simulating a bulk material using the direct method, one must use simulation cells of different lengths to account for size effects. The largest cells may contain tens or hundreds of thousands of atoms. Only one simulation is required for each configuration, which may need a few hundred thousand time steps to reach convergence.
- In both methods, the biggest challenge is overcoming noise inherent to the simulations. In the GK method, noise can lead to a HCACF integral that is difficult to specify. The challenge in the direct method is calculating the temperature gradient and the heat flux. In both approaches it helps to run longer simulations (or simulations from different initial conditions for GK).

- The GK method provides the full thermal conductivity tensor. One can also use the simulations to investigate phonon dynamics and to predict other properties, like the specific heat. In the direct method, the thermal conductivity is only predicted in one direction, and the imposed nonequilibrium limits access to extra information.

Because of size effects and computational demands associated with the direct method, we believe the GK method should be used to predict bulk phase thermal conductivity. One exception is for materials whose HCACF does not converge neatly to zero (e.g., silicon [61]). The direct method is ideal for studying finite structures like thin films and superlattices; it allows for predictions of interfacial and boundary thermal resistances not possible in the GK method.

C. Expectations from Molecular Dynamics

The finite thermal conductivities predicted in MD simulations clearly indicate that inter-phonon interactions are present. The discussion in Section IV indicates that the nature of the phonon transport in small MD systems differs from that in a particle-based description. This does not mean, however, that one should expect large differences between experimental results, predictions from a quantum model, and/or predictions from MD simulations. The nonlocalized description is a more general approach to the problem that could be applied to the quantum system. What to seek when addressing the issue of phonon transport in the MD system is a consistent framework in which to interpret the results of the classical simulations and those from a quantum-particle model.

The classical-quantum issue may lead to poor agreement in some properties at low temperatures due to freezing out of the high-frequency modes in the quantum theory. This is evident from the specific heat predicted by MD, which remains finite as temperature approaches zero. Regarding thermal conductivity prediction, some have argued [61,69,71] that above the Debye temperature the difference should not be significant. That said, classical simulations, run at temperatures significantly lower than this value, have reproduced experimental results reasonably well.

While agreement between MD thermal conductivity predictions and experimental data has been found in some cases, success is not universal. With large computational cells that eliminate size effects, MD simulations of diamond at a temperature of 300 K with the GK method predict a thermal conductivity of 1,200 W/m K [77]. This is almost half the experimental value of 2300 W/m K. On the other hand, MD simulations of germanium with the GK method over-predict the experimental room-temperature thermal conductivity of 63 W/m K by a factor of 2 [83]. When MD results overpredict the experimental thermal conductivity, one often assumes it is because the MD system contains no impurities, removing a source of phonon scattering. When MD under-predicts the experimental results, one often assumes it is because the simulation cell was too small to capture long-wavelength phonons. One can take size effects into account, however, and very high-purity single crystals are available in experimental investigations. The question of the classical nature of the simulations also arises. Discrepancies between experimental data and MD predictions most likely result from the interatomic potential used. Because of this, many investigations still consider their efforts to be methodological in nature, or only good for qualitative predictions. To simulate real devices, MD must be applied as a predictive tool. As such, the development of new potentials must continue. The potential functions currently available are limited, and in some cases of dubious origin, especially when modeling thermal transport.

VIII. Concluding Remarks

We have presented a formulation for studying thermal transport in dielectric materials using MD simulations. The simulations allow for analysis in both the real and phonon spaces. The natural inclusion of anharmonic effects through the form of the interatomic potential presents a significant advantage over harmonic theories. We have described, examined, and compared the two major approaches for predicting thermal conductivity from MD simulations (the GK and direct methods). Each has advantages and disadvantages, and the method chosen strongly depends on the problem of interest. Generally, the GK method is superior for bulk phase simulations, while the direct method is best for finite structures. By no means are the simulations an exact representation of the real world; they are limited by their classical nature, the small systems sizes that must be considered for reasonable computational times, and the available interatomic potentials. Still, many investigations have found reasonable agreement with experimental data, and the simulations have allowed for atomic-level observations not otherwise possible – resulting in elucidation of interesting phenomena.

The exchange of energy between normal modes underlies all of the thermal transport behavior discussed. Without a fundamental knowledge of the nature of the transport, it may not be possible to understand how a complex crystal structure localizes energy, or to design a material with a specified thermal conductivity, or even to interpret the shapes of relaxation–time curves. To date, work in molecular simulation and thermal transport seems to have skirted this issue entirely [11]. While phonons are often mentioned, little effort has been taken to observe them, or to investigate how energy flows in a frequency space whose discretization cannot be ignored.

In terms of applying MD simulations to real systems, current computational resources cannot accurately model anything close to a micron in size on an atom-by-atom level. The upscaling of MD results to larger length-scale models is a promising and exciting avenue. For example, the results of MD simulations can be used to parameterize the BTE expression for the thermal conductivity, allowing for predictions without the use of fitting parameters [28]. Upscaling has been applied in a different context to phonon transport across material interfaces by Schelling and Phillpot [51].

Acknowledgements

This work has been supported by the United States Department of Energy, Basic Energy Sciences Division under grant DE-FG02-00ER45851, the Natural Sciences and Engineering Research Council of Canada, and the Rackham Graduate School at the University of Michigan.

Nomenclature

a	lattice constant, constant	KE	kinetic energy
A	constant	l	wave packet size
B	constant	L	MD simulation cell size
c_v	specific heat at constant volume, J/K (either per particle or per mode)	m	mass
		n	number of atoms in unit cell
C_p	volumetric specific heat at constant pressure, J/kg K	N	number of atoms
		p	probability distribution function
C_v	specific heat at constant volume, J/kg K	$\mathbf{p}$	particle momentum vector
		P	pressure, function
C	constant	$\mathbf{q}, q$	heat flux vector, heat flux
D	dynamical matrix	$\mathbf{r}, r$	particle position, inter-particle separation
$\mathbf{e}$	normal mode polarization vector		
E	energy (kinetic and potential)	R	radius
$\mathbf{F}$	force vector, forcing function	S	normal mode coordinate
g	radial distribution function	$\mathbf{S}$	heat current vector
$\mathbf{G}$	reciprocal space lattice vector	t	time
$\hbar$	Planck constant/2π, 1.0546 × 10^{-34} J s	T	temperature
		$\mathbf{u}, u$	particle displacement from equilibrium
k_B	Boltzmann constant, 1.3806 × 10^{-23} J/K		
		$\mathbf{v}, v$	particle or phonon velocity
k	thermal conductivity	V	volume
K	spring constant	x	$\hbar\omega/k_B T$

Greek Letters

α	thermal diffusivity
ε	energy
ε_{LJ}	Lennard-Jones energy scale
η	unit cells in linear dimension, Nose–Hoover thermostat parameter
$\kappa, \mathbf{\kappa}$	wave vector, wave number
λ	phonon wavelength
Λ	phonon mean free path
ν	polarization branch
ρ	density
σ_{LJ}	Lennard-Jones length scale
τ	time constant, relaxation time
ϕ, Φ	potential energy
$\mathbf{\Phi}$	force constant matrix
ω	angular frequency

Subscripts

ac	acoustic
am	amorphous
c	cutoff
D	Debye
eff	effective
equ	equilibrium
FD	first dip
G	center of mass
harm	harmonic
i	summation index, particle label
j	summation index, particle label
k	summation index, particle label, phonon mode label, conduction label
l	
lg	long range
LJ	Lennard-Jones
nn	nearest neighbor
o	self (referring to a particle), equilibrium
op	optical
r	relaxation
sh	short-range
sim	length of simulation
t	total
T	thermostat
u	convection
α, β, γ	x, y, or z direction
0	zero-temperature equilibrium
1	related to single exponential fit
∞	bulk

Superscripts

*	complex conjugate, dimensionless
~	deviation from average value
–	instantaneous average
'	new (velocity)

Abbreviations

BTE	Boltzmann transport equation
BZ	Brillouin zone
CP	Cahill–Pohl
EF	exponential fit
fcc	face-centered cubic
FD	first dip
GK	Green–Kubo
HCACF	heat current autocorrelation function
LJ	Lennard-Jones
MD	molecular dynamics
RDF	radial distribution function
RMS	root mean square

References

1. Cahill, D. G., Goodson, K. E., and Majumdar, A. (2002). Thermometry and thermal transport in micro/nanoscale solid-state devices and structures. *J. Heat Transfer* **124**, 223–241.
2. Cahill, D. G., Ford, W. K., Goodson, K. E., Mahan, G. D., Maris, H. J., Majumdar, A., Merlin, R., and Phillpot, S. R. (2003). Nanoscale thermal transport. *J. Appl. Phys.* **93**, 793–818.

3. Ashcroft, N. W. and Mermin, N. D. (1976). "Solid State Physics". Saunders College Publishing, Fort Worth.
4. Srivastava, G. P. (1990). "The Physics of Phonons". Adam Hilger, Bristol.
5. Dove, M. T. (1993). "Introduction to Lattice Dynamics". Cambridge University Press, Cambridge.
6. Ziman, J. M. (2001). "Electrons and Phonons". Oxford University Press, Oxford.
7. Roth, J., Gahler, F., and Trebin, H.-R. (2000). A molecular dynamics run with 5 180 116 000 particles. *Int. J. Mod. Phys. C* **11**, 317–322.
8. Kadau, K., Germann, T. C., and Lomdahl, P. S. (2004). Large-scale molecular-dynamics simulation of 19 billion particles. *Int. J. Mod. Phys. C* **15**, 193–201.
9. Touloukian, Y. (1970). "Thermophysical Properties of Matter Volume 2". Plenum, New York.
10. Touloukian, Y. (1970). "Thermophysical Properties of Matter Volume 3". Plenum, New York.
11. McGaughey, A. J. H. (2004). Phonon Transport in Molecular Dynamics Simulations: Formulation and Thermal Conductivity Prediction. PhD Thesis, University of Michigan, Ann Arbor, MI.
12. Omini, M. and Sparavigna, A. (1996). Beyond the isotropic-model approximation in the theory of thermal conductivity. *Phys. Rev. B* **53**, 9064–9073.
13. Yu, R. C., Tea, N., Salamon, M. B., Lorents, D., and Malhotra, R. (1992). Thermal conductivity of single crystal C_{60}. *Phys. Rev. Lett.* **68**, 2050–2053.
14. Poulikakos, D., Arcidiacono, S., and Maruyama, S. (2003). Molecular dynamics simulation in nanoscale heat transfer: a review. *Micro. Therm. Eng.* **7**, 181–206.
15. Allen, M. P. and Tildesly, D. J. (1987). "Computer Simulation of Fluids". Clarendon, Oxford.
16. Frenkel, D. and Smit, B. (1996). "Understanding Molecular Simulation: From Algorithms to Applications". Academic Press, San Diego.
17. Li, J. (2000). "Modeling Microstructural Effects on Deformation Resistance and Thermal Conductivity". PhD Thesis, Massachusetts Institute of Technology, Cambridge, MA.
18. Ding, H.-Q., Karasawa, N., and Goddard, W. A. (1992). The reduced cell multipole method for Coulomb interactions in periodic-systems with million-atom unit cells. *Chem. Phys. Lett.* **192**, 6–10.
19. Ding, H.-Q., Karasawa, N., and Goddard, W. A. (1992). Atomic level simulations on a million particles – the cell multipole method for Coulomb and London nonbond interactions. *J. Chem. Phys.* **97**, 4309–4315.
20. Wolf, D., Keblinski, P., Phillpot, S. R., and Eggebrecht, J. (1999). Exact method for the simulation of Coulombic systems by spherically truncated, pairwise r^{-1} summation. *J. Chem. Phys.* **110**, 8254–8282.
21. McQuarrie, D. A. (2000). "Statistical Mechanics". University Science Books, Sausalito.
22. Nose, S. (1984). A molecular dynamics method for simulations in the canonical ensemble. *Mol. Phys.* **52**, 255–268.
23. Nose, S. (1984). A unified formulation of the constant temperature molecular dynamics method. *J. Chem. Phys.* **81**, 511–519.
24. Hoover, W. G. (1985). Canonical dynamics: equilibrium phase-space distributions. *Phys. Rev. A* **31**, 1695–1697.
25. Anderson, H. C. (1980). Molecular dynamics simulations at constant pressure and/or temperature. *J. Chem. Phys.* **72**, 2384–2393.
26. Parrinello, M. and Rahman, A. (1981). Polymorphic transitions in single crystals: a new molecular dynamics method. *J. Appl. Phys.* **52**, 7182–7190.

27. McGaughey, A. J. H. and Kaviany, M. (2004). Thermal conductivity decomposition and analysis using molecular dynamics simulations. Part I. Lennard-Jones argon. *Int. J. Heat Mass Transfer* **27**, 1783–1798.
28. McGaughey, A. J. H. and Kaviany, M. (2004). Quantitative validation of the Boltzmann transport equation phonon thermal conductivity model under the single-mode relaxation time approximation. *Phys. Rev. B* **69**, 094303-1-12.
29. Kaburaki, H., Li, J., and Yip, S. (1998). Thermal conductivity of solid argon by classical molecular dynamics. *Mater. Res. Soc. Symp. Proc.* **538**, 503–508.
30. Tretiakov, K. V. and Scandolo, S. (2004). Thermal conductivity of solid argon from molecular dynamics simulations. *J. Chem. Phys.* **120**, 3765–3769.
31. Einstein, A. (1911). Elementare betrachtungen uber die thermische molekularbewegung in festen korpern. *Ann. Phys.* **35**, 679–694.
32. Cahill, D. G. and Pohl, R. O. (1989). Heat flow and lattice vibrations in Glasses. *Solid State Commun.* **70**, 927–930.
33. Cahill, D. G., Watson, S. K., and Pohl, R. O. (1992). Lower limit to thermal conductivity of disordered crystals. *Phys. Rev. B* **46**, 6131–6140.
34. McGaughey, A. J. H. and Kaviany, M. (2004). Thermal conductivity decomposition and analysis using molecular dynamics simulations. Part II. Complex silica structures. *Int. J. Heat Mass Transfer* **27**, 1799–1816.
35. Ohara, T. (1999). Contribution of intermolecular energy transfer to heat conduction in a simple fluid. *J. Chem. Phys.* **111**, 9667–9672.
36. Greegor, R. B. and Lyle, F. W. (1979). Extended X-ray absorption fine structure determination of thermal disorder in Cu: comparison of theory and experiment. *Phys. Rev. B* **20**, 4902–4907.
37. Porter, L. J., Yip, S., Yamaguchi, M., Kaburaki, H., and Tang, M. (1997). Empirical bond-order potential description of thermodynamic properties of crystalline silicon. *J. Appl. Phys.* **81**, 96–106.
38. Kittel, C. (1996). "Introduction to Solid State Physics", 7th ed. Wiley, New York.
39. Tschaufeser, P. and Parker, S. C. (1995). Thermal expansion behavior of zeolites and $AlPO_4$s. *J. Phys. Chem.* **9**, 10609–10615.
40. Tiano, W., Dapiaggi, M., and Artioli, G. (2003). Thermal expansion in cuprite-type structures from 10 K to decomposition temperature: Cu_2O and Ag_2O. *J. Appl. Crystallogr.* **36**, 1461–1463.
41. Pryde, A. K. A., Hammonds, K. D., Dove, M. T., Heine, V., Gale, J. D., and Warren, M. C. (1996). Origin of the negative thermal expansion in ZrW_2O_8 and ZrV_2O_7. *J. Phys.: Condens. Matter* **8**, 10973–10982.
42. Fujii, Y., Lurie, N. A., Pynn, R., and Shirane, G. (1974). Inelastic neutron scattering from solid ^{36}Ar. *Phys. Rev. B* **10**, 3647–3659.
43. Bernandes, N. (1958). Theory of solid Ne, A, Kr, and Xe at 0°K. *Phys. Rev.* **112**, 1534–1539.
44. Callaway, J. (1959). Model for lattice thermal conductivity at low temperatures. *Phys. Rev.* **113**, 1046–1051.
45. Holland, M. G. (1963). Analysis of lattice thermal conductivity. *Phys. Rev.* **132**, 2461–2471.
46. Ladd, A. J. C., Moran, B., and Hoover, W. G. (1986). Lattice thermal conductivity: a comparison of molecular dynamics and anharmonic lattice dynamics. *Phys. Rev. B* **34**, 5058–5064.
47. Chung, J. D., McGaughey, A. J. H., and Kaviany, M. (2004). Role of phonon dispersion in lattice thermal conductivity. *J. Heat Transfer* **126**, 376–380.

48. Allen, P. B. and Feldman, J. L. (1993). Thermal conductivity of disordered harmonic solids. *Phys. Rev. B* **48**, 12581–12588.
49. Feldman, J. L., Kluge, M. D., Allen, P. B., and Wooten, F. (1993). Thermal conductivity and localization in glasses: numerical study of a model of amorphous silicon. *Phys. Rev. B* **48**, 12589–12602.
50. Schelling, P. K., Phillpot, S. R., and Keblinski, P. (2002). Phonon wave-packet dynamics at semiconductor interfaces by molecular-dynamics simulations. *Appl. Phys. Lett.* **80**, 2484–2486.
51. Schelling, P. K. and Phillpot, S. R. (2003). Multiscale simulation of phonon transport in superlattices. *J. Appl. Phys.* **93**, 5377–5387.
52. Sinha, S., Schelling, P. K., Phillpot, S. R., and Goodson, K. E. (2005). Scattering of g-process longitudinal optical phonons at hotspots in silicon. *J. Appl. Phys.* **97**, 023702-1-9.
53. Green, M. S. (1954). Markoff random processes and the statistical mechanics of time-dependent phenomena. II. Irreversible processes in fluids. *J. Chem. Phys.* **22**, 398–413.
54. Kubo, R. (1957). Statistical mechanical theory of irreversible processes. I. General theory and simple applications to magnetic and conduction problems. *J. Phys. Soc. Japan* **12**, 570–586.
55. R. Zwanzig, (1965). Time-correlation functions and transport coefficients in statistical mechanics. In: "Annual Review of Physical Chemistry Volume 16", (H. Eyring, C. J. Christensen, and H. S. Johnston, eds.), pp. 67–102 Annual Reviews, Palo Alto.
56. Helfand, E. (1960). Transport coefficients from dissipation in a canonical ensemble. *Phys. Rev.* **119**, 1–9.
57. Lindan, P. J. D. and Gillan, M. J. (1991). A molecular dynamics study of the thermal conductivity of CaF_2 and UO_2. *J. Phys. Condens. Matter* **3**, 3929–3939.
58. Gillan, M. J. (1991). The molecular dynamics calculation of transport coefficients. *Phys. Scripta* **T39**, 362–366.
59. Paolini, G. V., Lindan, P. J. D., and Harding, J. H. (1997). The thermal conductivity of defective crystals. *J. Chem. Phys.* **106**, 3681–3687.
60. Li, J., Porter, L., and Yip, S. (1998). Atomistic modeling of finite-temperature properties of crystalline β-SiC. II. Thermal conductivity and effects of point defects. *J. Nucl. Mater* **255**, 139–152.
61. Schelling, P. K., Phillpot, S. R., and Keblinski, P. (2002). Comparison of atomic-level simulation methods for computing thermal conductivity. *Phys. Rev. B* **65**, 144–306.
62. Vogelsang, R., Hoheisel, C., and Ciccotti, G. (1987). Thermal conductivity of the Lennard-Jones liquid by molecular dynamics calculations. *J. Chem. Phys.* **86**, 6371–6375.
63. Volz, S. G., Saulnier, J.-B., Lallemand, M., Perrin, B., Depondt, B., and Mareschal, M. (1996). Transient Fourier-law deviation by molecular dynamics in solid argon. *Phys. Rev. B* **54**, 340–347.
64. Motoyama, S., Ichikawa, Y., Hiwatari, Y., and Oe, A. (1999). Thermal conductivity of uranium dioxide by nonequilibrium molecular dynamics simulation. *Phys. Rev. B* **60**, 292–298.
65. Picu, R. C., Borca-Tasciuc, T., and Pavel, M. C. (2003). Strain and size effects on heat transport in nanostructures. *J. Appl. Phys.* **93**, 3535–3539.
66. A. J. H. McGaughey, M. I. Hussein, M. Kaviany, and G. Hulbert, (2004) Phonon band structure and thermal transport correlation in a two-atom unit cell. ASME paper IMECE2004-62328, presented at 2004 ASME International Mechanical Engineering Congress and Exhibition, Anaheim, CA, USA, November 13–19.
67. Chen, Y., Lukes, J. R., Li, D., Yang, J., and Wu, Y. (2004). Thermal expansion and impurity effect on lattice thermal conductivity of solid argon. *J. Chem. Phys.* **120**, 3841–3846.

68. Tretiakov, K. V. and Scandolo, S. (2004). Thermal conductivity of solid argon at high pressure and high temperature: a molecular dynamics study. *J. Chem. Phys.* **121**, 11177–11182.
69. Lee, Y. H., Biswas, R., Soukoulis, C. M., Wang, C. Z., Chan, C. T., and Ho, K. M. (1991). Molecular-dynamics simulation of thermal conductivity in amorphous silicon. *Phys. Rev. B* **43**, 6573–6580.
70. Volz, S. G. and Chen, G. (1999). Molecular dynamics simulation of thermal conductivity of silicon nanowires. *Appl. Phys. Lett.* **75**, 2056–2058.
71. Volz, S. G. and Chen, G. (2000). Molecular-dynamics simulation of thermal conductivity of silicon crystals. *Phys. Rev. B* **61**, 2651–2656.
72. Volz, S. G. and Perrin, B. (2002). Si crystal thermal conductance in the THz frequency range by molecular dynamics. *Physica B* **316–317**, 286–288.
73. Murakawa, A., Ishii, H., and Kakimoto, K. (2004). An investigation of thermal conductivity of silicon as a function of isotope concentration by molecular dynamics. *J. Cryst. Growth* **267**, 452–457.
74. J. Li, and S. Yip, (2005) Spectral method in thermal conductivity calculation. Submitted.
75. Murashov, V. V. (1999). Thermal conductivity of model zeolites: molecular dynamics study. *J. Phys: Condens. Mater* **11**, 1261–1271.
76. Hirosaki, N., Ogata, S., Kocer, C., Kitagawi, H., and Nakamura, Y. (2002). Molecular dynamics calculation of the ideal thermal conductivity of single-crystal α- and β-Si_3N_4. *Phys. Rev. B* **65**, 134110-1-11.
77. Che, J., Cagin, T., Deng, W., and Goddard III, W. A. (2000). Thermal conductivity of diamond and related materials from molecular dynamics simulations. *J. Chem. Phys.* **113**, 6888–6900.
78. Che, J., Cagin, T., and Goddard III, W. A. (2000). Thermal conductivity of carbon nanotubes. *Nanotechnology* **11**, 65–69.
79. Berber, S., Kwon, Y.-K., and Tomanek, D. (2000). Unusually high thermal conductivity of carbon nanotubes. *Phys. Rev. Lett.* **84**, 4613–4616.
80. Grujicic, M., Cao, G., and Gersten, B. (2004). Atomic scale computations of the lattice contribution to thermal conductivity of single-walled carbon nanotubes. *Mater Sci. Eng. B* **107**, 204–216.
81. Shenogin, S., Bodapati, A., Xue, L., Ozisik, R., and Keblinski, P. (2004). Effect of chemical functionalization on thermal transport of carbon nanotube composites. *Appl. Phys. Lett.* **85**, 2229–2231.
82. Zhang, W., Zhu, Z., Wang, F., Wang, T., Sun, L., and Wang, Z. (2004). Chirality dependence of the thermal conductivity of carbon nanotubes. *Nanotechnology* **15**, 936–939.
83. Dong, J., Sankey, O. F., and Myles, C. W. (2001). Theoretical study of lattice thermal conductivity in Ge framework semiconductors. *Phys. Rev. Lett.* **86**, 2361–2364.
84. Ishii, H., Murakawa, A., and Kakimoto, K. (2004). Isotope-concentration dependence of thermal conductivity of germanium investigated by molecular dynamics. *J. Appl. Phys.* **95**, 6200–6203.
85. Yamada, K., Kurosaki, K., Uno, M., and Yamanaka, S. (2000). Evaluation of thermal properties of uranium dioxide by molecular dynamics. *J. Alloy Compd.* **307**, 10–16.
86. Kurosaki, K., Yano, K., Yamada, K., Uno, M., and Yamanaka, S. (2000). A molecular dynamics study of the thermal conductivity of uranium mononitride. *J. Alloy Compd.* **311**, 305–310.
87. Inoue, R., Tanaka, H., and Nakanishi, K. (1996). Molecular dynamics calculation of the anomalous thermal conductivity of clathrate hydrates. *J. Chem. Phys.* **104**, 9569–9577.

88. Keblinski, P., Phillpot, S. R., Choi, S. U. S., and Eastman, J. A. (2002). Mechanisms of heat flow in suspensions of nano-sized particles (nanofluids). *Int. J. Heat Mass Transfer* **45**, 855–863.
89. Volz, S. G., Saulnier, J.-B., Chen, G., and Beauchamp, P. (2000). Computation of thermal conductivity of Si/Ge superlattices by molecular dynamics techniques. *Microelectr. J* **31**, 815–819.
90. Tretiakov, K. V. Personal communication.
91. Luchnikov, V. A., Medvedev, N. N., Naberukhin, Y. I., and Novikov, V. N. (1995). Inhomogeneity of the spatial distribution of vibrational modes in a computer model of amorphous argon. *Phys. Rev. B* **51**, 15569–15572.
92. Wolfing, B., Kloc, C., Teubner, J., and Bucher, E. (2001). High performance thermoelectric Tl_9BiTe_6 with extremely low thermal conductivity. *Phys. Rev. Lett.* **86**, 4350–4353.
93. Konstantinov, V. A. (2001). Manifestation of the lower limit to thermal conductivity in the solidified inert gases. *J. Low Temp. Phys.* **122**, 459–465.
94. Domingues, G., Saulnier, J.-B., and Volz, S. G. (2004). Thermal relaxation times and heat conduction in β-cristobalite and α-quartz silica structures. *Superlattice Microstruct.* **35**, 227–237.
95. Evans, D. J. (1982). Homogeneous NEMD algorithm for thermal conductivity: application of non-canonical linear response theory. *Phys. Lett.* **91A**, 457–460.
96. Gillan, M. J. and Dixon, M. (1983). The calculation of thermal conductivity by perturbed molecular simulation. *J. Phys. C. Solid State* **16**, 869–878.
97. Evans, D. J. and Morriss, G. P. (1984). Non-Newtonian molecular dynamics. *Comp. Phys. Rep.* **1**, 297–344.
98. Evans, D. J. and Morriss, G. P. (1990). "Statistical Mechanics of Nonequilibrium Liquids". Academic Press, New York.
99. Hansen, J. P. and McDonald, I. R. (1986). "Theory of Simple Liquids". Academic Press, New York.
100. Maeda, A. and Munakata, T. (1995). Lattice thermal conductivity via homogeneous nonequilibrium molecular dynamics. *Phys. Rev. E* **52**, 234–239.
101. Osman, M. A. and Srivastava, D. (2001). Temperature dependence of the thermal conductivity of single-wall carbon nanotubes. *Nanotechnology* **12**, 21–24.
102. Majumdar, A. (1998). Microscale energy transport in solids. In "Microscale Energy Transport" (C.-L. Tien, A. Majumdar and F.M. Gerner, eds.). Taylor Francis, Washington.
103. Christen, D. K. and Pollack, G. L. (1975). Thermal conductivity of solid argon. *Phys. Rev. B* **12**, 3380–3391.
104. Maiti, A., Mahan, G. D., and Pantelides, S. T. (1997). Dynamical simulations of nonequilibrium processes heat flow and the Kapitza resistance across grain boundaries. *Solid State Commun.* **102**, 517–521.
105. Muller-Plathe, F. (1997). A simple nonequilibrium molecular dynamics method for calculating the thermal conductivity. *J. Chem. Phys.* **106**, 6082–6085.
106. Schelling, P. K. and Phillpot, S. R. (2001). Mechanism of thermal transport in zirconia and yttria-stabilized zirconia by molecular-dynamics simulation. *J. Am.Ceram. Soc.* **84**, 2997–3007.
107. Lukes, J. R., Li, D. Y., Liang, X.-G., and Tien, C.-L. (2000). Molecular dynamics study of solid thin-film thermal conductivity. *J. Heat Transfer* **122**, 536–543.
108. Chantrenne, P. and Barrat, J.-L. (2004). Finite size effects in determination of thermal conductivities: comparing molecular dynamics results with simple models. *J. Heat Transfer* **126**, 577–585.

109. Mountain, R. D. and MacDonald, R. A. (1983). Thermal conductivity of crystals: a molecular dynamics study of heat flow in a two-dimensional crystal. *Phys. Rev. B* **28**, 3022–3025.
110. Michalski, J. (1992). Thermal conductivity of amorphous solids above the plateau: molecular-dynamics study. *Phys. Rev. B* **45**, 7054–7065.
111. Oligschleger, C. and Schon, J. C. (1999). Simulation of thermal conductivity and heat transport in solids. *Phys. Rev. B* **59**, 4125–4133.
112. Noya, E. G., Srivastava, D., Chernozatonskii, L. A., and Menon, M. (2004). Thermal conductivity of carbon nanotube peapods. *Phys. Rev. B* **70**, 115416-1-5.
113. Cummings, A., Osman, M. A., Srivastava, D., and Menon, M. (2004). Thermal conductivity of Y-junction carbon nanotubes. *Phys. Rev. B* **70**, 115405-1-6.
114. Poetzsch, R. H. H. and Bottger, H. (1994). Interplay of disorder and anharmonicity in heat conduction: molecular-dynamics study. *Phys. Rev. B* **50**, 15757–15763.
115. Imamura, K., Tanaka, Y., Nishiguchi, N., Tamura, S., and Maris, H. J. (2003). Lattice thermal conductivity in superlattices: molecular dynamics calculations with a heat reservoir method. *J. Phys.: Condens. Matter* **15**, 8679–8690.
116. Tenenbaum, A., Ciccotti, G., and Gallico, R. (1982). Stationary nonequilibrium states by molecular dynamics Fourier's law. *Phys. Rev. A* **25**, 2778–2787.
117. Maruyama, S. (2000). Molecular dynamics method for microscale heat transfer. *Adv. Numer. Heat Transfer* **2**, 189–226.
118. Maruyama, S. (2003). A molecular dynamics simulation of heat conduction of a finite length single walled nanotube. *Microscale Therm. Eng.* **7**, 41–50.
119. Jund, P. and Jullien, R. (1999). Molecular-dynamics calculation of the thermal conductivity of vitreous silica. *Phys. Rev. B* **59**, 13707–13711.
120. Yoon, Y.-G., Car, R., Srolovitz, D. J., and Scandolo, S. (2004). Thermal conductivity of crystalline quartz from classical simulations. *Phys. Rev. B* **70**, 012302-1-4.
121. Heino, P. and Ristolainen, E. (2003). Thermal conduction at the nanoscale in some metals by MD. *Microelectr. J* **34**, 773–777.
122. Ikeshoji, T. and Hafskjold, B. (1994). Nonequilibrium molecular dynamics calculation of heat conduction in liquid and through liquid–gas interface. *Mol. Phys.* **81**, 251–261.
123. Lukes, J. R. and Tien, C.-L. (2004). Molecular dynamics simulation of thermal conduction in nanoporous thin films. *Microscale Therm. Eng.* **8**, 341–359.
124. Liang, X.-G. and Shi, B. (2000). Two-dimensional molecular dynamics simulation of the thermal conductance of superlattices. *Mater Sci. Eng. A Struct.* **292**, 198–202.
125. Abramson, A. R., Tien, C.-L., and Majumdar, A. (2002). Interface and strain effects on the thermal conductivity of heterostructures: a molecular dynamics study. *J. Heat Transfer* **124**, 963–970.
126. Chen, Y., Li, D., Yang, J., Wu, Y., Lukes, J. R., and Majumdar, A. (2004). Molecular dynamics study of the lattice thermal conductivity of Kr/Ar superlattice nanowires. *Physica B* **349**, 270–280.
127. Konashi, K., Ikeshoji, T., Kawazoe, Y., and Matsui, H. (2003). A molecular dynamics study of thermal conductivity of zirconium hydride. *J. Alloy. Compd.* **356–357**, 279–282.
128. Kotake, S. and Wakuri, S. (1994). Molecular dynamics study of heat conduction in solid materials. *JSME Int. J. B – Fluid T* **37**, 103–108.
129. Bedrov, R. and Smith, G. D. (2000). Thermal conductivity of molecular fluids from molecular dynamics simulations: application of a new imposed-flux method. *J. Chem. Phys.* **113**, 8080–8084.
130. Daly, B. C., Maris, H. J., Imamura, K., and Tamura, S. (2002). Molecular dynamics calculation of the thermal conductivity of superlattices. *Phys. Rev. B* **66**, 024301-1-7.

131. Daly, B. C., Maris, H. J., Tanaka, Y., and Tamura, S. (2003). Molecular dynamics calculation of the in-plane thermal conductivity of GaAs/AlAs superlattices. *Phys. Rev. B* **67**, 033308-1-3.
132. Allen, P. B., Feldman, J. L., Fabian, J., and Wooten, F. (1999). Diffusions, locons and propagons: character of atomic vibrations in amorphous Si. *Philos. Mag. B* **79**, 1715–1731.
133. Anderson, C. V. D. R. and Tamma, K. K. (2004). An overview of advances in heat conduction models and approaches for prediction of thermal conductivity in thin dielectric films. *Int. J. Numer. Method Heat* **14**, 12–65.

Heat and Mass Transfer in Fluids with Nanoparticle Suspensions

G.P. PETERSON and C.H. LI

Rensselaer Polytechnic Institute, Troy, NY 12180, USA

I. Introduction

The fundamental laws of thermodynamics govern the transfer of heat and state that when a temperature gradient exists in a body, there is an energy transfer from the high-temperature region to the low-temperature region or from a region of high potential to a lower energy state. The French mathematical physicist Joseph Fourier [1] simplified this phenomenon and described it for a semi-infinite solid between two parallel planes, where the lower plane was maintained at a constant temperature, $T1$, and was heated uniformly by a constant heat source and the upper plane was maintained at a constant temperature, $T2$, and cooled uniformly such that $T2$ was less than $T1$. After an infinite period of time, the heat transferred from the lower plane would propagate until it reaches the upper plane, where it is rejected to the heat sink, thus maintaining the upper plane at a temperature of $T2$. The final and fixed state results in a temperature distribution in the intermediate solid that has a monotonically decreasing value or a straight line extending from $T1$ on the lower plane to $T2$ on the upper plane. If z represents the height of one point in the intermediate solid, and h the distance between the two planes, the temperature of the point at z can be calculated as

$$T(z) = T1 - (T1 - T2)\frac{z}{h} \quad (1)$$

which leads to the definition of "effective thermal conductivity".

To understand how much heat would be transferred during a given time under this type of imposed temperature distribution, assume $T1$ is maintained at a higher value during the entire heat transfer process, and $T2$ is maintained at a lower value. During time t, a specific quantity of heat, q, is transferred. This amount of heat can be calculated with what is now known as Fourier's equation,

$$q'' = k\frac{T1 - T2}{h} \tag{2}$$

where k is a measure of the thermal conductivity of the solid.

While seemingly quite elementary, this equation is a macrostatistical expression for the relationship between thermal potential and heat flux, where the coefficient, k, is a thermophysical property that varies from material to material and governs how fast and how large the heat transfer will be. In order to clarify the phenomena of heat conduction, it is extremely important to clarify precisely what is meant by the thermal conductivity, k.

The micromechanisms that impact the thermal conductivity of different materials are quite varied and play differing roles for different materials. For solid materials, the thermal conductivity is principally the result of two types of heat transport mechanisms: lattice waves (which always exist) and free electrons (which exist in metals and semiconductors). In special situations, electromagnetic waves, spin waves, and other excitation mechanisms must also be considered. Fundamentally, the thermal conductivity can be expressed as

$$k = \frac{1}{3}\sum_i C_i v_i l_i \tag{3}$$

where C is the contribution of the carriers to the specific heat per unit volume, v the mean velocity, and l the mean free path. The two main carriers described above serve to determine the thermal conductivity of solids and may provide a large range of values and properties, due to the scattering processes of these carriers in the solid, and the interaction between them. In addition, the influence of this scattering varies with the corresponding variations in both bulk temperature and temperature gradient.

In solid bodies, there are two principal effects that govern the heat transfer, one from the electrons and the other from the phonons. In most metals, the contribution of phonon heat transport to the thermal conductivity is less than 5%, implying that electrons dominate the thermal conductivity of these materials. The phonon mean free path is influenced by two processes: U-processes and phonon–electron scattering. There are also two kinds of scattering mechanisms for electrons: scattering from imperfections and scattering by phonons. The type and concentration of imperfections are not dependent on temperature, and the scattering by phonons is independent of imperfections. However, with a decrease of temperature, the thermal resistivity due to mechanical imperfections will increase, while the thermal resistivity due to phonons effects will decrease. In the very low-temperature region, the thermal conductivity of a metal will vary approximately linearly with respect to the temperature. In the higher-temperature region, the thermal conductivity of metals will reach a maximum value and then decrease slightly, until reaching a limiting value. For sufficiently impure metals, there is no such maximum value in the high-temperature region, due to the high-defect concentration. For less highly conducting metals, the intrinsic electronic thermal conductivity contribution and the lattice wave thermal conductivity contribution may be comparable in magnitude. For a dielectric solid with a crystalline structure, the thermal conductivity will vary with $1/T$ in the high-temperature region, while in the moderate-temperature region it will vary with respect to $1/T^2$. In the low-temperature region, scattering on the external boundary of a specimen will result in a mean free path that is comparable to the shortest linear dimension of the specimen, and the thermal conductivity will change in proportion to T^3 or with respect to the change of specific heat, $k = \frac{1}{3}C(T)vl$. As a result, for a crystalline, nonmetallic solid, scattering due to the density differences and surface boundaries plays a dominant role in determining the value of the thermal conductivity. In a highly disordered and noncrystalline solid, the phonon mean free path is even shorter, and the thermal conductivity, even less. For example, the phonon mean free path in glass is of the order of 1 nm. Therefore solids, which depend on phonon transfer to transmit thermal energy, typically have a much lower thermal conductivity than those that depend on electrons, which explains why good electrical conductors are also typically good thermal conductors.

The mechanism of thermal conduction in a gas is much simpler and normally only two-body interactions and binary collisions need be considered. The kinetic energy of a gas molecule can be related to the bulk temperature in the equilibrium state as

$$\frac{1}{2}mv^2 = \frac{3}{2}KT \tag{4}$$

where m is the molecular mass, v the molecular velocity, K the Boltzmann constant, and T the absolute temperature. The intermolecular pair potential or Lennard–Jones potential can be described as

$$p(R) = 4\varepsilon\left[\left(\frac{\sigma}{R}\right)^{12} - \left(\frac{\sigma}{R}\right)^6\right] \tag{5}$$

where R is the distance between two molecules, ε the emissivity, and σ the distance value when the potential $p(R) = 0$. Because gas has no fixed

TABLE I

Comparison of the Calculated Value and Experimental Data for the Thermal Conductivities of Gaseous Argon [2]

Temperature (K)	90.2	194.7	273.2	373.2	491.2	579.1
Calculated value $\times 10^4$ (W/cm K)	0.57	1.22	1.64	2.11	2.59	2.91
Experimental data $\times 10^4$ (W/cm K)	0.59	1.23	1.65	2.12	2.57	2.87

microstructure, during the transport process of mass, momentum, and energy of gas molecules, diffusion will dominate. However, the molecules in both the high- and low-temperature regions are in a continuous random motion, colliding with one another and exchanging both energy and momentum. In the high-temperature region, the molecules have a higher velocity or higher energy than the molecules in the lower-temperature regions. As a result, the overall thermal conductivity of the gas depends on the temperature, which governs how fast and how far the molecules are transported as well as the thermophysical properties of the fluid, and is approximately proportional to the square root of the absolute temperature. As the temperature increases, the thermal conductivity of the gas will also increase. For the simple case of spherically symmetric molecules, i.e., an inert gas, this reflects reality. As an example, the calculated values for gaseous argon, match the experimental data quite well as illustrated in Table I [2].

For liquids, however, the situation is significantly more complicated. The intermolecular potential is not only a two-body problem, but involves three- or multibody potential, and is directly dependent on the molecular structure of the liquid. The molecular structure of a liquid is believed to be very similar to that of a solid, in that the volume only changes approximately 10% as a material undergoes a phase change from the solid to liquid state. Since the liquid has a structure similar to that of the solid, the lattice model is quite suitable for modeling the structure of liquids. In the lattice model, liquid molecules are believed to be confined to cells created by the potential barriers of the molecules adjacent to them. However, the lattice structure of a liquid is not quite the same as it is for solids, in that the molecules in the cells of liquid molecule potential barriers can escape from the cells with relative ease, while in a solid state, it is considerably more difficult. For this reason, the heat conduction in a liquid comprises two distinct mechanisms, one is the vibration of the molecules in the center of the cells and collisions with molecules adjacent to them, and the other is the transport of the molecules from the center of the cells to the region outside of the cells. The first mechanism is similar to the effect of phonons or elastic waves in a solid, except that the collisions and in-cell oscillations are not as violent as in a solid. The second mechanism results from the random transport of molecules from one location to another location in a gas, except that the mean free path of the molecules is not nearly as long as the mean free path for molecules in a gaseous state. For liquids, the contribution of both mechanisms can be described by the Boltzmann equation.

The study of isotopes has provided experimental verification of the two mechanisms described above. In the following, the mechanisms that govern the thermal conductivities of these three different phases of materials are presented, contrasted, and compared. First, metals in the solid state have the highest conductivity. For example, the thermal conductivity of pure silver can be as high as 410 W/(m K) at the ice-melt temperature; the alloy of chrome–nickel steel has a thermal conductivity of 16.3 W/(m K) at the same temperature. Second, nonmetallic solids have a large range of thermal conductivity values due to their different structures. For example, with quartz, the thermal conductivity parallel to the principal axis can be as high as 41.6 W/(m K) at the ice-melt temperature, while the thermal conductivity of glass wool can be as low as 0.038 W/(m K). This is even lower than the thermal conductivity of most of liquids and is comparable to the thermal conductivity of some gases. Third, liquid metals have a relatively high thermal conductivity that is comparable to the thermal conductivity of some metals and is higher than the values for most nonmetallic solids. For example, the thermal conductivity of mercury is 8.21 W/(m K) at the ice-melt temperature, while the thermal conductivity for nonmetallic liquids is comparably very low. Finally, gases typically have the lowest thermal conductivity of all materials.

Within these gaseous materials, single-atom molecule gases have the highest thermal conductivities, such as the thermal conductivity of hydrogen, which is 0.175 W/(m K), followed by multi-atom gases, such as air and carbon dioxide, which have thermal conductivities as low as 0.024 and 0.0146 W/(m K), respectively. This variation is primarily due to the shorter mean free path of molecules and the relative variation within the molecular structure.

With the exception of metals, the thermal conductivity variation of different phase materials indicates that when the molecular collision and oscillation inside a lattice cell dominates the principal mechanism of thermal conduction in the heavy-density materials with a fixed material structure, the materials will have higher thermal conductivities. With a decrease in the density and a relaxation of the material structure, the mechanisms of oscillation and collision inside the lattice cell and the mechanisms of inter-cell random transport will have an effect on the thermal conduction that has approximately equal magnitude, decreasing the thermal conductivity. In the gaseous state, when the mechanism of random transport of molecules dominates the thermal conduction, the value of the thermal conductivity goes even lower.

More than a century ago, Maxwell [3] predicted that adding high thermal conductivity particles to a liquid would enhance the effective thermal conductivity and result in a composite material with an increased effective thermal conductivity. Because many engineering heat transfer applications could benefit from a material that could combine the comparably high thermal conductivity of a solid with the flow properties of liquids, there have been a considerable number of experimental investigations conducted with milli- and micrometer-sized particle suspensions since that initial hypothesis. These investigations have, for the most part, supported the initial theoretical prediction. More recently, these investigations have been directed at the evaluation of the effects of variations in the size of the particles and the relative response time.

The original ideas that form the basis of the field of nanotechnology were initially formulated in two articles by Richard Feynman: "There's Plenty of Room at the Bottom" [4] delivered in 1959, which was republished 33 years later in the *Journal of Microelectromechanical Systems* (1992) [5], and "Infinitesimal Machinery" republished again in 1993 [6].

The study of the thermal behavior of nanomaterials is rapidly increasing in importance and is currently one of the most exciting and significant fields of study within thermal science. At present, the study of the thermal behavior in nanometer size particles is primarily focused on experimental research: tests to measure the specific heat, thermal conductivity, thermal expansion, etc; the study of individual nanoparticles, nanometer membranes, nanotubes, and nanoparticle suspensions; and the thermal behaviors of many compositions or structures formed by nanometer size particles. Meanwhile, theoretical and molecular simulation methods that utilize molecular simulation and the Boltzmann transport equation, have been devised and are still under development to try to understand the fundamental thermophysical phenomenon that governs the behavior of these materials.

The study of heat transfer in nanoparticle suspensions has increased in importance and achievements in this area, allied with achievements in other areas of nanotechnology, offer tremendous promise for engineering applications.Water with 3 nm Cu nanoparticle suspensions was found to have an effective thermal conductivity that was almost twice that of pure water. In order to understand the fundamental reasons why nanoparticle suspensions could cause this type of dramatic increase in thermal conductivity, a number of experimental and theoretical investigations have been conducted. While informative, none of these have provided information that can clearly reveal the nature of this seemingly anomalous thermal behavior.

Building on the theoretical analyses of the different mechanisms of thermal conduction in solids, liquids, and gaseous materials, Wang *et al.* [7] developed a numerical simulation of the effective thermal conductivity of

SiO_2 nanoparticle/water suspensions that focused on the nature of this phenomenon. SiO_2 nanoparticles were selected because the mean free path of the phonon carrier of SiO_2 is smaller than 1 nm, [8] which is much smaller than the diameter of the SiO_2 nanoparticles used (25 nm). Hence there would be essentially no temperature jump at the surface of the SiO_2 nanoparticles. With the elimination of the temperature jump at the surface and the relatively low thermal conductivity of the SiO_2 particles, there should be essentially no difference in the thermal conductivity enhancement between suspensions with millimeter-sized particles and those with nanometer-sized particles, except for the nonsedimentation effects in the nanoparticle suspension. To verify this hypothesis, a numerical simulation was performed in which the SiO_2 nanoparticles oscillated around their initial location. This oscillation was modeled, based on experimental observations of particle suspensions in sessile drops [9] in which the particles in the drops were observed to have a kind of oscillating Brownian motion. The objective of this simulation was to determine the significance of the movement of the nanoparticles and to what effect it would enhance the effective thermal conductivity of the suspensions.

The results of the simulation indicated that the oscillation of the SiO_2 nanoparticles resulted in a significant enhancement in the effective thermal conductivity when compared to pure water. This enhancement appears to be the result of the oscillating Brownian motion of the nanoparticles in the fluid that works much like the lattice cell vibration occurring in solids, which as explained previously, gives rise to an increased thermal conductivity. When the particles are sufficiently small, the oscillation or Brownian motion of the nanoparticles in the liquid solution causes a "microconvection" induced by the random transport of nanoparticles and the heat conduction in the fluid.

More recently, experimental investigations of the effective thermal conductivity of various nanoparticle suspensions indicates that the effect of this oscillating motion and the random transport of nanoparticles will vary in magnitude for the same fluid, depending upon the volume ratio, shape, size, and material composition of the nanoparticles.

In the following, the experimental and theoretical aspects of the effective thermal conductivity of nanoparticle suspensions are reviewed and examined separately. Section II begins with the manufacture and preparation of nanoparticle suspensions and proceeds from a discussion of the various novel properties of nanoparticles and how these property variations affect the overall behavior, through the different experimental methods and facilities used to measure the effective thermal conductivity of nanoparticle suspensions. Section III reviews the development of the thermal conductivity equations from Maxwell's initial formulation to the most recent developments and includes a discussion of the effects of the Brownian motion and thermal phoresis occurring in nanoparticles, and finally describes other heat transfer aspects of nanoparticle suspensions. A summarizing discussion and conclusion is presented in the last section.

II. Experimental Studies and Results

A. Preparation of the Nanoparticle Suspensions

Techniques for manufacturing ultrafine particles with unique physical and chemical properties have been under development since the first technique was identified at the beginning of the 20th century. Since that time, a number of methods and techniques have been developed, capable of manufacturing a wide variety of nanometer size particles for engineering applications and scientific study. Of the many methods of production, the most common is the so-called "inert gas evaporation" method. In this procedure, a crucible or a resistive filament is heated in a vacuum to a predetermined temperature. Following the introduction of an inert gas, the metal material in the crucible or the filament depending upon the method employed is evaporated or in this case sublimated in the gas. The particles sublimate, combine in the inert gas, and then deposit as nanoparticles on a cooled surface. The distribution of the particle diameters obtained through this process is determined by the evaporation rate, the inert gas pressure, the atomic mass of the gas, and the evaporation temperature [10]. Granqvist and Buhrman [11] managed to produce nanometer size particles by evaporation in a temperature-regulated oven containing a reduced atmosphere of an inert gas with nanometer size crystal particles having a diameter less than 20 nm and a spherical shape. Larger nanometer size crystal particles typically assumed a crystalline structure that distorted this spherical shape. This early work, demonstrated that nanometer size particles with a diameter below 20 nm might exhibit noncrystalline properties. In this study, the logarithm size distribution of the particle diameters was reported to assume a Gaussian distribution for the small nanoparticles, with the diameter of nanoparticles directly dependent upon the vapor pressure in which they were produced. To date, a variety of nanometer size particles can be made from common metal or metal oxide materials, as well as from rarely seen ZnO, ZnS, CdS, CdSe, Cd_3P_2, Zn_3P_2, PbI_2, HgI_2, BiI_3, and GaN. The diameters of the particles range from several nanometers to several hundred nanometers and can be produced in a highly reliable process that can accurately control the composition, shape, and size of the particles.

Regardless of the method used to produce the nanoparticles, determination of the mean diameter and the diameter distribution can be made using a variety of techniques, ranging in complexity from a relatively simple technique that utilizes an electron microscope to much more sophisticated methods. In the first of these, samples are made by placing a microscope grid at the center of the cooling plate on which the nanoparticles deposit, or into a colloid made of nanoparticles. After the grid has been dried, the nanoparticles can be readily observed using an electron microscope. In other, more complicated methods described by Jani et al. [12] it was observed that the Fourier transform of the autocorrelation function of a predetection optical signal, contains all of the necessary information on particle size. Here it was demonstrated that by measuring the ratio of the contents of two specific channels of the histogram containing the Fourier transform of the autocorrelation function, the radius of the particle can be calculated as a monotonous function of this ratio.

Hummes et al. [13] applied a differential mobility analyzer (DMA) to determine the size distribution of nanoparticles with great success and to measure this distribution using an atomic force microscope and transmission electron microscopy (TEM). Figures. 1a and b illustrate the results and show TEM images of Al_2O_3 and SiO_2 nanoparticles, respectively.

The size effect of the nanoparticles has resulted in a number of rather unique properties. Buhrman and Halperin [14] measured the effects of thermodynamic fluctuations on the superconducting transition of mostly spherical, single crystal, zero-dimensional Al particles, with an average radius ranging from 400 nm to < 12.5 nm under different temperatures and magnetic fields. In this investigation, zero-dimensional superconductivity was found to be valid both inside and outside the critical region. Meier and Wyder [15] measured the static magnetic moment of a collection of small metallic spherical indium particles with a mean diameter of approximately 5 nm. A saturating paramagnetic moment was found, which indicated that the previous models of electronic quantum size effects are not alone sufficient to explain the experimental results. For this reason, it was assumed that the particles should be treated as having full rotational symmetry, which resulted in a qualitative match with the observed behavior. Novotny and Meincke [16] measured the specific heat of lead particles with diameters of 2.2, 3.7, and 6.0 nm and indium particles with a diameter of 2.2 nm at low temperatures. For the lead particles, the maximum surface-specific heat enhancement existed at temperatures below 15 K, and for indium particles, the maximum surface-specific heat enhancement occurred above 15 K and was significantly greater than that measured for the lead particles. In addition, an anomalously large specific heat discontinuity was observed at the superconducting transition temperature of the indium particles. All of these phenomena were assumed to be the result of the size effect.

Tanner and Sievers [17] measured the far infrared adsorption of small metallic particles of Cu, Al, Sn, and Pb, with diameters ranging from 6.5 to

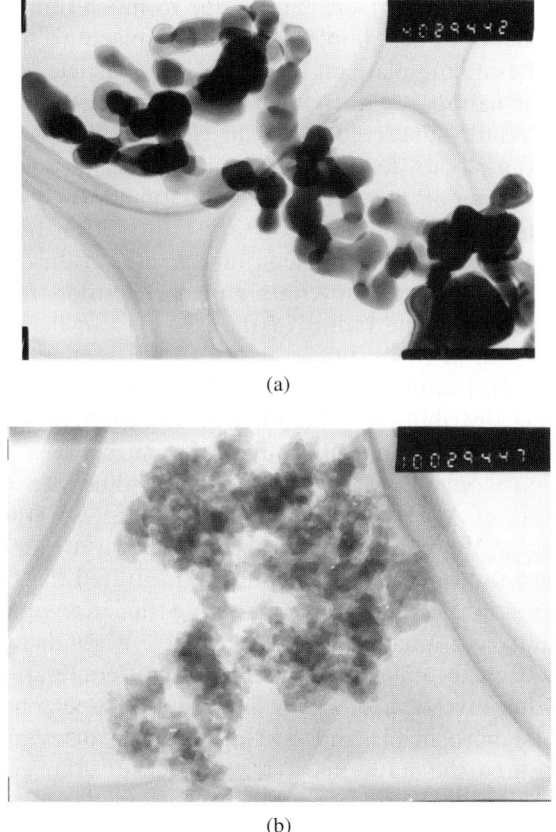

Fig. 1. TEM pictures of nanoparticle of (a) Al_2O_3 and (b) SiO_2.

35 nm. The results indicated that the mean energy-level spacing at the Fermi surface is strongly dependent on the diameter. The adsorption ability of the particles had a linear relation to the frequency in the low-frequency region and a temperature range from 1.2 to 20 K. This investigation also indicated that the assumption of constant internal field would fail at a diameter of 20 nm, when calculating the absorption. Yee and Knight [18] conducted nuclear magnetic resonance line shift experiments on copper particles having diameters ranging from 2.5 to 45 nm. The experiments indicated that the quantum size effects exist for nanoscale particles, and that either the effect of a narrowed conduction electron spin resonance line would occur only in particles of the order of possibly 1 nm or less, or that the theoretical predictions must be modified.

In the early 1980s, Ho [19] and Ben-Jacob et al. [20] discussed the theoretical possibility that the Coulomb charging energy of the particles would suppress the junction conductances that were observed in the (DC) experimental investigation of Giaever and Zeller [21]. The experiments indicated that small isolated Sn particles with diameters down to 2.5 nm would still be superconducting and suggested that there should be no lower size limit for superconductivity. Instead, it was predicted that when the energy-level spacing in a small particle is of the same order as the energy gap, the particle would not become superconducting. For instance, small particles, those with radii below 10 nm could not be superconducting and would deviate from the bulk superconductive properties. Later, Cavicchi and Silsbee [22] applied alternating current (AC) measurements to a "tunnel capacitor" and revealed the details of the tunneling transfer of electrons, from the bulk metal to small metallic particles with an average radius of less than 10 nm. The experiments confirmed the physical theories of Ho [19] and Ben-Jacob et al [20].

In the early 1990s, Kamat and Dimitrijevic [23,24] studied the size quantization effects, nonlinear optical and enhanced photoredox properties of nanometer size particles as semiconductors. Ball and Garwin [25] investigated the optical effects of nanometer size particles and found that the electronic structure does not necessarily acquire the true band-like characteristics of the bulk solid and that the molecular orbitals might remain valid. As a result, it was determined that the discrete highest occupied molecular orbital (HOMO), and the lowest unoccupied molecular orbital (LUMO), with a greater energy gap than that between the fully broadened bulk bands might exist, as well as a slight blue shift of the absorption spectrum. Apart from the physical property variations of these individual nanometer size particles, some distinct physical properties of nanometer size particles were also observed.

In addition, the influence of size on the dielectric behavior of nanoparticles has been studied. In experiments conducted by Xu et al. [26], the relative dielectric constant, ε_r, of 20 nm, 32 nm, and 1 μm particles of GaN were tested with frequencies of 100 Hz to 10 MHz. The results indicated that GaN coarse-grain powders have a lower relative dielectric constant when compared to nanometer size particles. It was hypothesized that this was due to the existence of interfaces with a large volume fraction of nanosized particles. Nalwa [8] found that at nanolength scales, the electrons and phonons by which heat is conducted would be significantly influenced by the conduction dimensions and the interfaces between the different regions. When the mean free path of the heat energy carriers is comparable to, or larger than the size of the particles, the heat energy carriers will be scattered at the interface and, as a result, the heat conduction will be reduced.

This phenomenon is a classic example of the size effect. If the size of the particles is less than the thermal wavelength, the quantum effects will be an important factor and this quantum effect will reduce the thermal conductivity even more. These size effects are especially strong at low temperatures and have been studied in detail by Guczi et al. [27], who measured the morphology, electron structure, and catalytic activity of Au, FeO_x, SiO_2, and Si nanoparticle samples in CO oxidation with X-ray photoelectron spectroscopy, UV photoelectron spectroscopy, and transmission electron microscopy.

As is the case for the individual or groups of nanoparticles discussed above, nanoparticle suspensions also demonstrate some rather unique and novel thermal properties when compared to traditional heat transfer fluids. Nanoparticle suspensions are produced by mixing nanoparticles with a fluid, such that the nanoparticles remain in suspension for long periods of time. There are two principal methods of producing these nanoparticle suspensions: The one-step method in which nanoparticles are produced directly in the base fluid to obtain the suspension, and the two-step method where the nanoparticles are produced independently and then mixed with a base fluid to obtain the suspension. Because of the ease with which the concentration and size distribution can be controlled, a majority of the experimental investigations conducted to date have utilized the two-step method. In the most common approach, the nanoparticles are produced in a vessel by evaporation and fall into the base fluid to form the suspension in a single process. This technique is called the Vacuum Evaporation on Running Oil Substrate (VEROS) technique. Changing the pH value of the suspension, adding surfactants or a suitable surface activator, or using ultrasonic or microwave vibration, are all techniques that have been used with the two-step method to better disperse and more evenly distribute the nanoparticles in the base fluid and maintain the stability of the suspension. When the nanoparticles are small enough, the weight/volume ratio is suitable, and the dispersion method is applied correctly, the nanoparticles will be very well dispersed and the suspension will be stable for several days. However, in a majority of the investigations studied, the literature reports that the suspension samples can typically be maintained in a homogeneous stable state for not more than 24 h.

Like nanoparticles, because of the extremely small size of the particles used, nanoparticle solutions have a number of interesting and seemingly unusual properties; and a number of recent investigations using well-prepared suspensions have been conducted to investigate the different physical aspects of nanoparticle suspensions, such as viscosity, flow behavior, and effective thermal conductivity, etc. Among the most interesting property variations is the effect that the addition of nanoparticles to a base fluid can have on the effective thermal conductivity of the nanoparticle suspension. Several of these experimental investigations have demonstrated that nanoparticle suspensions have remarkably high effective thermal conductivities compared with traditional heat transfer fluid media, like water, engineering oil, and organic fluids and that the increase in the effective thermal conductivity is greater than what might be predicted by the conventional expressions developed for larger particles discussed earlier. In order to better utilize the novel thermal properties of these suspensions or nanofluids, such as the SiO_2 nanoparticles in suspension illustrated in Fig. 2, it is necessary to more fully understand the fundamental mechanisms that govern the behavior of these nanoparticle suspensions.

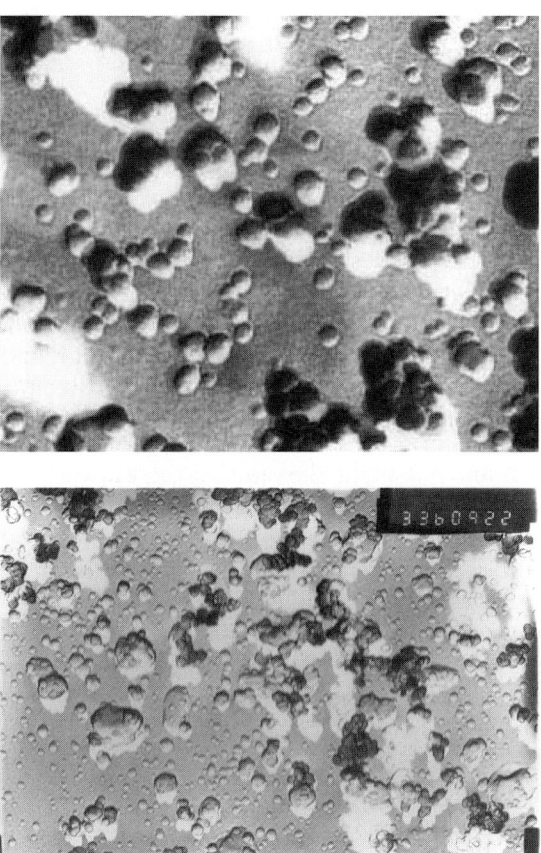

FIG. 2. TEM pictures of SiO_2 nanoparticles in suspension.

B. Experimental Methods and Test Facilities

There are two principal experimental methods that have been used to measure the effective thermal conductivity of nanoparticle suspensions: the transient approach, which typically utilizes a hot-wire method; and steady-state methods, which utilize a guarded hot-plate, a cut-bar apparatus, or a temperature oscillation technique.

The most commonly used apparatus in the measurement of effective thermal conductivity of fluids and nanoparticle suspensions is the transient hot-wire system. Nagasaka and Nagashima [28] first applied this method to measure the thermophysical properties of electrically conducting liquids. In this approach, a coated platinum hot wire is suspended symmetrically in a liquid contained within a vertical cylindrical container. This hot wire serves as both a heating element, through electrical resistance heating and as a thermometer, by measuring the electrical resistance of the fluid. The thermal conductivity can be calculated from the relationship between the electrical and thermal conductivity as

$$T(t) - T_{ref} = \frac{q}{4\pi k} \ln\left(\frac{4K}{a^2 C} t\right) \qquad (6)$$

where $T(t)$ is the temperature of the platinum hot wire in the fluid at time t, T_{ref} the temperature of the test cell, $\dot{q}$ the applied electric power applied to the hot wire, k the thermal conductivity, K the thermal diffusivity of the test fluid, a the radius of the platinum hot wire, and $\ln C = g$, where g is Euler's constant. This relationship between δT and $\ln(t)$ is linear, and the data of δT was valid only over a valid range of $\ln(t)$, namely between time t_1 and time t_2, the thermal conductivity of the fluid can be calculated as

$$k = \frac{q}{4\pi(T_2 - T_1)} \ln\left(\frac{t_2}{t_1}\right) \qquad (7)$$

where $T_1 - T_2$ is the temperature difference of the platinum hot wire between times t_1 and t_2.

Since this system has a wire coated with a layer of electrically insulating material, a number of problems that may arise have been analyzed. These problems may include (1) the effects of the insulation layer on the temperature rise of the metallic wire, (2) the effects of the insulation layer on the reference temperature, (3) the effects of the thermal contact resistance between the metallic wire and the insulation layer, and (4) the effects of the finite length of the wire. Experimental calibrations using NaCl solutions, verified that the whole system could be used with an overall accuracy of $\pm 0.5\%$.

Following the publication of the manuscript by Nagasaka and Nagashima [28] a number of investigators applied the transient hot-wire method to measure the effective thermal conductivity of different nanoparticle suspensions. These investigators included those of Choi [29], Eastman et al. [30] Lee et al. [31], Eastman et al. [32], Xuan and Li [33,34], Xie et al. [35,36] and the accuracy of this method has been found to be excellent, as shown by the following experimental data (see Figs. 3 and 4).

The second method for measuring effective thermal conductivity, the steady-state method, utilizes a number of different approaches. The first of these was used by Wang and Xu [37] and designed by Challoner and Powell [38]. As illustrated in Fig. 5, this one-dimensional, parallel-plate method, utilizes a guarded hot plate that surrounds two round copper plates positioned parallel to each other and set compactly in an aluminum cell. The nanoparticle suspension is placed between the two plates and by measuring the temperature of the plates and the cell, the overall thermal conductivity of the two copper plates and the sample fluid can be calculated using the one-dimensional heat conduction equation relating the heat flux, q, the temperature difference ΔT, and the geometry of the liquid cell.

$$k = \frac{qL_g}{S\Delta T} \quad (8)$$

$$k_e = \frac{kS - k_g S_g}{S - S_g} \quad (9)$$

where L_g is the thickness of the glass spacer between the two copper plates, S the cross-sectional area of the top copper plate, and k_g and S_g are the thermal conductivity and the total cross-sectional area of the glass spacers, respectively. The accuracy of this type of system has been reported to be greater than $0.02°C$ for the temperature gradient, which yields an overall experimental uncertainty of less than $\pm 3\%$.

Another steady-state measurement system that utilizes the same fundamental approach is the cut-bar apparatus used extensively by Peterson and Fletcher [39–41] to measure the thermal conductivity of both saturated and unsaturated dispersed ceramics and the contact resistance of various interfacial materials [42]. This experimental facility, which was first described by Miller and Fletcher [43] is illustrated in Fig. 6, and consists of a vertical column composed of a heat source; two test fixtures typically fabricated from stainless steel 304, copper or aluminum; the test cell; a load cell, to determine the contact pressure; and a heat source and sink. The fixtures and test cell are arranged so that a cylindrical column of spherical particles is

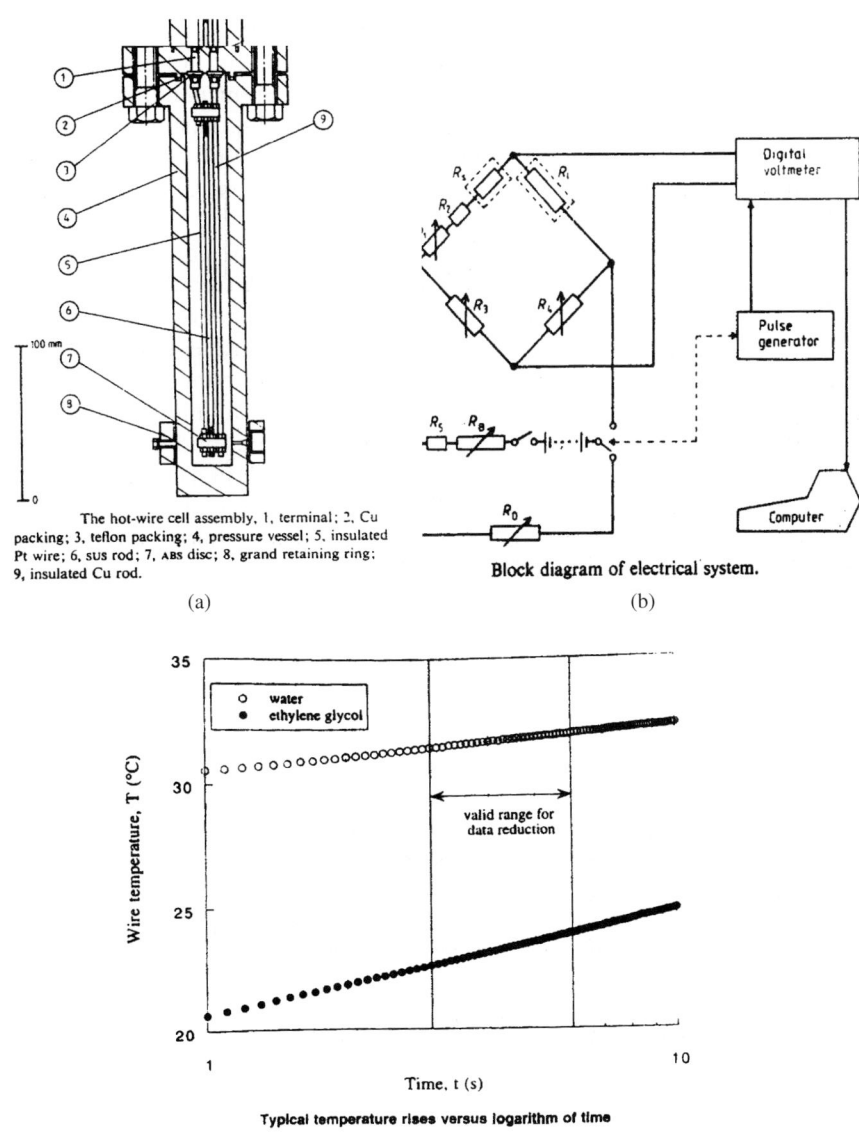

Fig. 3. Transient hot-wire method and data-logging method: (a), (b) [28], (c) [31].

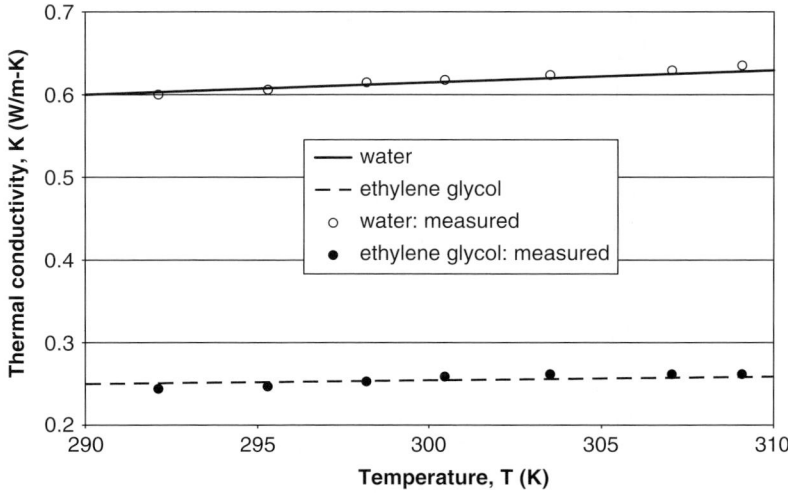

Fig. 4. Calibrating of transient hot-wire experimental setup (measuring thermal conductivity of basefluids of water and ethylene glycol) [31].

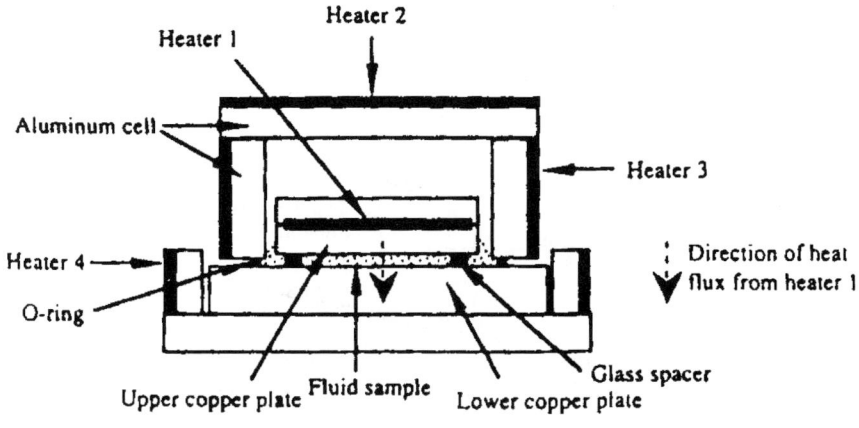

Fig. 5. The steady-state one-dimension method appatatus [37].

sandwiched between the two test fixtures. The test bed particles are contained within an insulated tube to reduce the heat transferred through the sleeve. The top test fixture is placed in a heat source, heated using electrical resistance heaters and the other test fixture is placed in direct contact with a liquid-cooled heat sink. A series of thermocouples evenly spaced in the test

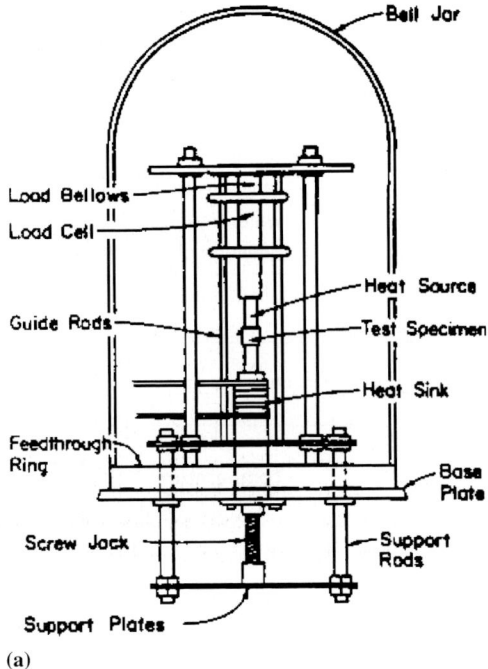
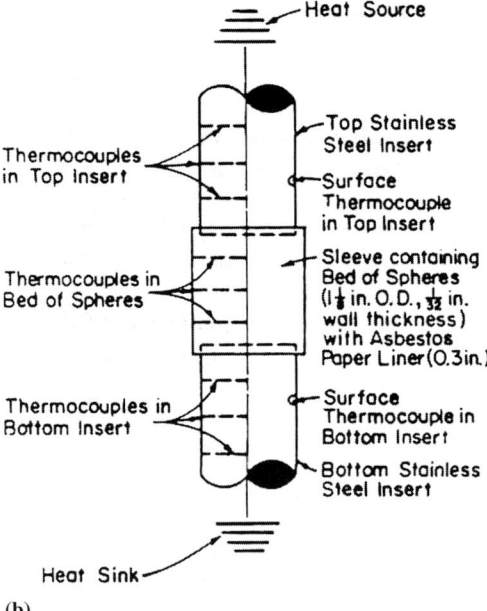

FIG. 6. The vacuum steady-state thermal conductivity test rig [42]; (a) the experimental test facillity, (b) the packed-bed test specimens.

fixtures are used to determine the average heat flux above and below the test fixture, and hence the temperature distribution in the test cell as illustrated in Fig. 6. The accuracy of the system is enhanced by placing the entire system in a vacuum environment of less than 10^{-5} Torr and surrounding the test sample with a passive or active radiation shield to reduce radiation losses.

In this system, the effective thermal conductivity can be obtained as

$$K_{eff} = \frac{Qh}{A\Delta T} \qquad (10)$$

where Q is the heat flux through the test cell, h the height corresponding to the temperature difference ΔT, and A the cross-sectional area in the test cell. This method, while used extensively for measuring the thermal contact resistance, has also been applied with good success to the measurement of the effective thermal conductivity of materials and saturated porous materials and has been shown to have an accuracy of nearly 0.02°C and an overall experimental uncertainty of less than $\pm 3\%$.

A third technique, the temperature oscillation technique, was first reported in 1995 by Czarnetzki and Roetzel [44] and is illustrated in Fig. 7.

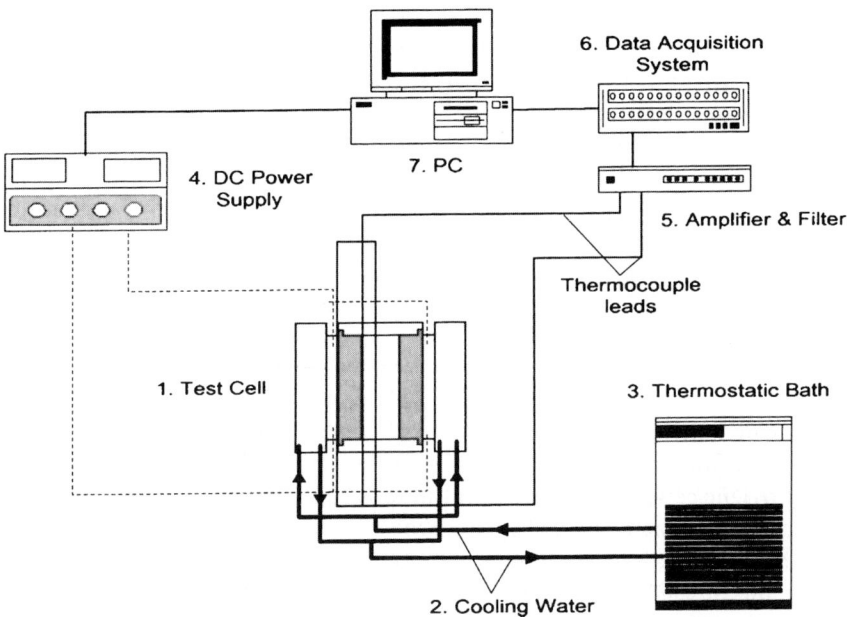

FIG. 7. The temperature oscillation system [45].

This approach has been applied to measure effective thermal conductivity of a variety of materials [45] and utilizes the temperature oscillation in a test section through a number of thermocouples to determine the thermal diffusivity of the nanoparticle suspension. Then, by comparison with a reference layer, the thermal conductivity can be calculated. The thermal diffusivity and thermal conductivity are related using the one-dimensional form of energy equation

$$\frac{\partial^2 T}{\partial \xi^2} = \frac{\partial T}{\partial \tau} \qquad (11)$$

where $\xi = x\sqrt{\omega/\alpha}$ and represents the dimensionless space coordinate, α the thermal diffusivity, and ω the constant angular frequency. Using the assumption that the periodic temperature oscillations are generated with the same frequency in the boundary regions, but different amplitude and phases than at the two surfaces, the thermal diffusivity of the nanoparticle suspensions can be obtained by measuring the phase shift and amplitude of the temperature oscillation at these two surfaces and the center point of the fluid sample. The drawback of this method is that the accuracy of the thermal conductivity measurement of the nanoparticle suspensions depends directly upon the accuracy of the measured thermal diffusivity of the reference layer. The average deviation of the measured thermal diffusivity from the standard handbook values [45] has been shown to be 2.7% in the temperature range of 20–50 °C and less than 2.11% in the range of 20–30 °C.

C. Measurements of the Effective Thermal Conductivity

Over the past 45 years, there have been a number of reports documenting the results of experimental investigations conducted to determine the effective thermal conductivity of different types of particle suspensions, with particles ranging in size from millimeters to nanometers. As early as 1961, Meredith and Tobias [46] investigated the thermal conductivity of water/ propylene carbonate emulsions with volume fractions ranging from 0% to 50%. The thermal conductivity ratios between the continuous and discontinuous phases of these emulsions, ranged from 17.2% to 10,100%. As observed in this investigation, when water was the discontinuous phase and the propylene carbonate was the continuous phase, that is, when the thermal conductivity of the dispersed phase, where propylene carbonate was higher than that of the continuous phase, water, the effective thermal conductivity of the mixture was very sensitive to the volume ratio between the two phases. However, when water was the continuous phase and the propylene was the dispersed phase, the effective thermal conductivity of the emulsion was relatively insensitive to the volume ratio of the two phases. In addition, when the water was the dispersed phase and the volume ratio was below 20%, the experimental data were found to fall below the data line calculated using Maxwell's equation, but when the volume ratio was greater than 20%, the experimental data fell above the values calculated using Maxwell's equation.

In 1962, Hamilton and Crosser [47] measured the effective thermal conductivity of mixtures containing millimeter-sized aluminum and balsa particles with silastic rubber at a mean temperature of approximately 95°F using an apparatus that consisted of an electrically heated sphere surrounded by a spherical shell of the mixture. The particles used in the experiments were different in shape, i.e., spheres, cylinders, parallelepiped, disks, and cubes. The thermal conductivity enhancement of the suspensions with different shapes of particles varied considerably as shown in Table II, where the thermal conductivity ratio between the copper spheres and the base fluid was more than 1000 times greater and the volume ratio ranged from 5% to 20%. In the experiments using aluminum, the cylindrical shape with a length-to-diameter ratio of 10 to 1, had nearly 1.5 times the thermal conductivity of the mixture containing spheres. However, the thermal conductivity of the mixture containing nonsphere-shaped balsa wood particles showed no shape effect.

TABLE II

THERMAL CONDUCTIVITY OF MIXTURES OF RUBBER AND PARTICLES OF ALUMINUM OR BALSA AS THE DISCONTINUOUS PHASE [47]

Material	Shape	Particle dimensions (mm)	Volume of particles (%)	Measured conductivity of mixture (B.t.u./Sq. Ft. Hr. °F./Ft.)
Aluminum	Spheres	1.2 diameter.	27.0	0.234
Aluminum	Spheres	1.2 diameter	27.0	0.235
Aluminum	Spheres	1.2 diameter	15.5	0.173
Aluminum	Spheres	1.2 diameter	27.5	0.168
Aluminum	Spheres	0.012 diameter	27.5	0.238
Aluminum	Cylinders	5 × 1	15.5	2.04
Aluminum	Parallelepipeds	1.6 × 1.6 × 0.4	15.5	0.216
Aluminum	Cylinders	2.7 × 0.27	15.5	0.236
Balsa	Disks	0.8 × 7.2	25.0	0.087
Balsa	Disks	0.8 × 7.2	25.0	0.086
Balsa	Cubes	0.8 × 0.8	25.0	0.084
Balsa	Disks	0.8 × 7.2	14.0	0.096

Eastman et al. [30] measured the effective thermal conductivities of CuO/water nanoparticle suspensions with a mean diameter of the CuO nanoparticles of 36 nm and Al_2O_3/water nanoparticle suspensions with a mean diameter of the Al_2O_3 particles of 33 nm. The results indicated that for volume ratios of 0%, 1%, 2%, 3%, 4%, and 5%, the CuO/water nanoparticle suspensions demonstrated an effective thermal conductivity enhancement of as much as 60% and as much as 30% for the Al_2O_3/water nanoparticle suspension. The effective thermal conductivity enhancement had a linear relationship to the volume ratio of the nanoparticle to water as shown in Fig. 8. Further experiments indicated that both Cu/Duo-Seal oil nanoparticle suspensions and Cu/HE-200 oil nanoparticle suspensions, exhibited even higher thermal conductivity enhancements than those observed in the Al_2O_3/water or the CuO/water nanoparticle suspensions, and that the linear relationship between the enhancement of the effective thermal conductivity and volume ratio did not change as shown in Fig. 9.

In 1999, Lee et al. [31] measured the thermal conductivity enhancement of Al_2O_3/ethylene glycol and CuO/ethylene glycol nanoparticle suspensions. The results are illustrated in Fig. 10, and are compared with the experimental data obtained by Masuda et al. [31] for Al_2O_3/water and CuO/water nanoparticle suspensions, in Fig. 11. The Al_2O_3 nanoparticles used by Masudu et al. [31] had a diameter of 13 nm. The thermal conductivity ratios observed by Lee et al. [31] were lower than those of Masuda et al. [31] by

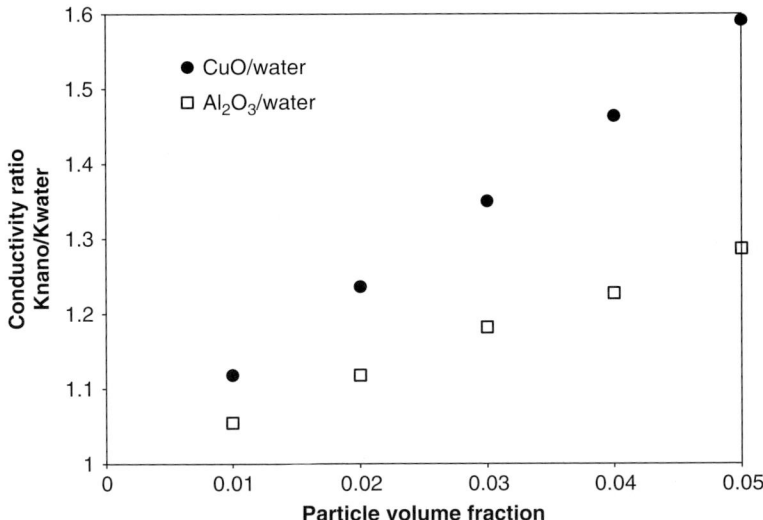

FIG. 8. Effective thermal conductivity in water basefluid vs. particle volume fraction [30].

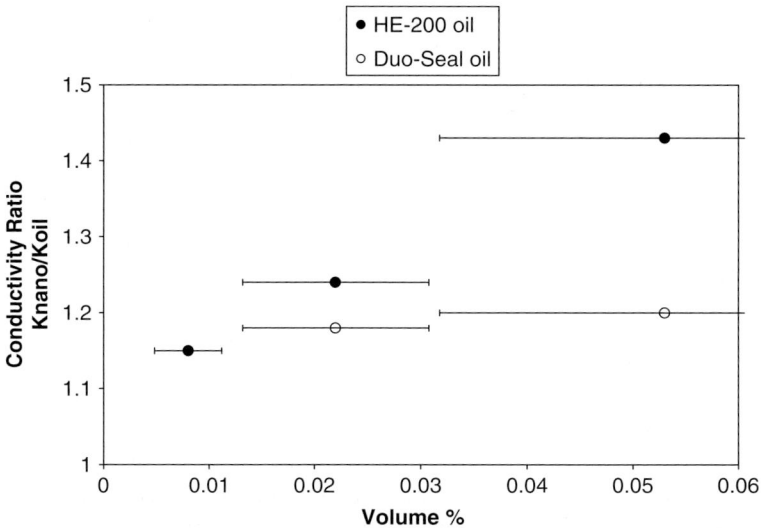

FIG. 9. Effective thermal conductivity in oil base fluid vs. particle volume fraction [30].

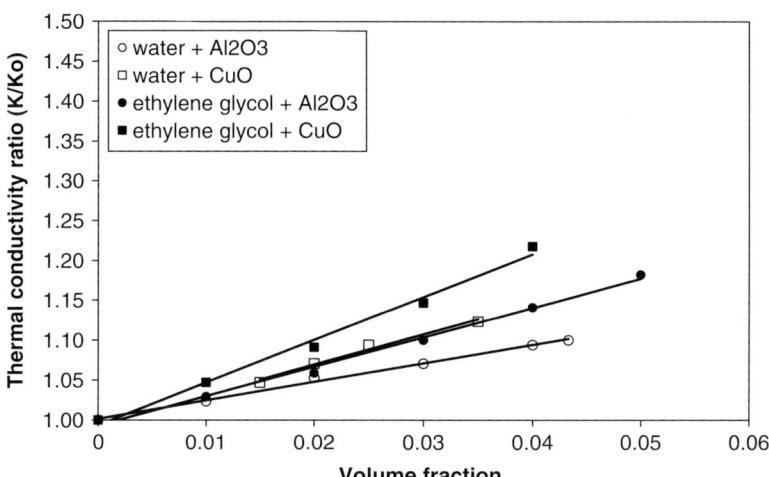

FIG. 10. Thermal conductivity vs. volume ratio [31].

more than 20%. This variation was attributed to the agglomeration of the nanoparticles in the nanoparticle suspensions of Le et al. [31] As illustrated in Fig. 12, the data were also compared with the prediction of the Hamilton and Crosser model [47]. Using this comparison, Lee et al. [31] concluded

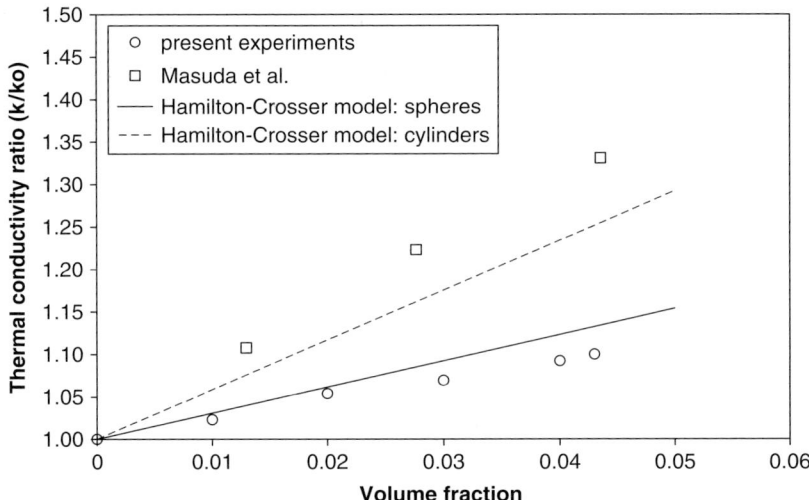

FIG. 11. Comparison of experiments by S. Lee, S. U.S. Choi et al. [31]. with data from Al$_2$O$_3$/water nanoparticle suspension experiments by Masuda et al.

that the particle shape and size were both dominant in the thermal conductivity enhancement of nanoparticle suspensions of this type, a result that is consistent with that of Hamilton and Crosser [47].

Meanwhile, nanoparticle suspensions comprise CuO and Al$_2$O$_3$ nanoparticles with differing base fluids were tested by Wang and Xu . In this investigation, the average diameter of the CuO and Al$_2$O$_3$ nanoparticles were 23 and 28 nm, respectively. The base fluids were pump fluid, engine oil, ethylene glycol and water, as illustrated in Figs. 13–16, respectively. In this investigation, three different methods were utilized to produce the nanoparticle suspensions. The first of these was to disperse nanoparticles into the base fluid with an ultrasonic bath; method 2 was to add polymer, coatings (styrene–maleic anhydride, 5000 mol.wt., 2.0% by weight) and keep the pH value at 8.5–9.0; method 3 was to filtrate and remove particles larger than 1 μm in diameter. The thermal conductivities of Al$_2$O$_3$/water nanoparticle suspensions are compared in Fig. 17. As is apparent from the experimental data presented, the nanoparticle suspensions produced using the third method resulted in the greatest enhancement in the thermal conductivity.

The results were also compared with a number of different theoretical models. The effect of the Brownian motion of the nanoparticles was considered, as were the effects of electric double layer and the van der Waals force. Micromotion caused by the electric double layer, van der Waals force, and the chain structure of the nanoparticles were considered to be the

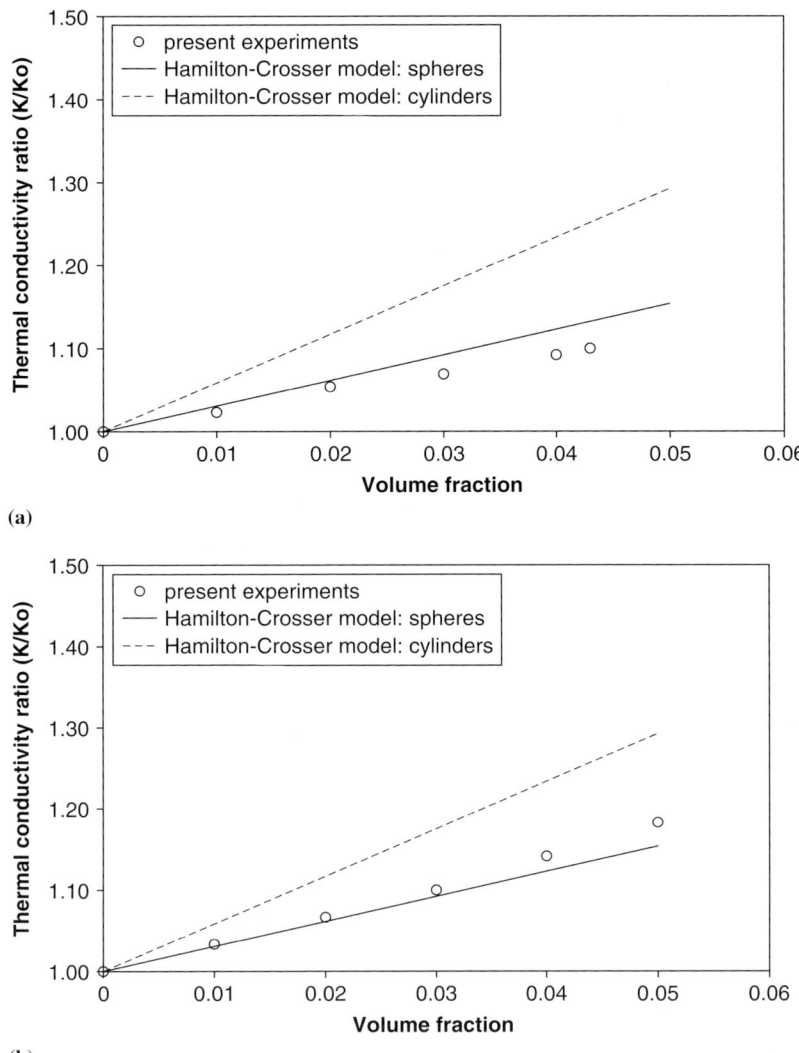

Fig. 12. Comparison of experimental data with Hamilton–Crosser model [31]: (a) Al_2O_3/water nanoparticle suspension, (b) Al_2O_3/ethylene glycol nanoparticle suspension, (c) CuO/water nanoparticle suspension, and (d) CuO/ethylene glycol nanoparticle suspension.

principal reasons and provided the main explanation for the enhancement in this particular case.

Xuan and Li [33,34] used the hot-wire method to evaluate suspensions of copper nanoparticles in transformer oil and copper nanoparticles in water.

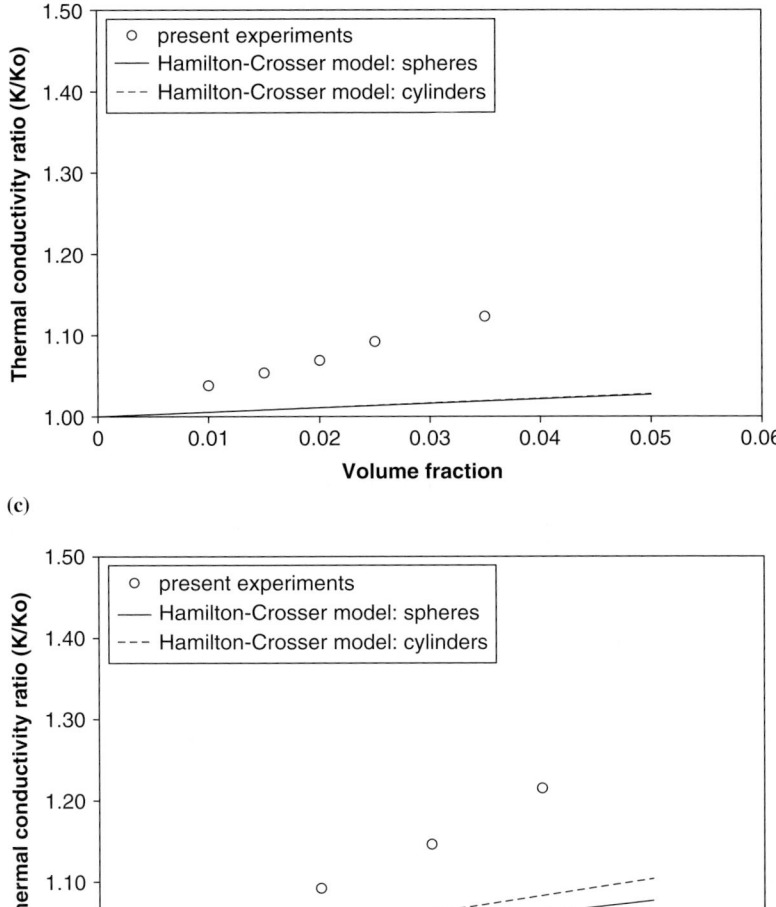

FIG. 12. Continued

The diameter of the copper nanoparticles ranged from 30 to 100 nm and the volume ratio between the nanoparticle and the base fluid varied from 2.5% to 7.5%. In this experiment, surface activators were added to the suspensions. For the copper nanoparticle/transformer oil suspension, an oleic acid was applied, and for the copper nanoparticle/transformer oil suspension, a

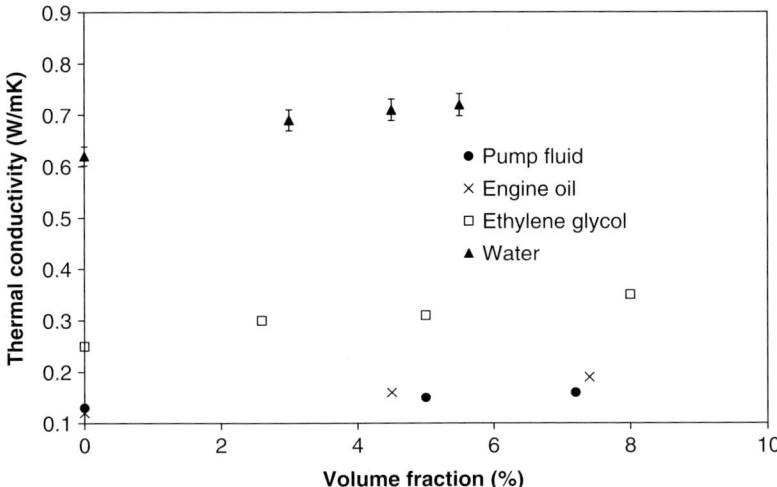

FIG. 13. Thermal conductivity as a function of volume fraction of Al_2O_3 powders in different fluids [37].

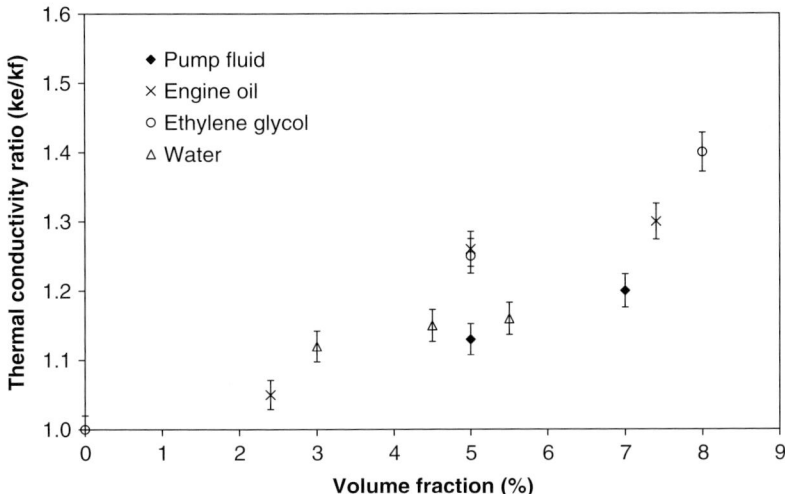

FIG. 14. Thermal conductivity ratio as a function of volume fraction of Al_2O_3 powders in different fluids [37].

laurate salt was used. Both suspensions were subjected to ultrasonic waves to enhance the mixing and disperse the nanoparticles. The final effective thermal conductivity data were compared with the data obtained by Eastman *et al.* in 1997, as shown in Figs. 18 and 19.

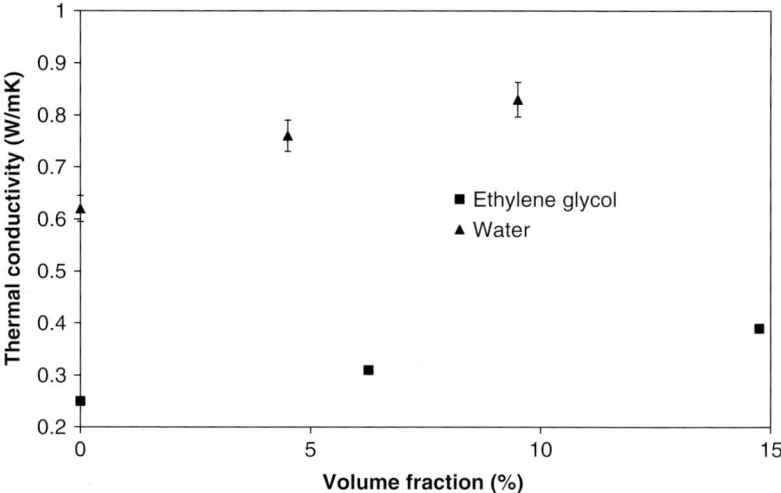

FIG. 15. Thermal conductivity as a function of volume fraction of CuO powders in ethylene glycol and water [37].

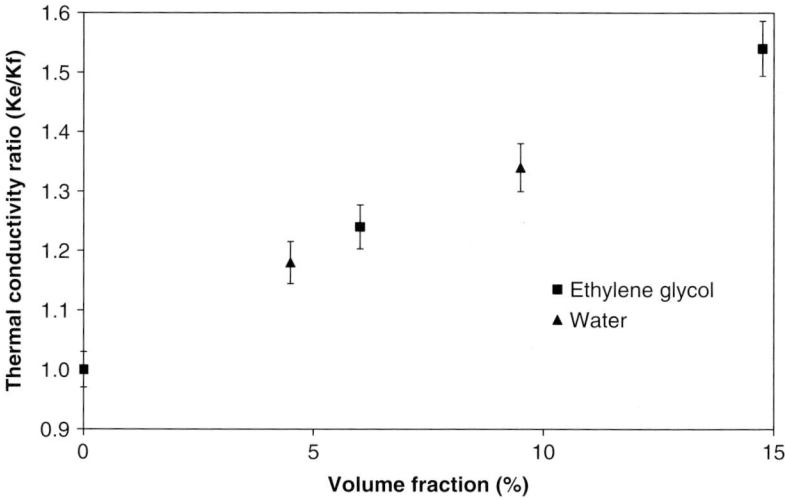

FIG. 16. Thermal conductivity ratio as a function of volume fraction of CuO powders in ethylene glycol and water [37].

In a different investigation, the thermal conductivity of several nanoparticle suspensions using different stabilizing agents and Cu/ethylene glycol combinations were evaluated experimentally. The diameter of the Cu nanoparticles was less than 10 nm [32]. As illustrated in Fig. 20, the nanoparticle

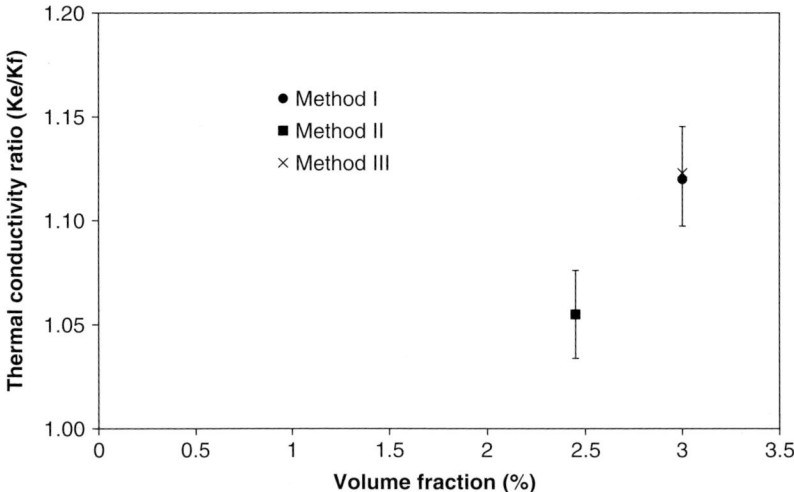

FIG. 17. Thermal conductivity of Al_2O_3/water mixtures prepared by three different methods [37].

suspensions containing thioglycolic acid as a stabilizing agent showed improved behavior when compared to nonacid samples containing the nanoparticle suspensions, while fluids containing thioglycolic acid, but no particles, showed no improvement in thermal conductivity. Fresh nanoparticle suspensions tested within two days of preparation exhibited slightly higher conductivities than fluids that were stored up to two months prior to measurement.

The measured effective thermal conductivities of metallic and oxide nanoparticle suspensions are shown in Fig. 21. The results indicate that nanoparticle suspensions with metallic nanoparticles have higher thermal conductivities than nanoparticle suspensions with oxide nanoparticles, and imply that the higher intrinsic thermal conductivity of the metallic nanoparticles was the principal cause of this increase over the oxide nanoparticle suspensions. However, it is important to note that the metallic nanoparticles used in this experiment had a considerably smaller diameter than the oxide nanoparticles evaluated.

Experimental data from still another set of Al_2O_3 nanoparticle suspensions were reported by Xie et al. [35,36], in which the Al_2O_3 nanoparticles had a specific surface area ranging from 5 to $124 \, m^2/g$. In this investigation, it was found that with the same type of nanoparticles, the thermal conductivity enhancement of nanoparticle suspension was reduced with the increasing value of thermal conductivity of the base fluid. The particle size ranged from 30.2 to 12.2 nm and had different crystalline phases. When the

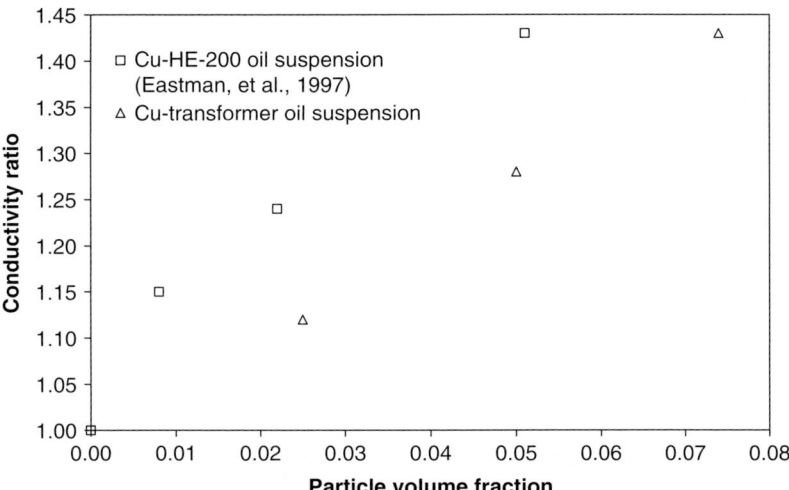

FIG. 18. Thermal conductivity enhancement of copper/transformer oil nanoparticle suspension vs. volume fraction of the nanoparticle [34].

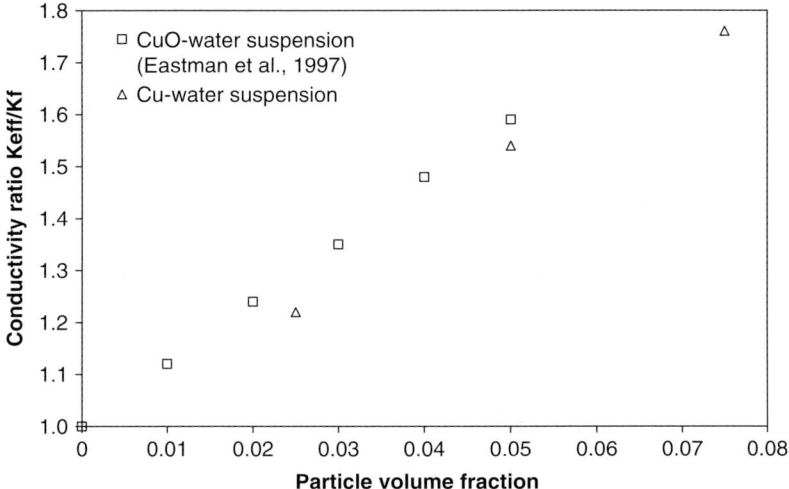

FIG. 19. Thermal conductivity enhancement of copper/water nanoparticle suspension vs. volume fraction of the nanoparticle [34].

same base fluid was used, however, the enhancement was shown to be dependent on the pH value of the suspension and the specific surface area of the nanoparticles. Fig. 22 indicates that the effective thermal conductivity increases with different slopes for different pH values, while Fig. 23

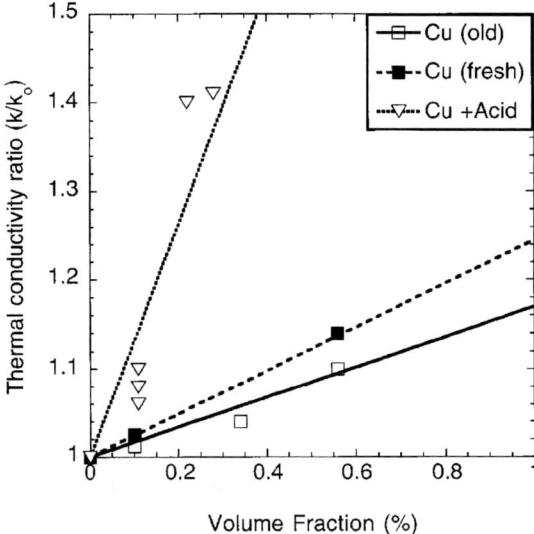

FIG. 20. Effective thermal conductivities of different Cu/ethylene glycol nanoparticle suspensions [32].

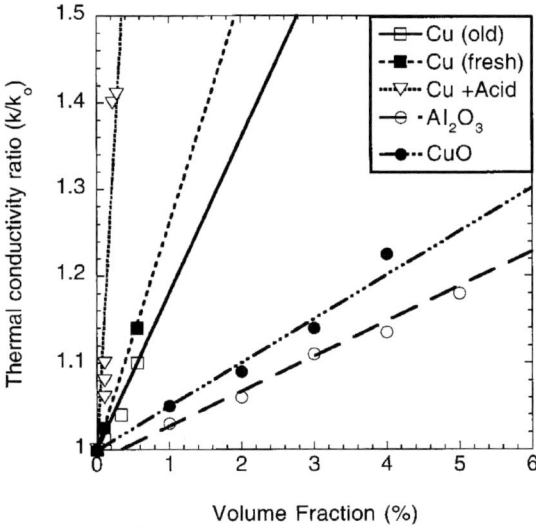

FIG. 21. Effective thermal conductivities of metallic and oxide nanoparticle suspensions [32].

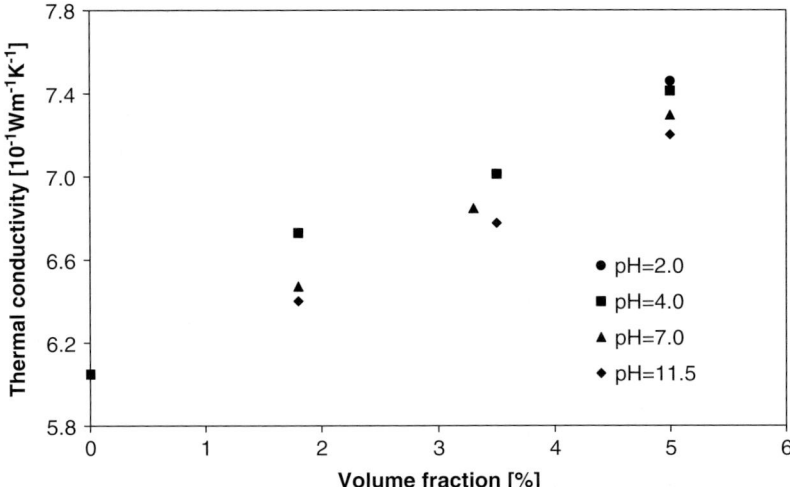

FIG. 22. Thermal conductivities of $\alpha A - 25/H_2O$ suspension as a function of solid volume fraction [36].

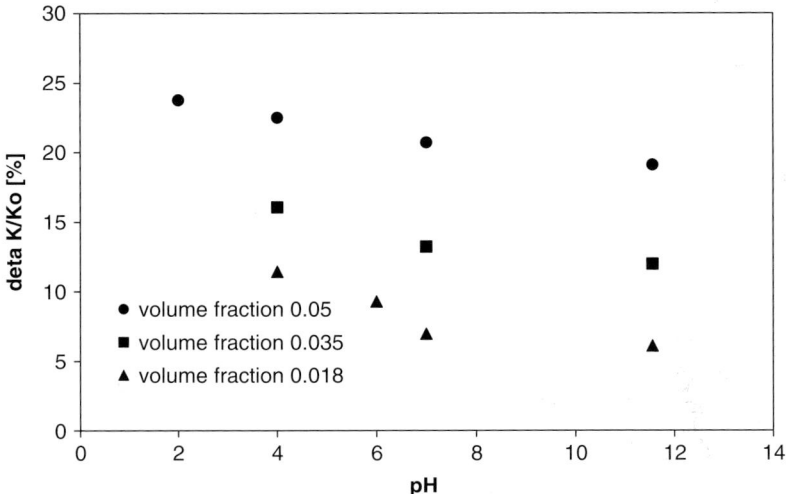

FIG. 23. Thermal conductivity enhancement ratios of $\alpha A - 25/H_2O$ suspension at different pH values [36].

indicates that the enhanced thermal conductivity ratio decreased with an increase in the pH values.

Previous results have indicated that the thermal conductivity enhancement increases nearly linearly with the volume fraction, but the rate of the enhancement ratios as a function of the volume fraction also depends on the type of dispersed nanoparticles as shown in Fig. 24.

The results presented in Fig. 25 indicate that the thermal conductivity ratio increases initially to a maximum at approximately $30\,m^2/g$ and then decreases with an increase in the specific surface area. The mean free path of phonons in polycrystalline nanoparticles was estimated to be around 35 nm. When the size of the polycrystalline nanoparticle was larger than 35 nm the interfacial area of the nanoparticle and base fluid increased with a decrease in the size of the nanoparticle and, hence, enhanced the effective thermal conductivity. When the size of the polycrystalline nanoparticle was smaller than 35 nm, the effect of the intrinsic thermal conductivity of the nanoparticle was reduced due to the scattering of the primary carriers of the energy at the interface between the nanoparticle and the base fluid.

The experimental results shown in Fig. 26 indicate that the thermal conductivity enhancement is higher in a base fluid with a low thermal conductivity than for base fluids with higher thermal conductivities. These results also demonstrate that the equations currently used for calculating the effective thermal conductivity cannot accurately predict the measured thermal conductivity enhancement.

The effect of different fluids used in the manufacture of nanoparticle suspensions has also been investigated. With α-Al_2O_3 nanoparticles, which have a specific surface area of $25\,m^2/g$, four different fluids with significantly

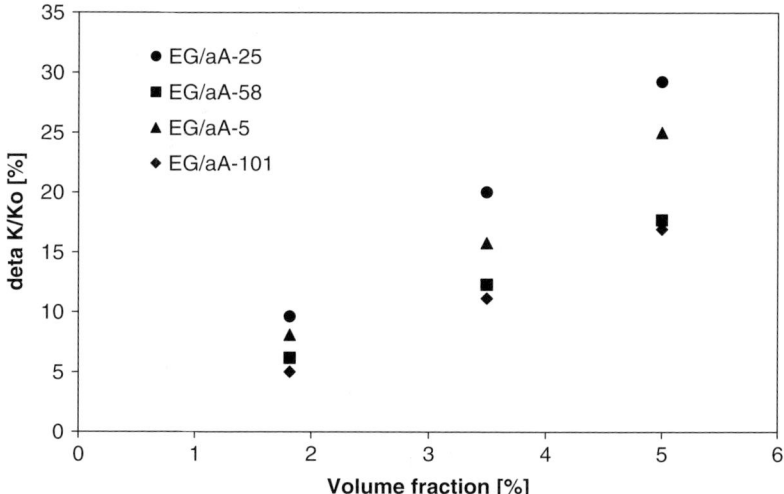

FIG. 24. Enhanced thermal conductivity ratios of the suspensions with nanoparticles in EG [36].

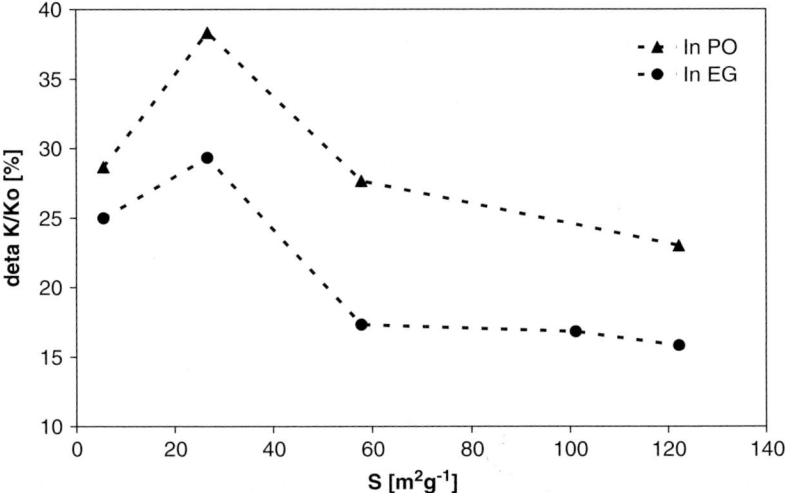

FIG. 25. Enhanced thermal conductivity ratios as a function of the specific surface area with solid volume fraction of 5% [36].

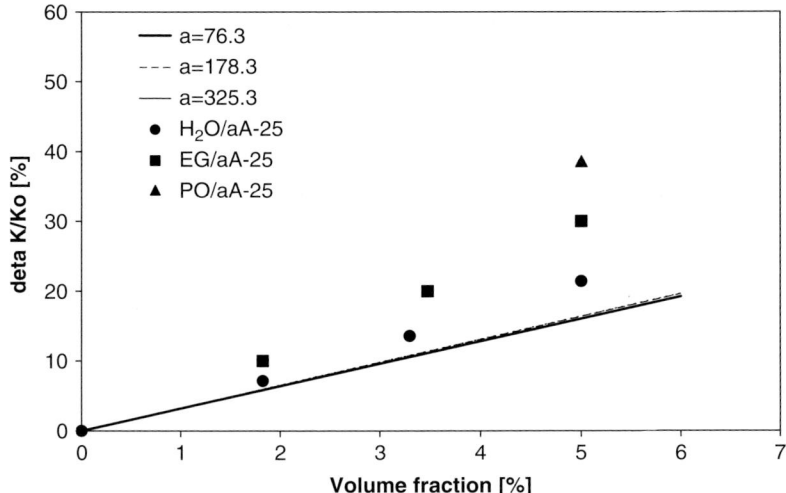

FIG. 26. Comparison between predicted values and the experimental results [36].

different thermophysical properties were used to make the suspensions. The fluids investigated were deionized water, glycerol, ethylene glycol, and pump oil. The volume fraction of the α-Al_2O_3 nanoparticles in the suspensions was held constant at 5.0%, and the experiments were initially conducted at a mean temperature of 25 °C. The results of this investigation are shown in Fig. 27. The base fluids used in these experiments were not pure materials, but rather fluid–fluid mixtures. The volume fractions of water in these fluid–fluid mixtures were defined as Φ_E and Φ_G for ethylene glycol and glycerol, respectively. The results, illustrated in Figs. 28–31, clearly indicate that the enhancement of the nanoparticle suspensions increases with a corresponding decrease in the thermal conductivity of the base fluid and that this trend is not dependent upon the particular base fluid used.

In 2003, nanoparticle suspensions of Al_2O_3/water and CuO/water were measured by Das et al. [45] at various nanoparticle volume ratios and different temperatures, using the temperature oscillation method described previously. Figure 32 shows the enhancement of thermal conductivity of Al_2O_3/water nanoparticle suspensions as a function of the temperature. As illustrated, for CuO/water nanoparticle suspensions with a 1% volume fraction, there is a considerable increase in the enhancement over the temperature range from 21 °C to 51 °C, i.e.; at room temperature, 20 °C, the enhancement is only approximately 2%, while at 51 °C the enhancement increases to approximately 10.8%. With a 4% volume fraction and CuO/water suspensions, the enhancement increases from 9.4% to 24.3% as the temperature increases from 21 °C to 51 °C.

The enhancement ratio relationship was also studied and the experimental data were compared with the prediction of the Hamilton–Crosser model

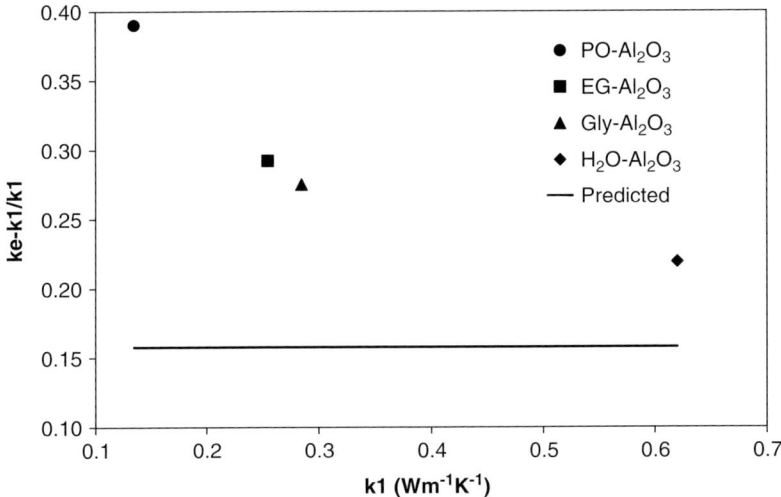

FIG. 27. The enhancement ratios of thermal conductivity between suspensions and the basefluids vs. the basefluids' thermal conductivities [35].

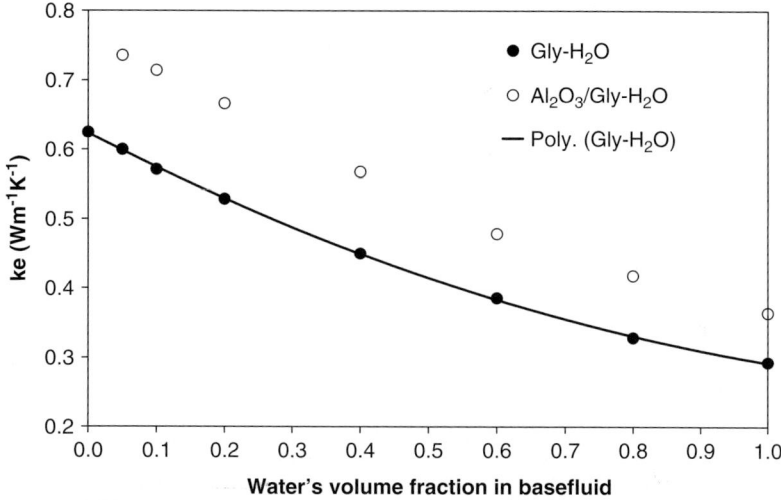

FIG. 28. Effective thermal conductivities of suspensions vs. the water volume fraction in water–glycerol mixture as the basefluid [35].

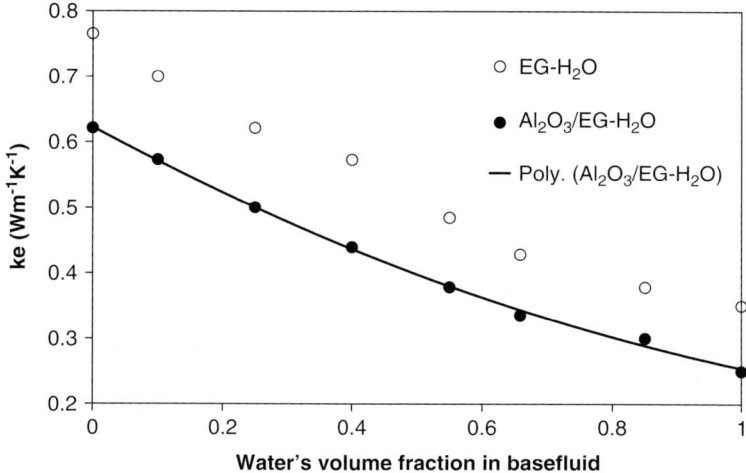

FIG. 29. Effective thermal conductivities of suspensions vs. the water volume fraction in water–ethylene glycol mixture as the basefluid [35].

[47], as shown in Fig. 33. Comparison of these results with those shown in Figs. 32 and 33 for Al_2O_3/water nanoparticle suspensions clearly indicate that the increasing slope of the data lines is governed by a combination of the volume ratio and temperature. Figures 34 and 35 show the same

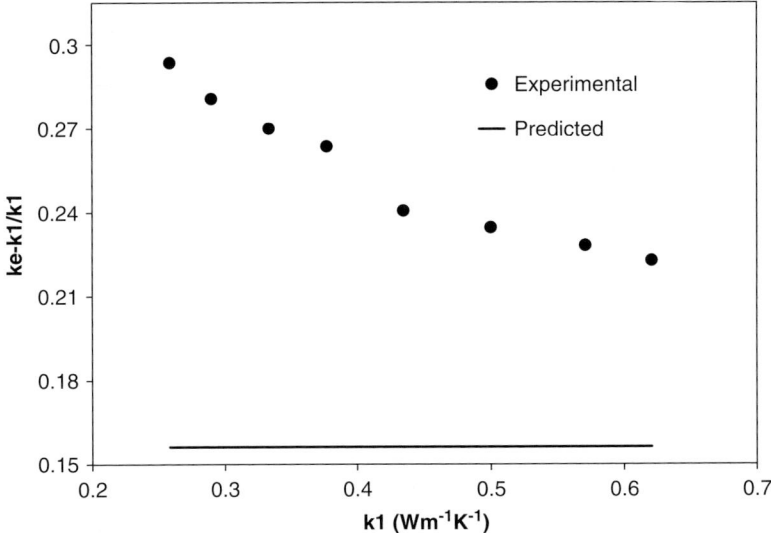

FIG. 30. Enhancement ratio of thermal conductivity between suspension and basefluid vs. the thermal conductivity of basefluid of water–ethylene glycol mixture [35].

phenomena for CuO/water nanoparticle suspensions. These experimental data when coupled with other previous investigations indicate that in addition to the relative volume fraction, thermal conductivity and particle size, a stochastic motion of nanoparticles may be a probable explanation for the difference between the prediction of Hamilton and Crosser theory and the experimental data.

A more recent article on nanoparticle suspensions published in 2003 by Patel et al. [48] found that the enhancement of the effective thermal conductivity did not increase linearly with the increase of temperature, for the same volume fraction of nanoparticle suspensions composed of Au–thiolate nanoparticles with 10–20 nm diameters and a base fluid of water and toluene. The enhancement ratio in this investigation was found to be 5–21% in the temperature range of 30–60 °C at a volume ratio of 0.00026 in toluene, as shown in Fig. 36 and 7–14% for Au particles stabilized with a monolayer of octadecanethiol, even for a loading of 0.001% in water, as shown in Fig. 37. This nonlinear finding is quite similar to findings by Wang et al. [49] who utilized SiO_2 nanoparticle suspensions (see Fig. 38). In both of these investigations, the transient hot-wire method was used to measure the effective thermal conductivity (see Fig. 39).

Li and Peterson [50] applied a steady-state method to measure the effective thermal conductivities of 36 nm diameter Al_2O_3/water and 29 nm

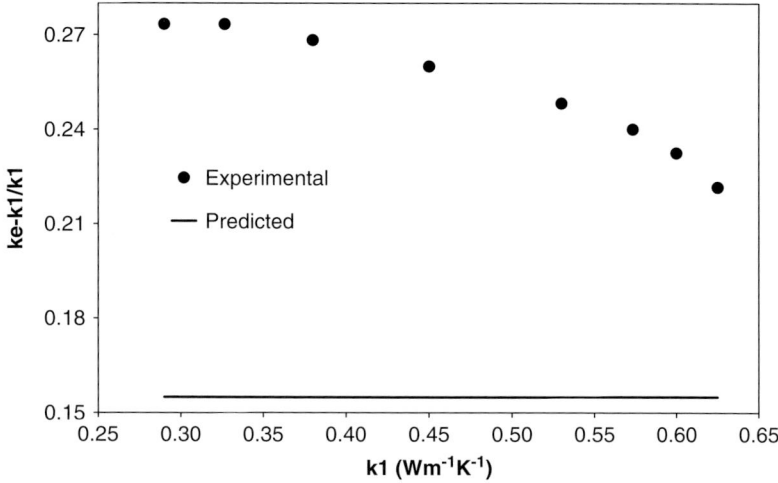

FIG. 31. Enhancement ratio of thermal conductivity between suspension and basefluid vs. the thermal conductivity of basefluid of water–glycerol mixture [35].

diameter CuO/water nanofluids as a function of temperature and volume fraction. In this investigation, the authors hypothesized that since the bulk thermal conductivity of Al_2O_3 decreases with increasing temperature, while the bulk thermal conductivity of water increases with increasing temperature, it is difficult to explain the enhanced effective thermal conductivity solely by the contribution of the higher thermal conductivity of the nanoparticles in the low thermal conductivity base fluid. In this paper, empirical equations for the effective thermal conductivities of both Al_2O_3/water nanofluid and CuO/water nanofluid are developed and used to demonstrate that the Brownian motion of the nanoparticles may contribute to the increase in the effective thermal conductivity, albeit not necessarily in the manner proposed by other investigators.

Most recently, driven by both a desire for a more complete understanding of the fundamental behavior of these nanoparticle suspensions and also of carbon nanotubes themselves, the application of carbon nanotubes in nanoparticle suspensions has begun to attract the attention of a number of researchers [51–61]. Hone *et al.* [51] found that bulk samples of single-walled nanotube bundles had very large thermal conductivities and that the thermal conductivity increased linearly with the temperature as illustrated in Figs. 40–42.

Che *et al.* [57] conducted a molecular dynamic numerical simulation of the thermal conductivity of carbon nanotubes and found that the thermal conductivity of the individual nanotubes was dependent on the structure,

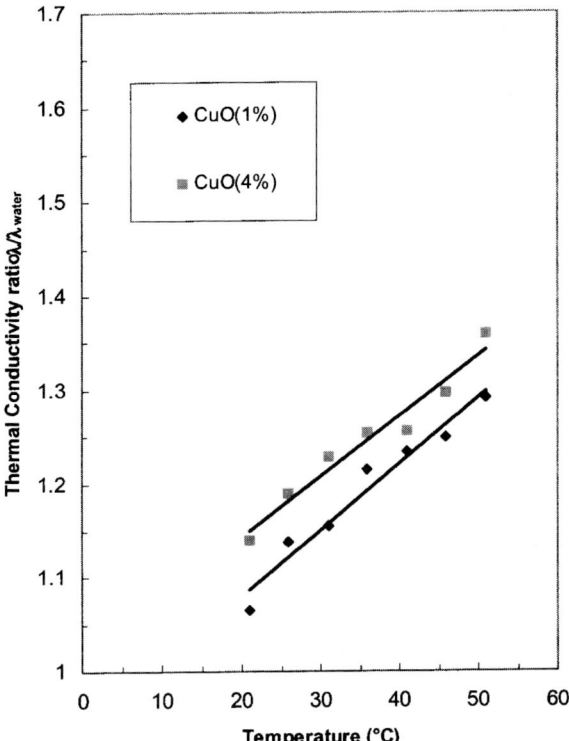

FIG. 32. Temperature dependence of thermal conductivity enhancement for Al_2O_3/water nanoparticle suspensions [45].

defects, and vacancies, and that the anisotropic character of the highly thermal conductive nanotubes is very similar to diamond crystal and in-plane graphite sheets. A number of other researchers have conducted theoretical studies of the thermal conductivity of carbon nanotubes and found that nanotubes have very high thermal conductivity. Kim and Shi *et al.* [59,60] measured the thermal conduction in 148 and 10 nm diameter single-wall carbon nanotube bundles. These materials showed a 1.5 power relation between the thermal conductance and the temperature, in the range of 20–100 K, as shown in Fig. 43, which is quite different from the quadratic relationship between thermal conductance and temperature for individual multi-wall carbon nanotubes (MWCNs).

The effective thermal conductivity of this kind of nanotube suspension has been shown to have an anomalously larger value than for other more classical nanoparticle suspensions and has demonstrated a nonlinear relationship with the nanotube volume fraction [55]. The volume ratio of

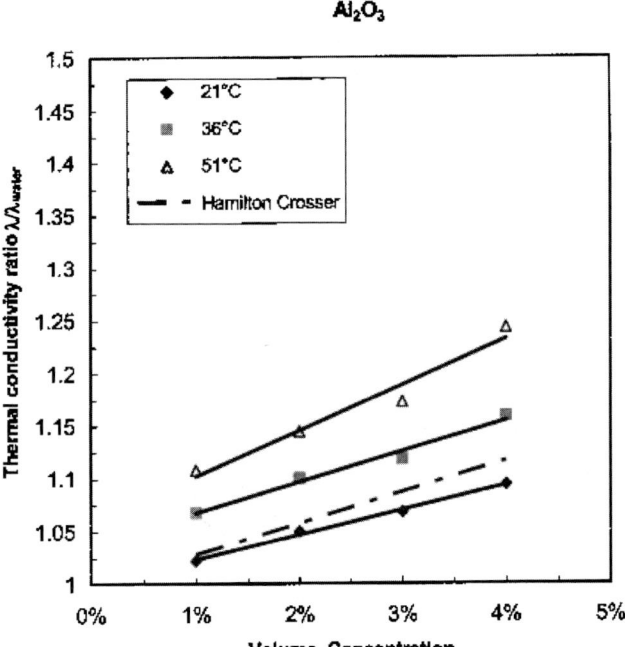

FIG. 33. Enhancement of thermal conductivity of Al_2O_3/water nanoparticle suspensions at different temperatures vs. volume fraction of nanoparticle [45].

nanotubes to base fluids ranged from only 2.5% to 1%, while the enhancement varied from 40% to 160%, as shown in Figs. 44–46. The results when compared with the current theoretical equations deviate even further from the experimental data for nanotube suspensions than the difference between experimental data and theoretical predictions of nanoparticle suspensions as shown in Fig. 47.

Kim and Peterson [62] employed a steady-state method to study the effects of the morphology of carbon nanotubes on the thermal conductivity of carbon nanotubes in suspension. In this investigation, the enhancement of three different types of carbon nanotube suspensions was evaluated as a function of temperature and volume fraction. The single-walled carbon nanotube suspension exhibited a 10% increase in the effective thermal conductivity at a volume fraction of 1%, and the MWCN suspension enhancement was 37% at a volume fraction of 1%. In general, the results demonstrate that carbon nanotube suspensions have a significantly greater effect on the effective thermal conductivity than metal oxide nanoparticles could have.

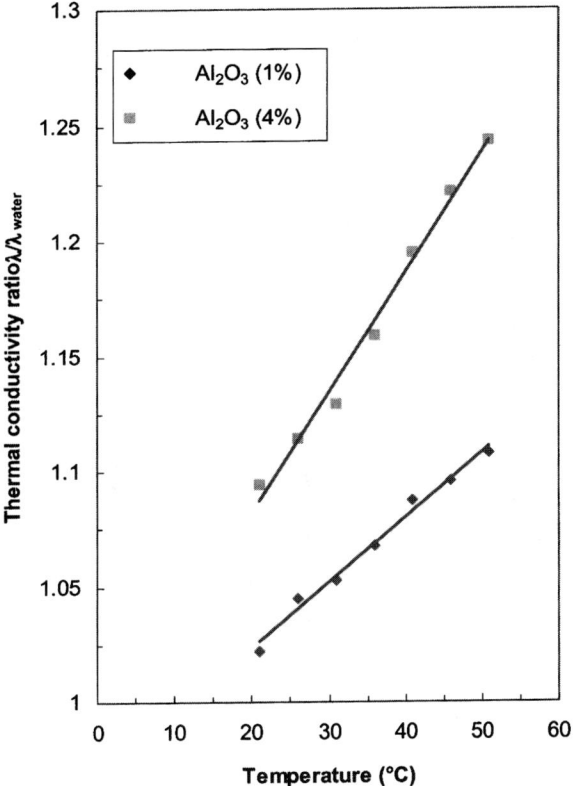

FIG. 34. Temperature dependence of thermal conductivity enhancement for CuO/water nanofluids [45].

As demonstrated in the previous experimental investigations, there appears to be an unusual phenomena occurring in these nanoparticle suspensions that has particular relevance in the application of these materials to both scientific study and engineering applications. This is particularly true for nanoparticle suspensions that utilize carbon nanotubes. However, no single method has yet been developed that is capable of accurately accounting for or predicting the effect of the size, volumetric ratio, and thermophysical effects on the effective thermal conductivity of these nanoparticle suspensions. The equations currently in use are only capable of predicting the effective thermal conductivities for the limited type of suspensions from which the equations were developed. Even before researchers began to focus on nanoparticle suspensions, experimental studies of the thermal conductivity in micrometer or millimeter particle suspensions or emulsions of

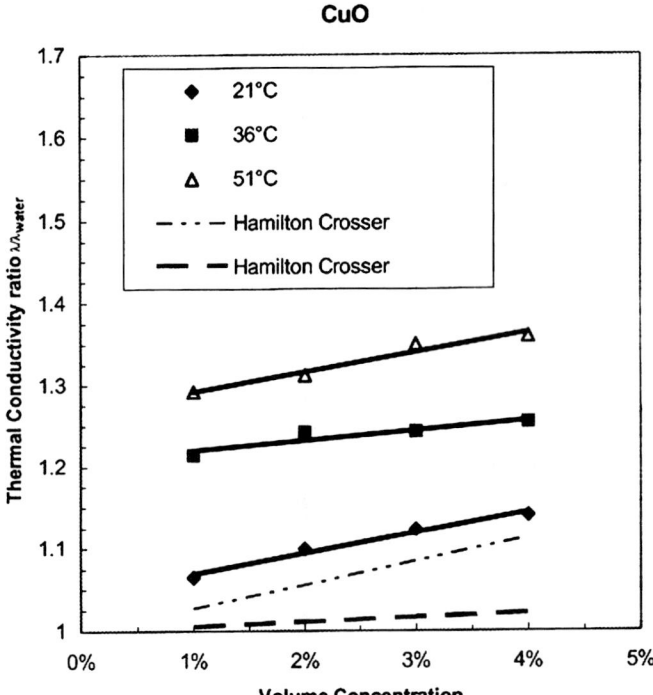

FIG. 35. Enhancement of thermal conductivity of CuO/water nanoparticle suspensions at different temperatures vs. volume fraction of nanoparticle [45].

water–propylene carbonate materials [46], found that neither Maxwell's equation nor the Bruggemann equation would yield satisfactory predictions of the effective thermal conductivity for micrometer or millimeter size particle suspensions. The preceding experimental data also indicated that a new and more generally satisfactory equation is necessary in order to completely understand and better predict the heat transfer in nanoparticle suspensions [37].

D. DISCUSSION AND CONCLUSIONS

The currently available experimental data for the effective thermal conductivity enhancement for different combinations of nanoparticle suspensions has demonstrated a significant variation in the thermal enhancement.

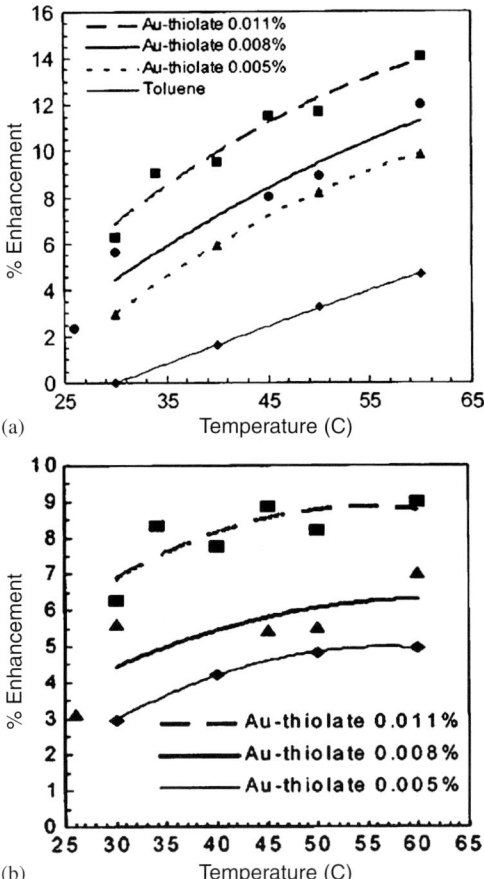

FIG. 36. Percentage of enhancement in thermal conductivity vs. the temperature of Au–thiolate in toluene with reference to the conductivity of toluene at (a) 30 °C and (b) respective temperatures [48].

While some variation has occurred for similar combinations of nanoparticles and fluids and similar experimental conditions, implying that the variation in the experimental data may be caused by the uncertainty associated with the experimental test facilities and methods, this alone cannot explain the wide variations observed. Aside from this experimental uncertainty, there appears to be two sources of error, those resulting from the conduction heat transfer and others resulting from the influence of the convection. Both of these appear to have impacted the accuracy of the

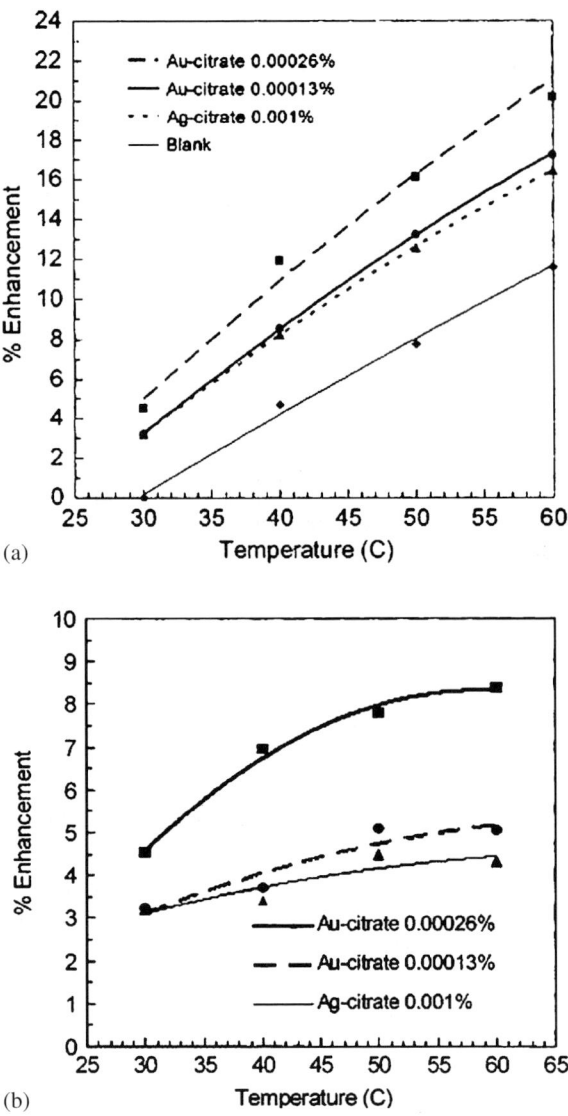

FIG. 37. Percentage of enhancement in thermal conductivity vs. the temperature of Au–citrate and Ag–citrate with reference to the conductivity of blank water (with 5 nM trisodium citrate) at (a) 30 °C and (b) respective temperatures [48].

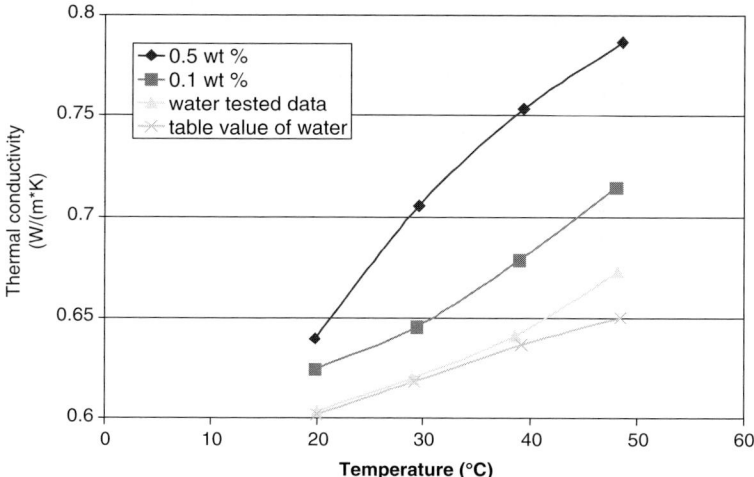

FIG. 38. Effective heat conductivity of SiO_2 nanoparticle suspension.

experimental test facilities and methodologies, regardless of the type of apparatus or experimental methods employed.

The influence of conduction depends on the molecular properties of the test fluid and the confining solid surfaces or nanoparticles and will always accompany the other heat transfer phenomena throughout all of the experimental measurements. The convective contribution, however, is a hydrodynamic phenomenon caused by the bulk movement of portions of the fluid in or around the conductivity cell through density gradients and, as such, may or may not occur, depending upon the type of test facility, size of the particles, and thermophysical properties of either the fluid and/or the nanoparticles. In addition, the convective contribution may introduce errors into the measurements, particularly for transient methods and because of the complexity of this convection-induced error it is difficult to develop and apply accurate correction factors.

In order to obtain accurate experimental data for the effective thermal conductivity of nanoparticle suspensions, three things must occur: first, the amount of heat transferred by conduction must be determined or measured as accurately as possible; second, the data must be obtained within the limited operating conditions for which convection is not likely to occur or for which the absence of this phenomenon can be proven experimentally; and third, a method by which the contribution resulting from microconvection must be developed and verified experimentally.

As indicated previously, the experimental uncertainty introduced by measurement of the pure conduction component will not usually be of great

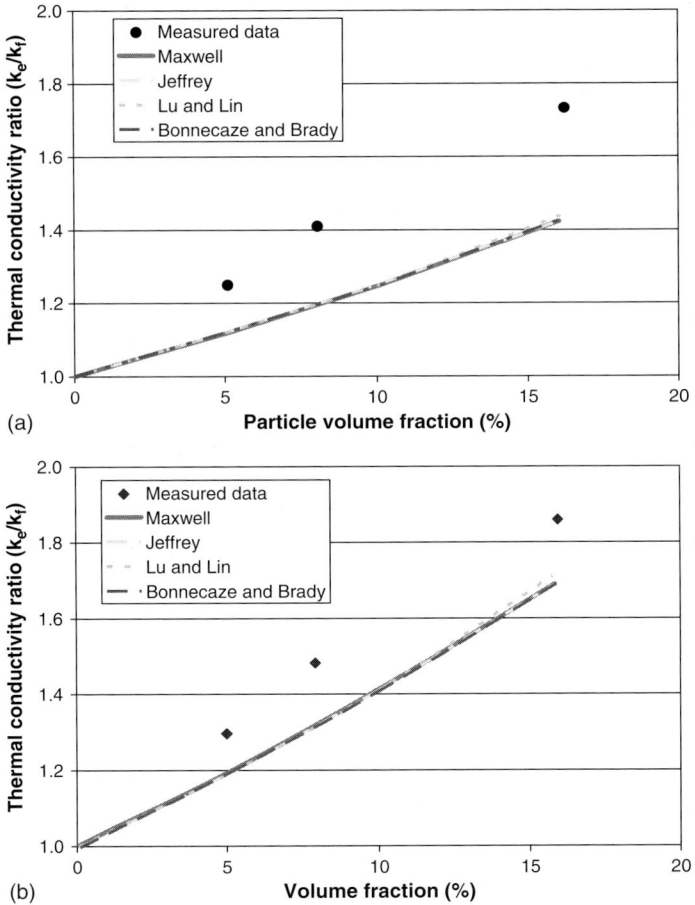

FIG. 39. Measured thermal conductivities of Al_2O_3/ethylene glycol mixtures vs. effective thermal conductivities calculated from theories: (a) $\alpha = 10$ and (b) $\alpha = \infty$ [37].

significance when measuring the effective thermal conductivity, particularly when compared with the effect of the convection induced errors. For this reason, the principal contribution to the errors in the measurement of the effective thermal conductivity of nanoparticle suspensions appears to be in the determination and quantification of the convection heat transfer during the experimental evaluation.

According to the common theoretical treatment of the coupled equations of flow and energy, if the change of pressure could be regarded as negligible, the density changes of the fluid in natural convection could be calculated from the Navier–Stokes and energy equations. The resulting solution can

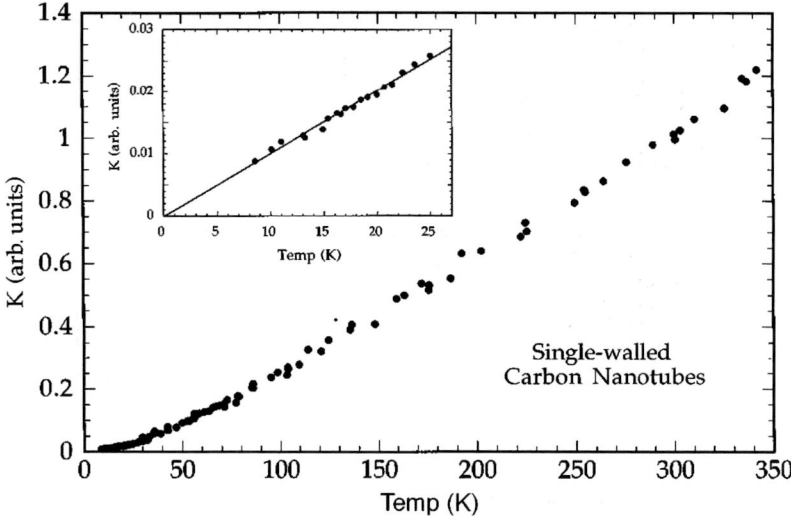

FIG. 40. Thermal conductivity of single-wall nanotubes vs. temperature [51].

be expressed as

$$\nabla d = \left(\frac{\partial d}{\partial T}\right)_p \nabla T = -\alpha d \nabla T \qquad (12)$$

where α is the volumetric expansion coefficient.

Again, due to the small velocity and large heat flux occurring as a result of conduction, the viscous dissipation portion is relatively small and negligible compared to the other components of the energy equation. For this reason, the equations were nondimensionalized and two parameters appeared, the Prandtl number

$$Pr = \frac{\eta c_p}{\lambda} \qquad (13)$$

where η is the dynamic viscosity of the fluid, and the Grashof number

$$Gr = l^3 d_1^2 \frac{(d_2/d_1) - 1}{\eta^2 g} \qquad (14)$$

where d_1, d_2 are the densities of the fluid at the two isothermal surfaces with respective temperatures T_1 and T_2, and g is the gravitational constant.

Rewriting Eq. (12) with respect to the fluid density and temperature at the two surfaces, this can be expressed as

$$d_2 = d_1[1 + \alpha(T_1 - T_2)] \qquad (15)$$

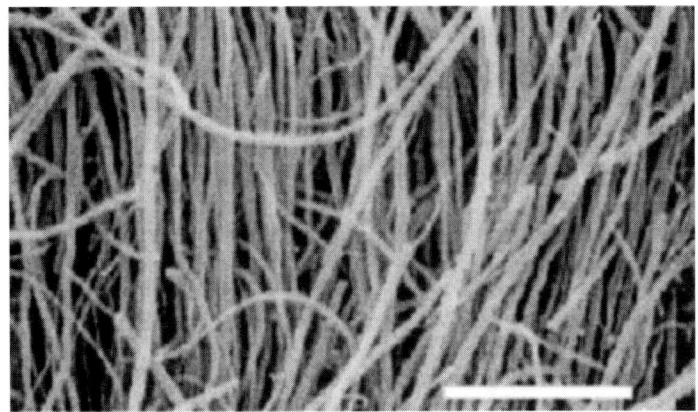

FIG. 41. Scanning electron microscopy image of multiwall nanotubes (scale bar is 800 nm) [52].

which yields the expression $\alpha(T_1 - T_2) = (d_1/d_2) - 1$. Substituting this expression into Eq. (14) for the Grashof number, Gr, yields the more familiar expression for the Grashof number

$$Gr = \frac{l^3 d_1^2 \alpha(T_1 - T_2)}{\eta^2 g} \qquad (16)$$

Here, the Grashof number can be interpreted as the ratio between the buoyancy forces and the viscous forces acting on the fluid during natural convection.

The error introduced by natural convection for the foregoing methods used to measure the effective thermal conductivity of nanoparticle suspensions can vary depending upon the methodology utilized. A thorough analysis of the steady-state convection problem for a horizontal flat-plate apparatus has been conducted [2] and indicated that the overall accuracy of the measurement of the thermal conductivity must be less than 1% for convection to be neglected, i.e.,

$$\gamma \frac{(Gr\ Pr\ \sin\ \phi)l}{180\pi 2 r_1} < 0.01 \qquad (17)$$

where ϕ is the angle of the plate surface with the horizontal plane and γ a factor that depends on $r \sin \phi$, which represents the fraction of the heat flowing away from the upper plate. This provides an indication of how large the applied temperature difference should be in order to accurately measure the thermal conductivity of nanoparticle suspensions experimentally. One additional problem with the steady-state methods is that it takes a relatively

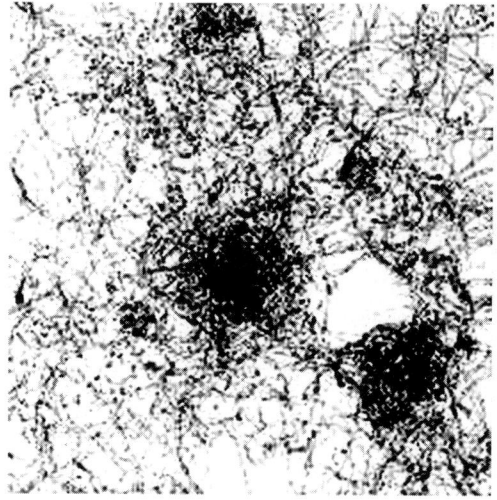

(a) TEM images of PCNTs, scale bar 200 nm

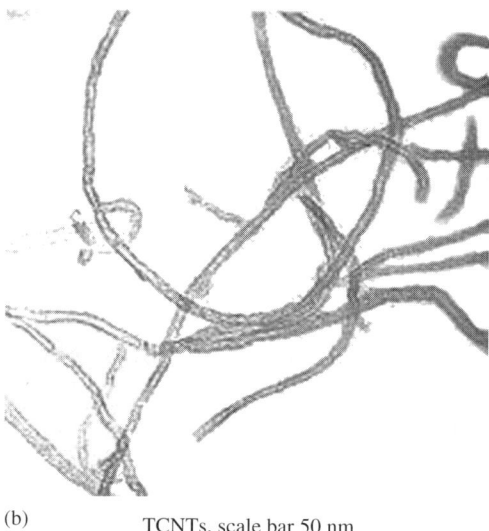

(b) TCNTs, scale bar 50 nm

FIG. 42. Transmission electron microscopy images of carbon nanotubes [55]: (a) TEM images of PCNTs (scale bar 200 nm), (b) TCNTs (scale bar is 50 nm).

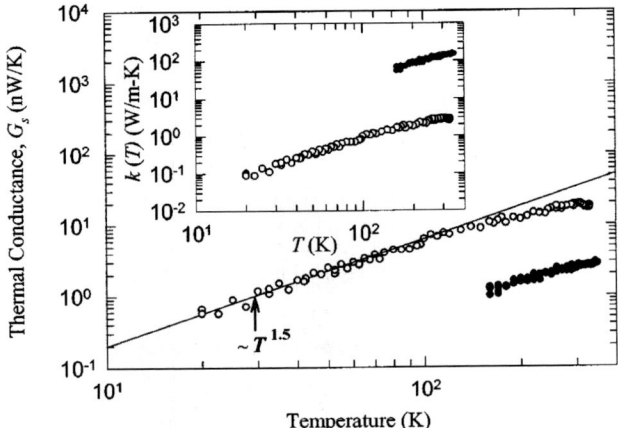

FIG. 43. Thermal conductance of two single-walled carbon nanotube (SWCN) bundles as a function of temperature. Solid and open circles represent the measurement results of the 10 and 148 nm SWCN, respectively; the line indicates thermal conductivity as a function of temperature [59].

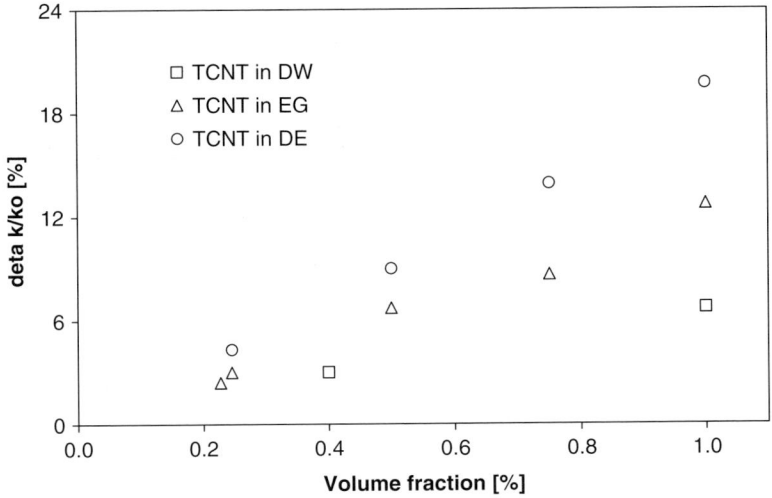

FIG. 44. Thermal conductivities of TCNT suspensions as a function of nanotube loadings [55].

long time for the test facility to reach steady-state, which for the case of nanoparticle suspensions and repeated tests over a reasonable temperature range, could allow the nanoparticles to settle on the bottom surface and thereby affect the measurement accuracy [2].

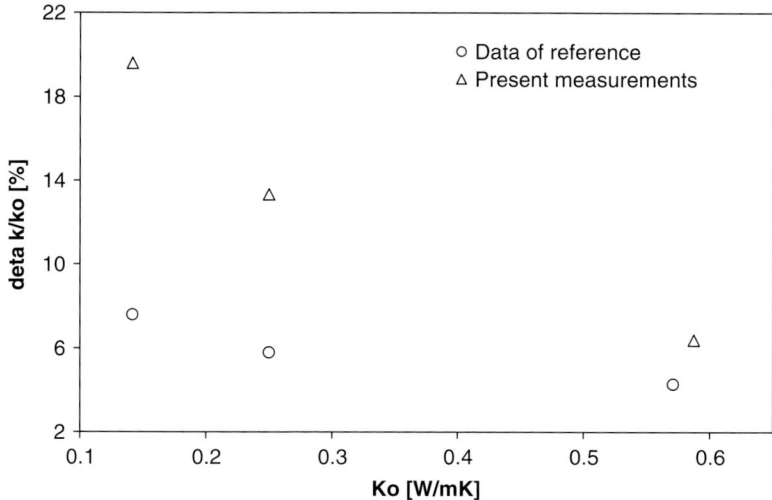

FIG. 45. Effective thermal conductivities of TCNT suspensions as a function of nanotube volume ratio [55].

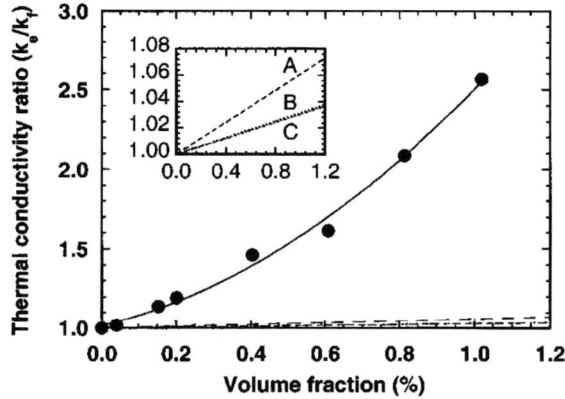

FIG. 46. Effective thermal conductivities of TCNT suspensions as a function of nanotube volume ratio, and (A) the prediction of Hamilton–Crosser equation; (B) the Boonecaze and Brady equation; and (C) Maxwell's equation [52].

Although the transient hot-wire test facility reduces the impact of this settling and the convective effects in the nanoparticle suspensions, these latter effects may still be quite large since the experimental data are typically taken approximately 1 min after the initial heat flux is applied. During this time interval, there is sufficient time for natural convection cells to form

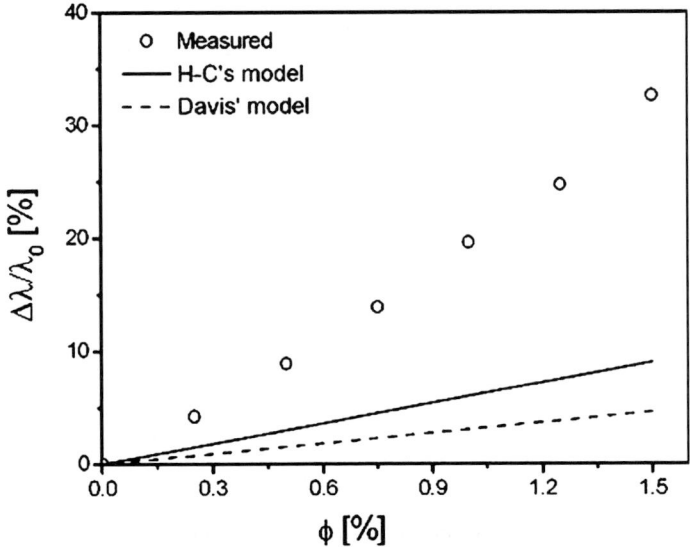

FIG. 47. Comparison between the calculated values and the experimental results for TCNT in DE (decene) [55].

around the hot wire and generate microconvection-induced temperature gradients.

Once the accuracy of the different experimental test facilities and methodologies are clear and the error sources understood, it is important to determine what actually contributes to the enhancement of the effective thermal conductivity.

(1) Foremost among these is the influence of the base fluid. For the same base fluid, and size and type of nanoparticle, suspensions using ethylene glycol and water as the base fluid, the enhancement will increase with increasing nanoparticle volume fraction. The dependence of the effective thermal conductivity enhancement with increasing nanoparticle volume fraction is significantly less pronounced for suspensions with engine oil or pump fluid as the base fluid, due to the higher conductivity of these fluids. Among these four base fluids, the thermal conductivity of water at 25°C is 6.07 W/(m K), ethylene glycol is 0.256 W/(m K), and pump fluid and engine oil are approximately 0.12 and 0.14 W/(mK), respectively. Using these data, it appears that the lower the thermal conductivity of the base fluid, the greater the relative enhancement of the effective thermal conductivity of the suspension for the same nanoparticle volume fraction (see Fig. 48).

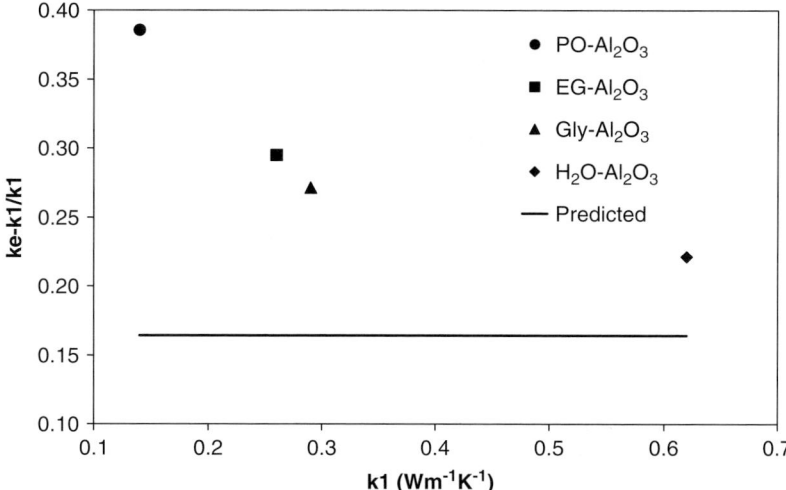

FIG. 48. The enhancement of effective thermal conductivity of suspension vs. the thermal conductivity of basefluid [35].

(2) The influence of a surfactant or acid, both of which serve to more evenly disperse the particles and provide long-lasting nanoparticle suspensions. For the same nanoparticle volume fraction, the addition of a surfactant will result in greater enhancement than that typically achieved in a suspension without a surfactant. The addition of an acidic solution can also impact the enhancement with higher pH values resulting in a reduction in the enhancement.

(3) The third factor that influences the magnitude of the enhancement is the thermal conductivity of the nanoparticle itself. For a constant nanoparticle diameter, the same base fluid, and an identical volume fraction, increases in the nanoparticle thermal conductivity will result in an increase in the enhancement factor. However, the experimental data that support this type of comparison are limited and additional studies are necessary to determine the effect of the thermal conductivity of the nanoparticle on the overall enhancement.

(4) The shape of the nanoparticle also plays an important role in the enhancement of the effective thermal conductivity. Increases in the nanoparticle diameter result in a decrease in the enhancement of the effective thermal conductivity of the suspension. In addition, the experimental data currently available for millimeter diameter particle suspensions indicates that suspensions with cylindrical particles having a length-to-diameter ratio of 10 will have an effective thermal

conductivity around 1.5 times that of a suspension with spherical particles at the same temperature and volume fraction. However, there are no such experimental data available for nanoparticle suspensions.

(5) The bulk temperature of the suspension also influences the effective thermal conductivity. Experimental evidence indicates that increases in the temperature result in corresponding increases in the relative enhancement. While this increase in the enhancement does not always have a linear relationship with temperature increase, the general trend is clearly apparent.

Although the experimental data are quite limited, suspensions of carbon nanotubes exhibit similar trends with respect to the effective thermal conductivity enhancement with respect to variations in the temperature, the type of the base fluid, and the volume fraction and for the type of particles being investigated.

III. Theoretical Investigations

As discussed above, the mechanisms that govern the enhanced effective thermal conductivity of nanoparticle suspensions may be associated with many factors, including, but not limited to the volume ratio of the nanoparticle to the base fluid, the ratio of the thermal conductivity of the nanoparticle to that of the base fluid, the molecular level adsorption of the liquid layer on the surface of the nanoparticle, the nonlocalized heat transport within the individual nanoparticles, the nanoparticle clustering, and the Brownian motion of the individual nanoparticles. As described above, the first two of these have been extensively investigated experimentally, and the predictive expressions for the variations in the effective thermal conductivity of particle suspensions have been shown to be reasonably accurate. As the particle size decreases, however, the other factors listed above become increasingly important and must be considered in order to fully explain the unusually high effective thermal conductivity of these suspensions.

A. Development of Effective Thermal Conductivity Equations

The first fundamental equation capable of predicting the effective thermal conductivity of solid particles and fluid mixtures was initially developed by Maxwell [3] in 1873 and is given below as Eq. (19) In 1924, Fricke [63] developed the following expression for the effective thermal conductivity:

$$\frac{(k/k_1 - 1)}{(k/k_1 + x)} = \frac{\phi(k_2/k_1 - 1)}{(k_2/k_1 + x)} \tag{18}$$

where k, k_1, and k_2 are the specific conductivities of the suspension, the suspending medium, and the suspended spheroids, respectively; φ is the volume concentration of the suspended spheroids; and x a function of the ratio of k_2/k_1 and the ratio a/b of the axis of symmetry of the spheroid to the other axis. Fricke compared the predicted results with experimental data for the conductivity of blood ($k_2 = 0$, $a/b = 1/4.25$, $x = 1.05$). The results showed excellent agreement for concentrations ranging from 10% to 90%. The influence of the geometric factors of the suspended particles was considered through a parameter, x. In this expression, the conductivity of a suspension is independent of the size of the suspended particles for a constant volume concentration and is also nearly independent of the shape of the particles when the difference between the conductivities of the suspended and the suspending phases is small. This is especially true for suspensions of prolated spheroids that are less conducting than the suspending medium.

This approach represents a purely mathematical treatment, and builds upon the previous work of Maxwell [3] who studied conduction through heterogeneous media. In this initial work, the potentials at the interface were assumed to be continuous and equal, thus the current passing through the interface of the two media would be the same. Examples of two and three different physical properties with interfaces that were spherical were conducted and the mathematical treatment was applied to introduce the potential expanded in solid harmonics and the surface harmonic. By analyzing the algebraic sign of the constants in the harmonics, it became clear that if the conduction of one of the materials were higher than the rest, the material with the higher conduction would tend to equalize the potential throughout the material. Otherwise, the current was prevented from reaching the inner of the lower conduction materials.

By examining the case of n spheres of radius a_1 and resistance R_1, placed in a medium of resistance material, R_2, where all the spheres and media were contained in a sphere of radius a_2, and assuming no interaction between the spheres or medium, the effective resistance of the sphere of radius a_2 could be calculated as follows:

$$R = \frac{2R_1 + R_2 + \phi(R_1 - R_2)}{2R_1 + R_2 - 2\phi(R_1 - R_2)} R_2 \qquad (19)$$

This equation serves as the original formula, which was subsequently modified by other researchers and has been extensively used to predict the behavior of these types of systems, with the limitation that the volume fraction, ϕ, should be very small in order to avoid the interaction between spheres and that the magnitude of $(R_1 - R_2)/(2R_1 + R_2)$ should not be large.

Based on the relationship between the refractive index and density conducted by Reyleigh [64] this problem was further restricted to cases where the spheres or bodies were not small compared to the distances between them and that they were arranged in rectangular or square orderly patterns. The first such case was that of two-dimensional thermal conductivity for heat in a uniform medium consisting of cylindrical obstacles in an ordered, rectangular arrangement. This investigation was later extended to the application of thermal conductivity of uniform media interrupted by cylindrical obstacles in three dimensions, under the assumption that the spheres were arranged in a cubic arrangement and that volume ratio was in the moderate range. In this analysis, two extreme conditions were evaluated, the case where the conductivity of the cylinders were infinite and the case where this value was zero. Using the same initial assumptions as Maxwell, i.e., the potential was continuous at the interface and the current remained the same, the effective thermal conductivity was expressed as

$$k_e = 1 - \frac{2\pi B_1}{SH} = 1 - \frac{2\pi a^2}{S(v' + a^2 \Sigma_2)} \qquad (20)$$

where B_1 is the constant in the expanded series of the potential external to the sphere, S the area of the rectangular cell around the sphere in two-dimensional condition, $h = B_1(v'a^{-2} + \Sigma_2)$ the unit potential, and $v' = (1+v)/(1-v)$, v the ratio of the conductivity of the sphere to that of the medium. By introducing the volume fraction for the spheres, $\phi = \pi a^2/S$, where a is the radius of the sphere, this expression could be written as

$$k_e = 1 - \phi \frac{2B}{a^2 H} = 1 - \frac{2\phi}{v' + \phi - \phi^4 \Sigma_2} \qquad (21)$$

This particular study was quite valuable in that it gave researchers an indication of how variations in the thermal conductivity might occur and provided a method by which the effect of variations in the different thermophysical and geometric properties of the constituents could be determined, but it had little value in terms of accurately predicting the actual experimental results.

Fricke [63] proposed Eq. (22) to predict the effective thermal conductivity of suspensions based on a review of previous research.

$$\frac{(k_e/k_1) - 1}{(k_e/k_2) + 2} = \phi \frac{k_2/(k_1 - 1)}{(k_2/k_1) + 2} \qquad (22)$$

It was noted that this equation was only suitable for calculating the effective thermal conductivity of suspensions that had spherical particles dispersed in base fluids for which the volume fraction of the spherical particles was small. For other particle geometries, such as ellipsoid or spheroid particles, the geometrical arrangement of the suspended particles was introduced and used to calculate β, which is a factor that represents the influence of the geometry of suspended particles to the effective thermal conductivity of the suspension material. By examining the expression for β, it was concluded that for a constant volume concentration, the effective thermal conductivity of a suspension was independent of the size of the suspended particles and also nearly independent of the shape of the particles, when the difference between the thermal conductivities of the suspended and the suspending phases was not large [61]. For suspensions of nonspherical particles, the effective thermal conductivity could be calculated using a form somewhat analogous to the equation for a suspension with spherical particles or

$$\frac{(k_e/k_1) - 1}{(k_e/k_2) + x} = \phi \frac{k_2/(k_1 - 1)}{(k_2/k_1) + x} \tag{23}$$

where $x = -(k_2/k_1 - 1) - (k_2/k_1)\beta/(k2/k1 - 1) - \beta$ and represents the geometrical influence on the effective thermal conductivity of the suspension. Comparing the experimental data of blood and a sand suspension with the values predicted by this equation demonstrates reasonably good correlation between the measured and predicted results. Because many suspensions have particles that either are or can be polarized, and the interaction due to this polarization can be quite large and could thereby significantly influence the effective conductivity of the suspension, it was suggested that the mean value of the forces due to the charges on the suspended particles throughout the medium should be added to the original potential field to derive an equation for the effective conductivity of the suspension [63].

Meredith and Tobias [46] used the expression for the effective conductivity proposed by Bruggeman [65] to compare predicted values with the experimental data obtained by Rue and Tobias, and found relatively good agreement. However, the experimental data used in this comparison were for suspended particles with sizes that varied by as much as three orders of magnitude. For experimental data in which the particle size was constrained within a relatively narrow range, neither the Bruggeman equation nor the Maxwell equation provided a good correlation. In these comparisons, the volume fraction of the surrounding material was defined as

$$1 - \phi = \frac{k_e/k_f - k_p/k_f}{(k_e/k_f)^{1/3}(1 - k_p/k_f)} \tag{24}$$

where, k is the conductivity and φ the volume fraction of particles in suspension, and the subscripts p represent the particle, f the fluid, and e the effective value, respectively.

Following this comparative evaluation, a new expression for predicting the effective conductivity of a suspension was developed by Meredith and Tobias [46] (Eq. (25)), which accurately predicted the experimental data as,

$$\frac{k_e}{k_f} = \left(\frac{2 + 2X\phi}{2 - X\phi}\right)\left(\frac{2 + (2X-1)\phi}{2 - (X+1)\phi}\right) \tag{25}$$

where $X = k_p/k_f - 1/k_p/k_f + 2$. Here, the assumption was made that only one size of particle was dispersed in the suspension.

This work was followed by an extensive theoretical investigation conducted in an attempt to develop a set of equations that could be used to calculate the effective thermal conductivity of various types of dispersed composites. In this investigation, Hamilton and Crosser [47] proposed an expression for predicting the effective thermal conductivity of heterogeneous systems consisting of a continuous phase with a discontinuous phase dispersed within it, based on the work of Maxwell and Fricke. In this investigation, it was found that a parameter n, which could be treated as a constant was dependent on the shape of the particle and the ratio of the thermal conductivities of the discontinuous and continuous phases. When the particle was spherical in shape, n did not exhibit this dependence. As a result of this observation, the effect of the shape of the particles was evaluated and a shape factor was added to the equation that related the variation of the particle shape being investigated to that of a spherical particle. This shape factor was based on experiments of both spherical and nonspherical particle mixtures. A second coefficient, ψ, was then introduced, which was defined as the ratio of the surface area of a sphere to the surface area of a nonspherical particle having the same volume, i.e., $n = 3/\psi$ for spherical particles, $n = 3/\psi^2$ for prolated ellipsoids, and $n = 3/\psi^{1.5}$ for oblate ellipsoids. For these cases, the equation for the effective thermal conductivity was expressed as

$$k = k_1 \left[\frac{k_2 + (n-1)k_1 - (n-1)\phi(k_1 - k_2)}{k_2 + (n-1)k_1 + \phi(k_1 - k_2)}\right] \tag{26}$$

With this expression, it was found that if the continuous phase had the higher conductivity, the shape had little effect on the effective thermal conductivity of the mixture. However, if the discontinuous phase had the higher conductivity, there was a strong shape effect. Application of Eq. (26) to situations where n was expressed in terms of the sphericity and the conductivity of the discontinuous phase particles was at least 100 times larger than the conductivity of the continuous phase, indicated good agreement with the experimental results of silastic rubber and different shaped particles of 1 mm aluminum and balsa at various volume compositions and a temperature of 95°F. It was also noted in this work that the agglomeration of the particles that would change the shape, size, and orientation of particles, could result in an increased deviation between the predicted and measured experimental data.

This work was followed by an investigation conducted by Keller [66] in which the extreme situations of spheres and circular cylinders with infinite conductivity were embedded in a medium of conductivity k_1 in a dense cubic array, and nonconducting cylinders were embedded in a square array. The results of this investigation indicated that for the nonconducting embedded cylinders contained in the medium, the resistance of the gap between two adjacent cylinders could be expressed as $R = 1/k_e c$, where c is the distance between the axes of the two adjacent cylinders. For this situation, the maximum fractional volume was determined to be $\pi/6$ for spheres and $\pi/4$ for cylinders. With this information, the equation for calculating the effective conductivity could be expressed as:

$$k/k_1 = -(\pi/2)\log[(\pi/6) - \phi] + \cdots + (\pi/6) - \phi \ll 1 \qquad (27)$$

When $\phi = 0.5161$, the results yield $k/k_1 = 7.65$, which agreed with the measured result of $k/k_1 = 7.6$. Using a similar procedure, the equation for a medium containing cylinders was obtained as

$$k/k_1 = -(\pi^{3/2}/2)\log[(\pi/4) - \phi]^{1/2} + \cdots + (\pi/4) - \phi \ll 1 \qquad (28)$$

which for the square array of perfectly conducting circular cylinders, agreed well with the numerical results obtained by Keller [66]. Finally, it was concluded that the expressions for the effective thermal conductivity of nonconducting cylinders and perfectly conducting cylinders were inversely related and had the following relationship:

$$\frac{k_e^\infty}{k_1} = \frac{k_1}{k_e^0} \qquad (29)$$

where k_e^∞, k_e^0 are the effective thermal conductivities for the case of the perfectly conducting cylinders and the nonconducting cylinders, respectively, and k_1 is the conductivity of the continuous medium.

Leal [67] analyzed the effective conductivity of a dilute suspension in a simple shear flow and studied the role of the mechanical and thermal properties of the particles to the bulk thermal transport characteristics of the suspension. The bulk heat flux of the suspension was calculated by averaging the microscale heat flux, and the local temperature field was considered under the assumption that the suspension was experiencing a simple shear flow with a linear bulk temperature field across the shear gradient in which the Reynolds number and Peclet number were very small. The results of this investigation indicated that the effective conductivity was influenced by the thermal conductivity of the suspended particles and the shear flow disturbance. It was also noted that while the flow always contributed positively, the particle thermal conductivity could have either a positive or negative influence, depending upon the conductivity ratio between the particle and the base fluid. This investigation provided an alternative hypothesis and indicated that the enhancement of the effective thermal conductivity of suspensions was not only a function of the physical and thermal properties of the suspended particles, but that the effective thermal conductivity of these suspensions were or could be strongly influenced by the motion of the particles as well. While presenting a new and somewhat radically different approach, because it lacked information on the effects of the particle size, shape, volume fraction, and other factors, this expression was not capable of accurately predicting the experimental data. The expression was given as

$$k = k_1 \left[1 + \phi \left\{ \frac{3(k_2 - k_1)}{k_2 + 2k_1} + \left(\frac{\frac{1.176(k_2-k_1)^2}{(k_2+2k_1)^2}}{+ \frac{2\mu_1+5\mu_2}{\mu_1+\mu_2}} \begin{pmatrix} 0.12 \frac{2\mu_1+5\mu_2}{\mu_1+\mu_2} \\ -0.028 \frac{k_2-k_1}{k_2+2k_1} \end{pmatrix} \right) \times Pe_1^{3/2} + 0(Pe_1^2) \right\} \right] \quad (30)$$

where μ is the viscosity and the subscript 1 represents the base fluid and 2 represents the particle.

Jeffrey [68] used a probability approach and suggested that the probability $P(r|o)$ for only one sphere within a distance of $O(r)$ was $O(\phi)$ and the probability of two spheres within a distance of $O(\phi^2)$ was $P(r|o)$. The choice used for the probability $P(r|o)$ satisfied the conditions that

$$P(r|o) = 0, \quad r \leq 2a$$
$$P(r|o) = n, \quad r \gg a \quad (31)$$

Two conditions were required here, the first of which was that the spheres did not overlap and the second was that all of the spheres were far away from the reference sphere. In solving the problems involving two spheres, twin spherical expansions were applied. After obtaining the expressions for the heat flux and temperature both in and outside of the spheres, the authors considered the interactions between the various pairs of spheres using a second-order expression of the volume ratio between the particles and the fluid. This equation was dependent on the way the various pairs of spheres were distributed. Convergence was addressed by averaging the interactions between the spheres to obtain the bulk properties of the suspension. The resulting equation was

$$k/k_1 = 1 + 3\beta\phi + 3\beta\phi^2 \left(\beta + \sum_{p=6}^{\infty} \frac{B_p - 3A_p}{(p-3)2^{p-3}} \right) \tag{32}$$

where $\beta = (\alpha - 1)/(\alpha - 2)$, volume fraction $\phi = 4/3\pi a^3 n$, a is the radius of the particle, $\alpha = k_2/k_1$, p the probability density for a second sphere located at r from the center of the reference sphere and A_p and B_p are known coefficients.

Rocha and Acrivos [69,70] assumed that no significant interactions existed in the dilute suspension and using Fourier's law, developed a general expression for the heat conduction in a statistically homogeneous suspension

$$q + k_1 \frac{\partial T}{\partial x_i} = (k_1 - k_2)c \int_{S_p} Tn_i \, dS \tag{33}$$

where q is the heat flux, k_1 the thermal conductivity of continuous matrix, k_2 the thermal conductivity of the dispersed particle, c the number of dispersed particles per unit volume, S_p the surface of the particle and n_i the normal to the surface. The resulting effective thermal conductivity of the suspension was then derived as

$$\frac{k_e}{k_1} = \{1 + 3[(\alpha - 1)/(\alpha + 2)]\phi\}\delta \tag{34}$$

where α is the ratio of the thermal conductivity of the dispersed particles to that of the continuous matrix, ϕ the volume fraction of the particles, and δ the Kronecker delta. This equation was only applicable for spherical particles, since for slender particles or ellipsoids, the effective thermal conductivity was anisotropic and had different values along different particle axes in the suspension. Finally, the effective thermal conductivity equations for suspensions with arbitrary shaped particles, axially symmetrical particles, and slender particles were obtained under the steady-state, noninteraction conditions. The general equation was

$$\frac{k_{e,ij}}{k_1} = \delta_{ij} + 2[(\alpha-1)/(\alpha+1)]\phi$$
$$\times \left\{ \delta_{ij} + V_2^{-1} \sum_{m=1}^{\infty} (2\pi)^m [(\alpha-1)/(\alpha+1)]^m \int_\Omega \right\} \quad (35)$$

Here, it was noted that the effective thermal conductivity of the suspension could be isotropic if the function were truncated to the first order of $(\alpha-1)/(\alpha+1)$, and was only a function of the volume fraction of the dispersed particles.

Batchelor [71] investigated the transport properties of heterogeneous mixtures composed of two phases that were each individually homogeneous. In this investigation, probability theory was used to determine the bulk properties of the resulting two-phase dispersion system. Beginning with the Fourier theory that $F = K \cdot G$, where F was the flux, K the bulk transport coefficient characteristic of the medium, and G the gradient of intensity, a table of the physical phenomenon that represented what was called an effective transport coefficient was used as the basis for the theoretical analysis. The resulting values are listed in Table III.

In this analysis, the applicability of the equations and theories to the various volume fractions were developed, along with the upper and lower limits of the effective thermal conductivity for arbitrary volume fractions, as determined by Hashin and Shtrikman [72] for the effective magnetic permeability of the multiphase materials. Calculations of the mean flux, F, in two-phase statistically homogeneous mixtures, in which the local heat flux always had a linear relation to the local intensity gradient were performed and a new parameter, the dipole strength S, was introduced, which served as a measure of the values obtained when the matrix material was replaced by the particle material.

Finally, three conditions for dilute suspensions with non-interaction between particles, first-order effects of the interaction between particles, and the strong interaction effects in small volume fraction suspensions, were all discussed. An expression for the effective thermal conductivity for the non-interaction between particles in dilute suspensions was given as

$$k_e = k_1 \{ 1 + [2(\alpha-1)/(\alpha+2)]\phi + [3(\alpha-1)^2/(\alpha+2)^2]\phi^2 \} \quad (36)$$

where k_e is the effective thermal conductivity, k_1 the thermal conductivity of the continuous matrix, α the ratio of the thermal conductivity of the particle

TABLE III

THE EFFECTIVE TRANSPORT COEFFICIENT OF DIFFERENT DISPERSE SYSTEMS [71]

Nature of the two uniform media of which the two-phase disperse system is composed	Quantity represented by **F**	Quantity represented by **G**	Transport coefficient (or scalar components in case of isotropic structure)	Local differential equation satisfied in each phase (in steady state)
(1) Thermal conductor	Heat flux	Temperature gradient	Thermal conductivity	**F** = **KG**
(2) Electrical conductor	Electric current	Electric field	Electrical conductivity	∇**F** = 0
(3) Electrical insulator	Electric displacement	Electric field	Dielectric constant	
(4) Dia- or para-magnetic material	Magnetic induction	Magnetic field	Magnetic permeability	
(5) Small fluid or rigid particles sedimenting through Newtonian incompressible fluid (a "fluidized bed")	Flux of particle number relative to zero-volume flux axes	Gravitational force on particles in unit volume of mixture (minus the buoyancy force)	Mobility	
(6) "Porous medium" consisting of a fixed array of small rigid particles through which incompressible Newtonian fluid is passing	Force on particles in unit volume of mixture (= pressure gradient calculated from pressure drop between distant parallel planes)	Flux of fluid volume relative to particles	Permeability (Darcy constant) divided by μ	$\nabla p = \mu \nabla^2 \mathbf{u}$ $\nabla \mathbf{u} = 0$ where **u** = velocity, p = pressure, μ = viscosity
(7) Small fluid or rigid particles suspended in incompressible Newtonian fluid	Deviatoric stress	Rate of strain	Share viscosity	As immediately above; or $F = 2\mu \mathbf{G}$, $\nabla \mathbf{F} = 0$ G = trace **G** = 0
(8) Elastic inclusions embedded in an elastic matrix	Stress	Strain	Lamé constants (for rigidity and bulk moduli)	$F = 2\mu \mathbf{G} + \lambda \mathbf{GI}$ $\nabla \mathbf{F} = 0$ where μ, λ = local Lamé constants

Note: Cases of two-phase disperse systems with random structure in which the problem is to determine an effective transport coefficient.

material to that of the matrix material, and ϕ the volume fraction of the particles. For the second and third conditions, a broader, more complicated analysis was conducted that considered the case where there was a bulk movement in the suspension. This latter case was considerably more complicated in that it included the interactions between particles plus the influence of other external forces.

In 1977, Wills [73] studied the bounds and self-consistent estimates for the effective thermal conductivity of a composite material comprising a matrix with a random volume fraction of dispersed particles with bounds of the Hashin–Shtrikman type and self-consistent estimate methods. The simple case of spherical particles aligned along one axis was discussed, where both the thermal conductivities of the matrix and the particles were assumed to be isotropic. Then assuming that the thermal conductivity of the dispersed particles was zero, the lower bound was found to be zero and the upper bound was equivalent to the thermal conductivity of the matrix. Using this approach, the effective thermal conductivity is given as

$$k_e \leq \frac{2(1-\phi)}{(2+\phi)}k_1 \qquad (37)$$

and for the self-consistent estimates of the effective thermal conductivity, is given as

$$k_e = \left(1 - \frac{3}{2}\phi\right)k_1 \qquad (38)$$

If the dispersed particles were assumed to be perfectly conducting, the upper bound of the effective thermal conductivity was infinite and the lower bound of the effective thermal conductivity was equal to the thermal conductivity of matrix. Here, the effective thermal conductivity is given as

$$k_e \geq \frac{(1+2\phi)}{(1-\phi)}k_1 \qquad (39)$$

and the self-consistent estimates of the effective thermal conductivity was given as

$$k_e = \frac{k_1}{1-3\phi} \qquad (40)$$

As indicated above, both of the self-consistent estimates for these two extreme situations suggested that when the volume fraction was 1/3, the effective thermal conductivity was zero for the composite consisting of nonconducting spherical particles, and when the volume fraction was 2/3, the effective thermal conductivity was infinite for the composite consisting of perfectly conducting spherical particles [73].

The special case of infinitely long, thin cylindrical particles and disc-shaped particles were also investigated and the critical volume fractions of the dispersed particles varied. O'Brien [74], who considered the case where the volume fraction of the particles was quite high and the particle interactions needed be considered, utilized an approach similar to that of Batchelor, and assumed that the mean heat flux of the statistical homogeneous suspension consisted of two parts, one for the continuous matrix material, and the other for the volume occupied by the particles, expressed in terms of dipole strength. By retaining the assumption that the probability of a particle having m particles within the effective distance of the strong interaction was of the order of ϕ^m, and that the interaction between the particles would decrease rapidly with increases in distance, the dipole strength in dilute suspension could be expressed as

$$F = -k_e \nabla T \tag{41}$$

$$F = -k_1 \nabla T + n \bullet S \tag{42}$$

$$S = S^0 + \int S'(r) p(r|0) \, dV(r) \tag{43}$$

where n is the number density of particles, S^0 the dipole strength in the absence of particle interactions, S' the amount of dipole strength altered by the presence of another particle at r, and $p(r|0) \, dV(r)$ the probability that the center of a particle lay with a distance r.

Using the divergence theorem applied to this integral, a term that represented the field strength due to a continuous distribution of dipoles over the volume contained in a large enclosure was obtained. The contribution to the fluctuation of the temperature field, at a point x distance from the particles, which lay in a distant volume, was then cancelled by the contribution from the continuous distribution of dipoles contained within that volume. Using this approach, the common integral convergence problem encountered by other investigators, was avoided and the expression for the fluctuation of the temperature field at point x converged. The cases, in which particles were arranged in both an ordered and random array, were also discussed and the influence of the interaction between the two particles was considered. Here, the effective thermal conductivity was corrected as $O(\phi^2)$, and the average dipole strength was corrected as $O(\phi)$.

Given that the calculations available for short fiber suspensions could not be applied to a condition in which fibers were randomly oriented in a continuous matrix, Hatta and Taya [75] proposed an equation for calculating the effective thermal conductivity of two- and three-dimensional cases with randomly oriented short fibers in a continuous matrix, based on the equivalent inclusion method for steady-state conditions. In order to describe the orientation of the fibers, two-coordinate systems were established as shown in Fig. 49, and the distribution functions for the random oriented short fibers were of two kinds: uniform and cosine type as shown in Fig. 50.

The results of this approach are illustrated in Figs. 51–53, where the thermal conductivity ratio of the fiber and the continuous matrix was 20, the aspect ratio of the fiber was 100, and the constant was $a = 90°$. The assumption that the fibers were completely surrounded by the continuous matrix required that the volume fraction of the short fibers in the suspension conform to $\phi \leq 40\%$.

Davis [76], fully aware of the difficulty of integral convergence encountered in earlier investigations, considered a situation where inside the composite material, the undisturbed temperature gradient varied inversely with

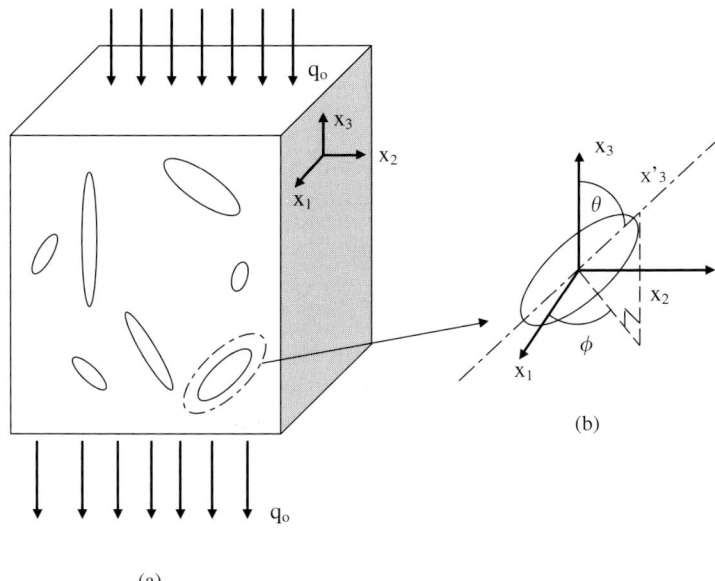

FIG. 49. Illustration of the two-coordinate systems used in the model developed by Hatta and Taya [75]: (a) The randomly oriented short fiber suspension, (b) the coordinate system of suspension and the coordinate system of fiber.

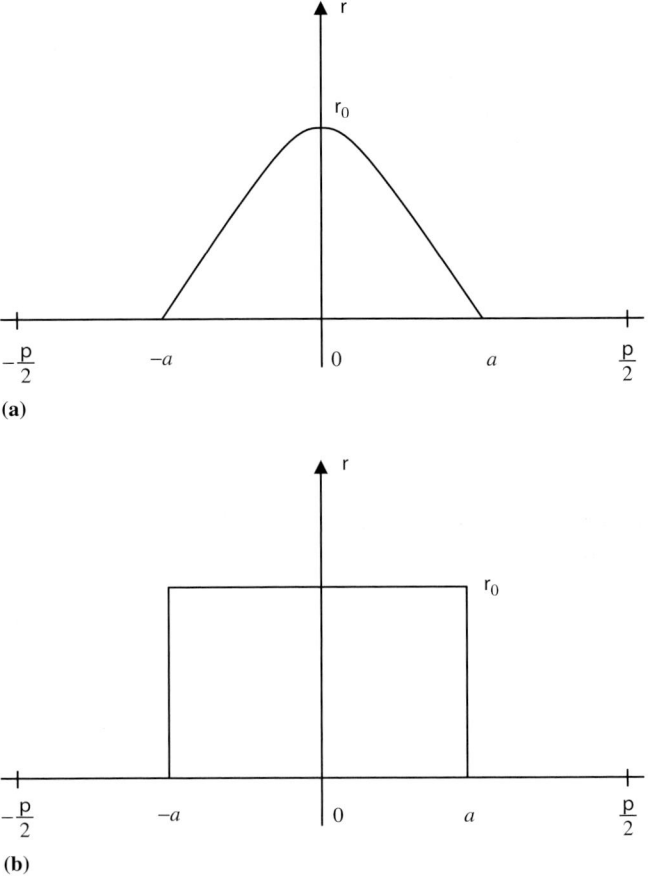

FIG. 50. Distribution function of fiber orientation [75]: (a) The uniform distribution $\rho = \rho_0$, (b) the cosine-type distribution $\rho = \rho_0 \cos a\theta$.

the square of the distance from the heated body. The resulting correction to the dipole strength of the reference sphere due to the presence of a second sphere, varied as $(a/r)^5$ instead of $(a/r)^3$, and only convergent integrals were encountered. Later Eq. (44), which was based on Green's theorem and is similar to Eq. (32) was used to calculate the rate of heat transfer from the heated body to the composite material.

$$k/k_1 = 1 + \frac{3(\alpha - 1)}{[\alpha + 2 - (\alpha - 1)\phi]} \{\phi + f(\alpha)\phi^2 + 0(\phi^3)\} \qquad (44)$$

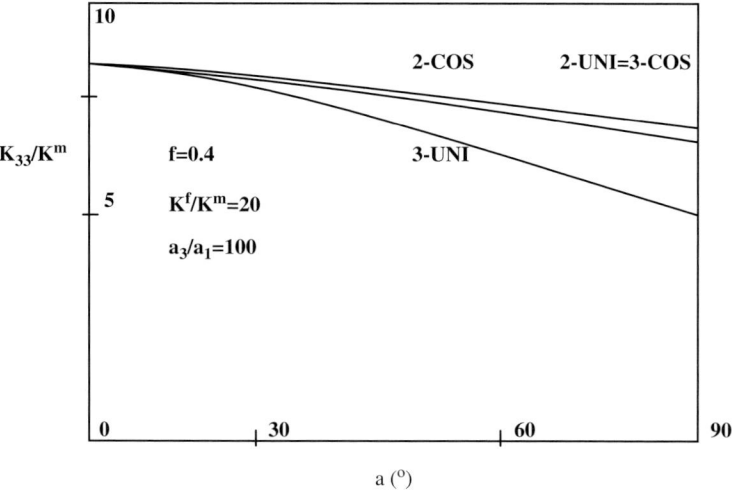

Fig. 51. The effective thermal conductivity of various types of short fiber suspension as a function of α [73].

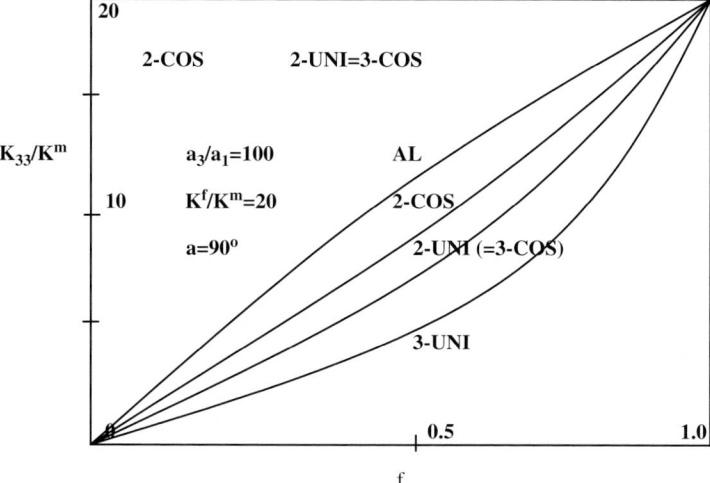

Fig. 52. The effective thermal conductivity of various types of short fiber suspension as a function of φ [73].

Here, $f(\alpha) = \sum_{p=6}^{\infty} B_p - 3A_p/(p-3)2^{p-3}$ is used to represent the decaying temperature field. In developing this equation, researchers had previously considered it a composite material of infinite extent, on which an undisturbed linear temperature field was imposed. The contribution of the term

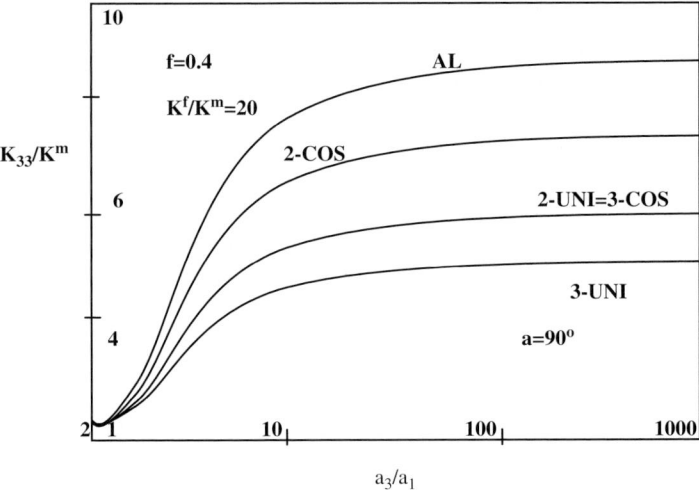

FIG. 53. The effective thermal conductivity of various types of short fiber suspension as a function of fiber aspect ratio [75].

$O(\phi^2)$ to the effective thermal conductivity was not entirely described by the solution for two spheres placed in the undisturbed temperature field, but the actual environment of the these two spheres was the undisturbed temperature plus an $O(\phi)$ term from all of the remaining spheres. It was also pointed out that contributions of $O(\phi^3)$ and higher orders were even more difficult to obtain and not generally needed for the dilute suspension cases. Here, in order to prevent the spheres from touching one another, the maximum volume ratio was 62%.

Based on the experimental results of spherical nickel dispersions in a sodium borosilicate glass matrix and carbon fiber-reinforced glass ceramic, Hasselman and Johnson [77] noted that the thermal barrier resistance at the interface between the dispersed particle and the continuous matrix could have an important influence on the effective thermal conductivity of the composite. As Maxwell had proposed, the idea that the potential was continuous and the current was the same at the two sides of the interface between dispersed particles and a continuous matrix was also adopted in this study.

$$k_1 \left(\frac{\partial T_1}{\partial r} \right) = k_2 \left(\frac{\partial T_2}{\partial r} \right) \tag{45}$$

$$T_1 - T_2 = -\frac{k_2}{h_c} \frac{\partial T_2}{\partial r} \tag{46}$$

where T_1 and T_2 are the temperatures in the continuous matrix and the dispersed particle, respectively; k_1 and k_2 the thermal conductivity of the continuous matrix dispersed particle, respectively; r is the radius of the interfacial surface; and h_c the interfacial thermal barrier resistance. If this latter term, the interfacial thermal barrier resistance, was zero, then $T_1 = T_2$ at the interface, which agreed with the original solution as proposed by Maxwell.

However, it is clear that physically the interfacial thermal barrier resistance, h_c, could not be zero, so the effective thermal conductivity of the composite was affected by the cumulative effect on T_1, and the radius of the large sphere, which included n dispersed spheres of radius a. When the larger sphere was sufficiently large, $b \to \infty$, and the equation for the effective thermal conductivity could be expressed as

$$k_e = k_1 \frac{[2(k_2/k_1 - k_2/ah_c - 1)\phi + k_2/k_1 + 2k_2/ah_c + 2]}{[(1 - k_2/k_1 + k_2/ah_c)\phi + k_2/k_1 + 2k_2/ah_c + 2]} \quad (47)$$

The equations for two kinds of particles, circular cylinders and flat plates, were also developed

$$k_e = k_1 \frac{[(k_2/k_1 - k_2/ah_c - 1)\phi + k_2/k_1 + 2k_2/ah_c + 1]}{[(1 - k_2/k_1 + k_2/ah_c)\phi + k_2/k_1 + 2k_2/ah_c + 1]} \quad (48)$$

$$k_e = \frac{k_2}{[(1 - k_2/k_1 + 2k_2/ah_c)\phi + k_2/k_1]} \quad (49)$$

and were found to be in good agreement with the previous solutions when the interfacial thermal barrier resistance was zero.

Brady et al. [78] presented a method for calculating the hydrodynamic interactions among particles in an infinite suspension, but with an infinitely small particle Reynolds number. The Stokesian dynamics method was applied as part of the numerical calculation for the spatially periodic suspension problem. Based on this work, Bonnecaze and Brady [79,80] developed a method for determining the individual dipoles for a finite number of particles in order to avoid the convergence problem, in integration over the infinite extent of the matrix. This method included the particle interactions in both long- and short-range fields. The individual particle potentials and the potential gradient were both related to the sum of the particle moments, to generate a "potential" matrix, which when inverted, formed a capacitance matrix. The potential invert produced the many-body reflections of the particle moments that were included in the initial formulation and the far-field interactions of the particles. By adding the exact two-body interactions into the potential invert, an accurate capacitance matrix was obtained. With this capacitance matrix, the dipole strength of the particles could be calculated directly under the given external potential field. The equation developed was given as

$$k_e = k_1 - n\left(\frac{S}{G^E}\right) \qquad (50)$$

where S is the dipole strength and G^E the potential gradient applied.

This method was applied to infinite suspensions using the method of O'Brien and periodic boundary conditions that reduced the system to a finite number of linear equations. In this way, the effective thermal conductivity of three different suspensions with a cubic array of spheres of (a) an infinite thermal conductivity, (b) a thermal conductivity equal to 10 times the thermal conductivity of the matrix, and (c) a thermal conductivity 0.01 times the thermal conductivity of the matrix, were all calculated. The results are shown in Fig. 54. The authors later used this simulation method to calculate the effective thermal conductivity of an infinite, statistically homogeneous suspension of hard spherical particles in a disordered configuration where the random pack configurations based on the computed radial distribution function were generated with the Monte Carlo method. In this approach, particles leaving one side of the periodic cell reappeared on

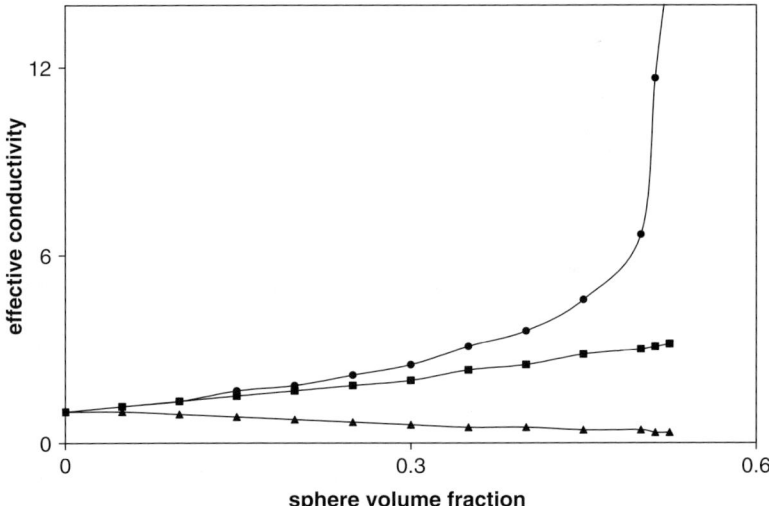

FIG. 54. Calculation results of the simple cubic lattice with sphere to matrix thermal conductivity ratio of infinite (dot), 10 (square), and 0.01 (triangle) [79].

the opposite side, but did not overlap or come in contact with others in the cell or in the neighboring periodic cells. The Ewald method was used to form rapidly converging lattice sums of particle interactions. In this study, the effective conductivity of the spherical particles in random arrays was calculated for particle-to-matrix conductivity ratios of infinity, 10, and 0.01, and the volume ratio was restricted to within 60% to insure no contact or agglomeration. The rapid increase in the effective conductivity near the densest packing for each array was explained as the result of the lubrication or "short-circuiting" of material flux between the particles. Cases of nonperfect conductors with conductivity ratios of 10 and 0.01, demonstrated that using the potential matrix invert to form the capacitance matrix provided a reasonably accurate estimation when compared to the results that included the influence of two-body particle interactions (see Table IV).

Lu and Kim [81] investigated the relationship between the microstructure of the dispersed particles and the effective thermal conductivity of anisotropic composites consisting of a homogeneous matrix with dispersed spherical particles. The systems with fixed aligned dispersed particles were the only ones considered and the variation of the particle positions influenced the effective thermal conductivity in this study. As had been done in previous investigations, the dipole strength and renormalization approach to the integration were used to calculate the dipole strength. A separate

TABLE IV

EFFECTIVE THERMAL CONDUCTIVITY IN THE RANDOM DISPERSED HARD SPHERE SUSPENSION [81]

Thermal conductivity ratio of particle-to-matrix material	Volume fraction of particles in suspension	Without the two-body interaction	With the two-body interaction	Jeffrey's results
Infinite	0.00	1.000	1.000	1.000
	0.05	1.160	1.162	1.61
	0.30	2.341	2.529	2.306
	0.60	5.100	8.853	4.424
10	0.00	1.000	1.000	1.000
	0.10	1.246	1.247	1.240
	0.30	1.889	1.944	1.812
	0.60	3.589	3.966	2.900
0.01	0.00	1.000	1.000	1.000
	0.10	0.859		0.858
	0.30	0.600		0.608
	0.60	0.308		0.297

model that included the assumption that dispersed particles would take nonoverlapping positions with equal probability was developed, and indicated that neglecting the potential energy between particles would result in the introduction of errors. Following this, the hard-sphere fluid model was adopted such that the pair distribution function was only a function of the distance between the two axes. The results obtained from this latter model demonstrated good agreement with both previous investigations and the experimental data. Here

$$k_e = \widehat{I} + \left(-\frac{M}{V_2 r_1}\right)\phi + \left(\frac{-\Xi + \Omega}{V_2 r}\right)\phi^2 \quad (51)$$

where $\widehat{I}$ is the unit tensor, V_2 the volume of the dispersed particles, M the second-order tensor for one-body contributions, Ξ the second-order tensor for two-body contributions, and Ω the renormalization quantity.

Kim and Torquato [82] utilized a computer simulation to calculate the effective thermal conductivity of aligned spheroid particle suspensions using a "first passenger time" algorithm. Here, the dispersed particles had a wide range of aspect ratios, ranging from 0.1 to 10 and the ratio of the thermal conductivities of the dispersed particle and the continuous matrix ranged from zero to infinity. This method was principally concerned with the Brownian motion of dispersed particles and utilized a totally different approach to the calculation of the effective thermal conductivity of suspensions when compared to previously methodologies, which focused on the use of derivatives of the Maxwell equation for varying volume ratios to include the far and near range interactions between two or more particles. This new approach addressed the motion instead of the statistic force field, to consider the mechanisms of the enhancement of the effective thermal conductivity of the suspension.

Lu and Lin [83] developed an expression for the effective conductivity tensor to calculate the effective conductivity in the form of a virial expansion in the dispersed particle volume fraction, ϕ, which was truncated at the $o(\phi^2)$ term. The particle shape discussed was ellipsoid and its mathematical expression was

$$\frac{x^2}{a^2} + \frac{y^2}{b^2} + \frac{z^2}{c^2} = 1, \quad a \geq b \geq c \quad (52)$$

where a, b, and c are the three axes of the particles.

In this pair interaction study, a boundary collocation scheme was used to analyze two aligned spheroids and solve the boundary value problem to evaluate the second-order tensor for the one-body contribution, M, the second-order tensor for the two-body contribution, Ξ, and the renormalization quantity, Ω, for the two aligned spheroids. This boundary collocation scheme was derived based on the premise that the Laplace equation was the governing equation, and that the system was limited to a two-particle system. The temperature fields outside the particles and inside of the two particles were expressed, first, with the general solution for the Laplace equation, then, they were forced to satisfy the boundary conditions at the particle surface, and finally the relevant thermal dipole strength. The series ratio inferred from the virial expansion of the equivalent inclusion estimate served as a reasonably good check on the validity of the test parameter for the second-order virial approximation. The magnitude of this series ratio was also a reflection of the intensity of the thermal interactions within the system. Two models for the pair distribution function were studied; one was the well-stirred model, and the other was the hard spheroid model. The equivalent inclusion estimates method of Hatta and Taya [75] was applied and derived as

$$\frac{k_{e,ij}}{k_1} = 1 + \frac{\alpha - 1}{(\alpha - 1)h_i + 1}\phi + \frac{(\alpha - 1)^2 h_i}{[(\alpha - 1)h_i + 1]^2}\phi^2 + \frac{(\alpha - 1)^3 h_i}{[(\alpha - 1)h_i + 1]^3}\phi^3 + \cdots \qquad (53)$$

where $\alpha = k_2/k_1$, and the equation would converge with the condition $\alpha - 1/(\alpha - 1)h_i + 1 < 1$.

The results of this analysis are compared with the results obtained by other investigators in Tables V–VIII.

The equation developed by Lu and Lin [83] indicated that the smaller the value of the series ratio, the better the agreement of the virial expansion with the results of the previous simulation. The magnitude of the ratio also provided a strong indication of the intensity of the interaction between the particles in the suspension.

In the development of the previously discussed methods for calculating the effective thermal conductivity of the suspensions, there was little information that related to real situations or applications of dispersed particles in suspension, or the effect of size variations of these dispersed particles on the suspension thermophysical properties. For the dispersed particles with a size ranging from millimeters to micrometers, this size effect was so small that it could be ignored in the calculations. However, when the size of the particles dispersed in the suspension is of the order of several nanometers or several tens of nanometers, the mean free path of the heat carriers for most nanoparticle materials is smaller than the size of the dispersed nanoparticles themselves. For example, the heat carrier mean free path of gold at room

TABLE V

The k_e/k_1 Value for the Well-Stirred Model: (a) Prolated Particle, (b) Oblate Particle [83]

(a)

a/b	σ	$\sigma_{\text{eff}}^{\parallel}$	$\sigma_{\text{eff}}^{\perp}$
10	0	$1-1.02f+0.00316f^2$	$1-1.96f+1.35f^2$
2		$1-1.21f+0.1716f^2$	$1-1.70f+0.94f^2$
1[a]		$1-1.50f+0.588f^2$	$1-1.50f+0.588f^2$
10	0.1	$1-0.917f+0.00497f^2$	$1-1.617f+0.977f^2$
2		$1-1.07f+0.143f^2$	$1-1.43f+0.694f^2$
1[a]		$1-1.291f+0.450f^2$	$1-1.29f+0.450f^2$
10	2	$1+0.981f+0.0265f^2$	$1+0.671f+0.243f^2$
2		$1+0.947f+0.0638f^2$	$1+0.679f+0.240f^2$
1[a]		$1+0.852f+0.146f^2$	$1+0.708f+0.228f^2$
		$1+0.766f+0.199f^2$	$1+0.742f+0.211f^2$
		$1+0.750f+0.208f^2$	$1+0.750f+0.208f^2$
10	10	$1+7.61f+2.85f^2$	$1+1.66f+1.77f^2$
5		$1+5.99f+3.98f^2$	$1+1.71f+1.85f^2$
2		$1+3.51f+3.24f^2$	$1+1.91f+2.01f^2$
10/9		$1+2.40f+2.39f^2$	$1+2.18f+2.22f^2$
1[b]		$1+2.25f+2.27f^2$	$1+2.25f+2.27f^2$
10	100	$1+32.9f+98.6f^2$	$1+2.00f+3.03f^2$
5		$1+15.2f+35.3f^2$	$1+1.171f+1.85f^2$
2		$1+5.45f+8.84f^2$	$1+1.91f+2.01f^2$
10/9		$1+3.17f+4.59f^2$	$1+2.18f+2.22f^2$
1[b]		$1+3.17f+4.59f^2$	$1+2.25f+2.27f^2$

(b)

a/c	σ	$\sigma_{\text{eff}}^{\parallel}$	$\sigma_{\text{eff}}^{\perp}$
10	0	$1-7.18f+34.8f^2$	$1-1.07f-0.0730f^2$
2		$1-2.12f+1.92f^2$	$1-1.31f+0.278f^2$
1a		$1-1.50f+0.588f^2$	$1-1.50f+0.588f^2$
10	0.1	$1-4.00f+11.6f^2$	$1-0.960f+0.00449f^2$
2		$1-1.71f+1.31f^2$	$1-1.14f+0.228f^2$
1a		$1-1.29f+0.450f^2$	$1-1.29f+0.450f^2$
10	2	$1+0.537f+0.261f^2$	$1+0.935f+0.0869f^2$
2		$1+0.571f+0.264f^2$	$1+0.889f+0.125f^2$
1a		$1+0.655f+0.247f^2$	$1+0.809f+0.176f^2$
		$1+0.734f+0.215f^2$	$1+0.758f+0.203f^2$
		$1+0.750f+0.208f^2$	$1+0.750f+0.208f^2$
10	10	$1+1.03f+1.15f^2$	$1+5.53f+5.25f^2$
5		$1+1.16f+1.32f^2$	$1+4.24f+4.19f^2$
2		$1+1.57f+1.70f^2$	$1+2.88f+2.83f^2$
10/9		$1+2.11f+2.16f^2$	$1+2.32f+2.33f^2$
1[b]		$1+2.25f+2.27f^2$	$1+2.25f+2.27f^2$
10	100	$1+1.15f+1.78f^2$	$1+12.5f+40.6f^2$
5		$1+1.31f+1.98f^2$	$1+7.72f+16.2f^2$
2		$1+1.86f+2.65f^2$	$1+4.06f+6.34f^2$
10/9		$1+2.69f+3.83f^2$	$1+3.04f+4.38f^2$
1[b]		$1+3.91f+4.19f^2$	$1+2.91f+4.19f^2$

TABLE V. (Continued)

	(a)				(b)	
a/b	σ	$\sigma_{\text{eff}}^{\parallel}$	$\sigma_{\text{eff}}^{\perp}$	a/c	σ	$\sigma_{\text{eff}}^{\parallel}$	$\sigma_{\text{eff}}^{\perp}$
10	10,000	$1+49.1f+271f^2$	$1+2.04f+3.40f^2$	10	10,000	$1+1.16f+1.92f^2$	$1+14.3f+57.9f^2$
5		$1+17.9f+52.6f^2$	$1+2.12f+3.38f^2$	5		$1+1.33f+2.10f^2$	$1+8.01f+19.7f^2$
2		$1+5.76f+10.1f^2$	$1+2.42f+3.64f^2$	2		$1+1.90f+2.80f^2$	$1+4.23f+7.00f^2$
10/9		$1+3.27f+4.19f^2$	$1+2.88f+4.30f^2$	10/9[b]		$1+2.76f+4.10f^2$	$1+3.13f+4.73f^2$
1[b]		$1+3.00f+4.51f^2$	$1+3.00f+4.51f^2$			$1+3.00f+4.51f^2$	$1+3.00f+4.51f^2$
10[c]	∞	$1+49.3f+275f^2$	$1+2.04f+3.42f^2$	10[c]	∞	$1+2.25f+2.27f^2$	$1+14.4f+58.3f^2$
5[c]		$1+17.9f+52.9f^2$	$1+2.212f+3.39f^2$	5[c]		$1+1.33f+2.10f^2$	$1+8.02f+19.8f^2$
2[c]		$1+5.76f+10.1f^2$	$1+2.42f+3.65f^2$	2[c]		$1+1.90f+2.80f^2$	$1+4.23f+7.03f^2$
10/9[c]		$1+3.27f+4.97f^2$	$1+2.88f+4.31f^2$	10/9[c]		$1+2.76f+4.12f^2$	$1+3.13f+4.74f^2$
1[a]		$1+3.00f+4.51f^2$	$1+3.00f+4.51f^2$	1[a]		$1+3.00f+4.51f^2$	$1+3.00f+4.51f^2$

[a] cited from Jeffrey [68].
[b] calculated based on Jeffrey's equation [68].
[c] cited from Lu and Kim [81].

TABLE VI

The Comparison of $k_{e,ij}/k_1$ to Previous Works for $k_2/k_1 = 10$ [83]

$a/b\ (>1)$ $c/a\ (<1)$	Models	$\sigma_{\text{eff}}^{\perp}$			$\sigma_{\text{eff}}^{\parallel}$		
		$f=0.1$	$f=0.3$	$f=0.5$	$f=0.1$	$f=0.3$	$f=0.5$
0.1	K–T	1.60	3.26	5.00	1.12	1.45	2.05
	Well stirred	1.61	3.13	5.08	1.11	1.41	1.80
	Hard spheroid	1.61	3.19	5.37	1.11	1.42	1.82
	[Series ratio]	[0.039]	[0.12]	[0.19]	[0.089]	[0.27]	[0.44]
0.2	K–T	1.47	2.78	4.39	1.13	1.48	2.09
	Well stirred[a]	1.47	2.65	4.17	1.13	1.47	1.91
	Hard spheroid	1.47	2.71	4.47	1.13	1.48	1.96
	[Series ratio]	[0.053]	[0.16]	[0.26]	[0.087]	[0.26]	[0.44]
0.5	K–T	1.37	2.11	3.35	1.18	1.63	2.35
	Well stirred	1.32	2.12	3.15	1.17	1.62	2.21
	Hard spheroid	1.32	2.16	3.39	1.17	1.64	2.32
	[Series ratio]	[0.068]	[0.20]	[0.34]	[0.083]	[0.25]	[0.41]
1	K–T	1.25	1.93	3.02	1.25	1.93	3.02
	Well stirred[a]	1.25	1.88	2.69	1.25	1.88	2.69
	Hard sphere[b]	1.25	1.91	2.88	1.25	1.91	2.88
	[Series ratio]	[0.075]	[0.225]	[0.375]	[0.075]	[0.225]	[0.375]
2	K–T	1.23	1.78	2.65	1.41	2.53	3.85
	Well stirred	1.21	1.75	2.46	1.38	2.35	3.57
	Hard spheroid	1.21	1.78	2.61	1.39	2.40	3.85
	[Series ratio]	[0.079]	[0.24]	[0.39]	[0.061]	[0.18]	[0.30]
5	K–T	1.22	1.74	2.61	1.71	3.21	4.87
	Well stirred	1.19	1.68	2.32	1.64	3.16	4.99
	Hard spheroid	1.19	1.70	2.43	1.64	3.21	5.28
	[Series ratio]	[0.081]	[0.24]	[0.40]	[0.033]	[0.10]	[0.17]
10	K–T	1.19	1.73	2.59	1.81	3.48	5.20
	Well stirred	1.18	1.66	2.27	1.79	3.54	5.52
	Hard spheroid	1.18	1.67	2.36	1.79	3.57	5.67
	[Series ratio]	[0.082]	[0.24]	[0.41]	[0.015]	[0.046]	[0.077]

[a]cited from Jeffrey [68].
[b]calculated based on Jeffrey's equation [68].

temperature is around 31 and for silicon is 43 nm. Hence, the traditional methods discussed previously, which are primarily built upon Fourier's law, may no longer fully describe the fundamental process occurring in these suspensions, and may not fully incorporate the heat exchange between the dispersed particle and the continuous phase immediately adjacent to the particle.

Chen [84] examined this problem along with a single nanoparticle immersed in fluid, assuming that there was a heat source in the nanoparticle. By evaluating the nonlocal and nonequilibrium heat conduction in the

TABLE VII

The Comparison of $k_{e,ij}/k_1$ to Previous Works for $k_2/k_1 = \infty$ [83]

$a/b\ (>1)\ c/a\ (<1)$	Models	$\sigma_{\text{eff}}^{\perp}$			$\sigma_{\text{eff}}^{\parallel}$		
		$f=0.1$	$f=0.3$	$f=0.5$	$f=0.1$	$f=0.3$	$f=0.5$
0.1	K–T	3.61	19.44	53.75	1.18	2.29	3.62
	Well stirred	3.01	10.5	22.7	1.13	1.46	1.88
	Hard spheroid	3.05	11.5	27.9	1.14	1.54	2.16
	[Series ratio]	[0.1]	[0.3]	[0.5]	[0.1]	[0.3]	[0.5]
0.2	K–T	2.30	7.11	22.69	1.24	2.32	3.70
	Well stirred[a]	2.00	5.18	9.93	1.15	1.59	2.19
	Hard spheroid	2.01	5.59	12.2	1.16	1.62	2.35
	[Series ratio]	[0.1]	[0.3]	[0.5]	[0.1]	[0.3]	[0.5]
0.5	K–T	1.51	3.35	6.88	1.28	2.36	3.81
	Well stirred	1.49	2.90	4.87	1.22	1.82	2.65
	Hard spheroid	1.50	3.05	5.74	1.22	1.87	2.94
	[Series ratio]	[0.1]	[0.3]	[0.5]	[0.1]	[0.3]	[0.5]
1	K–T	1.34	2.48	4.78	1.34	2.48	4.78
	Well stirred[a]	1.35	2.31	3.63	1.35	2.31	3.63
	Hard sphere[b]	1.35	2.40	4.18	1.35	2.40	4.18
	[Series ratio]	[0.1]	[0.3]	[0.5]	[0.1]	[0.3]	[0.5]
2	K–T	1.27	2.06	3.50	1.68	3.60	8.26
	Well stirred	1.28	2.05	3.12	1.68	3.63	6.40
	Hard spheroid	1.28	2.12	3.53	1.68	3.86	7.68
	[Series ratio]	[0.1]	[0.3]	[0.5]	[0.1]	[0.3]	[0.5]
5	K–T	1.25	1.99	3.46	3.13	10.12	25.42
	Well stirred	1.25	1.94	2.90	3.31	11.1	23.1
	Hard spheroid	1.25	2.00	3.24	3.35	12.1	28.6
	[Series ratio]	[0.1]	[0.3]	[0.5]	[0.1]	[0.3]	[0.5]
10	K–T	1.23	1.90	3.33	9.03	36.36	99.64
	Well stirred	1.24	1.92	2.87	8.61	40.1	93.2
	Hard spheroid	1.24	1.97	3.18	8.74	43.8	113.
	[Series ratio]	[0.1]	[0.3]	[0.5]	[0.1]	[0.3]	[0.5]

[a]cited from Jeffrey [68].
[b]calculated based on Jeffrey's equation [68].

vicinity of the nanoparticles, as shown in Fig. 55, the spherical particle with uniform heat generation embedded in a host medium could be modeled and the Boltzmann transport equation applied to predict the effective temperature rise at the particle surface. A temperature jump at the surface of the spherical nanoparticle was observed and Eq. (53) was developed to predict the effective thermal conductivity

$$\frac{k}{k_1} = \frac{3\tau_1/4}{3\tau_1/4 + 1} \tag{54}$$

TABLE VIII

The Comparison of $k_{e,ij}/k_1$ to Previous Works for $k_2/k_1 = 0$ [83]

a/b (>1) c/a (<1)	Models	$\sigma_{\text{eff}}^{\perp}$			$\sigma_{\text{eff}}^{\parallel}$		
		$f=0.1$	$f=0.3$	$f=0.5$	$f=0.1$	$f=0.3$	$f=0.5$
0.1	K–T	0.887	0.653	0.415	0.496	0.183	0.068
	Well stirred	0.892	0.671	0.444	0.630	1.98	6.12
	Hard spheroid	0.892	0.668	0.429	0.624	1.83	5.43
	[Series ratio]	[0.0075]	[0.02]	[0.037]	[0.62]	[1.9]	[3.1]
1	K–T	0.856	0.602	0.381	0.856	0.602	0.381
	Well stirred[a]	0.856	0.603	0.397	0.856	0.603	0.397
	Hard sphere[b]	0.856	0.595	0.351	0.856	0.595	0.351
	[Series ratio]	[0.05]	[0.15]	[0.25]	[0.05]	[0.15]	[0.25]
2	K–T	0.819	0.526	0.308	0.898	0.692	0.489
	Well stirred	0.817	0.533	0.357	0.898	0.694	0.490
	Hard spheroid	0.817	0.510	0.231	0.898	0.694	0.489
	[Series ratio]	[0.096]	[0.29]	[0.48]	[0.0021]	[0.0062]	[0.01]

[a]cited from Jeffrey [68].
[b]calculated based on Jeffrey's equation [68].

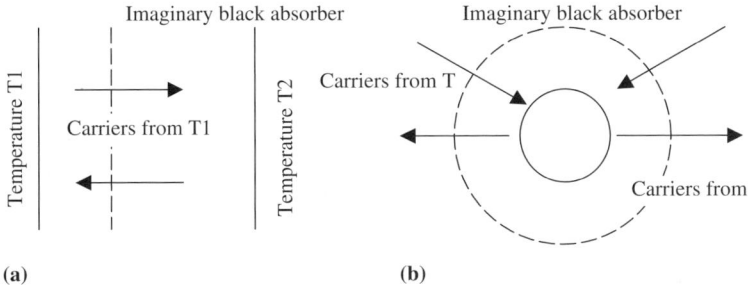

Fig. 55. Non-local and non-equilibrium heat transfer processes [84] (a) across a one-dimensional bar, (b) around a particle.

where $\tau_1 = a/\Lambda$ is the nondimensional radial distance, a the radius of the particle, and Λ the mean free path of the heat carrier. The results indicated that the nondimensional equivalent temperature distribution was a function of the radial distance normalized to the particle radius for different particle size parameters [84]. The temperature distribution calculated from the Boltzmann transport method approached the prediction of the Fourier law when the particle radius was much larger than the heat-carrier mean free path. For the condition $\tau_1 > 21$, the difference between the values predicted by Fourier's law and the value obtained using this nonlocal and nonequilibrium heat transfer method was less than 5% [84]. This investigation provided information on the interaction between the dispersed particle and the continuous matrix around it, as well as an idea of how the particle size might influence the effective thermal conductivity.

Xue [85] proposed an equation based on Maxwell's equation, to calculate the effective thermal conductivity by considering the interface effect between the solid particles and the base fluid. Utilizing polarization theory, Xue proposed that the molecules of base liquid close to the solid surface should be organized into layered structures much like a solid. This was expressed as

$$9\left(1 - \frac{v}{\lambda}\right)\frac{k - k_1}{sk + k_1} + \frac{v}{\lambda}\left[\frac{k - k_{c,x}}{k + B_{2,x}(k_{c,x} - k)} + 4\frac{k - k_{c,y}}{sk + (1 - B_{2,x})(k_{c,y} - k)}\right] = 0 \quad (55)$$

where $B_{2,j}$ are the depolarization factor components of the elliptical particle along the j-symmetrical axis which was dependent on the shape, v and v/λ the volume fraction of the nanoparticles and the complex nanoparticles, respectively, and $k_{c,j}$ is the effective dielectric constant component [85]. As shown in Figs. 56 and 57, when appropriate values for the special thickness

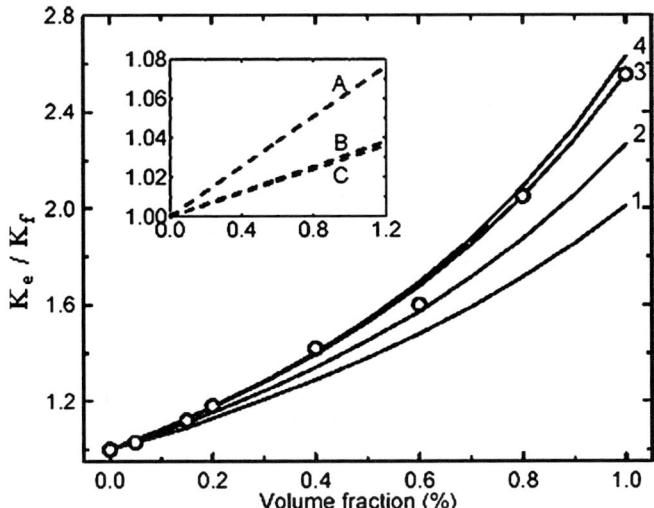

FIG. 56. Experimental data of carbon nanotube/oil suspension (open circles) and predictions with the data for the thickness and thermal conductivity of the interfacial shell [85]: (1) 1 nm and 5 W/mK, (2) 2 nm and 5 W/mK, (3) 3 nm and 5 W/mK, (4) 3 nm and 20 W/mK; (A)Hamilton–Crosser, (B) Bonnecaze and Brady, and (C) Maxwell.

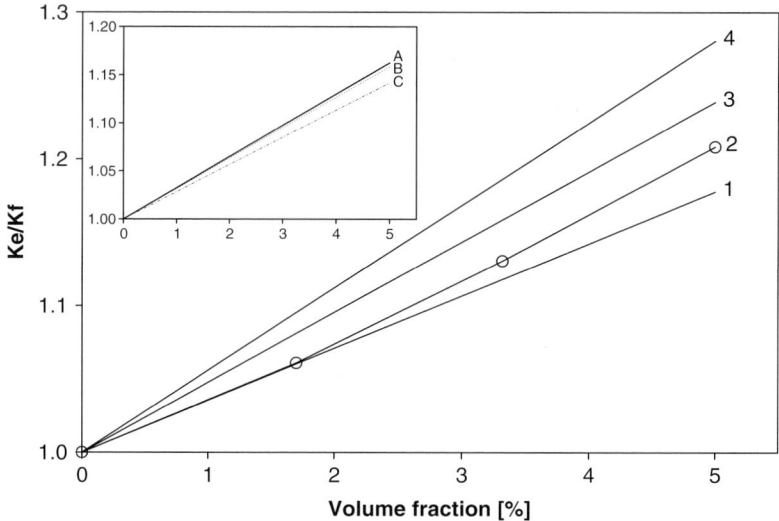

FIG. 57. Experimental data of Al2O3/water suspension (open circles) and predictions with the data for the thickness and thermal conductivity of the interfacial shell [85]: (1) 1 nm and 2.1 W/mK, (2) 3 nm and 2.1 W/mK, (3) 5 nm and 2.1 W/mK, (4) 5 nm and 10 W/mK; (A) Hamilton–Crosser, (B) Bonnecaze and Brady, and (C) Maxwell.

of the base liquid molecular layer and its thermal conductivity are used, this model yields good agreement when compared with the previous experimental data and predictions resulting from some of the expressions discussed previously.

Wang et al. [86] examined the effect of the agglomeration of nanoparticles, using fractal theory to describe the radial distribution of the nanoparticles, and developed a method to calculate the effective thermal conductivity of the nanoparticle suspension that included the effect of size and the adsorbed liquid layer on the particle surface. The resulting expression was given as

$$k/k_1 = \frac{(1-\phi) + 3\phi \int_0^\infty k_{cl}(r)n(r)/k_{cl}(r) + 2k_1 \, dr}{(1-\phi) + 3\phi \int_0^\infty k_1 n(r)/k_{cl}(r) + 2k_1 \, dr} \qquad (56)$$

where $k_{cl}(r)$ is the effective thermal conductivity of the cluster and $n(r)$ the radius distribution function. As shown in Fig. 58, comparison of the predicted values both with and without adsorption shows good agreement with the experimental data.

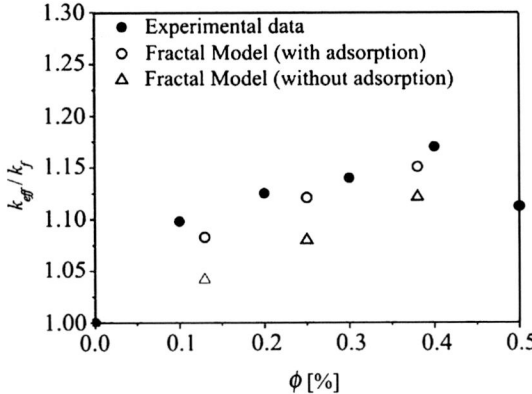

FIG. 58. Comparison of predicted data with/without the adsorbed liquid layer on the surface of the nanoparticle to the CuO/water nanoparticle suspension experiemental data of Wang et al. [86].

B. The Effects of Brownian Motion Coupled with Thermal Phoresis

Based upon these previous investigations it is apparent that the Brownian motion of the nanoparticles in these suspensions has not been adequately considered in the development of the equations to predict the thermal behavior and may be at least, partially responsible for the unusually high effective thermal conductivity of these suspensions. The omission of the potential impact of the Brownian motion is understandable when one realizes and understands the way in which the initial equations were developed and the fact that, as described in the previous sections, much of the early work occurred before the emergence of nanotechnology and focused on millimeter- or micrometer-sized particles, which because of their size, would not experience Brownian motion. For nanoparticles or nanotubes, however, the level of Brownian motion will be significant and could play an important role in the enhancement of the thermal conductivity.

In 1827, Robert Brown examined the form of *Clarkia pulchella* pollen particles immersed in water and observed that many of these particles were in continual motion that arose from neither currents in the fluid, nor from its gradual evaporation, but rather that appeared to belong to the particles themselves. This constant and irregular motion increased as the size of the particle decreased and continued for the entire time that the particle remained in suspension. Further observation of this "Brownian motion" illustrated the irregular energy and momentum exchanges of molecules caused by the collision between particles and molecules, and indicated a strong temperature dependence. This temperature dependence was thought to be a result of the mean kinetic energy of the fluid molecules; the viscosity of the fluid, which is also influenced by the temperature of the fluid; the size of the particles; and the nature of the fluid itself.

In 1909, Jean Perrin, as described by Haw and Mazo [87,88] successfully developed experiments to indirectly prove quantitatively, that Brownian motion is, in fact, the reflection of molecular kinetic theory. In this investigation, the sedimentation speed of the particles was measured for different heights inside the suspension. While demonstrating the existence of Brownian motion, the problem of direct measurement of the displacement of these particles remained and a number of other investigators have attempted to measure the displacement of individual particles due to this Brownian motion.

In 1995, Grasselli and Bossis [89] applied a three-dimensional micrometer size particle tracking technique that utilized microscopy and computer digitized imaging. In this investigation, individual particles were tracked simultaneously and measurements of both the Brownian motion and the sedimentation were made along with the particle hydrodynamic diameter and density. By analyzing the trajectory of individual particles on an X–Y plane with 22,000 different positions and a time interval of 0.1 s, as shown in Fig. 59, the diffusion coefficients for the particles were determined to be $DX = 1.438\mathrm{e}-13\,\mathrm{m}^2/\mathrm{s}$ and $DY = 1.412\mathrm{e}-13\,\mathrm{m}^2/\mathrm{s}$ in the X and Y directions, respectively, and are shown as the slope in Figs. 60 and 61, respectively. In addition, the vertical sedimentation velocity of the micrometer-sized particle was observed to be $\sim 0.23 \pm 0.02\,\mu\mathrm{m/s}$, as shown in Fig. 62.

In 1998, Gaspard et al. [90] verified the hypothesis of microscopic chaos by observing the Brownian motion of a 2.5 µm diameter particle suspended in deionized water at 22°C. In this experiment, the diffusion coefficient was calculated to be $D \approx 0.124\,\mu\mathrm{m}^2/\mathrm{s}$ as illustrated in Fig. 63. In 2001, Briggs et al. [91] introduced an experimental test facility and measurement procedure for tracking the Brownian motion of single particles, as small as 0.5 µm in diameter, while suspended in water. This work proved that these particles could, as a result of undergoing Brownian motion, behave as predicted by Stokes–Einstein diffusion theory. The desktop experimental facility is illustrated in Fig. 64. Data describing the Brownian motion displacements for 1.0 and 2.5 µm diameter particles are given in Figs. 65 and 66, which illustrates the mean-squared displacement as a function of time.

Goodson and Kraft [92] developed and studied the numerical properties of a stochastic algorithm for the modeling of nanoparticle dynamics in the free molecular regime. A new majorant kernel and the notion of fictitious jumps were developed to describe the formation and coagulation of fumed silica in the process of flame synthesis. The introduction of this new majorant kernel resulted in a simulation technique that was very efficient when

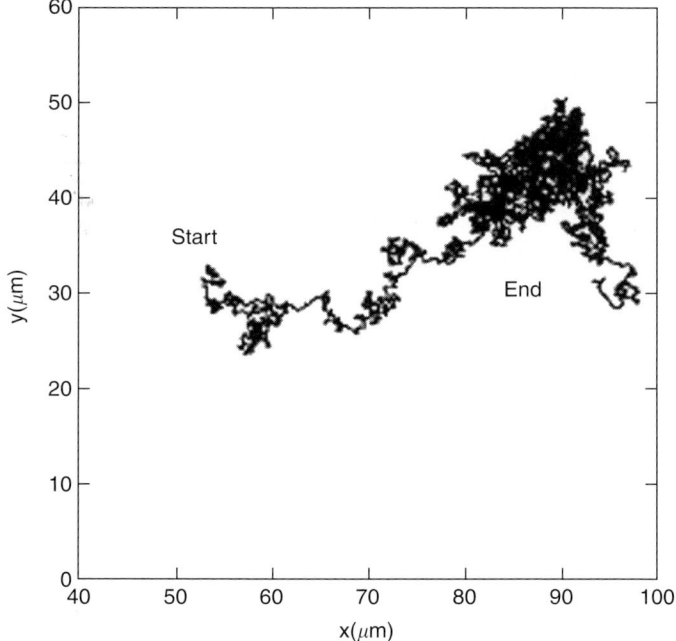

FIG. 59. Brownian diffusion of a latex particle (radius, 1.5 μm) in the horizontal plane. The particle was tracked for 40 min and 22,000 different positions of the particles were recorded [89].

compared to the linear majorant kernel algorithm and direct simulation Monte Carlo algorithms used prevously.

Particles in suspension and at equilibrium i.e., particles independent of external influences, such as electric fields, temperature fields, light, and gravity, will experience a dynamic equilibrium between the colloidal diffusive pressure and applied force from the fluid molecules, and the Stokesian particle flux produced by the applied force and the diffusive "current" of particles. Following this logic, Einstein obtained an expression for the diffusion constant of particles in suspension [88], which was later proven to be accurate by the experiments of Perrin [87,88]. Assuming that the friction constant was given by Stokes' law, the diffusion constant could be expressed as

$$D = \frac{kT}{6\pi\eta a} \qquad (57)$$

where k is the Boltzmann constant, η the viscosity of the medium, and a the radius of the particle. A similar expression for the rotational diffusion of

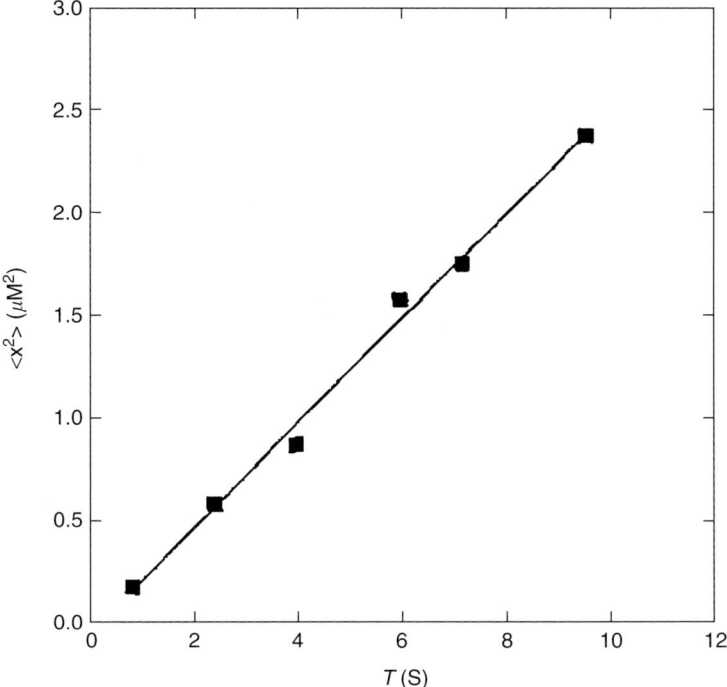

FIG. 60. The diffusion coefficient of X direction [89].

particles was derived as

$$D_R = \frac{kT}{8\pi\eta a^3} \tag{58}$$

For the translational diffusion, the displacement for each unit time is

$$l = \sqrt{\frac{kT}{\pi\eta a}} \tag{59}$$

However, the suspension is not always at the equilibrium state. When there are other potential fields present, such as a temperature field, a magnetic field, or an electric field, the particles will move along the direction of the field. In 1982, Batchelor [93,94], considering the effect of interactions between pairs of particles in Newtonian fluid and the Brownian diffusion, derived a formula to predict the mean sediment velocity of various particles. In this work, Batchelor demonstrated that when the pair distribution

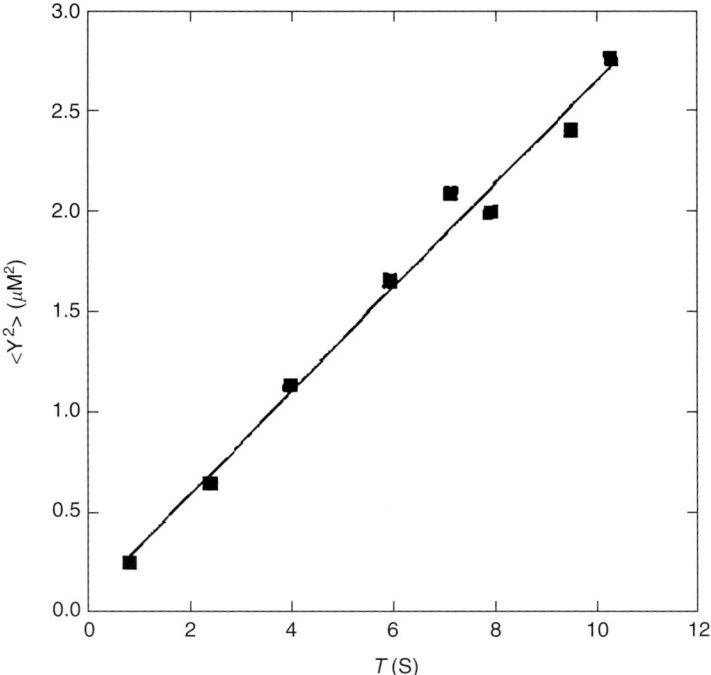

FIG. 61. The diffusion coefficient of Y direction [89].

function is not isotropic, the Brownian diffusion and the interparticle forces would significantly influence the mean sediment velocity. For small Peclet numbers, the Brownian diffusion contributed directly to a relative mass flux of the different size particles and to an absolute mass flux in the calculation of the velocity, the magnitude of which was comparable to that of gravity. This was followed by a numerical simulation of the sedimentation of interacting spheres [95], which considered gravity, Brownian diffusion, and interparticle forces, based on the assumption that a high Coulomb barrier and a van der Waals attractive force existed between the large separated particles. The results indicated that while gravity had a larger influence than either the Brownian diffusion or the interaction force between two adjacent particles, the influence of these two factors must be considered when calculating the pair-distribution function and sedimentation coefficients.

Torquato [96] investigated the effective electrical conductivity of two-phase composites, and derived a formula for an arbitrary dimensional electrical conductivity utilizing a three-point probability function of the composite by employing Brown's three-dimensional perturbation expansion

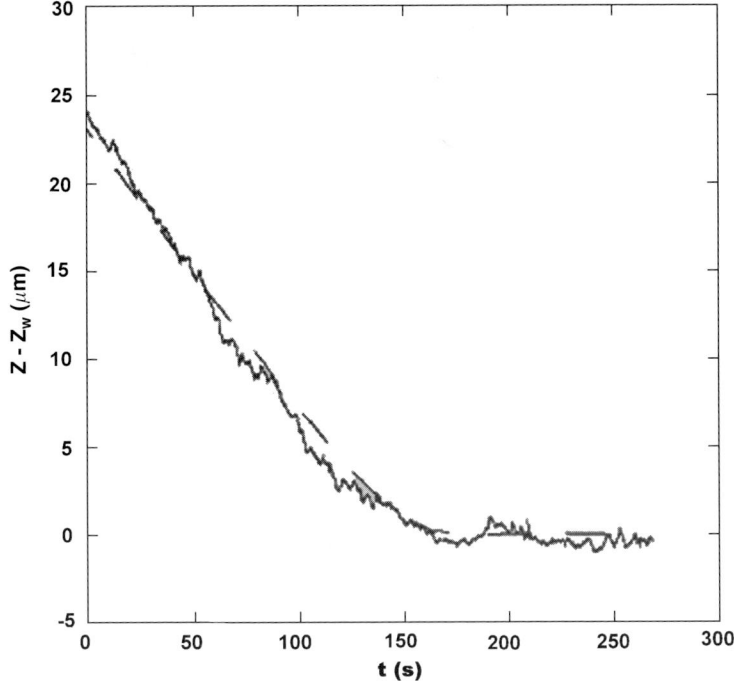

FIG. 62. The sedimentation curve of a latex particle of radius 1.5 μm representing the relative distance of the particle of the lower wall. z–z_w as function of time (thin line is the experimental curve; dash line is the theoretical curve) [89].

of effective conductivity for two-phase mixtures. The fundamental concept employed here was that by using Green's function to describe the Maxwell electric field and the induced polarization field, the effective conductivity could be included in the expression that defined the induced polarization field and the Lorentz electric field. The resulting expansion was described as

$$(\beta_{ij}\phi_i)^2 \left(\frac{k_e + (d-1)k_j}{k_i - k_j} \right) = \phi_i \beta_{ij} - \sum_{n=3}^{\infty} A_n^{(i)} \beta_{ij}^n \qquad (60)$$

where $i \neq j$ and $A_n^{(i)}$ are the integrals over a set of n-point probability functions. The Pade approximation was used to get the low-order bounds for k_e/k_i [96]. Finally, an expression that describes the ratio between k_e and k_i in three dimensions was derived, in which ζ_2 is the three-point parameter and

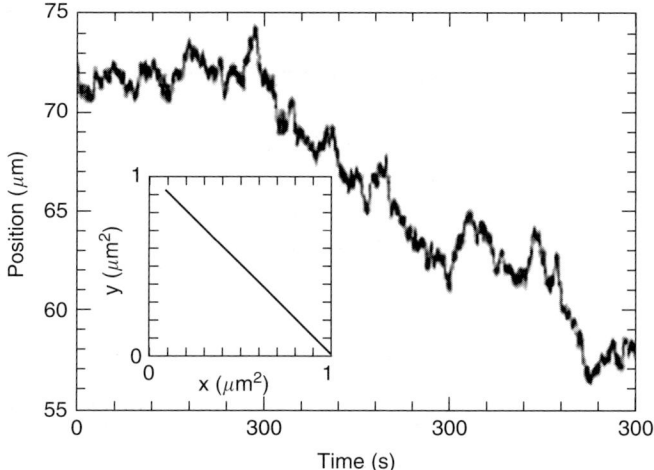

FIG. 63. Brownian motion of a particle of 2.5 μm diameter [88].

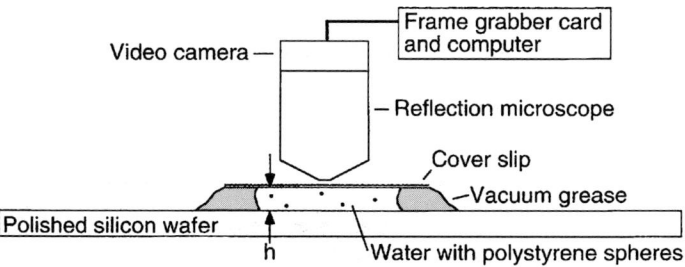

FIG. 64. The experimental setup for the observing Brownian motion of spherical particles in water [101].

ϕ_2 the volume fraction of the sphere.

$$\frac{k_e}{k_i} = \frac{1 + 2\phi_2\beta_{21} - 2\phi_1\zeta_2\beta_{21}^2}{1 - \phi_2\beta_{21} - 2\phi_1\zeta_2\beta_{21}^2} \qquad (61)$$

And for the permeable-sphere particle model, the final equation was

$$\frac{k_e}{k_i} = 1 + K_1\phi_2 + K_2\phi_2^2 \qquad (62)$$

Here, $K_1 = 3\beta_{21}$ and $K_2 = 3\beta_{21}^2 + 6\beta_{21}^3[0.21068 + 0.35078(1 - \lambda)]$, and λ is the impenetrability parameter for the particle which varied from zero to one.

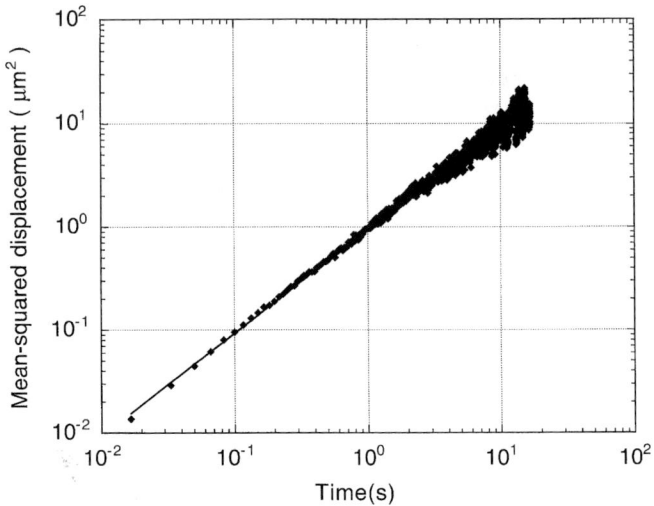

FIG. 65. The displacement vs. time of a 1.0 μm diameter particle in water [101].

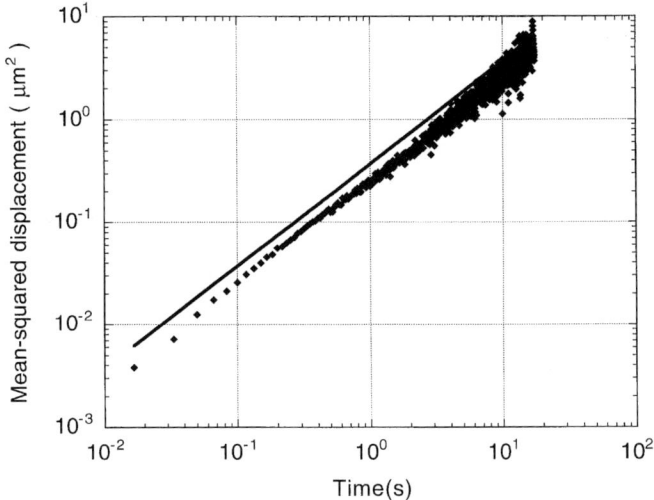

FIG. 66. The displacement vs. time of a 2.5 μm diameter particle in water [101].

Following this initial work, Torquato and Kim [97–100] developed a simulation technique to compute the effective transport properties of disordered heterogeneous media. This expression applied the concept of Brownian motion of the diffusing particle to the determination of the effective thermal conductivity and was consistent with the steady-state diffusion equation. It included the concept of the "first passage time" probability distribution and so-called "grid method." Based on this work, a Brownian motion simulation technique for computing the effective conductivity of an isotropic n-phase composite of arbitrary dimension d and microgeometry was developed. Equations were derived and the effective conductivities of a two-phase composite of high conductivity cylinders and low matrix conductivity with all volume fractions were computed. The general effective conductivity equation was

$$k_e = \frac{X^2}{2d\left\langle \sum_i R_i^2/2dk_i \right\rangle} \quad (63)$$

where X is the radius of the sphere cell, R the random walk path segment, and d the dimension number. The angular brackets indicated an ensemble average.

Later, the simulations at the extremes of the conductivity ratio were executed with this technique for cases in which inclusions were sphere, prolated, and oblate over a wide range of volume fractions and conductivity ratios. The data were compared with that developed by Brown [101] and were found to lie within the second-order bounds.

Using an approach somewhat different from the momentum approach utilized by Einstein and Smoluchowski *et al.*, Langevin applied the stochastic equations of motion in which the systematic forces used were linear with the different system state variables, to derive an expression for the Brownian motion velocity, expressed as Eq. (64). Uhlenbeck and Ornstein [102] also calculated the mean value of all the powers of the velocity and displacement of Brownian motion. Using the square of the displacement of the particles, expressions were obtained for the square deviation in Brownian motion as a function of the time and the initial deviation, in which a Gaussian process was assumed for the random force.

$$\frac{d}{dt}y(t) = -B(y(t)) + f(t) \quad (64)$$

where $B(y)$ is the systematic force, and $f(t)$ the random force. For the velocity, the distribution law was

$$G(u_0, u, t) = \left(\frac{m}{2\pi kT(1 - e^{-2\beta t})}\right)^{1/2} \exp\left[\frac{m}{2kT}\frac{(u - u_0 e^{-\beta t})^2}{1 - e^{-\beta t}}\right] \quad (65)$$

with β being the unit friction coefficient and u_0 the velocity at $t = 0$. For the frequency distribution of the displacement (the probability) was given as

$$F(x_0, x, t) = \left(\frac{m\beta^2}{2\pi k t(2\beta t - 3 + 4e^{-\beta t} - e^{-2\beta t})}\right)^{1/2}$$
$$\times \exp\left\{\frac{m\beta^2}{skT} \frac{[x - x_0 - u_0(1 - e^{-\beta t})/\beta]^2}{2\beta t - 3 + 4e^{-\beta t} - e^{-2\beta t}}\right\} \quad (66)$$

where x_0 is the initial position at $t = 0$. With these results, the case of periodic, aperiodic, and overdamped situations were treated.

Mazur and Bedeaux [103] studied the stochastic properties of the random force in Brownian motion in which a nonlinear Langevin equation was applied.

$$\frac{dp(t)}{dt} = -\beta(p(t))p(t) + f(t) \quad (67)$$

Here, $p(t)$ is the motion for the stationary process, $\beta(p(t))$ the momentum-dependent friction coefficient and $f(t)$ the random force due to the collisions with the particles of the fluid. In this approach, the assumption that the equilibrium distribution for the systematic force is a Maxwell distribution, was made

$$P_{eq}(p) = (2\pi mkT)^{-1/2} \exp\left(\frac{-p^2}{2mkT}\right) \quad (68)$$

where m is the mass of the particle, k the Boltzmann constant, and T the temperature. In this investigation, it was apparent that the dependence of the friction coefficient on the momentum must be considered in the description of Brownian motion.

Based upon the preceding analyses of the effective thermal conductivity, it appears that the parameters affecting the observed enhancement to the thermal conductivity of nanoparticle suspensions are principally a function of the result of the volume fraction, the size and shape, the conductivity ratio of the nanoparticle and the base fluid, the adsorption base fluid layer on the surface of the nanoparticle, and the presence and influence of the Brownian motion of the nanoparticle. The first of these factors has been studied in depth and a number of theoretical models have been developed. However, detailed investigations and development of an indepth understanding of the last two of these factors, the impact of the adsorption base fluid layer on the surface of the nanoparticle and the presence and influence of the Brownian motion of the nanoparticle, has just begun. Based upon this review, it is apparent that the equations that do not consider the adsorption layer and/or the Brownian motion of the nanoparticles do not accurately predict the experimental data for nanoparticle and/or nanotube suspensions hence, it is necessary to further investigate these two factors.

Although the mechanism of how the Brownian motion influences the various chemical and physical properties of nanoparticle or nanotube suspensions, is not clear or fully understood, it is quite apparent that this Brownian motion can induce micromovement to the fluid around small particles, and consequently, this micromovement will play an important role in the increased effective thermal conductivity of these suspensions.

Typically, when a nanoparticle suspension is used to transfer heat or is evaluated in order to determine the effective thermal conductivity, there is a temperature field acting on the suspension. In these cases, the Brownian motion will be accompanied by thermal phoresis, which will serve to increase the effective thermal conductivity of the nanoparticle suspensions. Thermal phoresis is introduced by the temperature difference around a solid particle. The fluid molecules in different temperature regions around the particle will have different kinetic energies, the molecules in the high temperature region will have a higher kinetic energy and the molecules in the lower temperature regions will have lower kinetic energy. Hence, the total force that the particle experiences from the surrounding molecules will be biased toward the direction of the lower temperature region. It is this phenomenon that researchers often ignore in the study of enhanced effective thermal conductivity of nanoparticle suspensions. The specific direction of the movement of the thermal phoresis is the same as that of the heat flux, and as a result, will yield the maximum heat transfer enhancement from the perspective of the vector dot product between the particle velocity and temperature field gradient. The combination of this thermal force, acting together with the Brownian motion has some similarity to the investigations conducted previously, in which the combined effect of the gravitational field and the Brownian motion were investigated. As described earlier, investigations of this latter problem, i.e., the combined effect of all the forces acting on a particle in fluid, can be traced as far back as the beginning of 1900s. Hence, the study of sedimentation of particles in suspension can be very helpful in the development of a better understanding of both, the Brownian motion, as it was to Perrin, and the other second-order effects that relate to or contribute to the other forces acting on the particles.

Batchelor [93] studied the sedimentation phenomena of dilute single-size particle suspensions and developed a method, Eq. (69), to determine the velocity of particles under the influence of gravity, the existence of the other particles, small Reynolds number (no inertia forces), and Brownian motion.

In this expression

$$U = U_0\left[1 + S\phi + O(\phi^2)\right] \tag{69}$$

and ϕ is the volume fraction of the particles, U_0 the descent velocity of a particle in isolation, and the sedimentation coefficient, S, is -6.55 for rigid spherical particles without any interaction between each other except for direct contact. Because it is a dilute dispersion, $\phi \ll 1\%$, the effect of the particle interaction for three or more particles is proportional to ϕ^2, so it is entirely reasonable to expect that the velocity difference due to the existence of other particles is $S\phi$, which means that only two-particle interactions should be considered. In this study, it was explained that this reduced settling velocity was largely caused by the diffuse upward fluid current, expressed as

$$U_0 = \frac{2a^2(\rho_p - \rho_f)}{9\mu} g \tag{70}$$

The statistical structure of a pair of particles can be expressed as the Boltzmann pair distribution, and the Brownian motion will play an indirect role in supporting the assumption of a uniform pair-distribution function of particles in a base fluid. Stated differently, the probability density of the location of one sphere center to another

$$P(x + r|x) = \begin{cases} n, & r \geq 2a \\ 0, & r < 2a \end{cases}$$

, n can be expressed as the mean number density of the particles, where μ is the fluid viscosity, and a the radius of the particle.

The relative velocity, effect of Brownian motion, interaction effects, density difference, and variations in the volume fraction of species particles, were explored in detail for polydisperse suspensions [94], with particular emphasis on two-particle interactions. The results indicated that the magnitude of the effect of the Brownian motion could be expressed as the inverse of the Peclet number and that it increased as the particle radius decreased. Because of the relative velocity of two neighboring particles, the effect of the Brownian motion was significant and was found to be directly related to the analysis of sedimentation. The resulting velocity equation for multi-species particles in suspension was found to be

$$U_i = U_i^0 \left(1 + \sum_{j=1}^{m} S_{ij}\phi_j\right), \quad (i = 1, 2, \ldots, m) \tag{71}$$

From the expression for a pair-distribution function, P_{ij}, the Peclet number is shown to be comparable to the ratio of the effect of gravity and the diffusion, i.e., the Brownian motion, which was expressed as

$$p_{ij} = \frac{\frac{1}{2}(a_i + a_j)V_{ij}^0}{D_{ij}^0} \qquad (72)$$

where V_{ij}^0 and D_{ij}^0 are the relative velocity and relative diffusivity tensor, respectively, for two separated particles. When the Peclet number was small, the diffusive effects were found to be dominant and the pair-distribution function could be approximated by two spherically symmetric particles. The specific Brownian motion contribution to the velocity of a particle of a specific species is given by

$$\Delta U_i = kT \sum_{j=1}^{m} \frac{1}{4}(a_i + a_j)^2 n_j \int_{s \geq 2} (b_{11} - b_{12}) \nabla_s P_{ij} \, ds \qquad (73)$$

where b_{11}, b_{12} are the mobility tensors depending on the geometry of the two-sphere configuration.

Numerical simulation of the multi-species particles suspensions was conducted and based on previous theoretical analyses [94,95]. In these, the assumption was made that a high Coulomb barrier existed at a certain sphere separation as did van der Waals attractive force at larger separation distances. The sedimentation coefficient for these cases was calculated, and the influence of gravity, the particle interaction, and the Brownian motion were both verified and found to be significant. While the interparticle forces were the focus of this numerical simulation [95], the case of both single-specie particle suspension and multi-specie particle suspensions were studied. Different values for the ratio of the radii of the particles, the ratio of densities, small and large values of the Peclet number, and different forms of the potential mutual force were all calculated. It was concluded that the pair-distribution function should consider the interparticle force and Brownian motion effects, and that the approximation, S_{ij}, could be used to determine the role of the gravitational body force.

In a similar study that used a somewhat different approach, Brady *et al.* [78] performed a dynamic simulation of hydrodynamically interacting single-specie particle suspensions using a modification of the simulation method developed by O'Brien. The hydrodynamic mobility and resistance matrices included all far-field, but no convergent or near-field interactions. The fluid motion was defined by the Stokes' equations and the spatially periodic simulation results were compared with the exact results, and shown to be in good agreement. The velocity of sedimentation was defined as

$U = M \cdot F$, where M is the entire mobility matrix and F the interacting force.

Gupte *et al.* [104] investigated the role of microconvection induced by the sedimentation of particles in stationary fluid, numerically. The particles were identical and evenly dispersed in the form of regular periodic arrays in the fluid. A subdomain was assumed to exist around each particle and all the calculations and conditions were applied within this subdomain. The unit cell model and subdomain are shown in Fig. 67. In this investigation, two sets of boundary conditions were compared. The first of these was that the fluid within the spherical subdomain was defined such that at the surface of a spherical unit cell, the fluid had the same particle velocity, while at the surface of the particle, the velocity was zero and the particle was fixed in the center of this cell. The second set of boundary conditions, is essentially the converse of the first, and was defined in such a way that at the particle surface, the velocity was the same as that of the particle, while at the surface of the spherical unit cell, the velocity was zero. The shear stresses at the surface of a spherical cell were zero under the second boundary condition, and were nonzero under the first boundary condition, which meant that

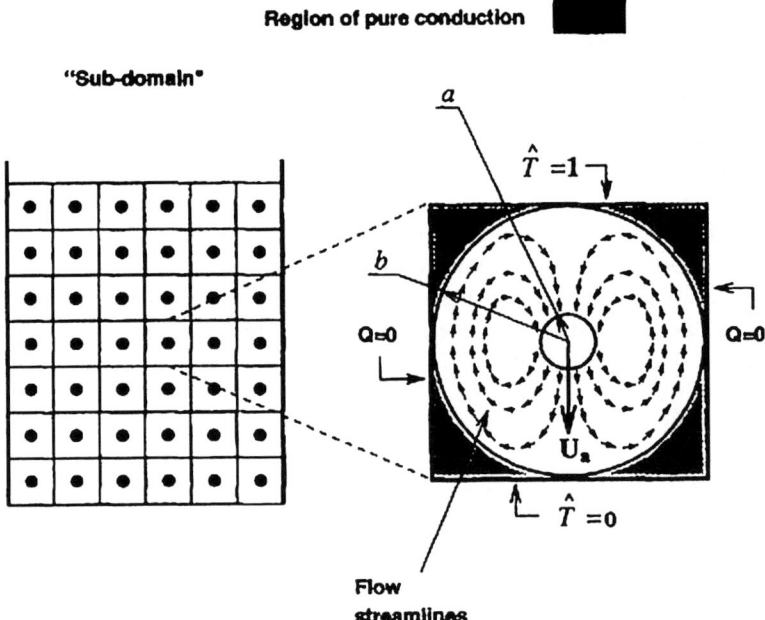

FIG. 67. The simulation model of nanoparticle suspension (schematic of unit cell model) [104].

there was a mass or momentum transfer across the outer envelope. As a result, the second boundary yielded better results than did the first. These results indicated that the microconvection induced by the sedimentation of particles would greatly enhance the overall heat transfer at a macroscopic scale. One significant problem remained with the two boundary conditions used in this simulation, and this was related to the pressure distribution within the spherical unit cell. Physically, the fluid right in front of the falling particle should experience a positive pressure as well as the same velocity as the particle, and the fluid immediately behind the falling particle should experience a negative pressure as well as the same velocity of the particle. This pressure distribution, in front or behind the falling particle did not exist under either of the two sets of boundary conditions evaluated in this simulation.

To further investigate this effect, Waldmann [105] studied the forces on a particle in a static binary diffusing gas mixture with a temperature gradient and particle number density gradient. When the suspended particles were considered to be small, it was assumed that the existence of these particles did not disturb the distribution function of the gas or the law of reflection at the surface. When the particles were considered large, it was assumed that the distribution on the surface wall for the approaching molecules was simply "hydrodynamic," as proposed by Maxwell. The assumption was remedied by stating that on the wall surface, the conservation laws should be fulfilled as far as possible for the concentrations, temperature, and mean velocities. The experiments were executed to verify the theoretical coefficients selected in the development of the theory, and slip boundary conditions on the wall were applied.

Brock [106] attempted to explain the discrepancies between the experimental observations and theoretical predictions of the thermal force acting on particles in a slip-flow regime. In this analysis, the assumption was made that the contribution of the thermal stress to the thermal force was small; thus, the Navier–Stokes equations could be applied to the slip-flow boundary regime. Using a first-order slip flow boundary condition, an equation for the thermal force acting on the particles was derived as follows:

$$F = -9\pi \frac{\mu^2}{\rho T} a \left(\frac{1}{1 + 3c_m l/a} \right) \left(\frac{k_f/k_p + c_t l/a}{1 + 2k_f/k_p + 2c_t l/a} \right) \nabla T_\infty \qquad (74)$$

where l is the mean free path of the molecule, a the radius of the particle, μ the coefficient of shear viscosity, k the thermal conductivity, T the temperature, and c_t and c_m are the temperature jump and velocity slip constants, depending on the nature of the fluid–particle interaction.

Rosner et al. [107] studied the dynamics and transport of small particles suspended in gases while exposed to a temperature gradient and undergoing heat transfer. Thermal phoresis of small particles suspended in gas was demonstrated, using a small spherical particle, suspended in a radiation-free fluid, exposed to a temperature gradient. As observed by Maxwell, whenever the gas–solid momentum accommodation coefficient was nonzero, the temperature gradient in the vicinity of the particle induced a local gas motion, directed from the cold region to the hotter region and, in turn, acted to propel the particle toward the cooler gas. In cases where the Knudson number was much >1, the thermal phoretic diffusivity was expressed as

$$(\alpha_T D)_p = \left(\frac{3}{4}\right) v_g \left\{1 + \left(\frac{\pi}{8}\right) \alpha_{mom}\right\}^{-1} \tag{75}$$

where v_g is the local gas kinematic viscosity and α_{mom} the gas/solid momentum accommodation coefficient. When the Knudsen number was much <1, the thermal phoretic diffusivity was expressed as

$$(\alpha_r D) = \frac{2c_s v \left(k_g/k_p + c_t l_g/a\right)\left[1 + l_g/a\left(A + be^{-ca/l_g}\right)\right]}{\left(1 + 3c_m l_g/a\right)\left(1 + 2k_g/k_p + 2c_t l_g/a\right)} \tag{76}$$

In many respects, the photo phoresis of particles is analogous to the thermal phoresis of particles. The ratio between these as a function of the particle diameter was determined and is shown in Fig. 68. Based upon these results and the preceding analysis, it was concluded that phoresis could cause dramatic changes in the rates at which suspended particles will be deposited or will drift along in the direction of the local energy flux.

Huisken and Stelzer [108] conducted an experimental investigation of the photo phoretic force. A weakly focused Gaussian laser beam was used to levitate absorbing nontransparent metal oxides and Mie particles against gravity in the experimental facility illustrated in Fig. 69. In this investigation, the beam of an Ar^+ laser was expanded and directed upwards and a lens focused the beam into a sealed glass chamber, in which convection was minimized. Particles were then dropped into the chamber through a small hole at the top and a stereomicroscope was used to image the levitated samples in a plane parallel to the optical axis. As predicted by Maxwell, the forces on an absorbing particle were due to the thermal creep, resulting in a higher pressure on the warm side than on the cold side of the particle. The principal contribution to the force on an absorbing particle was that the warmer medium exerted a higher pressure than the cold medium above the particle did. [108] As a result, the surface of the particle was unevenly heated by its orientation in the laser beam and the medium closest to the

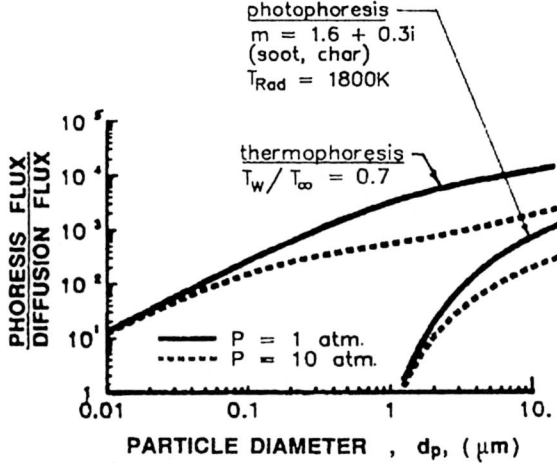

FIG. 68. Mass transfer due to thermo- and photophoresis vs. particle diameter (relative importance of particle mass transport by the mechanisms of Brownian diffusion, thermophoresis, and photophoresis for equal radiative and Fourier energy flux across laminar gaseous boundary layers. Size dependence for absorbing particles. $T_\infty = 1000\,\text{K}$, $k_g/k_P = 0.1$) [107].

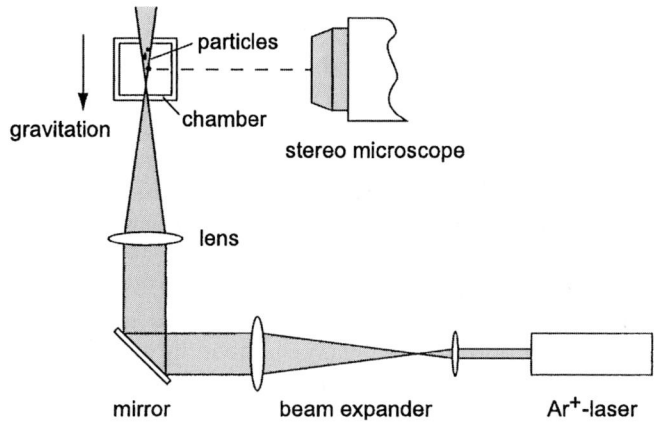

FIG. 69. Levitation apparatus [108].

surface was subsequently heated as well. Correspondingly, the density of the medium with higher temperature would decrease, and the medium molecules diffused in the direction of increasing temperature (thermal creep). Hence, the increased local pressure on the hot surface of a particle existed and the resulting force was expressed as

$$F_z = -\int_S P\cos\theta \, dA \qquad (77)$$

This expression was integrated over the particle surface, S, where θ is assumed to be the angle between the direction of the incoming ray and the normal to particle's surface.

Buevich et al. [109] utilized the momentum conservation equation to describe the diffusion process in flows for particle suspension and introduced the thermodynamic forces. Initially, only the isotropic Brownian motion effects were considered and the chemical potential was calculated from the thermodynamic force. The case in which both Brownian motion and pseudoturbulence effects were considered simultaneously was then investigated and because the physical mechanisms responsible for generating these two effects were considered in different ways and independently from each other, the Brownian motion and pseudoturbulence effects were examined separately. Two kinds of diffusion coefficients, self-diffusion, and gradient diffusion, were evaluated for both the Brownian motion and pseudoturbulence.

More recently, Jang and Choi [110] studied the effect of Brownian motion of nanoparticles to the effective thermal conductivity of nanofluids. By using it they proposed that using a three-diameter thickness base fluid molecule adsorption layer on the nanoparticle and the microconvection due to the Brownian motion of the nanoparticle, the enhancement could be well represented. Prasher et al. [111] derived another equation, which incorporates a similar idea, i.e., that the Brownian motion of the nanoparticle could induce a microconvection of the base fluid and thereby enhances the effective thermal conductivity of the nanofluids. The experimental results presented were in good agreement with the predicted values. However, both of these investigations [110,111], assume that the nanoparticles would absorb the thermal energy at the high-temperature region and release the thermal energy in the lower-temperature region through microconvection between the nanoparticles and base fluid. Li and Peterson [112] have suggested that instead of this microconvection, it is in fact a highly localized mixing of the base fluid in the region around each and every nanoparticle caused by the Brownian motion of the nanoparticles that is responsible for this enhancement and it is this mixing that results in a reduced temperature gradient throughout the nanofluid suspension at both micro- and macroscale.

In addition to the transportation effects caused by the Brownian motion and other potential fields, the enhancement of effective thermal conductivity of particle suspensions may also be the result of other factors, such as the adsorption of the base fluid layer on the surface of the nanoparticle, and the hydrodynamic flow and heat transfer effects of nanoparticle suspensions.

Yu et al. [113] observed a thin film of tetrakis silane film that had been spread on a silicon substrate, which was cleaned in strong oxidizer, using a specular X-ray beam. The results indicated that the density of the liquid film changed along the liquid film thickness. A model-independent analytical method was used to determine this density variation and the structure of the thin film was clearly observed to deviate from the ideal, isotropic liquid structure anticipated. At about the same time, Wang et al. [114] used a transmission electron microscope to observe a sample of 30 nm silicon oxide nanoparticles in distilled water. Here, it was found that there were traces of a thin film of liquid adsorbed to the surface around the nanoparticles and their clusters. By applying the Gibbs energy analysis method, it was determined that when the nanoparticle and the liquid around it reached a state of equilibrium and the Gibbs energy of this system reached a minimum, the adsorbed liquid film would be 1–2 nm thick.

Keblinski et al. [115,116] applied nonequilibrium molecular dynamic simulations to the study of the thermal resistance of liquid–solid interfaces. In this simulation, the model for both liquid and solid used the Lennard–Jones interatomic potential. The results indicted that the liquid with a weak atomic bond at the particle–matrix interfaces, exhibited high thermal resistance, while for wetting systems, the interfacial resistance was small.

The effective thermal conductivity of nanoparticle suspensions in flows has also been investigated and has demonstrated an enhanced heat transport capacity of nanoparticle suspensions in flow situations. Ahuja [117,118] conducted an experimental investigation of the effective thermal conductivity of suspensions in flow by using the apparatus pictured in Fig. 70. In this investigation, the experimental system was calibrated using tap water, and the overall accuracy was estimated to be within 10%. Following some preliminary calibration experiments, the following assumptions were made: (i) the polystyrene sphere particle suspension was homogeneous, (ii) the suspension behaved as a Newtonian liquid, and (iii) the Poiseuille law was applicable. The suspension samples were 88–105 µm in diameter and 44–53 µm diameter polystyrene spherical particles with a weight fraction of 1.2%, 1.9%, 3.1%, 4.6%, and 8.8% were placed in a base liquid of aqueous sodium chloride (5.2 wt.%) and aqueous glycerine (20 wt.%). The total heat transfer coefficient of the system was measured and the following expressions were then used to calculate the effective thermal conductivity:

$$h = \frac{(\dot{m}c/\pi d_m L)[(t_{sus,e} - t_{sus,i})}{(\Delta t_i - \Delta t_e)] \ln(\Delta t_i/\Delta t_e)} \tag{78}$$

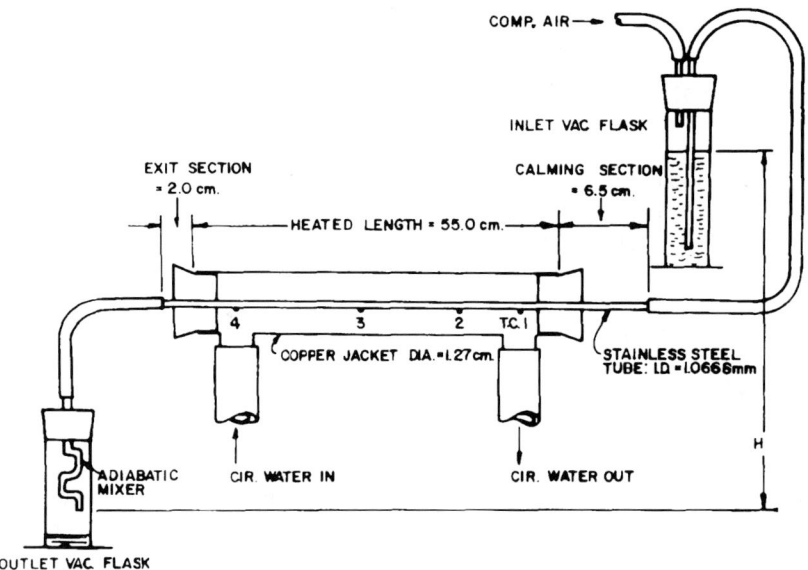

FIG. 70. Picture of experimental setup [117].

$$\frac{1}{h} = \frac{1}{h_1} + \frac{b}{k_{tube}} + \frac{1}{h_2} \quad (79)$$

$$h_1 D = 3.65 k_{sus} + \frac{0.27 \dot{m} c/\pi L}{1 + 0.04(4\dot{m}c/\pi L k_{sus})^{2/3}} \quad (80)$$

where $b = 1/2\, d_m \ln(D_0/D)$ and c is the specific heat.

In this experiment [117,118], the effective thermal conductivity of the suspension in flow was two times higher than that of the nonflow nanoparticle suspension and varied with different shear rates, particle concentrations, particle sizes, tube sizes, and base liquids (see Fig. 71). Following this experimental investigation, an analytical model, Eq. (81), was developed to describe the relationship between the enhancement of the effective thermal conductivity of the flowing suspensions and the various parameters mentioned above, and was shown to be

$$\frac{k_{mov}}{k_{stat}} - 1 = C \left[\frac{\phi\left(\frac{\omega a^2}{v_f}\right)\left(\frac{\omega a^2}{\alpha_f}\right)\left(\frac{R}{a}\right)^2\left(\frac{L}{2a}\right) \times 10^{-8}}{\times \text{doublet} - \text{collision} - \text{frequency} - \text{ratio}} \right]^n \quad (81)$$

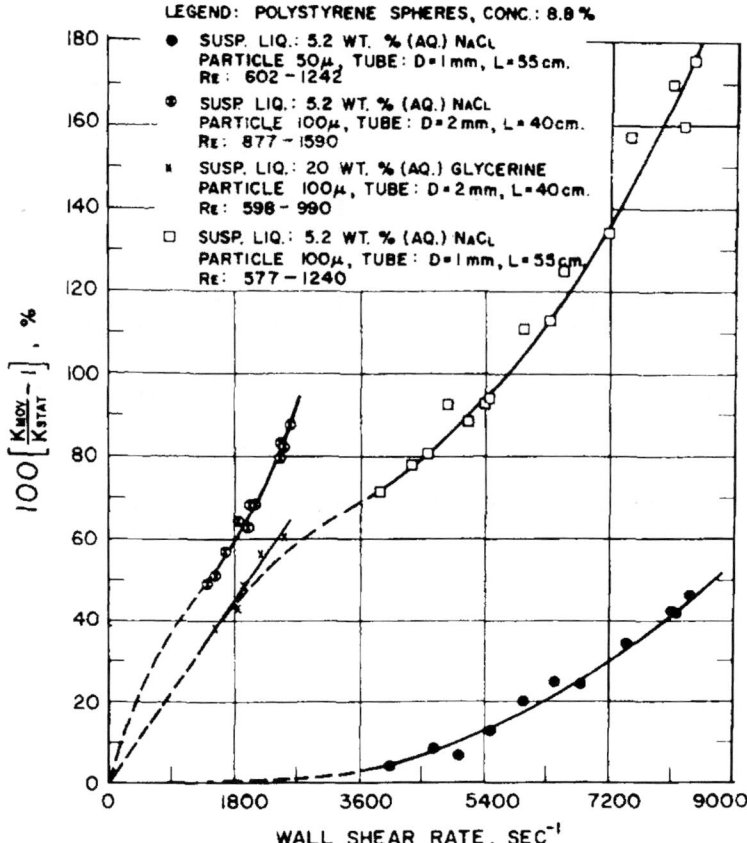

Fig. 71. Augmentation in thermal conductivity increases with the increasing shear rate of suspension [117].

where C is a constant and n greater than, equal to, or less than 1 for particle concentrations, ϕ, less than, equal to, or greater than 4.6%, respectively. The parameters a, ω, α_f, ν_f, L, and R are the radius of the particle, the angular velocity, the thermal diffusivity of the fluid, the kinematic viscosity of the fluid, the length of the tube, and the radius of the tube, respectively. The results indicated that the angular Reynolds and Peclet numbers both played an important role in the enhancement of the effective thermal conductivity of the suspension in a flow situation.

The adsorbed base fluid layer is another aspect that may be important in the determination of the effective thermal conductivity of nanoparticle suspensions. For a long time, this was believed to be well ordered at the surface and the thickness was typically believed to be approximately the molecular diameter of the base fluid. This thin layer of base fluid has a number of varying impacts on the effective thermal conductivity of the nanoparticle suspension. First, it will perform like a solid and have higher thermal conductivity than the bulk base fluid. Second, with this layer, the effective volume of the nanoparticle increases and may produce a concomitant effect, while at the same time prevent the agglomeration of the nanoparticles. Third, any clusters that are formed by the agglomeration of these nanoparticles will be well connected by this layer, allowing the clusters to have thermal properties different from the nanoparticles and approaching that of a solid of similar material. Although there has been little theoretical research conducted on the effect of the adsorbed base fluid layer, this and other previously discussed experimental investigations have demonstrated its importance.

C. OTHER TRANSPORT PHENOMENA IN NANOPARTICLE SUSPENSIONS

Although determination of the effective thermal conductivity has been the focus of most of the investigations of the thermophysical properties of nanoparticle suspensions, a number of researchers have recently begun to examine the other parameters, such as the heat transfer coefficient of nanoparticle suspensions. Lee and Choi [119] observed nanoparticle suspensions using an X-ray beam mirror cooling system that was cooled using a micro channel system with water and liquid nitrogen as the coolant. In this investigation, little information was provided regarding the kinds of nanoparticles and fluids used to make the suspensions, other than the notation that the effective thermal conductivities were two or three times greater than that of water. The experimental results demonstrated that the nanoparticle suspension with an effective thermal conductivity of three times greater than that of water, had a performance curve similar to that of liquid nitrogen.

Xuan and Roetzel [120] described two different methods used to study the convective heat transfer occurring in nanoparticle suspensions. The first was the traditional method in which the nanoparticle suspension was treated as a single-phase fluid and the second included the features of the dispersed nanoparticles using a multi-phase fluid model. Using this latter technique, Xuan and Li [121] developed the experimental test facility illustrated in Fig. 72 to determine the heat transfer coefficient of nanoparticle suspensions consisting of Cu particles, which have diameters below 100 nm and volume fractions of 0.3%, 0.5%, 0.8%, 1.0%, 1.2%, 1.5%, and 2.0%. The system uncertainty of this test facility was less than 3%. The results were presented in terms of the heat transfer coefficient as a function of velocity and are shown in Fig. 73. As shown, the results clearly indicate that the convective

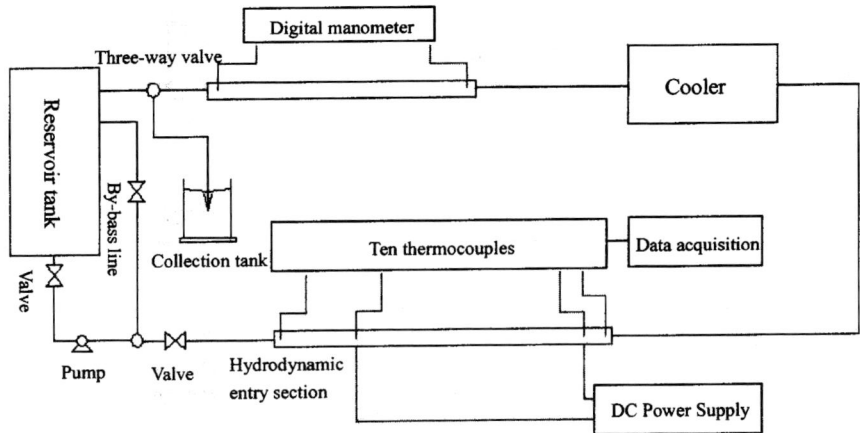

FIG. 72. The experimental system diagram of the convective heat transfer feature of nanoparticle suspensions [121].

heat transfer coefficient of the suspension increased with the flow velocity as well as with the volume fraction.

Based on this experiment, an equation to calculate the Nusselt number of this Cu nanoparticle suspension was developed. The resulting expression

$$Nu = 0.005991.0 + (7.6286\phi^{0.6886} Pe_{particle}^{0.001}) Re^{0.9238} Pr^{0.4} \quad (82)$$

was able to predict the Nusselt number of different volume fractions of Cu nanoparticle suspensions to within 8% when compared with the experimental data. In this expression, ϕ represents the volume fraction and $Pe_{particle}$ is the Peclet number, based on the diameter of the Cu nanoparticle. The experimental data and the trends predicted by the equation are shown in Fig. 74. The increase in the enhancement of the convective heat transfer coefficient for the nanoparticle suspensions were attributed to the increase in the effective thermal conductivity of the nanoparticle suspension and the chaotic movement of the nanoparticle, which accelerated the energy exchange process in the fluid.

In addition to the study of the forced convection effects of nanoparticle suspensions, a number of researchers have studied the natural convection occurring in these suspensions. Okada and Suzuki [122] investigated the natural convection occurring in two types of suspensions with micro soda glass beads in water with an average diameter of 4.75 and 6.51 μm, respectively. The weight fractions of both suspensions were less than 5%. The suspensions were heated in a rectangular cell at the center of the horizontal

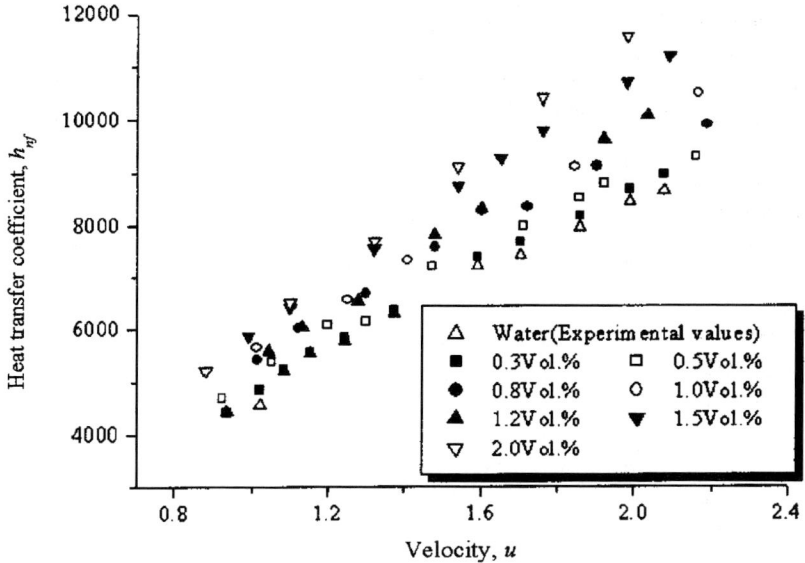

FIG. 73. The convective heat transfer coefficient vs. velocity [121].

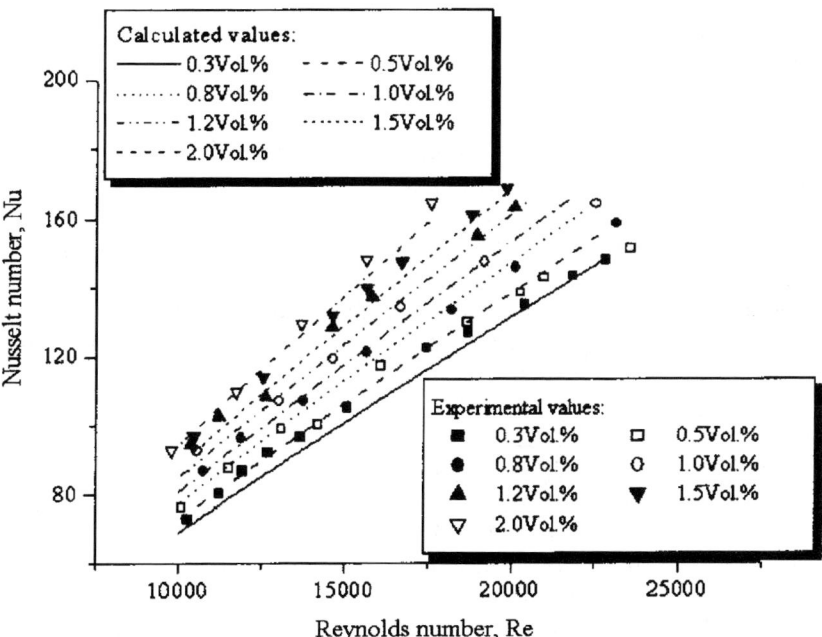

FIG. 74. The measured Nu data and predicted Nu values vs. Re of flow [121].

bottom wall. During the experiments, it was clear that several layers were formed in the test cell along the direction of the gravitational vector, and that each layer had several circular flows that were constrained to that cell and did not move or extend into the neighboring layers. Kang *et al.* [123] conducted a similar experiment on SiO_2 particles having a mean diameter of 2.97 µm with a 0.03 µm standard deviation in water. The difference between these two experiments was the method of heating. In the first experiment, the heat was added to the bottom and in the other it was added from one of the vertical walls, opposite the cold vertical wall. In the latter experiment, it was found that the natural convection of the suspension could be classified into one of the five distinguishable patterns, and that the critical wall temperature difference that would determine the natural convection pattern, decreased with a decrease of the particle concentration in the suspension.

Putra *et al.* [124] conducted an experimental study of the steady-state natural convection of two Al_2O_3 and CuO nanoparticle suspensions, one with a mean diameter of 131.2 and the other 87.3 nm in distilled water, under different conditions. The experimental test facility used in this investigation is shown in Fig. 75. The results indicated that temperature measured at the midpoint of the cold and hot walls was higher than the arithmetic mean value of the corresponding temperatures. This phenomenon was assumed to be due to the increase in the effective thermal conductivity of the nanoparticle suspensions with an increase in the temperature. It was also observed that a systematic degradation of natural convective heat transfer, i.e., that the Nusselt number decreased with the increase in the nanoparticle volume

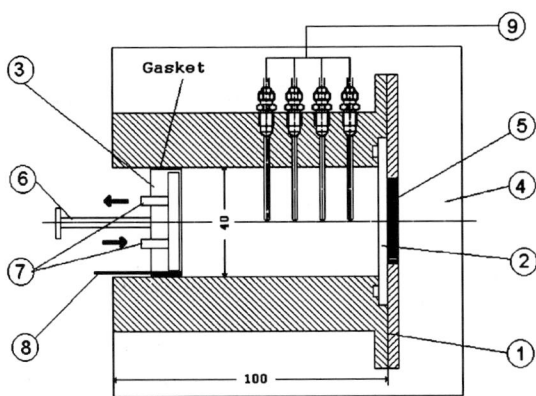

FIG. 75. Sketch of experimental rig [124]: (1) cylindrical block, (2) end cover as heating surface, (3) end cover as cooling surface, (4) cap, (5) resistance heating elements, (6) the piston shaft, (7) cooling water inlet and outlet, (8) narrow tube, (9) thermocouples.

fraction in the suspension, and that the decreases observed in the CuO nanoparticle suspension were more severe than in the Al_2O_3 nanoparticle suspensions for the same volume fraction. Here, a correlation between the Nusselt number and the Rayleigh number was given as $Nu = CRa^n$, where n was weakly dependent on the nanoparticle volume fraction in the suspension and the constant, C, was strongly dependent upon the volume fraction.

Khanafer et al. [125] developed a model to analyze the natural convection heat transfer performance of nanoparticle suspensions inside a two-dimensional enclosure. A numerical method using a finite-volume approach was employed along with an alternating direct implicit procedure. The physical model of the enclosure is illustrated in Fig. 76. In this investigation, the numerical simulation was conducted for Grashof numbers between 1,000 and 10,000, and nanoparticle volume fractions between 0% and 25%. The thermophysical properties used in this investigation are shown in Table IX.

Here, it was observed that the increase in the nanoparticle volume fraction and irregular and random movements of the particles increased, the energy exchange rates in the suspension increased, as did the thermal dispersion.

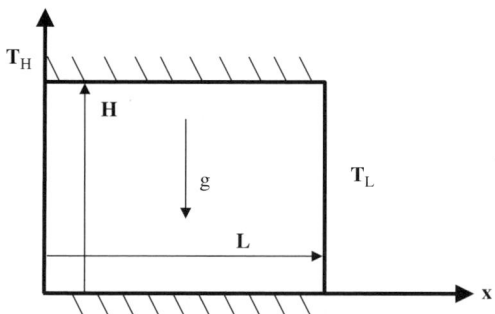

FIG. 76. The schematic for the physical model [125].

TABLE IX
THERMOPHYSICAL PROPERTIES OF DIFFERENT PHASES

Property	Fluid phase (water)	Solid phase (copper)
c_p (J/kg K)	4179	383
ρ (kg/m^3)	997.1	8954
k (W/m K)	0.6	400
β (K^{-1})	2.1e-4	1.67e-5

The correlation given for the Nusselt number of the suspension was

$$Nu = 0.5164(0.4436 + \phi^{1.0809})Gr^{0.3123} \tag{83}$$

where ϕ was the volume fraction of the nanoparticles in the suspension.

Another aspect of the heat transfer performance of nanoparticle suspensions is the effect of these suspensions on boiling heat transfer. You and Kim [126] experimentally investigated the enhancement of the critical heat flux (CHF) in pool boiling for Al_2O_3 nanoparticle suspensions. The experimental test facility used in this investigation is shown in Fig. 77 and the results indicted that in the nucleate boiling regime, the boiling heat transfer coefficient was not influenced by the addition of the Al_2O_3 nanoparticles. However, there was a dramatic increase in the CHF for different concentrations of the nanoparticle suspension that ranged as high as 300% of that measured for pure water. This increase in the CHF of Al_2O_3 nanoparticle suspensions varied with the concentration of nanoparticles as illustrated in Fig. 78. Unfortunately, the size and size distribution of the Al_2O_3 nanoparticles used in this investigation were not given.

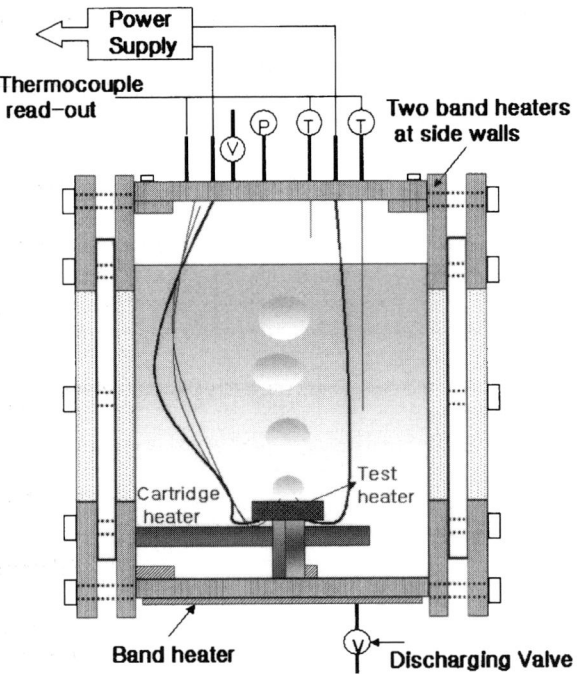

FIG. 77. The schematic of the pool boiling test facility [126].

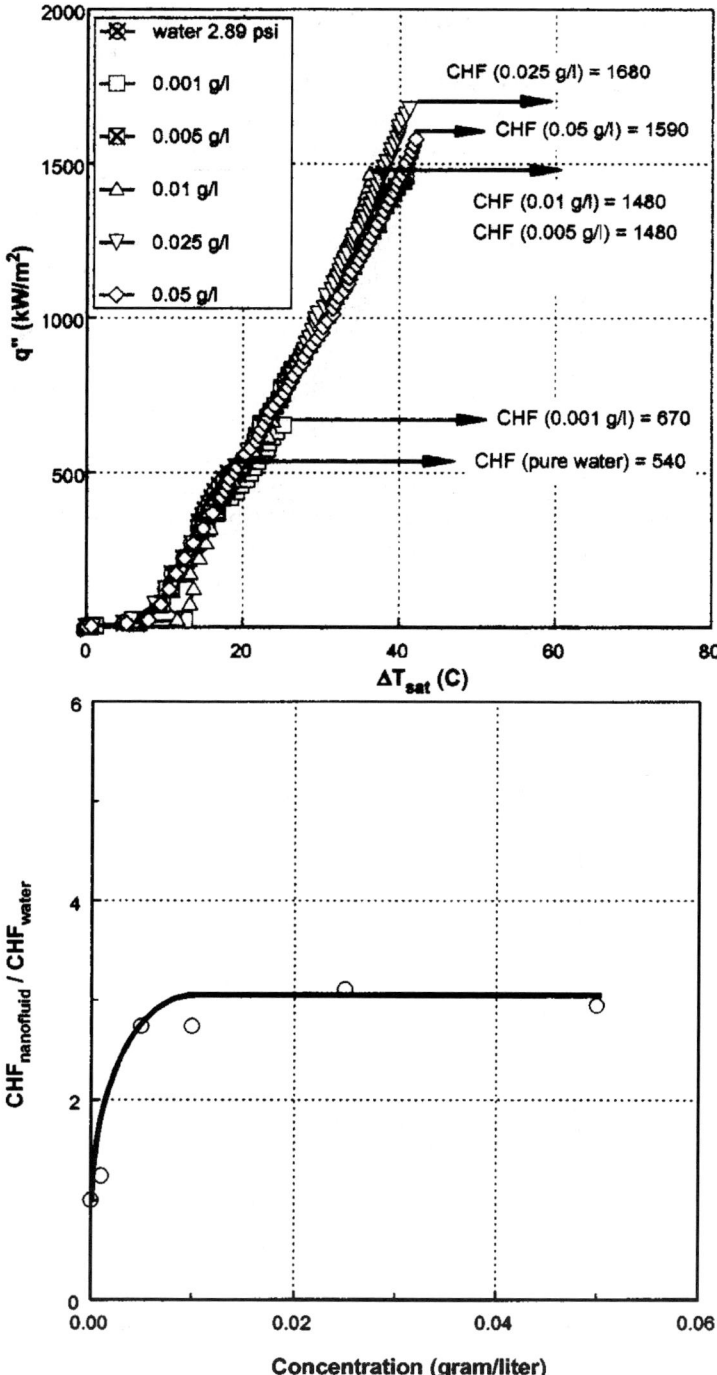

FIG. 78. The variance of Al_2O_3 nanoparticle suspensions' CHF with different concentration of nanoparticles in suspension. [126].

In the experimental investigation of Das *et al.* [127], the boiling performance of the water was found to deteriorate with the addition of the nanoparticles, and that a higher wall superheat existed for the same heat flux. With the increase of the nanoparticle concentration in the suspension, the deterioration was more apparent. The changes of heat flux along the superheat temperature are shown in Fig. 79. Here, it was observed that the nanoparticles with a diameter of 20–50 nm filled the cavities of an uneven heating surface and changed the surface characteristics by effectively making it smoother. With the increase of nanoparticle concentration, a layer of sedimentation of the nanoparticles developed, which deteriorated the boiling performance even further.

Vassallo *et al.*[128] studied the boiling characteristic of silica nanoparticles and microparticle suspensions with diameters of 15, 50, and 3000 nm in boiling situations, ranging from nucleate boiling to film boiling under atmospheric pressure. The heat flux was generated in a horizontally fixed, NiCr wire and the water level was observed to decrease ~6 mm during the experiment. All of the suspensions expressed no difference in the boiling

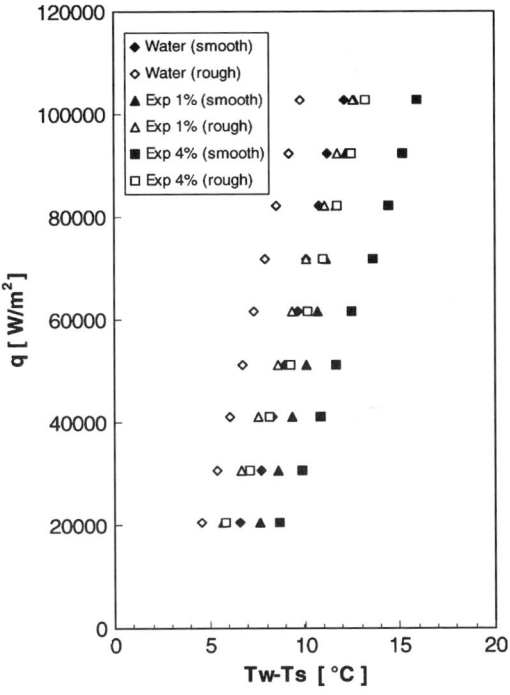

Fig. 79. The heat flux with different nanoparticle concentrations and heat wall roughness vs. superheat temperature [127].

performance of water, below the CHF point, but increased the CHF point dramatically. The difference between the performance of the nanoparticle suspensions and the microparticle suspensions was that the wire always failed prior to entering the film boiling regime for microparticle suspensions, while for the nanoparticle suspension, it was able to sustain operation. Silica coating on the wire was also found after the test and was determined to be part of the reason for the increase in the CHF. The experimental data obtained in this investigation are shown in Fig. 80.

This work was followed shortly thereafter by Zhou [129] who conducted an investigation of the boiling performance of copper nanoparticle suspensions with an average diameter of 80–100 nm, under conditions of acoustic cavitation. In this investigation, it was found that, without acoustic cavitation in the copper nanoparticle suspension, the boiling hysteresis with a wall superheat of 8.74 K did not occur as it did in conventional pool boiling, and that increases in the nanoparticle concentration in the suspension from 0.133 to 0.267 g/l did not change the heat transfer performance. This result was quite different from the observations of Das et al. [127] who observed even further enhancement of the heat transfer capacity with acoustic cavitation.

D. DISCUSSION AND CONCLUSIONS

The reasons given for the enhanced thermal conductivity of nanoparticle suspensions have been primarily focused on the variation in the volume

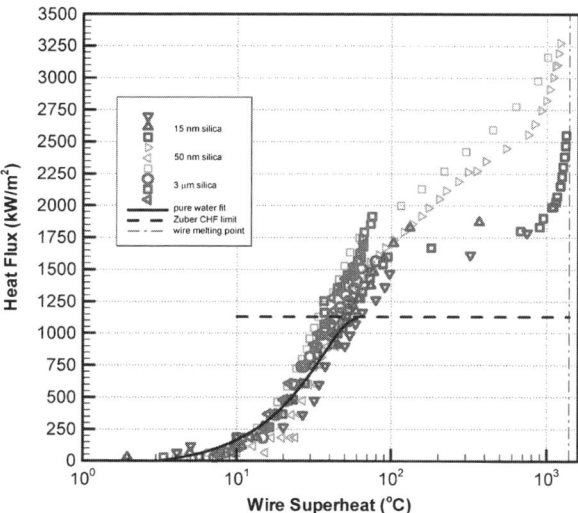

FIG. 80. Boiling curves of NiCr wire ($D = 0.4$ mm) in silica–water suspension [128].

fraction of the nanoparticles, the size of the nanoparticles, the shape of the nanoparticles, the conductivity ratio of the nanoparticle and the base fluid, the adsorption base fluid layer on the surface of the nanoparticle, and the Brownian motion and microconvection induced by the nanoparticles. The first three of these have been investigated in considerable detail and while there is still much to learn and some expressions that describe the effect of these parameters that have been established, there is still much to do in determining the effect of variations in the adsorption base fluid layer and the microconvection induced by Brownian motion. Previous investigations have shown that equations that do not consider the adsorption layer or the effect of the Brownian motion of nanoparticles are capable of accurately predicting the experimental data for nanoparticle and/or nanotube suspensions. Hence, it is very important to move the focus of research into these two arenas.

Although the impact of these two factors, especially the Brownian motion of nanoparticles, will influence the effective thermal conductivity of nanoparticle or nanotube suspensions, it is quite certain that Brownian motion can induce microconvection around these particles, and that this will work together with the thermal phoresis to enhance the thermal conductivity of these nanoparticle suspensions. In addition, when the direction of the thermal phoresis is the same as that of the heat flux, the maximum heat transferred per unit of time will be generated. Also, due to the viscosity of the base fluid and the temperature gradient in this transient process, the Brownian motion and thermal phoresis will form a kind of wave front that will enhance the microconvection.

Since the equations that take into consideration the volume fraction of the nanoparticles reflect the interaction between the dispersed particles with second-order effects and the shape of the nanoparticles, the ratio between thermal conductivity of nanoparticles does not successfully predict the experimental data satisfactorily. The question arises as to why the equations that include the adsorption base fluid layer, microconvection induced by the Brownian motion, and thermal phoresis of nanoparticles have not been developed? Just as the fluctuation of turbulent flow will dramatically increase the heat transfer coefficient over that of laminar flow, might not the Brownian motion of nanoparticles have the same effect on the thermal conductivity of suspensions?

IV. Summary

The area of nanoparticle suspension research is growing rapidly as is the developing field of nanotechnology and nanoscience. The effective thermal conductivity of nanoparticle suspensions is being studied both experimentally and theoretically. Experimental results of this research, followed by theoretical explanations, are attempting to develop a firm foundation on which a well understood explanation of this phenomenon can be built. The experimental results have identified several areas in which there is significant deviation from the theories developed to predict the effective thermal conductivity of micro- or millimeter size particle suspensions. However, because the experimental data are limited, theoretical studies have not as yet been verified to the extent that they can provide the basis for a well-defined set of equations that could inform subsequent experimental research.

The theoretical studies that describe the effective thermal conductivity of suspensions occur at three levels: the macroscopic level, the mesoscopic level, and the microscopic level. The macroscopic level is based on the conservation laws of mass, momentum and energy and experiments to measure and calculate the effective thermal conductivity with Fourier's equation. The study at the mesoscopic level deals not only with the average behavior of a suspension, but also with its stochastic behavior. This level of study may be complex compared to the one at the macroscopic level, in which probability theories are used to describe the system, and were rarely explored before the appearance of nanometer size particle suspensions. The microscopic level of study, which applies the dynamics of molecular motion in phase space, also requires further study. And although a basic understanding of the physical and chemical properties of nanoparticle suspensions is still not complete, this lack of clarity has not diminished the efforts to utilize these suspensions in various academic and industrial applications.

Evidence of these efforts is given in the work of Lieber *et al.* [130,131] who reported on a novel way to build electronic and optoelectronic devices with nanowires in suspension. Gorman [132] commented on a novel cleaning agent consisting of nanoparticle suspensions that were composed of uniform 8 nm surfactant spheres and water. This study identified a way to produce powerful cleaning agents and yet another application of nanoparticle suspension is in the biotechnical field, in which it has been reported that using nanoparticles in drug delivery could reduce severe disruptions in vessel walls with balanced infusion pressure, particle concentration, particle size, and administrated volume specific to the catheter system used. [133]

At this point, it is very clear that the experimental and theoretical investigation of the physical and chemical properties of nanoparticle suspensions will continue to attract interest for the foreseeable future, as will the exploration of many highly promising applications. With the further development of nanotechnology and nanoscience, and the advanced developments in biotechnology, the study of nanoparticle suspensions will continue to be an attractive subfield of the larger area of fluid dynamics and thermal sciences.

References

1. Fourier, J. (1878). "The Analytical of Heat". The University Press, Cambridge, UK.
2. Tye, R. P. (1969). "Thermal Conductivity". Academic Press, London and New York.
3. Maxwell, J. C. (1892). "A Treatise on Electricity and Magnetism", 3rd edn. Oxford University Press, Oxford.
4. Feynman, R. P. (1961). There's plenty of room at the bottom, "Miniaturization" (H. D. Gilbert, eds.). Reinhold, New York 282–296.
5. Feynman, R. P. (1992). There's plenty of room at the bottom. *J. Microelectromech. Systems* **1**(1), 60–66.
6. Feynman, R. P. (1993). Infinitesimal machinery. *J. Microelectromech. Systems* **2**(1), 4–14.
7. B. X. Wang, H. Li, X. F. Peng, G. P. Peterson, Numerical simulation for microconvection around Brownian motion moving nanoparticles, *In* Third International Symposium on Two-Phase Flow Modeling and Experimentation, Pisa, September (2004) 22–24.
8. Nalwa, H. S. (2004). *Encyclopedia of Nanoscience and Nanotechnology* **X**, 1–30.
9. J. Sterling, Evaporation history of sessile drops: effect of surface property and liquid composition, "56th Annual Meeting of the Division of Fluid Dynamics", The American Physical Society, New Jersey, November, 2003, 23–25.
10. Gleiter, H. (1969). Theory of grain boundary migration rate. *Acta Metall* **17**(7), 853–862.
11. Granqvist, C. G. and Buhrman, R. A. (1976). Ultrafine metal particles. *J. Appl. Phys.* **47**(5), 2200–2219.
12. Jani, P., Nagy, A., and Czitrovszky, A. (1996). Aerosol particles size determination using a photo correlation laser Doppler anemometer. *J. Aerosol Sci.* **27**(1), S531–S532.
13. Hummes, D., Neumann, S., Schmidt, F., Drotboom, M., and Fissan, H. (1996). Determination of the size distribution of nanometer-sized particles. *J. Aerosol Sci.* **27**(1), S163–S164.
14. Buhrman, R. A. and Halperin, W. P. (1973). Fluctuation diamagnetism in a "zero-dimensional" superconductor. *Phys. Rev. Lett.* **30**(15), 692–695.
15. Meier, F. and Wyder, P. (1973). Magnetic moment of small indium particles in the quantum size-effect regime. *Phys. Rev. Lett.* **30**(5), 181–184.
16. Novotny, V. and Meincke, P. P. M. (1973). Thermodynamic lattice and electronic properties of small particles. *Phys. Rev. B* **8**(9), 4186–4199.
17. Tanner, D. B. and Sievers, A. J. (1975). Far-infrared absorption in small metallic particles. *Phys. Rev. B* **11**(4), 1330–1341.
18. Yee, P. and Knight, W. D. (1975). Quantum size effect in copper: NMR in small particles. *Phys. Rev. B* **11**(9), 3261–3267.
19. Ho, T. L. (1983). Effect of quantum voltage fluctuations on the resistance of normal junctions. *Phys. Rev. Lett.* **51**(22), 2060–2063.
20. Ben-Jacob, E., Mottola, E., and Schon, G. (1983). Quantum shot noise in tunnel junctions. *Phys. Rev. Lett.* **51**(22), 2064–2067.
21. Giaever, I. and Zeller, H. R. (1968). Superconductivity of small tin particles measured by tunneling. *Phys. Rev. Lett.* **20**(26), 1504–1507.
22. Cavicchi, E. and Silsbee, R. H. (1984). Coulomb suppression of tunneling rate from small metal particles. *Phys. Rev. Lett.* **52**(16), 1453–1456.
23. Kamat, P. V. and Dimitrijevic, N. M. (1990). Colloidal semiconductors as photocatalysts for solar energy conversion. *Sol. Energy* **44**(2), 83–98.
24. Kamat, P. V. and Dimitrijevic, N. M. (1990). Picosecond charge transfer processes in ultrasmall CdS and CdSe semiconductor particles. *Mol. Cryst. Liquid Cryst.* **183**, 439–445.
25. Ball, P. and Garwin, L. (1992). Science at the atomic scale. *Nature* **355**(27), 761–766.

26. Xu, Y. P., Wang, W. Y., Zhang, D. F., and Chen, X. L. (2001). Dielectric properties of GaN nanoparticles. *J. Mater. Sci.* **36**, 4401–4403.
27. Guczi, L. D., et al. (2000). Modeling gold nanoparticles: morphology, electron structure, and catalytic activity in CO oxidation. *J. Phys. Chem. B* **104**, 3183–3193.
28. Nagasaka, Y. and Nagashima, A. (1981). Absolute measurement of the thermal conductivity of electrically conducting liquids by the transient hot-wire method. *J. Phys. E: Sci. Instrum.* **4**, 1435–1439.
29. Choi, S. U. S. (1995). Enhancing thermal conductivity of fluids with nanoparticles. *In* "Developments and Applications of Non-Newtonian Flows" (D.A. Siginer and H.P. Wang, eds.). American Society of Mechanical Engineers, New York.
30. Eastman, J. A., Choi, S. U. S., Li, S. Thompson, L. J., and Lee, S. (1997). Enhanced thermal conductivity through the development of nanofluids. *In* "Material Research Society Symposium Proceedings" (S. Komarnei, J.C. Parker and H.J. Wollenberger, eds.), Vol. 457, pp. 9–10, Warrendale, PA.
31. Lee, S., Choi, S. U. S., Li, S., and Eastman, J. A. (1999). Measuring thermal conductivity of fluids containing oxide nanoparticles. *Trans. ASME* **121**, 280–289.
32. Eastman, J. A., Choi, S. U. S., Li, S., Yu, W., and Thompson, L. J. (2001). Anomalously increased effective thermal conductivities of ethylene glycol-based nanofluids containing copper nanoparticles. *Appl. Phys. Lett.* **78**(6), 718–720.
33. Xuan, Y. M. and Li, Q. (2000). Heat transfer enhancement of nanofluids. *J. Eng. Thermophys.* **20**(4), 465–470, (Chinese).
34. Xuan, Y. M. and Li, Q. (2000). Heat transfer enhancement of nanofluids. *Int. J. Heat Fluid Flow* **21**, 58–64.
35. Xie, H., Wang, J., Xi, T., Liu, Y., and Ai, F. (2002). Dependence of the thermal conductivity of nanoparticle-fluid mixture on the base fluid. *J. Mater. Sci. Lett.* **21**, 1469–1471.
36. Xie, H., Wang, J., Xi, T., Liu, Y., and Ai, F. (2002). Thermal conductivity enhancement of suspensions containing nanosized alumina particles. *J. Appl. Phys.* **91**(7), 4568–4572.
37. Wang, X. and Xu, X. (1999). Thermal conductivity of nanoparticle–fluid mixture. *J. Thermophys. Heat Transfer* **13**(4), 474–480.
38. Challoner, A. R. and Powell, R. W. (1956). Thermal conductivities of liquids: new determinations for seven liquids and appraisal of existing values. *Proc. Royal Soc. London, Series A (Math. Phys. Sci.)* **238**(1212), 90–106.
39. Peterson, G. P. and Fletcher, L. S. (1987). Effective thermal conductivity of sintered heat pipe wicks. *J. Thermophys.* **1**(4), 343–347.
40. Peterson, G. P. and Fletcher, L. S. (1988). Thermal contact conductance of packed beds in contact with a flat surface. *J. Heat Transfer* **110**(1), 38–41.
41. Peterson, G. P. and Fletcher, L. S. (1989). On the thermal conductivity of dispersed ceramics. *J. Heat Transfer* **111**, 824–829.
42. Duncan, A. B., Peterson, G. P., and Fletcher, L. S. (1989). Effective thermal conductivity within packed beds of spherical particles. *J. Heat Transfer* **111**(4), 830–836.
43. R. G. Miller, L. S. Fletcher (1974). A facility for the measurement of thermal contact conductance. *In* "Proceeding of the 10th Southeastern Seminar on Thermal Sciences", New Orleans, LA, pp. 263–285.
44. Czarnetzki, W. and Roetzel, W. (1995). Temperature oscillation techniques for simultaneous measurement of thermal diffusivity and conductivity. *Int. J. Thermophys.* **16**(2), 413–422.
45. Das, S. K., Putra, N., Thiesen, P., and Roetzel, W. (2003). Temperature dependence of thermal conductivity enhancement for nanofluids. *J. Heat Transfer* **125**, 567–574.

46. Meredith, R. E. and Tobias, C. W. (1961). Conductivities in emulsions. *J. Electrochem. Soc.* **108**(3), 286–290.
47. Hamilton, R. L. and Crosser, O. K. (1962). Thermal conductivity of heterogeneous two-component systems. *I&EC Fundam* **1**(3), 187–191.
48. Patel, H. E., Das, S. K., and Sundararajan, T. (2003). Thermal conductivities of naked and monolayer protected metal nanoparticle based nanofluids: manifestation of anomalous enhancement and chemical effects. *Appl. Phys. Lett.* **83**(14), 2931–2933.
49. B. X. Wang, H. Li, and X. F. Peng (2003). Research on the effective thermal conductivity of nano-particle colloids. *In* "The Sixth ASME–JSME Thermal Engineering Joint Conference". Hawaii, USA.
50. Li, C. H. and Peterson, G. P. (2006). Experimental investigation of temperature and volume fraction variations on the effective thermal conductivity of nanoparticle suspensions. *J. Appl. Phys.* **99**, 084314.
51. Hone, J., Whitney, M., and Zettl, A. (1999). Thermal conductivity of single-walled carbon nanotubes. *Synth. Metals* **103**, 2498–2499.
52. Choi, S. U. S., Zhang, Z. G., Yu, W., Lockwood, F. E., and Grulke, E. A. (2001). Anomalous thermal conductivity enhancement in nanotube suspensions. *Appl. Phys. Lett.* **79**(14), 2252–2254.
53. W. Yu, S. U.S. Choi (2002). Analysis of thermal conductivity and convective heat transfer in nanotube suspensions. *In* "Proceedings of IMECE2002 ASME International Mechanical Engineering Congress and Exposition", New Orleans, LA.
54. Huxtable, S., Cahill, D. G., Shenogin, Xue, S., Ozisik, L. R., Barone, P., Usrey., M., et al. (2003). Interfacial heat flow in carbon nanotube suspensions. *Nat. Mater. Lett.* **2**, 731–734.
55. Xie, H., Lee, H., Youn, W., and Choi, M. (2003). Nanofluids containing multiwalled carbon nanotubes and their enhanced thermal conductivities. *J. Appl. Phys.* **94**(8), 4967–4971.
56. Shenogin, S., Xue, L., Ozisik, R., and Keblinski, P. (2004). Role of thermal boundary resistance on the heat flow in carbon-nanotube composites. *J. Appl. Phys.* **95**(12), 8136–8144.
57. Che, J., Cagin, T., and Goddard, W. A. III. (2000). Thermal conductivity of carbon nanotubes. *Nanotechnology* **11**, 65–69.
58. Dresselhaus, M. S. and Eklund, P. C. (2000). Phonons in carbon nanotubes. *Adv. Phys.* **49**(6), 705–814.
59. Shi, L., Li, D., Yu, C., Jang, W., Kim, D., Yao, Z., Kim, P., and Majumdar, A. (2003). Measuring thermal and thermoelectric properties of one-dimensional nanostructures using a microfabricated device. *J. Heat Transfer* **125**, 881–888.
60. Kim, P., Shi, L., Majumdar, A., and McEuen, P. L. (2001). Thermal transport measurements of individual multiwalled carbon nanotubes. *Phys. Rev. Lett.* **87**(21), 215502-1–215502-4.
61. Maruyama, S. (2003). A molecular dynamics simulation of heat conduction of a finite length single-walled carbon nanotubes. *Microscale Thermophys. Eng.* **7**, 41–50.
62. B. H.Kim, G. P. Peterson, Effect of morphology of carbon nanotubes on thermal conductivity enhancement of aqueous nanofluids, *AIAA J. Thermodynam. Heat Transfer*, in press.
63. Fricke, H. (1924). A mathematical treatment of the electric conductivity and capacity of disperse systems. *Phys. Rev.* **24**, 575–587.
64. Rayleigh, L. (1892). On the influence of obstacles arranged in rectangular order upon the properties of a medium. *Philos. Mag.* **34**, 481–502.
65. Bruggeman Von, D. A. G. (1935). Berechnung verschiedener physikalisher konstanten von heterogenen substanzen. *Ann. Phys.* **5**(24), 636–664, (German).

66. Keller, J. B. (1963). Conductivity of a medium containing a dense array of perfectly conducting spheres or cylinders or nonconducting cylinders. *J. Appl. Phys.* **34**(4), 991–993.
67. Leal, L. G. (1973). On the effective conductivity of a dilute suspension of spherical drops in the limit of low particle Peclet number. *Chem. Eng. Commun.* **1**, 21–31.
68. Jeffrey, D. J. (1973). Conduction through a random suspension of spheres. *Proc. Royal Soc. London, Series A (Math. Phys. Sci.)* **335**(1602), 355–367.
69. Rocha, A. and Acrivos, A. (1973). On the effective thermal conductivity of dilute dispersions: highly conducting inclusions of arbitrary shape. *Quart. J. Mech. Appl. Math.* **26**(4), 441–455.
70. Rocha, A. and Acrivos, A. (1973). On the effective thermal conductivity of dilute dispersions. *Quart. J. Mech. Appl. Math.* **26**(2), 217–233.
71. Batchelor, G. K. (1974). Transport properties of two-phase materials with random structure. *Annu. Rev. Fluid Mech.* **6**, 227–255.
72. Hashin, Z. and Shtrikman, S. (1962). A variational approach to the effective magnetic permeability of multiphase materials. *J. Appl. Phys.* **33**, 3125–3129.
73. Wills, J. R. (1977). Bounds and self-consistent estimates for the overall properties of anisotropic composites. *J. Mech. Phys. Solids* **25**(3), 185–202.
74. O'Brien, R. W. (1979). A method for the calculation of the effective transport properties of suspensions of interacting particles. *J. Fluid Mech.* **9**(1), 17–39.
75. Hatta, H. and Taya, M. (1985). Effective thermal conductivity of a misoriented short fiber composite. *J. Appl. Phys.* **58**(7), 2478–2486.
76. Davis, R. H. (1986). The effective thermal conductivity of a composite material with spherical inclusions. *Int. J. Thermophys.* **7**(3), 609–620.
77. Hasselman, D. P. H. and Johnson, L. F. (1987). Effective thermal conductivity of composites with interfacial thermal barrier resistance. *J. Comp. Mater.* **21**, 508–515.
78. Brady, J. F., Phillips, R. J., Lester, J. C., and Bossis, G. (1988). Dynamic simulation of hydrodynamically interacting suspensions. *J. Fluid Mech.* **195**, 257–280.
79. Bonnecaze, R. T. and Brady, J. F. (1990). A method for determining the effective conductivity of dispersions of particles. *Proc. Royal Soc. London (Math. Phys. Sci.)* **430**(1879), 285–313.
80. Bonnecaze, R. T. and Brady, J. F. (1991). The effective conductivity of random suspensions of spherical particles. *Proc. Royal Soc. London, Series A (Math. Phys. Sci.)* **432**, 445–465.
81. Lu, S. Y. and Kim, S. T. (1990). Effective thermal conductivity of composites containing spheroidal inclusions. *AIChE J* **36**(6), 927–938.
82. Kim, I. C. and Torquato, S. (1993). Effective conductivity of composites containing spheroidal inclusions: comparison of simulations with theory. *J. Appl. Phys.* **74**(3), 1844–1854.
83. Lu, S. Y. and Lin, H. C. (1996). Effective conductivity of composites containing aligned spheroidal inclusions of finite conductivity. *J. Appl. Phys.* **79**(9), 6761–6769.
84. Chen, G. (1996). Nonlocal and nonequilibrium heat conduction in the vicinity of nanoparticles. *J. Heat Transfer* **118**, 539–545.
85. Xue, Q. Z. (2003). Model for effective thermal conductivity of nanofluids. *Phys. Lett. A* **307**, 313–317.
86. Wang, B. X., Zhou, L. P., and Peng, X. F. (2003). A fractal model for predicting the effective thermal conductivity of liquid with suspension of nanoparticles. *Int. J. Heat Mass Transfer* **46**, 2665–2672.
87. Haw, M. D. (2002). Colloidal suspensions, Brownian motion, molecular reality: a short history. *J. Phys.: Condens. Matter* **14**, 7769–7779.

88. Mazo, R. M. (2002). "Brownian Motion Fluctuations, Dynamics and Applications. (International series of monographs on physics)". Oxford Science Publications, Clarendon, Oxford, UK.
89. Grasselli, Y. and Bossis, G. (1995). Three-dimensional particle tracking for the characterization of micrometer-size colloidal particles. *J. Colloid Interf. Sci.* **170**, 269–274.
90. Gaspard, P., Briggs, M. E., Francis, M. K., Sengers, J. V., Gammon, R. W., Dorfman, J. R., and Calabrese, R. V. (1998). Experimental evidence for microscopic chaos. *Nature* **394**, 865–868.
91. Briggs, M. E., Sengers, J. V., Francis, M. K., Gaspard, P., Gammon, R. W., Dorfman, J. R., and Calabrese, R. V. (2001). Tracking a colloidal particle for the measurement of dynamic entropies. *Physica: A* **296**, 42–59.
92. Goodson, M. and Kraft, M. (2002). An efficient stochastic algorithm for simulating nanoparticle dynamics. *J. Comput. Phys.* **183**, 210–232.
93. Batchelor, G. K. (1972). Sedimentation in a dilute dispersion of spheres. *J. Fluid Mech.* **52**, 245–268.
94. Batchelor, G. K. (1982). Sedimentation in a dilute polydisperse system of interacting spheres, part 1. General theory. *J. Fluid Mech.* **119**, 379–408.
95. Batchelor, G. K. and Wen, C. S. (1982). Sedimentation in a dilute polydisperse system of interacting spheres, part 2. Numerical results,. *J. Fluid Mech.* **124**, 495–528.
96. Torquato, S. (1985). Effective electrical conductivity of two-phase disordered composite media. *J. Appl. Phys.* **58**(10), 3790–3797.
97. Torquato, S. and Kim, I. C. (1989). Efficient simulation technique to compute effective properties of heterogeneous media. *Appl. Phys. Lett.* **55**(18), 1847–1849.
98. Kim, I. C. and Torquato, S. (1993). Effective conductivity of composites containing spheroidal inclusions: comparison of simulations with theory. *J. Appl. Phys.* **74**(3), 1844–1854.
99. Kim, I. C. and Torquatol, S. (1991). Effective conductivity of suspensions of hard spheres by Brownian motion simulation. *J. Appl. Phys.* **69**(4), 2280–2289.
100. Kim, I. C. and Torquato, S. (1990). Determination of the effective conductivity of heterogeneous media by Brownian motion simulation. *J. Appl. Phys.* **68**(8), 3892–3903.
101. Brown, W. F., Jr. (1955). Solid mixture permittivities. *The J. Chem. Phys.* **23**(8), 1514–1517.
102. Uhlenbeck, G. E. and Ornstein, L. S. (1930). On the theory of the Brownian motion. *Phys. Rev.* **36**, 823–841.
103. Mazur, P. and Bedeaux, D. (1992). Nature of the random force in Brownian motion. *Langmuir* **8**, 2947–2951.
104. Gupte, S. K., Advani, S. G., and Huq, P. (1995). Role of micro-convection due to non-affine motion of particles in a mono-disperse suspension. *Int. J. Heat Mass Transfer* **38**(16), 2945–2958.
105. Waldmann, L. (1961). On the motion of spherical particles in nonhomogeneous gases. *In* "Rarefied Gas Dynamics. Section 4: Applications of Kinetic Theory" (L. Talbot, ed.). Academic Press, New York.
106. Brock, J. R. (1962). On the theory of thermal forces acting on aerosol particles. *J. Colloid Sci.* **17**, 768–780.
107. Rosner, D. E., Mackowski, D. W., Tassopoulos, M., Castillo, J., and Ybarra, P. G. (1992). Effects of heat transfer on the dynamics and transport of small particles suspended in gases. *Ind. Eng. Chem. Res.* **31**(3), 766–769.
108. Huisken, J. and Stelzer, E. H. K. (2002). Optical levitation of absorbing particles with a nominally Gaussian laser beam. *Opt. Lett.* **27**(14), 1223–1225.

109. Buevich, Y. A., Zubarev, A. Y., and Isaev, A. M. (1990). The hydromechanics of suspensions. *J. Eng. Phys.* **57**(3), 1030–1038, (English translation of *Inz.-Fiz. Z.* 57(3), 1989, 402–412.).
110. Jang, S. P. and Choi, S. (2004). U.S. role of Brownian motion in the enhanced thermal conductivity of nanofluids. *Appl. Phys. Lett.* **84**, 4316.
111. Prasher, R., Bhattacharya, P., and Phelan, P. E. (2005). Thermal conductivity of nanoscale colloidal solutions (nanofluids). *Phys. Rev. Lett.* **94**, 025901.
112. C. H.Li, G. P. Peterson (2005). Due role of nanoparticle in the thermal conductivity enhancement of nanoparticle suspension. In "Proceedings of IMECE 2005 ASME International Mechanical Engineering Congress and Exposition", November 5–11, Orlando, FL.
113. Yu, C. J., Richter, A. G., Datta, A., Durbin, M. K., and Dutta, P. (1999). Observation of molecular layering in thin liquid films using X-ray reflectivity. *Phys. Rev. Lett.* **82**(11), 2326–2329.
114. Wang, B. X., Li, H., and Peng, X. F. (2002). Research on the heat conduction enhancement for liquid with nanoparticle suspensions. *J. Therm. Sci.* **11**(3), 214–219.
115. Keblinski, P., Phillpot, S. R., Choi, S. U. S., and Eastman, J. A. (2002). Mechanisms of heat flow in suspensions of nano-sized particles (nanofluids). *Int. J. Heat Mass Transfer* **45**, 855–863.
116. Xue, L., Leblinski, P., Phillpot, S. R., Choi, S. U. S., and Eastman, J. A. (2003). Two regimes of thermal resistance at a liquid–solid interface. *J. Chem. Phys.* **118**(1), 337–339.
117. Ahuja, A. S. (1975). Augmentation of heat transport in laminar flow of polystyrene suspensions. I. Experiments and results. *J. Appl. Phys.* **46**(8), 3408–3416.
118. Ahuja, A. S. (1975). Augmentation of heat transport in laminar flow of polystyrene suspensions. II. Analysis of data. *J. Appl. Phys.* **46**(8), 3417–3425.
119. Lee, S. P. and Choi, S. (1996). U.S. application of metallic nanoparticle suspensions in advanced cooling systems. In "Recent advances in solids/structures and application of metallic materials PVP-Vol. 342/MD-Vol. 72" (Y. Kwon, D. Davis and H. Chung, eds.), pp. 227–234. The American Society of Mechanical Engineers, New York.
120. Xuan, Y. M. and Roetzel, W. (2000). Conceptions for heat transfer correlation of nanofluids. *Int. J. Heat Mass Transfer* **43**, 3701,3077.
121. Xuan, Y. M. and Li, Q. (2003). Investigation on convective heat transfer and flow features of nanofluids. *J. Heat Transfer* **125**, 151–155.
122. Okada, M. and Suzuki, T. (1997). Natural convection of water-fine particle suspension in a rectangular cell. *Int. J. Heat Mass Transfer* **40**(13), 3201–3208.
123. Kang, C., Okada, M., Hattori, A., and Oyama, K. (2001). Natural convection of water-fine particle suspension in a rectangular vessel heated and cooled from opposing vertical walls (classification of the natural convection in the case of suspension with a narrow-size distribution). *Int. J. Heat Mass Transfer* **44**, 2973–2982.
124. Putra, N., Roetzel, W., and Das, S. K. (2003). Natural convection of nano-fluids. *Heat Mass Transfer* **39**, 775–784.
125. Khanafer, K., Vafai, K., and Lightstone, M. (2003). Buoyancy-driven heat transfer enhancement in a two-dimensional enclosure utilizing nanofluids. *Int. J. Heat Mass Transfer* **46**, 3639–3653.
126. You, S. M. and Kim, J. H. (2003). Effect of nanoparticles on critical heat flux of water in pool boiling heat transfer. *Appl. Phys. Lett.* **83**(16), 3374–3376.
127. Das, S. K., Putra, N., and Roetzel, W. (2003). Pool boiling characteristics of nano-fluids. *Int. J. Heat Mass Transfer* **46**, 851–862.
128. Vassallo, P., Kumar, R., and D'Amico, S. (2004). Pool boiling heat transfer experiments in silica–water nano-fluids. *Int. J. Heat Mass Transfer* **47**, 407–411.

129. Zhou, D. W. (2004). Heat transfer enhancement of copper nanofluid with acoustic cavitation. *Int. J. Heat Mass Transfer* **47**, 3109–3117.
130. Duan, X., Huang, Y., Cui, Y., Wang, J., and Lieber, C. M. (2001). Indium phosphide nanowires as building blocks for nanoscale electronic and optoelectronic devices. *Nature* **409**, 66–69.
131. Duan, X., Huang, Y., Agarwal, R., and Lieber, C. M. (2003). Single-nanowire electrically driven lasers. *Nature* **421**, 241–245.
132. Gorman, J. (2003). Nanofluid flow, detergents may benefit from new insight. *Sci. News* **163**(19), 292–293.
133. Westedt, U., Barbu-Tudoran, L., Schaper, A. K., Kalinowski, M., Alfke, H., and Kissel, T. (2004). Effects of different application parameters on penetration characteristics and arterial vessel wall integrity after local nanoparticle delivery using a porous balloon catheter. *Eur. J. Pharm. Biopharm.* **58**, 161–168.

The Effective Thermal Conductivity of Saturated Porous Media

H.T. AICHLMAYR[1] and F.A. KULACKI[2]

[1]Sandia National Laboratories, 7011 East Ave., Livermore, CA 94550, USA
[2]Thermodynamics and Heat Transfer Laboratory, Department of Mechanical Engineering, The University of Minnesota, 111 Church St. SE, Minneapolis, MN 55455, USA

I. Introduction

Despite decades of experimental and theoretical work, predicting the effective thermal conductivity of saturated porous media continues to be one of the great unsolved problems in heat transfer science. The problem is unsolved because the effective conductivity is a phenomenological characterization of a solid–fluid medium rather than a thermo-physical property. Consequently, it is experiential and characterized in macroscopic terms such as the thermal conductivities and volume fractions of the constituent phases. The effective conductivity, however, also depends on the geometry and arrangement of the phases–information which is generally not available from experiments. Therefore, a discontinuity exists in our understanding of the effective conductivity and it must be bridged by models for heat conduction in the microstructure.

Although one can argue that the effective thermal conductivity problem is inherently intractable, many aspects of conduction in composite media are fairly well understood. For example, physical limits for the effective conductivity are established. In addition, the effective conductivity of small to modest solid–fluid conductivity ratio media can be predicted with reasonable accuracy. In contrast, our understanding of heat conduction in large solid–fluid conductivity ratio media is incomplete.

II. The Effective Thermal Conductivity

Consider a cylindrical chamber filled with uniform solid spherical particles and a liquid (Fig. 1). The solid particles and saturating fluid have thermal conductivities k_s and k_f. The chamber has length L_c, and the

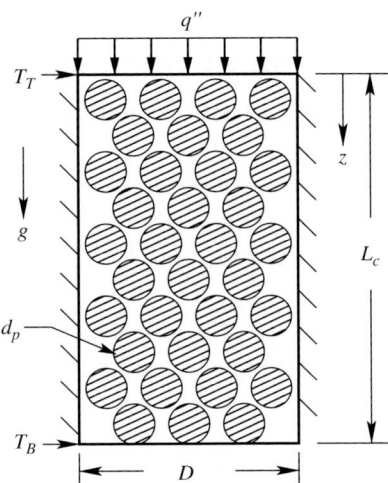

FIG. 1. The effective thermal conductivity problem. The chamber is filled with a medium comprised of solid particles and a liquid having thermal conductivities k_s and k_f. The chamber length is L_c, and the diameters of the particles and the chamber are d_p and D, respectively. The boundary heat flux, q'' and the boundary temperatures, T_T and T_B, are measured.

diameters of the particles and the chamber are d_p and D, respectively. The particles occupy volume V_p, and the total chamber volume is V_t. The porosity, ϕ is defined to be the fraction of the total chamber volume which is void space [1]. When the fluid phase completely occupies this space, the medium is said to be saturated, which implies that the porosity is given by

$$\phi = 1 - \frac{V_p}{V_t} \tag{1}$$

One should note that if the particles are spheres with $d_p \ll D$, the porosity is essentially uniform except in the vicinity of the wall [2].

To define the effective thermal conductivity of the material in the chamber, the following conditions are imposed: the chamber is heated from above to avoid the formation of buoyancy-induced convection currents; the saturating fluid is opaque to ensure that conduction is the dominant heat transfer mode; the vertical boundaries are adiabatic; and the surface temperatures of the horizontal boundaries are uniform. Consequently, when the chamber attains steady state, the overall heat flow is unidirectional. Therefore, applying Fourier's Law,

$$q'' = -k_e \frac{dT}{dz} \tag{2}$$

to the contents of the chamber implies that the effective thermal conductivity, k_e is given by

$$k_e = \frac{q'' L_c}{T_T - T_B} \qquad (3)$$

where q'', L_c, T_T, and T_B are measured quantities.

To model macroscopic heat conduction in a porous medium, one typically assumes that the effective conductivity is isotropic and that the solid and fluid phases are in thermal equilibrium. The former assumption applies generally, but the latter may not be satisfied during transient heating [3,4] or in the vicinity of a heated boundary [5]. When these assumptions are valid, conduction is described by

$$\frac{1}{\alpha_e}\frac{\partial T}{\partial t} = \nabla^2 T \qquad (4)$$

where

$$\alpha_e = \frac{k_e}{(\rho c_p)_e} \qquad (5)$$

is the effective thermal diffusivity. The effective volumetric heat capacity is defined by

$$(\rho c_p)_e = \phi(\rho c_p)_f + (1 - \phi)(\rho c_p)_s \qquad (6)$$

and it is the void-fraction-weighted arithmetic mean of the volumetric heat capacities of the constituent phases. One should note that Eq. (6) is independent of medium morphology because mass and energy are extensive properties.

A. EXPERIMENTAL DETERMINATION

Understanding the effective conductivity of porous media begins with careful measurements. The basic measurement strategy is to experimentally approximate a conduction problem with a well-known analytical solution. Consequently, the solution is used to interpret experimental data and to determine the effective conductivity. Experimental techniques appear in two varieties, steady state and transient.

1. *Steady-State Techniques*

Steady-state measurements are the traditional means to determine the thermal conductivity of a substance [6]. These methods have the advantage that the solid and fluid phases are in thermal equilibrium [4]. The salient measured quantities include heat flux and temperature. Heat flux can be measured directly, e.g., Ohmic dissipation in an electric heater, or indirectly by determining the temperature gradient through a material of known thermal conductivity. Unfortunately, contact resistance and imperfect thermal insulation limit the accuracy of the conduction problem approximation [6]. For instance, guard heaters are usually necessary to realize adiabatic boundary conditions. Additionally, porous systems have comparatively large heat capacities; hence the time required to achieve steady state can be considerable [7]. Despite these difficulties, steady-state experimental techniques are widely used. In general, they employ rectilinear or radial conduction.

For example, Prasad *et al.* [8] use axial conduction in a 17.8 cm cylindrical chamber similar to Fig. 1. They measure the effective conductivities of solid–fluid systems comprised of glass, steel, acrylic, water, and glycol. The chamber is heated from above by a thermofoil heater recessed within a 25.4 mm acrylic plate. The bottom boundary consists of aluminum and acrylic plates that are maintained at constant temperature by water circulating through internal passages.

Aside from measuring the top and bottom surface temperatures, Prasad *et al.* provide few details of their data collection and reduction technique. Presumably, they measure the power consumption of the heater and use an expression like Eq. (3) to compute the effective conductivity. In addition, they monitor the spatial temperature distributions of the axial boundaries to detect radial heat flow, but they do not compensate for heat losses at the boundaries.

More importantly, however, Prasad *et al.* fail to disclose whether or not a guard heater is incorporated in the top boundary. This is a crucial omission because without a guard heater, the fraction of energy that enters the test section cannot be determined. Consequently, if "heat leakage" is not taken into consideration, Eq. (3) will yield inaccurate results. This illustrates an important limitation of steady-state techniques: A thorough accounting of energy flows is necessary to ensure accuracy.

In contrast, Chen *et al.* [9] employ axial conduction in a cylindrical test chamber to determine the effective conductivity of mono- and bi-dispersed porous media. Their test cell is nearly identical to that of Prasad *et al.* [8], but they incorporate a guard heater in the top boundary.

In a similar manner, Calmidi and Mahajan [10] and Bhattacharya *et al.* [11] measure the effective conductivity of open-cell aluminum foams saturated with air and water. A rectangular sample is brazed to top and bottom aluminum plates that bound the test cell. The upper plate is heated by an electric heater and the bottom plate is cooled by immersion in a water bath.

The sides are insulated with polystyrene foam and the temperatures of the aluminum plates are measured. The effective conductivity is determined from the slope of a temperature-heat-flux line.

Ofuchi and Kunii [12] use axial conduction in a cylindrical chamber to measure the effective conductivity of porous systems comprised of various solid particles and saturating fluids that include water, air, carbon dioxide, helium, and hydrogen. Their test cell is depicted in Fig. 2. The test cell has an inside diameter of 20 cm and a height of 10 cm. Jacketed copper plates form the top and bottom surfaces of the test cell. The top boundary is heated by steam and the bottom boundary is cooled by water. Circumferential insulation is provided by an annular Dewar jacket. A marble disk having a known thermal conductivity occupies the lower half of the test cell. Hence, the axial heat flux is determined by estimating the temperature gradient through it. Several thermocouples placed inside the porous medium provide the axial temperature profile, which in turn, is used to estimate the effective conductivity.

Similarly, Hadley [13] employs axial conduction to determine the effective conductivity of pressed metal powder. The apparatus is depicted in Fig. 3, and one should note that a sample disk is sandwiched between copper and "standard" disks. Standard disks are made of Dynasil 4000 glass and 304 stainless steel; both have accurately measured thermal conductivities. The sample is heated from above by a nichrome wire embedded in an aluminum plate. A liquid-cooled brass plug constitutes the cold boundary.

Hadley determines the effective conductivity in a manner similar to that of Ofuchi and Kunii [12]. Specifically, the heat flux through the entire assembly is determined from the temperature gradient across the standard material. The thermal conductivity of the test specimen is then estimated from its surface temperatures and dimensions.

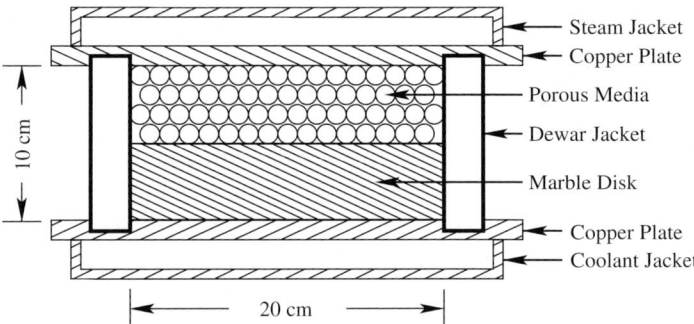

FIG. 2. The experimental setup of Ofuchi and Kunii [12]. The test section is axisymmetric and the overall direction of heat flow is axial. (Adapted from Ref. [12], Copyright (1965) with permission from Elsevier.)

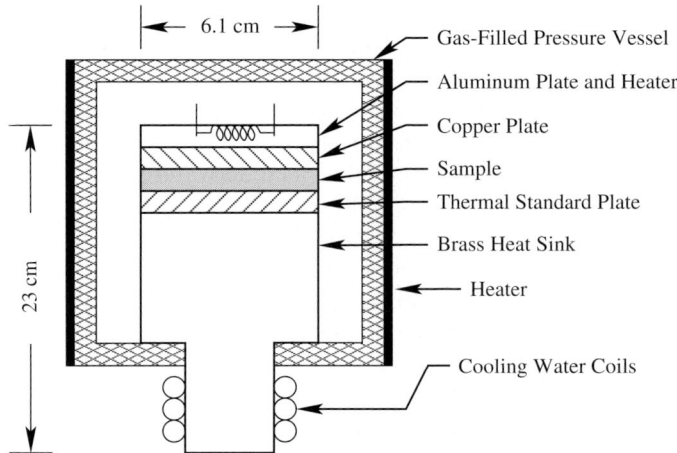

FIG. 3. The experimental setup of Hadley [13]. The test section is axisymmetric and heat flows in the axial direction. The samples consist of pressed metal powder disks. (Adapted from Ref. [13], Copyright (1986) with permission from Elsevier.)

Of note, Hadley employs calibrated thermistors to measure the surface temperatures of the disks. Thermistors have greater temperature sensitivity than thermocouples. Consequently, comparatively small temperature gradients can be used without compromising the overall measurement uncertainty [13]. He estimates the uncertainty in the effective conductivity measurements to be 10 percent. The major contributors to the overall uncertainty are non-uniform contact resistance and thermistor drift.

Widenfeld et al. [7] employ steady-state axial conduction in an experimental apparatus similar to that of Hadley [13] to determine the effective conductivity of particle beds subjected to axial compression. The specimen is bounded at the top and bottom surfaces by stainless steel rams. The top ram is heated by an electric heater and the bottom is cooled by water circulating in external tubes. Using guard heaters, however, is impractical because the upper boundary must move to compress the sample. Consequently, Widenfeld et al. use the the stainless-steel rams to determine the heat flux through the sample and to infer surface temperatures. This is accomplished by measuring the temperatures at several axial locations on the rams and estimating the temperature gradient.

Yagi and Kunii [14] employ steady-state radial conduction to measure the effective conductivity of gas–solid porous media. The annular test chamber is depicted in Fig. 4. One should note that locating the heater at the center of the medium rather than at the end reduces heat leakage considerably. Guard

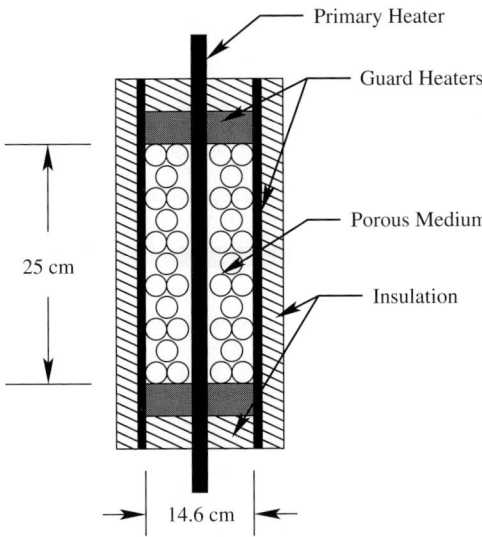

FIG. 4. The experimental configuration of Yagi and Kunii [14]. The overall direction of heat flow is radial. (Adapted with permission of the American Institute of Chemical Engineers.)

heaters, however, are located along the circumference and at the ends of the chamber for thermal management and to minimize heat loss.

Thermocouples are positioned at several radial locations within the chamber to obtain internal temperature measurements. Yagi and Kunii subsequently monitor the radial temperature profiles and use

$$k_e = \frac{Q \ln(D_o/D_i)}{2\pi L(T_i - T_o)} \qquad (7)$$

to determine the effective conductivity of the sample. The temperatures T_i and T_o are measured at the inner and outer diameters, D_i and D_o, of the annular chamber.

Deissler and Eian [15] and Deissler and Boegli [16] also employ radial conduction to determine the effective conductivity of magnesium oxide, stainless steel, and uranium oxide powders saturated with mixtures of helium, argon, neon, nitrogen, and air. A sketch of the experimental apparatus is presented in Fig. 5. The powder is located in the cylindrical test chamber and porous plugs provide a path for gas to enter the test section. Consequently, saturating fluids are changed by evacuating the gas case and refilling with the desired gas. The primary heater is located at the axis of the test section and nichrome–wire heaters are wrapped around the ends to

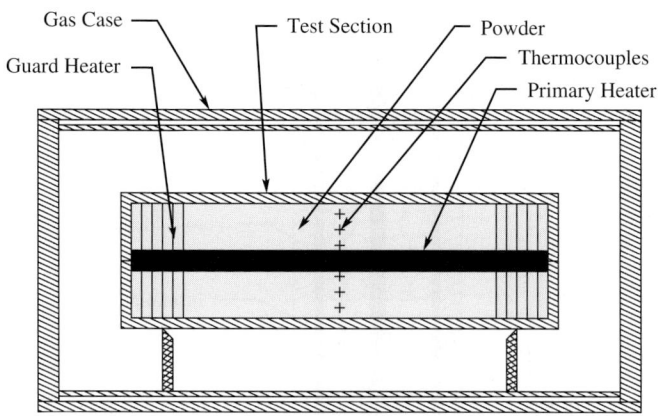

FIG. 5. The experimental setup of Deissler and Eian [15] and Deissler and Boegli [16]. The overall direction of heat flow is radial. (Adapted from Ref. [16], Copyright (1958) with permission from ASME.)

compensate for axial heat loss. An array of radially distributed thermocouples located at the midplane of the test section acquire the radial temperature profile. This profile is used to determine the effective conductivity.

Of note, they assume a temperature-dependent thermal conductivity. The radial heat flow is given by

$$Q = \frac{2\pi k_m L(T_1 - T)}{\ln(r/r_1)} \qquad (8)$$

where the temperatures T_1 and T are recorded at radii r_1 and r, respectively [15]. The mean effective conductivity, k_m, is evaluated at $(T_1 + T)/2$ and given by

$$k_m = \frac{k_1 + k}{2} \qquad (9)$$

for a linear temperature dependence, i.e., $k(T) = k_1 + c_1 T$. They determine the effective conductivity by rearranging Eq. (8) and substituting Eq. (9) to yield

$$T = T_1 - \frac{Q}{2\pi L k_1}\left[\frac{2k_1}{2k_1 + c_1 T}\ln\left(\frac{r}{r_1}\right)\right] \qquad (10)$$

Hence two parameters, i.e., k_1 and c_1, must be determined. Their data reduction procedure is to assume values for k_1 and c_1, plot the measured temperature profile versus the bracketed quantity in Eq. (10), revise the estimate of k_1 using the slope of the line, and replot. The procedure is repeated until convergence is attained.

Slavin et al. [17] also employ radial conduction to determine the effective conductivity of a medium consisting of alumina and helium. The experimental apparatus resembles that of Yagi and Kunii [14] and they employ a similar data reduction scheme. Notably, they describe their experimental apparatus and uncertainty analysis in detail.

2. Transient Techniques

Transient techniques offer several advantages relative to steady-state methods. First, experiments proceed comparatively quickly. Second, large thermal time constants are actually a benefit because they ensure that temperature changes are slow enough to be captured by typical data acquisition equipment. Third, the semi-infinite domain is a convenient alternative to adiabatic boundary conditions. Although, this condition must be implemented carefully [18]. In addition, conducting an energy balance in a transient experiment is a challenge because it necessitates the estimation of thermal energy accumulation in materials [18]. Fortunately, this problem can be mitigated to some extent with a carefully chosen test configuration or by employing data reduction techniques that yield the effective conductivity without determining the heat flux.

In general, two types of transient techniques exist. Both types approximate conduction problems involving the sudden heating of infinite domains, but they differ by the manner in which heat is applied. The first type employs a heated boundary and the second type employs a line heat source.

For example, Nozad et al. [19] employ the sudden end heating of a rectangular semi-infinite domain to determine the effective conductivity of glass, stainless-steel, bronze, urea-formaldehyde, and aluminum spheres saturated with air, glycerol, and water. The experimental apparatus is depicted in Fig. 6. The test section consists of an acrylic rectangular parallelepiped 10 cm wide, 10 cm deep, and 20 cm tall. The thermal boundary conditions are applied by heat exchangers located at the ends of the test section. Copper plates constitute the thermal interface between circulating water and test media. The water streams originate in hot or cold temperature-regulated reservoirs. Initially, the test section is maintained at a uniform temperature by circulating cold water in both heat exchangers. A test commences by admitting hot water to the upper heat exchanger. The internal temperature profile is acquired by eight thermocouples located at the centerline of the test section and sampled at 1 s intervals [19].

To determine the effective conductivity of the sample, the test cell is approximated by a semi-infinite domain. Consequently, Eq. (4) is solved

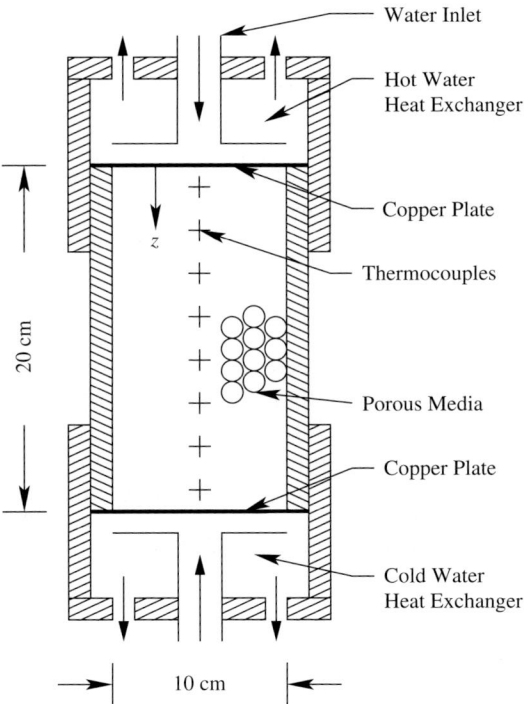

FIG. 6. The experimental apparatus used by Nozad *et al.* [19]. The test section is a rectangular parallelepiped 10 cm wide, 10 cm deep, and 20 cm tall. (Adapted from Ref. [19], Copyright (1985) with permission from Elsevier.)

subject to the initial and boundary conditions,

$$T(t=0,z) = T_0, \quad T(t>0, z=0) = T_1, \quad \text{and} \quad T(t, z \to \infty) = T_0 \tag{11}$$

The solution is

$$\frac{T - T_0}{T_1 - T_0} = 1 - \text{erf}\left(\frac{z}{2\sqrt{\alpha_e t}}\right) \tag{12}$$

which can be rearranged to yield

$$\text{erf}^{-1}(1 - \Theta) = \frac{1}{\sqrt{\alpha_e}} \left(\frac{z}{2\sqrt{t}}\right) \tag{13}$$

where

$$\Theta = \frac{T - T_0}{T_1 - T_0} \qquad (14)$$

Nozad et al. plot $\text{erf}^{-1}(1 - \Theta)$ versus $z/2\sqrt{t}$ and determine the effective thermal diffusivity from the slope of the line. They subsequently use this result and Eqs. (5) and (6) to estimate the effective conductivity. Nozad et al. estimate the accuracy of this method to be 5 percent. Natural convection is a source of experimental error and they note that it affects aluminum–air systems most.

In a similar manner, Aichlmayr [18] uses the sudden end-heating of a semi-infinite cylinder to determine the effective conductivity of solid–fluid systems comprised of glass, steel, air, and water. The cylindrical test chamber is depicted in Fig. 7, and it is 22.2 cm in diameter and 40.6 cm long. One should note that the salient difference between his study and that of Nozad et al. is the use of a constant heat flux boundary condition. Consequently, Eq. (4) is solved subject to the initial and boundary conditions,

$$T(t=0,z) = T_0, \quad -k_e \frac{\partial T}{\partial z}\bigg|_{z=0} = q'', \quad \text{and} \quad T(t, z \to \infty) = T_0 \qquad (15)$$

The result is

$$\theta(z,t) = \frac{2q''\sqrt{\alpha_e t}}{\sqrt{\pi}k_e} \exp\left(-\frac{z^2}{4\alpha_e t}\right) - \frac{q''z}{k_e} \text{erfc}\left(\frac{z}{2\sqrt{\alpha_e t}}\right) \qquad (16)$$

where $\theta(z, t) = T(z, t) - T_0$.

One should note that the effective conductivity can be determined by substituting temperature, time, position, and heat flux measurements into Eq. (16) and solving the transcendental equation for the effective thermal diffusivity. The effective conductivity is then found from Eq. (5) and the effective volumetric heat capacity (Eq. (6)). Lacroix et al. [20] employ this strategy to determine the effective conductivity of copper wool and duraluminum turnings.

Alternatively, the effective conductivity can be determined without measuring the heat flux. To illustrate, let θ_o and ξ represent the temperature of the boundary and the argument of the complimentary error function. That is,

$$\theta_o = \frac{2q''\sqrt{t}}{\sqrt{\pi k_e (\rho c_p)_e}} \qquad (17)$$

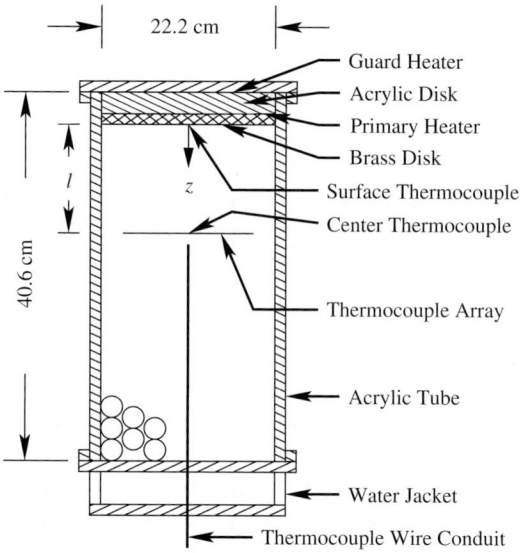

FIG. 7. Experimental setup of Aichlmayr [18]. The cylindrical test chamber is 22.2 cm in diameter and 40.6 cm long.

and

$$\xi = \frac{z\sqrt{(\rho c_p)_e}}{2\sqrt{k_e t}} \quad (18)$$

Incidentally, one should note that

$$\xi = \frac{Fo^{-1/2}}{2} \quad (19)$$

where

$$Fo = \frac{\alpha_e t}{z^2} \quad (20)$$

is the Fourier number at position z. Consequently, substituting Eqs. (17) and (18) into Eq. (16) and dividing by θ_o gives

$$\Theta(\xi) = \exp(-\xi^2) - \sqrt{\pi}\xi \, \text{erfc}(\xi) \quad (21)$$

where

$$\Theta = \frac{\theta(z,t)}{\theta_o} \quad (22)$$

is the temperature ratio.

The effective conductivity is determined from experimental data and Eq. (21) by the following procedure: (1) Two axial temperature measurements are recorded while the medium is heated. The first temperature, θ_o is located at the boundary and the second, θ_1 is located at a distance, l from the boundary. (2) The dimensionless boundary temperature, Θ is computed for each pair of measurements. (3) The corresponding value of ξ is obtained from Eq. (21). (4) The quantity $l/2(\rho c_p)_e^{1/2} t^{-1/2}$ is computed from ancillary experimental data. (5) The effective conductivity is obtained from a plot of ξ versus $l/2(\rho c_p)_e^{1/2} t^{-1/2}$.

Although this procedure is insensitive to the magnitude of the heat flux, one should note that it must be constant for Eq. (16) to be valid. To satisfy this requirement, Aichlmayr heats the chamber from above by a thermofoil heater sandwiched between brass and acrylic disks and employs a guard heater of identical design located between the acrylic disk and cover plate. The power applied to the guard heater is adjusted until the temperatures at opposite faces of the disk are equal, i.e., the heat flux through the disk is zero, to ensure that the heat flux applied to the medium is constant. The boundary temperature is measured by a thermocouple mounted in the brass plate that bounds the test cell. A thermocouple array is positioned at location l from the upper boundary. The center thermocouple provides the interior temperature measurement and the remaining thermocouples are used to estimate the radial temperature profile.

The "hot wire" and "thermal conductivity probe" methods are the second variety of transient effective conductivity measurement techniques. Both techniques employ the conduction problem of a continuous line heat source, e.g., a thin wire, located in an infinite medium [21, p. 261]. The heat source is oriented parallel to the axis of a cylinder of infinite extent and emits heat at a constant rate, q_w per unit length. After heating commences, the temperature distribution in the surrounding material is

$$\theta = \frac{q_w}{4\pi k_e} \int_0^t \exp\left[-\frac{r^2}{4\alpha_e(t-t')}\right] \frac{dt'}{t-t'} \qquad (23)$$

Next, substituting

$$u = \frac{r^2}{4\alpha_e(t-t')} \qquad (24)$$

into Eq. (23), yields

$$\theta = \frac{q_w}{4\pi k_e} \int_{r^2/4\alpha_e t}^{\infty} \frac{e^{-u}}{u} du \qquad (25)$$

which can be expressed in terms of the exponential integral,

$$\theta = \frac{q_w}{4\pi k_e} E_1\left(\frac{r^2}{4\alpha_e t}\right) \qquad (26)$$

Incidentally, one should note that $-Ei(-x)$ is often substituted for $E_1(x)$ in the literature [e.g., 3,6].

Typically, an approximate form of Eq. (26) is used to determine the effective conductivity. This form is obtained from the series representation of the exponential integral,

$$E_1(z) = -\hat{\gamma} - \ln z - \sum_{m=1}^{\infty} \frac{(-1)^m (z)^m}{mm!} \qquad (27)$$

where $\hat{\gamma} = 0.5772$ is Euler's constant and z is complex [22]. For small z, i.e., large times and small radii, the higher-order terms in Eq. (27) may be neglected. In which case, the temperature distribution is

$$\theta = \frac{q_w}{4\pi k_e} \ln\left(\frac{4\alpha_e t}{r^2}\right) - \frac{\hat{\gamma} q_w}{4\pi k_e} \qquad (28)$$

Alternatively, Eq. (28) may be expressed in the more convenient form,

$$\theta = \frac{q_w}{4\pi k_e} \ln\left(\frac{4\alpha_e t}{r^2 \hat{c}}\right) \qquad (29)$$

where $\hat{c} = \exp \hat{\gamma}$ [3]. To determine the effective conductivity using Eqs. (28) or (29), one can compute it directly from measurements of temperature, heat input, and thermal probe position; compute it using measurements of heat input and temperature rise; or determine it from a plot of temperature versus $\ln t$.

For example, Woodside and Messmer [6] employ a thermal conductivity probe 152 mm long and 1.7 mm in diameter to measure the thermal conductivity of quartz sand, glass beads, and lead shot saturated with air, water, and oil. The conductivity probe consists of an 18 μm heater wire located inside a stainless-steel sheath. A very thin wire is used to minimize the temperature resistance coefficient and therefore minimize the change in the energy dissipation rate. The temperature of the material adjacent to the probe is measured with a sheathed thermocouple junction located at the midpoint of the conductivity probe. The difference in junction temperatures at times t_2 and t_1 is

$$\theta_2 - \theta_1 = \frac{q_w}{4\pi k_e} \ln\left(\frac{t_2}{t_1}\right) \qquad (30)$$

which follows from Eq. (28). Consequently, they use Eq. (30) to determine k_e from the measured temperature rise and power dissipated in the wire. Incidentally, Woodside and Messmer adjust the thermocouple measurement circuit to yield the temperature rise directly.

Swift [23] employs a similar technique to measure the effective conductivity of uranium, zirconium, aluminum, copper, magnesium, nickel, and Pyrex powders saturated with nitrogen, hydrogen, helium, argon, methane, and Refrigerant-12. The powders are packed into a cylinder 51 mm long and 19 mm in diameter. The heater wire and two thermocouple junctions are sheathed within a 0.6 mm diameter tube located at the center of the test section. The thermocouple junctions are arranged in a thermopile configuration for greater temperature sensitivity. Typical test durations and temperature rises are 1 min and 5°C, respectively.

Tavman [24] also uses the hot-wire method to determine the effective conductivity of sand saturated with air. Tavman modifies the technique however, by sandwiching the wire and a thermocouple between rectangular samples of an insulating material and the medium. The effective conductivity is given by

$$k_e = F_c \frac{q_w \ln(t_2/t_1)}{\theta_2 - \theta_1} - H_c \qquad (31)$$

where F_c and H_c are constants determined by calibration. Tavman claims that the accuracy of the modified technique is 5 percent, but the calibration procedure is not disclosed.

Alternatively, Glatzmaier and Ramirez [3,25] employ a 25.4 μm diameter platinum wire to measure the effective conductivity and diffusivity of single- and two-phase materials. They employ a platinum wire because when incorporated in a bridge circuit, it is both a heat source and resistance temperature detector. They note that maintaining a uniform heat flux is a problem, but it can be mitigated by maintaining a constant voltage across the wire [25].

Glatzmaier and Ramirez obtain the effective conductivity and diffusivity from the wire temperature history. To determine the effective conductivity, they use

$$k_e = \frac{q_w}{4\pi} \frac{d(\ln t)}{d\theta} \qquad (32)$$

which is obtained by differentiating Eq. (29). Similarly, the diffusivity is

$$\alpha_e = \frac{r^2 \hat{c}}{4t} \exp\left[\theta \frac{d(\ln t)}{d\theta}\right] \qquad (33)$$

which results from substituting Eq. (32) into Eq. (29). The derivative, $d(\ln t)/d\theta$ is estimated from a least-squares fit of θ versus $\ln t$. In general, Glatzmaier and Ramirez find that the thermal conductivity can be measured with greater accuracy than the diffusivity. Also, they find that measuring the thermal conductivity of gases is an experimental challenge because convection currents develop near the wire.

Gori et al. [26] also determine the effective conductivity from a plot of temperature versus $\ln t$. They employ a thermal conductivity probe that integrates a heater wire and thermocouple junction in a 1.2 mm diameter tube. Additionally, they explore the design of the conductivity probe and its contribution to the measurement uncertainty.

In particular, Gori et al. explore the consequences of finite-conductance material between the wire, thermocouple junction, and surrounding material. Compensating for these effects however, may not be necessary. Specifically, Carslaw and Jaeger [21, p. 344] consider the case of a finite radius wire sheathed in an insulator. They note that although the large-time temperature distribution differs from Eq. (28), both cases asymptotically approach $\theta = q_w/k_e$. Consequently, the hot wire method is virtually insensitive to the properties of the wire and the thermal resistance between the wire and medium.

B. Results And General Trends

The generally accepted method for comparing effective conductivity data is to normalize the solid and effective conductivities by the thermal conductivity of the fluid. These ratios are assigned symbols, κ and η, i.e.,

$$\kappa = \frac{k_s}{k_f} \quad \text{and} \quad \eta = \frac{k_e}{k_f} \tag{34}$$

To illustrate, effective conductivity measurements reported by several investigators are plotted in Fig. 8. One should note that although the solid phases consist of both powders [16,23] and spheres [12,18,19], the data are grouped in a fairly coherent manner. In particular, the normalized effective conductivity, η, clearly depends on the solid–fluid conductivity ratio, κ.

Figure 8, however, also demonstrates that this dependence is complicated. For example, when $1 \leqslant \kappa < 10$, the data could be correlated by an expression such as,

$$\eta = a\kappa^b \tag{35}$$

where a and b are constants. In fact, most of the scatter could be attributed to experimental uncertainty if one were to assume that the uncertainty

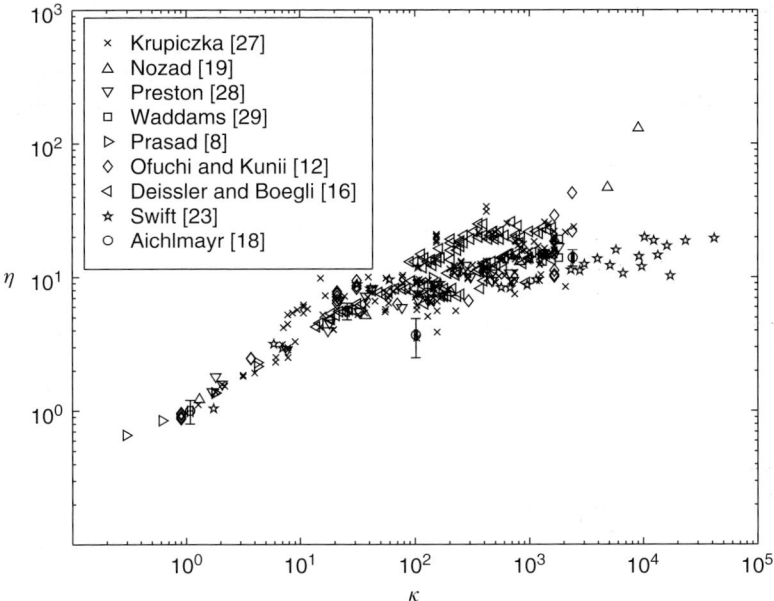

FIG. 8. Effective thermal conductivity measurements reported by several investigators [8,12,16,18,19,23,27–29]. The effective thermal conductivity data, k_e is plotted versus the solid-phase thermal conductivity, k_s; both are normalized by the fluid thermal conductivity, k_f. Although the composition of the media, e.g., shape, packing, and materials, vary considerably, the data are clearly grouped according to the solid–fluid conductivity ratio.

bounds reported by Aichlmayr [18] for the glass–water system ($\kappa \approx 1$) typify measurements within this range. On the other hand, the scatter increases considerably when the solid–fluid conductivity ratio is greater than 10. Moreover, the data appears to bifurcate when the solid–fluid conductivity ratio approaches 1000. This phenomena is most evident when the data of Nozad et al. [19] and Ofuchi and Kunii [12] are compared to that of Swift [23] and Aichlmayr [18].

These observations suggest that the relationship between the effective conductivity and the solid–fluid conductivity ratio is characterized by regimes that in turn, are defined by solid–fluid conductivity ratio intervals. Therefore to elucidate this relationship, the effective conductivity problem is examined within the context of small, intermediate, and large solid–fluid conductivity ratios. These ranges are delineated by $1 \leqslant \kappa < 10$, $10 \leqslant \kappa < 10^3$, and $\kappa \geqslant 10^3$, respectively.

We hypothesize that these solid–fluid conductivity ratio intervals demarcate the relative sensitivity of the effective conductivity to the geometry of the solid–fluid interface. Furthermore, we submit that successful modeling strategies are evidence of this property. For example, if interface geometry is a minor contributor when the solid–fluid conductivity ratio is small, then consideration of this feature is not a prerequisite for small solid–fluid conductivity ratio models to be successful. In contrast, interface geometry must be taken into account at modest conductivity ratios. But this can be accomplished in a variety of ways because the sensitivity is mild. Finally, the effective conductivity is most sensitive to interface geometry when the solid–fluid conductivity ratio is large. Hence, the success of these models is closely related to their representation of the microstructure.

III. Small Solid–Fluid Conductivity Ratios

We first consider the case of small solid–fluid conductivity ratios. Referring to Fig. 8, the region of interest is delimited by $1 \leqslant \kappa < 10$. Examples of solid–fluid systems that fit this description include glass-water and Pyrex-helium; additional examples are presented in Table I.

A. Maxwell's Formulae

Estimating the electrical conductivity, magnetic permeability, or permittivity of composite media is a classical problem in applied physics [30]. For example, Rayleigh [31] analyzes regular arrays of cylindrical and spherical objects surrounded by a fluid. In contrast, Maxwell [32] considers dilute suspensions of particles. He obtains

$$R_e = R_2 \frac{2R_1 + R_2 + p(R_1 - R_2)}{2R_1 + R_2 - 2p(R_1 - R_2)} \tag{36}$$

TABLE I
Systems with Small Solid–Fluid Conductivity Ratios

System	ϕ	κ	η	Source
Glass–water	0.34	0.90	0.90	[12]
Glass–water	0.369	0.90	0.96	[12]
Glass–water	0.376	1.08	1.00	[18]
Glass–water	0.396	1.79	1.36	[8]
Glass–glycol	0.349	4.25	2.16	[8]
Pyrex–hydrogen	0.40	5.88	3.10	[23]
Pyrex–helium	0.40	7.04	2.90	[23]

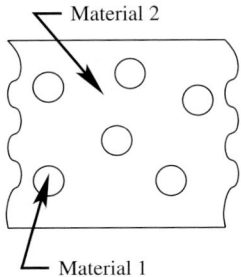

FIG. 9. The composite medium problem considered by Maxwell [32].

for the electrical resistance of a medium comprised of a random distribution of spherical particles in a continuous matrix (Fig. 9). The matrix has resistance R_2, and the spheres have resistance R_1. Of note, this analysis assumes that the spheres do not interact, i.e., the sphere radii are small relative to their separation distances. Thus, Eq. (36) assumes that the volume fraction of the spheres, p, is small. This solution however, can be extended to the case of interacting particles [33,34].

1. Effective Thermal Conductivity Problems

Two analogous effective thermal conductivity problems may be derived from the composite medium problem considered by Maxwell. The first problem consists of solid spherical particles suspended in a continuous fluid [13]. The corresponding effective conductivity is

$$\eta = \frac{2\phi + \kappa(3 - 2\phi)}{3 - \phi + \kappa\phi} \tag{37}$$

which follows from substituting

$$R_e = \frac{1}{k_e}, \quad R_1 = \frac{1}{k_s}, \quad R_2 = \frac{1}{k_f}, \quad \text{and} \quad p = 1 - \phi \tag{38}$$

into Eq. (36). One should note that Eq. (38) and the stipulation that the spheres not interact, limit Eq. (37) to large porosities. The second effective conductivity problem is that of a solid containing a random distribution of spherical fluid-filled voids [13]. In this case, one substitutes

$$R_e = \frac{1}{k_e}, \quad R_1 = \frac{1}{k_f}, \quad R_2 = \frac{1}{k_s}, \quad \text{and} \quad p = \phi \tag{39}$$

into Eq. (36) and obtains

$$\eta = \frac{2\kappa^2(1-\phi) + (1+2\phi)\kappa}{(2+\phi)\kappa + 1 - \phi} \qquad (40)$$

which is restricted to small porosities. Although these problems appear somewhat innocuous, the following discussion will demonstrate that Eqs. (37) and (40) are significant results.

2. *Statistical Approach*

Brown [30] investigates the influence of particle shape on the dielectric constant of a composite medium. Specifically, he considers the following question: Given a material comprised of phases A and B, can the dielectric constant of the composite medium be determined if only the dielectric constants ε_A and ε_B, and the phase volume fractions p and $1-p$, are known? Brown addresses this question by using a statistical description of a sample medium to predict the dielectric constant, ε, of the parent population. The result is

$$\frac{1}{K} = \frac{1}{pK_0} - \lambda K_0 + \cdots \qquad (41)$$

where

$$K = \frac{3(\varepsilon - \varepsilon_B)}{\gamma_c(\varepsilon + 2\varepsilon_B)} \qquad (42)$$

$$K_0 = \frac{3(\varepsilon_A - \varepsilon_B)}{\gamma_c(\varepsilon_A + 2\varepsilon_B)} \qquad (43)$$

λ is a parameter, and γ_c a constant. Brown then shows that λ and the higher-order terms in Eq. (41) depend on statistical properties of the phase geometry. Therefore, he concludes that detailed geometric information is necessary to determine the effective properties of a composite medium.

On the other hand, in the absence of such information, the best estimate of the dielectric constant is obtained by neglecting terms $-\lambda K_0$ and higher in Eq. (41) [30]. This yields

$$\frac{\varepsilon - \varepsilon_B}{\varepsilon + 2\varepsilon_B} = \frac{p(\varepsilon_A - \varepsilon_B)}{\varepsilon_A + 2\varepsilon_B} \qquad (44)$$

which is Maxwell's solution. One should note that Eq. (40) results when

$$\varepsilon = \frac{1}{k_e}, \quad \varepsilon_A = \frac{1}{k_f}, \quad \varepsilon_B = \frac{1}{k_s}, \quad \text{and} \quad p = \phi \tag{45}$$

are substituted into Eq. (44). Similarly, one obtains Eq. (37) by interchanging ε_A and ε_B and letting $p = 1-\phi$ in Eq. (45), and substituting the result into Eq. (44). Consequently, the Maxwell formulae, Eqs. (37) and (40), give the best estimate for the effective conductivity when the geometry of the solid–fluid interface is unknown.

This conclusion is supported by Hashin and Shtrikman [35], who employ variational analysis to estimate the physical limits for the magnetic permeability of a composite medium. Their analysis yields Eqs. (37) and (40) in terms of the magnetic permeability. Hence, they conclude that these expressions are the most stringent bounds for the composite permeability; which is supported by experimental evidence. Therefore, by analogy, Eqs. (37) and (40) are rigorous bounds for the effective thermal conductivity.

B. Mixture Rules

Although less stringent than the Maxwell formulae, mixture rules provide convenient predictions for the physical limits of the effective conductivity. These formulae are based on the volume-fraction-weighted arithmetic, harmonic, and geometric means of the phase conductivities. For example, the effective conductivity based on the arithmetic mean is given by

$$k_e = \phi k_f + (1 - \phi) k_s \tag{46}$$

or

$$\eta = \phi + (1 - \phi)\kappa \tag{47}$$

In addition, Eq. (46) has a useful physical interpretation. It is the effective conductivity of a composite medium comprised of parallel layers (Fig. 10(a)). Alternatively, the formula based on the harmonic mean is

$$\frac{1}{k_e} = \frac{\phi}{k_f} + \frac{1 - \phi}{k_s} \tag{48}$$

or

$$\eta = \frac{\kappa}{\kappa\phi + (1 - \phi)} \tag{49}$$

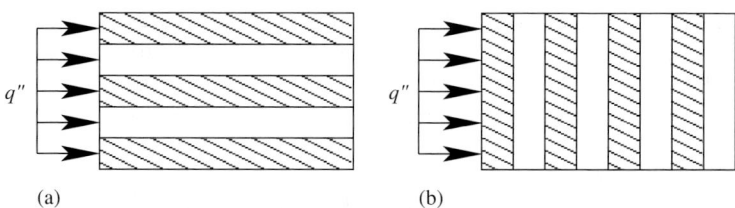

FIG. 10. Physical interpretation of the arithmetic (a) and harmonic (b) mean effective conductivity mixture rules. (a) Solid and fluid layers-in-parallel physical interpretation of Eq. (46). (b) Solid and fluid layers-in-series physical interpretation of Eq. (48).

in normalized form. Physically, the harmonic mean corresponds to the effective conductivity of a composite medium having layers arranged in series (Fig. 10(b)). One should note, however, that the physical interpretations of Eqs. (46) and (48) are somewhat fortuitous.

Lastly, the geometric mean is given by

$$k_e = k_f^\phi k_s^{1-\phi} \tag{50}$$

and

$$\eta = k^{1-\phi} \tag{51}$$

in normalized form. The geometric mean, however, does not have a meaningful physical interpretation. Thus, it is primarily used to correlate effective conductivity data [36,37].

Brown [30] explores the relationship between the mixture models and the statistical properties of phase geometry. To do this, Brown expands Eq. (41) in parameters derived from the mixture models. For example, the expansion of Eq. (41) in terms of the arithmetic mean is given by

$$\frac{\varepsilon}{\varepsilon'} = 1 - \frac{1}{3}pq\left(\frac{\delta'}{\varepsilon'}\right)^2 + \left[-\frac{2}{9}pq(p-q) + \frac{(p^2\lambda - q^2\mu)}{2\gamma_c^2}\right]\left(\frac{\delta'}{\varepsilon'}\right)^3 + \cdots \tag{52}$$

where,

$$\varepsilon' = p\varepsilon_A + q\varepsilon_B \tag{53}$$

$$\delta' = \varepsilon_A - \varepsilon_B \tag{54}$$

$$q = 1 - p \tag{55}$$

and λ and μ are parameters that depend on the statistics of the phase geometry. One should note that the arithmetic mean mixture rule is recovered when all terms in Eq. (52) except for the first, are discarded; hence the mixture rule is a first-order approximation for ε. A geometry-independent second-order approximation for ε is

$$\varepsilon = p\varepsilon_A + (1-p)\varepsilon_B - \frac{p(1-p)(\varepsilon_A^2 - 2\varepsilon_A\varepsilon_B + \varepsilon_B^2)}{3[p\varepsilon_A + (1-p)\varepsilon_B]} \qquad (56)$$

The analogous conductivity expression is

$$\eta = \phi + (1-\phi)\kappa - \frac{\phi(1-\phi)(1 - 2\kappa + \kappa^2)}{3[\phi + (1-\phi)\kappa]} \qquad (57)$$

Similarly, the second-order conductivity equations based on the harmonic and geometric means are given by

$$\eta = \frac{1}{\phi + (1-\phi)/\kappa} + \frac{2\phi(1-\phi)(1 - 1/\kappa)}{3[\phi + (1-\phi)/\kappa]} \qquad (58)$$

and

$$\eta = \exp\left[(1-\phi)\ln \kappa + \frac{\phi(1-\phi)}{6}(\ln \kappa)^2\right] \qquad (59)$$

Brown notes that the quotient δ'/ε' appearing in Eq. (52) must be small for Eqs. (57)–(59) to be valid. This implies that

$$\kappa < \frac{1+\phi}{\phi} \qquad (60)$$

is a criteria for mixture rule accuracy; it is plotted in Fig. 11.

C. Physical Bounds for the Effective Conductivity

The distinguishing feature of small solid–fluid conductivity ratio models is that they apply irrespective of the arrangement of the phases or the geometry of the solid–fluid interface. Consequently, these expressions define physical bounds for the effective conductivity; several are plotted in Fig. 12. In general, the expressions describing the first-order arithmetic and harmonic means (Eqs. (47) and (49)) yield the least-restrictive bounds for the effective conductivity. The second-order approximations (Eqs. (57) and (58)), on the other hand, are more restrictive and approach the Maxwell limits (Eqs. (37) and (40)). Interestingly, the second-order arithmetic mean (Eq. (57)) and the second Maxwell limit (Eq. (40)) virtually coincide.

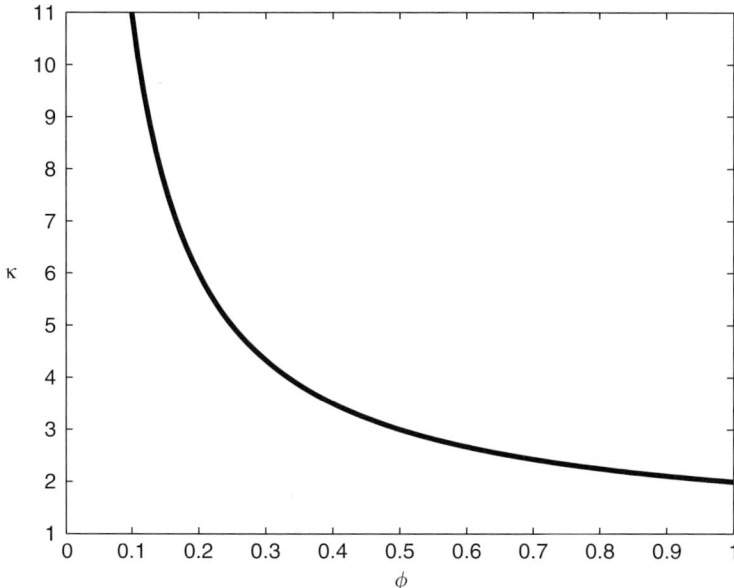

Fig. 11. The maximum solid–fluid conductivity ratio for which a mixture rule can accurately predict the effective conductivity of a saturated porous medium (Eq. (60)).

Similarly, the geometric mean (Eq. (51)) coincides with the first Maxwell limit (Eq. (37)) when $\kappa < 4$.

D. Closing Comments on Small Conductivity Ratio Media

Small solid–fluid conductivity ratios are a special case of the effective conductivity problem. Figure 12 demonstrates that the geometry-independent effective conductivity models presented in this section (Table II) are applicable when $\kappa < 10$. In fact, if one were to assume that the uncertainty limits reported by Aichlmayr [18] are typical, any of the plotted expressions give reasonable estimates for the effective conductivity when $\kappa < 4$. This suggests that the effective conductivity is insensitive to interfacial geometry when the solid–fluid conductivity ratio is small. Thus, it mostly depends on gross properties of the medium, e.g., volume fractions and conductivities of the constituent phases, rather than morphology.

IV. Intermediate Solid–Fluid Conductivity Ratios

We next consider the intermediate solid–fluid conductivity ratio case. Referring to Fig. 13, this category is approximately delimited by 10

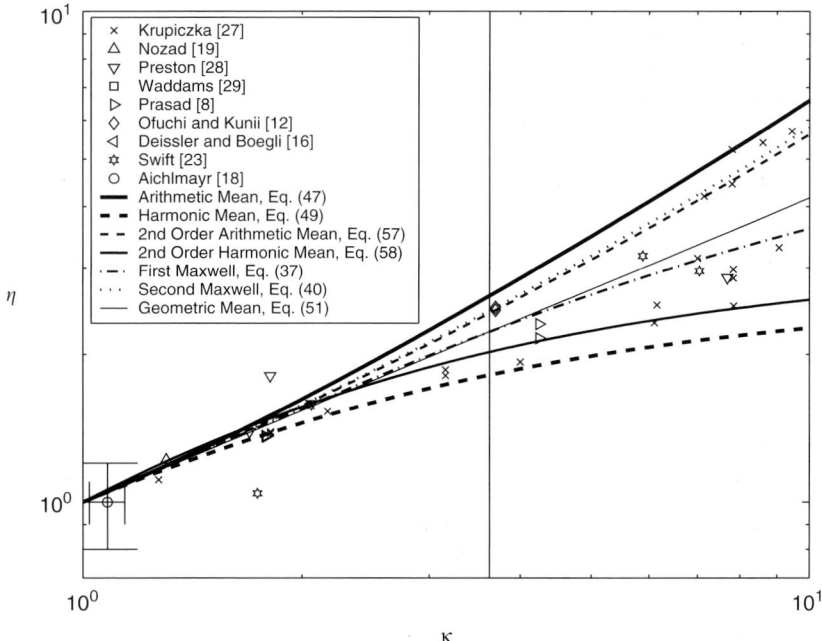

FIG. 12. Estimates for the effective conductivity of small solid–fluid conductivity ratio media. A porosity of 0.38 is used to perform the computations. The vertical line denotes the mixture rule limit ($\kappa = 3.63$) for this porosity. This limit is defined by Eq. (60).

$\leqslant \kappa < 10^3$ and it includes solid–fluid systems such as, glass–air, steel–water, and magnesium–hydrogen (Table III). The justification for these limits is provided by the geometry-independent Maxwell formulae. That is, these expressions bound the data, but they clearly fail to predict the effective conductivity when $\kappa > 10$. This failure is a consequence of neglecting the geometry of the solid–fluid interface.

A. CONDUCTION MODELS

Capturing the contribution of the interface geometry to the effective conductivity necessitates modeling heat conduction in the microstructure. Unfortunately, this requires a detailed description of the medium morphology, which in general, is not available.

The problem becomes manageable however when the microstructure is approximated by a regular arrangement of objects. For example, Fig. 14(a) depicts an array of spheres or a bundle of rods. The key feature of the array is that when the overall heat flow is unidirectional, isotherms are collinear

TABLE II
SUMMARY OF EFFECTIVE CONDUCTIVITY MODELS APPLICABLE TO SMALL SOLID–FLUID CONDUCTIVITY RATIO MEDIA

Name	Model description	Result	Reference
Lower Maxwell limit	Spherical particles suspended in a continuous matrix	Eq. (37)	[13]
Upper Maxwell limit	Solid containing a random distribution of spherical fluid-filled voids	Eq. (40)	[13]
Parallel layers	Arithmetic mean mixture rule	Eq. (47)	Unknown
Second-order Arithmetic mean	Second-order arithmetic mean mixture rule	Eq. (57)	[30]
Series layers	Harmonic mean mixture rule	Eq. (49)	Unknown
Second-order Harmonic mean	Second-order harmonic mean mixture rule	Eq. (58)	[30]
Geometric mean	Geometric mean mixture rule	Eq. (51)	[36]
Second-order geometric mean	Second-order geometric mean mixture rule	Eq. (59)	[30]

with the horizontal centerlines of the objects and the temperature distribution is symmetric about vertical centerlines. Consequently, the array and a unit cell having horizontal isotherms and vertical adiabats (Fig. 14(b)) are identical heat conduction problems; thus a difficult problem is replaced by a simpler one.

For example, Schumann and Voss [38] employ this technique to estimate the effective conductivity of a packed bed of spheres. They approximate the bed with a planar unit cell that resembles Fig. 14(b). The solid–fluid interface is represented by the hyperbola

$$xy = \hat{p}(\hat{p} + 1) \qquad (61)$$

where $\hat{p}$ is a shape parameter. The shape parameter is related to the porosity by

$$\phi = \hat{p}(\hat{p} + 1)\ln\left(\frac{1+\hat{p}}{\hat{p}}\right) - \hat{p} \qquad (62)$$

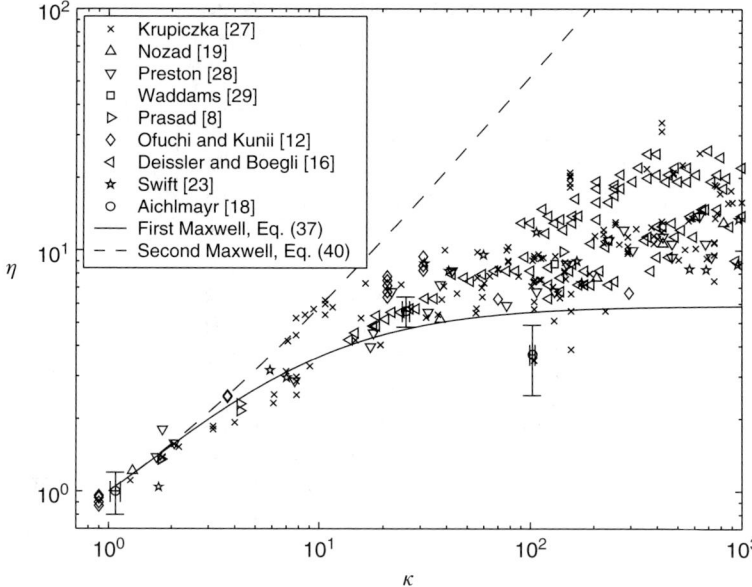

FIG. 13. Measurements and physical bounds for the effective conductivity of small and intermediate solid–fluid conductivity ratio media. A porosity of 0.38 is assumed when plotting Eqs. (37) and (40).

TABLE III

SYSTEMS WITH INTERMEDIATE SOLID–FLUID CONDUCTIVITY RATIOS

System	ϕ	κ	η	Source
Glass–air	0.34	21	6.73	[12]
Glass–air	0.465	21	9.05	[12]
Glass–air	0.376	25.7	5.6	[18]
Pyrex–argon	0.40	59.8	9.5	[23]
Steel–water	0.403	70	6.28	[12]
Pyrex–freon 12	0.40	106	11.9	[23]
Steel–glycol	0.416	143	9.9	[8]
Uranium–helium	0.40	166	9.0	[23]
Steel–helium	0.413	294	7.09	[12]
Magnesium–hydrogen	0.40	955	8.7	[23]

Schumann and Voss estimate the effective conductivity of the unit cell by dividing it into vertical slices of infinitesimal width and assuming unidirectional heat flow through the slice. Consequently, they compute the thermal resistance of the slice, integrate the result in the horizontal direction, and

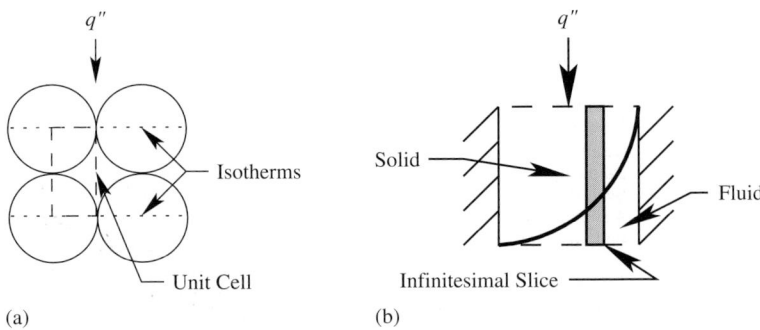

FIG. 14. Conduction model representation of a porous medium microstructure. The overall heat flow is unidirectional. Consequently the centerline of each horizontal layer is an isotherm and iso-heat flux lines are vertical. (a) An array of spherical particles or cylindrical rods of infinite axial extent. (b) A half-unit cell with isothermal horizontal boundaries and adiabatic vertical boundaries.

apply pre-determined boundary conditions; this gives

$$\eta = \phi^3 + \frac{\kappa(1-\phi^3)}{1+\hat{p}(1-\kappa)}\left[1 + \frac{\hat{p}(1+\hat{p})(1-\kappa)}{1+\hat{p}(1-\kappa)}\right]\ln\left(\frac{1+\hat{p}}{\hat{p}\kappa}\right) \quad (63)$$

Wilhelm et al. [39] argue that Eq. (63) does not account for conduction through particle contact points. Consequently, they propose an empirical correction factor, but it has limited applicability [27].

Similarly, Deissler and Eian [15] develop two unidirectional conduction models to correlate effective conductivity measurements of magnesium oxide powder saturated with air, argon, and helium. They first consider a square array of infinite cylinders. The unit cell resembles Fig. 14(b) and the effective conductivity is obtained by applying a procedure similar to that of Schumann and Voss [38]; the result is

$$\eta = \frac{\kappa}{\kappa-1}\left[\frac{2\kappa}{\sqrt{2\kappa-1}}\arctan(2\kappa-1) - \frac{\pi}{2}\right] \quad (64)$$

One should note that the shape of the solid–fluid interface is fixed. Consequently, $\phi = 0.214$ is tacitly assumed. Their second model consists of a cubic array of spheres. The unit cell is similar to Fig. 14(b) and the effective conductivity is given by

$$\eta = \frac{\pi}{2}\left(\frac{\kappa}{\kappa-1}\right)^2\left(\ln\kappa - 1 + \frac{1}{\kappa}\right) + 1 - \frac{\pi}{4} \quad (65)$$

This model also assumes a fixed solid–fluid interface; thus it applies when $\phi = 0.475$. Similar strategies are employed by Gemant [40], Woodside [41], and Swift [23] to derive unidirectional heat flow models for the effective conductivity of soil, snow, and powder, respectively.

Deissler and Eian find that the unidirectional sphere model (Eq. (65)) correlates their experimental data fairly well. This is somewhat unexpected because unidirectional models neglect the "bending" of heat flux lines, i.e., transverse conduction. To explore this effect, they determine the effective conductivity of the infinite-cylinder unit cell by numerically solving the two-dimensional conduction equation. They consider several solid–fluid conductivity ratios and generate a curve. When comparing this curve to experimental data, however, they find that the correspondence is worse than the unidirectional models.

Deissler and Boegli [16] extend the work of Deissler and Eian [15] by considering a three-dimensional cubic array of spheres (Fig. 14(a)). The unit cell consists of an eighth-sphere located in a cylinder. The unit cell is axisymmetric and it resembles Fig. 14(b). Also, $\phi = 0.475$ because the geometry of the solid–fluid interface is fixed. The conduction equation is numerically solved to obtain a curve of η versus κ. This curve is found to correlate experimental data well. Wakao and Kato [42] and Wakao and Vortmeyer [43] use similar techniques to investigate the effects of thermal radiation and gas pressure on the effective conductivity of gas-saturated packed beds.

Krupiczka [27] builds on the work of Deissler and Eian [15] and Deissler and Boegli [16] by obtaining analytical solutions to their respective multidimensional conduction models. Krupiczka first considers a bundle of semi-infinite cylinders. The two-dimensional conduction equation is solved and the result is a non-orthogonal infinite series. Krupiczka next considers a cubic array of spheres (Fig. 14(a)). Rather than solve the conduction equation in three dimensions, Krupiczka approximates the radial temperature distribution of the sphere with the temperature distribution of a semi-infinite cylinder. This approximation is shown to introduce minimal error.

Although Krupiczka demonstrates that the array of spheres solution is consistent with the curve generated by Deissler and Boegli [16], using the solution is impractical because the series converges slowly and the terms are difficult to evaluate. Consequently, Krupiczka proposes a correlation to represent the solution and literature data. The correlation is given by

$$\eta = \kappa^{[A^\circ + B^\circ \log_{10} \kappa]} \qquad (66)$$

where

$$A^\circ = 0.280 - 0.757 \log_{10} \phi \qquad (67)$$

and

$$B° = -0.057 \qquad (68)$$

This correlation applies for $0.215 \leqslant \phi \leqslant 0.476$.

Several conduction models are plotted in Fig. 15 and compared to effective conductivity measurements of magnesium oxide, stainless-steel, and uranium oxide powders saturated with air, helium, argon, nitrogen, and neon [16]. The three-dimensional sphere model of Deissler and Boegli [16] corresponds well to experimental data. In fact, the correspondence is remarkable when compared to the unidirectional cylinder (Eq. (64)) and the Schumann and Voss [38] models (Eq. (63)). On the other hand, the unidirectional sphere model (Eq. (65)) and the Krupiczka [27] correlation (Eq. (66)) yield fair estimates for the effective conductivity. The relative success of the sphere models versus the failure of the cylinder model suggests that although transverse conduction is noticeable, it is a minor contributor to the effective conductivity.

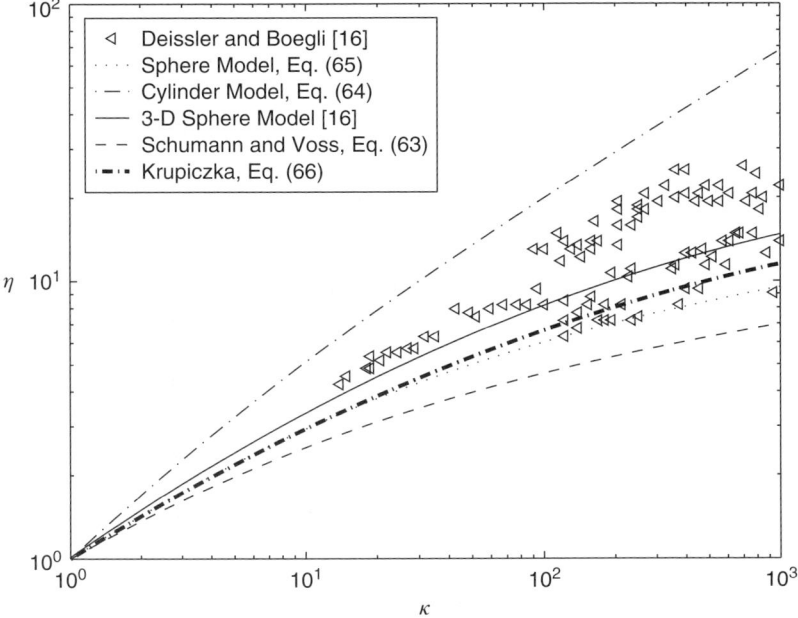

FIG. 15. A comparison of experimental data by Deissler and Boegli [16] and various conduction models. A porosity of 0.475 is assumed when evaluating the Schumann and Voss [38] expression (Eq. (63)) and the Krupiczka [27] correlation (Eq. (66)). The sphere models assume a porosity of 0.475 and the unidirectional cylinder model (Eq. (64)) assumes a porosity of 0.215.

B. Lumped Parameter Models

An alternative strategy to solving the conduction equation in a unit cell is to lump heat flow paths into thermal resistances acting in series and parallel. For example, Kunii and Smith [44] develop a lumped parameter model for heat transfer in packed beds of unconsolidated materials. They consider a unit cell comprised of two spherical particles (Fig. 16(a)). The overall heat flow is assumed to be unidirectional and they assume that heat transfer through the unit cell can be represented by two mechanisms in parallel: (1) conduction and radiation through the void space and (2) heat transfer through the solid phase. The latter mechanism consists of four processes: (2a) conduction through the contact surface; (2b) conduction through a fluid layer in the vicinity of the contact point; (2c) radiation between particle surfaces; and (2d) conduction through the solid. They further assume that mechanisms (2a), (2b), and (2c) are in parallel and that mechanism (2d) is in series with the combination of (2a), (2b), and (2c). The overall mechanism is illustrated in Fig. 16(b).

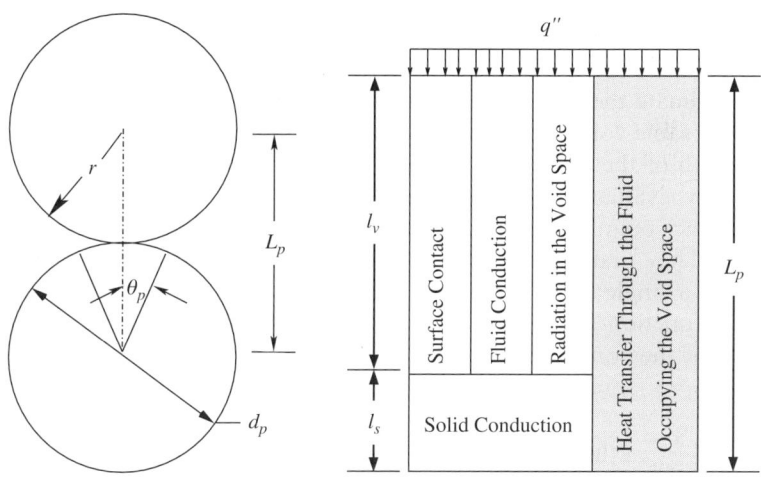

(a) Actual configuration of spherical particles [44] (Adapted with permission of the American Institute of Chemical Engineers).

(b) One-dimensional composite model used to estimate the effective conductivity.

FIG. 16. The Kunii and Smith [44] model for heat transfer near contact points of particles. The length l_v is the thickness of a slab of fluid having a thermal resistance equal to the fluid near the contact point, l_s is the thickness of a slab having a thermal resistance equal to a sphere, and L_p is the effective length between spherical particles. Note that because l_v, l_s, and L_p are effective thicknesses, $l_v + l_s \neq L_p$, contrary to what is indicated in Fig. 16(b).

To estimate the effective conductivity, Kunii and Smith assume that the aforementioned heat transfer mechanisms can be represented by equivalent thermal resistances. The effective lengths l_v, l_s, and L_p appearing in Fig. 16(b) are the thickness of a slab of fluid having a thermal resistance equal to the fluid near the contact point, the thickness of a slab having a thermal resistance equal to a sphere, and the effective length between spherical particles, respectively. One should note that $l_v + l_s \neq L_p$, because these are not physical lengths. Kunii and Smith next argue that radiation in the void space can be neglected. Consequently, the effective conductivity of the composite system is,

$$\eta = \phi + \frac{\beta(1-\phi)}{\Psi + \gamma/\kappa} \qquad (69)$$

where β, γ, and Ψ are dimensionless geometric parameters defined by L_p/d_p, l_s/d_p, and l_v/d_p respectively.

Kunii and Smith next estimate the geometric parameters β and γ. They first consider various packing arrangements to estimate the particle separation distance, β. They find that $\beta = 0.895$ for close packing and they argue that $\beta = 1.0$ for open packing. Consequently, they conclude that $0.9 \leqslant \beta \leqslant 1.0$ for any packed bed. Next, they consider a cylinder and a sphere of diameter d_p to estimate the effective length of a particle. Both objects are assumed to have the same volume, and the effective length of the particle is defined to be the length of the cylinder. This assumption yields $\gamma = 2/3$.

Finally, evaluation of the effective length of the fluid film in the vicinity of the contact point, l_v necessitates an analysis of conduction through contact points. They first assume that a fraction of the total heat transfer occurs through a single contact point. Consequently, the dimensionless parameter Ψ, is a function of the number of contact points and the thermal conductivities of the constituent phases. This function is

$$\Psi = \frac{[(\kappa-1)/\kappa]^2 \sin^2 \theta_p}{2\{\ln[\kappa - (\kappa-1)\cos\theta_p] - [(\kappa-1)/\kappa](1-\cos\theta_p)\}} - \frac{2}{3\kappa} \qquad (70)$$

where the contact angle, θ_p depends on the number of contact points in a particular packing arrangement, n. This relation is given by

$$\sin^2 \theta_p = \frac{1}{n} \qquad (71)$$

One should note that evaluating Eq. (70) requires knowledge of the packing arrangement, which in general, is not available. Consequently,

Kunii and Smith argue that Eq. (70) can be approximated by Ψ evaluated at the loose- and close-packing conditions. That is,

$$\Psi = \Psi_2 + (\Psi_1 - \Psi_2)\frac{\phi - 0.260}{0.216} \qquad (72)$$

where Ψ_1 and Ψ_2 indicate that Eq. (70) is evaluated with $n = 1.5$ and $4\sqrt{3}$, respectively. This approximation is valid for $0.260 \leqslant \phi \leqslant 0.470$.

Zehner and Schlünder [45] estimate the effective conductivity of packed beds by combining conduction and lumped parameter modeling techniques. Their axisymmetric unit cell consists of an eighth sphere in a cylindrical volume (Fig. 17). The inner cylinder, $0 \leqslant r \leqslant 1$ contains both solid and fluid phases. The annular region $1 < r \leqslant R$, however, contains only fluid.

The effective conductivity is found by assuming unidirectional heat flow throughout the unit cell. In addition, the inner cylinder and the outer annulus are assumed to be parallel thermal resistances. Hence, the effective conductivity of the unit cell is,

$$k_e = \left(1 - \frac{1}{R^2}\right)k_f + \frac{1}{R^2}k_{sf} \qquad (73)$$

where k_{sf} is the effective conductivity of the inner cylinder. Zehner and Schlünder use this expression and an analogy to mass transfer in packed beds to infer that

$$\frac{1}{R^2} = \sqrt{1 - \phi} \qquad (74)$$

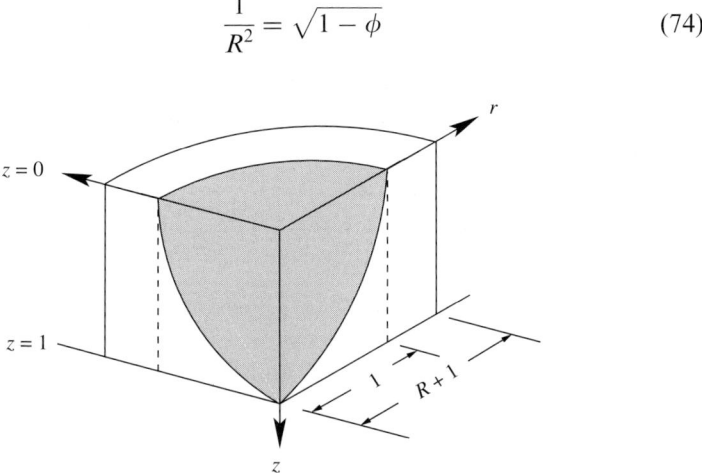

FIG. 17. The Zehner and Schlünder [45] unit cell with variable particle shape. (Adapted from Ref. [45], Copyright (1970) with permission from Wiley-VCH.)

To determine k_{sf}, the inner cylinder is first divided into annular slices similar to Fig. 14(b). Next, the solid and fluid phases are assumed to be thermal resistances in series. The combined resistance depends on the solid–fluid interface. The equation of this interface is

$$r^2 + \frac{z^2}{[B-(B-1)z]^2} = 1 \qquad (75)$$

where B is a shape parameter defined by

$$\phi = 1 - \left[\frac{B(3-4B+B^2+2\ln B)}{(B-1)^3}\right]^2 \qquad (76)$$

One should note that $B=0$, $B=1$, and $B\to\infty$ yield $r=0$, $r^2+z^2=1$, and $r^2=1$, which correspond to the z-axis, a sphere, and a cylinder, respectively. Next, the effective conductivity of the slice is determined and the result is integrated in the radial direction to give

$$\frac{k_{sf}}{k_f} = \frac{2\kappa}{\kappa-B}\left[\frac{\kappa(\kappa-1)B}{(\kappa-B)^2}\ln\left(\frac{\kappa}{B}\right) - \frac{B+1}{2} - \frac{\kappa(B-1)}{\kappa-B}\right] \qquad (77)$$

Lastly, Eqs. (77) and (74) are substituted into Eq. (73) to yield

$$\eta = 1 - \sqrt{1-\phi} + \frac{2\kappa\sqrt{1-\phi}}{\kappa-B}\left[\frac{\kappa(\kappa-1)B}{(\kappa-B)^2}\ln\left(\frac{\kappa}{B}\right) - \frac{B+1}{2} - \frac{\kappa(B-1)}{\kappa-B}\right] \qquad (78)$$

for the effective conductivity of the unit cell. Zehner and Schlünder report that $\phi=0.42$ yields the best correspondence to experimental data. In addition, they propose the correlation

$$B = C\left(\frac{1-\phi}{\phi}\right)^{m_c} \qquad (79)$$

where $C=1.25$ and $m_c=10/9$ to approximate Eq. (76). Hsu et al. [46], however, find that $C=1.364$ and $m_c=1.055$ is more accurate.

Prasad et al. [8] compare Kunii and Smith [44], Krupiczka [27], Zehner and Schlünder [45], and the series and parallel models (Eqs. (49) and (47)) to effective conductivity measurements of solid–fluid systems comprised of glass, steel, and acrylic spheres saturated with water and glycol. These solid–fluid combinations yield $0.349 \leqslant \phi \leqslant 0.427$ and $0.254 \leqslant \kappa \leqslant 143$. Their data is plotted in Fig. 18 and compared to these models. In general, Prasad

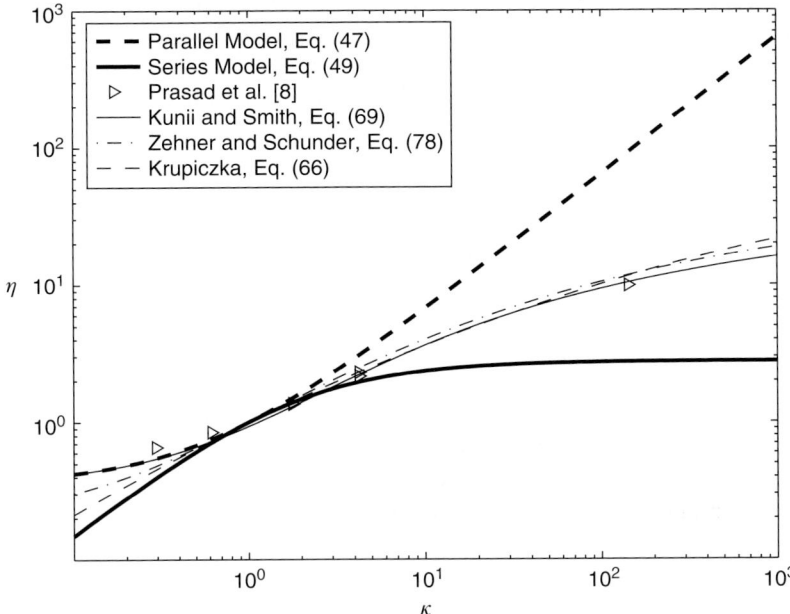

FIG. 18. A comparison of experimental data from Prasad *et al.* [8] and Kunii and Smith [44], Krupiczka [27], Zehner and Schlünder [45], and the series and parallel models (Eqs. (49) and (47)) with $\phi = 0.36$ assumed.

et al. find that the correlation of Krupiczka and the models of Zehner and Schlünder and Kunii and Smith adequately predict the effective conductivity when $\kappa > 1$.

Nield [37] however, questions the accuracy of this work. He notes that some of the measured effective conductivities of Prasad *et al.* do not fall within established physical bounds. He also argues that the model of Kunii and Smith is slightly more accurate than those of Krupiczka and Zehner and Schlünder. Nield recommends using the geometric mean (Eq. (50)) when $\kappa \sim 1$.

C. RESISTOR NETWORK MODELS

Resistor network models are an approximate method for capturing the influence of the interfacial geometry on the effective conductivity. In general, these models represent a solid–fluid medium by regions having geometries amenable to a one-dimensional analysis, e.g., the layers-in-series (Fig. 10(b)) and the layers-in-parallel (Fig. 10(a)) models. The dimensions and thermal resistances of these regions depend on the porosity and conductivities of the phases. The effective conductivity is subsequently determined by an electrical network analysis.

Crane and Vachon [47] provide physical insight into resistor network models. They argue that the layers-in-series and the layers-in-parallel models represent limiting cases for the lateral thermal resistance. For instance, the maximum network resistance occurs when the lateral resistance is infinite. On the other hand, the minimum network resistance occurs when the lateral resistance is zero. Physically, these cases correspond to uniform heat flux lines parallel to, and isotherms normal to, the overall direction of heat flow.

To illustrate, consider the electrical network representation of a porous medium depicted in Fig. 19. If the lateral resistance, R_L is assumed to be infinite, then the horizontal currents, e.g., I_{1-3} and I_{3-2}, will be equal. Consequently the network reduces to a circuit comprised of six resistors in three parallel branches (Fig. 20(a)). On the other hand, zero lateral resistance causes the intermediate temperatures T_3, T_4, and T_5 to be equal. Consequently, the network reduces to the circuit depicted in Fig. 20(b), which is essentially two resistors in series. One should recognize that the uniform heat flux circuit yields the maximum overall resistance. This result can be verified by assuming that R_f and R_s are equal and comparing overall

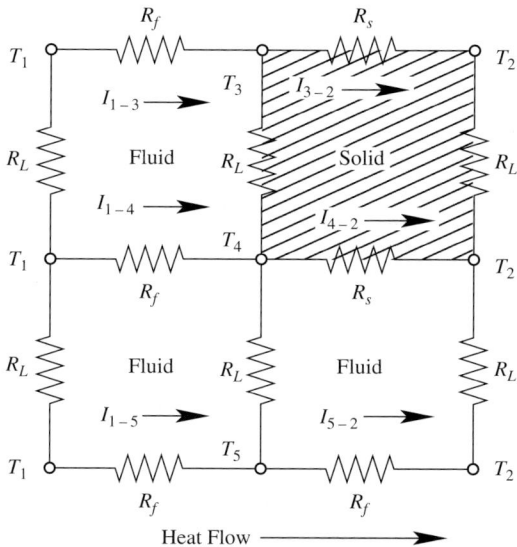

FIG. 19. Electrical network representation of one-dimensional heat conduction in a porous medium. The uniform heat flux and parallel isotherm limits result from assuming that the lateral resistances, R_L are infinite and zero, respectively.

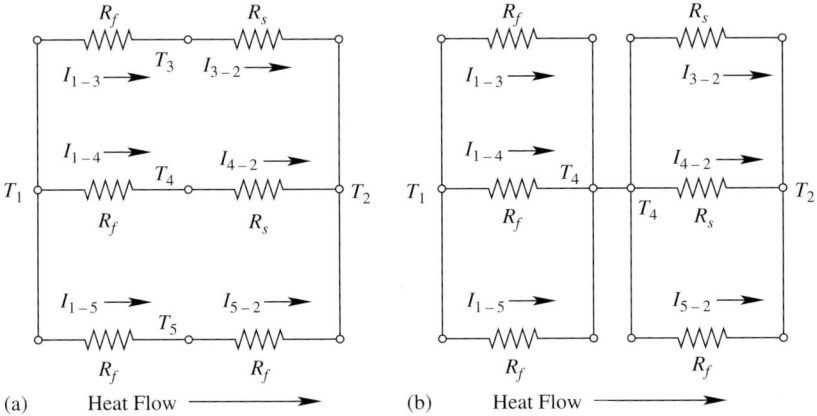

FIG. 20. Electrical circuit representations for the limiting cases of one-dimensional heat conduction in a porous medium. One should note that if R_f and R_s are equal, the overall resistance of the uniform heat flux circuit (a) is twice that of the parallel isotherm circuit (b). (a) Uniform heat flux and maximum overall resistance (R_L is infinite). (b) Parallel isotherm and minimum overall resistance (R_L is zero).

resistances. If this were the case, one would find that the effective resistance of the uniform heat flux circuit is twice that of the parallel isotherm circuit. Hence the uniform heat flux case corresponds to the lower limit for the effective conductivity.

Crane and Vachon [47] employ a resistor network model to predict the physical bounds for the effective conductivity. Their method entails taking an arbitrary cube of solid–fluid medium, splitting the cube into parallelepipeds consisting of solid and fluid phases in series, and rearranging the phases to obtain a phase-volume-fraction distribution. The effective conductivity is subsequently determined by the orientation of the parallelepipeds relative to the heat flux vector, e.g., one direction yields parallel resistances, and the phase-volume-fraction distribution. The problem then reduces to correlating the porosity to this distribution. Tsao [48] determines this function experimentally, Cheng and Vachon [49] assume a distribution function, and Crane and Vachon develop a stochastic model to estimate this function. The results of the model are plotted in Fig. 21.

Jaguaribe and Beasley [50] employ a resistor network model to predict the effective conductivity of porous media having arbitrary phase distributions. The basic strategy is to assume an irregular geometry for a medium and then transform it into a simpler geometry, i.e., one that can be treated with a one-dimensional analysis. Their prototype porous medium consists of a bundle of infinite cylinders. Hence, radial conduction is assumed. Following transformation to a one-dimensional system, the resulting expressions for the

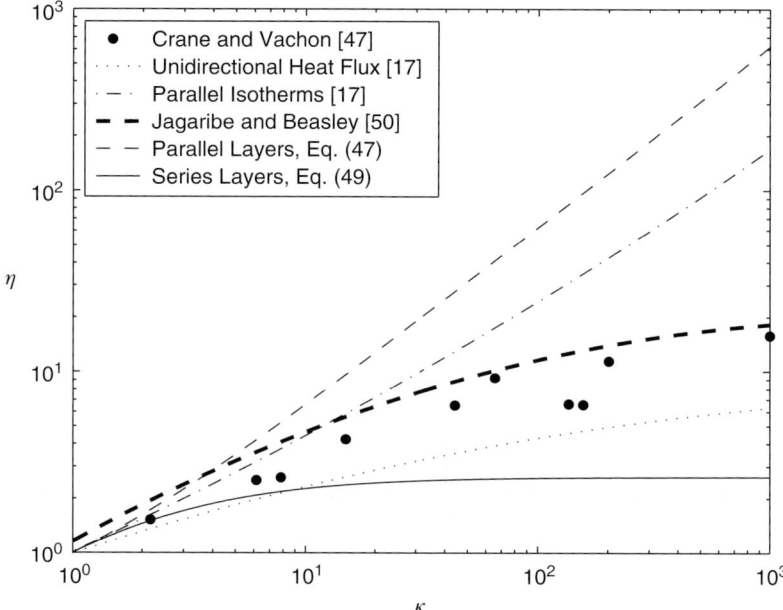

FIG. 21. The predicted bounds for the effective conductivity determined by Crane and Vachon [47]. A porosity of 0.38 is assumed to compute Eqs. (47) and (49). Also, the model of Jaguaribe and Beasley [50] is plotted for comparison.

effective conductivity are evaluated numerically. Their results are plotted in Fig. 21.

D. Closing Comments on Moderate Conductivity Ratios

The evidence presented in this section demonstrates that when the solid–fluid conductivity ratio is moderate, the effective conductivity is sensitive to the geometry of the solid–fluid interface. This effect however is comparatively mild because a variety of modeling techniques, e.g., unidirectional heat flow, lumped parameter, and resistor network, are able to capture it. The three-dimensional conduction and the lumped parameter models appear to capture this dependence best. On the other hand, the failure of the unidirectional cylinder models suggest that transverse conduction is less important than conduction in the direction of the overall heat flux. Although, transverse conduction is appreciable because the unidirectional models consistently under-predict the effective conductivity.

An advantage of the relative insensitivity of the effective conductivity to microstructure morphology is that the practitioner has a variety of models

TABLE IV
Summary of Effective Conductivity Models Applicable to Intermediate Solid–Fluid Conductivity Ratio Media

Name	Model description	Result	Reference
Schumann and Voss	Unidirectional heat flow with hyperbolic solid–fluid interface	Eq. (63)	[38]
Two-dimensional cylinders	Unidirectional heat flow through infinite cylinders	Eq. (64)	[15]
Two-dimensional spheres	Unidirectional heat flow through cubic array of spheres	Eq. (65)	[15]
Three-dimensional spheres	Three-dimensional heat flow through cubic array of spheres	Fig. 15	[16]
Krupiczka correlation	Correlation derived from two-dimensional analytical conduction solutions	Eq. (66)	[27]
Kunii and Smith model	Lumped parameter conduction model	Eq. (69)	[44]
Zehner and Schlünder model	Lumped parameter conduction model with variable particle shape	Eq. (78)	[45]
Crane and Vachon model	Resistance network model based on stochastic porosity distribution model	Fig. 21	[47]
Jaguaribe and Beasley model	Resistance network conductivity model	Fig. 21	[50]

from which to choose (Table IV). Most models predict the effective conductivity of moderate solid–fluid conductivity ratio media with reasonable accuracy. The Kunii and Smith [44] and Zehner and Schlünder [45] models offer the best combination of utility and accuracy; the Krupiczka [27] correlation is also acceptable. The conduction model of Deissler and Boegli [16] is more accurate, but less useful. The resistor network models of Crane and Vachon [47] and Jaguaribe and Beasley [50], however, appear to offer the least utility.

V. Large Solid–Fluid Conductivity Ratios

Lastly, we consider the case of large solid–fluid conductivity ratios. This category includes copper, aluminum, and uranium particles saturated with

TABLE V
Systems with Large Solid–Fluid Conductivity Ratios

System	ϕ	κ	η	Source
Uranium–argon	0.40	1.41×10^3	14.6	[23]
Steel–helium	0.342	1.65×10^3	15.0	[12]
Steel–carbon dioxide	0.413	2.37×10^3	12.5	[12]
Steel–carbon dioxide	0.352	2.37×10^3	42.7	[12]
Copper–helium	0.40	2.77×10^3	11.1	[23]
Bronze–air	0.39	4.34×10^3	45.7	[19]
Aluminum–air	0.41	8.08×10^3	145.0	[19]
Aluminum–nitrogen	0.40	9.20×10^3	14.3	[23]
Aluminum–argon	0.40	1.34×10^3	14.6	[23]
Copper–freon 12	0.40	4.18×10^3	19.6	[23]

helium, air, and argon (Table V). It is demarcated by $\kappa > 10^3$ and it is readily identified in Fig. 22 by the failure of the Kunii and Smith [44] and Zehner and Schlünder [45] models. The inability of these models to capture the bifurcation of effective conductivity data is a result of neglecting the details of particle contact.

A. The Method of Volume Averaging

The method of volume averaging [4] is a formal procedure for deriving spatially smoothed equations that govern transport phenomena in porous media. The objective of the method is to replace the intractable pore-scale problem with one that is valid on average throughout the macro-scale. This is accomplished by devising a continuum of averaged variables, e.g., concentration or temperature, and capturing pore-scale details in effective transport properties. The representative elementary volume (REV) (Fig. 23) is the basis for this continuum and it provides the morphological information to determine effective transport properties.

1. Development of the Volume-Averaged Conduction Equation

Nozad *et al.* [51] use the method of volume averaging to estimate the effective conductivity of two-phase porous media; a review of their method follows. First, they define the microscopic problem. Referring to Fig. 23, heat conduction in the REV is governed by

$$(\rho c_p)_f \frac{\partial T_f}{\partial t} = \nabla \cdot (k_f \nabla T_f) \tag{80}$$

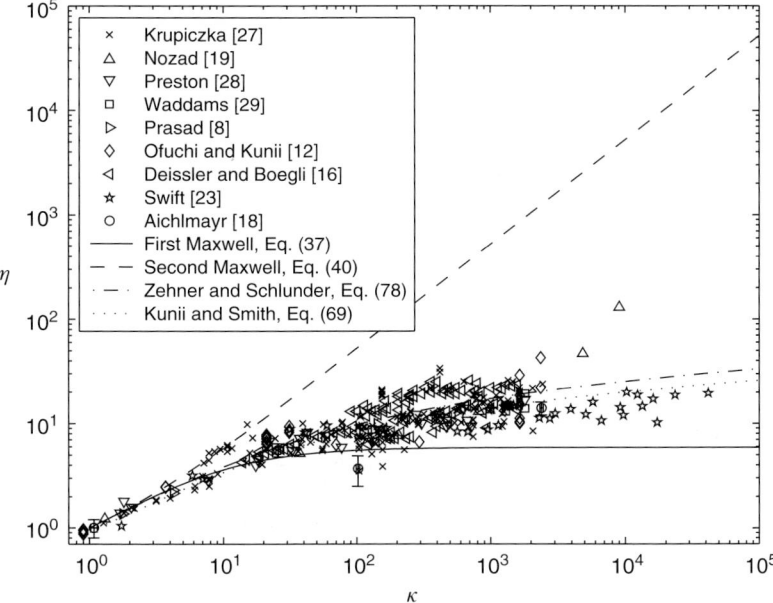

Fig. 22. Measurements and physical bounds for the effective conductivity of small, intermediate, and large solid–fluid conductivity ratio media. A porosity of 0.38 is assumed when plotting Eqs. (37), (40), (69), and (78).

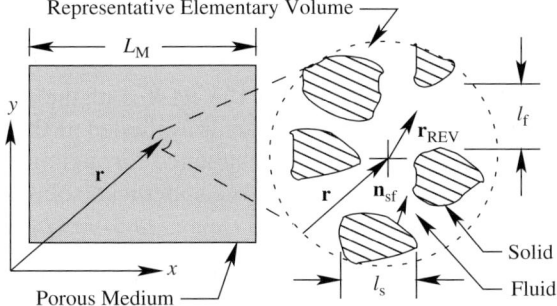

Fig. 23. The representative elementary volume (REV) and its relationship to the macroscopic medium. The vector $\mathbf{r}$, locates the centroid of the REV and indicates the location at which volume-averaged quantities are evaluated. The vector $\mathbf{r}_{REV}$, locates a point within the REV and $|\mathbf{r}_{REV}| = r_0$, is the boundary of the REV. The dimension L_M is a characteristic length of the macro-scale. Similarly, l_s and l_f are characteristic lengths of the solid and fluid phases. Broadly, the REV must satisfy $r_0 \gg l_s$, $r_0 \gg l_f$, and $r_0 \ll L_M$ [4].

and

$$(\rho c_p)_s \frac{\partial T_s}{\partial t} = \nabla \cdot (k_s \nabla T_s) \tag{81}$$

One should note that the fluid and solid temperatures T_f and T_s, vary according to position within the REV. Additionally, the boundary conditions at the solid–fluid interface are

$$T_f = T_s \tag{82}$$

and

$$\mathbf{n}_{fs} \cdot k_f \nabla T_f = \mathbf{n}_{fs} \cdot k_s \nabla T_s \tag{83}$$

where $\mathbf{n}_{fs}$ is the outward-directed surface normal. One should note that $\mathbf{n}_{fs} = -\mathbf{n}_{sf}$ (Fig. 23). Also, the initial conditions

$$T_f = g(t) \tag{84}$$

and

$$T_s = h(t) \tag{85}$$

apply at the bounding surface of the REV.

Next, Nozad *et al.* apply the averaging process to the variables and governing equations. For example, volume averaging the fluid temperature yields the superficial fluid temperature, i.e.,

$$\langle T_f \rangle|_\mathbf{r} = \frac{1}{V_{REV}} \int_{V_f(\mathbf{r})} T_f|_{\mathbf{r}+\mathbf{r}_{REV}} dV \tag{86}$$

where V_{REV} and V_f are the volumes of the REV and the fluid phase. One should note that the superficial temperature is evaluated at the centroid of the REV, whereas the phase temperature is evaluated throughout the REV. This is noted explicitly in Eq. (86). Similarly, the superficial solid temperature is

$$\langle T_s \rangle = \frac{1}{V_{REV}} \int_{V_s} T_s \, dV \tag{87}$$

with evaluation symbols omitted. The volume-averaged form of Eqs. (80) and (81) are

$$(\rho c_p)_f \frac{1}{V_{REV}} \int_{V_f} \frac{\partial T_f}{\partial t} dV = \frac{1}{V_{REV}} \int_{V_f} \nabla \cdot (k_f \nabla T_f) \, dV \tag{88}$$

and

$$(\rho c_p)_s \frac{1}{V_{REV}} \int_{V_s} \frac{\partial T_s}{\partial t} dV = \frac{1}{V_{REV}} \int_{V_s} \nabla \cdot (k_s \nabla T_s) dV \tag{89}$$

respectively. Using Eqs. (86) and (87), Eqs. (88) and (89) become

$$(\rho c_p)_f \frac{\partial \langle T_f \rangle}{\partial t} = \langle \nabla \cdot (k_f \nabla T_f) \rangle \tag{90}$$

and

$$(\rho c_p)_s \frac{\partial \langle T_s \rangle}{\partial t} = \langle \nabla \cdot (k_s \nabla T_s) \rangle \tag{91}$$

where angle brackets indicate volume averaged quantities.

Next, Nozad *et al.* apply the spatial averaging theorem [4] to Eqs. (90) and (91). The result is

$$(\rho c_p)_f \frac{\partial \langle T_f \rangle}{\partial t} = \nabla \cdot \langle k_f \nabla T_f \rangle + \frac{1}{V_{REV}} \int_{A_{fs}} \mathbf{n}_{fs} \cdot k_f \nabla T_f \, dA \tag{92}$$

and

$$(\rho c_p)_s \frac{\partial \langle T_s \rangle}{\partial t} = \nabla \cdot \langle k_s \nabla T_s \rangle + \frac{1}{V_{REV}} \int_{A_{fs}} \mathbf{n}_{sf} \cdot k_s \nabla T_s \, dA \tag{93}$$

where A_{fs} is the area of the solid–fluid interface. One should note that these equations are coupled by the integral of the interfacial heat flux vector (Eq. (83)). The superficial average however, is an unsuitable variable because it can yield erroneous results. For example, if the fluid temperature were a constant, the superficial average would differ from this constant [4]. On the other hand, intrinsic phase averages do not have this shortcoming. These averages are defined by

$$\langle T_f \rangle^f = \frac{1}{V_f} \int_{V_f} T_f \, dV \tag{94}$$

and

$$\langle T_s \rangle^s = \frac{1}{V_s} \int_{V_s} T_s \, dV \tag{95}$$

Also, intrinsic averages are related to superficial averages by

$$\langle T_f \rangle = \varepsilon_f \langle T_f \rangle^f \tag{96}$$

and

$$\langle T_s \rangle = \varepsilon_s \langle T_s \rangle^s \tag{97}$$

where ε_f and ε_s are the volume fractions of the fluid and solid phases. One should note that $\varepsilon_f = \phi$ and $\varepsilon_s = 1-\phi$ for a two-phase system.

Next, Nozad *et al.* substitute Eqs. (96) and (97) into Eqs. (92) and (93) and again apply the spatial averaging theorem; this gives

$$\varepsilon_f (\rho c_p)_f \frac{\partial \langle T_f \rangle^f}{\partial t} = \nabla \cdot \left\{ k_f \left[\nabla \langle T_f \rangle + \frac{1}{V_{REV}} \int_{A_{fs}} \mathbf{n}_{fs} T_f \, dA \right] \right\}$$
$$+ \frac{1}{V_{REV}} \int_{A_{fs}} \mathbf{n}_{fs} \cdot k_f \nabla T_f \, dA \tag{98}$$

and

$$\varepsilon_s (\rho c_p)_s \frac{\partial \langle T_s \rangle^s}{\partial t} = \nabla \cdot \left\{ k_s \left[\nabla \langle T_s \rangle + \frac{1}{V_{REV}} \int_{A_{fs}} \mathbf{n}_{sf} T_s \, dA \right] \right\}$$
$$+ \frac{1}{V_{REV}} \int_{A_{fs}} \mathbf{n}_{sf} \cdot k_s \nabla T_s \, dA \tag{99}$$

The quantities in square brackets are simplified by introducing the spatial decompositions

$$T_f = \langle T_f \rangle^f + \widetilde{T}_f \tag{100}$$

and

$$T_s = \langle T_s \rangle^s + \widetilde{T}_s \tag{101}$$

and by applying scaling arguments and the spatial averaging theorem. The result is

$$\nabla \langle T_f \rangle + \frac{1}{V_{REV}} \int_{A_{fs}} \mathbf{n}_{fs} T_f \, dA = \varepsilon_f \left[\nabla \langle T_f \rangle^f + \frac{1}{V_f} \int_{A_{fs}} \mathbf{n}_{fs} \widetilde{T}_f \, dA \right] \tag{102}$$

and

$$\nabla \langle T \rangle_s + \frac{1}{V_{REV}} \int_{A_{fs}} \mathbf{n}_{sf} T_s \, dA = \varepsilon_s \left[\nabla \langle T_s \rangle^s + \frac{1}{V_s} \int_{A_{fs}} \mathbf{n}_{sf} \widetilde{T}_s \, dA \right] \tag{103}$$

Hence substituting Eqs. (102) and (103) into Eqs. (98) and (99) yield

$$\varepsilon_f(\rho c_p)_f \frac{\partial \langle T_f \rangle^f}{\partial t} = \nabla \cdot \left\{ \varepsilon_f k_f \left[\nabla \langle T_f \rangle^f + \frac{1}{V_f} \int_{A_{fs}} \mathbf{n}_{fs} \widetilde{T}_f \, dA \right] \right\}$$
$$+ \frac{1}{V_{REV}} \int_{A_{fs}} \mathbf{n}_{fs} \cdot k_f \nabla T_f \, dA \qquad (104)$$

and

$$\varepsilon_s(\rho c_p)_s \frac{\partial \langle T_s \rangle^s}{\partial t} = \nabla \cdot \left\{ \varepsilon_s k_s \left[\nabla \langle T_s \rangle^s + \frac{1}{V_s} \int_{A_{fs}} \mathbf{n}_{sf} \widetilde{T}_s \, dA \right] \right\}$$
$$+ \frac{1}{V_{REV}} \int_{A_{fs}} \mathbf{n}_{sf} \cdot k_s \nabla T_s \, dA \qquad (105)$$

which are precursors to a two-equation model [52].

Lastly, if certain conditions are satisfied [4], the solid and fluid phases can be assumed to be in local thermal equilibrium. Consequently, the REV is represented by a single temperature, i.e.,

$$\langle T \rangle = \frac{1}{V_{REV}} \int_{V_{REV}} T \, dV = \varepsilon_f \langle T_f \rangle^f + \varepsilon_s \langle T_s \rangle^s \qquad (106)$$

Nozad et al. obtain a one-equation model by substituting the decompositions

$$\langle T_f \rangle^f = \langle T \rangle + \widehat{T}_f \qquad (107)$$

and

$$\langle T_s \rangle^s = \langle T \rangle + \widehat{T}_s \qquad (108)$$

into Eqs. (104) and (105), and adding the results. This yields

$$(\rho c_p)_e \frac{\partial \langle T \rangle}{\partial t} = \nabla \cdot \left\{ \varepsilon_f k_f \left[\nabla \langle T \rangle + \frac{1}{V_f} \int_{A_{fs}} \mathbf{n}_{fs} \widetilde{T}_f \, dA \right] \right.$$
$$+ \varepsilon_s k_s \left[\nabla \langle T \rangle + \frac{1}{V_s} \int_{A_{fs}} \mathbf{n}_{sf} \widetilde{T}_s \, dA \right] \right\}$$
$$- \left\{ \varepsilon_f (\rho c_p)_f \frac{\partial \widehat{T}_f}{\partial t} + \varepsilon_s (\rho c_p)_s \frac{\partial \widehat{T}_s}{\partial t} \right.$$
$$\left. - \nabla \cdot (\varepsilon_f k_f \nabla \widehat{T}_f) - \nabla \cdot (\varepsilon_s k_s \nabla \widehat{T}_s) \right\} \qquad (109)$$

By assumption, changes in the phase deviations are much smaller than changes in the spatially averaged temperatures, i.e.,

$$\nabla \langle T \rangle \gg \nabla \widehat{T}_f, \nabla \widehat{T}_s \quad \text{and} \quad \frac{\partial \langle T \rangle}{\partial t} \gg \frac{\partial \widehat{T}_f}{\partial t}, \frac{\partial \widehat{T}_s}{\partial t} \qquad (110)$$

Consequently, the last term in Eq. (109) can be neglected and the one-equation model is,

$$(\rho c_p)_e \frac{\partial \langle T \rangle}{\partial t} = \nabla \cdot \left\{ \varepsilon_f k_f \left[\nabla \langle T \rangle + \frac{1}{V_f} \int_{A_{fs}} \mathbf{n}_{fs} \widetilde{T}_f \, dA \right] \right.$$
$$\left. + \varepsilon_s k_s \left[\nabla \langle T \rangle + \frac{1}{V_s} \int_{A_{fs}} \mathbf{n}_{sf} \widetilde{T}_s \, dA \right] \right\} \qquad (111)$$

The utility of Eq. (111) however, rests upon evaluation of the integrals involving the spatial deviations $\widetilde{T}_f$ and $\widetilde{T}_s$.

2. Closure and Numerical Solution

To close Eq. (111), Nozad *et al.* derive governing equations for the spatial deviations. First, the intrinsic temperature decompositions given by Eqs. (100) and (101) are substituted into Eqs. (80) and (81) to give

$$(\rho c_p)_f \frac{\partial \langle T_f \rangle^f}{\partial t} - \nabla \cdot (k_f \nabla \langle T_f \rangle^f) = -\left[(\rho c_p)_f \frac{\partial \widetilde{T}_f}{\partial t} - \nabla \cdot (k_f \nabla \widetilde{T}_f) \right] \qquad (112)$$

and

$$(\rho c_p)_s \frac{\partial \langle T_s \rangle^s}{\partial t} - \nabla \cdot (k_s \nabla \langle T_s \rangle^s) = -\left[(\rho c_p)_s \frac{\partial \widetilde{T}_s}{\partial t} - \nabla \cdot (k_s \nabla \widetilde{T}_s) \right] \qquad (113)$$

Next, taking the intrinsic phase averages of Eqs. (112) and (113) yields

$$(\rho c_p)_f \frac{\partial \langle T_f \rangle^f}{\partial t} - \nabla \cdot (k_f \nabla \langle T_f \rangle^f) = -\left\langle (\rho c_p)_f \frac{\partial \widetilde{T}_f}{\partial t} - \nabla \cdot (k_f \nabla \widetilde{T}_f) \right\rangle^f \qquad (114)$$

and

$$(\rho c_p)_s \frac{\partial \langle T_s \rangle^s}{\partial t} - \nabla \cdot (k_s \nabla \langle T_s \rangle^s) = -\left\langle (\rho c_p)_s \frac{\partial \widetilde{T}_s}{\partial t} - \nabla \cdot (k_s \nabla \widetilde{T}_s) \right\rangle^s \qquad (115)$$

where one should note that when $(r_0/L_M)^2 \ll 1$, intrinsic averages are constant with respect to the averaging process. Next, substituting Eqs. (114) and (115) into Eqs. (112) and (113) gives

$$(\rho c_p)_f \frac{\partial \widetilde{T}_f}{\partial t} - \nabla \cdot (k_f \nabla \widetilde{T}_f) = \left\langle (\rho c_p)_f \frac{\partial \widetilde{T}_f}{\partial t} - \nabla \cdot (k_f \nabla \widetilde{T}_f) \right\rangle^f \quad (116)$$

and

$$(\rho c_p)_s \frac{\partial \widetilde{T}_s}{\partial t} - \nabla \cdot (k_s \nabla \widetilde{T}_s) = \left\langle (\rho c_p)_s \frac{\partial \widetilde{T}_s}{\partial t} - \nabla \cdot (k_s \nabla \widetilde{T}_s) \right\rangle^s \quad (117)$$

which are the governing equations for the spatial deviations, $\widetilde{T}_f$ and $\widetilde{T}_s$, respectively.

Next, Nozad et al. argue that because the phase length scales, l_s and l_f are small with respect to the macroscopic characteristic length, the spatial deviations are quasi-steady within the REV. Consequently, Eqs. (116) and (117) reduce to

$$\nabla \cdot \left(k_f \nabla \widetilde{T}_f \right) = \frac{1}{V_f} \int_{V_f} \nabla \cdot \left(k_f \nabla \widetilde{T}_f \right) dV \quad (118)$$

and

$$\nabla \cdot \left(k_s \nabla \widetilde{T}_s \right) = \frac{1}{V_s} \int_{V_s} \nabla \cdot \left(k_s \nabla \widetilde{T}_s \right) dV \quad (119)$$

Additionally, if phase conductivity gradients are smaller than gradients of deviations, i.e., $\nabla k_f \cdot \nabla \widetilde{T}_f \ll k_f \nabla^2 \widetilde{T}_f$ and $\nabla k_s \cdot \nabla \widetilde{T}_s \ll k_s \nabla^2 \widetilde{T}_s$, then Eqs. (118) and (119) become

$$\nabla^2 \widetilde{T}_f = \frac{1}{V_f} \int_{V_f} \nabla^2 \widetilde{T}_f \, dV \quad (120)$$

and

$$\nabla^2 \widetilde{T}_s = \frac{1}{V_s} \int_{V_s} \nabla^2 \widetilde{T}_s \, dV \quad (121)$$

The boundary conditions for Eqs. (120) and (121) at the solid–fluid interface are derived in a similar manner. They are given by

$$\widetilde{T}_f = \widetilde{T}_s \quad (122)$$

and
$$\mathbf{n}_{fs} \cdot \nabla \widetilde{T}_f = \kappa \mathbf{n}_{fs} \cdot \nabla \widetilde{T}_s + (\kappa - 1)\mathbf{n}_{fs} \cdot \nabla \langle T \rangle \qquad (123)$$

One should note that Eq. (122) incorporates the assumptions $\widetilde{T}_f \gg \widehat{T}_f$ and $\widetilde{T}_s \gg \widehat{T}_s$. Also, Eq. (123) results from the inequalities given by Eq. (110). The initial conditions are given by

$$\widetilde{T}_f = g'(t) \qquad (124)$$

and

$$\widetilde{T}_s = h'(t) \qquad (125)$$

but they do not affect the solution because fluctuations are assumed to be quasi-steady. Thus, Eqs. (120)–(123) define a local problem [4].

To solve Eqs. (120)–(123), Nozad et al. propose that

$$\widetilde{T}_f = \mathbf{b}_f \cdot \nabla \langle T \rangle + \xi_1 \qquad (126)$$

and

$$\widetilde{T}_s = \mathbf{b}_s \cdot \nabla \langle T \rangle + \xi_2 \qquad (127)$$

where $\mathbf{b}_f$ and $\mathbf{b}_s$ are vector fields that map gradients of the average temperature field to spatial fluctuations, and ξ_1 and ξ_2 are constants. The vector fields $\mathbf{b}_f$ and $\mathbf{b}_s$ are defined by the boundary value problem,

$$\nabla^2 \mathbf{b}_f = -\frac{\kappa - 1}{\varepsilon_f} \nabla \varepsilon_f \qquad (128)$$

in V_f,

$$\nabla^2 \mathbf{b}_s = \frac{\kappa - 1}{\varepsilon_s \kappa} \nabla \varepsilon_s \qquad (129)$$

in V_s, and

$$\mathbf{b}_f = \mathbf{b}_s \qquad (130)$$

and

$$\mathbf{n}_{fs} \cdot \nabla \mathbf{b}_f = \kappa \mathbf{n}_{fs} \cdot \nabla \mathbf{b}_s + (\kappa - 1)\mathbf{n}_{fs} \qquad (131)$$

at the solid–fluid interface. Also, the initial conditions

$$\mathbf{b}_f = \mathbf{g}''(t) \qquad (132)$$

and
$$\mathbf{b}_s = \mathbf{h}''(t) \tag{133}$$

apply at the boundary of the REV. Additionally, the volume averages of the vector fields must be zero, i.e.,

$$\frac{1}{V_f} \int_{V_f} \mathbf{b}_f \, dV = 0 \tag{134}$$

and

$$\frac{1}{V_s} \int_{V_s} \mathbf{b}_s \, dV = 0 \tag{135}$$

Nozad et al. demonstrate that the constants in Eqs. (126) and (127) can be neglected when

$$|\kappa - 1|\left(\frac{l_f}{L_M}\right) \ll 1 \tag{136}$$

A further simplification is realized by assuming that the medium is spatially periodic. In this case, ξ_1 and ξ_2 are identically zero [51] and Eqs. (128)–(133) become,

$$\nabla^2 \mathbf{b}_f = 0 \tag{137}$$

$$\nabla^2 \mathbf{b}_s = 0 \tag{138}$$

$$\mathbf{b}_f = \mathbf{b}_s \tag{139}$$

$$\mathbf{n}_{fs} \cdot \nabla \mathbf{b}_f = \kappa \mathbf{n}_{fs} \cdot \nabla \mathbf{b}_s + (\kappa - 1)\mathbf{n}_{fs} \tag{140}$$

$$\mathbf{b}_f(\mathbf{r} + \mathbf{l}_i) = \mathbf{b}_f(\mathbf{r}); \quad i = 1, 2, 3, \ldots \tag{141}$$

and

$$\mathbf{b}_s(\mathbf{r} + \mathbf{l}_i) = \mathbf{b}_s(\mathbf{r}); \quad i = 1, 2, 3, \ldots \tag{142}$$

respectively. Therefore, substituting Eqs. (126) and (127) with $\xi_1 = 0$ and $\xi_2 = 0$ into Eq. (111) and neglecting spatial variations in k_f and k_s yields the closed expression,

$$(\rho c_p)_e \frac{\partial \langle T \rangle}{\partial t} = \left[(\varepsilon_f k_f + \varepsilon_s k_s)\mathbf{I} + \frac{(k_f - k_s)}{V_{REV}} \int_{A_{fs}} \mathbf{n}_{fs} \mathbf{b}_f \, dA\right] : \nabla \nabla \langle T \rangle \tag{143}$$

One should note that Eq. (143) defines the effective conductivity tensor,

$$\frac{\mathbf{K}_e}{k_f} = (\varepsilon_f + \varepsilon_s \kappa)\mathbf{I} + \frac{(1 - \kappa)}{V_{REV}} \int_{A_{fs}} \mathbf{n}_{fs} \mathbf{b}_f \, dA \qquad (144)$$

which can be evaluated following determination of the $\mathbf{b}_f$ field. An interesting feature of Eq. (144) is that it is the sum a geometry-independent and a geometry-dependent term. Thus it is reminiscent of the expressions derived by Brown [30], e.g., Eqs. (41) and (52), for the effective properties of composite media.

3. *Numerical Solution and Experimental Validation*

To solve the closure relation, Nozad *et al.* [51] consider two spatially periodic unit cells (Fig. 24) and numerically solve the boundary value problem defined by Eqs. (137)–(142). The first unit cell is depicted in Fig. 24(a) and it is a two-dimensional approximation of a porous medium comprised of particles that do not interact. The second unit cell is depicted in Fig. 24(b) and it is a two-dimensional approximation for contacting particles. In the latter case, particles are connected by solid "bridges" having width c. Contact is characterized by the parameter $R_c = c/a$. The vector fields $\mathbf{b}_f$ and $\mathbf{b}_s$ are determined and the effective conductivity tensor is computed. One should note that the geometry of the unit cell implies that

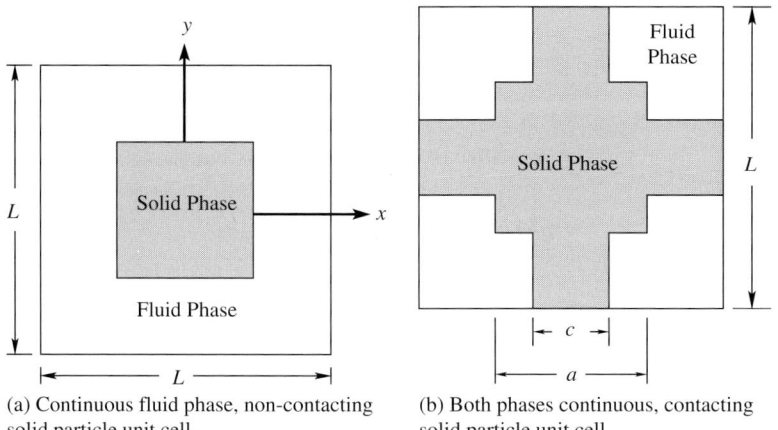

(a) Continuous fluid phase, non-contacting solid particle unit cell.

(b) Both phases continuous, contacting solid particle unit cell.

FIG. 24. Two-dimensional spatially periodic models of Nozad *et al.* [51]. The unit cell length is L, and the characteristic length of the particle is a. The contact region has characteristic length c and $R_c = c/a$, is the contact parameter. (Adapted from Ref. [51], Copyright (1985) with permission from Elsevier.)

the effective conductivity tensor is symmetric and that the effective conductivity is isotropic.

To validate the effective conductivity model, Nozad et al. [19] measure the effective conductivity of glass, urea–formaldehyde, stainless steel, bronze, and aluminum spheres saturated with water, air, and glycerol. These solid–fluid combinations yield porosities and conductivity ratios of $0.39 \leqslant \phi \leqslant 0.41$ and $1.23 \leqslant \kappa \leqslant 8077$. The experimental apparatus is depicted in Fig. 6 and except for the air–aluminum media, Nozad et al. estimate the relative uncertainty in the measurements to be 5 percent. In the case of air–aluminum media, lateral temperature variations are detected. This suggests that the fluid phase is in motion and that measurements of the air–aluminum effective conductivity are exaggerated.

The experimental and numerical results of Nozad et al. [19,51] are plotted in Fig. 25. The applicability of the continuous fluid-phase model is limited to small solid–fluid conductivity ratios. In fact, this model and the Maxwell solution (Eq. (37)) apparently have the same functional form. They differ

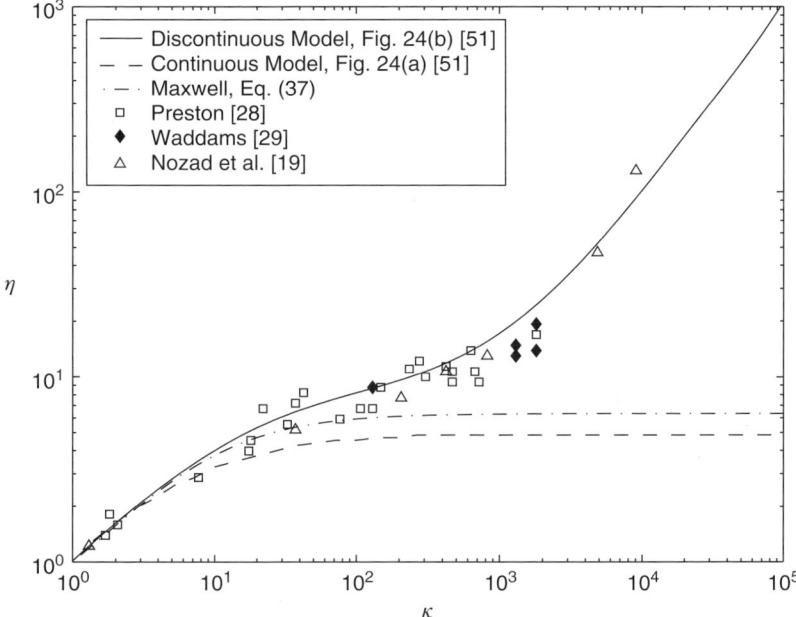

FIG. 25. Results of the numerical solution of Nozad et al. [51] and the experiments of Nozad et al. [19]. The numerical solutions and the Maxwell solution (Eq. (37)) assume a porosity of 0.36. Also, a contact parameter of 0.02 is assumed for the contacting-particle model. (Adapted from Ref. [51], Copyright (1985) with permission from Elsevier.)

however, because the former employs square particles [53]. On the other hand, the contacting-particle model (Fig. 24(b)) captures the relation between the effective conductivity and solid–fluid conductivity ratio quite well. Nozad et al. [51] report that a contact parameter of 0.02 fits experimental data best.

The value of the contact parameter however, is not clear. For example, Shonnard and Whitaker [54] report finding an error in the numerical analysis of Nozad et al. [51]; they recommend $R_c = 0.01$. On the other hand, Sahraoui and Kaviany [55] use a different numerical procedure and conclude that $R_c = 0.002$. Hence, the contact parameter is an adjustable constant rather than a property of large solid–fluid conductivity ratio media.

B. Empirical Contact Parameter Models

Although the introduction of an empirical contact parameter adds a degree of complication, models which employ this strategy are successful. This technique is used to introduce particle contact into correlations, lumped parameter models, and numerical conduction models.

1. Correlations

Hadley [13] develops a semi-empirical correlation to predict the effective conductivity of two- and three-phase porous media. First, he considers the volume-averaged temperature of a system comprised of n_p phases. He notes that this temperature, $\langle T \rangle$ is the sum of the superficial phase temperatures, i.e.,

$$\langle T \rangle = \sum_{i=1}^{n_p} \langle T_i \rangle \tag{145}$$

where $\langle T \rangle$ and $\langle T_i \rangle$ are defined by Eqs. (106) and (86). Next, the gradient of Eq. (145) is taken and the spatial averaging theorem [4] is applied to the result. Hadley, subsequently, argues that the surface integrals cancel to give

$$\nabla \langle T \rangle = \sum_{i=1}^{n_p} \langle \nabla T_i \rangle \tag{146}$$

Next, the same procedure is applied to the heat flux vector. The result is

$$\mathbf{k}_e \nabla \langle T \rangle = \sum_{i=1}^{n_p} \mathbf{k_i} \langle \nabla T_i \rangle \tag{147}$$

where $\mathbf{k}_e$ and $\mathbf{k}_i$ are thermal conductivity vectors. Next, the intrinsic phase average temperatures (Eq. (94)) are substituted for the superficial temperatures in Eqs. (146) and (147). This gives

$$\nabla \langle T \rangle = \sum_{i=1}^{n_p} \varepsilon_i \langle \nabla T_i \rangle^i \qquad (148)$$

and

$$\mathbf{k}_e \nabla \langle T \rangle = \sum_{i=1}^{n_p} \varepsilon_i \mathbf{k_i} \langle \nabla T_i \rangle^i \qquad (149)$$

where ε_i is the volume fraction of phase i. Next, Hadley restricts Eqs. (148) and (149) to two phases and isotropic conductivities. This yields

$$\nabla \langle T \rangle = \phi \langle \nabla T_f \rangle^f + (1-\phi) \langle \nabla T_s \rangle^s \qquad (150)$$

and

$$\eta \nabla \langle T \rangle = \phi \langle \nabla T_f \rangle^f + \kappa (1-\phi) \langle \nabla T_s \rangle^s \qquad (151)$$

which are simultaneous equations for $\langle T \rangle$ in terms of $\langle \nabla T_f \rangle^f$ and $\langle \nabla T_s \rangle^s$. A third relation, however, is needed to close the system.

Hadley hypothesizes that a closure relation can be derived from effective conductivity mixture rules. First, he notes that a choice of

$$\langle \nabla T_f \rangle^f = \kappa \langle \nabla T_s \rangle^s \qquad (152)$$

yields the layers-in-series mixture model, Eq. (49). Alternatively, a closure model given by

$$\langle \nabla T_f \rangle^f = \langle \nabla T_s \rangle^s \qquad (153)$$

results in the layers-in-parallel mixture model, Eq. (47). Hence Eqs. (152) and (153) are limiting possibilities for the closure relation. Therefore, Hadley proposes

$$\langle \nabla T_f \rangle^f = [f + \kappa(1-f)] \langle \nabla T_s \rangle^s \qquad (154)$$

where $f = f(\kappa, \phi)$ and $0 \leqslant f \leqslant 1$.

To further constrain f, Hadley next considers the effective conductivity of a dilute suspension of spheres (see Eq. (37)), which is valid for $\phi \to 1$. He notes that Eq. (37) results from a choice of $f = 2/3$. On the other hand, the thermal conductivity of a solid containing fluid-filled voids is shown in Eq. (40), which corresponds to

$$f = \frac{2\kappa}{2\kappa + 1} \tag{155}$$

Consequently,

$$\frac{2}{3} \leq f \leq \frac{2\kappa}{2\kappa + 1} \tag{156}$$

which implies that

$$\eta = \frac{\phi f + \kappa(1 - \phi f)}{1 - \phi(1 - f) + \kappa\phi(1 - f)} \tag{157}$$

where $f = f(\phi)$. Hence for unconsolidated porous media, i.e. non-touching particles, f is independent of κ and monotone increasing in $1-\phi$. Hadley notes, however, that the limited experimental data precludes further specification of the functional dependence. Consequently, f is assumed to be a constant, i.e., $f = f_o$.

For consolidated media, Hadley argues that the effective conductivity is proportional to Eq. (40). Hence, the effective conductivity can be represented by the sum of Eqs. (157) and (40). That is

$$\eta = (1 - \alpha_p)\frac{\phi f_o + \kappa(1 - \phi f_o)}{1 - \phi(1 - f_o) + \kappa\phi(1 - f_o)} + \alpha_p \frac{2\kappa^2(1 - \phi) + (1 + 2\phi)\kappa}{(2 + \phi)\kappa + 1 - \phi} \tag{158}$$

where α_p is a parameter representing the degree of consolidation.

To determine the empirical parameters f_o and α_p, Hadley considers the limiting cases for κ. The consolidation parameter is related to the effective conductivity of an evacuated medium. That is

$$\eta_{\text{VAC}} = 2\alpha_p \frac{1 - \phi}{2 + \phi} \tag{159}$$

which follows from taking the limit of Eq. (158) for $\kappa \to \infty$. Hadley subsequently uses Eq. (159) and experimental data to generate a curve of α_p versus percent-theoretical density. One should note that percent-theoretical density is equivalent to $1-\phi$ for two-phase media; several consolidation parameters taken from this curve are presented in Table VI. On the other

TABLE VI
Two-Phase Consolidation Parameters [13]

Solid	ϕ	α_p
Brass powder	0.60	1.0×10^{-3}
Stainless-steel powder	0.42	1.0×10^{-2}
Stainless-steel spheres	0.32	1.1×10^{-1}
Stainless-steel spheres	0.25	1.5×10^{-1}
Brass spheres	0.24	1.4×10^{-1}
Stainless-steel spheres	0.21	1.9×10^{-1}
Brass spheres	0.19	2.1×10^{-1}
Stainless-steel spheres	0.15	2.8×10^{-1}
Brass spheres	0.13	3.2×10^{-1}
Brass spheres	0.10	3.2×10^{-1}

hand, f_o is determined from intermediate κ, e.g., water-saturated, media; it is found to vary from 0.8 to 0.9.

Hadley reports that Eq. (158) captures data from the literature and experiments with 18 percent accuracy. This result is obtained with $f_o = 0.8$ and α_p determined from the consolidation parameter curve. Further, he finds that Eq. (158) with $\alpha_p = 0.02$ and $f_o = 0.8$ yields excellent correspondence with the touching square model of Nozad et al. [51] (Fig. 27). Hadley notes that the consolidation parameter equals the contact parameter and he speculates that it represents the fraction of heat conducted through the contact point. He acknowledges however, that the equivalence is likely fortuitous.

Lund et al. [56] also develop a correlation that incorporates particle contact. The correlation is derived from a finite element solution of a conduction problem consisting of two hemispheres. They introduce empirical parameters to represent the fractional contact area and the fractional gap between the hemispheres. In addition, the thermal conductivities of certain regions within the fluid-filled gap are equated to the solid conductivity to approximate asperites. The effective conductivity is determined for a variety of conductivity ratios, contact area fractions, and hemisphere-gap ratios. The correlation is derived from these results and it is shown to correspond well to the packed sand data of Tavman [24]. Unfortunately, little guidance is given for estimating the contact area fractions or gap ratios of practical systems.

2. Lumped Parameter Models

Hsu et al. [46] improve the accuracy of the Zehner and Schlünder [45] model at large solid–fluid conductivity ratios by incorporating conduction through a finite-area contact point. Referring to Fig. 26, this is accomplished

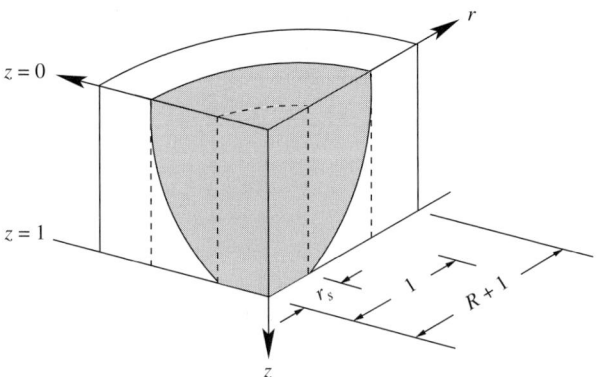

FIG. 26. The modified Zehner and Schlünder unit cell of Hsu *et al.* [46]. (Adapted from Ref. [46], Copyright (1994) with permission from ASME.)

by introducing a contact region having radius, r_s at the intersection of the solid and the plane $z = 1$.

To determine the effective conductivity of the unit cell, Hsu *et al.* parallel the analysis of Zehner and Schlünder. First, the equation of the solid–fluid interface is defined by

$$r^2 + \frac{z^2}{[(1+\alpha_o)B - (B-1)z]^2} = 1 \qquad (160)$$

where B and α_o are the shape and deformation parameters, respectively. These parameters are related to the contact radius by

$$r_s^2 = 1 - \frac{1}{(1+\alpha_o B)^2} \qquad (161)$$

which follows from substituting $z = 1$ into Eq. (160). Also, one should note that Eq. (160) reduces to Eq. (75) when $\alpha_o = 0$.

Hsu *et al.* then assume unidirectional heat flow through the unit cell and they divide it into three parallel thermal resistances. Referring to Fig. 26, the resistances consist of the axial projection of the contact area ($r \leqslant r_s$), the annular region encompassing the solid–fluid interface ($r_s < r \leqslant 1$), and the outer fluid region ($1 < r \leqslant R+1$). Hence the effective conductivity of the unit cell is

$$k_e = \left(1 - \frac{1}{R^2}\right)k_f + \left(\frac{1 - r_s^2}{R^2}\right)k_{sf} + \left(\frac{r_s}{R}\right)^2 k_s \qquad (162)$$

where k_{sf} is the effective conductivity of the region containing the solid–fluid interface. Also, R is related to the porosity by Eq. (74).

Next, k_{sf} is found by dividing the region $r_s \leqslant r \leqslant 1$ into annular slices that resemble Fig. 14(b). The conductivity of a slice is found by assuming that the solid and fluid are composite layers in series. Thus,

$$\pi^2(1-r_s)^2 k_{sf} = \int_{r_s}^1 \frac{2\pi r k_f}{(1-z)+(1/\kappa)z}\, dr \qquad (163)$$

where $z = z(r)$ by Eq. (160). Substituting Eqs. (160) and (161) into Eq. (163) gives

$$\frac{k_{sf}}{k_f} = 2(1+\alpha_o B)^2 \int_0^1 \frac{(1+\alpha_o)Bz}{[1+(1/\kappa-1)z][(1+\alpha_o)B-(B-1)z]^3}\, dz \qquad (164)$$

Consequently, computing the integral in Eq. (164) and substituting the result and Eq. (74) into Eq. (162) yields

$$\eta = 1 - \sqrt{1-\phi} + \kappa\sqrt{1-\phi}\left[1 - \frac{1}{(1+\alpha_o B)^2}\right]$$
$$+ \frac{2\sqrt{1-\phi}}{[1-(1/\kappa)B+(1-1/\kappa)\alpha_o B]} \left\{ \frac{(1-1/\kappa)(1+\alpha_o)B}{[1-(1/\kappa)B+(1-1/\kappa)\alpha_o B]^2} \right.$$
$$\times \ln\left[\frac{\kappa(1+\alpha_o B)}{(1+\alpha_o)B}\right] - \frac{B+1+2\alpha_o B}{2(1+\alpha_o B)^2}$$
$$\left. - \frac{(B-1)}{[1-(1/\kappa)B+(1-1/\kappa)\alpha_o B](1+\alpha_o B)} \right\} \qquad (165)$$

for the effective conductivity of the unit cell.

Finally, the contact parameter, α_o and the shape parameter, B are related to the porosity by

$$\phi = 1 - \frac{B^2}{(1-B)^6(1+\alpha_o B)^2}\{(B^2-4B+3)$$
$$+ 2(1+\alpha_o)(1+\alpha_o B)\ln\left[\frac{(1+\alpha_o)B}{1+\alpha_o B}\right]$$
$$+\alpha_o(B-1)(B^2-2B-1)\}^2 \qquad (166)$$

which follows from computing the fractional volume of solid in the unit cell. One should note that when $\alpha_o = 0$, Eqs. (165) and (166) reduce to Eqs. (78) and (76), respectively.

The modified Zehner and Schlünder model (Eq. (165)) is compared to the Zehner and Schlünder [45] model (Eq. (78)) and the touching-particle model of Nozad et al. [51] in Fig. 27. One should note that two empirical parameters must be specified to evaluate Eq. (165). Hsu et al. recommend $\alpha_o = 0.002$ and $\phi = 0.42$ to best fit the air–aluminum data of Nozad et al. [19]; which is confirmed in Fig. 27. They also note that the effective conductivity is very sensitive to the deformation parameter. Additionally, the divergence of Eqs. (165) and (78) at approximately $\kappa = 10^3$ indicates that particle contact is a significant effect and that it is greatest for large solid–fluid conductivity ratio media.

In a contemporaneous effort, Hsu et al. [57] develop three semi-empirical lumped parameter models for spatially periodic unit cells comprised of rectangles, cylinders, and cubes. They first consider a two-dimensional unit cell derived from an array of contacting parallelepipeds. The unit cell is

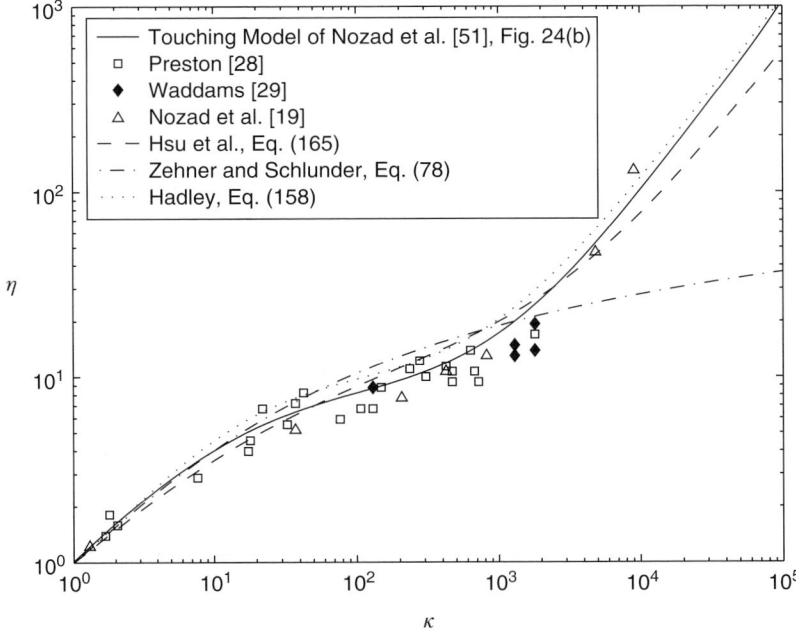

FIG. 27. A comparison of empirical contact models and the Zehner and Schlünder [45] model. One should note that the model of Hsu et al. [46] assumes $\phi = 0.42$ and $\alpha_0 = 0.002$. Also, $\alpha_p = 0.02$, $f_o = 0.8$, and $\phi = 0.36$ are used to evaluate the Hadley [13] correlation.

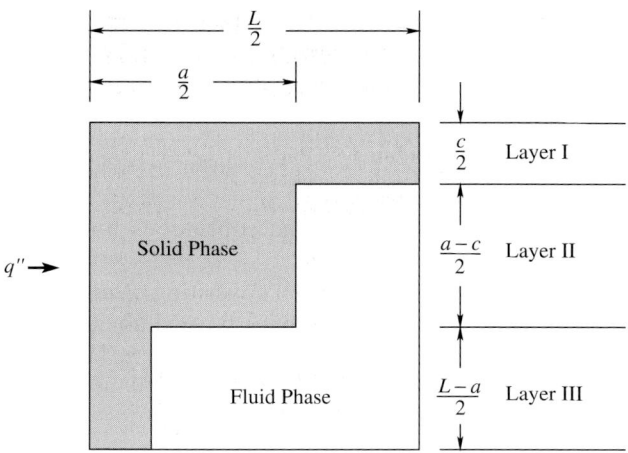

FIG. 28. Touching parallelepiped unit cell of Hsu *et al.* [57]. Layers I, II, and III represent parallel thermal resistances. (Adapted from Ref. [57], Copyright (1995) with permission from ASME.)

depicted in Fig. 28 and one should note that it is essentially a quadrant of the discontinuous fluid-phase model of Nozad *et al.* [51] (Fig. 24(b)).

Hsu *et al.* assume that the unit cell is a composite medium consisting of three layers. The layers represent parallel thermal resistances, hence the effective conductivity of the unit cell is,

$$k_e = \gamma_a \gamma_c k_s + \gamma_a(1 - \gamma_c)k_{sf1} + (1 - \gamma_a)k_{sf2} \qquad (167)$$

where $\gamma_a = a/L$ and $\gamma_c = c/a$. The conductivities k_{sf1} and k_{sf2}, are the thermal conductivities of Layers II and III. These conductivities are determined by assuming that the solid and fluid phases are thermal resistances in series. Substituting the equivalent expressions for k_{sf1} and k_{sf2} into Eq. (167) yields

$$\eta = \gamma_a \gamma_c \kappa + \frac{\gamma_a(1 - \gamma_c)}{1 + (1/\kappa - 1)\gamma_a} + \frac{(1 - \gamma_a)}{1 + (1/\kappa - 1)\gamma_a \gamma_c} \qquad (168)$$

The geometric parameters, γ_a and γ_c are related to the porosity by

$$1 - \phi = \gamma_a^2 + 2\gamma_c \gamma_a(1 - \gamma_a) \qquad (169)$$

which is equivalent to the area fraction of solid in the unit cell.

Additionally, Hsu et al. note that γ_c is equivalent to the contact parameter of Nozad et al. [51]. Consequently if $\gamma_c = 0$, Eq. (168) becomes

$$\eta = 1 - \sqrt{1-\phi} + \frac{\sqrt{1-\phi}}{1 + (1/\kappa - 1)\sqrt{1-\phi}} \quad (170)$$

which is presumed to correlate with the continuous fluid-phase model of Nozad et al. (Fig. 24(a)).

Hsu et al. [57] next consider a unit cell consisting of an array of touching circular cylinders (Fig. 29). The cylinders have diameter a and they are connected by plates having thicknesses, c and height, $\hat{h}$. The plate heights are given by

$$\frac{\hat{h}}{L} = 1 - \gamma_a \sqrt{1 - \gamma_c^2} \quad (171)$$

with $\gamma_a = a/L$ and $\gamma_c = c/a$. The contact angle is defined by

$$\theta_c = \arcsin \gamma_c \quad (172)$$

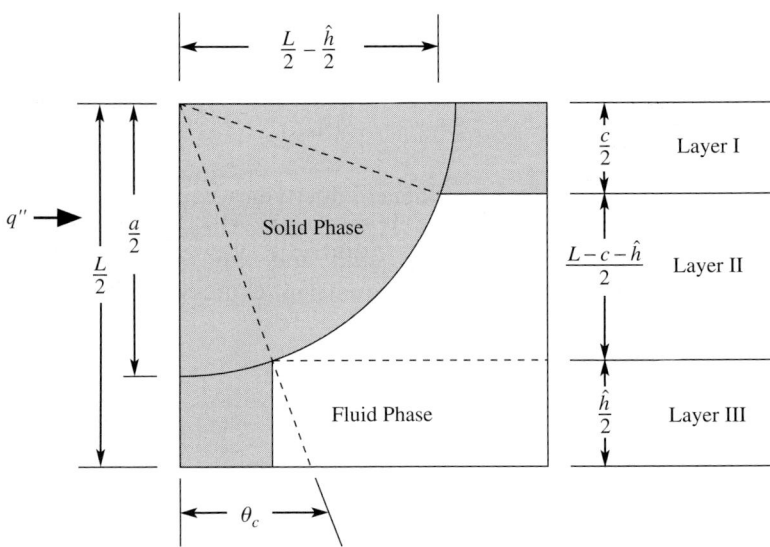

FIG. 29. Touching circular cylinder unit cell of Hsu et al. [57]. Layers I, II, and III represent parallel thermal resistances. (Adapted from Ref. [57], Copyright (1995) with permission from ASME.)

and one should note that the geometric parameters are related to the porosity by

$$\phi = 1 - \gamma_c\gamma_a - \frac{\gamma_a^2}{2}\left(\frac{\pi}{2} - 2\theta_c\right) \tag{173}$$

although Hsu et al. do not present this expression.

The effective conductivity of the unit cell is found by assuming that Layers I, II, and III in Fig. 29 are parallel thermal resistances. Hence,

$$k_e = \frac{c}{L}k_s + \left(1 - \frac{c}{L} - \frac{\hat{h}}{L}\right)k_{sf1} + \frac{\hat{h}}{L}k_{sf2} \tag{174}$$

where k_{sf1} and k_{sf2} are the effective conductivities of Layers II and III. The layer conductivities are given by

$$\frac{k_{sf1}}{k_f} = \frac{1}{\gamma_a(\sqrt{1-\gamma_c^2}-\gamma_c)}\left[\left(\frac{\kappa}{1-\kappa}\right)\left(\frac{\pi}{2}-2\theta_c\right)\right.$$

$$\left. - \left(\frac{\kappa}{1-\kappa}\right)\int_{\theta_c}^{\pi/2-\theta_c}\frac{d\theta}{1+(1/\kappa-1)\gamma_a\sin\theta}\right] \tag{175}$$

and

$$\frac{k_{sf2}}{k_f} = \frac{1}{1+(1/\kappa-1)\gamma_a\gamma_c} \tag{176}$$

respectively. One should note that when integrating Eq. (175), the result depends on $(1/\kappa-1)\gamma_a$. Consequently, computing Eq. (175) and substituting the result and Eq. (176) into Eq. (174) yields three expressions:

$$\eta = \gamma_c\gamma_a\kappa + \frac{1-\gamma_a\sqrt{1-\gamma_c^2}}{\gamma_a\gamma_c(1/\kappa-1)+1} + \frac{\kappa(\pi/2-2\theta_c)}{1-\kappa}$$

$$- \frac{2\kappa}{(1-\kappa)\sqrt{1-(1/\kappa-1)^2\gamma_a^2}}\left\{\arctan\left[\frac{\tan(\pi/4-\theta_c/2)+(1/\kappa-1)\gamma_a}{\sqrt{1-(1/\kappa-1)^2\gamma_a^2}}\right]\right.$$

$$\left. - \arctan\left[\frac{\tan(\theta_c/2)+(1/\kappa-1)\gamma_a}{\sqrt{1-(1/\kappa-1)^2\gamma_a^2}}\right]\right\} \tag{177}$$

for $(1/\kappa - 1)\gamma_a < 1$,

$$\eta = \gamma_c \gamma_a \kappa + \frac{1 - \gamma_a \sqrt{1 - \gamma_c^2}}{\gamma_a \gamma_c (1/\kappa - 1) + 1}$$
$$+ \frac{\kappa(\pi/2 - 2\theta_c)}{1 - \kappa} - \frac{\kappa}{(1 - \kappa)\sqrt{(1/\kappa - 1)^2 \gamma_a^2 - 1}}$$
$$\times \left\{ \ln \left[\frac{\tan(\pi/4 - \theta_c/2) + (1/\kappa - 1)\gamma_a - \sqrt{(1/\kappa - 1)^2 \gamma_a^2 - 1}}{\tan(\pi/4 - \theta_c/2) + (1/\kappa - 1)\gamma_a + \sqrt{(1/\kappa - 1)^2 \gamma_a^2 - 1}} \right] \right.$$
$$\left. - \ln \left[\frac{\tan(\theta_c/2) + (1/\kappa - 1)\gamma_a - \sqrt{(1/\kappa - 1)^2 \gamma_a^2 - 1}}{\tan(\theta_c/2) + (1/\kappa - 1)\gamma_a + \sqrt{(1/\kappa - 1)^2 \gamma_a^2 - 1}} \right] \right\} \quad (178)$$

for $(1/\kappa - 1)\gamma_a > 1$, and

$$\eta = \frac{\gamma_c \gamma_a^2}{\gamma_a + 1} + \frac{1 - \gamma_a \sqrt{1 - \gamma_c^2}}{\gamma_c + 1}$$
$$+ \gamma_a \left(\frac{\pi}{2} - 2\theta_c \right) - \left[\tan\left(\frac{\pi}{4} - \frac{\theta_c}{2}\right) - \tan\left(\frac{\theta_c}{2}\right) \right] \quad (179)$$

for $(1/\kappa - 1)\gamma_a = 1$.

Hsu *et al.* next consider a three-dimensional unit cell consisting of cubes. To determine the effective conductivity, they partition the unit cell into regions through which conduction is one-dimensional. The parallel paths are "columns" of solid and fluid. Otherwise, the analysis is identical to that of the two-dimensional models. Hsu *et al.* obtain

$$\eta = 1 - \gamma_a^2 - 2\gamma_c \gamma_a + 2\gamma_c \gamma_a^2 + \gamma_c^2 \gamma_a^2 \kappa$$
$$+ \frac{\gamma_a^2 - \gamma_c^2 \gamma_a^2}{[1 - \gamma_a + \gamma_a(1/\kappa)]} + \frac{2(\gamma_c \gamma_a - \gamma_c \gamma_a^2)}{[1 - \gamma_c \gamma_a + \gamma_c \gamma_a(1/\kappa)]} \quad (180)$$

for the effective conductivity of the unit cell. The geometric parameters γ_a and γ_c are related to the porosity by

$$1 - \phi = (1 - 3\gamma_c^2)\gamma_a^3 + 3\gamma_c^2 \gamma_a^2 \quad (181)$$

The rectangle, cylinder, and cube models are plotted in Fig. 30. One should note that $\phi = 0.36$ is assumed. Also, Hsu *et al.* recommend $\gamma_c = 0.01$ for the rectangle and cylinder models and $\gamma_c = 0.13$ for the cube model to

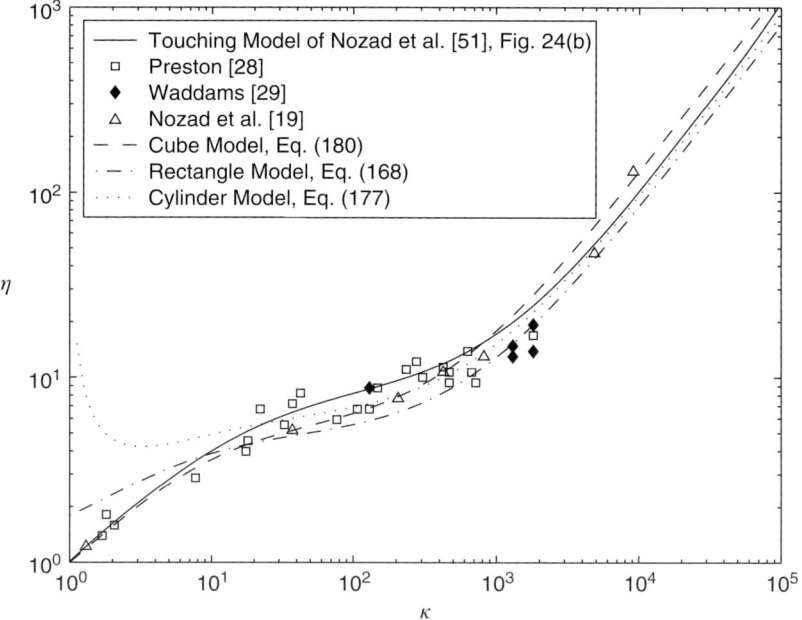

FIG. 30. A comparison of the lumped parameter models of Hsu *et al.* [57] and experimental data. One should note that $\phi = 0.36$ is assumed. Also, the rectangle and cylinder models assume $\gamma_c = 0.01$ and the cube model assumes $\gamma_c = 0.13$.

best approximate the numerical solution of Nozad *et al.* and experimental data.

Referring to Fig. 30, the rectangle and the cube models correspond well to the numerical solution of Nozad *et al.* and experimental data. But contrary to the claims of Hsu *et al.*, the cylinder model gives unrealistic estimates for the effective conductivity when $\kappa < 100$. Presumably, this indicates that Hsu *et al.* made an error in the derivation of Eq. (175). Specifically, Eq. (175) has a singularity at $\kappa = 1$, whereas $k_{\text{sf1}}/k_f = 1$ is clearly the expected limit. Consequently, Eq. (175) cannot represent the effective conductivity of Layer II (Fig. 29).

3. *Numerical Conduction Models*

Empirical contact parameters also appear in numerical conduction models. For example, Wakao and Kato [42] and Wakao and Vortmeyer [43] extend the conduction model of Deissler and Boegli [16] to include thermal radiation and conduction through finite contact areas. Contact is included by considering unit cells consisting of eighth-spheres that touch at flattened surfaces. Wakao and Vortmeyer numerically solve the three-dimensional conduction equation and obtain

$$\frac{k_c}{k_s} = 0.18 R_c^2 \qquad (182)$$

for the thermal conductivity of the contact point, k_c. This relation is restricted to $R_c < 0.17$. Additionally, Bauer and Schlünder [58] extend the model of Zehner and Schlünder [45] to include radiation, gas pressure, and particle contact. Heat flux through contact points is introduced by a correlation representing the numerical solution of Wakao and Kato. They adjust the contact parameter, R_c to match the model to experimental data; it is found to vary from 0.003 to 0.03 and to depend on the packing arrangement.

C. Contact Models

Although a connection between particle contact and the effective conductivity of large solid–fluid conductivity ratio media is clear, characterizing this phenomena continues to be both an experimental and a theoretical challenge. For example, Deissler and Boegli [16] attempt to investigate this phenomena with an electrical analog of a porous medium, i.e., carbon spheres saturated with an electrolyte. Unfortunately, the experiment is extremely sensitive to bed loading and gives inconsistent results. Nonetheless, a relationship between the effective conductivity and bed loading is confirmed; results of more rigorous efforts follow.

1. Point Contact

Batchelor and O'Brien [59] theoretically investigate the effective conductivity of granular material having large solid–fluid conductivity ratios. They first show that the volume-averaged heat flux through the medium is,

$$\langle \mathbf{q} \rangle = -k_f \langle \nabla T \rangle + n_c \langle \mathbf{s} \rangle \qquad (183)$$

where n_c is the number density of particles and $\mathbf{s}$ is the "thermal dipole strength" of a particle. The thermal dipole strength is defined by

$$\mathbf{s} = \left(1 - \frac{1}{\kappa}\right) \int_{A_p} \mathbf{x} \mathbf{q} \cdot \mathbf{n} \, dA \qquad (184)$$

where A_p is the surface area of a particle, $\mathbf{x}$ a vector locating a point on the particle surface, and $\mathbf{n}$ the surface normal.

Batchelor and O'Brien next argue that when the solid–fluid conductivity ratio is large, i.e., $\kappa \gg 1$, the following assumptions can be made. First, the second term on the right hand side of Eq. (184) is negligible. Second, temperature gradients are concentrated in the vicinity of contact points. Consequently if $\mathbf{x}_i$ is the position of a contact point and

$$q_{ci} = \int_{A_{ci}} \mathbf{q} \cdot \mathbf{n} \, dA \qquad (185)$$

is the heat flux through a contact region of approximate area A_{ci}, then

$$\mathbf{s} \approx \sum_i \mathbf{x}_i q_{ci} \qquad (186)$$

because contact points are discrete; their positions are determined by the packing arrangement. Third, temperature gradients removed from contact areas are small. Thus, Eq. (183) becomes

$$\langle \mathbf{q} \rangle \approx n_c \left\langle \sum_i \mathbf{x}_i q_{ci} \right\rangle \qquad (187)$$

which implies that the overall heat flux is essentially determined by properties of the contact regions.

Batchelor and O'Brien next consider two conduction problems to relate heat fluxes to contact region features. The first problem consists of spherical particles separated by a layer of fluid. The corresponding dimensionless heat flux is

$$H_1 = \ln\left(\frac{\hat{a}}{h}\right) - G(\tau) + \hat{K} \qquad (188)$$

where $\hat{a}$ is the mean particle radius, h the minimum particle separation distance, $\hat{K}$ a constant of order unity, and H the dimensionless heat flux through the contact region. The function G is related to the local temperature distribution and $\tau = \kappa^2 h/\hat{a}$ is the dimensionless particle separation distance. Batchelor and O'Brien compute this function and find that

$$\ln\left(\frac{\hat{a}}{h}\right) - G(\tau) \approx \ln \kappa^2 + \hat{K} - 3.9 - 0.1\tau \qquad (189)$$

when $\tau \leqslant 1$. Therefore, evaluating the limit, $h \to 0$ of Eq. (189) and substituting the result into Eq. (188) yields

$$H_2 = \ln \kappa^2 + \hat{K} - 3.9 \qquad (190)$$

which is the dimensionless heat flux through spherical particles in point contact.

The second conduction problem pertains to particles that contact at flat circular regions. A contact region is presumed to result from local particle deformation. According to Hertz theory [60], the radius of a region is

$$\rho_c = P_L^{1/3} \left[\frac{3(1-v^2)\hat{a}}{4E}\right]^{1/3} \quad (191)$$

where P_L, v, and E are the tangential compression force, Poisson ratio and elastic modulus, respectively. The corresponding dimensionless heat flux is

$$H_3 = H_c(\beta_r) + H_m(\beta_r) + \ln \kappa^2 + \hat{K} - 3.9 \quad (192)$$

where H_c is the dimensionless heat flux across the contact circle and H_m is the difference of the overall and point-contact heat fluxes. Both functions depend on the dimensionless contact radius, $\beta_r = \kappa \rho_c / \hat{a}$.

Next, Batchelor and O'Brien estimate the effective conductivity of the medium by relating the thermal dipole strength to dimensionless heat fluxes and packing arrangements. They argue that the volume-averaged thermal dipole strength is

$$\langle s \rangle = \frac{2}{3}\pi \hat{a}^3 k_s H n \langle \nabla T \rangle \quad (193)$$

where n is the average number of contact points for a particular packing arrangement (Table VII). Hence, the effective conductivity is

$$\eta = \frac{1-\phi}{2} nH \quad (194)$$

For example, the effective conductivity of a random distribution of spherical particles in point contact is obtained by substituting $\phi = 0.37$,

TABLE VII
SPHERICAL PARTICLE PACKING ARRANGEMENTS [59]

Packing arrangement	ϕ	Contact points (n)
Simple cubic	0.476	6
Body-centered cubic	0.32	8
Face-centered cubic	0.26	12
Random isotropic	0.37	6.5

$n = 6.5$, and Eq. (190) into Eq. (194); the result is

$$\eta = 4.0 \ln \kappa \tag{195}$$

when constants are neglected. Similar relations can be obtained for other packing arrangements and contact conditions, but Eqs. (188) and (192) do not yield simple algebraic expressions. In addition, Batchelor and O'Brien compare Eq. (195) to literature data and conclude that

$$\eta = 4.0 \ln \kappa - 11 \tag{196}$$

correlates literature data best.

Shonnard and Whitaker [54] attempt to experimentally verify the point-contact result of Batchelor and O'Brien [59]. The conduction cell is depicted in Fig. 31 and it consists of two 5 cm diameter machined hemispheres located in an acrylic tube. The hemispheres are brought into point contact by moving their relative positions until a sudden change in the electrical conductivity of the cell is detected. The test cell is heated from above and cooled from below. The heat flux through the test section is determined by measuring the

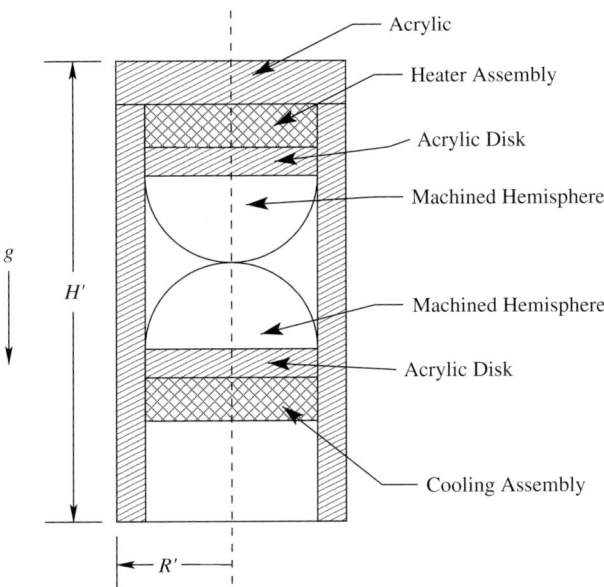

FIG. 31. Experimental apparatus used by Shonnard and Whitaker [54] to verify the point contact model of Batchelor and O'Brien [59]. (Adapted from Ref. [54], Copyright (1989) with permission from Elsevier.)

temperature gradient through acrylic disks located directly behind the hemispheres. The gradients are subsequently used to determine the effective conductivity of the test cell. Experiments are conducted with aluminum, copper, and stainless steel hemispheres saturated with air or helium. These solid–fluid combinations yield conductivity ratios of 100–12,000.

Shonnard and Whitaker find that accurately measuring point-contact effective conductivities is an experimental challenge. For example, the comparatively large dimensions of the test cell favor natural convection and thermal radiation. Consequently, they fill the void space with fiberglass insulation to suppress these modes, but achieve limited success. Additionally, the container and the reference disks have identical thermal conductivities, which compromises the unidirectional heat flow assumption. Shonnard and Whitaker estimate that these effects account for experimental uncertainties of 20 percent.

Despite these difficulties, Shonnard and Whitaker conclude that the functional form given by Batchelor and O'Brien [59] (Eq. (196)) is accurate. Furthermore, they report excellent correspondence with the effective conductivity measurements of Swift [23]. The latter finding is interesting because it implies that metal powders comprised of 100 μm particles and arrays comprised of 5 cm spheres have identical contact properties.

2. *Constriction Resistance*

The problem of conduction through arrays of particles that touch at flattened surfaces is considered in greater detail by Yovanovich [61], Kaganer [62], Chan and Tien [63], and Ogniewicz and Yovanovich [64]. For example, Yovanovich estimates the effective conductivity of elastic spheres bounded by plane surfaces and a rarefied gas. A conduction solution for cylinders of infinite extent bounded by isothermal planes is used to estimate the thermal resistance of the particles. The effective conductivity of the layer is estimated by combining the thermal resistance of the particles with thermal resistances corresponding to gas conduction and thermal radiation.

Kaganer [62] estimates the heat flux through a bed of spherical particles in a vacuum. He neglects thermal radiation and assumes that particles touch at flattened regions resulting from elastic deformation. To determine the thermal resistance of a particle, he solves the conduction equation and applies constant heat flux boundary conditions at the contact areas. The thermal analysis is simplified by approximating the contact regions by spherical segments. The segments are located at diametrically opposite positions on the sphere and given by

$$\theta_c = \arcsin R_c \tag{197}$$

where $R_c = \rho_c/\hat{a}$. The thermal resistance of the sphere is derived from the conduction solution; it is

$$R_{cc} = \frac{1}{\pi \rho_c k_s \sin \theta_c} \sum_{m=0}^{\infty} \frac{1}{2m+1} [P_{2m}(\cos \theta_c) - P_{2m+2}(\cos \theta_c)]$$
$$\times [1 + P_{2m+1}(\cos \theta_c)] \qquad (198)$$

where P denotes Legendre polynomials.

Kaganer next introduces a significant simplification. He notes that Carslaw and Jaeger [21, p. 216] consider the problem of conduction through a circular region into a semi-infinite half-space. Using their result for a constant heat flux boundary condition at the region, he derives

$$R_{cc} = \frac{16}{3\pi^2 \rho_c k_s} \qquad (199)$$

which is an analogous expression for the thermal resistance of a sphere. He compares this result to Eq. (198) and finds that it is an accurate approximation for $R_c < 0.02$. Consequently, he substitutes Eq. (199) for Eq. (198) when estimating the effective conductivity of the bed. The result is

$$\eta = \frac{3.12(1-\phi)^{4/3} P_c^{1/3} \kappa}{E^{1/3}} \qquad (200)$$

where P_c is a mean contact point compressive stress.

Chan and Tien [63] estimate the effective conductivities of packed beds comprised of solid, hollow, and coated spherical particles in a vacuum. Like Kaganer [62], they assume that particles contact at surfaces flattened by elastic deformation. They also determine thermal resistances by solving the conduction equation with constant heat flux boundary conditions at the contact regions. Moreover, contact regions are approximated by spherical segments in the thermal analysis.

Chan and Tien first estimate the thermal resistance of a solid particle in a simple-cubic packing arrangement. That is, the sphere has a pair of diametrically opposed contact points aligned with the overall heat flux vector. Using the conduction solution, they obtain

$$R_{cc} = \frac{2\hat{a}}{\pi k_s \rho_c^2 (1 - \cos \theta_c)} \sum_{m=1}^{\infty} \frac{[P_{2m-1}(\cos \theta_c) - P_{2m}(\cos \theta_c)]^2}{(2m-1)(4m-1)} \qquad (201)$$

for the thermal resistance of a sphere. One should note that Eq. (201) is based on the mean temperatures of the contact regions, which follows from the definition of thermal resistance.

Chan and Tien next consider solid particles in face- and body-centered packing arrangements. These arrays have three and four pairs of diametrically opposed contact points. Consequently, contact-point pairs are not parallel to the heat flux vector and the traditional definition of thermal resistance is inapplicable. A thermal resistance based on point temperatures of contact regions is used instead; it is given by

$$\hat{R}_{cc} = \frac{2\hat{a}}{\pi k_s \rho_c^2} \sum_{m=1}^{\infty} \frac{[P_{2m-2}(\cos \theta_c) - P_{2m}(\cos \theta_c)]}{(2m-1)} \qquad (202)$$

Also, Eq. (202) is related to Eq. (201) by $R_{cc} = S_r \hat{R}_{cc}$, where S_r is a parameter that depends on the contact parameter, R_c.

The thermal resistances of the face- and body-centered packing arrangements are obtained by superposing contact-pair heat fluxes; they are given by

$$R_{fc} = \frac{2\hat{a}}{3\pi k_s \rho_c^2} \sum_{m=1}^{\infty} \frac{[P_{2m-2}(\cos \theta_c) - P_{2m}(\cos \theta_c)]}{(2m-1)} \left[1 + 2P_{2m-1}\left(\frac{1}{2}\right)\right] \qquad (203)$$

and

$$R_{bc} = \frac{2\hat{a}}{4\pi k_s \rho_c^2} \sum_{m=1}^{\infty} \frac{[P_{2m-2}(\cos \theta_c) - P_{2m}(\cos \theta_c)]}{(2m-1)} \left[1 + P_{2m-1}\left(\frac{1}{3}\right)\right] \qquad (204)$$

respectively. Similar expressions are obtained for hollow and coated particles. Additionally, an investigation of the mathematical properties of the hollow particle result reveals that

$$R_{cc} = \frac{0.53}{k_s \rho_c} \qquad (205)$$

and

$$\hat{R}_{cc} = \frac{0.64}{k_s \rho_c} \qquad (206)$$

are excellent approximations to Eqs. (201) and (202) when $R_c < 0.1$.

TABLE VIII
Spherical Particle Packing Parameters [63]

Parameter	Simple cubic	Body-centered	Face-centered
ϕ	0.476	0.32	0.26
N_a	$1/4\hat{a}^2$	$3/16\hat{a}^2$	$1/2\sqrt{3}\hat{a}^2$
S_f	1	$\sqrt{3}/4$	$1/\sqrt{16}$
S_p	1.36	1.96	2.72
S_n	0.452	0.713	1.02

Chan and Tien next investigate limiting conditions for the effective conductivity of a packed bed in a vacuum. They first consider a bed subjected to an external load much greater than the weight of the constituent particles. The corresponding stress at a contact region is

$$P_c = S_f \frac{P_e}{N_a} \qquad (207)$$

where P_e is the externally applied compression stress. Also, the parameters S_f and N_a depend on the packing arrangement; they are presented in Table VIII. For small contact parameters, the thermal resistance of the contact region is described by Eqs. (205) or (206). The corresponding effective conductivity of the bed is

$$\eta = \kappa S_p \left(\frac{1-v^2}{E} P_c \right)^{1/3} \qquad (208)$$

where S_p is a parameter that captures the packing arrangement dependence (Table VIII).

The second case pertains to a packed bed loaded by particle weight. Consequently, the compressive stress is proportional to bed height and the thermal resistance of a layer of particles is inversely proportional to its height. Assuming uniform particles, the effective conductivity of the bed is

$$\eta = \kappa S_n \left[\frac{(1-v^2)\rho_s V_s L_b}{E \hat{a}^3} \right]^{1/3} \qquad (209)$$

where S_n is a parameter that depends on the packing arrangement and L_b is the height of the bed. Also, ρ_s and V_s are the mass density and volume of the particles.

Ogniewicz and Yovanovich [64] estimate the effective conductivity of regularly packed spherical particles subjected to compressive forces and surrounded by a rarefied gas. They first represent the thermal resistance of a contact region with

$$R_{cc} = \frac{1}{4k_s \rho_c} \quad (210)$$

which follows from approximating the particle by a semi-infinite half-space and assuming that the contact area is an isothermal circle [21, p. 215]. One should note that for small contact parameters, the isothermal and constant heat flux boundary conditions are essentially interchangeable [62]. They next estimate the thermal resistance of the gas in the vicinity of the contact region. This is accomplished by conducting a detailed analysis of one-dimensional heat conduction through a rarefied gas layer having a variable thickness.

Ogniewicz and Yovanovich use thermal resistances to estimate the effective conductivity of particles arranged in the simple-cubic, face-, and body-centered packing configurations. In contrast to Chan and Tien [63], Ogniewicz and Yovanovich vary the orientation of the heat flux and load vectors with respect to the packing pattern. Consequently the effective conductivity is determined by dividing the packing arrangement into unit cells featuring isotherms and adiabats that are perpendicular and parallel to the heat flux vector. The particle arrangement in a unit cell follows from assuming that only contact planes perpendicular to the heat flux and load vectors participate in conduction. The unit cells are subsequently divided into contact regions and gas layers. The effective conductivity of the unit cell is obtained from a thermal network analysis.

Lastly, Ogniewicz and Yovanovich compare model predictions to effective conductivity measurements of particles subjected to mechanical loads in the presence of rarefied gases. The first experiment consists of 25.4 mm spheres sandwiched between horizontal flat surfaces [65]. With the exception of light loads, Ogniewicz and Yovanovich find that model and experiment correspond well. They next compare model predictions to measurements of face-centered particle arrays in a vacuum. The test cell consists of 32 mm steel spheres sandwiched between vertical weighted rods. The upper rod is heated and the lower rod is cooled. Both rods have known thermal conductivities and they are instrumented to determine the heat flux through the test section. The pressure of the saturating gas is varied by locating the experimental apparatus in a vessel. Ogniewicz and Yovanovich again find excellent correspondence between model and experiment.

3. *Solid Mechanics*

The most detailed analyses of the effective conductivity of packed beds consider both solid mechanics and heat conduction problems. For example,

Buonanno and Carotenuto [66] consider a unit cell comprised of infinitely long elastic cylinders surrounded by a fluid. They assume that the cylinders contact and that the vertical force is proportional the height of the bed, i.e., resulting from particle weight. They consider beds comprised of glass, lead, bronze, aluminum, and stainless steel cylinders and employ the Galerkin finite element method to solve the corresponding solid mechanics problems. They demonstrate that the radii of both horizontal and vertical contact regions are proportional to bed height and that the latter is an order of magnitude smaller than the former. Incidentally, they compare the radii of horizontal contact regions to Hertz theory (Eq. (191)) and find that the difference is small, although it depends on the bed height.

Buonanno and Carotenuto next estimate the effective conductivity of the unit cell using the Galerkin finite element method. The same grid is employed, but it is adjusted to reflect particle deformation. They consistently find however, that this method significantly exaggerates the effective conductivity. They hypothesize that the over-estimation is a consequence of neglecting the thermal contact resistance that arises from irregular surfaces. They introduce contact resistance by assuming that particles are connected by rectangular slabs; hence the unit cells resemble Fig. 29. The lengths of the slabs are adjusted until the thermal conductance of the slabs equals the contact conductance predicted by a correlation [67].

Buonanno and Carotenuto subsequently obtain estimates for the effective conductivity of beds having various heights and particle roughness. They compare model predictions to literature data, but they note that insufficient information is available to yield meaningful comparisons. They successfully demonstrate, however, that bed height and surface roughness could account for significant variations in the measured effective conductivities of otherwise identical solid–fluid systems.

To employ a more realistic representation of a packed bed, Buonanno and Carotenuto [68] consider three-dimensional unit cells comprised of spherical particles. In contrast to Buonanno and Carotenuto [66], they employ Eq. (191) to estimate the dimensions of the contact area and incorporate contact resistance by modifying the thermal conductivity of the contact region. They also investigate the effect of an oxide layer surrounding the particles. In addition, they consider the simple-cubic and body-centered packing arrangements to estimate the limits for the effective conductivity when bed height and surface roughness are varied. They demonstrate that the model predictions bound much of the effective conductivity data found in the literature.

Buonanno et al. [69] investigate the effective conductivity of packed beds comprised of steel spheres and air. First, they experimentally determine the effective conductivity of four sets of 19.05 mm spheres arranged in the simple-cubic and face-centered packing arrangements. The bed is loaded by particle weight. Next, they develop a two-dimensional axisymmetric numerical model of the packed bed. This model neglects the packing arrangement, but incorporates an elasto-plastic model of local asperite deformation. Thus thermal contact resistance is captured without employing adjustable parameters. Finally, they measure the surface roughness of the four sets of spherical particles and use this information to compare model predictions to experimental results. They find that model predictions generally fall within the limits of experimental uncertainty.

In contrast to Buonanno et al. [69], Widenfeld et al. [7] experimentally investigate the effective conductivity of packed beds subjected to externally applied loads. The beds consist of 100 µm limestone particles and 1 and 0.5 mm steel spheres. The saturating fluid is air at atmospheric pressure. Experiments are conducted by increasing the axial load to a specified load and then decreasing it at a controlled rate. Their experiments demonstrate that the effective conductivity increases significantly with applied load. Additionally, they find that the relationship between the effective conductivity and load exhibits hysteresis. They note that this effect is greatest for the limestone powder and smallest for the large steel spheres. They attribute hysteresis to particle rearrangement and localized plastic deformation. Therefore, they conclude that the load history of a packed bed must be specified to predict its effective conductivity.

D. Closing Comments on Large Conductivity Ratio Media

The salient feature of large conductivity ratio media is that in the absence of thermal radiation, heat predominantly flows through small regions where particles meet, i.e., contact points. The sizes and distributions of these regions depend on: the mechanical properties of the solid phase; the load history of the bed, and the packing arrangement. Hence, the effective conductivity becomes extremely sensitive to microstructure morphology and predicting this property necessitates the development of novel modeling paradigms. The large solid–fluid conductivity ratio modeling strategies presented in this review include the method of volume averaging, empirical contact parameters, and contact models. The key results are summarized in Table IX. Some modeling strategies not addressed in this review include thermal particle dynamics [70,71], contact angle effects [72], and bed structure models [73,74].

The success of large solid–fluid conductivity ratio modeling strategies is mixed. For example, the method of volume averaging captures the medium morphology, but employs an adjustable parameter to incorporate particle contact. Adjustable parameters can also be used to adapt lumped-parameter

TABLE IX
Summary of Effective Conductivity Models Applicable to Large Solid–Fluid Conductivity Ratio Media

Name	Model description	Result	Reference
Discontinuous Fluid-phase model of Nozad et al.	Two-dimensional model featuring (empirical) contacting rectangular particles	Fig. 25	[51]
Hadley correlation	Correlation obtained from volume-averaged equations and mixture-rule based closure relation	Eq. (158)	[13]
Modified Zehner and Schlünder model	Lumped parameter conduction model with empirical contact relation	Eq. (165)	[46]
Contacting rectangle model	Two-dimensional lumped parameter conduction model with empirical contact relation	Eq. (168)	[57]
Contacting cylinder model	Two-dimensional lumped parameter conduction model with empirical contact relation (use with caution)	Eq. (177)	[57]
Contacting cube model	Three-dimensional lumped parameter conduction model with empirical contact relation	Eq. (180)	[57]
Point contact	Analytial result for two spheres making point contact	Eq. (196)	[59]
Flattened spheres	Analytical result for a bed of spherical particles in a vacuum	Eq. (200)	[62]
Externally loaded packed bed	Analytical result for spherical particles in a vacuum	Eq. (208)	[63]
Internally Loaded Packed bed	Analytical result for spherical particles in a vacuum	Eq. (209)	[63]

and conduction models to the large solid–fluid conductivity ratio case. More fundamental models of large solid–fluid conductivity ratio packed beds incorporate solid mechanics, but success is limited because little experimental information pertaining to the loading state or structure of practical systems is available.

VI. Conclusion

Although the effective thermal conductivity problem is unsolved, several aspects of the problem are understood. Specifically, the effective conductivity depends on the solid–fluid conductivity ratio, the volume fractions of the constituent phases, and the geometry of the solid–fluid interface. The latter dependency causes the most difficulty because it is the least precisely known feature of a porous medium. Additionally, the relationship between the effective conductivity and medium morphology depends on the solid–fluid conductivity ratio. To clarify this relationship, this article has examined the effective conductivity problem within the context of small ($1 \leqslant \kappa < 10$), intermediate ($10 \leqslant \kappa < 10^3$), and large ($\kappa \geqslant 10^3$) solid–fluid conductivity ratios; the key results follow.

First, the effective conductivity is least sensitive to interfacial geometry when the solid–fluid conductivity ratio is small. The success of geometry independent models such as the Maxwell formulae, provide ample evidence that the effective conductivity mostly depends on the volume fractions and conductivities of the constituent phases. These models also establish physical limits for the effective conductivity.

Second, the effective conductivity is mildly sensitive to morphology when the solid–fluid conductivity ratio is moderate. Consequently, interface geometry must be taken into account when predicting the effective conductivity, but this can be accomplished in several ways. For example, the unidirectional heat flow, lumped parameter, and resistor-network modeling strategies are able to capture this dependency to some degree. Three-dimensional conduction and lumped parameter models appear to be most successful.

Finally, the effective conductivity is most sensitive to interface geometry when the solid–fluid conductivity ratio is large. This property results from essentially all of the heat transfer occurring through particle contact regions. The state of the art is to model conduction through such regions empirically or, at best, semi-empirically. Although prediction of particle deformation and hence contact area have been attempted, the literature has yet to develop sufficiently to provide a definitive paradigm for overall modeling and interpretation of experiments. For very large κ and when the solid-phase morphology is unspecified, either averaging methods or statistical methods must be used. Unfortunately, averaging methods bring with them closure issues, much like the Reynolds averaging of the turbulent transport equations for energy and momentum in a single-phase fluid. Statistical methods are beyond the scope of the current review, and the reader is referred to Torquato [75] for an introduction to the framework of the subject.

Acknowledgment

Sandia is a multiprogram laboratory operated by Sandia Corporation, a Lockheed Martin Company, for the United States Department of Energy's National Nuclear Security Administration under contract DE-AC04-94AL85000.

Nomenclature

ROMAN SYMBOLS

A°	empirical parameter (unitless), Eq. (67)	c_p	constant pressure specific heat (kJ/kg K), Eq. (5)
A_p	particle surface area (m^2), Eq. (184)	D	chamber inside diameter (m)
		D_i	inner diameter (m), Eq. (7)
A_{ci}	contact point area (m^2), Eq. (185)	D_o	outer diameter (m), Eq. (7)
A_{fs}	area of the solid-fluid interface (m^2), Eq. (92)	d_p	particle diameter (m)
		E	elastic modulus (MPa), Eq. (191)
a	correlation constant (unitless), Eq. (35)	F_c	calibration constant (unitless), Eq. (31)
a	particle characteristic length (m)	f	closure function (unitless), Eq. (154)
$\hat{a}$	mean particle radius (m), Eq. (188)	f_o	unconsolidated parameter (unitless), Eq. (158)
B	shape parameter (unitless), Eq. (75)	G	contact point temperature distribution (unitless), Eq. (188)
B°	empirical parameter (unitless), Eq. (68)	g	gravitational acceleration constant (9.81 m/s^2)
b	correlation constant (unitless), Eq. (35)	$g'(t)$	initial condition for the fluid phase fluctuation (°C), Eq. (124)
$\mathbf{b}_f$	vector field that maps $\nabla\langle T\rangle$ to $\tilde{T}_f$ (m), Eq. (126)	$g(t)$	initial fluid temperature (°C), Eq. (84)
$\mathbf{b}_s$	vector field that maps $\nabla\langle T\rangle$ to $\tilde{T}_s$ (m), Eq. (127)	$\mathbf{g}''(t)$	vector field initial condition (m), Eq. (132)
C	curve fit parameter (unitless), Eq. (79)	H	dimensionless contact point heat flux (unitless), Eq. (188)
c	contact region characteristic length (m)	H_c	dimensionless heat flux across a contact circle (unitless), Eq. (192)
$\hat{c}$	constant equal to exp $\hat{\gamma}$ (unitless), Eq. (29)	H_m	dimensionless difference in overall and point contact heat fluxes (unitless), Eq. (192)
c_1	thermal conductivity temperature coefficient (W/m K^2), Eq. (10)	H_c	calibration constant (W/m K), Eq. (31)

H'	height of the point-contact test chamber (m), Eq. (196)	L_M	macroscopic characteristic dimension (m)
h	minimum particle separation distance (m), Eq. (188)	L_b	packed bed height (m), Eq. (209)
$h'(t)$	initial condition for the solid phase fluctuation (°C), Eq. (125)	l	thermocouple array position (m)
		l_f	fluid phase characteristic dimension (m)
$h(t)$	initial solid temperature (°C), Eq. (85)	l_s	solid phase characteristic dimension (m)
$\hat{h}$	height of plates that connect circular cylinders (m), Eq. (171)	l_s	effective slab thickness of the spherical solid phase (m), Eq. (69)
$\mathbf{h}''(t)$	vector field initial condition (m), Eq. (133)	l_v	effective slab thickness of the fluid film in the region of the contact point (m), Eq. (69)
I	thermal current (W)		
$\mathbf{I}$	identity tensor (unitless), Eq. (143)	$\mathbf{l}$	lattice vectors that define a spatially periodic medium (m), Eq. (141)
i	summation index (unitless), Eq. (145)	m	summation index (unitless), Eq. (27)
K	population electric susceptibility (unitless), Eq. (42)	m_c	curve fit parameter (unitless), Eq. (79)
K_0	sample electric susceptibility (unitless), Eq. (43)	N_a	particles per unit area (unitless), Eq. (207)
$\hat{K}$	constant (unitless), Eq. (188)	n	number of contact points (unitless), Eq. (71)
$\mathbf{K}_e$	effective conductivity tensor (W/m K), Eq. (144)	n_c	number density of particles (1/m^3), Eq. (183)
k	local effective thermal conductivity (W/m K), Eq. (9)	n_p	number of phases (unitless), Eq. (145)
k_e	effective thermal conductivity (W/m K), Eq. (2)		
k_f	fluid thermal conductivity (W/m K)	$\mathbf{n}$	particle surface normal vector (unitless), Eq. (184)
k_s	solid thermal conductivity (W/m K)	$\mathbf{n}_{fs}$	interfacial normal vector (unitless), Eq. (83)
k_{sf}	combined solid-fluid effective conductivity (W/m K), Eq. (73)	$\mathbf{n}_{sf}$	interfacial normal vector (unitless)
k_1	reference effective thermal conductivity (W/m K), Eq. (9)	P	legendre polynomial (unitless), Eq. (198)
k_m	mean effective thermal conductivity (W/m K), Eq. (9)	P_c	average contact point compression stress (N/m^2), Eq. (200)
k_c	thermal contact conductivity (W/m K), Eq. (182)	P_c	contact point compression stress (MPa), Eq. (207)
k_{sf1}	composite layer effective conductivity (W/m K), Eq. (167)	P_e	external compression stress (MPa), Eq. (207)
k_{sf2}	composite layer effective conductivity (W/m K), Eq. (167)	P_L	tangential compression force (N), Eq. (191)
$\mathbf{k}_i$	phase thermal conductivity vector (W/m K), Eq. (147)	p	volume fraction of particles (unitless), Eq. (36)
$\mathbf{k}_e$	effective thermal conductivity vector (W/m K), Eq. (147)	$\hat{p}$	shape parameter (unitless), Eq. (62)
L	unit cell length (m)	Q	total energy dissipated in the heater (W), Eq. (7)
L_c	chamber length (m), Eq. (3)		
L_p	effective length between centers of spherical particles (m), Eq. (69)	q	matrix volume fraction (unitless), Eq. (54)

q''	unidirectional heat flux (W/m²), Eq. (2)	T_B	bottom boundary temperature (°C), Eq. (3)
q_w	wire energy dissipation rate (W/m), Eq. (23)	T_T	top boundary temperature (°C), Eq. (3)
q_{ci}	contact point heat flux (W/m²), Eq. (185)	T_i	inner temperature (°C), Eq. (7)
q	heat flux vector (W/m²), Eq. (183)	T_o	outer temperature (°C), Eq. (7)
R	unit cell outer radius (m), Eq. (73)	T_f	point-wise fluid-phase temperature (°C), Eq. (80)
R_1	matrix electrical resistance (Ω), Eq. (36)	T_s	point-wise solid-phase temperature (°C), Eq. (81)
R_2	particle electrical resistance (Ω), Eq. (36)	$\tilde{T}_f$	spatial temperature fluctuation of the fluid phase (°C), Eq. (100)
R_c	contact parameter (unitless)	$\tilde{T}_s$	spatial temperature fluctuation of the solid phase (°C), Eq. (101)
R_e	electrical resistance of a composite medium (Ω), Eq. (36)	$\hat{T}_f$	fluid-phase temperature deviation (°C), Eq. (107)
R_{bc}	sphere thermal resistance for body-centered packing arrangement (K/W), Eq. (204)	$\hat{T}_s$	solid-phase temperature deviation (°C), Eq. (108)
R_{cc}	sphere thermal resistance for simple-cubic packing arrangement (K/W), Eq. (198)	$\langle T \rangle$	spatially averaged temperature (°C), Eq. (106)
		$\langle T_f \rangle$	superficial fluid-phase temperature (°C), Eq. (86)
$\hat{R}_{cc}$	modified sphere thermal resistance (K/W), Eq. (202)	$\langle T_s \rangle$	superficial solid-phase temperature (°C), Eq. (87)
R_{fc}	sphere thermal resistance for face-centered packing arrangement (K/W), Eq. (203)	$\langle T_f \rangle^f$	fluid intrinsic average temperature (°C), Eq. (94)
R_f	fluid thermal resistance (K/W)	$\langle T_s \rangle^s$	solid intrinsic average temperature (°C), Eq. (95)
R_s	solid thermal resistance (K/W)	t	time (s), Eq. (4)
R_L	lateral thermal resistance (K/W)	t'	integration variable (s), Eq. (23)
R'	outer radius of the point-contact test chamber (m), Eq. (196)	t_1	first reference time (s), Eq. (30)
		t_2	second reference time (s), Eq. (30)
r	radial coordinate (m), Eq. (68)	u	substitution variable (unitless), Eq. (24)
r	radial dimension (m), Eq. (75)		
r_0	REV radius (m)	V_f	fluid-phase volume (m³), Eq. (86)
r_s	contact area radius (m), Eq. (161)	V_p	particle volume (m³), Eq. (1)
		V_t	total sample volume (m³), Eq. (1)
r	macroscopic position vector (m)	V_{REV}	averaging volume (m³), Eq. (86)
r$_{REV}$	microscopic position vector (m)	V_s	solid-phase volume (m³), Eq. (89)
S_f	contact point load parameter (unitless), Eq. (207)	V_s	particle volume (m³), Eq. (209)
		x	position vector (m), Eq. (184)
S_n	internally loaded bed parameter (unitless), Eq. (209)	**x**$_i$	contact point position vector (m), Eq. (186)
S_p	externally loaded bed parameter (unitless), Eq. (208)	z	axial coordinate (m)
		z	complex variable (unitless), Eq. (27)
S_r	sphere thermal resistance parameter (unitless)		
s	thermal dipole strength vector (W m), Eq. (183)		
T	average temperature (°C), Eq. (2)		

GREEK SYMBOLS

α_e effective thermal diffusivity (m²/s), Eq. (5)

α_p	consolidation parameter (unitless), Eq. (158)	θ_1	first reference temperature (°C), Eq. (30)
α_0	particle deformation parameter (unitless), Eq. (160)	θ_2	second reference temperature (°C), Eq. (30)
β	geometrical parameter (unitless), Eq. (69)	θ_o	boundary temperature (°C), Eq. (17)
β_r	dimensionless contact radius (unitless), Eq. (192)	θ_c	particle contact angle (rad), Eq. (172)
γ	geometrical parameter (unitless), Eq. (69)	θ_p	contact point angle (unitless), Eq. (71)
γ_a	particle size parameter (unitless), Eq. (167)	κ	solid–fluid thermal conductivity ratio (unitless), Eq. (34)
γ_c	contact parameter (unitless), Eq. (167)	λ	geometry-dependent parameter (unitless), Eq. (41)
γ_c	electromagnetic unit constant (unitless), Eq. (42)	μ	geometry-dependent parameter (unitless), Eq. (52)
$\hat{\gamma}$	Euler's constant (unitless), Eq. (27)	ν	poisson ratio (unitless), Eq. (191)
δ'	difference in dielectric constants (F/m), Eq. (54)	ξ	complimentary error function argument (unitless), Eq. (18)
ε	composite medium dielectric constant (F/m), Eq. (42)	ξ_1	closure constant (°C), Eq. (126)
		ξ_2	closure constant (°C), Eq. (127)
ε_f	fluid-phase volume fraction (unitless), Eq. (96)	ρ	mass density (kg/m³), Eq. (5)
ε'	arithmetic mixture rule dielectric constant (F/m), Eq. (53)	ρ_c	radius of flattened contact region (m), Eq. (191)
ε_A	dielectric constant of material A (F/m), Eq. (42)	ρ_s	mass density of particles (kg/m³), Eq. (209)
ε_B	dielectric constant of material B (F/m), Eq. (42)	$(\rho c_p)_f$	fluid phase heat capacity (J/mK), Eq. (6)
ε_i	volume fraction of phase i (unitless), Eq. (148)	$(\rho c_p)_s$	solid phase heat capacity (J/mK), Eq. (6)
ε_s	solid-phase volume fraction (unitless), Eq. (97)	τ	dimensionless particle separation distance (unitless), Eq. (188)
η	effective-fluid thermal conductivity ratio (unitless), Eq. (34)	ϕ	porosity, volume fraction of void space (unitless), Eq. (1)
η_{VAC}	vacuum effective conductivity ratio (unitless), Eq. (159)	Ψ	geometrical parameter (unitless), Eq. (69)
Θ	temperature ratio (unitless), Eq. (21)	Ψ_1	loose packing limit geometrical parameter (unitless), Eq. (72)
θ	temperature (°C), Eq. (28)	Ψ_2	close packing limit geometrical parameter (unitless), Eq. (72)

References

1. Nield, D. A. and Bejan, A. (1999). "Convection in Porous Media", 2nd edn. Springer, New York.
2. Roblee, L. H. S., Baird, R. M., and Tierney, J. W. (1958). Radial porosity variations in packed beds. *AIChE J.* **4**(4), 460–464.

3. Glatzmaier, G. C. and Ramirez, W. F. (1988). Use of volume averaging for the modeling of thermal properties of porous materials. *Chem. Eng. Sci.* **43**(12), 3157–3169.
4. Whitaker, S. (1999). "The Method of Volume Averaging". Kluwer Academic Publishers, Dordrecht.
5. Lage, J. L. (1999). The implications of the thermal equilibrium assumption for surrounding-driven steady conduction within a saturated porous medium layer. *Int. J. Heat Mass Transfer* **42**(3), 447–485.
6. Woodside, W. and Messmer, J. H. (1961). Thermal conductivity of porous media. I. unconsolidated sands. *J. Appl. Phys.* **32**(9), 1688–1699.
7. Widenfeld, G., Weiss, Y., and Kalman, H. (2003). The effect of compression and preconsolidation on the effective thermal conductivity of particulate beds. *Powder Technol.* **133**(1–3), 15–22.
8. Prasad, V., Kladias, N., Bandyopadhaya, A., and Tian, Q. (1989). Evaluation of correlations for stagnant thermal conductivity of liquid-saturated porous beds of spheres. *Int. J. Heat Mass Transfer* **32**(9), 1793–1796.
9. Chen, Z. Q., Cheng, P., and Hsu, C. T. (2000). A theroetical and experimental study on stagnant thermal conductivity of bi-dispersed porous media. *Int. Commun. Heat Mass Transfer* **27**(5), 601–610.
10. Calmidi, V. V. and Mahajan, R. L. (1999). The effective thermal conductivity of high porosity fibrous metal foams. *J. Heat Transfer* **121**(2), 466–471.
11. Bhattacharya, A., Calmidi, V. V., and Mahajan, R. L. (2002). Thermophysical properties of high porosity metal foams. *Int. J. Heat Mass Transfer* **45**(5), 1017–1031.
12. Ofuchi, K. and Kunii, D. (1965). Heat transfer characteristics of packed beds with stagnant fluids. *Int. J. Heat Mass Transfer* **8**(5), 749–757.
13. Hadley, G. R. (1986). Thermal conductivity of packed metal powders. *Int. J. Heat Mass Transfer* **29**(6), 909–920.
14. Yagi., S. and Kunii, D. (1957). Studies on effective thermal conductivities in packed beds. *AIChE J.* **3**(3), 373–381.
15. Deissler, R. G. and Eian, C. S. (1952). Investigation of effective thermal conductivities of powders. Tech. Rep. NACA RM E52C05, National Advisory Committee for Aeronautics.
16. Deissler, R. G. and Boegli, J. S. (1958). An investigation of effective thermal conductivities of powders in various gases. *ASME Trans.* **80**(7), 1417–1425.
17. Slavin, A. J., Londry, F. A., and Harrison, J. (2000). A new model for the effective thermal conductivity of packed beds of solid spheroids: alumina in helium between 100 and 500°C. *Int. J. Heat Mass Transfer* **43**(12), 2059–2073.
18. Aichlmayr, H. T. (1999) The Effective Thermal Conductivity of Saturated Porous Media, Master's Thesis, The University of Minnesota.
19. Nozad, I., Carbonell, R. G., and Whitaker, S. (1985). Heat conduction in multiphase systems II experimental method and results for three-phase systems. *Chem. Eng. Sci.* **40**(5), 857–863.
20. Lacroix, C., Bala, P. R., and Feidt, M. (1999). Evaluation of the effective thermal conductivity in metallic porous media submitted to incident radiative flux in transient conditions,. *Energy Conv. Manag.* **40**(15–16), 1775–1781.
21. Carslaw, H. S. and Jaeger, J. C. (1959). "Conduction of Heat in Solids", 2nd edn. Clarendon Press, Oxford, UK.
22. Abramowitz, M. and Stegun, I. A. (1965). "Handbook of Mathematical Functions". Dover, New York, NY.
23. Swift, D. L. (1966). The thermal conductivity of spherical metal powders including the effect of oxide coating. *Int. J. Heat Mass Transfer* **9**(10), 1061–1074.

24. Tavman, I. H. (1996). Effective thermal conductivity of granular porous materials. *Int. Commun. Heat Mass Transfer* **23**(2), 169–176.
25. Glatzmaier, G. C. and Ramirez, W. F. (1985). Simultaneous measurement of the thermal conductivity and thermal diffusivity of unconsolidated materials by the transient hot wire method. *Rev. Sci. Instrum.* **58**(7), 1394–1398.
26. Gori, F., Marino, C., and Pietrafesa, M. (2001). Experimental measurements and theoretical predictions of the thermal conductivity of two phases glass beads. *Int. Commun. Heat Mass Transfer* **28**(8), 1091–1102.
27. Krupiczka, R. (1967). Analysis of thermal conductivity in granular materials. *Int. Chem. Eng.* **7**(1), 122–144.
28. Preston, F. W. (1957). Mechanism of Heat Transfer in Unconsolidated Porous Media at Low Flow Rates, Ph.D. Thesis, Pennsylvania State University.
29. Waddams, A. L. (1944). The flow of heat through granular material. *J. Soc. Chem. Ind.* **63**, 336–340.
30. Brown, W. F., Jr. (1955). Solid mixture permitivities. *J. Chem. Phys.* **23**(8), 1514–1517.
31. Rayleigh, L. (1892). On the influence of obstacles arranged in rectangular order upon the properties of a medium. *Phil. Magazine* **34**, 481–502.
32. Maxwell, J. C. (1891). "*A Treatise on Electricity and Magnetism*," Vol. 1, 3rd edn. Clarendon Press, Oxford.
33. Jeffrey, D. J. (1973). Conduction through a random suspension of spheres. *Proc. Roy Soc. Lond. A* **335**, 355–367.
34. O'Brien, R. W. (1979). A method for the calculation of the effective transport properties of suspensions of interacting particles. *J. Fluid Mech.* **91**(1), 17–39.
35. Hashin, Z. and Shtrikman, S. (1962). A variational approach to the theory of the effective magnetic permeability of multiphase materials. *J. Appl. Phys.* **33**(10), 3125–3131.
36. Lichteneker, K. (1926). *Phys. Z.* **27**, 115–158.
37. Nield, D. A. (1991). Estimation of the stagnant thermal conductivity of saturated porous media. *Int. J. Heat Mass Transfer* **34**(6), 1575–1576.
38. Schumann, T. and Voss, V. (1934). Heat flow through granular material. *Fuel Sci. Practice* **13**(8), 249–256.
39. Wilhelm, R. H., Johnson, W. C., Wynkoop, R., and Collier, D. W. (1948). Reaction rate, heat transfer, and temperature distribution in fixed-bed catalytic converters. *Chem. Eng. Prog.* **44**(2), 105–116.
40. Gemant, A. (1950). The thermal conductivity of soils. *J. Appl. Phys.* **21**, 750–752.
41. Woodside, W. (1958). Calculation of the thermal conductivity of porous media. *Can. J. Phys.* **36**, 815–823.
42. Wakao, N. and Kato, K. (1969). Effective thermal conductivity of packed beds. *J. Chem. Eng. Jpn.* **2**(1), 24–33.
43. Wakao, N. and Vortmeyer, D. (1971). Pressure dependence of effective thermal conductivity of packed beds. *Chem. Eng. Sci.* **26**(10), 1753–1765.
44. Kunii, D. and Smith, J. M. (1960). Heat transfer characteristcs of porous rocks. *AIChE J* **6**(1), 71–78.
45. Zehner, P. and Schlünder, E. U. (1970). Thermal conductivity of granular materials at moderate temperatures. *Chem. Ing. Tech.* **42**(14), 933–941.
46. Hsu, C. T., Cheng, P., and Wong, K. W. (1994). Modified Zehner–Schlunder models for stagnant thermal conductivity of porous media. *Int. J. Heat Mass Transfer* **37**(17), 2751–2759.
47. Crane, R. A. and Vachon, R. I. (1977). A prediction of the bounds on the effective thermal conductivity of granular materials. *Int. J. Heat Mass Transfer* **20**(7), 711–723.

48. Tsao, G. T. (1961). Thermal conductivity of two-phase materials. *Ind. Eng. Chem.* **53**(5), 395–397.
49. Cheng, S. C. and Vachon, R. I. (1969). The prediction of the thermal conductivity of two and three phase solid heterogeneous mixtures. *Int. J. Heat Mass Transfer* **12**(3), 249–264.
50. Jaguaribe, E. F. and Beasley, D. E. (1984). Modeling of the effective thermal conductivity and diffusivity of a packed bed with stagnant fluid. *Int. J. Heat Mass Transfer* **27**(3), 399–407.
51. Nozad, I., Carbonell, R. G., and Whitaker, S. (1985). Heat conduction in multiphase systems. I. Theory and experiment for two-phase systems. *Chem. Eng. Sci.* **40**(5), 843–855.
52. Quintard, M. and Whitaker, S. (1993). One- and two-equation models for transient diffusion processes in two-phase systems. *"Advances in Heat Transfer"*, Vol. 23. Academic Press, New York, pp. 369–464.
53. Ochoa-Tapia, J. A., Stroeve, P., and Whitaker, S. (1994). Diffusive transport in two-phase media: Spatially periodic models and Maxwell's theory for isotropic and anisotropic systems. *Chem. Eng. Sci.* **49**(5), 709–726.
54. Shonnard, D. R. and Whitaker, S. (1989). The effective thermal conductivity for a point-contact porous medium: An experimental study. *Int. J. Heat Mass Transfer* **32**(3), 503–512.
55. Sahraoui, M. and Kaviany, M. (1993). Slip and no-slip temperature boundary conditions at interface of porous, plain media: conduction. *Int. J. Heat Mass Transfer* **36**(4), 1019–1033.
56. Lund, K. O., Nguyen, H., Lord, S. M., and Thompson, C. (1999). Numerical correlation for thermal conduction in packed beds. *Can. J. Chem. Eng.* **77**(4), 769–774.
57. Hsu, C. T., Cheng, P., and Wong, K. W. (1995). A lumped parameter model for stagnant thermal conductivity of spatially periodic porous media. *J. Heat Transfer* **117**(2), 264–269.
58. Bauer, R. and Schlünder, E. U. (1978). Effective radial thermal conductivity of packings in gas flow. Part II. thermal conductivity of the packing fraction without gas flow. *Int. Chem. Eng.* **18**(2), 189–204.
59. Batchelor, G. K. and O'Brien, R. W. (1977). Thermal or electrical conduction through a granular material. *Proc. Roy Soc. Lond. A* **355**, 313–333.
60. Landau, L. D. and Lifshitz, E. M. (1970). "Theory of Elasticity". Pergamon, New York.
61. Yovanovich, M. M. (1967). Thermal contact resistance across elastically deformed spheres. *J. Spacecraft Rockets* **4**(1), 119–122.
62. Kaganer, M. G. (1966). Contact heat transfer in granular materials under vacuum. *J. Eng. Phys.* **11**(1), 19–22.
63. Chan, C. K. and Tien, C.-L. (1973). Conductance of packed spheres in vacuum. *J. Heat Transfer* **95**(3), 302–308.
64. Ogniewicz, Y. and Yovanovich, M. M. (1977). Effective conductivity of regularly packed spheres: Basic cell model with constriction. *In* "Heat and Thermal Control Systems" (L. S. Fletcher, ed.), Vol. 60 *Progress in Astronautics and Aeronautics*, pp. 209–228. American Institute of Aeronautics and Astronautics, Reston, VA.
65. Kitscha, W. W. and Yovanovich, M. M. (1975). Experimental investigation on the overall thermal resistance of sphere-flat contacts. *In* "Heat Transfer with Thermal Control Applications" (M. M. Yovanovich, ed.), Vol. 39 *Progress in Astronautics and Aeronautics*, pp. 93–100, American Institute of Aeronautics and Astronautics, Reston, VA.
66. Buonanno, G. and Carotenuto, A. (1997). The effective thermal conductivity of a porous medium with interconnected particles. *Int. J. Heat Mass Transfer* **40**(2), 393–405.
67. Mikic, B. B. (1974). Thermal contact conductance: Theoretical considerations. *Int. J. Heat Mass Transfer* **17**, 205–214.
68. Buonanno, G. and Carotenuto, A. (2000). The effective thermal conductivity of packed beds of spheres for a finite contact area. *Numer. Heat Transfer, Part A* **37**(4), 343–357.

69. Buonanno, G., Carotenuto, A., Giovinco, G., and Massarotti, N. (2003). Experimental and theoretical modeling of the effective thermal conductivity of rough steel spheroid packed beds. *J. Heat Transfer* **125**(4), 693–702.
70. Vargas, W. L. and McCarthy, J. J. (2002). Stress effects on the conductivity of particulate beds. *Chem. Eng. Sci.* **57**(15), 3119–3131.
71. Vargas, W. L. and McCarthy, J. J. (2002). Conductivity of granular media with stagnant interstitial fluids via thermal particle dynamics simulation. *Int. J. Heat Mass Transfer* **45**(24), 4847–4856.
72. Siu, W. W. M. and Lee, S. H. K. (2000). Effective conductivity computation of a packed bed using constriction resistance and contact angle effects. *Int. J. Heat Mass Transfer* **43**(21), 3917–3924.
73. Cheng, G. J., Yu, A. B., and Zulli, P. (1999). Evaluation of effective thermal conductivity from the structure of a packed bed. *Chem. Eng. Sci.* **54**(19), 4199–4209.
74. Kikuchi, S. (2001). Numerical analysis model for thermal conductivities of packed beds with high solid-to-gas conductivity ratio. *Int. J. Heat Mass Transfer* **44**(6), 1213–1221.
75. Torquato, S. (2001). "Random Heterogeneous Materials: Microstructure and Macroscopic Properties". Springer, New York.

Mesoscale and Microscale Phase-Change Heat Transfer

P. CHENG and H.Y. WU

School of Mechanical & Power Engineering, Shanghai Jiaotong University, Shanghai, China

I. Introduction

The miniaturization of heat exchangers is made feasible by the developments in aluminum extrusion and brazing processes for manufacturing of mesochannels (or minichannels) beginning more than 20 years ago. Typical tubes produced by this technology have parallel rectangular ports with hydraulic diameters on the order of 1–3 mm. These compact heat exchangers have been widely used in automotive and aerospace applications [1]. For this reason, much work on the study of boiling and condensation heat transfer in mesochannels has been carried out in recent years. Because heat transfer rates in these mesochannels (or minichannels) are considerably higher than those in conventional channels, meso (mini) heat exchangers are more compact and lighter in weight for a given heat transfer capacity.

Meanwhile, advances in microelectronic fabrication technology have led to the miniaturization of silicon components. This coupled with the improved processing speed in clock frequencies from megahertz (MHz) to gigahertz (GHz) has generated increasingly larger amount of heat in microprocessors. The large amount of heat, if not properly dissipated, will cause overheating of a chip, leading to its degrading performance and eventual damage. In the year 2000, heat generated in a microprocessor at a frequency of 70 MHz, was about 10 W [2]. By the year of 2004, the heat dissipation increased to about 50 W at the frequency range of 2–3 GHz. Based on processor's die surface area of 1 cm^2, the heat flux of the microprocessor has increased five times during the past 5 years from 10 to 50 W/cm^2 [2]. It is anticipated that this rate of increasing heat flux in microprocessors will continue in the future. To cope with the demand for improved cooling technology for the next generation of high-power electronics devices (such as high-power laser diodes, power transistors, digital mirror devices in projectors, etc.), various kinds of meso-/microheat pipes and two-phase microchannel heat sinks [3] utilizing latent heat of evaporation for more efficient heat transfer have been developed in recent years.

Advances in microfabrication not only led to the miniaturization of microelectronic components but also led to many MEMS products. One of the most successful MEMS products is the thermal inkjet printhead. The technology was developed by Canon [4] and Hewlett Packard (HP) [5] in the early 1980s. The thermal inkjet printhead uses a microheater under electrical pulse of several microseconds (μs) to generate a microbubble periodically. The microbubble will expel a small drop of ink through a nozzle at a high frequency to a specified position on a paper to compose the text and graphics. The quality of the print depends on the size of the drops and drop velocity which is closely related to bubble generation. Recently, there is a great deal of interest in using thermal bubbles as a thermal actuator [6] and for DNA hybridization enhancement [7,8]. For this reason, much attention has recently been given to microbubble generation and collapse under pulse heating conditions.

In this review, we will focus our attention to boiling and condensation heat transfer processes occurring in various meso-/microdevices such as heat exchangers, heat sinks, heat pipes, thermal inkjet printheads, and thermal microbubble actuators. Particular emphasis will be placed on the review of most recent literature on phase change heat transfer phenomena in these devices.

II. Meso-(Mini-)/Microdevices Involving Phase-Change Heat Transfer Processes

A. Classification of Meso-(Mini-) and Microsystems: The Bond Number

In the past, classifications of micro-, meso-(mini-), and macrochannels involving phase change heat transfer have been based on the value of its hydraulic diameter in comparison to some arbitrarily chosen thresholds values which have no physical significance [1,9,10]. In a recent paper, Li and Wang [11] carried out an analysis to study the gravity effect on the transition from symmetry flow (where the gravity influence can be ignored) to asymmetry flow (i.e., stratified flow due to gravity) patterns during condensing flow in meso-/microtubes. Based on the Young–Laplace equation, they obtained the following critical and threshold values of tube diameter, d_c, and d_{th}, in terms of the capillary length ℓ_c as

$$d_c = 0.224\ell_c \tag{1a}$$

and

$$d_{th} = 1.75\ell_c \tag{1b}$$

with ℓ_c given by

$$\ell_c = \left[\frac{\sigma}{(\rho_\ell - \rho_v)g}\right]^{1/2} \quad (1c)$$

where σ, g, ρ_ℓ, and ρ_v are the surface tension, gravitational acceleration, densities of the liquid and vapor at the saturated pressure, respectively. Note that the critical diameter given by Eq. (1a) is comparable in magnitude to the bubble departure diameter given by Fritz [12] as

$$d_{bub} = 0.0208\beta\ell_c \quad (2)$$

if $\beta = 10°$ is used for the contact angle in degree. Dupont and Thome [13] show that Fritz's correlation with $\beta = 10°$ agrees with other correlations for bubble departure very well.

Table I shows the values of d_c and d_{th} for various fluids [11] at different temperatures under normal gravity. It is worthy to note from this table that water has the largest values of d_c and d_{th}, and these critical and threshold values decrease with the increase of saturated temperature (or saturated pressure). For example, the critical and the threshold diameters of water at 300 K are $d_c = 600\,\mu\text{m}$ and $d_{th} = 4680\,\mu\text{m}$, respectively, while the

TABLE I

PREDICTED d_c AND d_{th} FOR FLOW CONDENSATION (LI AND WANG [11])

Liquid	Temperature (K)	Surface tension σ (N/m)	d_c (μm)	d_{th} (μm)
Water	300	0.0717	600	4680
	350	0.0632	560	4374
	400	0.0536	534	4170
	450	0.0429	490	3827
Glycol	300	0.0478	464	3624
	330	0.0451	454	3546
	360	0.0425	444	3468
	373	0.0413	442	3452
R22	283	0.0104	206	1609
	303	0.0076	183	1429
	323	0.0047	153	1195
	333	0.0034	136	1062
R134a	283	0.0103	207	1617
	303	0.0075	183	1429
	323	0.0050	156	1218
	333	0.0038	140	1093

corresponding values at 450 K are $d_c = 490\,\mu m$ and $d_{th} = 3827\,\mu m$, respectively. Li and Wang [11] also gave the following condensation flow regimes based on the tube diameter d:

(i) When $d \leq d_c$, the effect of gravity on the flow regime can be ignored completely, and flow is symmetric with possible flow patterns: annular, lengthened bubble, and bubble flow.

(ii) When $d_c \leq d \leq d_{th}$, the effect of gravity on flow condensation inside the tube will cause asymmetrical distributions of the condensate although the surface tension effect is dominate. The corresponding flow regimes are: asymmetrical annular and asymmetric elongated bubble flow.

(iii) When $d \geq d_{th}$, the condensation flow regimes will be similar to those in macrosize (conventional) tubes, and the effect of surface tension is small in comparison with the gravitational effect.

Kew and Cornwell [14] conducted a series of experiments on flow boiling of R141b in small tubes (with diameters ranging from 1.39 to 3.69 mm and a length of 500 mm) to study the effect of confinement on flow boiling in small tubes. For this purpose, they defined the Confinement number Co as the ratio of the capillary length ℓ_c and the tube diameter d, i.e.,

$$Co = \frac{\ell_c}{d} \tag{3}$$

Based on their experimental data, Kew and Cornwell [14] found that correlations for flow boiling in macrotubes are applicable to small tubes if

$$Co = \frac{\ell_c}{d} < 0.5 \quad \text{or} \quad \text{if} \quad d > 2\ell_c \tag{4a,b}$$

It is interesting to note that the threshold diameter given by Eq. (4b) determined from flow boiling experiments is in good agreement with the threshold diameter given by Eq. (1b) obtained theoretically by Li and Wang [11] for flow condensation.

For the convenience of subsequent discussion on phase change heat transfer, we will now classify the micro-, meso- (or mini-), and macrochannels in terms of the Bond number (Bo), which is defined as

$$Bo = \left(\frac{d}{\ell_c}\right)^2 = \left(\frac{1}{Co}\right)^2 = \frac{g(\rho_\ell - \rho_v)d^2}{\sigma} \tag{5}$$

It is relevant to point out that the Bond number is one of the most important dimensionless parameters in phase change heat transfer, which is a measure of the relative importance of the buoyancy force to surface tension force [15]. Based on the critical and thresholds diameters obtained by Li and Wang [11], we can classify phase change heat transfer in channels according to the Bond number as follows:

(i) *Microchannel*. If $Bo < 0.05$ where gravity effect can be neglected.
(ii) *Mesochannel (or minichannel)*. If $0.05 < Bo < 3.0$ where surface tension effect becomes dominant and gravitational effect is small.
(iii) *Macrochannel*. If $Bo > 3.0$ where surface tension is small in comparison with gravitational force. Note that this criterion is more stringent than the one given by Kew and Cornwell [14], i.e., Eq. (4b), which is equivalent to $Bo > 4.0$.

B. Meso-/Microchannel Heat Exchangers

As mentioned previously, mesochannels (minichannels) having hydraulic diameters in the range of 1–3 mm are used in compact heat exchangers, which are used in automotive, aerospace, and refrigeration industries. Figure 1 shows four types of multi-port extruded tubes, where types A and B consist of plane rectangular mesochannels, while types C and D consist of rectangular mesochannels with straight microfins [16]. Mesochannels, typically 0.5–2 mm in depth, are also used in the printed circuit heat exchanger [17], which was originally developed in Australia for applications in refrigeration and processing. These printed circuit heat exchangers are constructed from flat alloy plates with fluid flow passages photo-chemically machined (etched) into them. This process is similar to manufacturing electronic-printed circuit boards, which gives rise to the name of the exchanger.

Advances in microfabrication technology enable fabrication of microchannels for cooling at the very close proximity to the heat source of a semiconductor chip. After the publication of Tuckerman and Pease's pioneering work [18] in 1981, integrated microchannel heat sinks for electronic cooling at chip level has received a great deal of attention. In an experiment on forced convection of water (single phase) in a microheat sink of $1 \times 1\,cm^2$, consisting of an array of microchannels having rectangular cross sections of 50~56 µm in width and 302~320 µm in depth, Tuckerman and Pease found that a total of 790 W/cm^2 of heat (based on the area of the heat sink) could be dissipated at a Reynolds number of 730. These silicon microchannels (see Fig. 2) can have trapezoidal and triangular cross sections if the wet (chemical) etching method is used. The silicon microchannels can also have rectangular cross sections with greater depth if the dry (plasma)

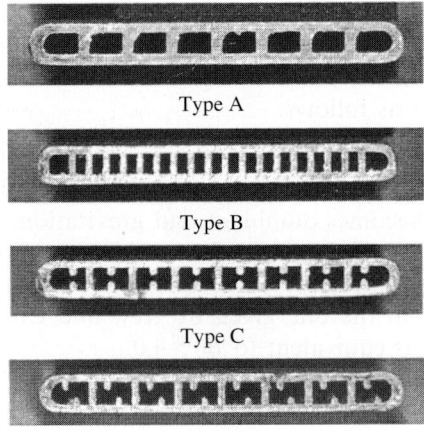

FIG. 1. Photographs of four types of multi-port extruded tubes. From Koyama et al. [16]. Copyright (2003), with permission from ASME.

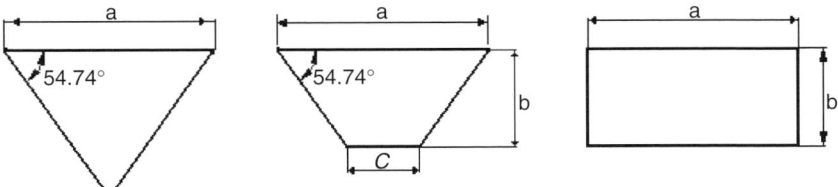

FIG. 2. Triangular, trapezoidal, and rectangular cross-sectional shapes of microchannels.

etching method is used. Most recently, increasing attention has been given to the study of two-phase microheat sinks for microelectronic cooling [19,20]. The advantages of using two-phase heat sinks over those of single-phase heat sinks are obvious: while single-phase heat sinks have large streamwise increase in coolant and wall temperatures at higher heat flux, two-phase heat sinks (utilizing latent heat exchange) maintain a constant temperature of the coolant and the heat sink. In particular, Koo et al. [20] have fabricated a microchannel two-phase flow heat sink with dimensions of $25 \times 25\,mm^2$. This two-phase heat sink, consisting of 100 microchannels with a rectangular cross section (150 μm wide and 200 μm deep), could dissipate 200 W of heat with a temperature rise of 40°C using water as a coolant.

C. Meso-(Mini-)/Microheat Pipes

A heat pipe, made of a metallic tube sealed on both ends, is a device that transfers heat from the evaporator section to the condenser section by latent heat of evaporation. The metallic tube is partially filled with a phase change liquid depending on the operation temperature range. It is a passive device since the motion of the working fluid is driven by capillary force in the wick as in conventional heat pipes, or in microgrooves as in meso heat pipes. Aluminum/ammonia heat pipes, with diameters up to a few centimeters, are used for satellite thermal control. Meso heat pipes, having diameters of a few millimeters, have been used in the cooling of CPUs in desktop and laptop computers. For electronic cooling where the temperature range is up to 100°C, water, acetone, methanol, or ethanol can be used as the working medium. Copper, having a high thermal conductivity, is the most common wall material for low-temperature heat pipes because of its compatibility with water and other low-temperature working fluids.

Recently, much work has been carried out on the study of microheat pipes, which are usually used as a heat spreader to eliminate hot spots in the chip owing to uneven distribution of heat flux. A microheat pipe is a non-circular channel with a hydraulic diameter of $10\sim500\,\mu m$ that was partially filled with a phase change fluid. The sharp-angled corners of the channels serve as liquid arteries. A microheat pipe, with water as a working fluid, typically can dissipate heat flux on the order of $10\sim16\,W/cm^2$. An extensive review on heat pipes prior to 1998 is given by Groll et al. [21]. For this reason, only representative work will be briefly discussed below.

1. Meso-(Mini-) Heat Pipes

Cao et al. [22] tested two meso copper heat pipes, having different sizes of axial rectangular microgrooves ($0.1 \times 0.25\,mm^2$ and $0.12 \times 0.25\,mm^2$) and using water as the working fluid. The maximum heat input and heat flux at the evaporator in the horizontal orientation were about 31 W and 20.6 W/cm^2, respectively. The effective thermal conductance of the heat pipe was estimated on the order of 40 times that of copper based on the external cross-sectional area of the miniature heat pipes. It was found that the capillary limit is the dominant heat transfer limitation for the meso heat pipe.

Hopkins et al. [23] presented a detailed experimental and theoretical analysis on heat transfer capabilities of copper flat meso heat pipes with microcapillary grooves (Fig. 3). Two kinds of capillary structure, namely, diagonal trapezoidal microcapillary grooves and axial rectangular microcapillary grooves were studied. The 120-mm-long axial-grooved heat pipe had a vapor channel cross-sectional area of approximately $1.5 \times 12\,mm^2$ with rectangular grooves of 0.20 mm in width and 0.42 mm in depth. It was

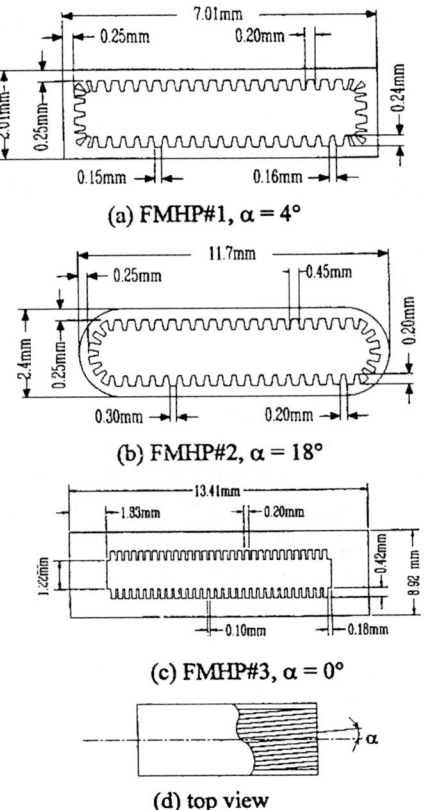

FIG. 3. Schematic cross sections of flat mini heat pipes used in Hopkins' experiments. From Hopkins et al. [23]. Copyright (1999), with permission from ASME.

found that the maximum heat flux of this axial-grooved heat pipe exceeded 90 W/cm^2 in the horizontal orientation and 150 W/cm^2 in the vertical orientation using water as the working fluid.

Take and Webb [24] designed a passive, keyboard-sized aluminum-integrated plate heat pipe for cooling CPUs in a notebook computer. The heat pipe had an array of mesochannels with an arc-shaped cross section of 1.0 mm in depth and 3.3 mm in width. An analysis was performed to estimate thermal resistances in the heat pipe, including the effect of the vapor pressure drop. The modified design using a heat spreader at the evaporator significantly reduced the heat pipe resistance. Test results showed that the integrated plate heat pipe, using R12 as the coolant, could dissipate 18 W of heat and maintained the CPU at 65°C above ambient temperature.

Lin et al. [25] performed an experimental study on two meso heat pipes with two different types of grooves made from a folded copper sheet fin. The meso heat pipes, having an internal cross section of $12.7 \times 6.35\,\text{mm}^2$ and 108 mm long, were partially filled with distilled water. The grooves were rectangular in shape having a dimension of $0.89 \times 0.203\,\text{mm}^2$. Heat fluxes from the two heat pipes with fully and partially opened grooves were measured. The results showed that a heat flux of $140\,\text{W}/\text{cm}^2$ was achieved in the heat pipe with partially opened grooves, which was higher than that with the fully opened grooves.

2. MicroHeat Pipes

The microheat pipe was proposed by Cotter in 1984 [26]. The operating principle of a microheat pipe is similar to that of a macroheat pipe without a wick. Heat is supplied to one end of the heat pipe that vaporizes the liquid in that region and forces it to move to the cooler end, where it condenses and gives up the latent heat of vaporization. Figure 4 shows this vaporization and condensation process in a triangular microheat pipe that causes the liquid–vapor interface in the corners to change continually along the pipe. This results in a capillary pressure difference between the evaporator and condenser regions, which drives the working fluid from the condenser back to the evaporator through the corner regions.

Peterson et al. [27] carried out an experimental investigation on two microheat pipe arrays fabricated in silicon wafers. The first one consisted of 39 parallel rectangular microchannels, 45 µm wide and 80 µm deep, which were machined in a $2 \times 2\,\text{cm}^2$-silicon wafer. The second one consisted of etched triangular microchannels (120 µm wide and 80 µm deep) in a silicon wafer of the same size. For the two microheat pipes with rectangular and triangular microchannels at an input power of 4 W and using methanol as the working fluid, it was found that reductions in the maximum chip temperature of 14.1

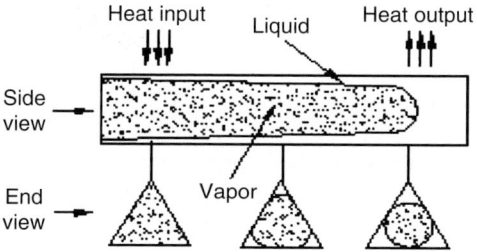

FIG. 4. Schematic of a microheat pipe. From Peterson et al. [27]. Copyright (1993), with permission from ASME.

and 24.9°C, and increases in the effective thermal conductivity of 31% and 81%, respectively.

Longtin *et al.* [28] developed a steady one-dimensional model for the evaporator and adiabatic sections of a microheat pipe. Separate one-dimensional conservation equations of mass and momentum for the vapor and liquid phase were applied. Since the vapor phase was assumed to be isothermal, only the conservation of energy in the liquid phase was needed. The resulting ordinary differential equations with the distance from the inlet of the evaporator x as the independent variable were solved numerically. Results are presented for pressure, velocity, and film thickness as a function of x. It was found that the maximum heat transport capability of a microheat pipe varied with the inverse of its length and the cube of its hydraulic diameter. This implies that a heat pipe with the largest cross section and shortest length has the highest heat transport capability.

To enhance heat transfer characteristics of microheat pipes, Kang and Huang [29] fabricated two microheat pipes with star and rhombus grooves in silicon wafers, as shown in Figs. 5 and 6, respectively. The heat-transfer performance of these microheat pipes was improved due to better capillarity provided by more acute angles and microgaps. The results show that for the silicon wafer with an array of 31 star grooves (having a hydraulic diameter of 340 µm) filled with 60% methanol at 20 W of power input, reduction in the maximum wafer temperature was 32°C. For the silicon wafer with an array of 31 rhombus grooves (having a hydraulic diameter of 55 µm) filled with 80% methanol at 20 W of power input, reduction in the maximum wafer temperature was 32°C. The best thermal conductivities of star grooves

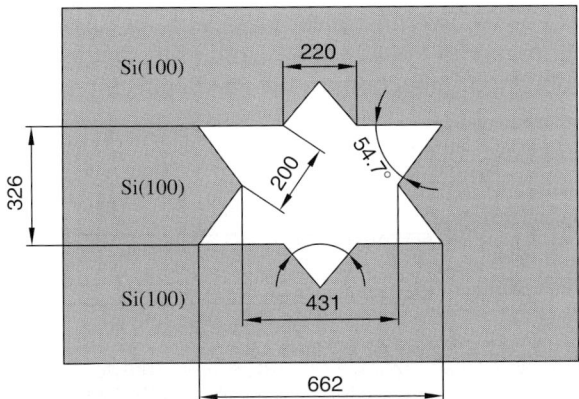

FIG. 5. Dimensions (µm) of a star groove. From Kang and Huang [29]. Copyright (2002), with permission from IOP Publishing Ltd.

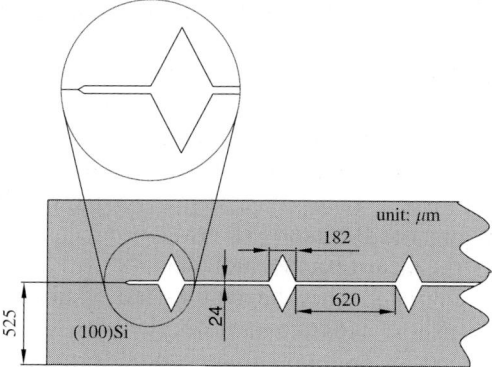

FIG. 6. Dimensions (μm) of a rhombus groove. From Kang and Huang [29]. Copyright (2002), with permission from IOP Publishing Ltd.

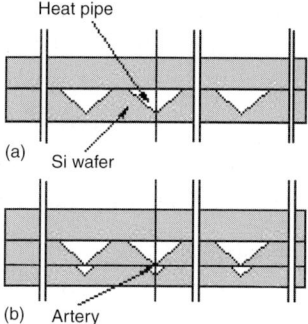

FIG. 7. Transverse cross-sections of two microheat pipe arrays: (a) trianglular microchannels in two silicon wafers and (b) triangular microchannels with arteries in three silicon wafers. From Berre et al. [30]. Copyright (2003), with permission from IOP Publishing Ltd.

micropipe and rhombus grooves microheat pipe were found to be 277.9 and 289.4 W/m K, respectively.

Berre et al. [30] fabricated two sets of microheat pipes arrays in silicon wafers. The first array, as shown in Fig. 7a, was made from two silicon wafers with 55 triangular parallel microchannels, 230-μm wide and 170-μm deep. Figure 7b shows the second set of array, which was made from three silicon wafers having two sets of 25 parallel microchannnels, with the larger ones placed on the top of the smaller ones. The smaller triangular channels were used as arteries to drain the liquid to the evaporator. Ethanol and methanol were used as the working fluids. Fill rate from 0% up to 66% were tested. The results show a maximum improvement of 300% in effective thermal conductivity at high heat flux, which demonstrates enhanced heat transfer in a prototype with liquid arteries.

Wang and Peterson [31] fabricated a wire-bonded microheat pipe by placing parallel aluminum wires (0.5–0.8 mm in diameter) sandwiched between two square aluminum sheets as shown in Fig. 8. Using acetone as the working fluid, a maximum heat flux of 31 W was obtained for a wire diameter of approximately 1 mm at a saturation temperature of 55°C. In another paper, Wang and Peterson [32] developed a steady one-dimensional model to predict the heat-transfer performance and optimum design parameters in a wire-bonded microheat pipe. The results indicated that the maximum heat transport capacity increased with increasing the wire spacing, and there existed an optimal configuration that yielded the maximum heat transport capacity.

Launay *et al.* [33] have carried out experimental and theoretical investigations on a copper/water wire plate microheat pipe. The plate microheat pipe had a dimension of $10 \times 78 \, \text{mm}^2$ with a wire diameter of 1.45 mm. With a hydraulic diameter greater than 1 mm, this heat pipe can be considered as

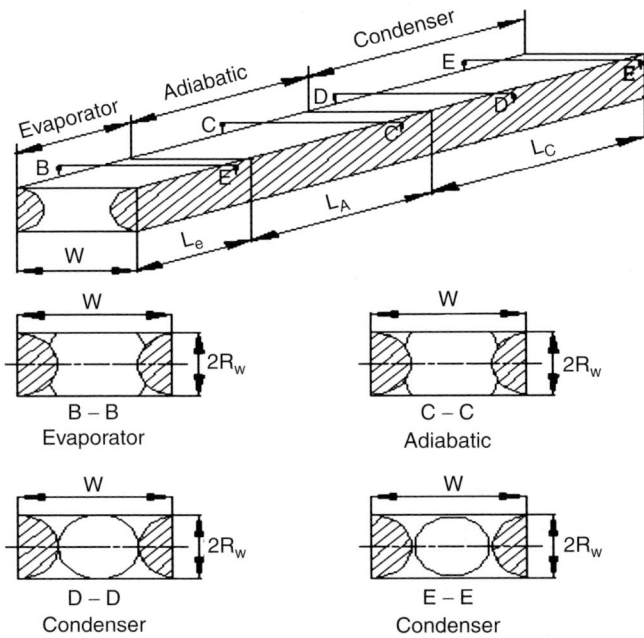

FIG. 8. Flat wire-bonded microheat pipes. From Wang and Peterson. [32]. Copyright (2003), with permission from IOP Publishing Ltd. Copyright (2002), with permission from American Institute of Aeronautics and Astronautics, Inc.

a meso heat pipe. Nevertheless, this system operates like a microheat pipe, using the sharp corners formed between the wires and plates as capillary structure. Following the approach of Longtin *et al.* [28], a model was developed to predict the velocity and temperature distributions, and the maximum heat flux. The temperature distribution obtained from experiments was found in good agreement with the numerical solution.

A radial-grooved microheat pipe was designed and fabricated in a silicon wafer by Kang *et al.* [34,35]. A schematic diagram of such a heat pipe is illustrated in Fig. 9, which shows that the radial-grooved microheat pipe consisted of a three-layer structure, with the middle layer serving as the interface between liquid and vapor phases flowing in the upper and bottom layers, respectively. The separation of the liquid and vapor flow was designed to reduce the viscous shear force. The microheat pipe was fabricated by bulk micromachining and eutectic bonding techniques. Both the vapor- and the liquid-phase grooves were 23 mm in length and trapezoidal in shape, with 70 grooves spreading in a radial manner from the center outward. For the vapor-phase grooves, the widths on the inner and outer ends of the grooves were 350 and 700 µm, respectively. The best heat transfer performance of 27 W at a filling rate of 70% was obtained for this microheat pipe. Most recently, Kang *et al.* [35] presented two wick designs of copper

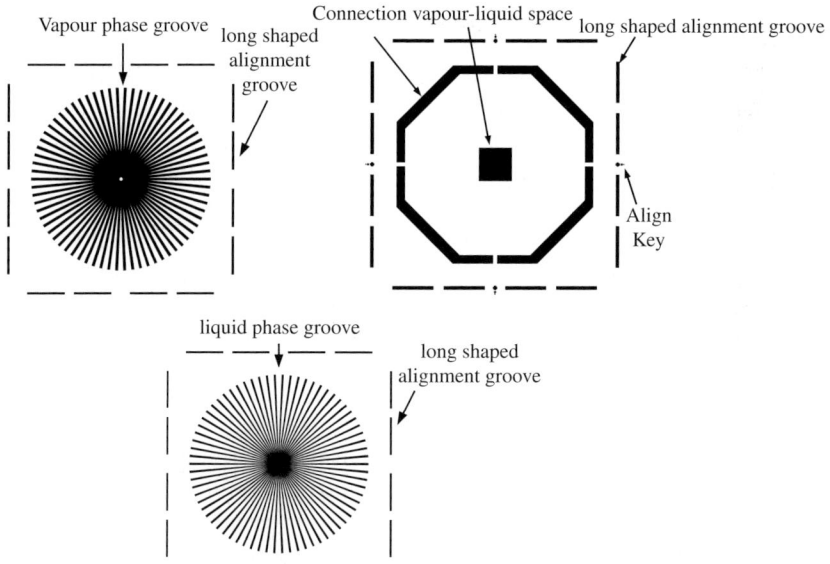

FIG. 9. A radial-grooved microheat pipe: vapor-phase grooves (top left); interface (top right); and liquid-phase grooves (bottom). From Kang *et al.* [34]. Copyright (2002), with permission from Elsevier.

microheat pipe. The first design had almost the same structure as depicted in Fig. 9(top), whereas the second one had 100-mesh copper screens as wick structure. It was found that the groove-shaped microheat pipe, filled with methanol at an 82% fill rate, achieved better performance at a heat flux of 35 W than the microheat pipe using mesh copper screens as wick structure.

3. Meso Pulsating Heat Pipes

The pulsating heat pipe (PHP) is a relatively new type of heat pipe that was first proposed by Akachi [36] in the early 1990s. A PHP is made from a long continuous capillary tube bent into many turns with the evaporator and condenser sections located at these turns. When a capillary tube is partially filled with a working fluid, it will break up into a number of vapor plugs and liquid slugs in the meso tube owing to the effect of surface tension. Hosoda et al. [37] carried out a numerical solution to determine the maximum inner diameter of a tube that can hold a vapor plug, which is

$$d < d_{th} = 1.84 \ell_c \qquad (6)$$

It is interesting to note that the above equation agrees with Eq. (1b).

There are two types of PHPs [38]: the looped and the unlooped PHPs. In a looped PHP, two ends of the pipe are connected to one another (Fig. 10) and the working fluid is circulating continuously within the tube. In the unlooped PHP, both ends are sealed. It has been reported that the looped heat pipe has better heat transfer performance than the unlooped heat pipe. This is because the working fluid in the looped heat pipe can circulate, whereas the working fluid in the unlooped heat pipes is unable to circulate and heat is transferred only by oscillatory motion [39].

When heat is supplied to the evaporator section of a PHP, the pressure in the vapor plugs is increased, while the pressure in the vapor plugs in the condenser section decreases because of the cooling. The uneven distribution of the vapor plugs leads to uneven distribution of pressures in the parallel tubes. These pressure differences become the driving force for the axial oscillation of vapor plugs and liquid slugs in the tubes. Thus, a self-sustained oscillating flow is maintained and heat is transferred from the evaporator section to the condenser section by the phase change fluid.

Tong et al. [40] carried out a flow visualization study on a closed-loop PHP using a CCD camera. The PHP had an internal diameter of 1.8 mm and was partially filled with methanol. The power input was set at 50 W and the PHP was placed both in horizontal and vertical orientation. It was found that the fluid oscillated with large amplitudes during the start-up period, whereas it circulated at steady state. The direction of circulation was

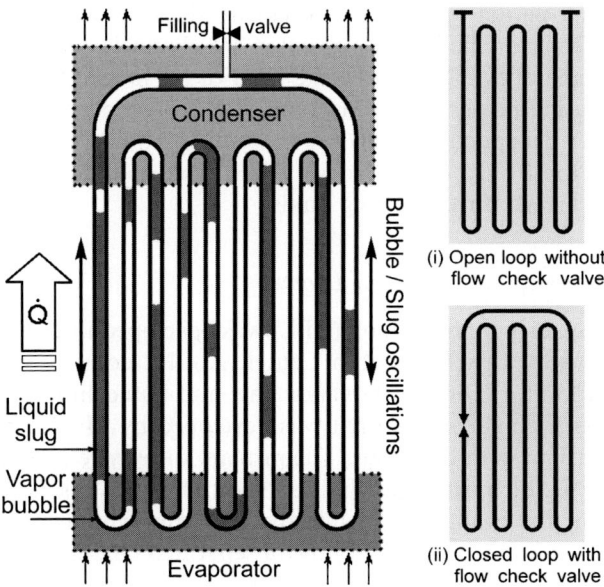

FIG. 10. A looped pulsating heat pipe. From Groll and Khandekar [38]. Copyright (2004), with permission from ASME.

consistent once circulation pattern was attained, but the direction of circulation could be different during the same experimental run.

Shafii et al. [41] developed a model for predicting the behavior of vapor plugs and liquid slugs in the PHP under constant wall temperature condition. The momentum equation was applied to a liquid slug and a vapor plug separately. The results showed that heat transfer in both looped and unlooped PHPs was due to the exchange of sensible heat. Higher surface tension resulted in a thinner liquid film, leading to a slight increase in the total heat transfer. The tube diameter, heating wall temperature, and charging ratio had significant effects on the performance of the PHP. Total heat transfer significantly decreased with a decrease in the heating wall temperature. Increasing the diameter of the tube resulted in higher total heat transfer. The results also showed that the PHP could not operate for charge ratios higher than 90%.

Zhang et al. [42] performed an experimental study on a PHP using FC-72, ethanol, and water as working fluids. The capillary copper tube used in this study had an inside diameter of 1.18 mm with fill ratios from 60% to 90%. Thermal oscillations in wall temperature were due to the passing of vapor plugs and liquid slugs. Thermal oscillation amplitudes were much smaller in FC-72 (possibly due to its lower surface tension) than for ethanol and water.

The oscillation period in FC-72 was shorter than in two other fluids, indicating the faster oscillation movement in the channels, possibly due to the lower latent heat of evaporation for FC-72. For the looped PHP, there was a minimum heating power that initiates the PHP working. Such a minimum heating power depended strongly on the working fluid, and was considerably smaller for FC-12 compared to water. The optimal filling ratio was around 70% for all three working fluids.

4. *Meso-/Microcapillary-Pumped Loop*

The first capillary pumped loop (CPL) was developed in 1960s by Stenger [43] at the NASA Lewis Research Center. A CPL consists of four parts: capillary evaporator, condenser, vapor line, and liquid line. The loop heat pipe (LHP) is a similar design that was developed by Maidanik [44] in the former Soviet Union in the 1980s. The primary difference between the CPL and the LHP is the location of the reservoir. The CPL reservoir is located remotely from the evaporator, while the LHP reservoir is coupled to the evaporator. The vapor and liquid flow in separated channels in a CPL or a LHP, which differs from those in a conventional heat pipe where the vapor and liquid flows are in contact and run countercurrent to each other. The CPL and LHP not only offer greater geometric freedom over traditional heat pipes, but can also carry much greater heat loads due to the co-current flow of vapor and liquid. These two types of heat pipes have been used as a thermal management device in satellites.

The basic operation of a CPL is illustrated in Fig. 11, where heat is added to a wick causing the liquid to evaporate. The evaporation results in a slightly higher pressure at the wick that forces the vapor to move to the cold side of the loop where the vapor is recondensed. The condensed liquid is pulled back to the evaporator by the capillary pressure difference between the condenser meniscus and the menisci in the evaporator wick. However, liquid sometimes accumulate in the curved portion of the vapor leg under certain conditions. The accumulation of liquid would continue until the

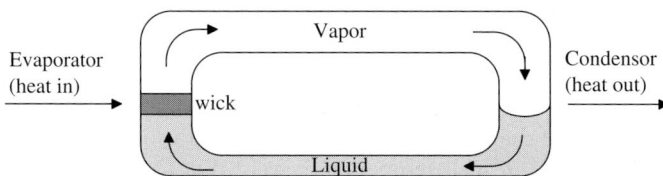

FIG. 11. Sketch of a capillary pump loop. From Allen and Hallinan [45]. Copyright (2001), with permission from Elsevier.

liquid lobe suddenly transform into a slug of liquid. The liquid slug may prevent the flow of vapor to the condenser leading to possible dryout of the condenser [45].

Most recently, some work has been done on the development of microCPL for cooling at chip level. Kirshberg et al. [46] fabricated the first microCPL, where the wicking structure for both the evaporator and condenser were etched into the glass although the whole system had been made with the wicking structure etched into the silicon. Liepmann [47] fabricated a second generation of microCPL based on Kirshberg's original design with an enlarged evaporator having a dimension of $1000 \times 500\,\mu m^2$. It was found that unsteady boiling occurred in the evaporator although the temperature of the evaporator was remarkably steady. At Reynolds numbers of 26 and 430 in the liquid and vapor lines, respectively, the microCPL could remove approximately 4 W of heat and the surface temperature of the CPL was at 30°C. The wick dried out at an input power of 8 W.

Kang et al. [48] fabricated a $60 \times 33 \times 0.8\,mm^3$ microloop heat pipe in a $<100>$ silicon wafer, consisting of an evaporator, vapor line, condenser, and two liquid lines. A glass wafer was bonded from the top to allow visualization. The area of the evaporator was $1\,cm^2$ and the vapor and liquid lines were 40 mm in length. Three wick structures, consisting of parallel V-grooves with hydraulic diameters of 47, 67, and 83 μm, were tested. Water and methanol were used as the working fluid. It was found that the noncondensable vapor had a significant effect on the fluid circulation in the microchannels. The test results showed that heat transfer performance increased as the hydraulic diameter of the microchannels decreased. In addition, it was found that deionized water had a wider heat load performance range (3.3–12.96 W) than that of methanol (1.2–5.85 W).

Cytrynowicz et al. [49] fabricated a silicon-loop heat pipe with a coherent porous silicon wick located inside the evaporator, using deionized water as the working fluid. The wick consisted of an array of micrometer size pores that were produced by a novel technique known as Coherent Porous Silicon Technology. Quartz fiber was used as a secondary wick to handle vapor that generated in the reservoir during the start-up stage of the operation. It was claimed that the theoretical cooling capability of $300\,W/cm^2$ is achievable in this device.

D. Thermal Bubble Actuators

In this section, we will discuss three types of vapor-bubble actuators, namely thermal inkjet printheads, vapor-bubble micropumps, and vapor-bubble perturbators.

1. *Thermal Inkjet Printheads*

As mentioned earlier, thermal inkjet printheads are the most successful products of MEMS technology to date. To be commercially viable, a thermal inkjet printhead must be reliable, with high performance, and at low manufacturing cost. High performance generally implies high-quality droplets (e.g., no satellite droplets), high-frequency response (faster than 10 kHz) and high resolution (droplet volume smaller than 10 pl and nozzle spatial resolution over 300 dpi). At the present time, thermal-type inkjet printers are dominant in the market over piezoelectric-type inkjet printers due to their low cost and high quality.

An inkjet printhead consists of a series of tiny ink firing chambers. Every firing chamber is a microejector, which is composed of a microheater (a thin film resistor), nozzle, barrier, and ink feed channel. Depending on the printhead's design, the ink-feed edge is either on a slot in the substrate or along the substrate's edge. The positions of thin-film resistors and nozzles in the firing chamber are the most important factors that determine the firing modes, which are categorized by the direction of ink ejection relative to the direction of vapor-bubble growth. There are three types of firing modes: (i) the topshooter (also called the roofshooter) where the two directions are the same (see Fig. 12a), which has been used in HP's products [4], (ii) the sideshooter where the two directions are at right angle with each other (see Fig. 12b), which is used in Canon printers [5], and (iii) the backshooter where a droplet is ejected in the opposite direction to bubble growth (see Fig. 12c), was first patented by Bhaskar and Leban [50].

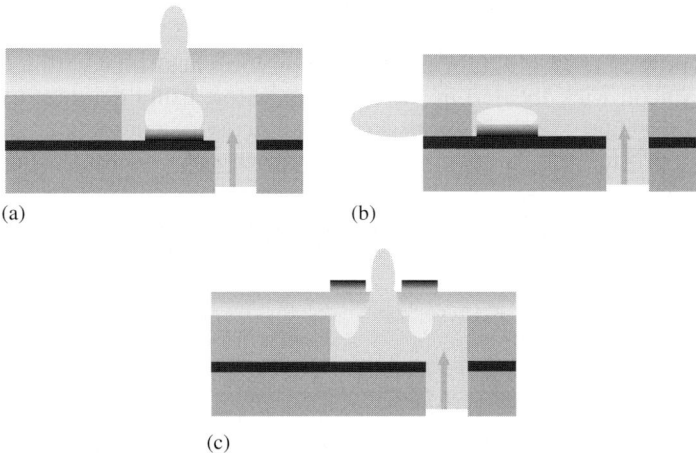

Fig. 12. Different designs of a thermal inject (TIJ) printhead: (a) topshooter, (b) sideshooter, and (c) backshooter.

Various designs of the topshooter, sideshooter, and backshooter types of thermal inkjet printheads have recently been proposed. Lee *et al.* [51] fabricated a monolithic thermal inkjet printhead, which employs the topshooter firing mode as shown in Fig. 13. The heater had an area of $40 \times 40\,\mu m^2$. The nozzles with a diameter of 50 μm had a resolution of 300 dpi and a nozzle-to-nozzle pitch of 170 μm. Experimental results show that ink ejection up to the operating frequency of 11 kHz with an average ink dot diameter of about 110 μm was achieved under current pulses of 0.3 Å and 5 μs. Thus, the dot diameter was close to the diameter (120–140 μm) of a conventional 300 dpi inkjet printhead, and the operating frequency was faster than that (8 kHz) of a conventional 600 dpi printhead.

Baek *et al.* [52] reported a novel sideshooter design with high nozzle density (NPI). They adopted two concepts to improve the NPI of the nozzle density in the existing topshooter design (Fig. 14a). First, the heater was placed on the sidewall of the ink inlet (see Fig. 14b) and thus, the area taken by the heater could be minimized. In order to realize this, indirect heating by heat transfer from chemical vapor deposition (CVD) diamond was used.

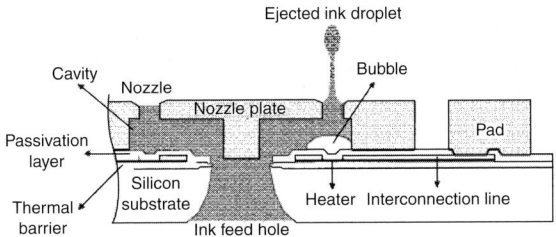

FIG. 13. Cross-sectional view of a monolithic "topshooter" inkjet printhead. From Lee et al. [51]. Copyright (1999), with permission from IEEE.

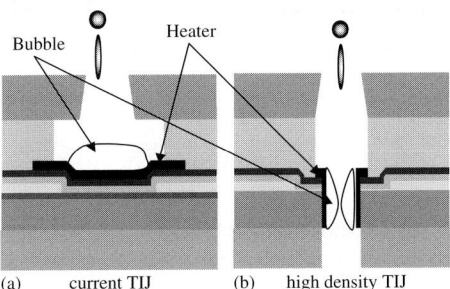

FIG. 14. Concept for a high-density "sideshooter" thermal inject (*a*) current topshooter design, (*b*) improved high-density "sideshooter" design. From Baek *et al.* [52]. Copyright (2004), with permission from IOP Publishing Ltd.

Second, the manifold, the inlet, pressure chamber, and the nozzle were arranged in a line vertically with respect to the substrate. The nozzle density was improved in this new design although it was not easy to fabricate using MEMS technology. To show the feasibility of the new design, the heater including SiO_2 and CVD diamond was fabricated. Bubble generation by heat transfer from the CVD diamond was demonstrated through a visualization test.

Tseng *et al.* [53,54] introduced a modified backshooter design, featuring two heaters (see Fig. 15): the heater at the manifold side was designed narrower than that at the chamber side. When a current pulse passed through these two heaters connected in series, the narrow one generated a bubble faster because of higher power dissipation (Fig. 15a). The bubble formed under the narrow heater blocks the liquid passage and isolated the chamber as the wide heater started to generate its own bubble (Fig. 15b). Expansion of the second bubble pushed the ink ejection. Blocking the leakage also prevented the hydraulic cross talk with neighboring chambers. As the two bubbles grew and eventually merged under the nozzle, the tail of the ejected droplet was cut off, preventing the formation of a long tail. Between firings, however, the virtual valve was opened, by collapsing the bubble, to reduce flow restriction for fast refilling of the microchamber. The use of bubble valves brought about fast and reliable device operation without the need of fabricating physical microvalves. A microinjector based on this design was fabricated [54], and test results showed that the droplet volume

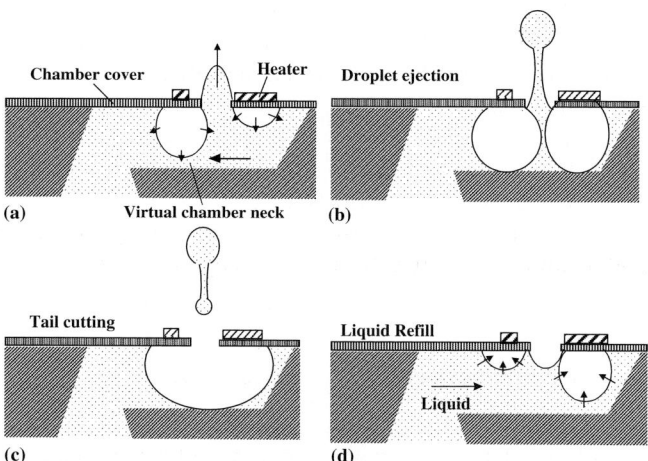

FIG. 15. An improved design of a "backshooter" thermal inject printhead with two unequal size heaters: (a) first set of bubbles generate for virtual neck formation. (b) Droplet ejection. (c) Bubbles meet to cut off long droplet column. (d) Virtual neck collapses for liquid fast refill. From Tseng *et al.* [53]. Copyright (2002), with permission from IEEE.

from the 10-μm-diameter nozzle was smaller than 1 pl (10 times smaller than those of commercial counterparts) and at a frequency of 35 kHz (three times higher than those of commercial counter parts).

Recently, Lee et al. [55] described a thermally driven monolithic printhead, DomeJet, based on "backshooter" design consisting of dome-shaped ink chambers and omega-shaped heaters with thin film nozzle guides. When the heater was activated by pulsed voltage, a doughnut-shaped bubble expanded in shape with the heater, followed by ejection of an ink droplet from the ink chamber. Fifty-six nozzles (18 μmin diameter) were embedded at two columns on each chip with 150 NPI. A pulse voltage of 2.6 μs was used in initial printing tests, resulting in 36 μm diameter dots. At a firing frequency of 1 kHz, the drop volume and velocity were 4 pl and 12 m/s, respectively. Most recently, Shin et al. [56] presented design modifications of the Dome-Jet, to enhance the firing frequency. As the firing frequency was increased, it was found that the residual heat accumulated in heater layer caused unstable droplet ejection. Thus, an efficient heat passage was necessary in order to increase the firing frequency. Contrary to a roofshooter architecture printhead, a backshooter-type printhead has the disadvantage in heat release since the heater is confined to a thin film that has a low thermal conductivity. Among the four modified designs under consideration, it was found that the best model could operate at a frequency of 40 kHz.

Baek et al. [57] reported a novel backshooting thermal inkjet printhead with monolithically fabricated nickel nozzle plate on a SOI (silicon on insulator) wafer. Twin horizontal heaters were adopted as the main design feature, and this inkjet printhead was named "T-jet". This new design has both advantages of the roofshooting, which has an excellent design freedom for flow structure, and of the backshooting which can be fabricated by monolithic process. This backshooter had a chamber and a restrictor with arbitrary shape by utilizing the silicon dioxide etch-stop layers in the bottom and sidewalls of chamber. The printhead had 56 nozzles in two columns with 600 NPI. The test results showed a drop velocity of 12 m/s, a drop volume of 30 pl, and a maximum firing frequency of 12 kHz for single-nozzle ejection.

2. Vapor Bubble Micropumps

Geng et al. [58] described a novel pumping device based on the periodic generation and collapse of a single vapor bubble in a tube. The device consisted of two stainless steel tubes with two different diameters (508 and 1067 μm, respectively). A divergent passage was connected from the smaller tube to the larger tubes (see Fig. 16). Pulse heating was applied at the narrow end of the divergent channel, where a vapor bubble was generated

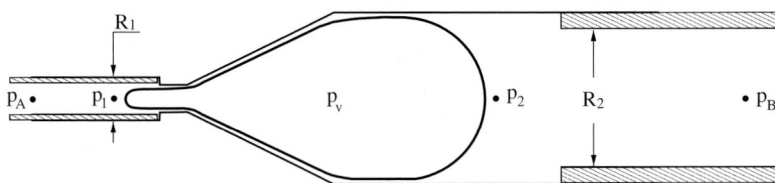

Fig. 16. Sketch of a microbubble drive micropump. From Geng et al. [58]. Copyright (2001), with permission from IOP Publishing Ltd.

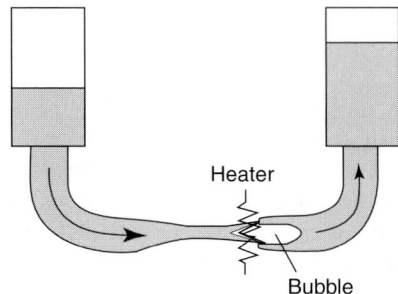

Fig. 17. Schematic diagram of a glass tube meso bubble pump. From Nakabeppu et al. [59]. Copyright (2003), with permission from ASME.

and collapsed periodically. Because of the difference in curvature of the bubble at the narrow tube and the large tube, a capillary force was created that drove the bubble toward the larger tube. At a pulse-heating period of 100~1000 ms (i.e., a frequency between 1 and ~10 Hz), the device could develop a head of a few centimeters of water with flow rates in the range of 100 μl/min.

Similarly, Nakabeppu et al. [59] constructed two micropumps made of glass tube (see Fig. 17) and silicon wafer, respectively, for electronic cooling. Each of these micropumps consisted of a convergent–divergent section with heating at the diverging section where an isolated vapor bubble was formed. The difference in curvature of the bubble resulted in a capillary force that droved the bubble toward the diverging direction. When the bubble expanded to the non-heating section, condensation began and the bubble disappeared at downstream. During the expansion of the bubble, it pushed the liquid to downstream. The diameter of the glass tube micropump was 400 μm. The convergent–divergent section had a throat diameter of 100~200 μm and 30~60 mm in length. Water or ethanol was used as a working fluid. At the power input of 5 W and using water as a working medium, a maximum pressure of 1100 Pa and a maximum flow rate of 800 μl/min were achieved. At the same power input, the maximum pressure and maximum flow rate were smaller if ethanol was used as a working medium. Silicon micropumps, having microchannels (500 μm wide and 250 μm deep), and converging–diverging sections (having throat widths ranging from 20 to 160 μm and 5 mm long), were also fabricated. The experimental results showed that the delivery pressure and flow rate in the silicon micropump were higher for smaller throat width and larger in heating rate.

Jun and Kim [60] used the growth and collapse of a single or multiple vapor bubble(s) to displace fluids in microchannels. The pumping action was provided by asymmetric heating along the microchannels that created a variation in vapor pressure and surface tension due to the heater-induced temperature gradient. A pumping device, consisting of a microchannel with a hydraulic diameter of 3.4 and 726 μm in length, was fabricated in a silicon wafer. Three heaters were turned on and off sequentially (Fig. 18) to actuate the movement of the vapor bubble(s). It was found that the pulse width (ranging from 0.25 to 1 s) and power input (ranging from 18 to 26 V) determined whether the micropump would operate in single or multiple bubble mode. Experimental results showed that pumping of isopropanol at velocities as high as 160 μm/s with a pressure head of approximately 800 Pa was achieved. A heat and mass transfer analysis was performed to understand the pumping mechanism and estimate its pumping capability.

Soong and Zhao [61] fabricated a micropump consisting of 12 short Pyrex glass tubes connected in series and using deionized water as a working fluid. The inner diameter of these glass tubes was 1 mm and a total of 12 miniature heaters were attached along the tubes. The displacement of fluids in such a

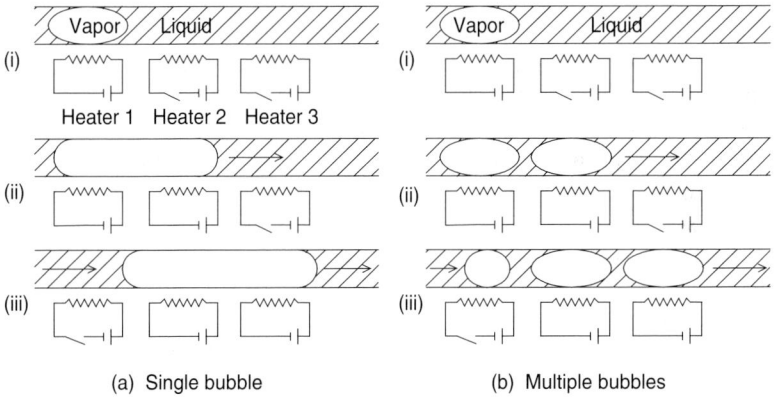

FIG. 18. Two modes of obtaining net fluid displacement: (a) single bubble and (b) multiple bubbles. From Jun et al. [60]. Copyright (1998), with permission from American Institute of Physics.

pump was realized by actuation of a moving vapor bubble generated by suitably phase heating elements. The experimental results indicated that this micropump, using deionized water as the working fluid, could achieve a maximum pressure head of 57 mm of water and a maximum volumetric flow rate of 300 µl/min when it was operated for a heating power ranging from 8 to 12 W and a heating time of about 3 s. A heat and mass transfer analysis was performed to understand the pumping mechanism and estimate its pumping capability. The results of the theoretical model were found in agreement with experimental data.

Xu et al. [62] presented an experimental study of two heat-driven mesopumps with different configurations that were made of glass tubes. The first one had a simple meso U-tube with two check valves, while the other had an additional horizontal branch across the heating section. The inner diameter of the glass tube was 3 mm and water was used as a working fluid. It was found that the pressure and temperatures fluctuated periodically and pumping motion was in a pulsating form. At a power input of 27.4 W and a flow rate of 280 g/h with an inlet water temperature of 24°C, the fluctuation had a cycle of 6.8 s. The heat-driven mesopump with the additional horizontal branch will shorten pulsating cycle periods and smooth temperature fluctuations. However, its discharge rate and the operation power range were not much different from the one without the horizontal branch. In another paper, Xu et al. [63] developed a numerical model to simulate the thermal hydraulic characteristics of the meso heat-driven pump. The formulation of the problem was divided into four stages: the method of characteristics was used to simulate the sensible heating during the initial stage. Moving boundaries formulation was used to model the remaining three stages: (i) phase change in horizontal branch with single-phase discharge stage, (ii) two-phase blow down stage, and (iii) liquid suction from the inlet check valve. The predicted variations of pressure and mass flow rate were in good agreement with experimental data.

3. Thermal Bubble Perturbators

Recently, Deng et al. [6,8] investigated the possibility of using a microthermal bubble, generated by a microheater under pulse heating at millisecond range, as an actuator for mixing and for DNA hybridization enhancement. A non-uniform width microheater (with the slim part having a dimension of $10 \times 3\,\mu m^2$) in a silicon wafer under pulse heating conditions was used to investigate microbubble growth and collapse in water and in single-stranded DNA (ssDNA) solutions with different concentrations. This non-uniform microheater can induce highly localized heat source, which can generate a single microbubble periodically at the narrow part of the heater

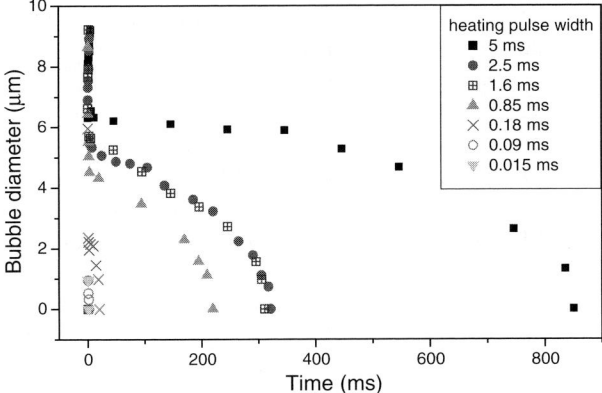

Fig. 19. Effect of heating pulse width on bubble diameter variation with time. From Deng *et al.* [6]. Copyright (2003) with permission from Elsevier.

under pulse heating. Figure 19 shows the effect of pulse width (ranging from 1.6 to 5 ms) on the bubble growth and collapse in the water adjacent to the heater at a power input of 28.2 mW. It is shown that the diameter of the bubble undergoes a rapid growth period and after reaching a maximum diameter, the diameter of the bubble suddenly drops and then the bubble begins to collapse slowly as a result of the condensation process. For short pulse widths (such as in the range of 0.015–0.18 ms), single bubble was barely observable on the heater surface and the growth and collapse of bubble was almost symmetric. When the heating pulse duration was in the millisecond range from 0.85 to 5 ms, bubble growth and collapse patterns became asymmetric. At the pulse width of 1.6 ms under a power input of 28.2 mW, the asymmetric growth and collapse of the water bubble led to a perturbation area about sixfolds of the maximum bubble diameter.

Deng *et al.* [8] carried out similar experiments to investigate bubble growth and collapse in ssDNA solutions with different concentrations under pulse heating conditions. Figure 20 shows the bubble diameter versus time in ssDNA solutions with different concentrations under a pulse width of 1.6 ms. As clearly shown in Fig. 20a, the overall bubble lifetime in ssDNA solutions decreased distinctly with the increase of ssDNA concentration. For all testing concentrations, the bubble growth period (which was comparable to the heating pulse width) was much shorter than the bubble collapse process. Since the bubble growth process was much shorter than the collapse process, a close-up view of the bubble growth process is given in Fig. 20b. In addition to the bubble lifetime, the bubble growth process was also highly affected by the DNA concentration. Figure 20b shows that the maximum bubble diameter decreases with the increase of the ssDNA

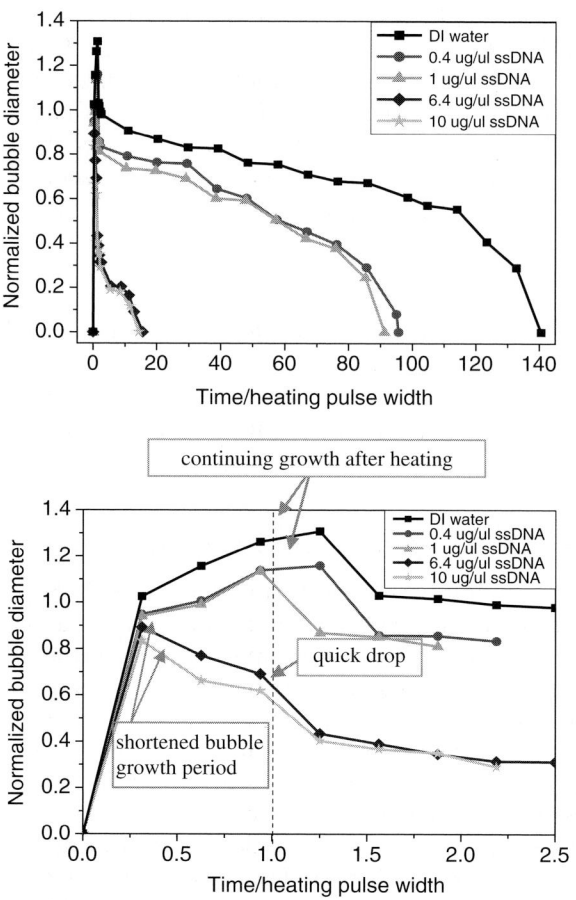

Fig. 20. Microbubble diameter versus time in ssDNA solutions: bubble-time history (top); close-up view of the bubble growth process (bottom). The heating pulse width was 1.6 ms. From Deng et al. [8]. Copyright (2004), with permission from IOP Publishing Ltd.

concentration. For the cases of deionized water and 0.4 µg/µl ssDNA solution, the maximum-size bubbles were both recorded at 2 ms after the beginning of the heating process, implying that the bubble kept on growing even though the heating pulse stopped at 1.6 m, which was due to the inertia effect. For the case of 1 µg/µl, the maximum-size bubble was recorded at 1.5 ms, implying that the bubble stopped growing at almost the end of the heating pulse. For high concentrations, 6.4 and 10 µg/µl, however, the maximum-size bubbles were both recorded at 0.5 ms, indicating that the bubble growing process was much shorter than that at moderate concentrations, and the bubble stopped growing even though heating from the heater was still going on. From this graph, it can be concluded that high concentrations of the DNA macromolecules in a solution can effectively retard the bubble growth by increasing the dissipation of bubble kinetic energy. As a result, both the bubble growth period and the recorded maximum bubble diameter are distinctively reduced at high DNA concentrations. For medium and low concentrations (1 and 0.4 μg/μl), the bubble growth period is comparable to the heating pulse width.

III. Meso- and Microscale Phase-Change Heat Transfer Phenomena

In this section, we will discuss some of boiling and condensation phenomena in meso-/microdevices, which have been discussed in Section II.B–D.

A. Homogeneous and Heterogeneous Nucleation in Pool Boiling

It has been well established that vaporization usually begins with an embryo in the superheated liquid [64]. Once the embryo is larger than a critical size, a vapor bubble will form. If the vapor embryo completely forms within a superheated liquid, it is called "homogeneous nucleation". On the other hand, if the vapor embryo forms at the interface between liquid and solid or gas, it is called "heterogeneous nucleation". The homogeneous nucleation is usually associated with high degrees of superheat and extremely rapid rates of vapor generation, while the heterogeneous nucleation can occur at lower degrees of superheat. The incipient nucleation temperature in pool boiling can be computed from the classical kinetics of nucleation [65] as follows:

(i) For homogeneous nucleation in superheated liquids, the bubble nucleation density J is given by

$$J = N_0 \left(\frac{kT_n}{h}\right) \exp\left[-\frac{16\pi\sigma^3}{3kT(p_v - p_\ell)^2}\right] \quad (7a)$$

where T_n, σ, k, and h are the nucleation temperature, the surface tension, the Boltzmann constant, and the Planck constant, respectively.

(ii) For heterogeneous nucleation on a smooth surface with no cavity, the bubble nucleation density is

$$J = N_0^{2/3} \psi \left(\frac{kT_n}{h}\right) \exp\left[-\frac{16\pi\sigma^3 \omega}{3kT_n(p_v - p_\ell)^2}\right] \quad (7b)$$

with

$$\psi = \frac{(1+\cos\theta)}{2} \quad \text{and} \quad \omega = \frac{(1+\cos\theta)^2(2-\cos\theta)}{4} \quad (8a,b)$$

where θ is the contact angle. With the aid of the Clausius–Clapeyron equation, Eq. (7) can be expressed in the following compact form [66]:

$$T_n - T_s = \frac{T_s}{h_{fg}\rho_v}\sqrt{\frac{16\pi\sigma^3\omega}{3kT_n \ell n\left(\frac{N_0^\gamma kT_n\psi}{Jh}\right)}} \quad (9)$$

which is applicable for both homogeneous nucleation with $\gamma = \psi = \omega = 1$ and heterogeneous nucleation with $\gamma = 2/3$ and ψ and ω given by Eqs. (8a, b). The nucleation temperatures for homogeneous and heterogeneous nucleation can be obtained from Eq. (9) iteratively if the value of J is known. Note that the nucleation temperature determined from Eq. (9) is not sensitive with the chosen value of J. After the nucleation temperature is obtained, the critical bubble radius for homogeneous and heterogeneous nucleation is given by

$$r_c = \frac{2\sigma}{P_s(T_n) - P_\ell} \quad (10)$$

For water at the atmospheric pressure, the homogeneous nucleation temperature and the critical bubble radius determined from Eqs. (9) and (10) are: $T_n = 303.7°C = 576.8\,K$ and $r_c = 3.15\,nm$.

Li and Cheng [66] also discussed the effects of dissolved gases and corners of the microchannels on the heterogeneous nucleation temperature of a smooth surface with no cavity such as those on a silicon wafer. Figure 21 shows the effect of dissolved air, contact angle, and the cross-sectional shape (triangular, rectangular, and trapezoidal) of the microchannel on the nucleation temperature of a smooth surface. It is shown that the heterogeneous nucleation temperature with dissolved air (Curve 2) is about 35°C lower than those without dissolved air (Curve 1). If the shape of the corner together with dissolved air is taken into consideration, the heterogeneous nucleation temperature can be lower to 200 K as shown in Fig. 21.

Brereton et al. [67] investigated both experimentally and theoretically the effect of confinement on nucleation temperature of water in a silica capillary tube with diameters ranging from 10 to 100 μm. Based on the modification of the classical theory, a model for heterogeneous nucleation temperature was obtained. Figure 22 is a comparison of the nucleation temperature obtained based on the theoretical model and experimental observation. It is shown that the nucleation temperature increased as the tube diameter

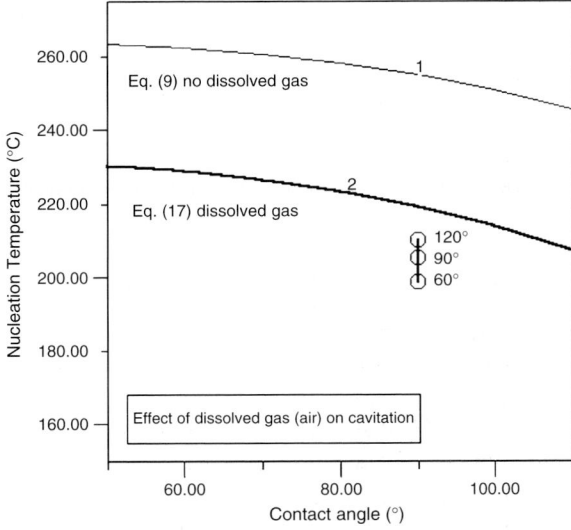

FIG. 21. Effects of dissolved air, corners, and contact angle on heterogeneous nucleation temperature of water. From Li and Cheng [66]. Copyright (2004), with permission from Elsevier.

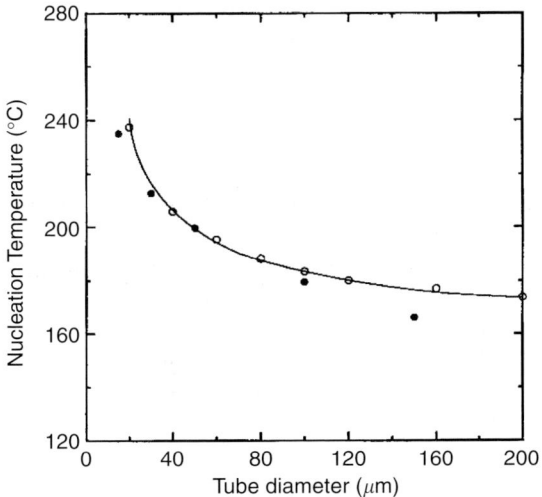

FIG. 22. Effect of tube diameter on the nucleation temperature of water precompressed to 100 MPa. From Brereton, *et al.* [67]. Copyright (1998), with permission from Elsevier.

decreased. It can be seen that the nucleation temperature of water was about 240°C in a 20 μm-diameter tube when the water was precompressed to 100 MPa. Peng *et al.* [68] performed an analysis on the bubble nucleation in microchannels. They assumed that the bubble embryo touches the walls of

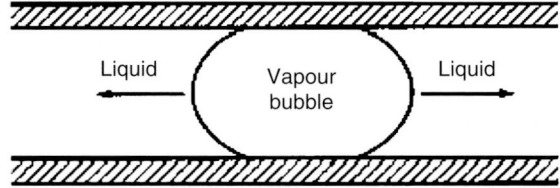

Fig. 23. Bubble nucleation model in a microchannel. From Peng et al. [68]. Copyright (1998), with permission from Elsevier.

the microchannel and the bubble growth is greatly restricted by the surrounding walls as shown in Fig. 23. From the phase stability condition in classical thermodynamics and with the aid of the Clausius–Clapeyron equation, Peng et al. [68] obtained a criteria for liquid nucleation in microchannels in terms of a dimensionless parameter N_{mb}

$$N_{mb} = \frac{h_{fg}\alpha}{c\pi(v_v - v_\ell)qd_h} \leq 1 \tag{11a}$$

where v_v and v_ℓ are specific volumes of vapor and liquid at saturation condition; h_{fg} the latent heat, α the thermal diffusivity, and d_h the hydraulic diameter. The above equation can be rewritten as

$$q \geq \frac{h_{fg}v\alpha}{c\pi(v_v - v_\ell)d_h} \tag{11b}$$

which is the minimum heat flux for liquid nucleation. From the Maxell's relation and with the aid of the Clausius–Clapeyron equation and the Young–Laplace equation, the wall superheat necessary for nucleation in microchannels is given by

$$\Delta T_{sup} \geq \frac{4AT_s(v_v - v_\ell)\sigma}{h_{fg}d_h} \tag{12}$$

where T_s is the saturation temperature and σ the surface tension. In Eqs. (11) and (12), $c = 1$ and $A = 280$ are empirical constants that were determined from the experiment. The variations of minimum heat flux for nucleation as well as the wall superheat for nucleation computed based on Eqs. (11) and (12) are presented in Figs. 24a and b, respectively. It is seen that the nucleation heat flux and wall superheat increase with the decrease in diameter, especially for microchannels with $d_h < 1$ mm. It is shown that a heat flux much higher than 10 W/cm^2 and a nucleation temperature much higher

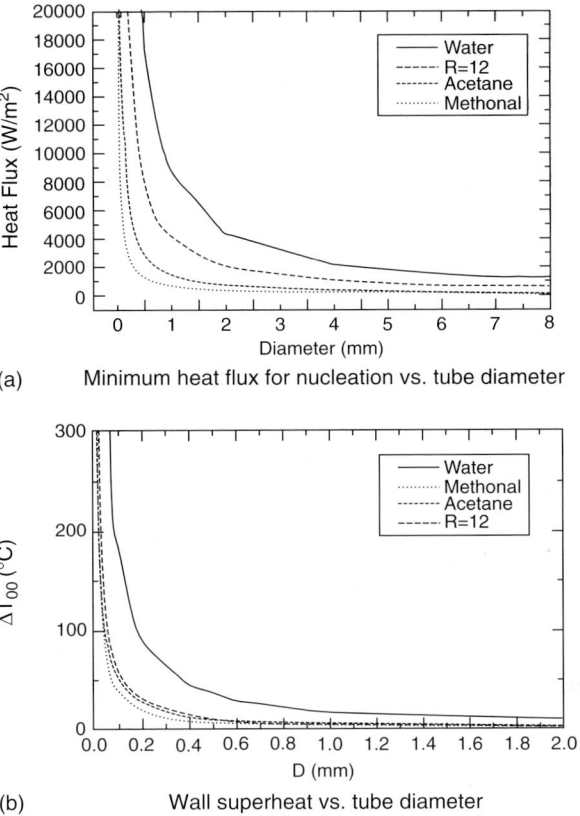

FIG. 24. Effects of tube diameter on (a) minimum heat flux for nucleation and (b) wall superheat for nucleation. From Peng et al. [68]. Copyright (1998), with permission from Elsevier.

than 300°C are needed for bubble nucleation in a microchannel with $d_h < 1$ mm.

B. Subcooled Pool Boiling Under Constant Heat Flux

In this section, we will discuss subcooled pool boiling about a microheater and about an ultra-thin wire under steady heating conditions.

1. Subcooled Pool Boiling About a Microfilm Heater

Tsai and Lin [69] studied transient microbubble formation on polysilicon microresistors $(95 \times 10 \, \mu m^2, \ 95 \times 5 \, \mu m^2)$ submerged in isopropyl alcohol under constant input currents from 15 to 40 mA. It was found that microbubble nucleation characteristics can be classified into three groups depending on the input current level: Group I: when the input current was less than 22 mA, no bubble was generated and the temperature of the heater remained nearly the same; Group II: when the input current was between 25 and 30 mA, a single microbubble was nucleated. The initial bubble growth rate was found proportional to the square root of time, which is similar to the heat diffusion-controlled mode during the initial stage of a macrobubble growth. The wall temperature increases initially and then drops before a bubble was formed. The slope of this temperature drop seems to be steeper as the magnitude of the input power is increased. After a bubble is nucleated, the wall temperature increases and goes above the initial temperature before reaching a steady state; Group III: when the input current was equal or higher than 40 mA, a bubble was formed as soon as the electrical current was applied. There was no detectable temperature drop when the thermal bubble was generated and the wall temperature continued to increase and the bubble continued to grow before a steady state was reached.

Tsai and Lin [69] pointed out the similarity on the temporal variations of wall temperature during bubble nucleation processes about a macroheater and those about a microfilm heater under medium current (i.e., Group II). Figure 25a is a sketch of the temporal wall temperature variation in a marco boiling experiment which can be divided into four stages. When bubbles begin to form in stage 1, the wall temperature begins to drop because of the evaporation of a microlayer liquid between the bubble and the heating wall. After the microlayer evaporates and dries out, the boiling process moves to stage 2, where bubbles continue to grow. The wall temperature rises in this stage owing to the reduction of the heat flux from the wall to the liquid due to the existence of vapor bubbles. When bubbles grow to a certain size and start to rise from the wall as shown in stage 3, cool liquid is sucked to the wall causing a quenching effect on the wall. The cool liquid is then heated up gradually in stage 4 to the original temperature, where a new bubble is nucleated and the process repeats itself. Figure 25b shows temporal variations of wall temperature on the microheater in responses to transient microbubble formation, which can be divided into three stages. In stage 1, the wall temperature of the microheater jumps immediately after an electric current is applied. It is believed that liquid is superheated locally in this stage and a microlayer is evaporated causing wall temperature drop in stage 2. As the temperature reaches the incipient bubble nucleation temperature, a single spherical bubble is generated. Subsequently, the wall temperature increases as shown in stage 3, in which a bubble grows and covers the entire microheater. The wall temperature increases in this stage because the vapor bubble is blocking the heat dissipation from the wall. This shows a

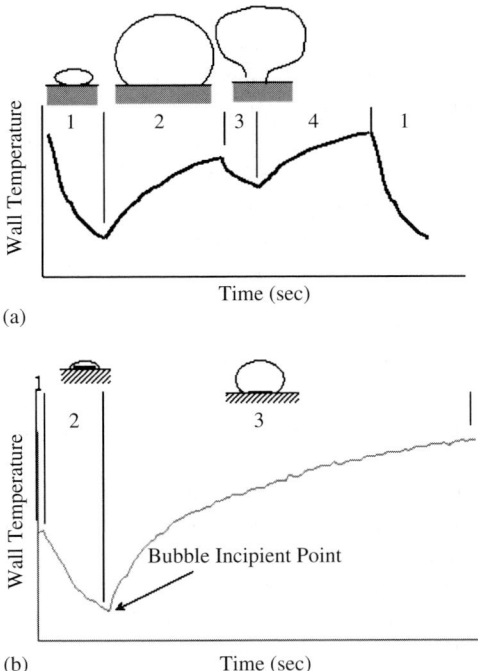

FIG. 25. Temporal wall temperature variation in response to bubble nucleation: (a) macro heater and (b) microheater. From Tsai and Lin [69]. Copyright (2002), with permission from ASME.

"V"-shaped wall temperature variation with time. The nucleated microbubble continues to grow slowly and sticks to the microheater due to Marangoni effect and the wall gradually reaches an equilibrium temperature.

Straub and co-workers [70,71] carried out a series of experiments to study thermocapillary flow around a bubble in subcooled pool boiling. When the top of a bubble has extended into the subcooled liquid, it gives rise to a temperature gradient along the interface, thus inducing a thermocapillary flow around the bubble as sketched in Fig. 26. This vapor bubble can grow, shrink, or remain constant depending on the heat and mass transfer process at the upper perimeter of the bubble. The thermocapillary flow around the bubble exerts a force on the bubble, preventing it to detach from the heated surface and increasing the time of bubble attachment. A numerical solution on thermocapillary flows around a bubble was carried out by Straub et al. [72].

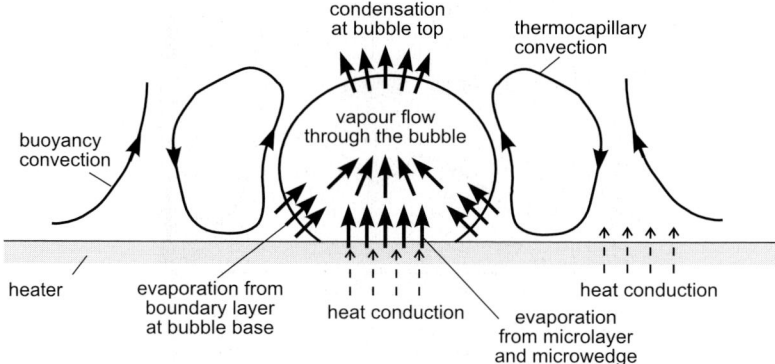

FIG. 26. Thermocapillary convection around a vapor bubble on a heating wall. From Marek and Straub [71]. Copyright (2001), with permission from Elsevier.

Kim et al. [73] performed an experiment on subcooled pool boiling of FC-72 on a microheater ($0.27 \times 0.27\,\text{mm}^2$) in low, normal, and high gravity environments. An array of 96 individual microheaters was used to measure the local heat transfer rate. It was found that the boiling behavior was dominated by the formation of a large primary bubble on the surface that acted as a "sink" for many smaller bubbles surrounding it. Dryout of the heater occurred under the primary bubble. Side view photo images showed that the Marangoni convection around the bubble formed a "jet" of liquid into the bulk liquid.

2. *Bubble Dynamics about an Ultra-Thin Heating Wire*

Wang et al. [74–78] performed a series of experiments on nucleate boiling on a horizontal thin heating platinum wire (0.1 mm in diameter) under steady heating that was submerged in subcooled alcohol or water. Using zoom routine and CCD camera, it was found that in addition to small and large bubbles, sweeping bubbles and different types of jet flows coexisted during nucleate boiling on the wire. Figure 27 shows the nucleation jet flow from the heating wire where vapor bubbles are formed and some jets evolve into meso bubbles. The black circle in the figure shows the process of the formation of meso bubbles at the end of nucleation jets. The occurrence of jet flows and sweeping bubbles has been attributed to the Marangoni effect [77,78].

Figure 28 shows the phenomenon of bubble forward-and-backward sweeping motion along the thin heating horizontal wire that was submerged in subcooled water at 80°C [75]. It was found that bubbles could move along

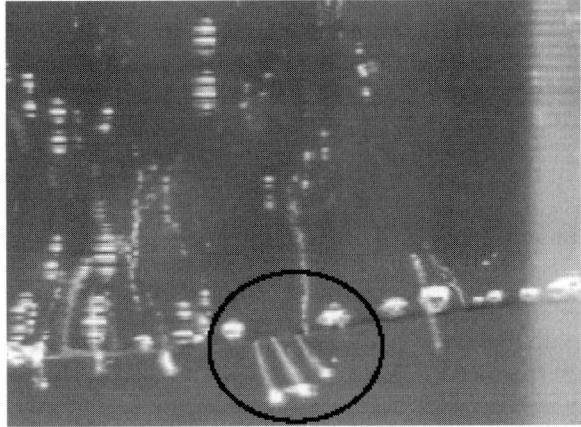

Fig. 27. Nucleation jets along a heating wire. From Wang et al. [74]. Copyright (2002), with permission from Elsevier.

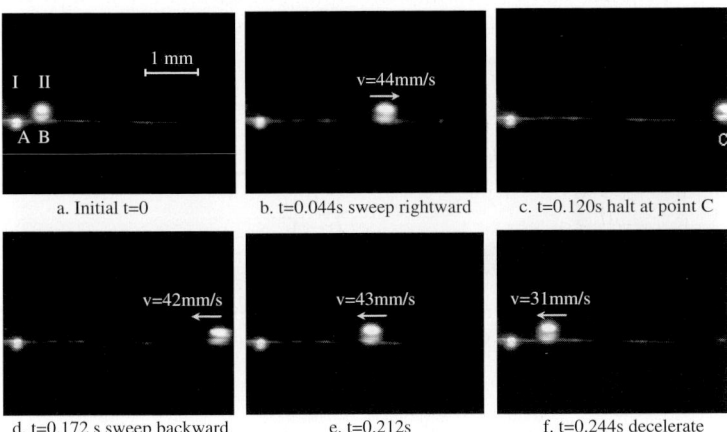

Fig. 28. Bubble forward-and-backward sweeping on a heating wire. From Wang et al. [75]. Copyright (2003), with permission from Elsevier.

the wire with a speed up to 44 mm/s and they could change their moving directions backward when encountering with another bubble. The observation of sweeping bubbles that could change direction along a thin heated wire are quite different from usual observations of nucleate boiling on a wall where the bubbles move in the direction of fluid flow or buoyancy force but cannot change their directions. This phenomenon of sweeping bubbles resulted in significant consequence on the liquid flow around bubbles and temperature of the liquid and the wire, which greatly altered the boiling behavior and heat transfer mode. The back-and-forth bubble sweeping motion of bubbles also occurred along a vertical thin wire [76].

Figure 29 is a sketch of the structure of a bubble-top jet flow which can be divided into four different regions: the pumped region, jet neck, expanding, and wake region [77]. In the pumped region, the hot liquid was pumped up from wire surface. The pumping effect was an important behavior of the jet flow, which caused highly efficient single-phase heat transfer near the bubble bottom, and also caused strong interactions between neighboring bubbles. The jet flows from downward-facing bubbles were as strong as the jet flows from-upward facing bubbles [76]. Bubble size greatly influenced the jet flow intensity and increased the complexity of the jet flow structure. The bubble size relative to the wire is an important factor affecting the jet flow structure. When the diameter of the bubble is close to or larger than the wire diameter, the outer edges of the bubble extending beyond the wire are exposed to cooler liquid. As a result, the lowest temperatures on the bubble surface occur on these two exposed sides, causing the two symmetrical jets to emanate from two sides of the bubble as shown in Fig. 30.

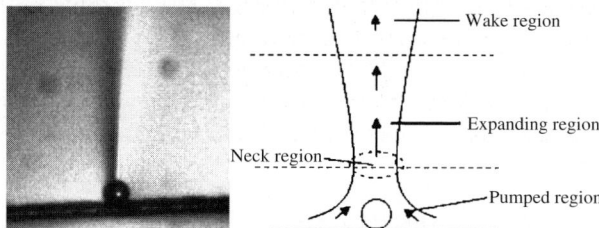

FIG. 29. Bubble-top jet flow in a vertical plan parallel to the axis of a heating wire. From Wang et al. [77]. Copyright (2004), with permission from Elsevier.

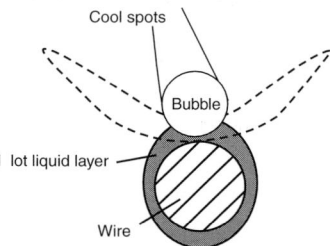

FIG. 30. Cross-sectional view of multi-bubble jet flow around a heating wire. From Wang et al. [77]. Copyright (2004), with permission from Elsevier.

A three-dimensional numerical simulation [78] was carried out to study the observed jet flow phenomena from a bubble (having prescribed diameters of 0.03 and 0.06 mm, respectively) on a thin wire (with a diameter of 0.1 mm). The Boussinesq approximation was applied to the governing equations. Phase change heat transfer and the Marangoni effect were taken into consideration at the bubble interface. The numerical solutions confirmed that the Marangoni effect is responsible for the appearance of bubble-top and multi-jet flows due to high temperature gradients near the wire.

C. Subcooled Pool Boiling under Pulse Heating or Transient Heating

Microbubble nucleation under pulse heating has important applications to thermal inkjet printheads, vapor bubble pumps, vapor bubble perturbators as discussed in Section II.D.1. Under rapid pulse heating, power densities equivalent to several megawatts per square meter can be generated. Bubble nucleation under an extremely high heat flux pulse differs from the usual nucleate boiling in many aspects. First, bubble nucleation is initiated at a higher temperature close to the theoretical superheat limit. Second, the boiling process is more explosive because the initial bubble pressure is very high. Third, the boiling process is more reproducible because its mechanism is governed by the property of the liquid (i.e., homogeneous nucleation) rather than by the surface characteristics (i.e., heterogeneous nucleation). Skripov [79] gave the following criterion for the homogeneous nucleation to occur under rapid heating

$$\frac{<dT/dt>}{<\varphi>^{1/k}[\pi\tilde{\Omega}_A]^{1/2k}} > T^* - T_s \tag{13}$$

where $<dT/dt>$ is the time-averaged heating rate, T^* and T_s the superheat limit of the liquid and the saturation temperature, respectively, $\tilde{\Omega}_A$ the number of pre-existing nuclei per unit area on the heater surface, k the Boltzmann constant, and $\langle\varphi\rangle$ a time average function depending on the superheat and the thermophysical parameter. In the following, we will discuss bubble growth and collapse about an ultra-thin heating wire and also on a microfilm heater under pulse heating.

1. *Bubble Growth and Collapse about an Ultra-Thin Heating Wire*

Fine platinum wires under pulse heating were used extensively for transient boiling study in early work where the temperature of the wire was determined by the principles of resistance thermometry. Skripov *et al.*

[79,80] performed transient boiling experiments on a fine wire under rapid heating in organic liquids, and demonstrated that it is possible to superheat a liquid to its homogeneous nucleation temperature, where the molecular energy fluctuations become the dominant mechanism for vapor nucleation. Their experiments [80] showed that a heating rate as high as $\langle dT/dt \rangle > 6 \times 10^6$ K/s was required for homogeneous nucleation around a platinum heating wire immersed in water.

Derewnicki [81] studied transient boiling in water using a thin platinum wire of 25 μm in diameter under slow and rapid pulse heating. Figure 31a shows the temperature variation of the platinum wire as a function of time under different pressures at a slow heating rate of about 9×10^5 K/s. Before nucleation takes place, temporal temperature variation was found to follow the theoretical solution for transient heat conduction and independent of pressure. The points of divergence from this curve indicate that onset of nucleation shifts to higher temperatures when liquid pressure is increased. At the atmosphere pressure, the bubble nucleation temperature was about 200°C, corresponding to a superheat of 100°C. In this case, only a few heterogeneous nucleation sites were active, leading to rapid growth and spreading of vapor bubbles along the heated wire. As a result, large cylindrical bubbles momentarily enclosing large portions of the wire were formed. On the other hand, at a high heating rate of 6×10^6 K/s as shown in Fig. 31b, the bubble nucleation temperature was about 300°C (i.e., 20 K below the predicted homogeneous nucleation temperature) and is relatively independent of pressure. Also, the value of this nucleation temperature was very repeatable for different wires and remained unchanged if the heating rate was increased further. This phenomenon was attributed to the onset of homogeneous nucleation on the wire.

Glod et al. [82] investigated microbubble generation in water from a platinum wire of 10 μm in diameter and 1 mm in length. At a slow heating rate of 10^5 K/s, Fig. 32a shows that nucleation was initiated by a single vapor bubble growing from a cavity on the wire surface, which subsequently triggered the boiling on the entire wire surface with heterogeneous nucleation to be the main mechanism. It was found that the nucleation temperature increased with heating rate until a maximum limit is reached. Figure 32b shows that the boiling process at a high heating rate of 8.6×10^6 K/s. It can be seen that the wire surface was almost instantaneously covered with a thin vapor film. A maximum nucleation temperature of 303°C was obtained at the maximum heating rate of 8.6×10^7 K/s when homogeneous nucleation occurred. The useful extractable mechanical work from the acoustic pressure wave as a function of time was calculated.

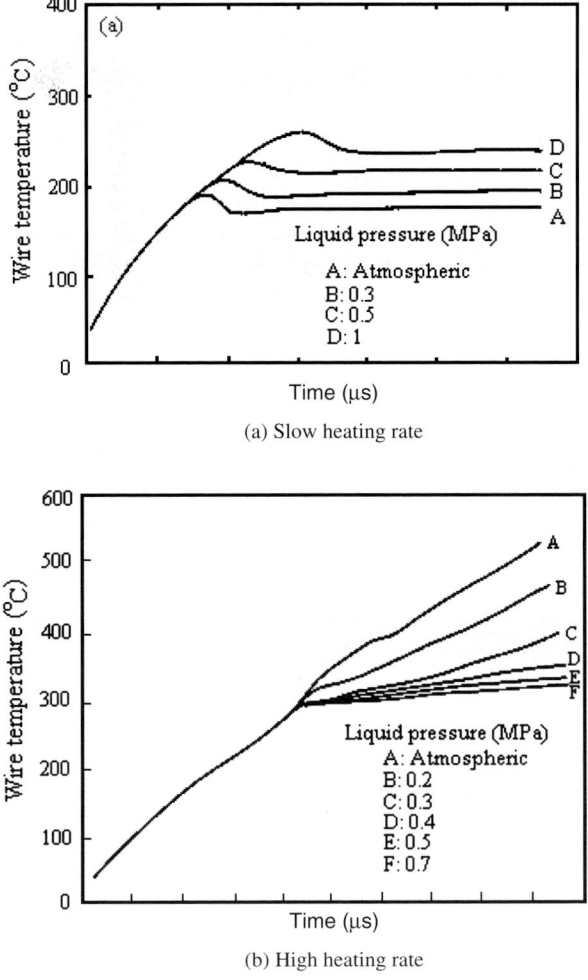

FIG. 31. Wire temperature versus time under different pressures: (a) at relatively slow heating rate and (b) at relatively high heating rate. From Derewnicki [81]. Copyright (1985), with permission from Elsevier.

2. Bubble Growth and Collapse on a MicroFilm Heater

Advances in MEMS technology have enabled the fabrication of planar film heaters of micrometer or even submicron in size, resulting in a much higher heat flux from the heater surface. Using the microfabrication

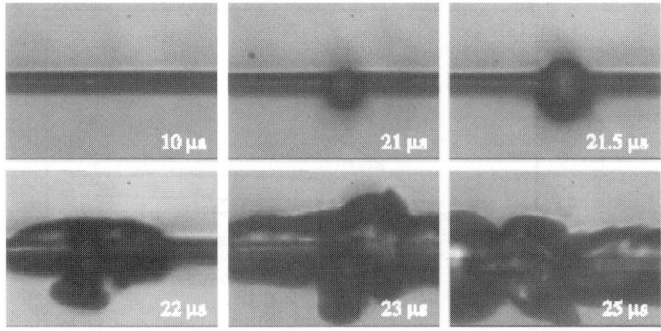

(a) Slow heating rate of 10^7 K/s

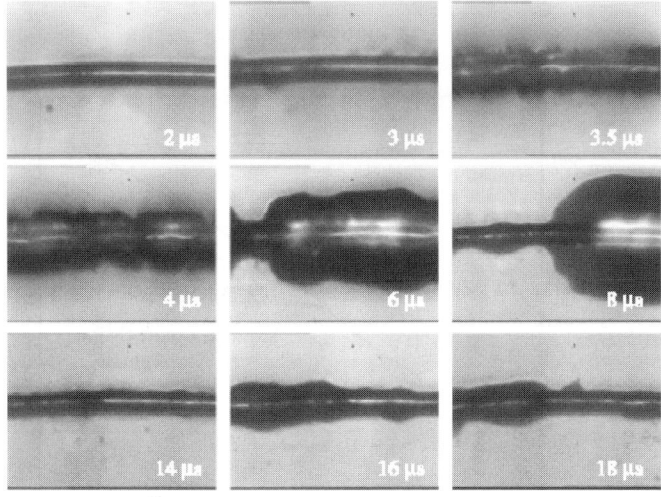

(b) Slow heating rate of 8.6×10^7 K/s

FIG. 32. Photo images on initial stage of explosive vaporization of water on an ultra-thin wire: (a) at a slow heating rate of 10^7 K/s (b) and at a high heating rate of 8.6×10^7 K/s (bottom). From Glod et al. [82]. Copyright (2002), with permission from Elsevier.

technology, a polysilicon microheater can be fabricated with its surface totally free of noticeable cavities.

Asai [83] carried out an experimental study on bubble nucleation on a microfilm heater $100 \times 100\,\mu m^2$ immersed in methanol under different power inputs from 16 to 51 MW/m^2 with pulse widths varying from 5.2 to 52 μs. A theoretical model was also carried out to predict the bubble growth. The model assumes nucleation in the superheated liquid followed by instantaneous formation of a vapor film, rapid bubble growth due to the pressure

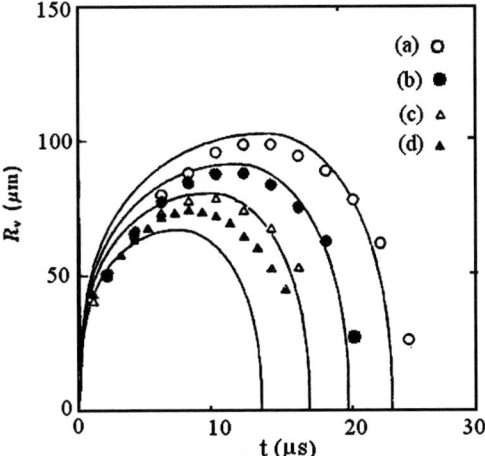

FIG. 33. Comparison of theory and experiments of bubble radius versus time on a microheater with pulse heating: (a) $q = 16\,\text{MW/m}^2$ and $\tau = 52\,\mu\text{s}$, (b) $q = 25\,\text{MW/m}^2$ and $\tau = 20.8\,\mu\text{s}$, (c) $q = 35\,\text{MW/m}^2$ and $\tau = 10.4\,\mu\text{s}$, and (d) $q = 51\,\text{MW/m}^2$ and $\tau = 5.2\,\mu\text{s}$. From Asai [83]. Copyright (1991), with permission from ASME.

impulse, and cavitation bubble collapse. Figure 33 is a comparison of theoretical and experimental data of bubble radius versus time that shows a nearly symmetric bubble growth and collapse period for pulse width in the microseconds range.

Lin et al. [84,85] investigated bubble formation mechanisms on a non-uniform width polysilicon microresistor (having two sections of $10 \times 3\,\mu\text{m}^2$ and $5 \times 2\,\mu\text{m}^2$) in an open environment with application to inkjet printer heads as well as in a closed environment with applications to microbubble pumps and microbubble valves. The average temperature of the heater could be obtained from the electrical resistance versus temperature curve. Distinct differences in boiling experiments by microheaters made from microfabrication technology and those of macroheaters were identified. While scattered bubbles are generated by macroheaters, a single, controllable microbubble can be generated by these silicon microheaters. Moreover, large cavities, existing on conventional macroheaters, do not exist on these silicon microheaters. Thus, established cavity theories for bubble nucleation in pool boiling on conventional heaters are not applicable to microheaters. It was found that individual, spherical vapor bubbles with diameters ranging from 2 to 500 μm could be generated by solid-state polysilicon microheaters, and that the size of the thermal microbubbles is controllable by a small power input. It was concluded that the microbubble formation on these microheaters is more likely quasi-homogeneous nucleation than heterogeneous nucleation.

Iida et al. [86] investigated bubble nucleation in ethyl alcohol, toluene, and water using a small Pt/Cr rectangular film heater ($100 \times 250\,\mu m^2$) sputtered on a quartz glass substrate at an extremely high heating rate of about 10^8 K/s. Photos were taken for boiling patterns on the microheater at different heating rates. At a low heating rate, the photos show that only a few bubbles were generated. At a moderate heating rate, more bubbles were generated and grew. When the heating rate was high, a large number of tiny bubbles generated concurrently on the heater in a short period of time. Under high heating rates, all the tiny bubbles observed in one frame were generated at different positions from those in the other frame, which were taken at a different time. Thus, it was concluded that the nucleation was not due to the pre-existing sites on the wall but due to fluctuation nucleation, which is a homogeneous boiling mechanism. The temperature data at boiling incipience versus the rate of temperature rise show that the temperature of boiling incipience increases with the temperature rise rate and approaches a constant value. The nucleation temperature, calculated based on classical nucleation theory, is higher than experimental data.

Using the same experimental apparatus as Iida et al. [86], Okuyama et al. [87] investigated the transient boiling in ethyl alcohol at a heating rate of 10^7 K/s, with particular emphasis on boiling dynamics and the corresponding heat transfer from the heater surface. The concurrently generated tiny bubbles coalesced to form a large bubble slightly after the boiling incipience and the heat flux decreased to about half of the value before boiling. It was deducted that the energy stored in the superheated liquid layer at the heater surface dominated the growth and collapse of the coalesced bubble.

Avedisian et al. [88] measured the average surface temperature of a Ta/Al heater ($65 \times 65\,\mu m^2$) on a silicon substrate of a commercial thermal inkjet printer (TIJ), which was immersed in subcooled water during pulse heating at microsecond duration. A SEM examination of the surface structure of the microheater showed that there were no obvious surface imperfections for nucleation sites on the heater surface. Figure 34 shows measurements of the temporal temperature variation of the microheater under various applied voltages with 5 μs duration. For very low voltage of less than 6 V, the surface temperature increased in an exponential fashion that is characteristics of a small heater. At an applied voltage of 7 V, the surface temperature increased in an exponential manner for the first 3 μs and thereafter oscillated, as shown in Fig. 34a. This is probably due to heterogeneous nucleation of vapor bubbles forming at one or more nucleation sites. When a growing bubble encounters subcooled water on its upper perimeter, condensation takes place and the bubble collapses. During collapse, cold water is drawn to

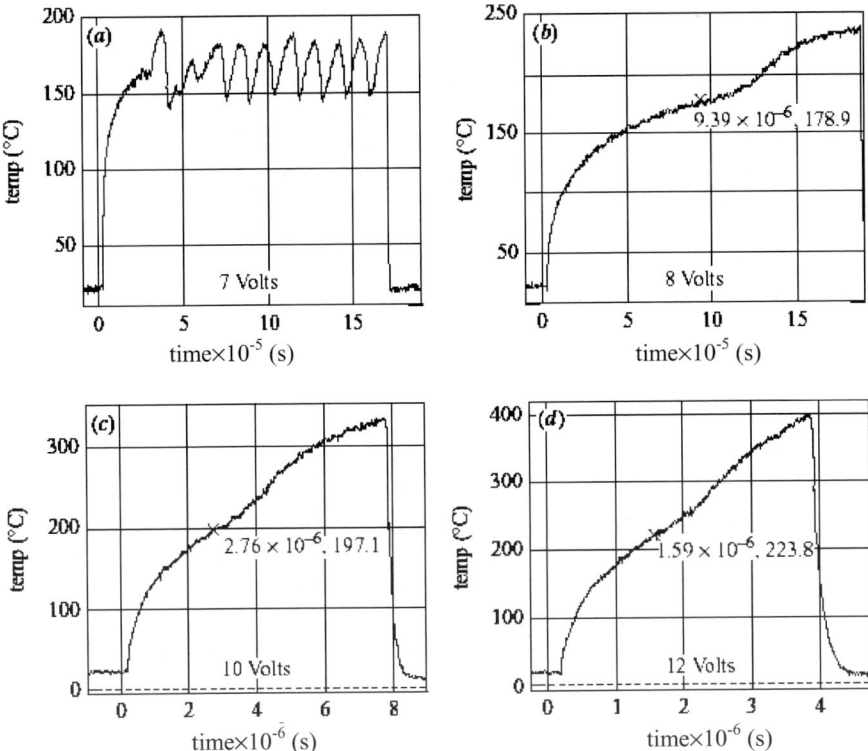

FIG. 34. Temporal variation of average surface temperature of a microheater at different input voltages. "x" in each figure indicates the nucleation temperature defined by the first inflection point, $d^2T/dt^2 = 0$. From Avedisian [88]. Copyright (1999), with permission from The Royal Society.

the surface which lowers the surface temperature. Since the surface of the microheater is smooth, the cyclic growth and collapse process probably occurred at few sites. At higher voltages, temperature oscillations were not measured (Figs. 34b–d). The nucleation temperature was identified by an inflection point on the temperature profile (where $d^2T/dt^2 = 0$) of the heater surface. It was found that the bubble nucleation temperature increased with the heating rate and approached a maximum value of 556 K, corresponding to a maximum heating rate of 2.5×10^{8}°C/s. This heating rate was higher than the 9.3×10^{7}°C/s value reported by Iida et al. [86] for pulse heating water using a thin platinum film heater ($100 \times 250\,\mu m^2$) as discussed above. Furthermore, the measured nucleation temperatures were in qualitative agreement with the prediction of the homogeneous boiling theory for an appropriate value of the contact angle. Figure 35 shows the variation of

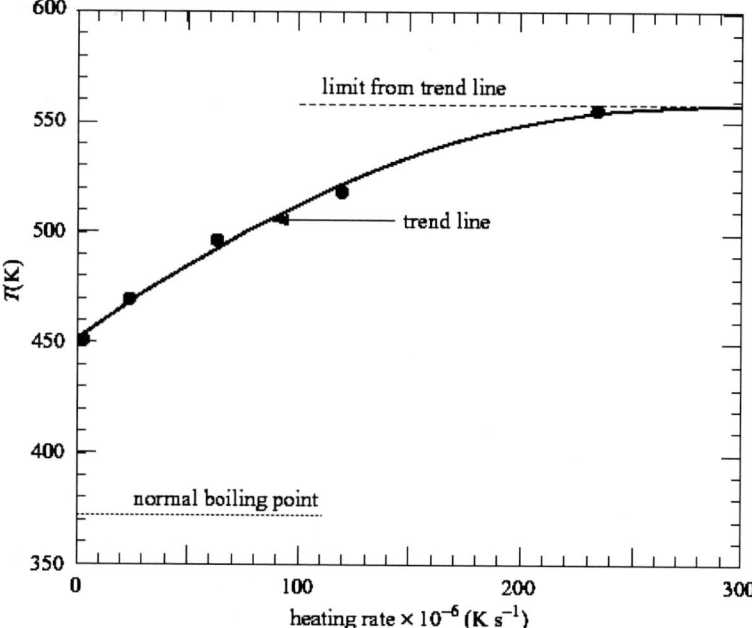

FIG 35. Variation of nucleation temperature with heating rate. Trend line shows an asymptote of 560 K. From Avedisian [88]. Copyright (1999), with permission from The Royal Society.

nucleation temperature with heating rate. Generally, the nucleation temperature increases with heating rate, and a limit appears to be approached in which the nucleation temperature is independent of heating rate. Note that this behavior is very similar to that observed by Iida et al. [86].

Zhao et al. [89] measured the acoustic pressure waves as the bubbles formed on a $100 \times 110\,\mu m^2$ thin film heater (under pulse heating in μs range) immersed in water. When a bubble undergoes expansion or contraction in the subcooled liquid, acoustic pressure waves are radiated from the bubble surface. The intensity of the acoustic pressure waves represents the intensity of the microscopic vapor explosion on the microheater surface. It was found that the acoustic pressure emission increased linearly with time shortly after the onset of explosive vaporization. The amount of extractable mechanical energy was approximately $0.3\,\mu J$ for the 37 V heating pulse. The corresponding extractable mechanical power was more than 0.2 W.

Yin et al. [90] studied the bubble nucleation in FC-72 on an impulsively powered square microheater ($260 \times 260\,\mu m^2$) for pulse widths of 1–10 ms duration at different heat fluxes between 3 and 44 MW/m^2. It was found that the bubble growth consisted of two steps, a relatively violent one at first, followed by a rapid shrinking of the vapor mass and a subsequent slower expansion. At a low heat flux of 3.43 MW/m^2, a single large bubble consistently nucleated near the center of the heater. It grew very dynamically overshooting its equilibrium size, shrinked, and then essentially stabilized over the heater. At high heat flux of 44 MW/m^2, nucleation occurred at several spots on the heater. The nucleation temperature was reached much earlier and there was insufficient time to heat up a substantial mass of liquid. In these conditions, the heat storing in the solid substrate played an important role in the secondary vapor growth after the heater had been turned off.

Jung et al. [91] investigated bubble nucleation in dielectric liquids (such as FC-72, FC-77, and FC-40) heated by polysilicon microheaters (50 × 3 μm^2, 50 × 5 μm^2) under steady or finite pulse of voltage input. It was found that the bubble nucleation temperature on the heater with 3 μm width was higher than the superheat limit, while the temperature on the broader width of 5 μm was considerably lower than the superheat limit.

Deng et al. [6,8,92,93] carried out a series of experiments to investigate microvapor bubble generation under pulse heating. Thin film planar platinum heaters of non-uniform width with the narrow part ranging from 0.5 up to 70 μm were fabricated on the same wafer to guarantee that the same fabrication process was applied to all of these heaters. A 1.66-ms-wide voltage pulse was imposed on each platinum heater, while the height of the voltage pulse was increased from zero until boiling was observed at a certain voltage. Thus, the minimum power input to the heater to induce a boiling process can be identified for each microheater at this pulse width. The onset bubble nucleation temperature was measured using each platinum heater as a self-sensing resistive temperature sensor. The bubble nucleation process was visualized by a high-speed CCD camera, and a 100 MHz high-speed digitizer was used for data acquisition. It was found that scaling effect of a heater could greatly affect the boiling pattern. Figure 36 is a map showing the effect of the heater size (slim part) on whether microbubble or film boiling is formed on the heater. It appears that the critical dimension of the microheater is 10 μm, and vapor bubble is formed when one side of the heater is less than 10 μm, while film boiling is achieved when one side of the heater is greater than or less than 10 μm. Note that the experimentally determined critical heater size (10 μm) is comparable to the thermal diffusion length for the present case ($\sqrt{Dt} = 7.1$ μm). Figure 37 shows that a single spherical bubble was generated at the slim part of the microheater (10 × 3 μm^2) at a pulse width of 1.6 ms under a power input of 28.2 mW. The maximum size of the bubble was about 9.2 μm that was comparable to the narrow part of the microheater. Figure 38 shows that an oblate vapor

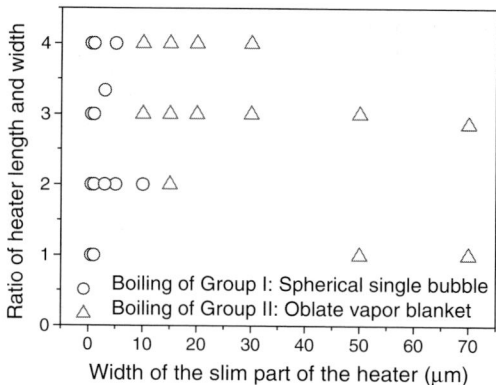

FIG. 36. Map of boiling patterns on microheaters: Group I: single vapor bubble; Group II: film boiling. (Deng [92]).

film was formed on the uniform width microheater surface with heater's feature size of $50 \times 10\,\mu m^2$ at a pulse width of 1.6 ms at 180.1 mW.

Deng et al. [92,93] used several non-uniform width microheaters with feature size less than 10 μm to study microbubble growth and collapse in a subcooled pool of water under pulse heating with a pulse width of 1.6 ms. Figure 39 shows temporal variations of surface temperatures of submicron heaters a and b (with dimensions of $0.5 \times 0.5\,\mu m^2$ and $1.0 \times 0.5\,\mu m^2$, respectively) and micron heaters c and d (with dimensions $2 \times 1\,\mu m^2$ and $10 \times 3\,\mu m^2$, respectively). As shown from this figure, immediately after the pulse heating was applied, the surface temperature of the heater increased exponentially with time for all the heaters (see the insets A–D for a close-up view), indicating lumped capacitance heat conduction characteristics. The heating rate was in the range of 10^6–$10^7\,K/s$, and the high heating rate was observed on submicron heaters due to their low thermal capacity. Note that this heating rate was similar to the experiment performed by Skripov [80], who showed that a heating rate as high as $6 \times 10^6\,K/s$ was required for homogeneous nucleation around a platinum heating wire in water. As shown in curves I–III, a temperature inflection point (indicated as "A," "B," and "C") was observed after the exponential increase of the heater temperature during the initial heating period. The temperature inflection point was also observed by other researchers [82,88] and the temperature at the inflection point was regarded to be the bubble nucleation temperature. According to Fig. 39, the onset bubble nucleation temperature (the temperature at the inflection point) on submicron heaters a, b, and micron heater c was 518, 480, and 487 K, respectively. Thus, the nucleation temperature is

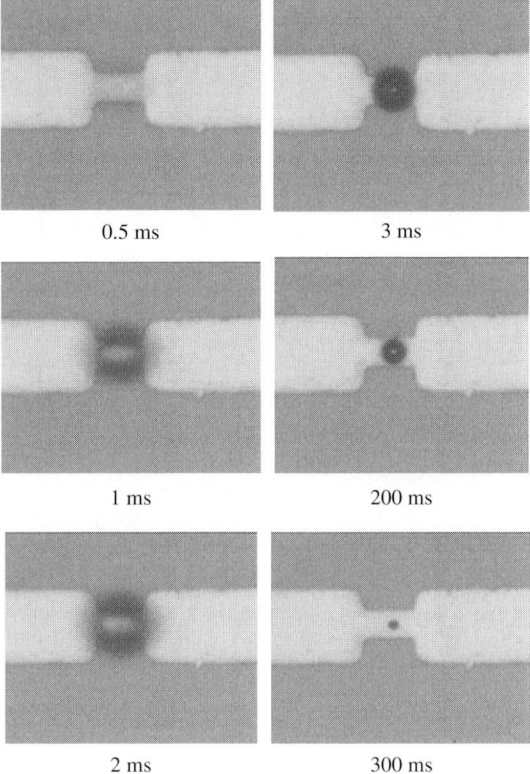

Fig. 37. Growth and collapse of a vapor bubble on a microheater ($3 \times 10\,\mu m^2$) submerged in deionized water. From Deng et al. [6]. Copyright (2003) with permission from Elsevier.

higher for smaller heater with a faster heating rate. The corresponding bubble nucleation time (the time of the occurrence of bubble nucleation after the beginning of the heating pulse) was 51 and 60 μs on submicron heaters (a) and (b), and 72 μs on micron heater (c) after the beginning of 1.6-ms-width pulse heating. As shown in Curves I–III, after the bubble nucleation, the temperature of the submicron heaters continue to increase till the end of the heating pulse, and the temperature rise during this period was about 110 K. For bubble nucleation on micron heater (d), however, the temperature inflection point was not observed during the boiling process. Instead, the temperature of the heater increase curve had a "V"-shaped change at about 240 μs after the beginning of the heating pulse. The temperature suddenly dropped from 485 to 476 K and then slowly increased to

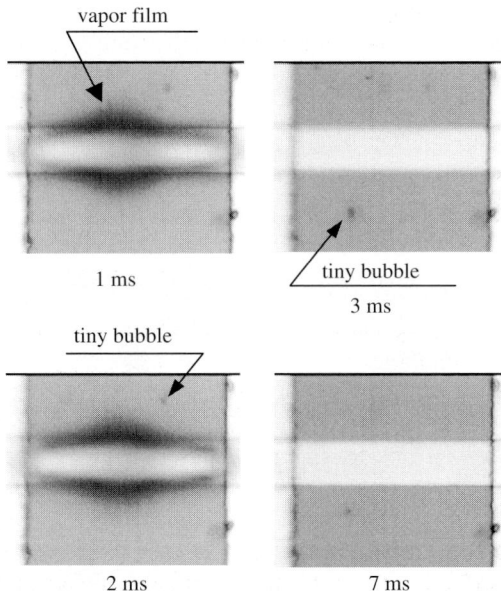

FIG. 38. Growth and collapse of a vapor film on a strip microheater ($50 \times 10\,\mu m^2$) submerged in deionized water. From Deng *et al.* [6]. Copyright (2003) with permission from Elsevier.

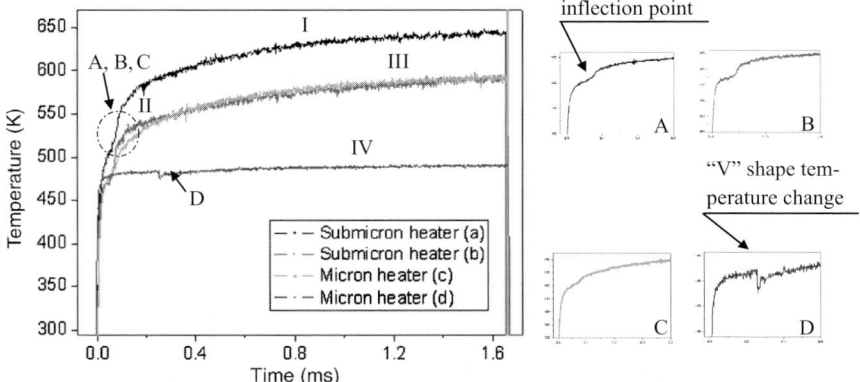

FIG. 39. Measured temporal variations of surface temperatures of the heaters. Curves I, II, III and IV are for heaters (A), (B), (C), and (D), respectively. (Deng et al. [93]).

491 K at end of the heating pulse (see D in Fig. 39), showing a discontinuity in temperature variation. As mentioned earlier, the "V"-shaped temperature variation was also observed by Tsai and Lin [69] in their experiment on transient bubble nucleation on a polysilicon microheater ($95 \times 10\,\mu m^2$) when a medium input current was employed. According to Tsai and Lin [69], the "V"-shaped temperature change in curve IV of Fig. 39 can be identified as the beginning of bubble nucleation on micron heater (d). Before bubble nucleation, the heat supplied by the heater was used to heat up the heater itself, the silicon nitride and oxide layer underneath the heater, and the water immediately adjacent to the heater by conduction heat transfer. During this period, the temperature of the heater increased exponentially indicating lumped capacitance heat conduction characteristics as shown during the initial time of curve IV in Fig. 39. As the surrounding water was heated up to the incipient bubble nucleation temperature, a highly localized near homogeneous boiling occurred in the water adjacent to the heater. A relatively large amount of heat was absorbed from the heater to cover the latent heat of evaporation of water. This resulted in a sudden drop of the heater temperature, as shown in the inset D of Fig. 39, if the power input to the heater was insufficient to cover the latent heat of evaporation. Therefore, the onset bubble nucleation temperature in the water on micron heater (d) could be identified at 485 K. It was found that the bubble nucleation occurred at approximately 240 μs after the beginning of a 1.66-ms-width pulse heating, and a temperature rise of 15°C was measured after bubble nucleation.

D. Flow Boiling in Meso-/Microchannels

Recently, a number of papers [9,10,94,95] have been published on the review of literature on heat transfer coefficients and pressure drop in flow boiling in meso-/microchannels. For this reason, early work in these topics will not be discussed further in this section. Instead, our attention will be focused on recent work on nucleation, flow patterns, and instability in meso-/microchannels.

1. *Nucleation in Meso-/Microchannels*

Peng and Wang [96] performed one of the early experiments on flow boiling of water in a microchannel with rectangular cross section of $0.6 \times 0.7\,mm^2$ that was made of stainless steel. It was found that boiling was initiated at very low superheats of only 2–8°C. However, without using a microscope and a high-speed video recording system, they claimed that no bubbles were generated at a high heat flux of $100\,W/cm^2$ in their experiments because of the confined space.

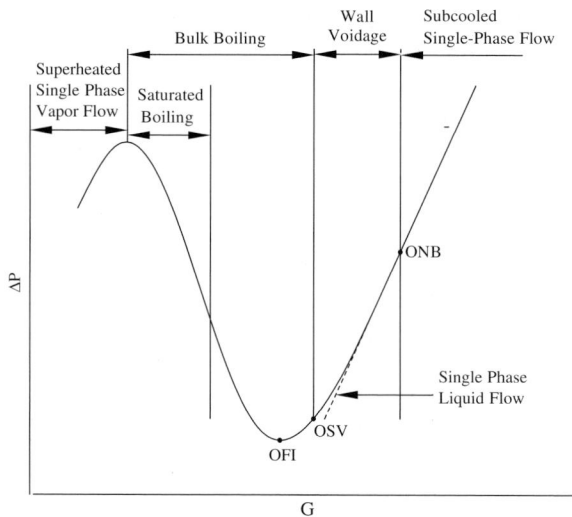

FIG. 40. Pressure drop versus mass flux characteristic curve for a uniformly heated channel. From Ghiaasiaan and Chedester [97]. Copyright (2002), with permission from Elsevier.

Figure 40 is a sketch of total pressure drop across a channel versus mass flux (ΔP–G) with a constant thermal load and a constant inlet temperature [97]. The onset of nucleate boiling (ONB) can be identified as the point where the calculated single-phase pressure drop line begins to deviate from the (ΔP–G) curve. Ghiaasiaan and Chedester [97] analyzed available heat flux data of water for the ONB (or boiling incipience) in microtubes with diameters in the 0.1–1 mm range, and found that the ONB heat flux were underpredicted by macroscale models and correlations. The higher ONB heat flux in microtubes was attributed to the dominance of thermocapillary force that tends to suppress microbubbles generated on wall cavities. For better prediction of the incipient boiling heat flux in microtubes, they presented a semi-empirical model. Figure 41 shows that the predicted ONB heat flux based on this new model increases monotonically with increasing Peclet number ($Pe = uD/\alpha$) and subcooling (dT), and with decreasing tube diameter, which is in good agreement with experimental data.

Yen et al. [98] performed an experiment on the flow boiling of HCFC123 and FC-72 in microtubes with diameters ranging from 0.19 to 0.51 mm. A twin plunge pump was used to keep a constant flow rate without introducing unwanted gas bubbles into the test section. It was found that the degree of superheat depended on the heat flux, mass flux, and tube diameter. Figure 42 shows that the maximum superheat temperature of HCFC123 decreases with the tube diameter: the maximum superheat temperatures of

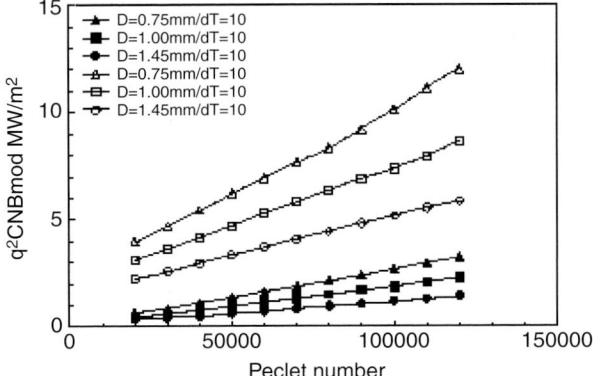

FIG. 41. Effects of tube diameter on incipient heat flux of water at different Peclet numbers at $P = 10$ atm. From Ghiaasiaan and Chedester [97]. Copyright (2002), with permission from Elsevier.

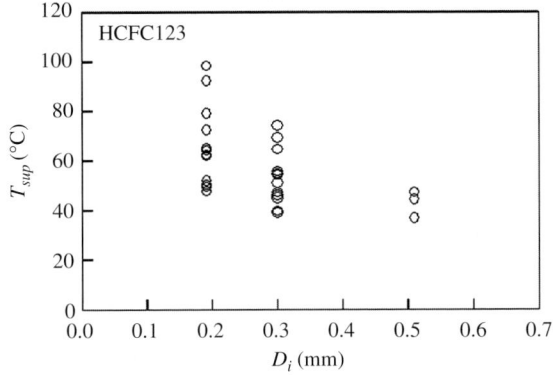

FIG. 42. Superheat temperature of HCFC123 versus tube diameter. From Yen et al. [98]. Copyright (2003), with permission from Elsevier.

HCFC123 in the 0.19 and 0.51 mm diameter tubes were 100°C and 45°C, respectively.

Zhang et al. [99] investigated the effects of wall roughness and surface tension of the liquid on nucleation in flow boiling in silicon microchannels. Figures 43a and b show the change of wall temperatures at different locations versus heating rate in two microchannels (50 μm wide and 44 μm deep) with the same hydraulic diameter of 47 μm having smooth and rough walls at the flow rate of 0.02 ml/min. A comparison of these two graphs

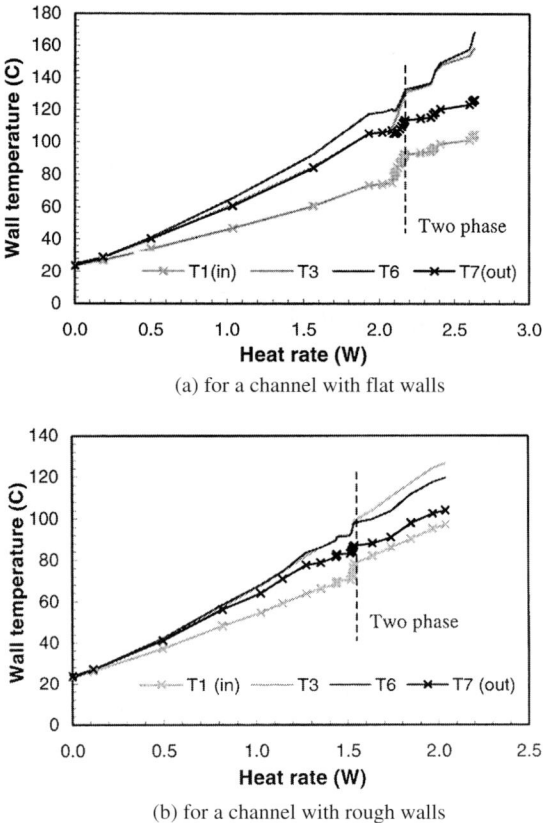

FIG. 43. Wall temperature versus heating rate in two microchannel at water flow rate of 0.02 ml/min: (a) with smooth walls and (b) with rough walls. From Zhang et al. [99]. Copyright (2002), with permission from IEEE.

shows that the wall superheat before the onset of boiling in a rough wall was more than 20°C lower than that in a smooth wall. However, no obvious difference in wall superheat was found when water was replaced by 200 ppm Trition X-100 solution that has a lower surface tension. Therefore, Zhang et al. [99] concluded that the increase of wall superheat before bubble nucleation in microchannels as reported in the literatures is primarily due to the lack of active nucleation sites rather than limited channel space or high surface tension of the liquid.

Lee et al. [100] investigated bubble dynamics of deionized water at the location of incipient boiling in a trapezoidal microchannel with a hydraulic diameter of 41.3 μm at flow rates ranging from 170 to 477 kg/m² s. With the

TABLE II

Bubble Growth Rate, Initial Bubble Radius, and Bubble Departure Radius in Single Microchannel under Various Conditions [100]

q″ (kW/m^2)	T_w (°C)	T_{sat}^a (°C)	$T_{sat,L}^b$ (°C)	Bubble growth rate (μm/ms)	Initial bubble radius (μm)	Bubble departure diameter (μm)	Remarks
$G = 170$ kg/m^2 s							
1.47±0.80	100	103	103	0.24±0.005	15.1±0.32	45.1	$T_w < T_{sat}$
57.6±20.0	108	105	105	2.62±0.21	12.6±1.38	47.3	
137±23.4	124	112	117	*95.3±3.07	21.3±0.40	30.6	
196±26.9	131	115	123	4.91±0.28	13.4±0.69	27.3	Reversed flow present
$G = 341$ kg/m^2 s							
6.94±2.13	98	106	107	0.16±0.002	9.74±0.23	33.6	$T_w < T_{sat}$
103±25.3	122	111	113	*94.63±3.74	1.06±0.70	28.4	
189±28.7	128	114	123	0.24±0.008	12.2±0.44	31.1	
264±30.1	134	117	127	0.32±0.02	17.2±0.43	26.9	
255±31.9	145	119	132	0.22±0.01	14.2±0.66	38.7	Reversed flow present
$G = 477$ kg/m^2 s							
15.7±1.87	100	108	108	0.19±0.003	5.56±0.29	40	$T_w < T_{sat}$
107±18.5	116	109	111	7.09±0.492	1.90±1.43	36.4	
218±21.1	130	114	118	*72.8±9.15	20.5±1.18	29.9	
353±24.3	144	120	130	0.54±0.036	13.1±1.18	38.9	
415±25.9	150	123	136	0.10±0.01	17.9±1.07	47	
449±26.7	157	124	137	0.10±0.01	14.1±0.67	19.2	Reversed flow present

*Extraordinary high growth rate

aids of a high-speed digital camera and Image-Pro, bubble nucleation, growth, departure size, and frequency were observed and recorded. Table II gives the bubble growth rate, initial bubble radius, and bubble departure radius in a single microchannel under various heat and mass fluxes conditions. It was found that the bubble nucleation in a microchannel can be predicted by the classical model with micro-sized cavities, but their growth rates are much smaller than that predicted by the classical Rayleigh equation due to the space constraint and the significant temperature gradient in the transversal directions. It was found that the bubble growth rate in a single microchannel can reach extraordinary high values (as indicated by * in Table II) for some cases, and subcooled nucleation boiling ($T_w < T_{sat}$) was observed at low heat fluxes due to the effect of dissolved gases. Figure 44 shows the comparison of experimental data with the results of an analysis

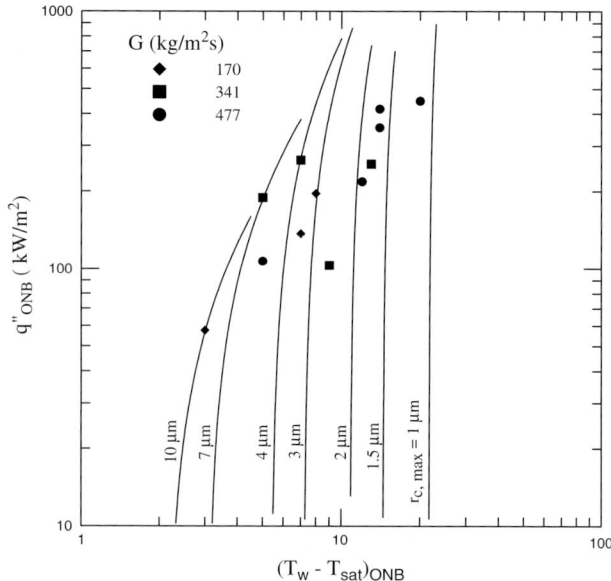

FIG. 44. Comparison of theory and measurements for incipient heat flux versus wall superheat. From Lee et al. [100]. Copyright (2004), with permission from Elsevier.

on the incipient boiling heat flux versus wall superheat for various cavity radii on the wall. It can be seen from this figure that at the same incipient heat flux, the wall superheat increases with the decrease of the cavity radius.

Using the same method as Lee et al. [100], Li et al. [101] investigated bubble dynamics in the two parallel trapezoidal microchannels with a hydraulic diameter of 47.7 μm at flow rates from 105 kg/m²s to 555 kg/m²s. Their results on bubble growth rate, initial bubble radius, and bubble departure radius are summarized in Table III. Similar to that of the single microchannel [100], subcooled nucleation boiling ($T_w < T_{sat}$) was also observed at low heat fluxes. However, extraordinary high values of bubble growth rate, which were observed by Lee et al. [100] in a single microchannel, did not exist in the parallel microchannels due to flow interaction in the channels.

Steinke and Kandlikar [102] performed an experimental investigation on the effect of dissolved oxygen contents (8.0, 5.4, and 1.8 ppm) on the nucleation temperature for flow boiling of water in six parallel copper microchannels having a diameter of 207 μm. They found that, for the dissolved oxygen content of 8.0 ppm, nucleation (boiling incipience) appeared at a subcooled surface temperature of 90.5°C at 1 atmospheric pressure and at

TABLE III
Bubble Growth Rate ($\dot{R}$) and Initial Bubble Radius (R_0) under Various Conditions [101]

q″ (kW/m²)	T_w (°C)	T_{sat}^a (°C)	$\dot{R}$ (μm/m s)	R_0 (μm)	Remarks
$G = 105\,\text{kg/m}^2\,\text{s}$					
17.8	106	102	4.29±0.88	~7	Bubble waiting period; reversed flow present
53.9	112	105	1.62±0.11	~7	Bubble waiting period and reversed flow present
121	124	108	2.27±0.39	8.70±1.59	Bubble waiting period occurs intermittently, reversed flow present
$G = 269\,\text{kg/m}^2\,\text{s}$					
12.3	97	104	0.087±0.001	4.32±0.29	$T_w < T_{sat}$; reversed flow present
30.1	103	104	0.175±0.003	7.27±0.32	$T_w < T_{sat}$; reversed flow present
65.4	108	106	1.32±0.080	8.15±1.16	Reversed flow present
143	118	110	0.334±0.008	6.40±0.44	Reversed flow present
190	132	111	0.193±0.014	8.69±0.41	Reversed flow present
245	146	114	0.211±0.02	8.64±0.51	Reversed flow present
303	157	115	0.147±0.002	6.16±0.20	Reversed flow present
$G = 555\,\text{kg/m}^2\,\text{s}$					
70.6	94	104	0.306±0.008	6.82±0.29	$T_w < T_{sat}$
123	112	103	4.12±0.23	8.48±0.28	
233	131	105	15.0±8.5	9.32±0.3	

the mass flow rate of 380 kg/m² s. However, for dissolved oxygen contents of 5.4 and 1.8 ppm, nucleation (boiling incipience) did not appear until the surface temperature reached 100°C at the same pressure and mass flow rate.

It can be concluded from the above experimental investigations that the wall roughness and the dissolved gas have great influence on boiling nucleation in microchannels. These two factors, plus the corner effect [66] in the channel cross section, cause the experimental observation of bubble nucleation at a much lower temperature than those predicted by Peng et al. [68] or by the classical homogeneous nucleation theory.

2. Boiling Flow Patterns and Instabilities

Jiang et al. [103] and Lee et al. [104] performed visualization and measurement investigation on flow boiling of water through silicon microchannels with hydraulic diameters as small as 26 and 24 μm, respectively. The heaters in these experiments were located at the inlet of microchannels, instead of locating along the microchannels as in other flow boiling experiments. Local nucleation was observed in their microchannels at low power inputs and low wall temperatures, which was attributed to the dissolved gas in the water. Their visualization studies showed that the size and shape of the microchannel have important effects on the two-phase flow patterns as well as flow transitions in the channels. Figure 45 gives the schematic illustrations of an annular flow in a triangular silicon microchannel as well as in two rectangular silicon microchannels with different depths. Consider the two-phase flow in Fig. 45, the forces across the vapor–liquid interface must satisfy the Young–Laplace equation

$$\Delta P = P_V - P_L = \frac{\sigma}{R} \qquad (14)$$

where P_V and P_L are the vapor and liquid pressure, while σ and R the surface tension coefficient and the interface curvature radius, respectively. For the triangular microchannel in Fig. 45a, if the vapor pressure decreases or the liquid pressure increases (i.e., $\Delta P \downarrow$), the interfaces move inward to the channel center forming a larger curvature radius ($R\uparrow$) that leads to a smaller surface tension force ($\sigma/R \downarrow$). Similarly, if the vapor pressure increases or the liquid pressure decreases ($\Delta P \uparrow$), the interfaces move outward further into the corners leading to a smaller curvature radius ($R \downarrow$), and consequently a higher surface tension ($\sigma/R \uparrow$). Therefore, there exists a restoring force that tends to damp out pressure perturbations and renders the annular flow stable in a triangular microchannel. However, this restoring force is absent in the nearly rectangular microchannel as shown in Fig. 45b. Since the interface movement due to pressure perturbations (either $\Delta P \uparrow$ or $\Delta P \downarrow$) does not cause changes in the curvature radius of the interface and the corresponding

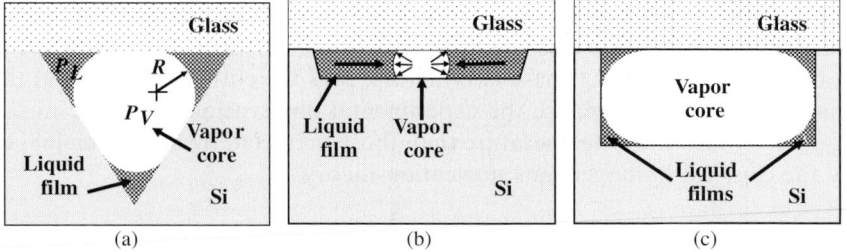

FIG. 45. Schematic illustrations of an annular flow mode in (a) triangular microchannel, (b) shallow rectangular microchannel, (c) deep rectangular microchannel. From Lee et al. [104]. Copyright (2003), with permission from IOP Publishing Ltd.

surface tension force, there exists no restoring force to damp out the pressure perturbations. Thus, the annular flow becomes less stable in this nearly rectangular microchannel. Figure 46 is a series of photos showing the development of unstable annular flow in a shallow rectangular microchannel with a depth of 14 μm and a width of 120 μm. It can be seen that within a 0.3 s of forming an annular flow (Fig. 46a), the vapor core became thinner and thinner as shown in Figs. 46b and c, and completely disappeared in Fig. 46d. Then, a new cycle of unstable flow begins. It should be noted that the size of the microchannel also affects the stability of annular flow in microchannels [103,104]. To illustrate this point, consider a rectangular microchannel as shown in Fig. 45c, which has the same width as in Fig. 45b but with a larger depth. In this deep microchannel, the liquid/vapor interface cannot extend across the entire channel height as in the shallow microchannel (Fig. 45b), and the liquid is confined in each corner similar to the triangular microchannel in Fig. 45a. Following the same restoring force analysis, it can be deduced that the annular flow in this deep rectangular microchannel (Fig. 45c) is more stable than that in the shallow rectangular microchannel (Fig. 45b).

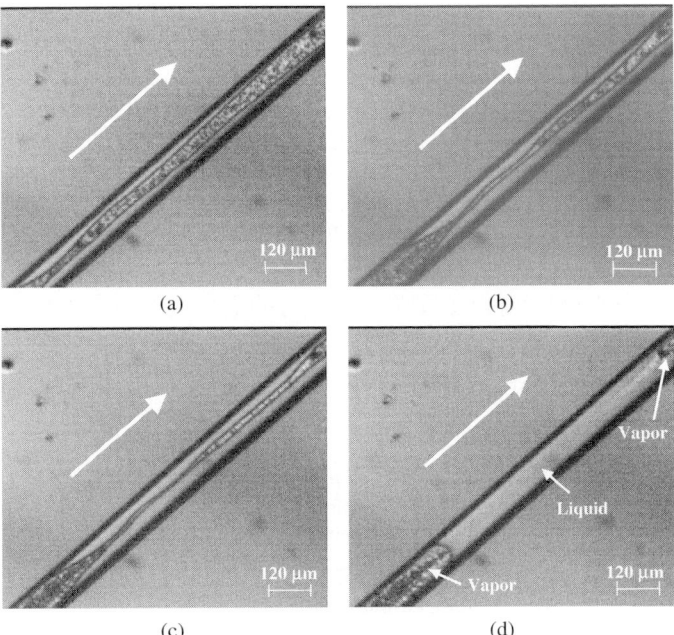

FIG. 46. Unstable annular flow pattern in a nearly rectangular microchannel. From Lee et al. [104]. Copyright (2003), with permission from IOP Publishing Ltd.

Sobierska et al. [105] investigated the flow patterns and their transitions for flow boiling of water in a vertical rectangular microchannel with a width of 0.86 mm and a depth of 2.0 mm, respectively. With a hydraulic diameter of 1.2 mm and a Confinement number greater than 0.5, the channel can be classified as a mesochannel. With the aids of visualization techniques, three basic flow patterns: bubbly, slug, and annular flow, were observed in this vertical mesochannel. The measurements presented in Fig. 47 show that the appearance of different flow patterns were dependent on the mass flux, heat flux, vapor quality, and axial locations. The transition boundaries for different flow patterns are presented in Fig. 48. By a comparison of their own data for flow boiling in microchannels with those in macrochannels, Sobierska et al. [105] found that the transition from slug to annular flow can be well described by the macrochannel criteria [106,107]. However, the macro models are unable to predict their bubbly slug transition data in the microchannels.

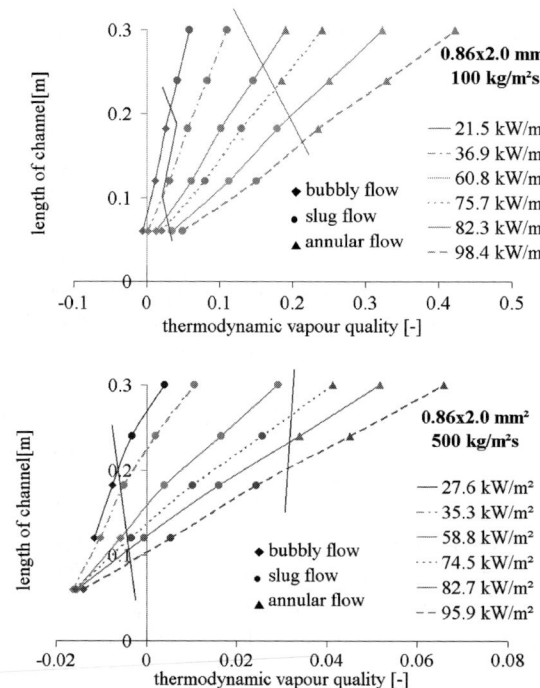

FIG. 47. Vapor qualities along the channel under different heat fluxes at two mass fluxes: 100 kg/m² s (top) and 500 kg/m² s (bottom). From Sobierska et al. [105]. Copyright (2004), with permission from ASME.

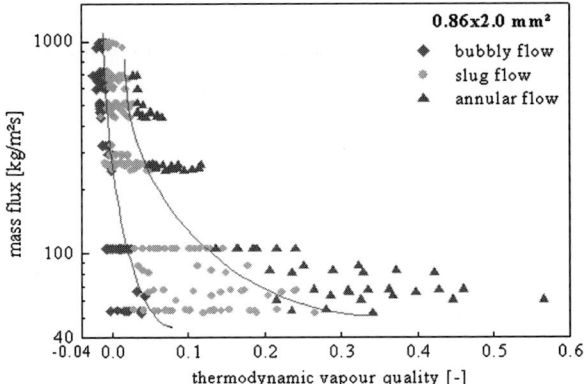

FIG. 48. Boiling flow regimes in terms of mass flux versus vapor quality. From Sobierska et al. [105]. Copyright (2004), with permission from ASME.

Wu and Cheng [108–110] carried out a series of visualization and measurement studies on boiling flow patterns and instabilities in deionized water flowing through parallel trapezoidal silicon microchannels (with hydraulic diameters ranging from 82.8 to 186 μm). The investigation was conducted with the combined use of a microscope, a high-speed video recording system (located at the top of the microchannels), and a multi-channel data acquisition system. Their experimental results showed that the heat and mass flux have great influences on boiling modes. With increasing heat flux and corresponding mass flux, three types of boiling flow modes were observed in the microchannels: liquid/two-phase alternating flow (LTAF), annular two-phase flow, and liquid/two-phase/vapor alternating flow (LTVAF). Figure 49 is a series of photos on the three boiling modes in the microchannels with a hydraulic diameter of 186 μm. The following is a brief description of these boiling modes and their associated fluctuations of various measurements.

Case 1: LTAF. At a heat flux of 13.5 W/cm^2 and mass flux of 14.6 g/cm^2 s, a LTAF appeared in the microchannels. As shown in Fig. 49a, after a period of liquid phase for about 5.4 s, a bubbly two-phase flow appeared. The bubbly flow lasted for about 10.0 s, and then, it was followed by a new cycle of liquid to two-phase alternating flow. Simultaneous measurements of temperature and pressure showed that these data fluctuated periodically with the same period as those obtained from the visualization study. The LTAF and its induced fluctuations in pressure drop and outlet temperature with small oscillation amplitudes were also reported earlier by Hetsroni

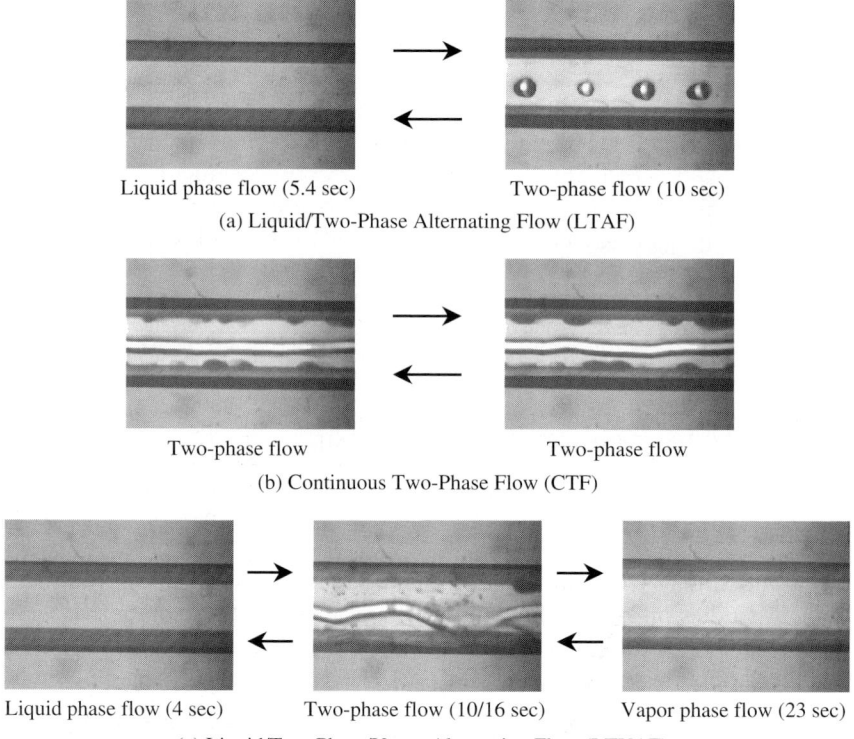

FIG. 49. Flow boiling modes at different heat fluxes and mass fluxes: (a) Liquid/two-phase alternating flow (LTAF): $q = 13.5\,\text{W/cm}^2$ and $m = 14.6\,\text{g/cm}^2\,\text{s}$; (b) continuous two-phase flow (CTF): $q = 18.8\,\text{W/cm}^2$ and $m = 11.9\,\text{g/cm}^2\,\text{s}$; (c) liquid/two-phase/vapor alternating flow (LTVAF): $q = 22.6\,\text{W/cm}^2$ and $m = 11.2\,\text{g/cm}^2\,\text{s}$. From Wu and Cheng [110]. Copyright (2004), with permission from Elsevier.

et al. [111] in parallel triangular silicon microchannels with a hydraulic diameter of 130 μm.

Case 2: Annular two-phase flow. As the heat flux was increased to 18.8 W/cm² and the corresponding mass flux decreased to 11.9 g/cm² s, an annular two-phase flow appeared in the microchannels. In this boiling mode as shown in Fig. 49b, no liquid phase period was observed and two-phase flow occupied the entire microchannel at all time. The oscillations of temperature and pressure measurements in this boiling mode were much less than those in LTAF mode, and this boiling mode can be regarded as a steady boiling mode. Note that in the two-phase flow period, a bright vapor core was formed along the middle of microchannels, while bubbles formed and collapsed in the corners at the same time.

Case 3: LTVAF. As the heat flux was increased to 22.6 W/cm^2 and the corresponding mass flux decreased to 11.2 g/cm^2 s, a most interesting boiling mode: LTVAF as shown in Fig. 49c was observed. In this boiling mode, the periods of single-phase liquid, two-phase mixture, and superheated steam during a cycle were 4, 26, and 23 s, respectively. Owing to the drastic alternation from subcooled liquid to superheated vapor, temperature and pressure measurements showed extraordinary large-amplitude oscillations, with more than 200°C in the wall temperature oscillation amplitude as shown in Fig. 50a. During the two-phase period in this LTVAF boiling mode, a bright band of vapor moving violently in different directions as shown in Fig. 49c was observed. The mechanism for causing large-amplitude oscillations in this boiling mode was discussed by Wu and Cheng [109,110].

Brutin *et al.* [112] carried out an experimental investigation on two-phase flow instabilities during the upward flow boiling of *n*-pentane in vertical rectangular aluminum microchannels having a hydraulic diameter of 889 μm. Two microchannels having a cross section of 0.5×4 mm^2 with lengths of 50 and 200 mm were used. To observe the flow pattern in the microchannel, each of the aluminum microchannel was covered with a transparent polycarbonate plate, and photos were taken by a high-speed video recording system. The inlet and outlet pressures were measured during the experiment. It was found that steady and unsteady thermohydraulic behaviors depended on the amount of heat flux and mass flow rate. For the 50-mm-long microchannel at a mass flow rate of 240 kg/m^2 s, different boiling behaviors were observed at different heat fluxes. At a moderate heat flux (under 20×10^4 W/m^2), a liquid zone at the bottom and a bubbly flow zone at the top coexisted in the microchannel, while steady temperature and pressure measurements were recorded. Above a critical value of the heat flux (38×10^4 W/m^2), an oscillatory regime appeared. A two-phase fluctuating flow can be clearly observed in the microchannel, with inlet and outlet pressures fluctuating at certain frequencies. Beyond a second critical heat flux (50×10^4 W/m^2), a steady mode (film boiling) appeared again, and a small fluid area at the bottom of the microchannel was surmounted by a two-phase flow with a high vapor quality. Figure 51 is a map of heat flux versus mass flow rate, showing stable and unstable flow regions in flow boiling in the 50-mm-long microchannel. Boiling fluctuations were also found in the 200-mm-long microchannel. Figures. 52a and b show the inlet and outlet pressure fluctuations in the 200-mm-long microchannel at the same mass flow rate but at different heat fluxes. Fluctuation amplitudes at the high heat flux of 9.6×10^4 W/m^2 were larger than those at the lower heat flux of 3.3×10^4 W/m^2. The visualization study showed that the larger fluctuations of inlet pressure were caused by the periodic back-flow of the vapor slug to the entrance.

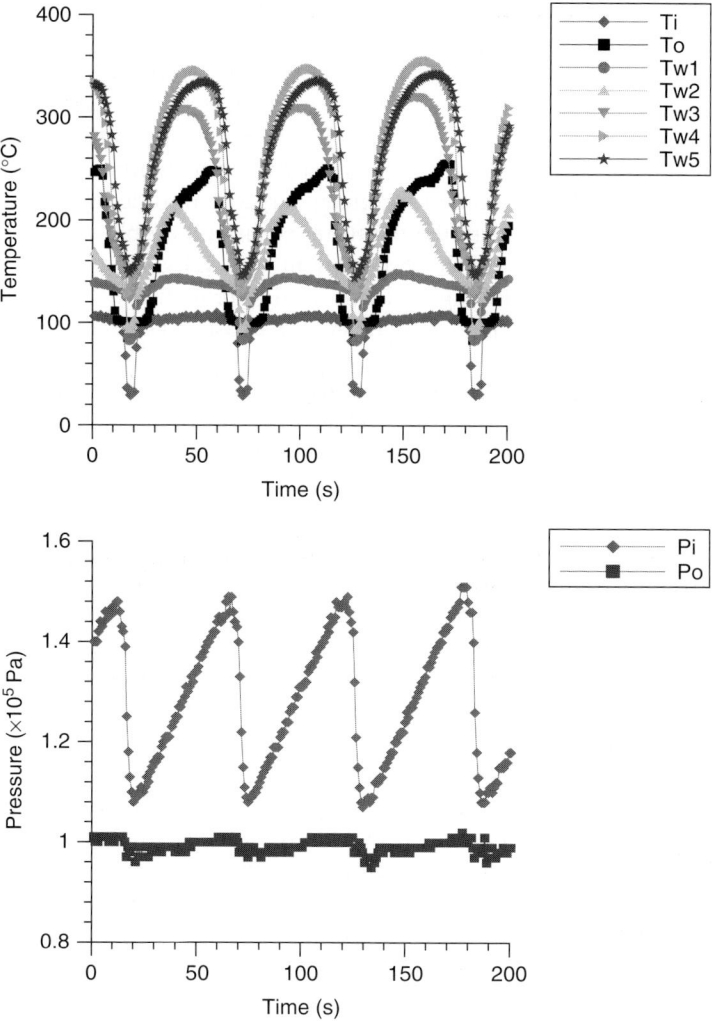

FIG. 50. Oscillations of temperature (top) and pressure (bottom) measurements for LTVAF at $q = 22.6\,\text{W/cm}^2$ and $m = 11.2\,\text{g/cm}^2\,\text{s}$. From Wu and Cheng [110]. Copyright (2004), with permission from Elsevier.

Qu and Mudawar [113] performed a flow boiling experiment of water in a two-phase microchannels heat sink made of 21 copper parallel microchannels with a cross-sectional area of $231 \times 713\,\mu\text{m}^2$. They identified two types of two-phase hydrodynamic instability in the microchannels: severe pressure drop oscillation and mild parallel channel instability. The pressure drop

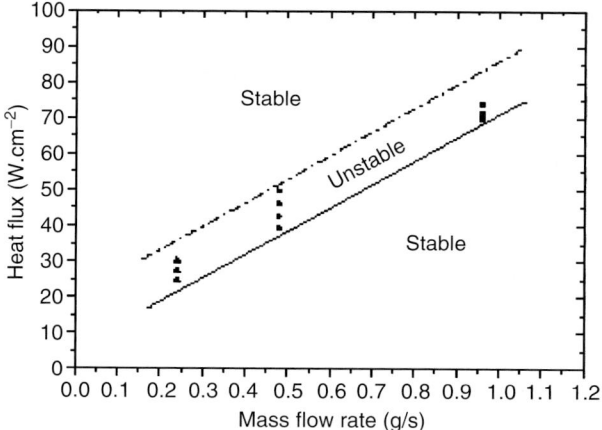

Fig. 51. Stability diagram of heat flux supplied versus mass flow rate in a $0.5 \times 4 \times 50\,\mathrm{mm}^3$ mesochannel. From Brutin et al. [112]. Copyright (2003), with permission from Elsevier.

oscillation often occurs when the upstream valve is fully opened, which allows the interaction between vapor generation in channels and compressible volume in the upstream of the heat sink. Figure 53a is a sketch showing the boundary between the single-phase liquid and two-phase mixture, oscillating back and forth in unison between the inlet and outlet in neighboring microchannels, during the severe pressure drop oscillation. This type of oscillation is usually so severe that the vapor in microchannels can enter the inlet plenums. Figure 54a shows the temporal measurements of inlet and outlet pressures of water during the severe pressure drop oscillation where large oscillation amplitudes were recorded. The severe pressure drop oscillation can be virtually eliminated by throttling the flow immediately upstream of the test section. With this added system stiffness, the boiling boundary was observed to oscillate at small amplitudes between neighboring microchannels in a random manner as illustrated in Fig. 53b. This type of instability has been classified as the "mild parallel channel instability". A comparison of Figs. 54a and b shows that the temporal fluctuations of inlet and outlet pressures of water during the "parallel channel instability" are much smaller and more irregular than those of "pressure drop oscillation".

The mild parallel channel instability due to the flow maldistribution in parallel multi-channels was also reported by Balasubramanian and Kandlikar [114]. The vapor slug in the parallel microchannels (having a hydraulic diameter of 333 μm) was observed to flow in different directions, which induced the boundary between the single-phase liquid and two-phase mixture oscillating in a random manner. Therefore, the pressure drop across the microchannels showed chaotic but mild fluctuations (with an amplitude

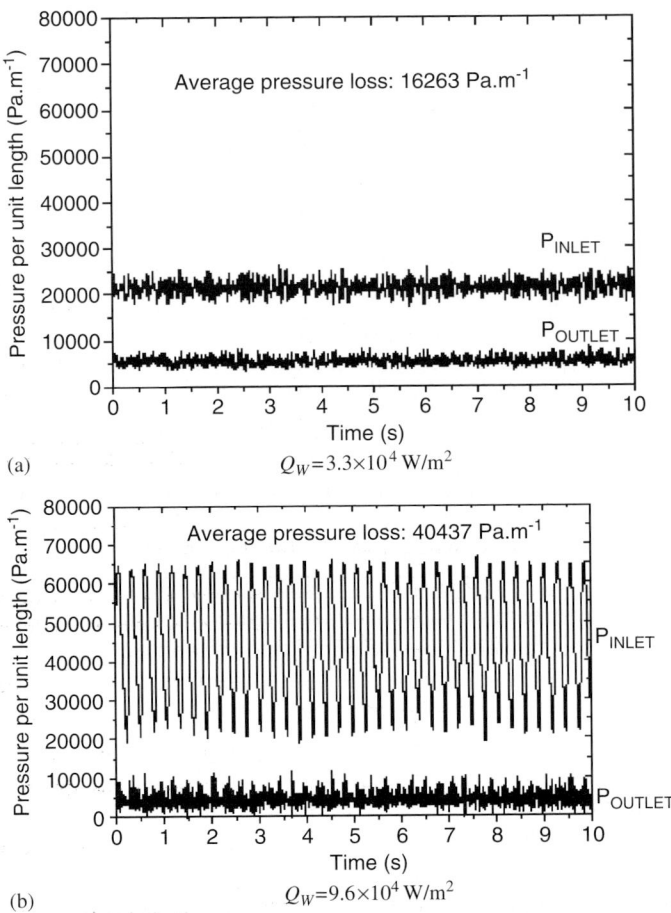

FIG. 52. Typical steady and unsteady boiling inlet and outlet pressure measurements in a mesochannels with $D_h = 888.9\,\mu m$, $L = 200\,mm$ at $G = 240\,kg/m^2\,s$ and at two different heat fluxes: (a) $Q_W = 3.3 \times 10^4\,W/m^2$ (top) and (b) $Q_W = 9.6 \times 10^4\,W/m^2$ (bottom). From Brutin et al. [112]. Copyright (2003), with permission from Elsevier.

of about 1.5 kPa). Based on a pressure drop signal analysis, it was found that the dominant frequency of the pressure drop fluctuation increases with an increase in the surface temperature due to the increase in bubble nucleation frequency. However, when the surface temperature was higher than 109°C, a decreasing trend of dominant frequency was found due to the formation of slug flow, which had a lower fluctuation frequency than bubble nucleation. The bubbly nucleate boiling and the slug flow were found to be

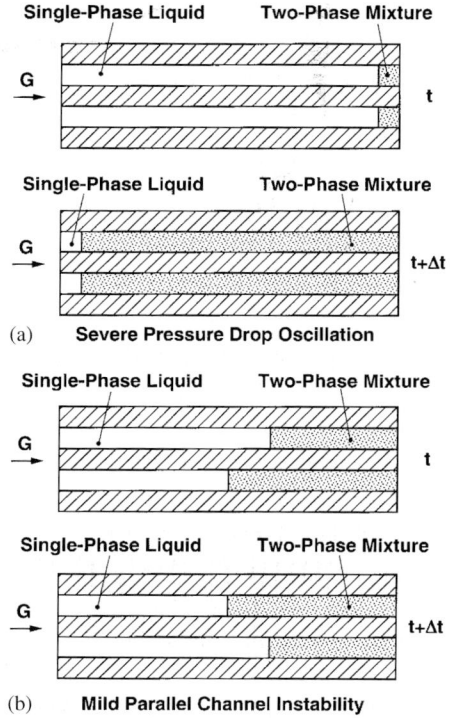

Fig. 53. Top view of two neighboring microchannels illustrating (a) pressure drop oscillation and (b) mild parallel channel instability. From Qu and Mudawar [113]. Copyright (2003), with permission from Elsevier.

the dominant flow patterns corresponding to the lower surface temperature range (103–109°C) and the higher surface temperature range (109–111°C), respectively.

The experimental study by Lee et al. [100] shows that boiling instabilities due to the reversed flow also occurred in the single microchannel at high heat fluxes, as shown in Table II. However, in the two parallel microchannels having the same size of cross section, reversed flow was observed for almost all mass flux and heat flux ranges, as shown in Table III. This indicates that boiling instabilities occurred more easily in parallel microchannels than in single microchannel due to the interaction of the parallel microchannels.

Most recently, Xu et al. [115] conducted an experiment to determine the onset of flow instability (OFI) in 26 parallel rectangular copper microchannels (300 μm wide and 800 μm deep), having a hydraulic diameter of

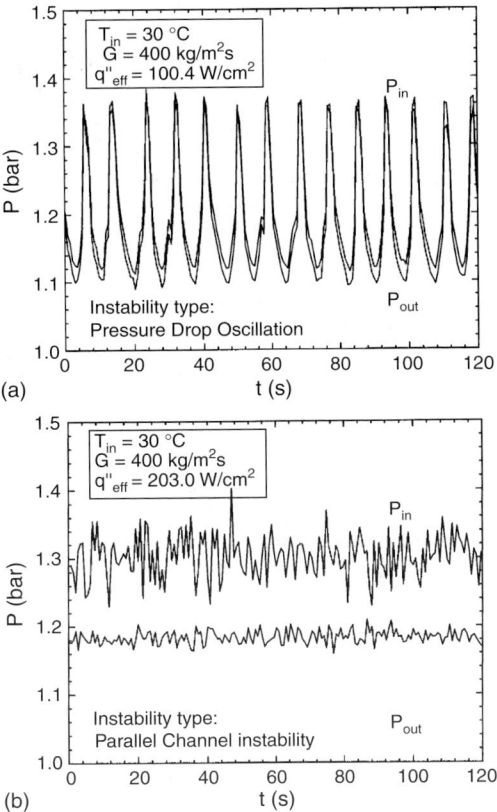

FIG. 54. Temporal variations of inlet and outlet pressures during (a) pressure drop oscillation and (b) parallel channel instability. From Qu and Mudawar [113]. Copyright (2003), with permission from Elsevier.

436 μm under different mass and heat fluxes. Deionized water and methanol were used as working fluids. Figure 55 gives the OFI at different inlet water temperatures, power inputs, and mass fluxes. It is shown that the OFI moved in the direction of larger mass flux with increasing inlet temperature and power input. When the mass flux increased beyond the OFI value, a stable liquid flow existed in the microchannels. However, when the mass flux decreased below the OFI value, unsteady boiling flow was observed. There are two types of oscillations depending on the level of inlet subcooling: large-amplitude/large-period oscillation (LALPO) and small-amplitude/short-period oscillation (SASPO). Figure 56a shows the LALPO of various measurements for the LALPO (which was superimposed with SASPO) at a large inlet subcooling ($T_{in} = 29.6°C$), while

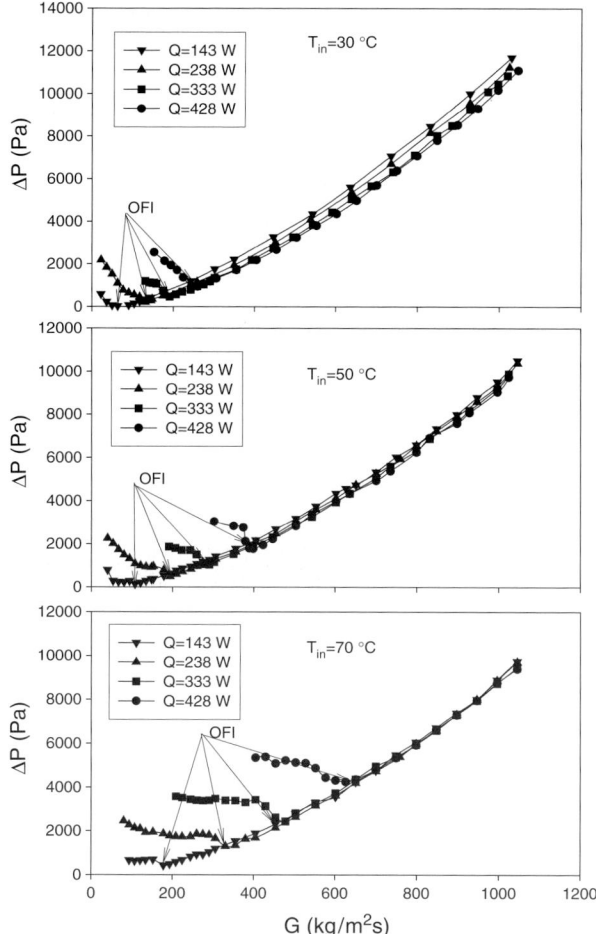

FIG. 55. Pressure drop versus mass flux for different inlet liquid temperatures and power inputs. From Xu et al. [115]. Copyright (2005), with permission from Elsevier.

Fig. 56b shows the SASPO of measurements for the SASPO at a smaller inlet subcooling ($T_{in} = 50.2°C$). With the aid of a microscope, it was found that temperature and pressure oscillations were caused by the liquid–vapor interface moving upstream and downstream in the microchannels. It was also found that the oscillation period of methanol was shorter than that of water under the same conditions, which was attributed to the small specific heat and small latent heat of methanol.

From the above studies, it can be concluded that the mass flux, heat flux, channel size, and shape have great influence on two-phase flow patterns in

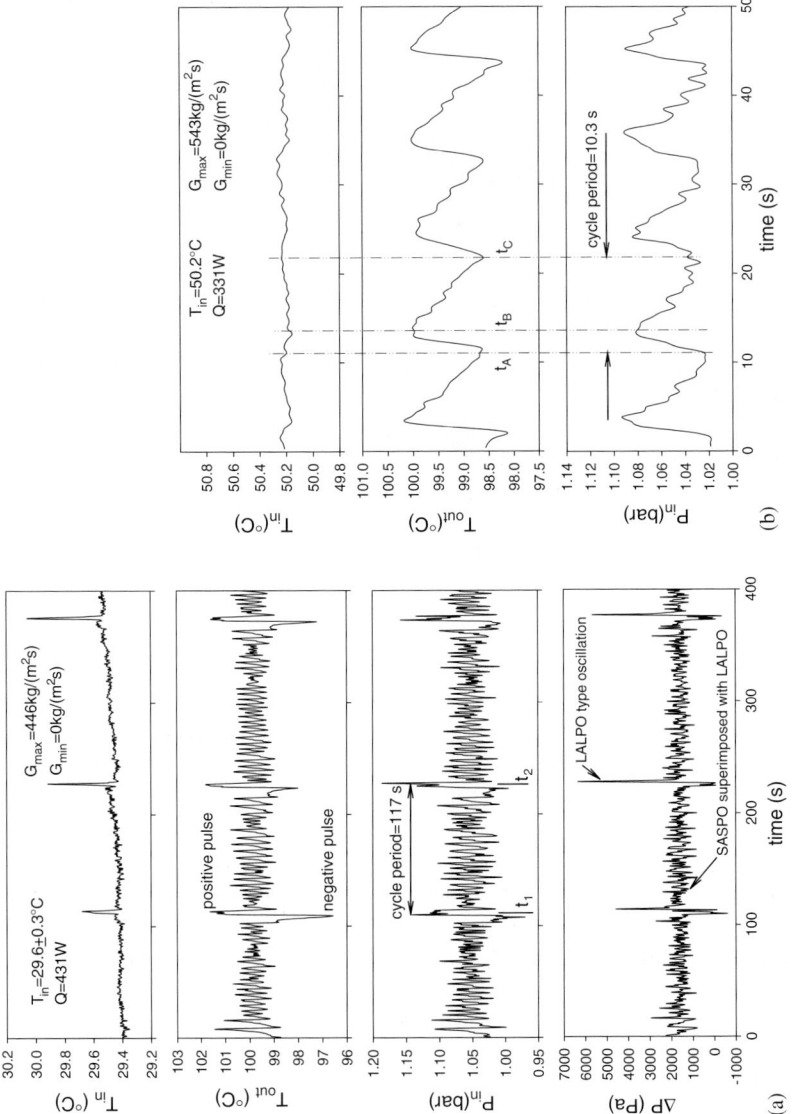

FIG. 56. Temperature and pressure oscillations during (a) LALPO-type oscillations at low inlet subcooling and (b) SASPO-type pressure oscillations at high inlet subcooling. From Xu et al. [115]. Copyright (2005), with permission from Elsevier.

microchannels. Bubbly flow, slug flow, annular flow, churn flow, and some new flow patterns [108–110] have been observed during flow boiling in microchannels. Under certain heat flux and mass flux conditions, single-phase flow and two-phase flow alternate periodically as seen from a fixed location. The period of alternation seems to be prolonged with the decrease in hydraulic diameter of the channel. Flow boiling was found prone to be unstable, especially under the conditions of low mass flow, high heat flux, large inlet subcooling, and large inlet compressibility. Also, boiling instabilities occur more easily in parallel microchannels than in a single microchannel.

E. Flow Condensation in Meso-/Microchannels

Condensation heat transfer in meso-/microchannels has important applications in compact condensers and CPLs. Compared with the large amount of work on boiling heat transfer in meso-/microchannels, the work on condensation in meso-/microchannel condensation is relatively limited. In this section, we will focus our attention on the observation of flow patterns, measurements of heat transfer coefficients and pressure drop as well as analytical studies of condensation process in meso-/microchannels. A summary of experimental investigations on flow patterns, heat transfer, and pressure drop characteristics is listed in chronological order in Table IV. As mentioned in Section II.A, the classification of meso- and microchannels in which condensation takes place can be based on whether the Bond number is greater than or less than 0.05.

1. *Condensation Flow Patterns*

Coleman and Garimella [116–118] have carried out a series of experimental investigations on flow condensation of refrigerant R134a in mesochannels having circular, square, and rectangular cross-sectional areas over a wide range of hydraulic diameters from 1 to 5 mm. Using visualization techniques, they observed four different flow regimes in condensing flow in these channels, as shown in Fig. 57. These four condensation flow regimes are: annular, wavy, intermittent, and dispersed flow. Within each flow regime, there exist several different flow patterns. Figure 57 shows that the annular regime can be divided into the mist, annular ring, wave ring, wave packet, and annular film flow patterns. The wavy flow regime includes the discrete- and disperse-wave flow patterns. The intermittent regime consists of the alternation of slug and plug flow patterns. The dispersed regime consists of bubbly flow. These complex flow regimes and patterns are the result of interaction of surface tension and gravitational force. Transitions lines between the respective flow patterns and regimes can be established based on experimental data. Figure 58 is the flow-regime map for flow condensation of R134a in a round

TABLE IV
EXPERIMENTAL INVESTIGATONS ON FLOW CONDENSATION IN MESO/MICROCHANNELS

Sources	Channel	Fluid	Flow pattern	Main results
Tengblad and Palm [126](1995)	Copper square: 2×2 mm, $d_h = 2$ mm	R142b, R134a, R22, Propane		All tested refrigerants were found to have higher condensation heat transfer coefficients than the classical Nusselt theory. The increase in heat transfer coefficients was attributed to the rapid drainage of the condensed liquid film induced by the sharp corners of the square cross section of the channels.
Wang and Du [127] (2000)	Copper circular $d = 1.94$, 2.8, 3.95, 4.98 mm	Water, steam		The effect of gravity (tube inclination angle) on condensation heat transfer coefficient decreases with the decrease in tube diameter. An analytical model has been used to give reasonable predictions of condensation heat transfer data.
Garimella et al. [116–119] (2000–2004)	Circular, square, rectangular, triangular, barrel-shaped, $d_h = 1$–5 mm	R134a	Intermittent, wavy, annular, disperse	Photos showing condensation flow patterns are presented. Flow maps for wavy flow, intermittent flow, and annular flow are obtained. Effects of hydraulic diameter and geometry of the channel on flow pattern are investigated.
Webb and Ermis [137] (2001)	Aluminum extruded tubes $d_h = 0.44$–1.56 mm	R134a		Both the condensation coefficient and pressure gradient were found to increase with the decrease in hydraulic diameter of the channel. Condensation heat transfer can be predicted by existing correlations of Akers et al. [130] and Moser et al. [134].

Wang et al. [120] (2002)	Aluminum, rectangular 5×1.4 mm $d_h = 1.46$ mm	HFC-134a	Slug, wavy, annular	The annular flow occurs over most of the quality range at higher mass fluxes, while slug flow exists in most of the quality range at low-mass flux of less than $100 \, \text{kg/m}^2\text{s}$. A weighted correlation is obtained for the overall condensation heat transfer coefficient in the mesochannels where the annular flow and stratified flow coexist.
Mederic et al. [121] (2003)	Borosilicate, circular $d = 560, 1100 \, \mu\text{m}$	n-pentane, unstable	Annular, annular-wavy, slug, bubbly	Photos show that the boiling flow pattern was asymmetric in the channel with 1100 μm hydraulic diameter but was nearly symmetric in the channel with 560 μm hydraulic diameter. Annular flow, slug flow, bubbly flow was observed along the flow direction in the channels. Liquid slugs were formed behind the annular flow. During the formation of slug, an elongated Taylor bubble is released which is transferred immediately into spherical bubbles. The interfacial instabilities were attributed to the differences of velocity between the liquid and the vapor phases.
Baird et al. [132] (2003)	Copper, circular $d = 0.92, 1.95$ mm	HCFC-123, R11	Annular	The mass flux and local vapor quality have a strong effect on the condensation heat transfer coefficient, while the influence of the system pressure and tube diameter on the heat transfer coefficient are small. A simple model of gas–liquid shear-driven annular flows with a turbulent liquid film was employed to give reasonable predictions of experimental data. However, the model became less accurate at low and high vapor qualities, presumably because the assumption of annular flow breaks down.

TABLE IV. (Continued)

Sources	Channel	Fluid	Flow pattern	Main results
Kim et al. [138] (2003)	Copper, round $d = 0.691$ mm	R134a, unstable	Bubbly, slug, annular, droplet	Pressure drop measurements are in good agreement with Friedel's correlation [139] except at low mass flux. Except for high quality flow, large pressure drop fluctuations in the experiment were observed, indicating that unstable condensation flow occurred under certain condition.
Koyama et al. [16] (2003)	Aluminum extruded channels $d_h \approx 1$ mm	R134a		Both the heat transfer data and the pressure drop data at low mass fluxes cannot be well predicted by existing correlations. Modified correlations for condensation heat transfer and pressure drop, taking into consideration of the surface tension and diameter effects, have been developed. The modified correlations are in satisfactory agreement with experimental data.
Cavallini et al. [133,136] (2003, 2004)	Aluminum-extruded channel $d_h = 1.4$ mm	R134a, R410a		Effects of vapor quality and mass flux on condensation heat transfer coefficient were observed. All available models for the annular flow regime in both macro and mesochannels, underestimate the experimental data at high mass velocities. The modified version of Moser et al. correlation [134] was found in better agreement with their data than other correlations.

Reference	Geometry	Fluid	Flow regimes	Remarks
Shin and Kim [128] (2004)	Copper, round: $d = 0.493$, $0.691, 1.067$ mm rectangular: $d_h = 0.494, 0.658, 0.972$ mm	R134a unstable	~	The Nusselt number data agree well with Shah's correlation as well as other correlations at low mass fluxes, but are considerably above most existing correlations at high mass fluxes. Effects of mass flux, channel geometry and diameter on condensation heat transfer coefficient have been investigated. Increase in heat transfer coefficients and pressure drop were observed as the hydraulic diameter was decreased. Large pressure drop fluctuations were noted, except in the high vapor quality flow regime, showing that the condensation flow in the channels was unstable.
Wu and Cheng [123,124] (2004, 2005)	Silicon trapezoidal $d_h = 82.8$ μm	Water, steam, unstable	Droplet, annular, injection, slug-bubbly	Depending on the mass flux, four condensation modes, i.e., fully droplet flow, droplet/annular/injection/slug-bubbly flow, annular/injection/slug-bubbly flow, and fully slug-bubbly flow were observed in the microchannels. The injection flow, a result of the breakup of the vapor core, was unstable. Owing to the unstable injection flow, large-amplitude fluctuations of wall temperatures, fluid temperatures and pressures at the inlet and outlet were measured.
Chen and Cheng [122] (2005)	Silicon trapezoidal $D = 75$ μm	Water, steam	Droplet intermittent	Stable droplet condensation was observed near the inlet of the microchannels, while an intermittent flow of vapor and condensate was observed at downstream of the microchannels.

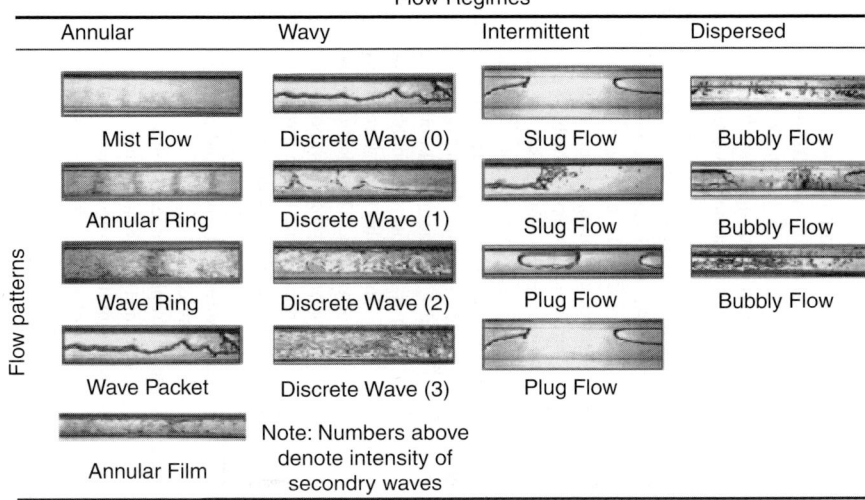

Fig. 57. Condensation flow patterns in mesochannels. From Coleman and Garimella [118]. Copyright (2003), with permission from Elsevier.

tube of 4.91 mm in diameter, which is plotted in terms of mass flux (greater than $100\,kg/m^2\,s$) versus quality. It can be seen that a major portion of this map is occupied by the wavy flow regime with a small region where the slug/plug, and discrete-wave flow patterns coexist. Both discrete- and disperse-wave patterns are present, with the waves become increasingly disperse as the quality and mass flux is increased. The annular flow regime includes both the mist flow pattern and the annular film pattern. At low mass fluxes, the wave packet flow pattern is found near the border between the mist-flow pattern and disperse-wave pattern. The annular ring flow pattern exists at high mass flux and high quality region. As shown in Figs. 59a and b, the intermittent and the annular regime became larger, with the wavy regime decreasing in size as the hydraulic diameter decreases. When the hydraulic diameter deceases to 1 mm, the wavy regime disappears completely (indicating a diminishing influence of gravity forces), and the annular and intermittent flow became the dominant flow [119]. The effect of tube shape on flow regime transitions was also investigated. It was found that the intermittent flow regime in the round tube is larger than that in the square tubes at low mass fluxes and approximately the same at high mass fluxes. The wavy flow regime is also larger in the round tube. However, the annular flow regime increases in the square and rectangular channels due to the retention effect of the corners. It was concluded that the hydraulic diameter is more significant than tube shape in determining the condensation flow pattern.

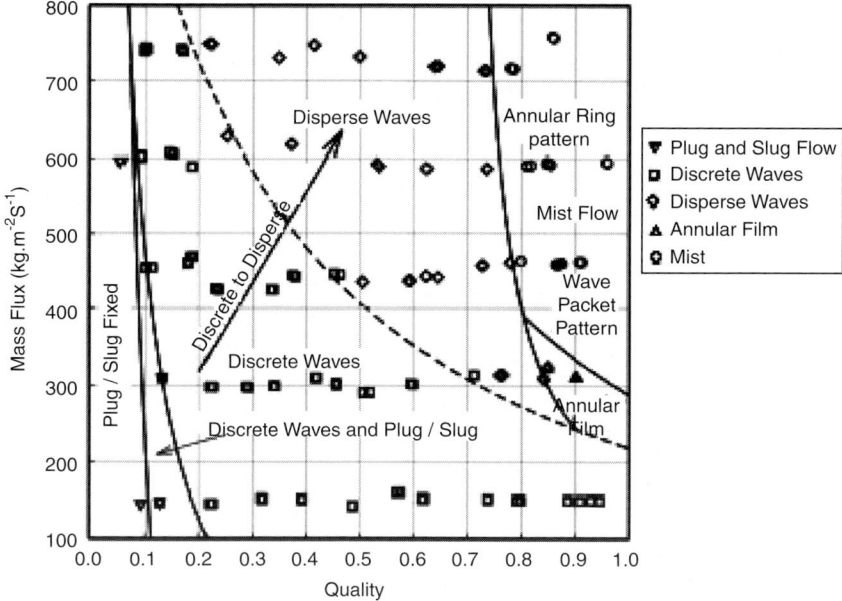

FIG. 58. Condensation flow regime map of R-134a in a 4.91 mm-round tube. From Coleman and Garimella [118]. Copyright (2003), with permission from Elsevier.

Wang et al. [120] observed flow patterns in their experiment on flow condensation of HFC-134a inside a horizontal multi-port aluminum tube with a rectangular cross section (1.5 × 1.4 mm²) having a hydraulic diameter of 1.46 mm, which can be classified as a mesochannel. They found that the quality and mass flux are major factors affecting condensation flow patterns in rectangular mesochannels. Figure 60 shows that the annular regime occurs over most of the quality range at higher mass fluxes, while slug flow exists at low mass flux of less than $100\,\text{kg/m}^2\,\text{s}$ over most of the quality range. Table V is a summary of the observed transitions of flow regimes at various mass fluxes and qualities.

Mederic et al. [121] carried out an experiment to study complete condensation flow of n-pentane in two small diameter horizontal borosilicate tubes with diameter of 1100 μm and 560 μm, respectively. Substituting the thermophysical properties of n-pentane into Eq. (1c) gives $l_c = 1250$ μm for the 1100 μm diameter tube. It follows that the gravity effect is important in the 1100 μm diameter tube, while its effect is small in the 560-μm-diameter tube. This is confirmed by the photo images shown in Fig. 61, which were taken by a high-speed camera. Figures 61a and b show the flow patterns at the entrance zone in the 1100 and 560 μm diameter tubes, respectively. An asymmetric annular flow with a thicker liquid film at the bottom and a

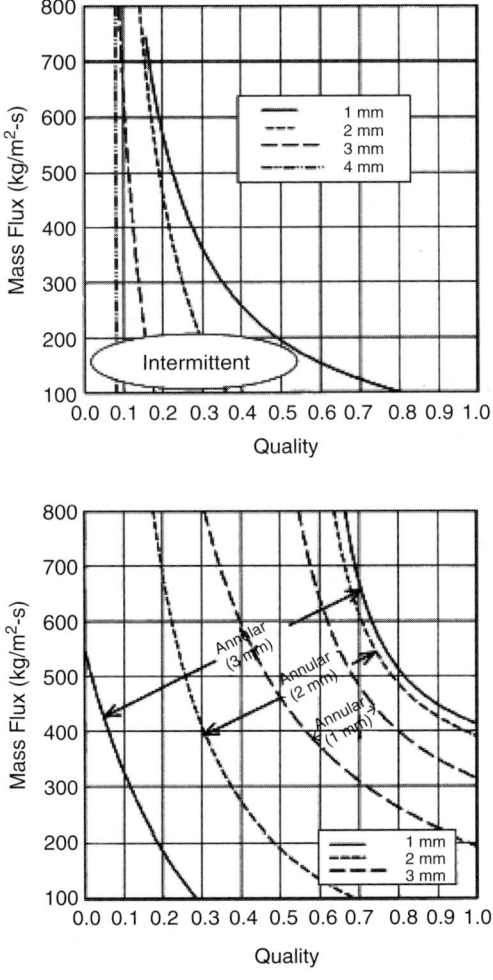

FIG. 59. Effects of hydraulic diameter on the intermittent flow regime (top) and on annular flow regime (bottom) of R-134a. From Garimella [119]. Copyright (2004), with permission from Taylor & Francis.

thinner film at the top (shown by circle) was observed in the 1100 μm-diameter tube due to the effect of gravity. On the other hand, an almost symmetric annular flow appeared in the 560 μm-diameter tube, showing that the effect of gravity is small in this smaller diameter tube. Figure 61c shows that there existed three flow regimes in the tube along the flow direction. In the entrance regime (i.e., Zone #1), the flow was annular which was characterized by high vapor quality. Behind Zone #1, the unstable

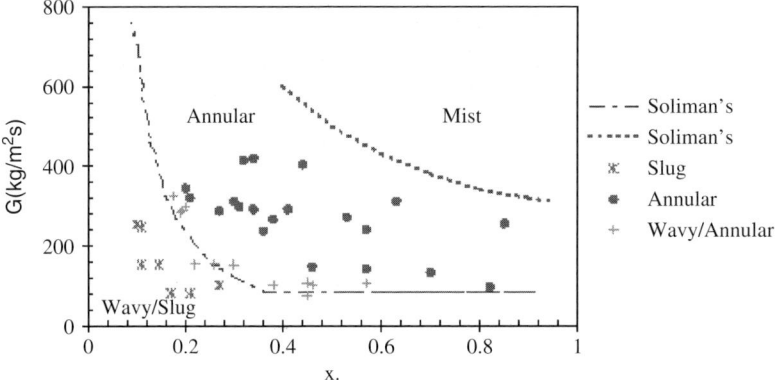

FIG. 60. Condensation flow regime visualization data plotted against Soliman transitions. From Wang et al. [120]. Copyright (2002), with permission from Elsevier.

TABLE V

Transitions of Condensation Flow Regimes of HFC-134a [120]

Mass flux g (kg/m²s)	Slug/wavy to wavy/annular Quality X	Wavy/annular to annular Quality X
75	0.45	0.88
100	0.38	0.70
150	0.21	0.42
250	0.14	0.30
300	0.12	0.22
350	0.10	0.17

condensing flow appeared. In this regime (Zone #2), a liquid slug (which releases an elongated Taylor bubble) was formed. This elongated bubble was in a non-equilibrium state and was transformed immediately into spherical bubbles in Zone #3. Mederic et al. [121] attributed the interfacial instabilities to the difference of velocity between the liquid and the vapor phases.

Chen and Cheng [122] performed a visualization study on condensation of steam in microchannels etched in a <100> silicon wafer bonded by a thin Pyrex glass plate from the top. The microchannels had a trapezoidal cross section with a hydraulic diameter of 75 μm. Saturated steam flowed through these parallel microchannels whose walls were cooled by natural convection of air at room temperature. The absolute pressure of the saturated steam at the inlet ranged from 127.5 to 225.5 kPa, and the outlet was at atmospheric

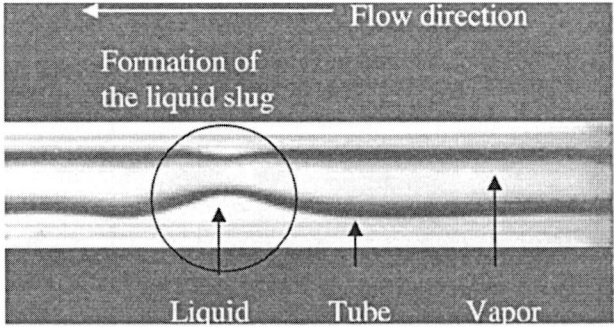

(a) Asymmetric annular flow in a 1100 μm-diameter tube just after the entrance

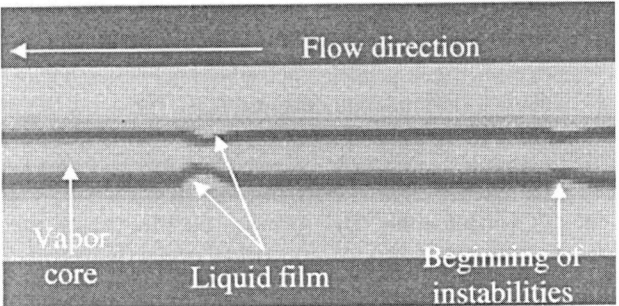

(b) Nearly symmetric annular flow in 560 μm-diameter tube just after the entrance

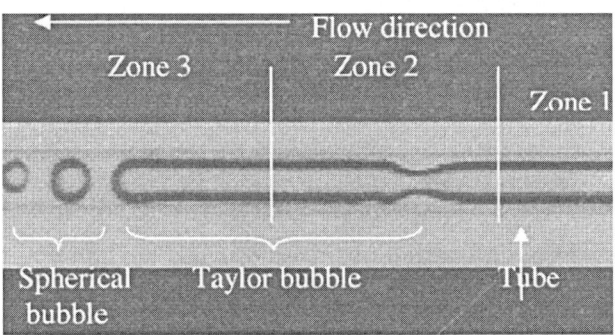

(c) Flow pattern transition

FIG. 61. Condensation flow patterns of n-pentane in tubes: (a) asymmetric annular flow in 1100 μm-diameter tube just after the entrance, (b) nearly symmetric annular flow in 560 μm-diameter tube just after the entrance, and (c) flow pattern transition in the flow direction. From Mederic et al. [121]. Copyright (2003), with permission from ASME.

pressure at approximately 101.3 kPa with the outlet temperature of the condensate ranging from 42.8 to 90°C. Stable droplet condensation was observed near the inlet of the microchannel. When the condensation process progressed along the microchannels, droplets accumulated on the wall. As the vapor core entrained and pushed the droplets, it became an intermittent flow of vapor and condensate at downstream of the microchannels.

Wu and Cheng [123,124] carried out a series of simultaneous visualization and measurement experiments on saturated steam condensing in parallel silicon microchannels, having a trapezoidal cross section with a hydraulic diameter of 82.6 μm. By changing the mass flux (through regulating the inlet valve, while maintaining other conditions in the system unchanged) from 47.5 to 19.3 g/cm², they observed four different condensation flow patterns in the microchannels, which are designated as Cases I, IIA, IIB, and III as follows:

Case I: Fully droplet flow. At the mass flux of 47.5 g/cm²s (with corresponding inlet pressure of 4.15×10^5 Pa), droplets were formed on the wall of the microchannel from the inlet to outlet. In this condensation flow pattern, the wall temperatures showed little fluctuations with time, revealing that the fully droplet flow was fairly steady.

Case IIA: Droplet/annular/injection/slug-bubbly flow. When the mass flux was decreased to 30.4 g/cm²s (with corresponding inlet pressure decreased to 2.15×10^5 Pa), the droplet flow zone shrank to the inlet, and the annular flow, injection flow, slug-bubbly flow appeared sequentially in the flow direction, as sketched in Fig. 62a. Figure 62b also gives the photos taken at different zones in this condensation flow pattern. The decrease in the vapor quality along the channel was responsible for different flow patterns. Of particular interest is the injection flow occurring at $0.5 < x/L < 0.75$, which is essentially the front end of the vapor core undergoing shrinking and detachment activities. The injection flow was found to appear and disappear periodically, and the appearance of slug-bubbly flow at downstream was actually the product of vapor core's detachment activity. The breakup of the vapor core was owing to the surface tension effect resulting in its instability. In this droplet/annular/injection/slug-bubbly flow pattern, the wall temperatures at the downstream ($x/L > 0.75$) of injection flow had large-amplitude fluctuations with time as shown in Fig. 62b, which confirmed that downstream had been affected by the unsteady injection flow. At the upstream ($x < L < 0.5$) of the injection flow, the wall temperatures show small fluctuations with occasional dips because of the more steady droplet flow in this zone.

Case IIB: Annular/injection/slug-bubbly flow. When the mass flux was further decreased to 23.6 g/cm²s (with corresponding inlet pressure decreased to 1.45×10^5 Pa), the droplet flow disappeared totally in the microchannel,

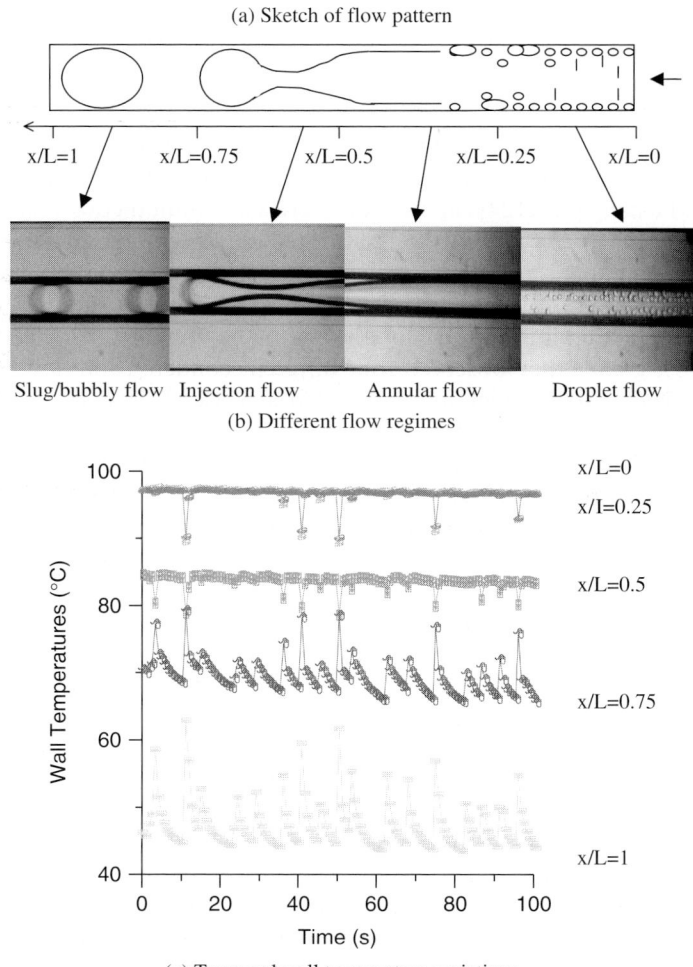

FIG. 62. Droplet/annular/injection/slug-bubbly condensation flow: (a) sketch of the flow pattern, (b) photos of the flow of steam pattern, and (c) wall temperature variations at different locations. From Wu and Cheng [124]. Copyright (2005), with permission from Elsevier.

and the annular flow, injection flow and slug-bubbly flow occupied the microchannel from the inlet to outlet successively as sketched in Fig. 63a. In this flow pattern, the injection flow occurring at $0.25 < x < 0.5$. This is evident by the fact that wall temperature variations for $x < 0.25$ and $x > 0.75$ had opposite trends. While the wall temperature decreased periodically at $x < 0.25$, the wall temperature increased periodically at $x > 0.75$. It was found that when the wall temperatures at the upstream of injection flow

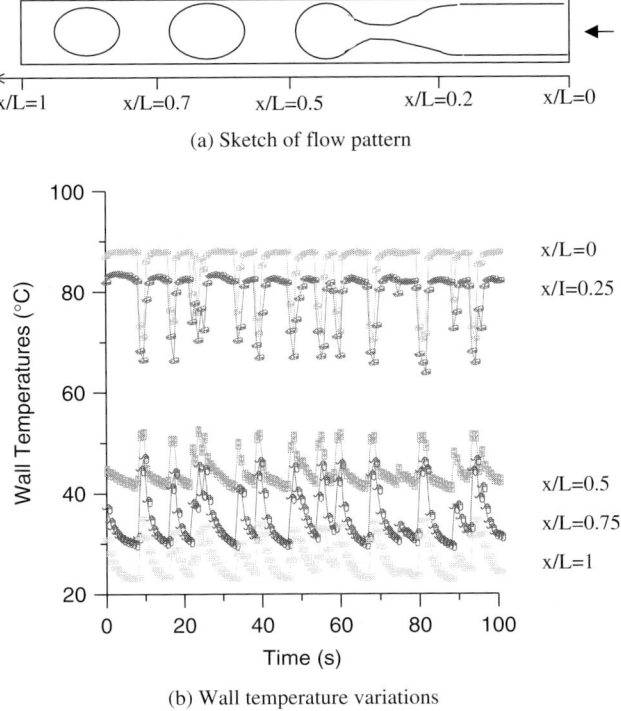

FIG. 63. Annular/injection/slug-bubbly flow condensation mode of steam: (a) sketch of the flow pattern, and (b) wall temperature variations at different locations. From Wu and Cheng [124]. Copyright (2005), with permission from Elsevier.

fluctuated from a high value (corresponding to vapor flow) to a low value (corresponding to annular flow), the wall temperatures at down stream fluctuated from a low value (corresponding to liquid flow) to a high value (corresponding to slug-bubbly flow). It was also found that the fluctuation amplitudes of wall temperatures in this boiling mode were larger than those in the previous two modes because of different flow pattern alternations with time at the same locations.

Case III: Fully slug-bubbly flow. When the mass flux was decreased further to $19.3 \, g/cm^2 \, s$ (with corresponding inlet pressure decreased to $1.25 \times 10^5 \, Pa$), the vapor injection flow moved to the inlet and the annular flow disappeared totally from the microchannel, while the slug-bubbly flow occupied everywhere in the microchannel as shown in Fig. 64a. In this condensation flow pattern, the vapor bubble and liquid slug appeared alternatively along the microchannels, which is usually referred to as the "intermittent flow" in the literature. It was found that due to the increased

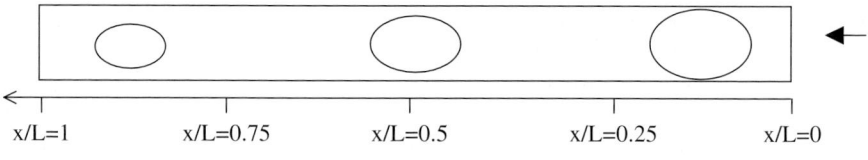

(a) Sketch of the flow pattern

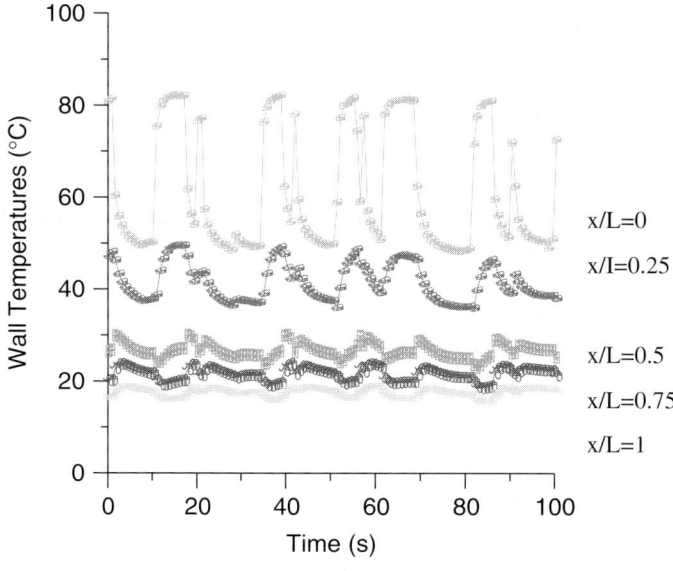

(b) Wall temperature variations

FIG. 64. Fully slug-bubbly condensation flow of steam: (a) sketch of the flow pattern and (b) wall temperature variations. From Wu and Cheng [124]. Copyright (2005), with permission from Elsevier.

amount of the condensate, the liquid slug became larger, while the vapor bubble became smaller along the flow direction. The fluctuation amplitudes of wall temperatures were greatly different, decreasing from the inlet to the outlet because it was farther away from the inlet unstable injection flow.

From Wu and Cheng's work [123,124], it can be concluded that (1) fully droplet flow exists at high mass flux and high quality, while slug-bubbly flow exists at low mass flux and low quality, which is consistent with Garimella's conclusion about flow patterns in a tube as mass flux and quality varied

[117–119] and also consistent with the observation by Wang et al. [120]; (2) at intermediate mass fluxes, droplet/annular/injection/slug-bubbly flow appear in the microchannel as the local vapor quality decreases along the flow direction, and (3) the appearance of the injection flow at low mass flux causes temperature fluctuations at upstream and downstream of the microchannels, revealing that condensation flow in microchannels is rather unstable at low mass flux.

The physical reasons for the breakup of the annular flow into slug/plug flow in a meso-/microchannel have been analyzed by Teng et al. [125] using an integro-differential approach. The resulting characteristic equation was used to analyze instability of the condensate film in a small-diameter-tube condenser. Their results show that there existed two different zones with different causes of film instability. In the zones of low relative vapor velocities, the surface tension is the cause of the film instability, and the disturbance wavelengths are approximately a function of the radius of the undisturbed inner condensate film, i.e., $\lambda_m \approx \sqrt{8\pi}a$. In the zones of high relative vapor velocities, the film instability is induced by the hydrodynamic force due to the difference in phase velocities, and the disturbance wave became shorter with the increase in hydrodynamic force. The prediction of the disturbance waves matched well with the measurements reported in the literature. Figure 65a shows that an annular liquid layer is initially formed on the inside of the microtube. The collars are formed either from condensation of the vapor on the liquid film in the microtube or from instability at the vapor/liquid interface as shown in Fig. 65b. Subsequently, the collar grows to form a liquid bridge as shown in Fig. 65c, and finally a liquid slug is formed between vapor bubbles as shown in Fig. 65d. Note that because of gravitational force, the bridge is not easily formed in a macrotube. Figures 66a–d are the sketches showing the formation of liquid blocking in a microchannel. It was pointed out that capillary instabilities that are encountered in small-diameter-thermosyphon condensers may occur in other small-channel condensers, such as those used in wickless microheat pipes.

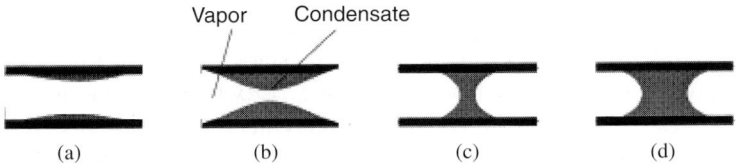

FIG. 65. Development of a condensate collar. From Teng et al. [125]. Copyright (1999), with permission from Elsevier.

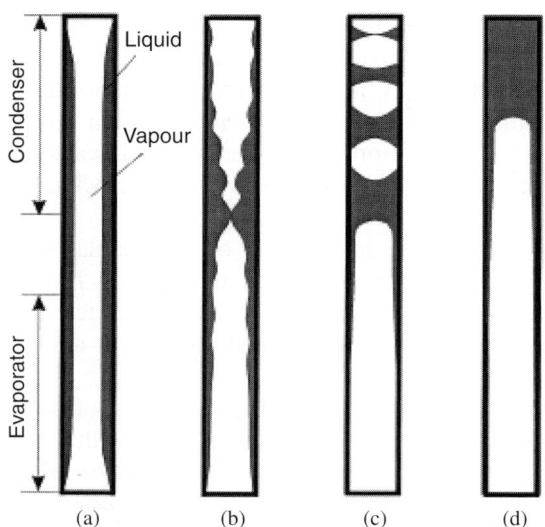

FIG. 66. Formation of liquid blocking as time advancing. From Teng et al. [125]. Copyright (1999), with permission from Elsevier.

2. *Heat Transfer Coefficients*

Tengblad and Palm [126] performed an experimental investigation on film condensation heat transfer in narrow and vertical channels of thermosyphons. The mesochannels used in their condensation experiments had a square cross section of $2 \times 2 \, \text{mm}^2$, through which four kinds of refrigerants (R142b, R134a, R22, and Propane) were tested. In comparison with Nusselt's classical theory of film condensation, it was found that all tested refrigerants have higher heat transfer coefficients, especially for R142b having a value in the range of $15 \sim 40 \, \text{kW/m}^2 \, \text{K}$, which is about twice as high as other refrigerants. They attributed the high condensation heat transfer coefficient to the rapid drainage of the condensed liquid film by the capillary force, which was induced by the sharp-edged corners of square cross sections of the channels.

Wang and Du [127] have carried out an experimental investigation on gravity effects on stream condensing in meso-inclined tubes with diameters of 1.94 and 3.95 mm, respectively. Figure 67 shows the variation of the flow condensation Nusselt number at four different inclination angles ($\beta = 0, 17°, 34°, 45°$) versus the liquid quality $(1-x)$ in the two tubes for the same inlet vapor Reynolds number of 4500. It is shown that the effect of inclination angle on the condensation heat transfer rate is small in the smaller diameter tube ($d = 1.94 \, \text{mm}$) indicating that the gravity effect is

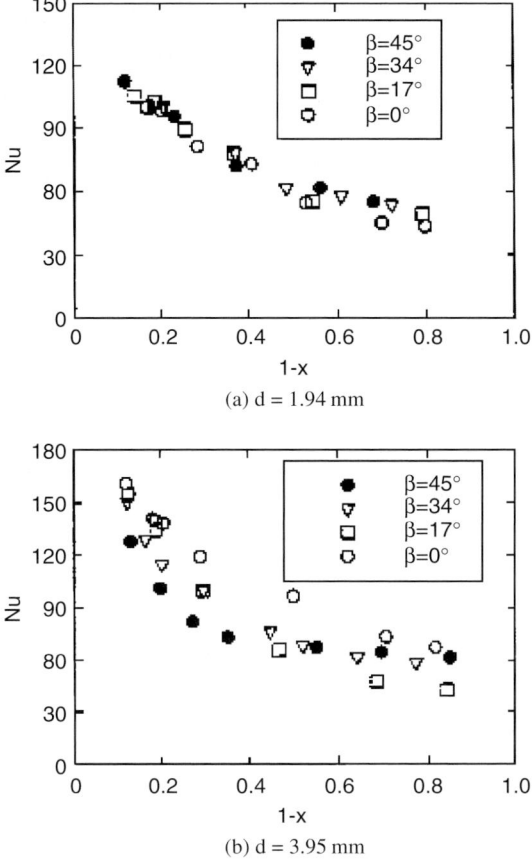

FIG. 67. Effect of inclination angle on Nusselt number of condensation flow of steam in two tubes with different diameters. From Wang and Du [127]. Copyright (2000), with permission from Elsevier.

small. On the other hand, the inclination angle is important in the larger diameter tube ($d = 3.95$ mm) indicating that the gravity effect is important.

Shin and Kim [128] investigated experimentally heat transfer characteristics during the condensation of R134a in horizontal single meso-/microchannels. Both circular ($d_h = 0.493$, 0.691, and 1.067 mm) and rectangular channels ($d_h = 0.494$, 0.658, and 0.972 mm) were used. The experiments were conducted at mass fluxes of 100, 200, 400, and 600 kg/m^2 s, heat fluxes ranging from 5 to 20 kW/m^2, and at a saturation temperature of 40°C. Figures 68a–d show the comparison of their Nusselt number data for $d_h = 0.493$ mm with the existing correlations. Except at low mass fluxes

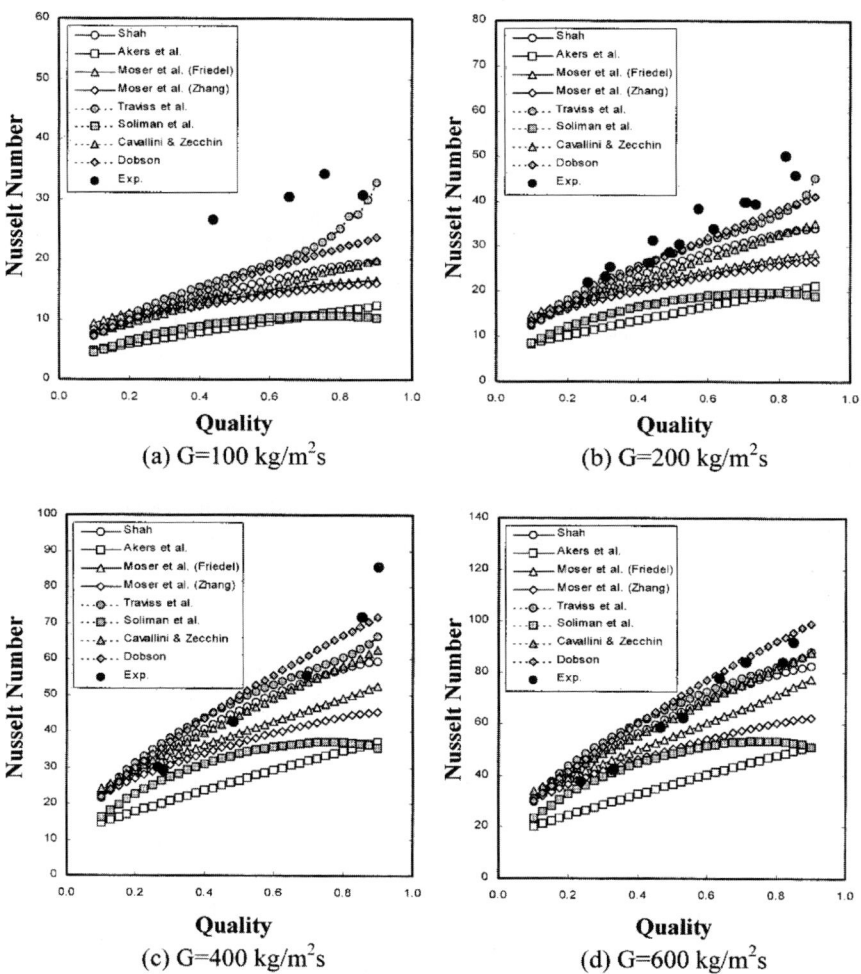

FIG. 68. Comparison of measured Nusselt number with existing correlations for condensations of R134a in a round tube with $D = 0.493$ mm. From Shin and Kim [128]. Copyright (2004), with permission from ASME.

(100 and 200 kg/m²s), their data agree well with Shah's correlation [129] as well as other correlations. However, their data are considerably above the prediction from the correlation by Akers and Rosson [130]. The effect of channel geometry and hydraulic diameter on the average condensation heat transfer coefficient is shown in Fig. 69. It was found that minichannels with a rectangular cross section have a higher condensation heat transfer coefficient than that of minichannels with a circular cross section having a

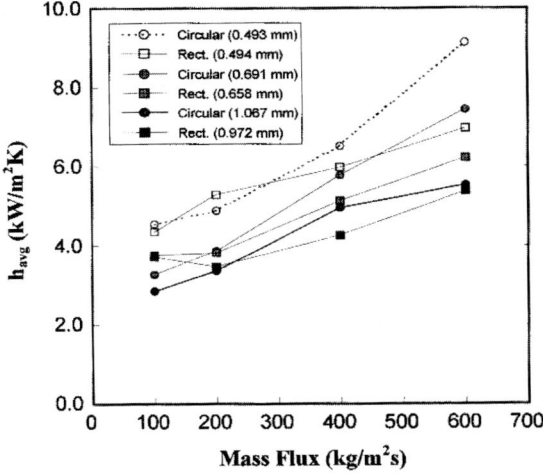

FIG. 69. Channel geometry effects on average condensation heat transfer coefficient of R134a. From Shin and Kim [128]. Copyright (2004), with permission from ASME.

similar hydraulic diameter at low mass fluxes. However, as the mass flux is increased, the minichannels with a circular cross section have a higher value of heat transfer coefficient than that of the rectangular ones.

Similarly, Koyama *et al.* [16] investigated experimentally condensation heat transfer of R134a in extruded aluminum tubes with hydraulic diameter of about 1 mm. They found that their experimental data do not agree with any existing correlations. A new correlation for condensation heat transfer in mesochannels based on an existing correlation is modified by taking into consideration of the surface tension effect (in terms of the Bond number) and the diameter effect. The modified correlation is shown to be in satisfactory agreement with their experimental data.

Wang *et al.* [120] reported heat transfer measurements of HFC-134a, condensing inside a horizontal rectangular multi-port aluminum condenser tube of 1.46 mm hydraulic diameter. Their data at high mass fluxes (assuming annular flow) are found in satisfactory agreement with Akers' correlation equation [130]. For low mass fluxes, a stratified flow pattern was observed at most qualities and the data were found in reasonably good agreement with Jaster and Kosky's correlation [131]. They obtained improved correlations for the Nusselt number in the annular flow regime (Nu_{anul}) and the stratified flow regime (Nu_{strat}), respectively. Since multiple flow regimes exist in the tube, a weighted correlation in terms of Nu_{anul}, Nu_{strat} and a weighted factor is obtained for predicting the overall condensation heat transfer coefficient in the mesochannels. This weighted correlation agreed well with their experimental data.

Baird et al. [132] measured the local heat transfer coefficient for flow condensation of HCFC-123 and R11 for a wide range of mass fluxes (7–60 kg/cm^2 s), heat fluxes (1.5–11 W/cm^2), vapor qualities (superheated to fully condensed), and pressure (120–410 kPa) in mini tubes with internal diameters of 0.92 and 1.95 mm. The data showed that the mass flux and local quality have a strong influence on the heat transfer coefficient, while the system pressure and tube diameter have a weaker or little influence on the heat transfer coefficient. To account for the major effects observed, a simple model of gas-liquid shear-driven annular flows with a turbulent liquid film was employed. Although the simple annular shear flow model is in better agreement with experimental data than other models and existing correlations, it became less accurate at low qualities ($x \leqslant 0.2$) and high qualities, presumably because the assumption of annular flow breaks down.

Cavallini et al. [133] introduced a new technique to measure the heat transfer coefficient during condensation of R134a inside a multi-port extruded mesochannel tube of 1.4 mm hydraulic diameter. Effects of vapor quality and mass flux on heat transfer coefficient were observed. The Moser et al. correlation [134], as modified by Zhang and Webb [135], was found in better agreement with their data than other correlations, with deviation of 20%. In another paper, Cavallini et al. [136] measured heat transfer coefficients during condensation of R134a and R410a inside multi-port mesochannels with a hydraulic diameter of 1.4 mm. It is shown that all the models available in the literature for the annular flow regime in both macro- and mesochannels, underestimate the experimental heat transfer coefficient at high mass velocities.

Webb and Ermis [137] performed an experimental study on the effect of hydraulic diameter on condensation heat transfer and pressure gradient for R134a in multi-port flat extruded aluminum tubes. The hydraulic diameters of the tested tubes ranged from 0.44 to 1.56 mm. It was found that the condensation coefficient and pressure gradient increase with the decrease in hydraulic diameter, and their Nusselt number data can be well predicted by the correlations by Akers et al. correlation [130] and Moser et al. [134].

3. *Pressure Drop Characteristics*

Kim et al. [138] obtained pressure drop measurements during the condensation of R134a in a horizontal round meso-tube with an inner diameter of 0.691 mm at a saturation temperature of 40°C. As shown in Fig. 70, pressure drop measurements are in good agreement with the well-known

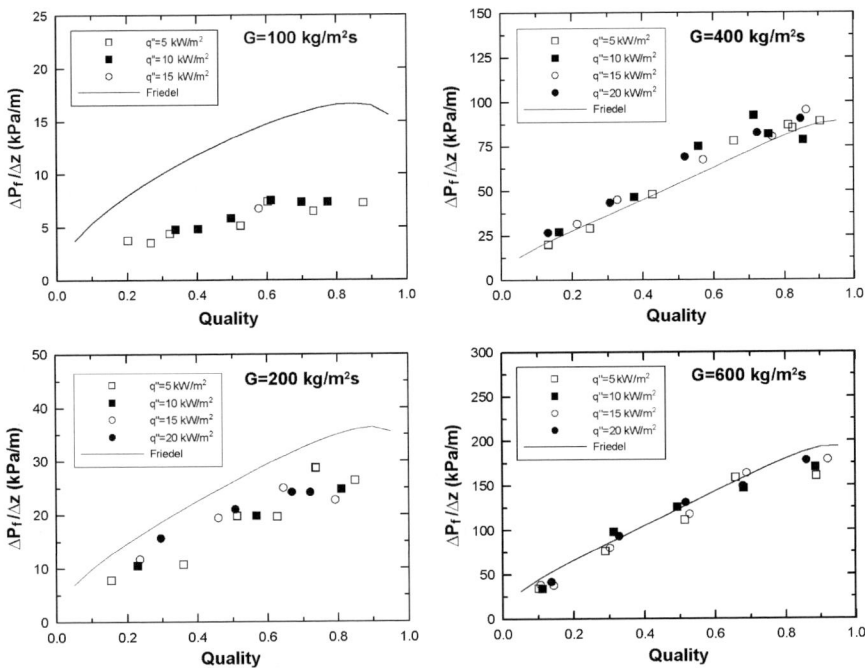

FIG. 70. Two-phase pressure drop versus quality at different mass fluxes of R-134a. From Kim *et al.* [138]. Copyright (2003), with permission from ASME.

Friedel's correlation [139] except at low mass flux ($G = 100 \, \text{kg/m}^2 \, \text{s}$). Except for high quality flow, large pressure drop fluctuations in the experiment were observed, inferring that unstable flow may occur under certain conditions during the condensation process in the meso-tube.

Similar experiments were performed by Koyama *et al.* [16] on pressure drop of R134a condensing in the extruded aluminum mesotubes with hydraulic diameter about 1 mm. They also found that their pressure data can be well predicted with Friedel's correlation [139] except at low mass flux, where their data are lower than that predicted by correlation equations. Taking into consideration of the surface tension effect in Mishima–Hibiki correlation [140] in terms of Bond number, Koyama *et al.* [16] found that their pressure drop measurements can be well predicted by the modified correlation equation.

Garimella *et al.* [141,142] developed experimentally validated pressure drop models for the intermittent flow regime of R134a in circular and non-circular mesochannels as well as the pressure drop model for the annular flow regime in circular mesochannels [143]. Figure 71 gives the predictions

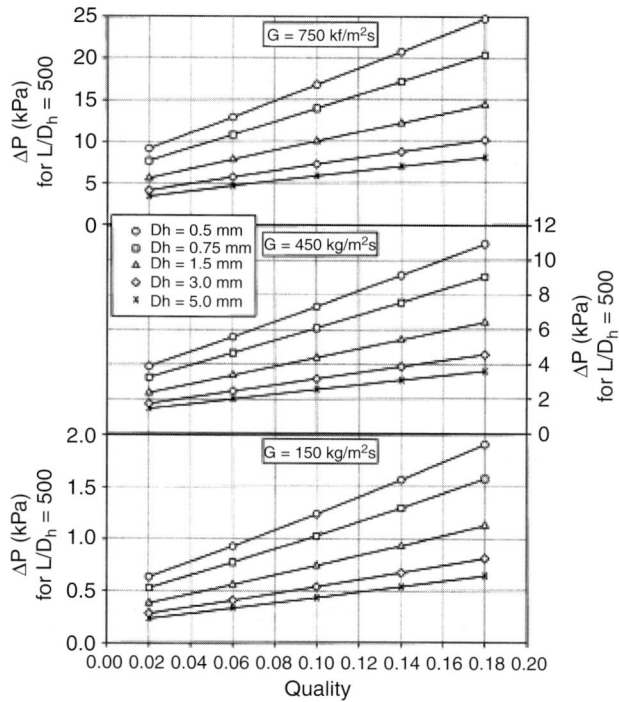

FIG. 71. Predicted effects of hydraulic diameter, mass flux, and quality on pressure drop in the intermittent condensation flow of R134a. From Garimella et al. [141]. Copyright (1999), with permission from ASME.

of condensation pressure drop as a function of quality in the intermittent flow of R134a for tube diameters from 0.5 to 5 mm [141]. It is seen that the effects of tube size on pressure drops are significant, especially at large vapor quality. By modifying and combining the intermittent and annular models, Garimella et al. [144] further developed a multiple flow-regime model, which can accurately predict condensation pressure drops of R134a in the annular, disperse wave, mist, discrete wave, and intermittent flow regimes as the condensation flow progresses from the vapor phase to the liquid phase. A comparison of calculated pressure drop from the model and the measured pressure drop in round channels with diameters ranging from 0.5 to 4.91 mm [144] shows that the predicted pressure drops are within ±13.5% of the measured values.

4. Theoretical Studies on Condensation in Meso-/Microchannels

Begg et al. [145] developed a mathematical model to study the complete condensation flow in a microtube. The complete condensation flow is

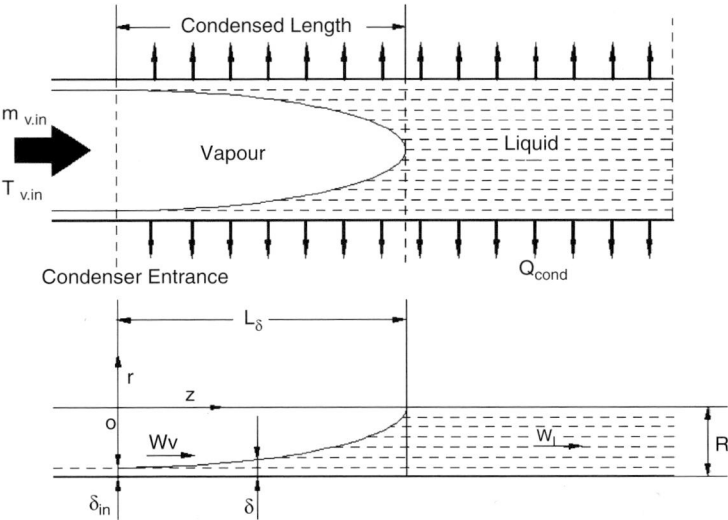

FIG. 72. The physical model for complete condensation in a meso tube. From Begg et al. [145]. Copyright (1999), with permission from ASME.

defined as the condensation of all incoming vapor in a film-wise manner in which the vapor flow terminates at a well-defined location forming a steady meniscus-like interface (see Fig. 72). Computations were carried out for water vapor condensing in a 3-mm-diameter condenser at a constant wall temperature of 340 K and a constant inlet vapor temperature of 363 K at a low mass flux of 0.01 g/s with two different inlet heat loads of $Q_{in} = 10$ and $Q_{in} = 12$ W, respectively. The results of the former case are represented by solid lines and the latter case are represented by dashed lines in Fig. 73. Figure 73a shows that the film thickness decreases in the downstream direction before converging due to the capillary forces acting at the liquid–vapor interface. The variation of vapor pressure drop is shown to be insignificant (Fig. 73b) compared to the liquid pressure drop (Fig. 73c). Figure 73d shows the cumulative heat flux (Q) that is removed from the vapor by condensation along the channel. For the case of $Q_{in} = 10$ W, calculation was terminated at the point $Q = 10$ W, i.e., where all incoming vapor is condensed that leads to complete condensation. To verify this model, a simple flow visualization experiment was performed to observe the condensation flow pattern of steam over the entire region of two-phase flow in the horizontal glass tubes having diameters of 3.4, 2.3, and 1.6 mm, respectively. Visual observations confirmed the existence of steady annular film condensation leading to the complete condensation in small circular tubes ($d < 3.4$ mm). It was shown that the condensation flow pattern depends on heat load and mass flux. In the tubes of 3.4 and 2.3 mm in diameters,

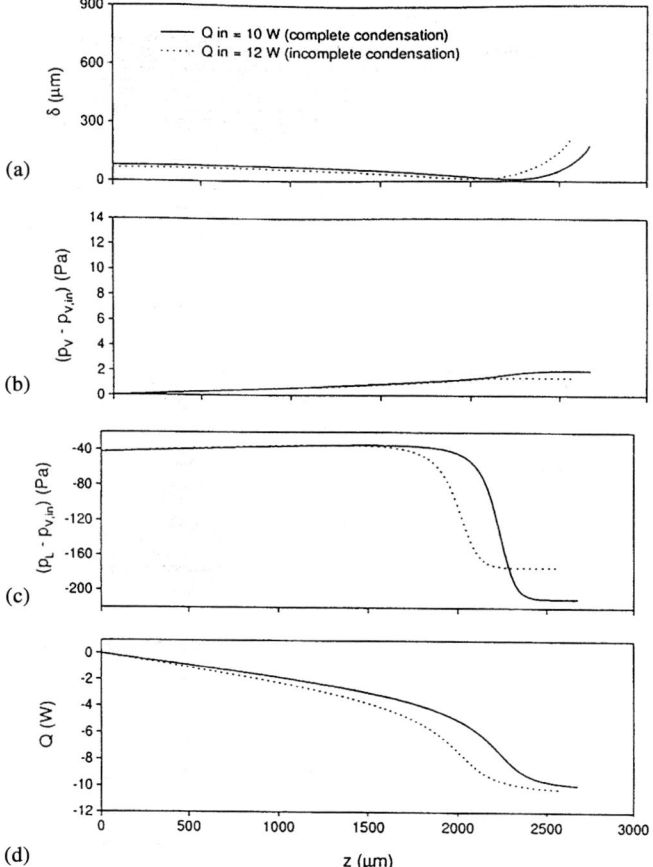

FIG. 73. Variations of different quantities in film condensation in a channel: (a) film thickness, (b) vapor pressure, (c) liquid pressure, and (d) cumulative heat rejected. From Begg et al. [145]. Copyright (1999), with permission from ASME.

the stratification of condensation film due to gravity effect was observed. This stratification effect was not obvious with the decrease in diameter ($d = 1.6$ mm). Note that although steady complete condensation at low heat loads was observed in small-diameter tubes, it was pointed out that unsteady slug flow may occur at higher heat load conditions.

Zhao and Liao [146] carried out an analytical study on the film condensation heat transfer of steam in vertical meso triangular channels. Their analysis was based on the assumption that the condensation two-phase flow in the mesochannel can be divided into three zones: the thin liquid film flow on the sidewall, the condensate flow in the corners, and the vapor core flow in the center. Effects of capillary force, interfacial shear stress, interfacial thermal resistance, gravity, axial pressure gradient, and saturation temperature were taken into account in the model. The cross-sectional average condensation heat transfer coefficient along the wall of the three equilateral triangular channels (with side lengths of 1.0, 1.5, and 2.0 mm) having

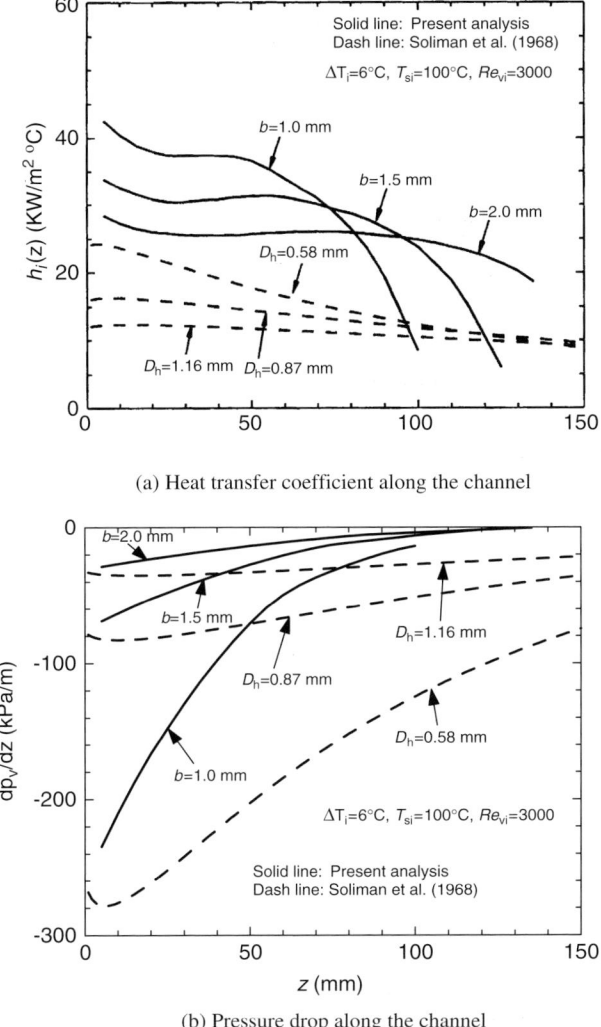

FIG. 74. Effects of channel diameter on film condensation of steam in a vertical triangular channel on heat transfer coefficient (top) and pressure drop (bottom). From Zhao and Liao [146]. Copyright (2002), with permission from Elsevier.

hydraulic diameters of 0.58, 0.87, and 1.16 mm at each z under the conditions of $Re_{vi} = 3000$ and $\Delta T_i = 6°C$ is presented in Fig. 74a. It is shown that the heat transfer coefficient in the entrance of a triangular channel is always higher than that computed from Soliman et al.'s correlation [147] for a round tube with the same hydraulic diameter. Zhao and Liao [146] attributed the heat transfer enhancement by the extremely thin liquid film on the sidewall that results from the liquid flow toward the channel corners due to surface tension. Figure 74b shows that the two-phase pressure drop in a triangular channel is higher than that in a round tube, and the pressure drop increases with decreasing channel diameter.

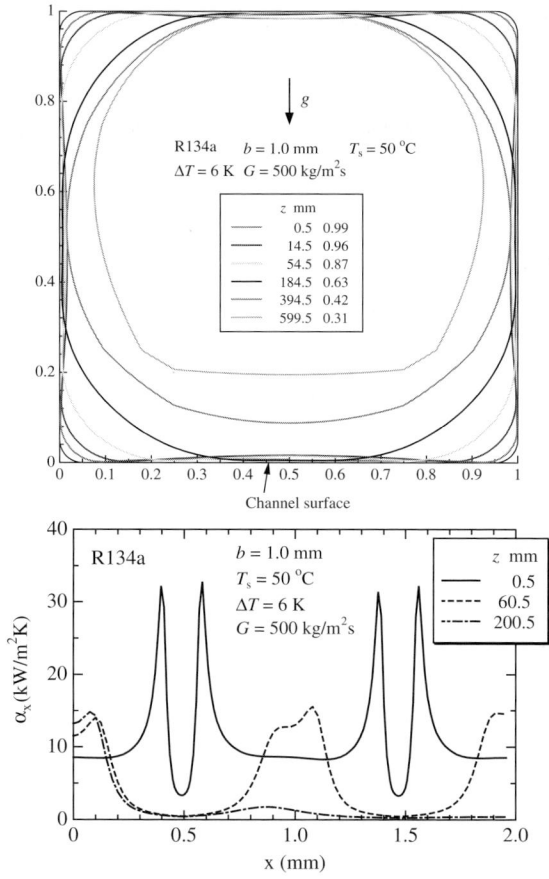

FIG. 75. Variations of film profiles (top) and local film condensation heat transfer coefficient (bottom) at different distances in a horizontal minichannel. From Wang et al. [148]. Copyright (2004), with permission from Institution of Chemical Engineers.

Based on a similar approach, Wang et al. [148,149] carried out analyses on the effects of surface tension, vapor shear stress, and gravity on laminar condensation heat transfer of R134a in horizontal minichannels with a square cross section of $1 \times 1\,\text{mm}^2$. Figure 75a shows the calculated condensate surface profiles at different distances z along the channel with shear stress, surface tension and gravity taken into consideration. The asymmetric film profile with respect to top and bottom of the channel shows that the film condensation flow is stratified due to gravity. Figure 75b shows the spanwise (measured clockwise from the top center of the tube) non-uniform distributions of the local heat transfer coefficient at different distances z along the channel. It is shown that the lowest heat transfer coefficient appears at the corners ($X = 0.5, 1.5$), while the highest local heat transfer coefficient appears at locations near the corners where the film is very thin.

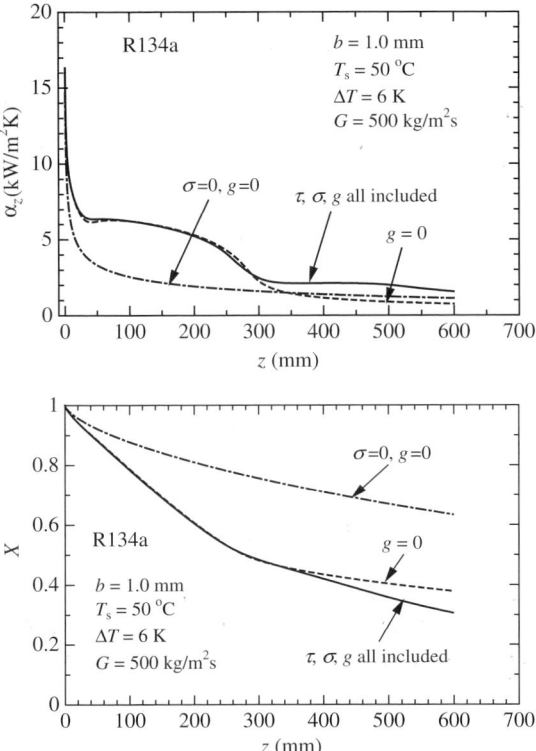

FIG. 76. Variations of mean heat transfer coefficient (top) and vapor mass quality (bottom) with distance long the channel in film condensation in a horizontal channel. From Wang et al. [148,149]. Copyright (2004), with permission from Institute of Chemical Engineers.

The effects of shearing stress, surface tension, and gravity on the variations of spanwise average heat transfer coefficient and the vapor mass quality are shown in Figs. 76a and b respectively. In these figures, the solid line includes all three mechanisms, the dash line represents the case where gravity effect is neglected, and chain dot line represents the case where the effects of surface tension and gravity are neglected. Figure 76a shows that the heat transfer coefficient over the first 300 mm of the channel is enhanced because of the surface tension effect. The variation of vapor mass quality with distance along the channel z is shown in Fig. 76b, where it can be seen that surface tension effect is important everywhere in the channel, while the gravity effect is negligibly small in the first 300 mm of the channel.

IV. Concluding Remarks

In this paper, an up-to-date review of literature on meso- and microscale phase change heat transfer processes occurring in heat exchangers, heat pipes, and thermal bubble actuators has been presented. Many new phase change heat transfer phenomena such as bubble growth and collapse under pulse heating, effects of dissolved gas, wall roughness, corners, and tube diameter of the channels and size of the heater on the nucleation temperature in microchannels have been discussed. New flow patterns and instabilities in flow boiling and flow condensation in microchannels due to surface tension effects have also been illustrated. In view of the continued trend in miniaturization of mechanical systems and the explosive growth of MEMS products, meso- and phase change heat transfer will find increasing applications in engineering and medical fields. Much research work is needed to provide rational tools for optimal designs of these meso and microdevices.

Acknowledgments

The authors would like to acknowledge the support of this work through the National Natural Science Foundation of China through Grant No. 50536010 and 50476017. They would also like to thank Drs. Peigang Deng, Jie Yi, and Jun Zhao for their contributions in the preparation of this paper.

References

1. Mehendale, S. S., Jacobi, A. M., and Shah, R. K. (1999). Heat exchangers at micro-and meso-scales. *In* "Compact Heat Exchangers and Enhancement Technology for the Process

Industries" (R. K. Shah, K. J. Bell, H. Honda and B. Thonon, eds.), pp. 55–74. Begell House, New York.
2. Ektummakij, P., Kumthonkittikun, V., Kuriyama, H., Mashiko, K., Mochizuki, M., Saito, Y., and Nguyen, T. (2004). New composite wick heat pipe for cooling personal computer. In "Proceedings of 13th International Heat Pipe Conference," September 21–25. Shanghai, China.
3. McGlen, R. J., Jachuck, R. Lin, S. (2004). Integrated thermal management techniques for high power electronic devices. Appl. Therm. Eng. **24**, 1143–1156.
4. Nielsen, N. J. (1985). History of thinkjet printhead development-preventing hydraulic crosstalk. Hewlett-Packard J. **36**, 4–10.
5. Kobayashi, H., Koumura, N., and Ohno, S. (1981). Canon Kabushiki Kaisha Liquid recording medium. US Patent Specification 4243994.
6. Deng, P. G., Lee, Y. K., and Cheng, P. (2003). The growth and collapse of a micro-bubble under pulse heating. Int. J. Heat Mass Transfer **46**, 4041–4050.
7. Okamoto, T., Suzuki, T., and Yamamoto, N. (2000). Microarray fabrication with covalent attachment of DNA using bubble jet technology. Nat. Biotechnol. **18**, 438–441.
8. Deng, P. G., Lee, Y. K., and Cheng, P. (2004). Micro bubble dynamics in DNA solutions. J. Micromech. Microeng. **14**, 693–701.
9. Kandlikar, S. G. (2002). Fundamental issues related to flow boiling in minichannels and microchannels. Exp. Therm. Fluid Sci. **26**, 389–407.
10. Thome, J. R. (2004). Boiling in microchannels: a review of experiment and theory. Int. J. Heat Fluid Fl. **25**, 128–139.
11. Li, J. M. and Wang, B. X. (2003). Size effect on two-phase regime for condensation in micro/mini tubes. Heat Transfer-Asian Res. **32**, 65–71.
12. Fritz, W. (1935). Berechnung des maximal volume von dampfblasen. Phys. Z **36**, 379–388.
13. Dupont, V. and Thome, J. R. (2004). Evaporation in microchannels: influence of the channel diameter on heat transfer. In "Proceedings of 2nd International Conference on Microchannels and Minichannels (ICMM2004)", pp. 461–468.
14. Kew, P. A. and Cornwell, K. (1997). Correlations for the prediction of boiling heat transfer in small-diameter channels. Appl. Therm. Eng. **17**, 705–715.
15. Kandlikar, S. G. (2004). Heat transfer mechanisms during flow boiling in microchannels. ASME J. Heat Transfer **126**, 8–16.
16. Koyama, S., Kuwahara, K., and Nakashita, K. (2003). Condensation of refrigerant in a multi-port channel. In "Proceedings of First International Conference on Microchannels and Minichannels,", pp. 193–205. New York, USA.
17. Guide to Compact Heat Exchangers (http://www.hw.ac.uk/mec www/hexag/CHX.pdf.).
18. Tuckerman, D. B. and Pease, R. F. W. (1981). High-performance heat sinking for VLSI. IEEE Electron Rev. Lett. EDL-2, 126–129.
19. Jiang, L. N., Wong, M., and Zohar, Y. (1999). Phase change in microchannel heat sinks with integrated temperature sensors. J. Microelectromech. Syst. **8**, 358–365.
20. Koo, J. M., Jiang, L. N., Zhang, L., Zhou, P., Banerjee, S. S., Kenny, T. W., Santiago, J. G., and Goodson, K. E. (2001). Modeling of two-phase microchannel heat sinks for VLSI chips. IEEE, pp. 422–426.
21. Groll, M., Schneider, M., Sartre, V., Zaghdoudi, M. C., and Lallemand, M. (1998). Thermal control of electronic equipment by heat pipes. Rev. Gen. Therm **37**, 323–352.
22. Cao, Y., Gao, M., Bea, J. E., and Donovan, B. (1997). Experimental and analyses of flat miniature heat pipes. J. Thermophys. Heat Transfer **11**, 158–163.
23. Hopkins, R., Fagri, A., and Khrustalev, D. (1999). Flat miniature heat pipes with micro capillary grooves. J. Heat Transfer **121**, 102–109.

24. Take, K. and Webb, R. L. (2001). Thermal performance of integrated plate heat pipe with a heat spreader. *ASME J. Electron. Packaging* **123**, 189–195.
25. Lin, L. C., Ponnappan, R., and Leland, J. (2002). High performance miniature heat pipe. *Int. J. Heat Mass Transfer* **45**, 3131–3142.
26. Cotter, T. P. (1984). Principles and prospect for microheat pipes. *In* "Proceedings of the 5th International Heat Pipe Conferance," Vol. 4, pp. 328–334. Tsukuba, Japan.
27. Peterson, G. P., Duncan, A. B., and Weichold, M. H. (1993). Experimental investigation of micro heat pipes fabricated in silicon wafers. *J. Heat Transfer* **115**, 751–756.
28. Longtin, J. P., Badran, B., and Gerner, F. M. (1994). A one-dimensional model of a micro heat pipe during steady-state operation. *J. Heat Transfer* **116**, 709–715.
29. Kang, S. W. and Huang, D. (2002). Fabrication of star grooves and rhombus grooves micro heat pipe. *J. Micromech. Microeng.* **12**, 525–531.
30. Berre, M. L., Launay, S., Sartre, V., and Lallemand, M. (2003). Fabrication and experimental investigation of silicon micro heat pipes for cooling electronics. *J. Micromech. Microeng.* **13**, 436–441.
31. Wang, Y. X. and Peterson, G. P. (2001). Investigation of the temperature distribution on radiator fins with micro heat pipes. *J. Theromphys. Heat Transfer* **15**, 42–49.
32. Wang, Y. X. and Peterson, G. (2002). Analysis of wire bonded micro heat pipes arrays. *J Thermophys. Heat Transfer* **16**, 346–355.
33. Launay, S., Sartre, V., Mantelli, B. H., de Paiva, K. V., and Lallemand, M. (2004). Investigation of wire plate micro heat pipe array. *Int. J. Therm. Sci.* **43**, 499–507.
34. Kang, S. W., Tsai, S. H., and Chen, H. C. (2002). Fabrication and test of radial grooved micro heat pipes. *Appl. Therm. Eng.* **22**, 1559–1568.
35. Kang, S. W., Tsai, S. H., and Ko, M. H. (2004). Metallic micro heat pipe heat spreader fabrication. *Appl. Therm. Eng.* **24**, 299–309.
36. Akachi, H. (1990). Structure of a heat pipe. US Patent 4921041.
37. Hosoda, M., Nishio, S., and Shirakashi, R. (1999). Meadering closed-loop heat transfer tube. *In* "Proceedings of the 5th ASME/JSME Joint Thermal Engineering Conference." San Diego, CA.
38. Groll, M., and Khandekar, S., (2004). State of the art on pulsating heat pipes. *In* "Proceedings of the 2nd International Conference on Minichannels and Microchannels, ICMM2004,", pp.33–44. Rochester, NY.
39. Lee, W. H., Jung, H. S., Kim, J. H., and Kim, J. S. (1999). Flow visualization of oscillating capillary tube heat pipe. *In* "Proceedings of 11th International Heat Pipe Conference", pp.131–136. Tokyo, Japan.
40. Tong, B. Y., Wong, T. N., and Ooi, K. T. (2001). Closed-loop pulsating heat pipe. *Appl. Therm. Eng.* **21**, 1845–1862.
41. Shafii, M. B., Faghri, A., and Zhang, Y. W. (2002). Analysis of heat transfer in unlooped and looped pulsating heat pipes. *Int. J. Numer. Meth. Heat Fluid Fl.* **12**, 585–609.
42. Zhang, X. M., Xu, J. L., and Zhou, Z. Q. (2004). Experimental study of a pulsating heat pipe using FC-72, ethanol, and water as working fluids. *Exp. Heat Transfer* **17**, 47–67.
43. Stenger, F. J. (1966). "Experimental Feasibility Study of Water-Filled Capillary-Pumped Heat Transfer Loops. NASA TM-X-1310". NASA Lewis Research Center, Cleveland, OH.
44. Maidanik, Y., Fershtater, Y., and Goncharov, K. (1991). Capillary pumped loop for the systems of thermal regulation of spacecraft. *In* 25th International Conference on Environmental Systems, SAE Paper.
45. Allen, J. S. and Hallinan, K. P. (2001). Liquid blockage of vapor transport lines in low Bond number systems due to capillary-driven flows in condensed annular films. *Int. J. Heat Mass Transfer* **44**, 3931–3940.

46. Kirshberg, J., Yerkes, K., Trebotich, D., and Liepmann, D. (2000). Cooling effect of a MEMS-based capillary pumped loop for chip-level temperature control. Proceedings of 2000 Int. Mech. Eng. Congress & Exposition **3**, 1–8, Orlando, Florida.
47. Liepmann, D. (2001). Design and fabrication of a micro-CPL for chip-level cooling. In "Proceedings of 2001 ASME International Mechanical Engineering Congress and Exposition", New York, USA.
48. Kang, S. W., Hsu, C. C., and Hou, T. F. (2004). Micro loop heat pipes. *Proceedings of 13th International Heat Pipe Conference* **1**, 78–84.
49. Cytrynowicz, D., Hamdan, M., Medis, P., Shuja, A., Henderson, H. T., Gerner, F. M., and Golliher, E. (2002). MEMS loop heat pipe based on coherent porous silicon technology. In "Proceedings of Space Technology and Applications International Forum (STAIF-2002)" (M. El-Genk, ed.), pp. 220–232. AIP Conference Proceedings 608, New York.
50. Bhaskar, E. V. and Leban, M.A. (1989). Integrated thermal inkjet printhead and method of manufacture, US Patent 4 847 630.
51. Lee, J. D., Yoon, J. B., Kim, J. K., Chung, H. J., Lee, C. S., Lee, H. D., Lee, H. J., Kim, C. K., and Han, C. H. (1999). A thermal inkjet printhead with monolithically fabricated nozzle plate and self-aligned ink feed hole. *J. Microelectromech. Syst.* **8**, 229–236.
52. Baek, S. S., Choi, B. Y., and Oh, Y. S. (2004). Design of a high-density thermal inkjet using heat transfer from CVD diamond. *Micromech. Microeng.* **14**, 750–760.
53. Tseng, F. G., Kim, C. J., and Ho, C. M. (2002). A high-resolution high-frequency monolithic top-shooting microinjector free of satellite drops. Part I. Concept, design, and model. *J. Microelectromech. Syst.* **11**, 427–436.
54. Tseng, F. G., Kim, C. J., and Ho, C. M. (2002). A high-resolution high-frequency monolithic top-shooting microinjector free of satellite drops. Part II. Fabrication, implementation, and characterization. *J. Microelectromech. Syst.* **11**, 437–447.
55. Lee, S. W., Kim, H. C., Keon, K., and Oh, Y. S. (2002). A monolithic inkjet print head: Domejet. *Sensors Actuators A* **95**, 114–119.
56. Shin, S. J., Kuk, K., Shin, J. W., Lee, C. S., Oh, Y. S., and Park, S. O. (2004). Thermal design modifications to improve firing frequency of back shooting inkjet printhead. *Sensors Actuators* **A 114**, 387–391.
57. Baek, S. S., Lim, H. T., Song, H., Kim, Y. S., Bae, K. D., Cho, C. H., Lee, C. S., Shih, J. W., Shin, S. J., Kuk, K., and Oh, Y. S. (2004). A novel back-shooting inkjet printhead using trench-filling and SOI wafer. *Sensors Actuators A* **114**, 392–397.
58. Geng, X., Yuan, H., Oguz, H. N., and Prosperetti, A. (2001). Bubble-based micropump for electrically conducting liquids. *J. Micromech. Microeng.* **11**, 270–276.
59. Nakabeppu, O., Seki, K., and Ohko, K. (2003). Operating Characteristics of Micro Machined Bubble Drive Micropumps. 6th ASME-JSME Thermal Engineeering Joint Conference.
60. Jun, T. K., Jin, C., and Kim, C. J. (1998). Valveless pumping using traversing vapor bubbles in microchannels. *J. Appl. Phys.* **83**, 5658–5664.
61. Soong, Y. J. and Zhao, T. M. (2001). Modelling and test of a thermally driven phase-change nonmechanical micropump. *J. Micromech. Microeng.* **11**, 713–719.
62. Xu, J. L., Huang, X. Y., and Wong, T. N. (2003). Study on heat driven pump. Part 1-experimental measurements. *Int. J. Heat Mass Transfer* **46**, 3329–3335.
63. Xu, J. L., Wong, T. N., and Huang, X. Y. (2003). Study on heat driven pump. Part 2-mathematical modeling. *Int. J. Heat Mass Transfer* **46**, 3337–3347.
64. Carey, V. P. (1992). "Liquid-Vapor Phase Phenomena". Hemisphere, Wasington, DC.
65. Cole, R. (1974). Boiling nucleation. *Adv. Heat Transfer* **10**, 86–166.

66. Li, J. and Cheng, P. (2004). Bubble cavitation in a microchannel. *Int. J. Heat Mass Transfer* **47**, 2689–2698.
67. Brereton, G. J., Crilly, R. J., and Spears, J. R. (1998). Nucleation in small capillary tubes. *Chem. Phys.* **230**, 253–265.
68. Peng, X. F., Hu, H. Y., and Wang, B. X. (1998). Boiling nucleation during liquid flow in microchannels. *Int. J. Heat Mass Transfer* **41**, 101–106.
69. Tsai, J. H. and Lin, L. W. (2002). Transient thermal bubble formation on polysilicon micro-resisters. *J. Heat Transfer* **124**, 375–382.
70. Straub, J. (2000). Microscale boiling heat transfer under 0 g and 1 g conditions. *Int. J. Therm. Sci.* **39**, 490–497.
71. Marek, R. and Straub, J. (2001). The origin of thermocapillary convection in subcooled nuclear pool boiling. *Int. J. Heat Mass Transfer* **44**, 619–632.
72. Straub, J., Betz, J. and Marek, R. (1994). Enhancement of heat transfer by thermocapillary convection around bubbles – a numerical study. Numerical Heat Transfer **A25**, 501–518.
73. Kim, J. H., Benton, J. F., and Wisniewski, D. (2002). Pool boiling heat transfer on small heaters: effect of gravity and subcooling. *Int. J. Heat Mass Transfer* **45**, 3919–3932.
74. Wang, H., Peng, X. F., Wang, B. X., and Lee, D. J. (2002). Jet flow phenomena during nucleate boiling. *Int. J. Heat Mass Transfer* **45**, 1359–1363.
75. Wang, H., Peng, X. F., Wang, B. X., and Lee, D. J. (2003). Bubble sweeping and jet flows during nucleate boiling of subcooled liquids. *Int. J. Heat Mass Transfer* **46**, 863–869.
76. Wang, H., Peng, X. F., Wang, B. X., Lin, W. K., and Pan, C. (2005). Experimental observations of bubble dynamics on ultra thin wires. *Experimental Heat Transfer* **18**, 1–11.
77. Wang, H., Peng, X. F., Lin, W. K., Pan, C., and Wang, B. X. (2004). Bubble-top jet flow on microwires. *Int. J. Heat Mass Transfer* **47**, 2891–2900.
78. Wang, H., Christopher, D. M., Peng, X. F., and Wang, B. X. (2005). Jet flows from a bubble during subcooled pool boiling on micro wires. Science in China Sero.E Engineering & Materials Science **48**, 385–402.
79. Skripov, V. P. (1974). "Metastable Liquids". Wiley, New York.
80. Skripov, V. P. and Pavlov, P. A. (1970). Explosive boiling of liquids and fluctuation nucleus formation. *High Temperature (USSR)* **8**, 833–839, 782–787.
81. Derewnicki, K. P. (1985). Experimental studies of heat transfer and vapor formation in fast transient boiling. *Int. J. Heat Mass Transfer* **28**, 2085–2092.
82. Glod, S., Poulikakos, D., Zhao, Z., and Yadigaroglu, G. (2002). An investigation of microscale explosive vaporization of water on an ultrathin Pt wire. *Int. J. Heat Mass Transfer* **45**, 367–379.
83. Asai, A. (1991). Bubble dynamics in boiling under high heat flux pulse heating. *J. Heat Transfer* **113**, 973–979.
84. Lin, L., Udell, K. S., and Pisano, A. P. (1994). Liquid vapor phase transition and bubble formation in micro structures. *Therm. Sci. Eng.* **2**, 52–59.
85. Lin, L. (1998). Microscale thermal bubble formation: thermophysical phenomena and applications. *Microscale Thermophys. Eng.* **2**, 71–85.
86. Iida, Y., Okuyama, K., and Sakurain, K. (1994). Boiling nucleation on a very small film heater subjected to extremely rapid heating. *Int. J. Heat Mass Transfer* **37**, 2771–2780.
87. Okuyama, K. and Iida, Y. (1990). Transient boiling heat transfer characteristics of nitrogen (bubble behavior and heat transfer rate at stepwise heat generation). *Int. J. Heat Mass Transfer* **33**, 2065–2071.
88. Avedisian, C. T., Osborne, W. S., McLeod, F. D. and Curley, C. M. (1999). Measuring bubble nucleation temperature on the surface of a rapidly heated thermal ink-jet heater immersed in a pool of water. *Proc. Royal Soc. Lond.* **A 455**, 3875–3899.

89. Zhao, Z., Glod, S., and Poulikakos, D. (2000). Pressure and power generation during explosive vaporization on a thin-film microheater. *Int. J. Heat Mass Transfer* **43**, 281–296.
90. Yin, Z., Prosperetti, A., and Kim, J. (2004). Bubble growth on an impulsively powered microheater. Int. J. Heat Mass Transfer **47**, 1053–1067.
91. Jung, J. Y., Lee, J. Y., Park, H. C., and Kwak, H. Y. (2003). Bubble nucleation on micro line heaters under steady or finite pulse of voltage input. *Int. J. Heat Mass Transfer* **46**, 3897–3907.
92. Deng, P. G., Lee, Y. K., and Cheng, P. (2006). An experimental study of heater size effect on micro bubble generation. *Int. J. Heat Mass Transfer* **49**, 2535–2544.
93. Deng, P. G. (2003). Micro Bubble Actuator for DNA Hybridization Enhancement, PhD Dissertation, The Hong Kong University of Science and Technology.
94. Ghiaasiaan, S. M. and Abdel-Khalik, S. I. (2001). Two-phase flow in microchannels. *Adv. Heat Transfer* **34**, 145–253.
95. Garimella, S. V. and Sobhan, C. B. (2003). Transport in microchanels – a critical review. *Annu. Rev. Heat Transfer* **13**, 1–50.
96. Peng, X. F. and Wang, B. X. (1993). Forced convection and flow boiling heat transfer for liquid flowing through microchannels. *Int. J. Heat Mass Transfer* **36**, 3421–3427.
97. Ghiaasiaan, S. M. and Chedester, R. C. (2002). Boiling incipience in microchannnels. *Int. J. Heat Mass Transfer* **45**, 4599–4606.
98. Yen, T. H., Kasagi, N., and Suzuki, Y. (2003). Forced convective boiling heat transfer in microtubes at low mass and heat fluxes. *Int. J. Multiphase Fl.* **29**, 1771–1792.
99. Zhang, L., Wang, E. N., Koo, J. M., Jiang, L., Goodson, K. E., Santiago, J. G., and Kenny, T. W. (2002). Enhanced nucleating boiling in microchannels. *15th IEEE International Conference on Micro Electro Mechanical Systems*, 89–92.
100. Lee, P. C., Tseng, F. G., and Pan, C. (2004). Bubble dynamics in microchannels. Part I: single microchannel. *Int. J. Heat Mass Transfer* **47**, 5575–5589.
101. Li, H. Y., Tseng, F. G., and Pan, C. (2004). Bubble dynamics in microchannels. Part II: two parallel microchannels. *Int. J. Heat Mass Transfer* **47**, 5591–5601.
102. Steinke, M. E. and Kandlikar, S. G. (2004). Control and effect of dissolved air in water during flow boiling in microchannels. *Int. J. Heat Mass Transfer* **47**, 1925–1935.
103. Jiang, L., Wong, M., and Zohar, Y. (2001). Forced convection boiling in a microchannel heat sink. *J. Microelectromech. Syst.* **10**, 80–87.
104. Lee, M., Wong, Y. Y., Wong, M., and Zohar, Y. (2003). Size and shape effects on two-phase flow patterns in microchannel forced convection boiling. *J. Micromech. Microeng.* **13**, 155–164.
105. Sobierska, E., Shuai, J., Mertz, R., Kulenovic, R., and Groll, M. (2004). Visualization and flow pattern map for flow boiling of water in a vertical micro channel. *Proc. Second Int. Conf. Microchannels Minichannels, ICMM 2004*, 491–497.
106. Mishima, K. and Ishii, M. (1984). Flow regime transition criteria for upward two-phase flow in vertical tubes. *J. Heat Mass Transfer* **27**, 723–736.
107. Taitel, Y., Bornea, D., and Duckler, A. E. (1980). Modelling flow pattern transitions for steady upward gas–liquid flow in vertical tubes. *AICHE J.* **26**, 345–354.
108. Wu, H. Y. and Cheng, P. (2003). Visualization and measurements of periodic boiling in silicon microchannels. *Int. J. Heat Mass Transfer* **46**, 2603–2614.
109. Wu, H. Y. and Cheng, P. (2003). Liquid/two-phase/vapor alternating flow during boiling in microchannels at high heat flux. *Int. Comm. Heat Mass Transfer* **30**, 295–302.
110. Wu, H. Y. and Cheng, P. (2004). Boiling instability in parallel silicon microchannels at different heat flux. *Int. J. Heat Mass Transfer* **47**, 3631–3641.
111. Hetsroni, G., Mosyak, A., Segal, Z., and Ziskind, G. (2002). A uniform temperature heat sink for cooling of electronic devices. *Int. Heat Mass Transfer* **45**, 3275–3286.

112. Brutin, D., Topin, F., and Tadrist, L. (2003). Experimental study of unsteady convective boiling in heated minichannels. *Int. J. Heat Mass Transfer* **46**, 2957–2965.
113. Qu, W. and Mudawar, I. (2003). Measurement and prediction of pressure drop in two-phase micro-channel heat sinks. *Int. J. Heat Mass Transfer* **46**, 2737–2753.
114. Balasubramanian, P. and Kandlikar, S. G. (2004). Experimental study of flow patterns, pressure drop and flow instabilities in a parallel rectangular minichannels. Proc. of 2nd International Conference on Microchannels and Minichannels, ICMM2004-2371, pp. 475–481, Rochester, NY.
115. Xu, J. L, Zhou, J., and Gan, Y. (2005). Static and dynamic flow instability of a parallel microchannels heat sink at high heat fluxes. *Energy Convers. Manage.* **46**, 313–334.
116. Coleman, J. W. and Garimella, S. (2000). Visualization of refrigerant two-phase flow during condensation. Proceedings of the 34th National Heat Transfer Conference, Vol. 1, NHTC2000–12115.
117. Coleman, J. W. and Garimella, S. (2000). Two-phase flow regime transitions in microchannel tubes: the effect of hydraulic diameter. *Proc. ASME Heat Transfer Div.-2000 HTD* **366**(4), 71–83.
118. Coleman, J. W. and Garimella, S. (2003). Two-phase flow regimes in round, square and rectangular tubes during condensation of refrigerant R134a. *Int. J. Refrig.* **26**, 117–128.
119. Garimella, S. (2004). Condensation flow mechanisms in microchannels: basis for pressure drop and heat transfer models. *Heat Transfer Eng.* **25**, 104–166.
120. Wang, W. W. W., Radcliff, T. D., and Christensen, R. N. (2002). A condensation heat transfer correlation for millimeter-scale tubing with flow regime transition. *Exp. Therm. Fluid Sci.* **26**, 473–485.
121. Mederic, B., Miscevic, M., Platel, V., Lavieille, P., and Joly, J. J. (2003). Complete convection condensation inside small diameter horizontal tubes. In "Proceedings of 1st International Conference on Microchannels and Minichannels", pp.707–712. Rochester, NY, USA.
122. Chen, Y. P. and Cheng, P. (2005). Condensation of steam in a silicon microchannel. *Int. Comm. Heat Mass Transfer* **32**, 175–183.
123. Wu, H. Y. and Cheng, P. (2004). Alternating condensation flow patterns in microchannels. In "Proceedings of Second International Conference on Microchannels and Minichannels", pp. 657–660. Rochester, NY, USA.
124. Wu, H. Y. and Cheng, P. (2005). Condensation flow patterns in silicon microchannels. *Int. J. Heat Mass Transfer* **48**, 2186–2197.
125. Teng, H., Cheng, P., and Zhao, T. S. (1999). Instability of condensate film and capillary blocking in small-diameter-thermosyphon condensers. *Int. J. Heat Mass Transfer* **42**, 3071–3083.
126. Tengblad, N. and Palm, B. (1995). Flow boiling and film condensation heat transfer in narrow channels of thermosiphones for cooling of electronic components. In "Proceedings of EuroHiam Sem. UoH5". Leuveu, Belgium.
127. Wang, B. X. and Du, X. Z. (2000). Study on laminar film-wise condensation for vapor flow in an inclined small/mini-diameter tube. *Int. J. Heat Mass Transfer* **43**, 1859–1868.
128. Shin, J. S. and Kim, M. H. (2004). An experimental study of flow condensation heat transfer inside circular and rectangular mini-channels. In "Proceedings of Second International Conference on Microchannels and Minichannels", pp. 633–640. New York, USA.
129. Shah, M. M. (1979). A general correlation for heat transfer during film condensation inside pipes. *Int. J. Heat Mass Transfer* **22**, 547–556.
130. Akers, W. W. and Rosson, H. F. (1960). Condensation inside a horizontal tube. *Chem. Eng. Prog. Symp. Serb* **30**, 145–149.

131. Jaster, H. and Kosky, P. G. (1976). Condensation in a mixed flow regime. *Int. J. Heat Mass Transfer* **19**, 95–99.
132. Baird, J. R., Fletcher, D. F., and Haynes, B. S. (2003). Local condensation heat transfer rates in fine passages. *Int. J. Heat Mass Transfer* **46**, 4453–4466.
133. Cavallini, A., Censi, G., Col, D. D., Doretti, L., Longo, G. A., Rossetto, L., and Zilio, C. (2003). Experimental investigation on condensation heat transfer coefficient inside multi-port minichannels. *In* "Proceedings of 1st International Conference on Microchannels and Minichannels", pp. 691–698. New York, USA.
134. Moser, K. W., Webb, K. L., and Na, B. (1998). A new equivalent Reynolds number model for condensation in smooth tubes. *J. Heat Transfer* **120**, 410–417.
135. Zhang, M. and Webb, R. L. (2001). Correlation of two-phase friction for refrigerants in small-diameter tubes. *Exp. Therm. Fluid Sci.* **25**, 131–139.
136. Cavallini, A., Col, D. D., Doretti, L., Matkovic, M., Rossetto, L., and Zilio, C. (2004). Condensation heat transfer inside multi-port minichannels. *In* "Proceedings of 2nd International Conference on Microchannels and Minichannels", pp. 625–632. Rochester, NY, USA.
137. Webb, R. L. and Ermis, K. (2001). Effect of hydraulic diameter on condensation of R-134a in flat, extruded aluminum tubes. *J. Enhanced Heat Transfer* **8**, 77–90.
138. Kim, M. H., Shin, J. S., Huh, C., Kim, T. J., and, Seo, K. W. (2003). A study of condensation heat transfer in a single mini-tube and a review of Korean micro- and min-channel studies. *In* "Proceedings of First International Conference on Microchannels and Minichannels", pp. 47–58. Rochester, NY, USA.
139. Friedel, L. (1979). Improved friction pressure drop correlation for horizontal and vertical two-phase pipe flow, European Two-phase Flow Group Meeting, Paper No.2, Ispra, Italy.
140. Mishima, K. and Hibiki, T. (1996). Some characteristics of air-water two-phase flow in small diameter vertical tubes. *Int. J. Multiphase Flow* **22**, 703–714.
141. Garimella, S., Killion, J. D., and Coleman, J. W. (2002). An experimentally validated model for two-phase pressure drop in the intermittent flow regime for circular microchannels. *ASME J. Fluids Eng.* **124**, 205–214.
142. Garimella, S., Killion, J. D., and Coleman, J. W. (2003). An experimentally validated model for two-phase pressure drop in the intermittent flow regime for noncircular microchannels. *ASME J. Fluids Eng.* **125**, 887–894.
143. Garimella, S., Agarwal, A., and Coleman, J. W. (2003). Two-pressure drops in the annular flow regime in circular microchannels. 21st IIR International Congress of Refrigeration, Washington, DC, International Institute of Refrigeration.
144. Garimella, S., Agarwal, A., and Killion, J. D. (2004). Condensation pressure drops in circular microchannels. *In* "Proceedings of Second International Conference on Microchannels and Minichannels", pp. 649–656. Rochester, NY, USA.
145. Begg, E., Khrustalev, D., and Faghri, A. (1999). Complete condensation of forced convection two-phase flow in miniature tube. *ASME J. Heat Transfer* **121**, 904–915.
146. Zhao, T. S. and Liao, Q. (2002). Theoretical analysis of film condensation heat transfer inside vertical mini triangular channels. *Int. J. Heat Mass Transfer* **45**, 2829–2842.
147. Soliman, M., Schuster, J. R., and Berenson, P. J. (1968). A general heat transfer correlation for annular flow condensation. *J. Heat Transfer* **90**, 267–276.
148. Wang, H. S., Rose, J. W., and Honda, H. (2004). A theoretical model of film condensation in square section horizontal microchannels. *Chem. Eng. Res. Des.* **82**, 430–434.
149. Wang, H. S. and Rose, J. W. (2005). A theory of film condensation in horizontal noncircular section microchannels. *J. of Heat Transfer* **127**, 1096–1105.

Jet Impingement Heat Transfer: Physics, Correlations, and Numerical Modeling

N. ZUCKERMAN and N. LIOR

Department of Mechanical Engineering and Applied Mechanics, The University of Pennsylvania, Philadelphia, PA, USA; E-mail: zuckermn@seas.upenn.edu; lior@seas.upenn.edu

I. Summary

The applications, physics of the flow and heat transfer phenomena, available empirical correlations and values they predict, and numerical simulation techniques and results of impinging jet devices for heat transfer are described. The relative strengths and drawbacks of the k–ε, k–ω, Reynolds stress model, algebraic stress models, shear stress transport, and $v^2 f$ turbulence models for impinging jet flow and heat transfer are compared. Select model equations are provided as well as quantitative assessments of model errors and judgments of model suitability.

II. Introduction

We seek to understand the flow field and mechanisms of impinging jets with the goal of identifying preferred methods of predicting jet performance. Impinging jets provide an effective and flexible way to transfer energy or mass in industrial applications. A directed liquid or gaseous flow released against a surface can efficiently transfer large amounts of thermal energy or mass between the surface and the fluid. Heat transfer applications include cooling of stock material during material forming processes, heat treatment [1], cooling of electronic components, heating of optical surfaces for defogging, cooling of turbine components, cooling of critical machinery structures, and many other industrial processes. Typical mass transfer applications include drying and removal of small surface particulates. Abrasion and heat transfer by impingement are also studied as side effects of vertical/short take-off and landing jet devices, for example in the case of direct lift propulsion systems in vertical/short take-off and landing aircraft.

General uses and performance of impinging jets have been discussed in a number of reviews [2–5].

In the example of turbine cooling applications [6], impinging jet flows may be used to cool several different sections of the engine such as the combustor case (combustor can walls), turbine case/liner, and the critical high-temperature turbine blades. The gas turbine compressor offers a steady flow of pressurized air at temperatures lower than those of the turbine and of the hot gases flowing around it. The blades are cooled using pressurized bleed flow, typically available at 600°C. The bleed air must cool a turbine immersed in gas of 1400°C total temperature [7], which requires transfer coefficients in the range of 1000–3000 W/m^2 K. This equates to a heat flux on the order of 1 MW/m^2. The ability to cool these components in high-temperature regions allows higher cycle temperature ratios and higher efficiency, improving fuel economy, and raising turbine power output per unit weight. Modern turbines have gas temperatures in the main turbine flow in excess of the temperature limits of the materials used for the blades, meaning that the structural strength and component life are dependent upon effective cooling flow. Compressor bleed flow is commonly used to cool the turbine blades by routing it through internal passages to keep the blades at an acceptably low temperature. The same air can be routed to a perforated internal wall to form impinging jets directed at the blade exterior wall. Upon exiting the blade, the air may combine with the turbine core airflow. Variations on this design may combine the impinging jet device with internal fins, smooth or roughened cooling passages, and effusion holes for film cooling. The designer may alter the spacing or locations of jet and effusion holes to concentrate the flow in the regions requiring the greatest cooling. Though the use of bleed air carries a performance penalty [8], the small amount of flow extracted has a small influence on bleed air supply pressure and temperature. In addition to high-pressure compressor air, turbofan engines provide cooler fan air at lower pressure ratios, which can be routed directly to passages within the turbine liner. A successful design uses the bleed air in an efficient fashion to minimize the bleed flow required for maintaining a necessary cooling rate.

Compared to other heat or mass transfer arrangements that do not employ phase change, the jet impingement device offers efficient use of the fluid, and high transfer rates. For example, compared with conventional convection cooling by confined flow parallel to (under) the cooled surface, jet impingement produces heat transfer coefficients that are up to three times higher at a given maximum flow speed, because the impingement boundary layers are much thinner, and often the spent flow after the impingement serves to turbulate the surrounding fluid. Given a required heat transfer coefficient, the flow required from an impinging jet device may be two orders of magnitude smaller than that required for a cooling approach using a free wall-parallel flow. For more uniform coverage over larger surfaces multiple jets may be used. The impingement cooling approach also offers a compact hardware arrangement.

Some disadvantages of impingement cooling devices are: (1) For moving targets with very uneven surfaces, the jet nozzles may have to be located too far from the surface. For jets starting at a large height above the target (over 20 jet nozzle diameters) the decay in kinetic energy of the jet as it travels to the surface may reduce average Nu by 20% or more. (2) The hardware changes necessary for implementing an impinging jet device may degrade structural strength (one reason why impinging jet cooling is more easily applied to turbine stator blades than to rotor blades). (3) In static applications where very uniform surface heat or mass transfer is required, the resulting high density of the jet array and corresponding small jet height may be impractical to construct and implement, and at small spacings jet-to-jet interaction may degrade efficiency.

Prior to the design of an impinging jet device, the heat transfer at the target surface is typically characterized by a Nusselt number (Nu), and the mass transfer from the surface with a Schmidt number (Sc). For design efficiency studies and device performance assessment, these values are tracked vs. jet flow per unit area (G) or vs. the power required to supply the flow (incremental compressor power).

A. IMPINGING JET REGIONS

The flow of a submerged impinging jet passes through several distinct regions, as shown in Fig. 1. The jet emerges from a nozzle or opening with a velocity and temperature profile and turbulence characteristics dependent upon the upstream flow. For a pipe-shaped nozzle, also called a tube nozzle or cylindrical nozzle, the flow develops into the parabolic velocity profile common to pipe flow plus a moderate amount of turbulence developed upstream. In contrast, a flow delivered by application of differential pressure across a thin, flat orifice will create an initial flow with a fairly flat velocity profile, less turbulence, and a downstream flow contraction (vena contracta). Typical jet nozzles designs use either a *round jet* with an axisymmetric flow profile or a *slot jet*, a long, thin jet with a two-dimensional flow profile.

After it exits the nozzle, the emerging jet may pass through a region where it is sufficiently far from the impingement surface to behave as a free submerged jet. Here, the velocity gradients in the jet create a shearing at the edges of the jet which transfers momentum laterally outward, pulling additional fluid along with the jet and raising the jet mass flow, as shown in Fig. 2. In the process, the jet loses energy and the velocity profile is widened in spatial extent and decreased in magnitude along the sides of the jet. Flow interior to the

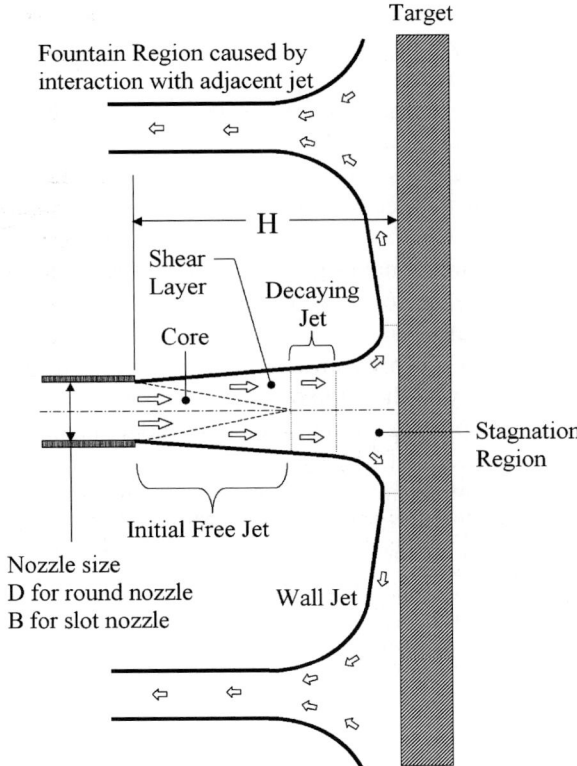

Fig. 1. The flow regions of an impinging jet.

progressively widening shearing layer remains unaffected by this momentum transfer and forms a core region with a higher total pressure, though it may experience a drop in velocity and pressure decay resulting from velocity gradients present at the nozzle exit. A free jet region may not exist if the nozzle lies within a distance of two diameters ($2D$) from the target. In such cases, the nozzle is close enough to the elevated static pressure in the stagnation region for this pressure to influence the flow immediately at the nozzle exit.

If the shearing layer expands inward to the center of the jet prior to reaching the target, a region of core decay forms. For purposes of distinct identification, the end of the core region may be defined as the axial position where the centerline flow dynamic pressure (proportional to speed squared) reaches 95% of its original value. This decaying jet begins four to eight nozzle diameters or slot-widths downstream of the nozzle exit. In the decaying jet, the axial velocity component in the central part decreases, with the radial

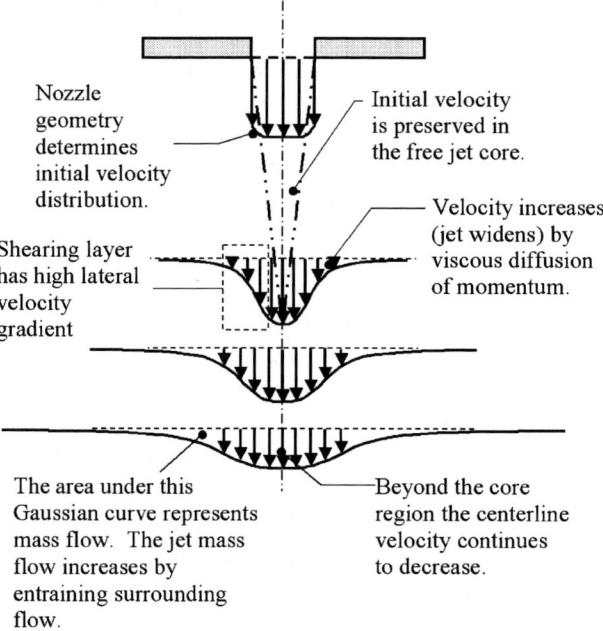

FIG. 2. The flow field of a free submerged jet.

velocity profile resembling a Gaussian curve that becomes wider and shorter with distance from the nozzle outlet. In this region, the axial velocity and jet width vary linearly with axial position. Martin [2] provided a collection of equations for predicting the velocity in the free jet and decaying jet regions based on low Reynolds number flow. Viskanta [5] further subdivided this region into two zones, the initial "developing zone," and the "fully developed zone" in which the decaying free jet reaches a Gaussian velocity profile.

As the flow approaches the wall, it loses axial velocity and turns. This region is labeled the stagnation region or deceleration region. The flow builds up a higher static pressure on and above the wall, transmitting the effect of the wall upstream. The nonuniform turning flow experiences high normal and shear stresses in the deceleration region, which greatly influence local transport properties. The resulting flow pattern stretches vortices in the flow and increases the turbulence. The stagnation region typically extends 1.2 nozzle diameters above the wall for round jets [2]. Experimental work by Maurel and Solliec [9] found that this impinging zone was characterized or delineated by a negative normal-parallel velocity correlation ($\overline{uv}<0$). For their slot jet this region extended to 13% of the nozzle height H, and did not vary with Re or H/D.

After turning, the flow enters a wall jet region where the flow moves laterally outward parallel to the wall. The wall jet has a minimum thickness within 0.75–3 diameters from the jet axis, and then continually thickens moving farther away from the nozzle. This thickness may be evaluated by measuring the height at which wall-parallel flow speed drops to some fraction (e.g. 5%) of the maximum speed in the wall jet at that radial position. The boundary layer within the wall jet begins in the stagnation region, where it has a typical thickness of no more than 1% of the jet diameter [2]. The wall jet has a shearing layer influenced by both the velocity gradient with respect to the stationary fluid at the wall (no-slip condition) and the velocity gradient with respect to the fluid outside the wall jet. As the wall jet progresses, it entrains flow and grows in thickness, and its average flow speed decreases as the location of highest flow speed shifts progressively farther from the wall. Due to conservation of momentum, the core of the wall jet may accelerate after the flow turns and as the wall boundary layer develops. For a round jet, mass conservation results in additional deceleration as the jet spreads radially outward.

B. Nondimensional Heat and Mass Transfer Coefficients

A major parameter for evaluating heat transfer coefficients is the Nusselt number,

$$Nu = hD_h/k_c \tag{1}$$

where h is the convective heat transfer coefficient defined as

$$h = \frac{-k_c \partial T / \partial \vec{n}}{T_{0jet} - T_{wall}} \tag{2}$$

where $\partial T/\partial n$ gives the temperature gradient component normal to the wall.

The selection of Nusselt number to measure the heat transfer describes the physics in terms of fluid properties, making it independent of the target characteristics. The jet temperature used, T_{0jet}, is the adiabatic wall temperature of the decelerated jet flow, a factor of greater importance at increasing Mach numbers. The non-dimensional recovery factor describes how much kinetic energy is transferred into and retained in thermal form as the jet slows down:

$$recovery\ factor = \frac{T_{wall} - T_{0jet}}{U_{jet}^2 / 2c_p} \tag{3}$$

This definition may introduce some complications in laboratory work, as a test surface is rarely held at a constant temperature, and more frequently held at a constant heat flux. Experimental work by Goldstein et al. [10] showed that the temperature recovery factor varies from 70% to 110% of the full theoretical recovery, with lowered recoveries in the stagnation region of a low-H/D jet ($H/D = 2$), and 100% elevated stagnation region recoveries for jets with $H/D = 6$ and higher. The recovery comes closest to uniformity for intermediate spacings around $H/D = 5$. Entrainment of surrounding flow into the jet may also influence jet performance, changing the fluid temperature as it approaches the target.

The nondimensional Sherwood number defines the rate of mass transfer in a similar fashion:

$$Sh = k_i D / D_i \tag{4}$$

$$k_i = D_i [\partial C / \partial n] / [C_{0jet} - C_{wall}] \tag{5}$$

where $\partial C / \partial n$ gives the mass concentration gradient component normal to the wall.

With sufficiently low mass concentration of the species of interest, the spatial distribution of concentration will form patterns similar to those of the temperature pattern. Studies of impinging air jets frequently use the nondimensional relation:

$$Nu/Sh = (Pr/Sc)^{0.4} \tag{6}$$

to relate heat and mass transfer rates.

The nondimensional parameters selected to describe the impinging jet heat transfer problem include the fluid properties such as Prandtl number Pr (the ratio of fluid thermal diffusivity to viscosity, fairly constant), plus the following:

- H/D : nozzle height to nozzle diameter ratio;
- r/D : nondimensional radial position from the center of the jet;
- z/D : nondimensional vertical position measured from the wall;
- Tu : nondimensional turbulence intensity, usually evaluated at the nozzle;
- Re_0 : Reynolds number $U_0 D / v$;
- M : Mach number (the flow speed divided by speed of sound in the fluid), based on nozzle exit average velocity (of smaller importance at low speeds, i.e. $M < 0.3$);
- p_{jet}/D : jet center-to-center spacing (pitch) to diameter ratio, for multiple jets;

- A_f: free area ($= 1-$[total nozzle exit area/total target area]);
- f: relative nozzle area ($=$ total nozzle exit area/total target area).

The fluid properties are conventionally evaluated using the flow at the nozzle exit as a reference location. Characteristics at the position provide the average flow speed, fluid temperature, viscosity, and length scale D. In the case of a slot jet the diameter D is replaced in some studies by slot width B, or slot hydraulic diameter $2B$ in others.

A complete description of the problem also requires knowledge of the velocity profile at the nozzle exit, or equivalent information about the flow upstream of the nozzle, as well as boundary conditions at the exit of the impingement region. Part of the effort of comparing information about jet impingement is to thoroughly know the nature and magnitude of the turbulence in the flow field.

The geometry and flow conditions for the impinging jet depend upon the nature of the target and the fluid source (compressor or blower). In cases where the pressure drop associated with delivering and exhausting the flow is negligible, the design goal is to extract as much cooling as possible from a given air mass flow. Turbine blade passage cooling is an example of such an application; engine compressor air is available at a pressure sufficient to choke the flow at the nozzle (or perhaps at some other point in the flow path). As the bleed flow is a small fraction of the overall compressor flow, the impinging jet nozzle pressure ratio varies very little with changes in the amount of airflow extracted. At high pressure ratios the jet emerges at a high Mach number. In the most extreme case, the flow exits the nozzle as an underexpanded supersonic jet. This jet forms complex interacting shock patterns and a stagnation or recirculation "bubble" directly below the jet (shown in Fig. 3), which may degrade heat transfer [11].

The details of the impingement device design affect the system pressure drop and thus the overall device performance. In the case of a device

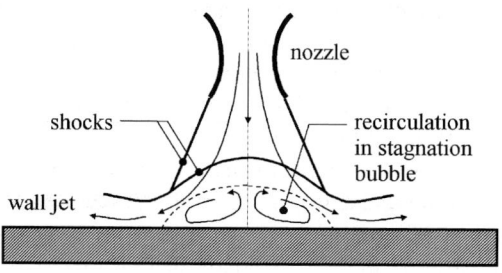

FIG. 3. Supersonic jet flow pattern.

powered by a blower or compressor, the blower power draw can be predicted using the required pressure rise, flow, and blower efficiency including any losses in the motor or transmission. For incompressible duct flow one can then estimate the power by multiplying the blower pressure rise Δp by the volumetric flow Q and then dividing by one or more efficiency factors (e.g., using a total efficiency of 0.52 based on a 0.65 blower aerodynamic efficiency times 0.80 motor efficiency). This same approach works for calculating pump power when dealing with liquid jets, but becomes more complex when dealing with a turbine-cooling problem where compressibility is significant.

The blower pressure rise Δp depends on the total of the pressure losses in the blower intake pathway, losses in the flow path leading to the nozzle, any total pressure loss due to jet confinement and jet interaction, and any losses exiting the target region. In cases where space is not critical the intake pathway and nozzle supply pathway are relatively open, for there is no need to accelerate the flow far upstream of the nozzle exit. When possible, the flow is maintained at low speed (relative to U_{jet}) until it nears the nozzle exit, and then accelerated to the required jet velocity by use of a smoothly contracting nozzle at the end of a wide duct or pipe. In such a case, the majority of the loss occurs at the nozzle where the dynamic pressure is greatest. For a cylindrical nozzle, this loss will be at least equal to the nozzle dump loss, giving a minimum power requirement of ($0.5 \, \rho \, U_{jet}^2 \, Q$).

Jet impingement devices have pressure losses from the other portions of the flow path, and part of the task of improving overall device performance is to reduce these other losses. For this reason, one or more long, narrow supply pipes (common in experimental studies) may not make an efficient device due to high frictional losses approaching the nozzle exit. When orifice plate nozzles are used the upstream losses are usually small, but the orifices can cause up to 2.5 times the pressure drop of short, smooth pipe nozzles (at a set Q and D). This effect is balanced against the orifice nozzle's larger shear layer velocity gradient and more rapid increase in turbulence in the free-jet region [12]. Such orifice plates take up a small volume for the hardware, and are relatively easy and inexpensive to make. A thicker orifice plate (thickness from $0.3D$ to $1.5D$) allows the making of orifice holes with tapered or rounded entry pathways, similar to the conical and bellmouth shapes used in contoured nozzles. This compromise comes at the expense of greater hardware volume and complexity, but reduces the losses associated with accelerating the flow as it approaches the orifice and increases the orifice discharge coefficient (effective area). Calculation of nozzle pressure loss may use simple handbook equations for a cylindrical nozzle [13,14], but for an orifice plate the calculations may require more specialized equations and test data (cf. [15,16]).

TABLE I
Comparison of Nozzle-Type Characteristics

Nozzle type	Initial turbulence	Free jet shearing force	Pressure drop	Nozzle exit velocity profile
Pipe	High	Low	High	Close to parabolic
Contoured contraction	Low	Moderate to high	Low	Uniform (flat)
Sharp orifice	Low	High	High	Close to uniform (contracting)

Table I compares characteristics of the most common nozzle geometries in a qualitative fashion.

C. Turbulence Generation and Effects

Jet behavior is typically categorized and correlated by its Reynolds number $Re = U_0 D/\nu$, defined using initial average flow speed (U_0), the fluid viscosity (ν) and the characteristic length that is the nozzle exit diameter D or twice the slot width, $2B$ (the slot jet hydraulic diameter). At $Re < 1000$ the flow field exhibits laminar flow properties. At $Re > 3000$ the flow has fully turbulent features. A transition region occurs with $1000 < Re < 3000$ [5]. Turbulence has a large effect on the heat and mass transfer rates. Fully laminar jets are amenable to analytical solution, but such jets provide less heat transfer at a given flow rate than turbulent ones, and therefore much more literature exists for turbulent impinging jets.

For example, an isolated round jet at $Re = 2000$ (transition to turbulence), $Pr = 0.7$, $H/D = 6$ will deliver an average Nu of 19 over a circular target spanning six jet diameters, while at $Re = 100,000$ the average Nu on the same target will reach 212 [2]. In contrast, laminar jets at close target spacing will give Nu values in the range of 2–20. In general, the exponent b in the relationship $Nu \propto Re^b$ ranges from $b = 0.5$ for low-speed flows with a low-turbulence wall jet, up to $b = 0.85$ for high Re flows with a turbulence-dominated wall jet. As an example of the possible extremes, Rahimi et al. [17] measured local Nu values as high as 1700 for a under-expanded supersonic jet at $Re = (1.028) \times 10^6$.

Typical gas jet installations for heat transfer span a Reynolds number range from 4000 to 80,000. H/D typically ranges from 2 to 12. Ideally, Nu increases as H decreases, so a designer would prefer to select the smallest tolerable H value, noting the effects of exiting flow, manufacturing capabilities, and physical constraints, and then select nozzle size D accordingly. For small-scale turbomachinery applications jet arrays commonly have D values of 0.2–2 mm, while for larger scale industrial applications, jet diameters are commonly in the range of 5–30 mm. The diameter is heavily influenced by manufacturing and assembly capabilities.

Modeling of the turbulent flow, incompressible except for the cases where the Mach number is high, is based on using the well-established mass, momentum, and energy conservation equations based on the velocity, pressure, and temperature:

$$\frac{\partial \overline{u_i}}{\partial x_i} = 0 \tag{7}$$

$$\rho \frac{\partial \overline{u_i}}{\partial t} + \rho \overline{u_i} \frac{\partial \overline{u_j}}{\partial x_j} = -\frac{\partial \overline{p}}{\partial x_i} + \frac{\partial \sigma_{ij}}{\partial x_j} + \frac{\partial \tau_{ij}}{\partial x_j} \tag{8}$$

$$\rho \frac{\partial \overline{u_i}}{\partial t} + \rho \overline{u_i} \frac{\partial \overline{u_j}}{\partial x_j} = -\frac{\partial \overline{p}}{\partial x_i} + \frac{\partial}{\partial x_j}\left[\mu\left(\frac{\partial \overline{u_i}}{\partial x_j} + \frac{\partial \overline{u_j}}{\partial x_i}\right)\right]$$
$$+ \frac{\partial}{\partial x_j}\left(-\rho\overline{u'_i u'_j}\right)(alternate\ form) \tag{9}$$

$$\rho c_p \frac{\partial \overline{T}}{\partial t} + \rho c_p \overline{u_j} \frac{\partial \overline{T}}{\partial x_j} = \sigma_{ij} \frac{\partial \overline{u_i}}{\partial x_j} + \frac{\partial}{\partial x_j}\left(\frac{\mu c_p}{Pr} \frac{\partial \overline{T}}{\partial x_j}\right) + \frac{\partial}{\partial x_j}\left(-\rho c_p \overline{u'_j T'}\right)$$
$$+ \mu \left(\frac{\partial u'_i}{\partial x_j} + \frac{\partial u'_j}{\partial x_i}\right) \frac{\partial u'_i}{\partial x_j} \tag{10}$$

$$\sigma_{ij} = \mu\left(\frac{\partial \overline{u_i}}{\partial x_j} + \frac{\partial \overline{u_j}}{\partial x_i}\right) \tag{11}$$

$$\tau_{ij} = -\rho\overline{u'_i u'_j} \tag{12}$$

where an overbar above a single letter represents a time-averaged term, terms with a prime symbol (') represent fluctuating values, and a large overbar represents a correlation.

The second moment of the time variant momentum equation, adjusted to extract the fluctuating portion of the flow field, yields the conservative transport equation for Reynolds stresses, shown for an incompressible fluid [18]:

$$\frac{\partial \tau_{ij}}{\partial t} + \bar{u}_k \frac{\partial \tau_{ij}}{\partial x_k} = \left[-\tau_{ik} \frac{\partial \bar{u}_j}{\partial x_k} - \tau_{jk} \frac{\partial \bar{u}_i}{\partial x_k} \right] + \left[\overline{\frac{p'}{\rho} \left(\frac{\partial u'_i}{\partial x_j} + \frac{\partial u'_j}{\partial x_i} \right)} \right]$$
$$+ \left[\frac{\partial}{\partial x_k} \left(-\overline{u'_i u'_j u'_k} - \frac{\overline{p'}}{\rho} \left\{ u'_i \delta_{jk} + u'_j \delta_{ik} \right\} \right) \right]$$
$$+ \left[-2\nu \overline{\frac{\partial u'_i}{\partial x_k} \frac{\partial u'_j}{\partial x_k}} \right] + \left[\nu \frac{\partial^2 \tau_{ij}}{\partial x_k \partial x_k} \right] \quad (13)$$

Each term of this equation has a specific significance.

- The term $\frac{\partial \tau_{ij}}{\partial t} + \bar{u}_k \frac{\partial \tau_{ij}}{\partial x_k}$ represents convective transport of Reynolds stresses.
- The term $-\tau_{ik} \frac{\partial \bar{u}_j}{\partial x_k} - \tau_{jk} \frac{\partial \bar{u}_i}{\partial x_k}$ measures turbulent production of Reynolds stresses.
- The term $\overline{\frac{p'}{\rho} \left(\frac{\partial u'_i}{\partial x_j} + \frac{\partial u'_j}{\partial x_i} \right)}$ measures the contribution of the pressure-strain rate correlation to Reynolds stresses.
- The term $\frac{\partial}{\partial x_k} \left(-\overline{u'_i u'_j u'_k} - \frac{\overline{p'}}{\rho} \left\{ u'_i \delta_{jk} + u'_j \delta_{ik} \right\} \right)$ gives the effects of the gradient of turbulent diffusion.
- The term $-2\nu \overline{\frac{\partial u'_i}{\partial x_k} \frac{\partial u'_j}{\partial x_k}}$ represents the effects of turbulent dissipation.
- The term $\nu \frac{\partial^2 \tau_{ij}}{\partial x_k \partial x_k}$ represents the effects of molecular diffusion.

The specific turbulent kinetic energy k, gives a measure of the intensity of the turbulent flow field. This can be nondimensionalized by dividing it by the time-averaged kinetic energy of the flow to give the turbulence intensity, based on a velocity ratio:

$$Tu = \sqrt{\frac{\overline{u'_j u'_j}}{\bar{u}_i \bar{u}_i}} \quad (14)$$

In addition to generation in the impinging jet flow field itself, turbulence in the flow field may also be generated upstream of the nozzle exit and convected into the flow. This often takes place due to the coolant flow distribution configuration, but can also be forced for increasing the heat transfer coefficients, by inserting various screens, tabs, or other obstructions in the jet supply pipe upstream of or at the nozzle. Experimental work has shown that this decreases the length of the jet core region, thus reducing the

H/D at which the maximal Nu_{avg} is reached [19]. The downstream flow and heat transfer characteristics are sensitive to both the steady time-averaged nozzle velocity profile and fluctuations in the velocity over time. Knowledge of these turbulent fluctuations and the ability to model them, including associated length scales, are vital for understanding and comparing the behavior and performance of impinging jets.

In the initial jet region the primary source of turbulence is the shear flow on the edges of the jet. This shear layer may start as thin as a knife-edge on a sharp nozzle, but naturally grows in area along the axis of the jet. At higher Reynolds numbers, the shear layer generates flow instability, similar to the Kelvin–Helmholtz instability. Figure 4 presents in a qualitative fashion the experimentally observed pattern of motion at the edges of the unstable free jet. At high flow speeds ($Re > 1000$) the destabilizing effects of shear forces may overcome the stabilizing effect of fluid viscosity/momentum diffusion. The position of the shear layer and its velocity profile may develop oscillations in space, seemingly wandering from side to side over time. Further downstream, the magnitude and spatial extent of the oscillations grow to form large-scale eddies along the sides of the jet. The largest eddies have a length scale of the same order of magnitude as the jet diameter and persist until they either independently break up into smaller eddies or meet and interact with other downstream flow features. The pressure field of the stagnation region further stretches and distorts the eddies, displacing them laterally until they arrive at the wall.

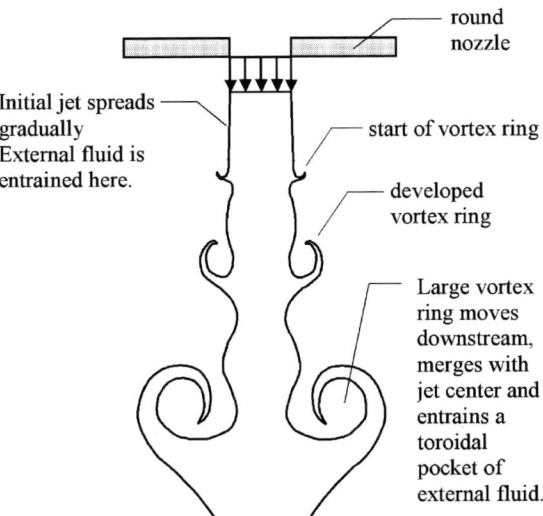

FIG. 4. Instability in the turbulent free jet.

Experiments by Hoogendorn [20] found that the development of turbulence in the free jet affected the profile of the local Nu on the target stagnation region as well as the magnitude. For pipe nozzles and for contoured nozzles at high spacing ($z/D > 5$) the Nu profiles had a peak directly under the jet axis. For contoured nozzles at $z/D = 2$ and 4 with low initial turbulence ($Tu \sim 1\%$), the maximum Nu occurred in the range $0.4 < r/D < 0.6$ with a local minimum at $r = 0$, typically 95% of the peak value.

In the decaying jet region the shear layer extends throughout the center of the jet. This shearing promotes flow turbulence, but on smaller scales. The flow in the decaying jet may form small eddies and turbulent pockets within the center of the jet, eventually developing into a unstructured turbulent flow field with little or no coherent structures in the entire jet core.

In the deceleration region, additional mechanisms take part in influencing flow field turbulence. The pressure gradients within the flow field cause the flow to turn, influencing the shear layer and turning and stretching large-scale structures. The deceleration of the flow creates normal strains and stresses, which promote turbulence. Numerical models by Abe and Suga [21] showed that the transport of heat or mass in this region is dominated by large-scale eddies, in contrast to the developed wall jet where shear strains dominate.

The flow traveling along the wall may make a transition to turbulence in the fashion of a regular parallel wall jet, beginning with a laminar flow boundary layer region and then reaching turbulence at some lateral position on the wall away from the jet axis. For transitional and turbulent jets, the flow approaching the wall already has substantial turbulence. This turbulent flow field may contain large fluctuations in the velocity component normal to the wall, a phenomenon distinctly different than those of wall-parallel shear flows [22].

Large-scale turbulent flow structures in the free jet have a great effect upon transfer coefficients in the stagnation region and wall jet. The vortices formed in the free jet-shearing layer, categorized as primary vortices, may penetrate into the boundary layer and exchange fluids of differing kinetic energy and temperature (or concentration). The ability of the primary vortex to dynamically scrub away the boundary layer as it travels against and along the wall increases the local heat and mass transfer.

The turbulent flow field along the wall may also cause formation of additional vortices categorized as secondary vortices. Turbulent fluctuations in lateral/radial velocity and associated pressure gradient fluctuations can produce local flow reversals along the wall, initiating separation and the formation of the secondary vortices, as shown in Fig. 5. Secondary vortices cause local rises in heat/mass transfer rates and like the primary vortices

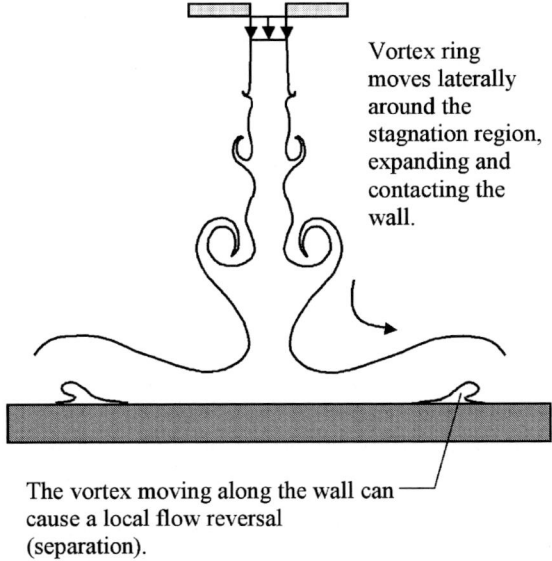

FIG. 5. Vortex motion in the impinging jet.

they result in overall loss of flow kinetic energy and after they disperse downstream may cause local regions of lower transfer rate. Gardon and Akfirat [23] noted a secondary peak in the transfer coefficient and attributed this to boundary layer transition along the wall. More recent studies at various Reynolds numbers concluded that large-scale vortex activity along the wall may generate a secondary peak in transfer coefficients and causes most of the variation in Nu over time [24]. Some investigations suggest that the turbulence in this region is generated by increased shear forces in the thin accelerating region immediately outside the stagnation region [25]. Time-averaged numerical modeling by the authors for $H/D = 2$ showed that the shear layer in the upper portion of the wall jet generates the majority of turbulence in the flow field. This high-turbulence region grows streamwise and also spreads in the wall-normal direction. The location of the secondary peak coincides with the location of highest turbulent kinetic energy adjacent to the wall. This numerically predicted effect correlates well with the findings of Narayanan et al. [26], who found that maximum Nu occurred in regions with high outer wall-jet region turbulence, rather than in regions exhibiting high turbulence only in the near-wall portion of the wall jet. Their specific conclusion was that the outer region turbulence caused an unsteadiness in the thermal boundary layer outside of the stagnation region.

D. Jet Geometry

These flow and turbulence effects, and the heat or mass transfer rates, are strongly influenced by the geometry of the impinging jet device. These include tubes or channels, and orifices (frequently as a perforated flat plate). The use of an enlarged plenum upstream of the orifice (Fig. 6) serves to dampen supply pressure oscillations, to smooth supply velocity and temperature profiles, and may form an important part of the structure of the jet impingement device. Tests by Lee and Lee [12] demonstrated that orifice nozzles produce higher heat transfer rates than a fully developed pipe flow at all radial positions, with local Nu increases of up to 65% at $H/D = 2$ and up to 30% at $H/D = 10$. This difference between the nozzle types becomes larger at decreasing H/D values.

A slot jet, shown in Fig. 7, provides a heat or mass transfer pattern that varies primarily in one spatial dimension on the target wall, an advantage when uniformity of transfer coefficient is desired, but presents some structural disadvantages relative to a round nozzle array orifice plate. The use of multiple nozzles to cover a target surface offers some improvements in efficiency and uniformity of transfer properties with both two-dimensional (slot) and three-dimensional nozzle geometries. For a typical single-round impinging jet the Nu values can vary by a factor of 4 or 5 from $r/D = 0$ to 9. The incorporation of a nozzle array can reduce this variation to a factor of 2. An array of pipe nozzles requires more effort to manufacture, but can also provide useful pathways for exiting flow.

E. The Effect of Jet Pitch: The Jet–Jet Interaction

The pitch p_{jet}, or center-to-center positioning of jets in an array, determines the degree of jet interaction. For jets spaced at pitch-to-diameter

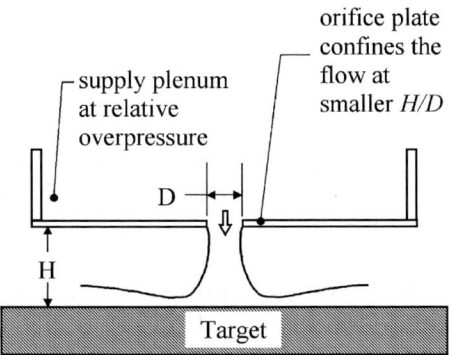

Fig. 6. Orifice plate nozzle and supply plenum.

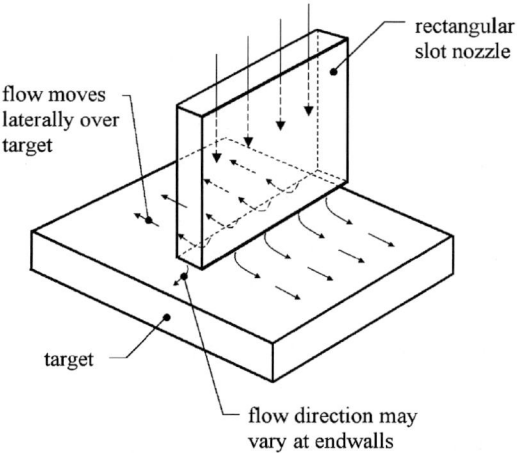

FIG. 7. Slot jet schematic.

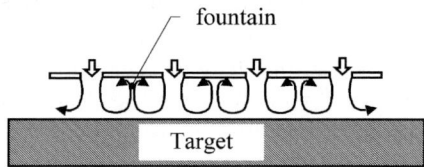

FIG. 8. Typical circulation pattern in the confined jet array.

ratio $p_{jet}/D < 4$, the jets show significant interaction. San and Lai [27] showed that for $H/D = 2$, the interference persisted up to spacings of $p_{jet}/D = 8$ or 10, and the maximal Nu occurred at $p_{jet}/D = 8$.

1. *Shear Layer Interference*

At the small p_{jet}/D for moderate-to-large jet lengths (such as $p_{jet}/D < 2$, $H/D \geqslant 6$), the growing shear-layer jet boundaries may influence each other. If the two neighboring shear layers grow and combine then the velocity gradient at the edge of the jet decreases in magnitude, reducing further turbulence generation and interfering with the generation of large-scale eddies. At the target impingement plate, the wall jets of two adjacent flows may collide, resulting in another local stagnation region or boundary layer separation, and a turning of the flow away from the wall into a "fountain" shape, shown in Fig. 8. This fountain effect can alter transfer rates in the location of colliding wall-jets, and for the highly constrained jets ($H/D < 2$) it may influence the free jet-shearing layer. If the fountain flow exchanges momentum with the free jet-shearing layer, the surface transfer rates are found to decrease [28]. In cases where the fountain is far from the free jet (large p_{jet}/D), the highly turbulent region beneath the fountain may have higher heat transfer rates than the upstream wall jet, with transfer rates of the same order of magnitude as those produced in the stagnation region.

2. Small Jet Lengths and Highly Confined Jets

Jet interaction in an array is also affected by the ratio of jet nozzle diameter to target wall spacing, H/D. At $H/D < 0.25$, the fountain effect may not occur, but other undesirable flow patterns may develop. The mild fountain effects occurring for $H/D > 1$ have a minimal effect on heat transfer, so the region in which fountain effects may degrade transfer is in the range $0.25 \leqslant H/D \leqslant 1$ [25]. At low values of H/D and low values of p_{jet}/D, the flow delivered by a single jet has limited space in which to exit the impingement region. As the flow travels along its exit pathway(s), the wall jets form a crossflow that encounters other jets. The presence of the crossflow causes asymmetric jet flow fields, moves stagnation points, disturbs other wall jets, and in general results in thicker boundary layers and reduced average transfer rates. The jet-to-jet interaction may not have a large influence on the peak Nu value, but the averaged Nu value shows a decrease due to the interference [29]. In experimental work with confined jet arrays, Huber and Viskanta [30] found that the major causes for the decrease in Nu due to jet–jet interaction take place before impingement, rather than in the wall jet or fountain region. With sufficiently small H/D, the flow pattern changes to give peak transfer coefficients off the central axis. For a round jet in a confined flow at $H/D = 2$, Huber and Viskanta found the primary Nu peak in a ring-shaped region at $r/D = 0.5$, as well as a secondary peak at $r/D = 1.6$.

3. Spacing and Interaction

As a rule of thumb, the jet interaction plays a minor role for $p_{jet}/D > 8$ and $H/D > 2$, and the undesired interference will increase as p_{jet}/D and H/D decrease from these values. Experiments have shown that the interference is much less sensitive to Re than to these two geometric ratios [27]. Table II summarizes the expected interaction effects at various H/D ranges.

4. Crossflow and Target Motion

The presence of a crossflow tends to disturb the impinging jet pattern, thicken wall boundary layers, and degrade transfer rates. Experimental work by Chambers et al. [31] modeled a turbine blade cooling channel

TABLE II
Jet Height and Spacing Effects

H/D	Effect upon jet array
Up to 0.25	Highly constrained flow, may have strong crossflow and high additional backpressure (on the order of magnitude of the nozzle exit dynamic pressure). Additional flow acceleration expected to shift peak Nu laterally by 0.5–$1.5D$
0.25–1.0	Fountain flow may greatly affect heat transfer in confined arrays
1–2	Mild fountain effects may occur. Minor turbulence generation. Flow will be affected by confinement wall, need to ensure a clear exit pathway
2–8	Shear layers may interact, need to maintain sufficient p_{jet}. Best performance tends to lie in this range
8–12	Minimal confinement effect is overshadowed by nozzle type. Need to ensure that neighboring jets remain separate
12+	Confining wall does not influence flow, instead nozzle type and jet spacing dominate the flow field. Nu affected by jet energy loss approaching the wall. Need to ensure that neighboring jets remain separate

including the crossflow due to spent flow passing thorough a confined channel. They found that as crossflow ratio (channel flow speed divided by jet speed) increased, the heat transfer patterns approached those of a film cooling design, with no impingement heat transfer enhancement at high crossflows. Centrifugal acceleration in rotating turbomachinery generates nonuniform supply pressures and generates secondary flows which add to crossflow effects in addition to creating radial variations in jet Re and Nu.

A series of experiments by L.W. Florschuetz et al. [32–34] mapped out the effects of jet nozzle spacing and the resulting crossflows in jet arrays. The results showed the immediate benefits of decreasing D and allowing space between jets for the channeling of spent flow. They also measured downstream displacement of heat transfer maxima caused by crossflow.

In some applications, the jet impinges on a moving target, which makes the heat or mass transfer in the direction of motion more uniform. Continuous industrial processes such as drying of paper or rolling of sheet stock or external heat transfer to rotating parts require the solid material to move. Selection of an effective speed depends on the jet spacing or spatial extent of nonuniformities as well as a time constant associated with the rate at which heat or mass can be transferred to or from the target. For small lateral translation speeds compared to the jet speed, target wall speed $<0.2U$, the motion has little effect on the fluid flow. When the wall velocity is higher, the effect is like that of superposing a crossflow. In a jet array this tends to decrease overall heat transfer, but with a single jet some overall benefit has been noted with target motion. Numerical investigation by Chattopadhyay et al. [35] found that for wall speeds greater than U, the turbulence production in the stagnation region is no longer dominated by wall-normal turbulent stresses but rather by wall-parallel velocity gradients and fluctuations in the wall-parallel fluid velocity component. Though industrial processes may use target speeds up to 10 times the jet speed, it was found that for a single jet a maximum averaged Nu occurred for target speeds at about 1.2 times the jet speed, with the maximum averaged Nu up to 25% higher than that with a stationary target [35].

F. ALTERNATE GEOMETRIES AND DESIGNS

Creative designers have added features to the simple impingement designs described thus far, primarily to obtain higher heat transfer coefficients, at an acceptable pressure drop and energy consumption, constrained by strength, space, and cost considerations. Many nozzle types were tested, including cross- and star-shaped cross-sections, fluted, scarfed, tabbed, and angled nozzles. Some nozzles were structured to **swirl** the flow before allowing it to hit the wall, resulting in a higher flow speed at a given nozzle mass flow [36]. At $H/D>6$, the beneficial effects of swirl tend to be lost, but for $H/D\leqslant 2$ the swirl can make the Nu distribution more uniform, at the cost of a lower peak value [37]. At high swirl the stagnation region heat transfer may decrease due to the formation of a recirculatory region on the target immediately under the swirling jet [38]. Wen [39] coupled a swirling jet flow with an axial vibration of the target plate, thereby increasing the fluid turbulence, promoting separation, and impeding the development of a steady boundary layer. This produced increases in Nu of up to 20% using vibratory frequencies of up to 10 Hz.

The **jet impingement angle** has an effect on heat transfer and was studied often. Impingement at an angle may be needed due to some unique feature of the hardware design, or is motivated by desire to reduce the penalties of jet interaction or to reduce losses in the approaching or spent-flow exit pathway. Inclination of the jet distorts the heat transfer contours, generating elliptical isoclines of Nu. It was found that Nu decreases as the impingement incidence angle becomes smaller than 90° (the normal direction) [40]. The maximum Nu (Nu_0) occurs downstream of the intersection of the nozzle axis and the target. Experiments by Sparrow and Lovell [41] found a displacement of up to 2.25 D for nozzle angles of 60° off perpendicular, with a corresponding decrease in Sh_0 or Nu_0 of 15–20%. Although the transfer rate pattern was affected by the jet inclination, the area-averaged transfer coefficient decreased by only 15–20%.

Jet arrays with **pulsed jets** generate large-scale eddy patterns around the exit nozzle, resulting in unsteady boundary layers on the target that may produce higher or lower heat transfer coefficients, depending primarily on frequencies, dimensions, and jet Reynolds number. Bart *et al.* [42] demonstrated an increase of up to 20% in Nu for lower pulse frequencies (200–400 Hz) at moderate jet spacings ($H/D = 4$–6). The large scale, unsteady disturbance of the boundary layer can cause local minima and maxima in the heat or mass transfer profile, in general an improvement over a steady more insulative layer. This effect is balanced against the effect of disturbing or breaking up the large-scale eddies before they reach the wall.

Göppert *et al.* [43] investigated the effects of an **unstable precessing jet** over a fixed target plate. As with a pulsed jet, the variation in local fluid velocity over the target prevented the development of a steady boundary layer. This effect was counteracted by an increased tendency of the precessing jet to mix with the surrounding fluid, lose energy, and reach the target at lower velocities than would be found with a stationary jet ($H/D = 25$), resulting in a 50% decrease in wall-jet Nu values.

Nozzles structures may be more complex than the single flow channels described previously. Hwang *et al.* [44] altered the flow pattern in the initial shearing layer by using **coaxial jets**. The entry velocity in a second layer of jet flow surrounding the primary jet was varied to control vortex shedding rate and the persistence of large-scale vortices. Experimental results showed that with high flow speed in the secondary nozzle, the onset of vortex formation was delayed, which increased Nu_0 by up to 25% for higher H/D ($9 < H/D < 16$). Some similar phenomena occur in the case of an annular jet, which produces Nu maxima spaced laterally outward from the center of the jet axis. It also creates a local Nu minimum on the target in the center of the annulus, where the flow washes upwards as a fountain and gets entrained into the surrounding annular jet [45]. The annular nozzle provides a means of widening the jet, and hence the stagnation region where Nu is highest, without increasing the required mass flow. This effect comes at a price, as the annular nozzle will have a higher wetted surface area and thus higher frictional losses and pressure drop. Compared to a cylindrical nozzle, the annular jet nozzle promotes additional turbulence downstream of the nozzle due to the internal and external shearing layers.

Alternate configurations may be dictated by the **target surface** or manufacturing process design, for example when cooling a rounded cylindrical object. Jet impingement on a cylindrical target generates a heat transfer pattern similar to that of a cylinder in crossflow. As expected, the greatest transfer coefficients occur in the stagnation region, and the lowest occur far along the cylinder surface downstream from the impingement point. In turbulent jet flows the minimum transfer rate occurs at cylinder azimuthal angles about 90 degrees from the impingement point with the transfer rate rising again when the boundary layer becomes turbulent further along the cylinder circumference, and often yet again in the turbulent wake on the back half of the cylinder. In slower, transitional flows Nu decreases continually along the cylinder surface, resulting in a minimum Nu directly opposite to the impinging point, in the center of the flow wake. Gau and Chung [46] found that for slot jets on convex targets, Nu_0 was proportional to $(d_{target}/B)^{-0.14}$ for $2 \leqslant H/B \leqslant 8$ and Nu_0 was proportional to $(d_{target}/B)^{-0.15}$ for $8 \leqslant H/B \leqslant 16$. For concave targets they determined Nu_0 was proportional to $(d_{target}/B)^{-0.38}$ for $2 \leqslant H/B \leqslant 16$. At higher curvatures the cylinder in crossflow experiences flow separation on the surface and earlier breakdown of surface eddies, though the wall jet tends to follow the curvature of the wall (stabilizing Coanda effect). Fleischer et al. [47] studied the effects of a jet on a cylinder and found separation angle to depend on not on Re but on cylinder curvature relative to jet diameter. The flow separation from the target surface results in a rapid decrease in heat transfer rates. Slot jet experiments by Gori and Bossi [48] using a cylindrical target found that the maximum average Nu occurred at $H/B = 8$. Convex target curvature reduces turbulence adjacent to the wall as the turbulence of the upper shear layer of the wall jet must spread farther laterally. Purely convective transport will carry the shear layer around the curved target but will not contribute to the increase in turbulence in the lower portion of the wall jet.

Cooling by impinging jet yields higher heat transfer rates than immersing a cylindrical target in a uniform flow field having the same flow speed (same Re). McDaniel and Webb [49] found that the average transfer rate on the surface was up to 40% higher than that of the parallel flow case for a rounded nozzle, and up to 100% higher with a sharp orifice nozzle. Compared to a uniform freestream flow, the impinging jet nozzle accelerates the flow on approach to the target cylinder, and adds turbulence in the approach to the stagnation region.

Concave target surfaces, such as those found in the front of an impingement-cooled turbine blade, present different flow fields. In general, the resulting pressure gradients tend to reduce the boundary layer growth and improve heat transfer. As summarized by Han et al. [7], the shape of the Nusselt number profile in the concave leading edge of a turbine blade has only weak dependence on H or Re. Nu/Nu_0 drops to only 0.4 at a lateral position of $x/D = 4$, with a flatter profile shape (smaller dNu/dx) than that seen on convex or flat target surfaces. Surface-curvature-forced recirculation of the jet flow in and around the stagnation region tends to disperse large-scale turbulent structures more rapidly, leading to a more unsteady turbulent wall flow without large coherent structures [50].

Experimental work with a **slot jet** showed the peak Nu moved outwards to 0.4–0.6 jet widths off of the central axis, with a local depression in Nu at the stagnation point of 7–10% below maximum Nu [51]. A series of experiments by Taslim et al. [52] studied the effects on Nu of geometric changes to a concave target representing the leading edge of a turbine blade. They found that the inclusion of conical bumps on the target surface did not improve the heat transfer per unit area, but increased the target heat transfer area while maintaining the same Nu, which generated an overall increase in heat transfer rate of up to 40%.

Impinging jets are used with other target modifications such as **ribbed walls**. The inclusion of ribs on the target surface disturbs the wall jet and increases turbulence. The ribs may also function as fins to increase the effective surface area for transfer of energy. While the ribs increase the transfer rates outside the stagnation region, the heightened drag of the ribs causes the wall jet to decelerate and disperse more rapidly, decreasing Nu far from the stagnation region. Experiments by Gau and Lee [53] found that in regions of low jet turbulence the space between the ribs can fill with an "air bubble" (a local region of low speed flow), which reduced Nu by 20–50%. By reducing rib height to 15% of the nozzle width and raising the nozzle to $H/D > 6$ they were able to set up a recirculating flow in these spaces and improve local Nu in the stagnation region. The resulting value of Nu_0 at $H/D = 10$ increased by up to 30% compared to a flat target.

A series of **additional holes in the fluid supply plate** of an orifice array, designed for the spent flow, can provide benefits in cases with restrictive exit pathways. These **effusion holes** vent to exit ducting or the surroundings to provide a lower-restriction exit pathway for spent air. The addition of effusion holes increases average transfer rates for $H/D < 2$ [54]. Rhee et al. [54] showed a 50% improvement in jet array average transfer rate at $H/D = 0.5$ by adding effusion holes, but noted minimal influence for $2 < H/D < 10$. The integration of a low-loss exit pathway into an impingement cooling device poses a design problem for which some simple solutions exist, but many more efficient approaches are possible. In a turbine blade the preferred effusion pathways are either through holes in the target wall itself (the blade exterior) to form a film cooling layer on the opposing surface, or through the confined flow region leading to aerodynamically favorable exit holes on or near the trailing edge of the blade.

Impinging jets are not limited to single-phase flows. Impinging **gas jets containing liquid droplets**, for example in the case of engine fuel injection, deliver a large increase in heat transfer resulting from the inclusion of the droplets. Numerical modeling by Li et al. [55] showed conduction from the wall to the droplets served as the major heat transfer path in the apparatus, with heat transfer rates raised by up to 50% at a liquid–gas mass ratio of

10%. In the extreme case, a high enough liquid content in a gaseous environment will result in a free surface jet, rather than a submerged jet, and only a small influence due to shearing forces between the gas and the liquid surfaces. The inclusion of small **solid particles** in an impinging jet flow will improve heat transfer as the particles impact the target [56]. In the stagnation region, the rebounding particles will increase the turbulence level. This technique has been shown to raise Nu_0 by up to 2.7 times that seen with a single-phase flow. In the wall jet, the particles will disturb the boundary layer and can thus increase heat transfer. High volume fractions of large particles can have a detrimental effect as they disrupt the formation and transport of large-scale eddies in the flow field [57].

The case of **flame impingement** forms a special subclass of jet impingement [5]. A jet carrying a reacting fuel transfers heat very effectively and tends to have higher turbulence. If some fuel travels through the stagnation region without complete combustion, further reaction in the wall jet will release additional thermal energy and improve the uniformity of the heat transfer. In addition to the convective mechanism, a flame impingement device also transfers heat to the surface by radiation from the flame. Experiments by Malikov et al. [58] demonstrated a flame impingement device in which convective heat transfer accounted for 60–70% of the heat transfer, with the remaining heat transfer primarily by radiation. For typical hydrocarbon fuels such as methane the concentration of fuel and of any supplemental oxygen in the flow (fuel–air equivalence ratio) must be controlled to limit accumulation of soot on the target surface, which will ultimately impede heat transfer. Interaction of neighboring jets may also alter the fuel concentration pattern and ultimately degrade heat transfer, including in wall jet fountain flows [59]. A number of analytical tools for predicting flame impingement heat transfer were published by Baukal and Gebhart [60,61].

III. Research Methods

Current research in improving impingement jet performance and predicting jet behavior falls into two categories, experimental and numerical. Where possible, researchers use experimental results to assist in the development and validation of numerical tools for predicting flow and heat transfer behavior.

A. EXPERIMENTAL TECHNIQUES

Impingement heat transfer experiments focus on measuring the flow field characteristics and the surface heat transfer coefficients. In an experiment a single jet or jet array is constructed and positioned above a solid target such as a plate or cylindrical surface. A pump or blower forces fluid onto the target plate while instrumentation collects information about fluid properties and target surface properties.

The target surface used is often supplied by a controlled constant heat flux (q'') using electrically resistive embedded heating strips or a resistive film in the target plate. To reduce lateral heat transfer within the target plate, which would make measurements more complicated, low thermal conductivity and/or thin target materials are chosen. The back side of the target is insulated to minimize energy loss to the surroundings and resultant uncertainty in the heat flux into the fluid. The end result is that the energy generated travels into the fluid in a one-dimensional pathway normal to the surface, with a uniform heat flux.

The surface heat transfer coefficient is evaluated using the equation

$$h = \frac{q''}{(T_s - T_f)} \quad (15)$$

where T_s and T_f are the measured surface and far-field fluid temperatures, respectively. The temperature measurement is performed either by conventional temperature sensors at discrete locations, or for the entire surface at once by non-contact optical devices such as infrared (IR) radiometers, thermally sensitive paints, and frequently by thermochromic liquid crystals (TLC) [62–64] that change their color with temperature.

When using thermocouples and other direct temperature sensors, they must be embedded in or carefully bonded to the target surface to give a proper reading. Rather than install a large array of thermocouples, some experiments use a single sensor mounted on the target, and move the target, thus positioning the sensor at different lateral positions relative to the jet axis. Such an approach is good only for large targets with small variation in their surface heat flux and small lateral thermal conductance.

TLC or other "temperature paints" change color with temperature. As part of the experiment, reference measurements are made to calibrate the paint response and build calibration curves of surface temperature vs. paint color [65].

This method allows the researcher to rapidly capture a complete high spatial resolution temperature distribution using photography, and process the images during or after the experiment. A drawback of the method is that most TLC materials undergo a full color transition over a narrow range of temperatures (e.g. 2–4 °C), which is often smaller than the full range of temperature on the target. As a solution, many researchers sequentially vary the heat flux in the experiment, causing the indicator color transition to occur at different locations on the target, and then collecting the results in one spatial plot of Nu. The accuracy of this method depends on the assumption that Nu does not vary significantly with heat flux. This method effectively turns an experiment at constant heat flux into an experiment with small, almost constant wall temperature, with the intent of producing equally valid results. Alternately, the experiment may use multiple closely spaced temperature paints with different transition ranges, sacrificing some spatial resolution [66]. Ultimately, the TLC must be selected based on ease of application and use, color transition sensitivity, and transition temperature range, balancing this against the difficulty in carefully maintaining fluid supply temperature and the error associated with making measurements at a small local heat flux or various heat fluxes (a potential consequence of using a TLC with a narrow temperature range).

Temperature patterns on the surface may be determined by measuring the intensity and spectral characteristics of the radiation emitted from the target. Many experimenters have successfully collected temperature data using long-wave IR cameras. For such an application, the surface is coated with a paint of known emissivity, and the instrumentation setup is calibrated using thermocouples, optical targets, or other sensors to define a calibration curve of temperature vs. radiation intensity. Commercial IR cameras come with software to collect, store, and process images as well as provide temperature scales given information about target emissivity. With the continuing improvements in digital photography, both this approach and the temperature paint/TLC methods are well-suited to digital post-processing to preserve a high level of spatial resolution in the data.

For mass transfer experiments the target surface is usually maintained at a set mass concentration, and the experiment tracks the amount of material removed over time. The target surface has a coating of a chemical film such as naphthalene. After running the apparatus for a set period of time, measurements are made to determine the change in the thickness of the chemical film, and therefore the rate of mass transfer. This yields a two-dimensional map of the Sherwood number on the target surface. Other coarser approaches have used a porous target material containing a known level of chemical, such as cloth strips soaked in water, and weighed the target material to determine the mass change over time.

A great number of techniques for measuring the flow field have been applied, including visualization by using smoke, paint, or other tracers, by Pitot and static probes, and hot-wire anemometers for discrete location measurements, and by the essentially noninvasive particle image velocimetry (PIV), particle tracking velocimetry (PTV), and laser doppler anemometry (LDA) for the entire viewed flow field.

For simple flow field visualization, the use of smoke, paint, and tracer particles provide clear visual pictures of initial streaklines, large-scale eddies, regions of flow separation and reversal, and onset of turbulence. Coarser time-averaged data providing fluid static and dynamic pressure can be tracked using Pitot and static probes, but these are used less due to the availability of smaller and more precise and sensitive devices.

For detailed, local measurements of fluid velocity, the hot-wire anemometer (HWA) remains the sensor of choice. The wire may be traversed through the flow field to construct a three-dimensional field of data. By utilizing one or more wires at different orientations, one can measure multiple velocity components. With a thin wire and a circuit delivering rapid time response, the anemometer can capture the rapid velocity variations associated with a turbulent flow. Instrumentation programs take the "jittery" time histories recorded by the HWA and process the signal, producing turbulent velocity (r.m.s.), turbulence intensity, and even frequency–amplitude plots of turbulent flow.

PIV and PTV techniques use small tracer particles to determine fluid velocity. By illuminating a thin layer of fluid using a sheet of laser light, the particles in a planar slice of the flow can be photographed at small time intervals. The change in position of a discrete particle between successive frames gives an indication of its velocity within the two-dimensional reference frame of the photograph. The velocity field through the whole flow field is measured by moving the light sheet to different positions within the flow. A velocity field with three components may be established by taking "slices" in multiple directions, but this requires a more complex experimental apparatus and is not performed as frequently as 2-D PIV. Experiments also use other tracer-based techniques such as Laser Doppler Anemometry.

The experimental results are typically correlated by the form

$$Nu = C Re^n Pr^m f(H/D) \qquad (16)$$

where $f(H/D)$ is an empirically determined function and C, n, and m are constants determined by experiment. Additional dimensionless parameters may be added to the correlation to account for other important effects, such as jet pitch, angle of incidence, surface curvature, pressure loss, etc.

B. Modeling

The designer of an impinging jet device needs to predict the transfer coefficient profile (Nu), necessary fluid flow (G), and pressure drops in advance of manufacturing the hardware. Highly accurate models or calculation methods are desirable as they minimize the amount of testing required. A reliable set of models provides the designer with a rapid, inexpensive, and flexible alternative to conducting a series of hardware tests.

1. Empirical Correlations

First, simple correlations such as those supplied by Martin [2] (with a summary in [1]) predict Nu as a function of the governing parameters in cases where the fluid has a continuously laminar flow ($Re_{jet} < 1,000$, $Re_{wall} < 10,000$) over the entire fluid and target region of interest. A survey of available impingement heat transfer correlations is collected in the appendix of this review.

2. Laminar Impingement

For laminar flows in many geometries, the governing equations may be reduced to analytical solutions, such as that for a stagnating flow field placed above a wall boundary layer [67]. Numerical modeling of steady laminar flows is fairly straightforward, using the mass, momentum and energy conservation equations in time-invariant forms. This simulation approach may even yield useful results for flows which are laminar over most but not all of the domain. Kang and Greif [68] successfully predicted flow field properties, separation locations, and heat transfer coefficients for impinging jets on cylinders for $100 \leq Re \leq 1000$, including exploration of buoyancy effects.

3. Turbulent Impingement Models

Most impinging jet industrial applications involve turbulent flow in the whole domain downstream of the nozzle, and modeling turbulent flow presents the greatest challenge in the effort to rapidly and accurately predict the behavior of turbulent jets. Numerical modeling of impinging jet flows and heat transfer is employed widely for prediction, sensitivity analysis, and device design. Finite element, finite difference, and finite volume computational fluid dynamics (CFD) models of impinging jets have succeeded in making rough predictions of heat transfer coefficients and velocity fields. The difficulties in accurately predicting velocities and transfer coefficients stem primarily from modeling of turbulence and the interaction of the turbulent flow field with the wall.

The computation grid must resolve both the upstream and downstream flow around the nozzles or orifices and must extend sufficiently far to the side of a single jet or array (typically eight to ten diameters) to provide realistic exit conditions. Zero-gradient and constant-static-pressure conditions have been used at the far-field model boundaries. Successful, stable modeling using both of these conditions can depend on properly shaping the boundary at the edge of the model domain. Turbulent impinging jet CFD employs practically all available numerical methods that will be critically reviewed in the following sections.

An earlier critical review of this topic was conducted by Polat et al. in 1989 [69]. Since that date, the variety of numerical models has grown and computational research has taken on a larger importance in predicting the physical behavior of impinging jets. The continuing increase in computing power has enabled more rapid computation including optimization by parametric variation. An inexpensive desktop computer may solve precise, high-resolution two-dimensional models within a day. Three-dimensional models and unsteady models are now possible without the use of super-computers, and have execution times ranging from several days to several weeks. The examples in the following review are primarily from impinging jet numerical modeling conducted since the original review by Polat et al.

Useful as a theoretically simple approach, the direct numerical simulation (DNS) method is the most complete and physically exact numerical method employed to predict the impinging jet flow field and transfer rates. This method solves the full Navier–Stokes, continuity, and energy/mass diffusion equations using discrete units of time and space, but requires an extremely small grid to fully resolve all the turbulent flow properties, because the microscopic turbulent length scales involved in jet impingement are far smaller than the macroscopic lengths involved (e.g. D_0 or H). The consequently long computation time practically limits the use of DNS to Reynolds numbers much lower than those in the gas turbine impingement heat transfer application. Since the DNS computational time to resolve turbulent eddies grows with the local turbulent Reynolds number (R_t) to the third power, this modeling method may be of academic interest for laminar flows but will remain impractical for turbulent jets for the foreseen future. Typical DNS CFD studies, using supercomputers, were limited to Reynolds numbers of the order of 10,000. To represent practical application successfully, the majority of DNS computations were limited to $Re < 1,000$, with even lower limits for highly complex flows.

In an attempt to remedy this situation, some CFD models use Large Eddy Simulation (LES). The time-variant LES approach tracks flow properties with the full equations down to some user-defined length scale (typically the grid spacing), and then uses additional sub-grid-scale equations to describe turbulent flow behavior at smaller scales. The LES method has shown encouraging results and clarified the understanding of formation, propagation, and effects of flow eddies upon the velocity fields and jet transfer characteristics [19,70–72], but it requires high resolution in space for accuracy, may require high resolution in time for stability and accuracy, and therefore still needs a great amount of computing power or time to produce satisfactory solutions for the transitional and turbulent flows of interest here

($Re > 1000$). Modeling by Cziesla et al. [73] demonstrated the ability of LES to predict local Nu under a slot jet within 10% of experimental measurements. The use of LES does not necessarily have an upper or lower limit on Re (though particular codes may be limited to $M \leqslant 1$), but for laminar flows ($Re < 1000$) the influence of turbulence is small enough that the DNS approach offers little improvement in accuracy over the time-averaged techniques detailed in the following sections. For those cases where computational cost is not a primary concern, the LES method offers the greatest information about the impinging jet flow field.

Steady-state time-averaged solution techniques, typically Reynolds-averaged Navier–Stokes (RANS) models, use some version of the Navier–Stokes equations adjusted for the presence of turbulent flow. The majority of RANS models used for jet flows fit into one of two categories, the two-equation eddy-viscosity models and the computationally more costly full second moment closure (SMC) models. Eddy viscosity models treat the turbulent viscosity as a scalar quantity, assuming or forcing an isotropy in the normal stresses [74]. The various full SMC models track all Reynolds stresses or track the various components of an anisotropic turbulent viscosity. These models approximate the Reynolds stresses and heat fluxes using semi-empirical equations based on expected physical trends rather than direct derivations. The semi-empirical equations provide approximations of undetermined terms within the second-moment equations, typically two-parameter correlations. With further manipulation, a series of higher-order-moment equations can be generated, but these more complex models have even more correlation terms and unknowns which require approximate modeling.

4. Near-Wall Treatment

In addition to the portions of the CFD model describing the fluid flow inside the computational domain, the steady and transient models require a description of how the flow behaves next to the wall (the target surface). This part of the model typically plays the major role in properly predicting both the flow and the heat transfer [75]. The fundamental difficulty comes from the need to describe how the turbulent regions of a decelerating flow field interact with the wall, including in the wall boundary layer. A variety of often very different wall reflection terms have been implemented. Numerical solutions have shown that heat transfer rates within the viscous sublayer are of a larger magnitude than outside the layer. The spatial region in which the turbulence models have the greatest difficulty approximating the flow is the same region in which the largest heat and mass gradients occur, and so this region cannot be neglected.

Numerical models of turbulence near the wall commonly feature one of two approaches. In the first approach, the grid near the wall is constructed at sufficiently high resolution to properly resolve flow in the entire viscous sublayer and turbulent boundary layer with turbulence equations intended for use at low cell Reynolds numbers. This requires a model capable of resolving turbulent behaviors very close to the wall, and a large computation effort.

The alternate method uses algebraic equations to relate steady and fluctuating velocity and scalar profiles to wall distance and surrounding fluid properties. These wall functions predict the flow properties in and above the viscous sublayer. This method requires only a single cell in the sublayer, and thus requires less computational time. Relations for high Re parallel flows such as the "Law of the Wall" are based upon flows in different geometry than that of the impinging jet, and may not produce a correct velocity profile near the wall, especially in cases where the flow separates or reverses on the target surface. The standard Law of the Wall is based upon the absence of pressure gradients near or along the wall, clearly a different flow field than that seen in the stagnation region of an impinging jet. The Non-equilibrium Law of the Wall is based upon differing turbulent energy generation and destruction rates and accounts for pressure gradients. Bouainouche et al. [76] performed modeling with various wall equations and concluded that the standard logarithmic Law of the Wall poorly predicted shear stresses (errors of up to -30% in the stagnation region) and that a generalized Non-equilibrium Law of the Wall performed well in the stagnation region but underpredicted wall shear stress in the wall-jet region (errors of up to -12%). Their "Hybrid Law of the Wall" model produced improved results by using the non-equilibrium Law in the stagnation region and switching to the logarithmic law in the wall-jet region.

Esch et al. [77] used a scalable wall function to resolve this problem, enforcing a limit on the length used for calculating the wall shear stress to force the first grid node off of the wall to lie outside the sublayer. At higher resolutions, this method effectively defines the wall surface of the CFD model as the outside of the viscous sublayer. This flexible wall function approach is recommended for the impinging jet application as it reduces computational effort without changing the governing equations.

Specific difficulties arise with the numerical modeling of impinging jets. A number of models reviewed in the following sections, such as k–ε, have been optimized for free-shear flows such as submerged jets. Some models, such as k–ω, perform best in boundary-layer flows such as the wall-jet region. Unfortunately, the impinging jet problem contains both of these as well as significant pressure gradients in the stagnation region. The normal strain, and rise in fluid pressure in the stagnation region, affect the turbulent flow through distinct terms in the second-moment RANS equations. The pressure plays a part in the turbulent diffusion term. The effects of changing pressure play an even greater role in the pressure–strain rate correlation term. Unlike the turbulent diffusion term, which most models focus on approximating, the pressure-strain correlation was usually of secondary interest. As a result, most models produce less accurate predictions for turbulent effects in the stagnation region. The pressure–strain rate correlation is typically divided into two parts, a "slow" pressure strain term tied to the turbulence dissipation rate or gradients of the velocity fluctuations, and a "rapid" term based on the gradient of the time-averaged velocity [18]. A wide variety of equation sets have been implemented to model these terms, with varying success. The two equation eddy-viscosity models, such as k–ε, contract the rank-2 tensors in the equations to eliminate terms, and thus drop these terms. That is, the two-equation models are based around assumptions about the low importance of pressure gradients and the minimal anisotropy of the Reynolds stresses, and experiments have shown that these modeling assumptions do not apply in the stagnation region.

5. The Boussinesq Approximation

The simplified RANS models need some approximation to determine the Reynolds stresses. An equation known as the Boussinesq approximation (or hypothesis) describes a simple relationship between turbulent stresses and mean strain rate. Given a strain rate tensor S_{ij}, where

$$S_{ij} = \frac{1}{2}\left(\frac{\partial \bar{u}_i}{\partial x_j} + \frac{\partial \bar{u}_j}{\partial x_i}\right) \quad (17)$$

the approximation gives a formula for the Reynolds stress tensor:

$$-\rho\overline{u'_i u'_j} = 2\mu'\left(S_{ij} - \frac{1}{3}S_{kk}\delta_{ij}\right) - \frac{2}{3}\rho k \delta_{ij} \quad (18)$$

By itself, the Boussinesq approximation does not constitute a complete turbulence model, as the value of μ' is unknown and depends on turbulence scales unique to each problem.

6. The k–ε Model

The commonly tested "k–ε" eddy viscosity model is widely acknowledged as producing poor results in the impinging jet problem, but remains a benchmark against which to compare better models. The k–ε model remains in use due to its common implementation and comparatively low computational cost. The model uses the Boussinesq hypothesis to calculate the Reynolds stresses as a direct function of the velocity gradients and is based on flow behavior at higher Reynolds numbers (fully turbulent fluid flow). It independently tracks turbulent energy k and turbulence destruction or dissipation rate ε, with a dissipation equation based upon expected trends. As with other RANS models it requires experimentally determined constants to fully close the equations. The k–ε model can produce acceptable results for free-shear flows but provides poor simulation of wall-jet flows. The model requires the user to specify ε at each boundary, but at the walls ε has a finite, non-zero value, which is not known in advance. For the impinging jet problem it gives useful results in the free-jet region, but poor results in the stagnation region and wall jet region, as detailed in the following discussion of experimental work. It gives poor predictions of the location of separation points on solid boundaries and for the impinging jet problem it may fail to predict the occurrence of secondary peaks in Nu. The standard k–ε model is formulated for flows at high Reynolds number. It does not apply in regions where viscous effects on the flow field are comparable in magnitude to turbulent effects (such as in the sublayer next to a wall). In many cases, the model uses wall functions to determine the velocity profiles. Alternately, k–ε models have been built with additional terms and damping functions to allow the model to simulate portions of the flow at low Reynolds numbers.

The following equations form the Launder and Sharma low Reynolds number model used by Craft et al. [75] in a comparative CFD study with only minor notation changes. It incorporates conservation equations for k and ε as well as a simple equation to set the velocity–temperature correlation (heat flux) proportional to the temperature gradient:

$$\frac{Dk}{Dt} = \frac{\partial}{\partial x_j}\left[\left(v + \frac{v'}{\sigma_k}\right)\frac{\partial k}{\partial x_j}\right] + v'\left(\frac{\partial \overline{u}_i}{\partial x_j} + \frac{\partial \overline{u}_j}{\partial x_i}\right)\frac{\partial \overline{u}_i}{\partial x_j} - \varepsilon \qquad (19)$$

$$\frac{D\tilde{\varepsilon}}{Dt} = c_{\varepsilon 1}\frac{\tilde{\varepsilon}v'}{k}\left(\frac{\partial \overline{u}_i}{\partial x_j} + \frac{\partial \overline{u}_j}{\partial x_i}\right)\frac{\partial \overline{u}_i}{\partial x_j} - c_{\varepsilon 2}f_\varepsilon\frac{\tilde{\varepsilon}^2}{k} + 2vv'\left(\frac{\partial^2 \overline{u}_i}{\partial x_j \partial x_k}\right)^2$$
$$+ \frac{\partial}{\partial x_j}\left[\left(v + \frac{v'}{\sigma_\varepsilon}\right)\frac{\partial \tilde{\varepsilon}}{\partial x_j}\right] + Y_c \qquad (20)$$

$$\overline{u_i T} = -\frac{v'}{\sigma_\theta}\frac{\partial \overline{T}}{\partial x_i} \qquad (21)$$

$$Y_c = c_w \left(\frac{k^{3/2}}{c_1 \tilde{\varepsilon} y} - 1\right) \left(\frac{k^{3/2}}{c_1 \tilde{\varepsilon} y}\right) \frac{\tilde{\varepsilon}^2}{k} = \text{Yap correction} \quad (22)$$

$$\varepsilon = \tilde{\varepsilon} + 2\nu \left(\frac{\partial [k^{1/2}]}{\partial x_j}\right)^2 = \text{turbulent kinetic energy dissipation rate}$$
$$(23)$$

$$\nu' = \frac{c_\mu f_\mu k^2}{\tilde{\varepsilon}} = \text{turbulent viscosity} \quad (24)$$

$$f_\mu = \exp\left(\frac{-3.4}{[1+(R_t/50)]^2}\right) = \text{damping function} \quad (25a)$$

$$f_\varepsilon = 1 - 0.3\,\exp(-R_t^2) \quad (25b)$$

$$R_t = \frac{k^2}{\nu \tilde{\varepsilon}} = \text{turbulent Reynolds number} \quad (26)$$

where c_n and σ_n values are empirical constants, given as $c_{\varepsilon 1} = 1.44$, $c_{\varepsilon 2} = 1.92$, $c_\mu = 0.09$, $c_w = 0.83$, $c_1 = 2.5$, $\sigma_k = 1.0$, $\sigma_\varepsilon = 1.3$, and $\sigma_\theta = 0.9$.

This example model includes the Yap correction term to adjust the dissipation rate $\tilde{\varepsilon}$ as a function of k, $\tilde{\varepsilon}$, and distance from the wall y. At low Re the damping function f_μ adds an adjustment to the turbulent viscosity used in the conservation equations. It increases the dissipation to reduce the turbulent length scale. Without the correction, the model will overpredict turbulent length scale and overpredict turbulent viscosity. From the equations and set of constants one can see that the model depends highly on empirical data, and that the correction terms and associated constants are therefore somewhat arbitrary. The adjustment and constants provided incorporate the best knowledge available at the time, but engineers continually invent alternate adjustment terms with different closure coefficients.

Heyerichs and Pollard [78] conducted a numerical comparison of three different wall function and five different wall damping functions with an impinging jet test case and concluded that the selected k–ε models with wall functions gave consistently poor results, with Nu errors in the range of

−21.5 to −27.8% in the stagnation region, and +32 to +38.4% at the secondary peak. Somewhat better matches were produced using models with damping functions, but those models still produced errors in Nu of up to 50% and misplaced the secondary peak. They concluded that basing the damping functions on wall position y^+ caused the poor results, as the damping functions using y^+ were based upon simple wall-parallel flows with simple boundary layers, rather than the flow found in the stagnation region of the impinging jet.

Craft et al. [75] presented a comparison of a 2-D implementation of the k–ε model vs. test data. For the test case at $Re = 23{,}000$ the model predicted centerline wall-normal r.m.s. velocity levels up to four times larger than those measured in the experimental work of Cooper et al. [22]. A specific problem noted in the k–ε model was that the model equation relating turbulent kinetic energy to turbulent viscosity caused increasing and erroneous turbulent kinetic energy levels in the stagnation region (increasing turbulent viscosity caused increasing turbulence intensity). The model similarly overpredicted wall normal r.m.s. velocity at $r/D = 0.5$, corresponding to the edge of the jet. Wall-parallel velocity errors were in the range of 15–20%, with errors of up to 50% in the $y/D < 0.05$ region very close to the wall. The model overpredicted Nu in the center of the impingement region by up to 40% and failed to predict the secondary Nu peak at $r/D = 2$. Craft et al. [79] continued work with this type of model, developing an alternate k–ε model which produced greatly improved impingement centerline wall-normal fluctuating velocity values and better Nu predictions in the $r/D < 2$ region. The largest errors in Nu were typically 15%, occurring in the range of $1 < r/D < 3$. Turgeon and Pelletier [80] built adaptive k–ε models which succeeded in generating a solution with minimal grid dependence, showing that the difficulties with applying the k–ε model are independent of grid resolution and persist for small mesh sizes. Merci et al. [81] devised and tested an altered non-linear variation of the k–ε model, yielding improved results over the standard model but an underprediction of Nu/Nu_0 of up to 25% (alternately interpreted as an overprediction of Nu_0). Souris et al. [82] showed that the upstream errors in low Reynolds number k–ε model predictions resulted in large downstream errors, giving wall jet thicknesses up to double that of experiment, and wall jet peak velocity as much as 44% below experimental results.

Tzeng et al. [83] compared seven low-Re modifications of the k–ε model using a confined turbulent slot jet array problem with three adjacent jets at $H/B = 1$. Each model used different adjustments to the ε equation to account for damping effects near the target wall, based on functions of $k^2/(\nu\varepsilon)$, $y(k^{0.5})/\nu$, and/or $(\nu\varepsilon)^{0.25} y/\nu$. The models were each run with three separate finite differencing formulas, specifically the power law, second-order upwind, and QUICK schemes. Out of this large array of combinations, no single set of results consistently improved the Nu prediction when compared to the standard k–ε model. Instead, various models produced local improvements at the cost of increased error at other positions. In the region around the central jet ($x/B<7$) the standard k–ε model produced better results than any of the low-Re models, matching Nu within 7%. In the regions where exhausting fluid placed impinging jets in a crossflow, the results varied a great deal. The standard k–ε model (QUICK) errors ranged as high as 20% under the impinging slot jet in crossflow. The worst models had Nu prediction errors as great as -65%. The favored model was the low-Re model of Abe, Kondoh, and Nagano [84] plus the QUICK scheme [85], which matched the Nu profile within 15% for crossflow regions of the target, with an error of up to -25% in the central wall-jet region prior to the crossflow.

From the various studies conducted, we can conclude that the even the best k–ε models and associated wall treatments will yield Nu profiles with local errors in the range of 15–30%, and the standard k–ε model is not recommended for use in the impinging jet problem. These shortcomings are attributed to the assumption of isotropic turbulence and the use of wall functions that poorly approximate near-wall velocity fluctuation and associated transport properties.

7. The k–ε RNG Model

Other variations of the model have been applied, such as the Renormalization Group Theory k–ε model (RNG). The RNG model incorporates an additional term in the turbulent energy dissipation equation based on strain rates, and includes adjustments for viscous effects at lower Re and a calculation of turbulent Prandtl number. Heck *et al.* [86] showed the RNG model provided a close match of Nu in the wall-jet region but an error up to 10% in the stagnation region. This is in part due to the RNG model's tendency to predict jet-spreading rates that are as high as twice that found in experiment [87]. This flaw on the upstream end of the model leads one to question how the downstream results did not stray as far from measured values. It offers some improved performance over the standard k–ε at a slightly higher computational cost and is recommended when only moderate accuracy is required.

8. The k–ω Model

The k–ω model solves for turbulence intensity (k) and dissipation rate per unit of turbulent kinetic energy (ω), where ω is determined through a conservation equation including experimentally determined functions, rather than direct calculation from the velocity field. The equations for ω treat it as a vorticity level or vortex fluctuation frequency. The model then produces turbulent viscosity as a function of k and ω:

$$\rho \frac{\partial k}{\partial t} + \rho U_j \frac{\partial k}{\partial x_j} = \tau_{ij} \frac{\partial U_i}{\partial x_j} - \beta^* \rho k \omega + \frac{\partial}{\partial x_j} \left[(\mu + \sigma^* \mu') \frac{\partial k}{\partial x_j} \right] \qquad (27)$$

$$\rho \frac{\partial \omega}{\partial t} + \rho U_j \frac{\partial \omega}{\partial x_j} = \alpha \frac{\omega}{k} \tau_{ij} \frac{\partial U_i}{\partial x_j} - \beta \rho \omega^2 + \frac{\partial}{\partial x_j} \left[(\mu + \sigma \mu') \frac{\partial \omega}{\partial x_j} \right] \qquad (28)$$

$$\mu' = \frac{\rho k}{\omega} \qquad (29)$$

$$\varepsilon = \beta^* \omega k \qquad (30)$$

The symbols α, β, β^*, σ, and σ^* represent constants set at $\alpha = 5/9$, $\beta = 3/40$, $\beta^* = 9/100$, $\sigma = 1/2$, and $\sigma^* = 1/2$. A wide variety of k–ω models have been generated and tested, with many different closure coefficients and corrections.

As with the k–ε model, the latest versions of the k–ω model include correction terms to improve predictions in the low Reynolds number flow regions. The k–ω model typically produces Nu profiles with a local error of up to 30% of the experimental Nu value. It can produce better predictions of the turbulent length scale than the k–ε model. The k–ω model can generate good predictions of flow properties in the wall jet, both in the sublayer and logarithmic region, without the need for damping functions. For a flow near a wall the boundary conditions are known – turbulent viscosity and the turbulent time scale are set to 0. The value of ω at or near the wall cell may be set proportional to v/y^2, meaning the user can fully specify the turbulence conditions at the wall, unlike in the k–ε model. Unfortunately, the k–ω model is sensitive to far-field boundary conditions, much more so than the k–ε model. Park et al. [88] demonstrated some improved results using the k–ω equations but noted that at higher Re (25,100) the secondary Nu peaks appeared too far inward, as low as 50% of the experimentally measured value of x/B. The local levels of Nu were overpredicted by as much as 100% as the result of misplacing this peak. A comparative study by Heyerichs and Pollard [78] found that the k–ω model overpredicted Nu by up to 18% and generated a secondary peak closer to the jet center than found in experiment, but concluded that for the impinging jet problem it clearly outperformed the nine different implementations of the k–ε model used in the study. The low-Re k–ω model gave good results by matching the shape of the experimental curves, but alternate formulations of the impinging jet CFD model using k–ω with wall functions gave poor results – they replaced the k–ω model with a cruder approximation in the very region where it gives the best results, overpredicting wall jet Nu by as much as 40%. Chen and Modi [89] successfully applied the k–ω model for mass transfer at high Sc, and claimed agreement within 10% of experimental results, given very high grid densities. The addition of cross-diffusion terms in various k–ω models have succeeded in reducing its sensitivity to far-field ω boundary conditions, a problem known to arise during use of the k–ω model for unconfined or partially confined flows.

9. Realizability Limits

In cases of high strain rates, the simple Boussinesq approximation may predict negative normal Reynolds stresses ($\overline{u_i^2} < 0$ [no summation], not physically possible) or excessively high Reynolds shear stresses ($\frac{\overline{u_i u_j}}{\overline{u_i^2}\,\overline{u_j^2}} > 1$, no summation). The k–ε, k–ω, and $v^2 f$ models described herein have been commonly modified to use realizability limits to prevent these problems. A common fix is to allow variation in the model constant found in the turbulent viscosity equation, for example using

$$C_\mu = \frac{1}{4.04 + \sqrt{6}\cos\left(\frac{1}{3}\cos^{-1}\left(\sqrt{6}\frac{S_{ij}S_{jk}S_{ki}}{\sqrt{S_{ij}S_{ij}}}\right)\right)\frac{k\sqrt{S_{ij}S_{ij}+\Omega_{ij}\Omega_{ij}}}{\varepsilon}} \quad (31)$$

in a non-rotating reference frame, where $\Omega_{ij} = \frac{1}{2}\left(\frac{\partial U_i}{\partial x_j} - \frac{\partial U_j}{\partial x_i}\right)$ is the fluid rotation rate [90]. This value is then used directly in the equation $v' = C_\mu(k^2/\varepsilon)$. Physical measurements have demonstrated variation in this "constant" in differing fluid flows. Other approaches put simple limits on time scales, length scales, strain rates, and/or terms including strain rates.

Abdon and Sunden [91] used nonlinear k–ε and k–ω models with realizability constraints to model impinging jets. These model adjustments produced results closer to experimental data, with the realizable k–ε model predicting Nu_0 within 10% (within the experimental data scatter) and the realizable k–ω model overpredicting Nu_0 by 20%. Further studies with nonlinear versions of the k–ε and k–ω models produced Nu profiles with errors equal to or greater than those of the standard linear models. The nonlinear models captured a secondary peak in Nu in the proper location at $r/D = 2$, but overpredicted the Nu value by up to 50%. Park and Sung [92] constructed a k–ε–f_μ model for low Re flows, where the turbulent viscosity damping function f_μ incorporated terms to describe damping near the wall and terms to describe the equilibrium flow farther from the wall. With the inclusion of realizability limits on eddy viscosity they were able to improve the Nu profile predictions for $r/D < 1.5$ to within 10–20% of experimental results, primarily by limiting overprediction of turbulent kinetic energy in the jet center. For the region of $r/D < 1$ the model was tuned to predict the Nu profile within 15%, giving a flat profile matching the experimental results. Given the slightly higher computational cost but potentially better results, realizability constraints are recommended for use in impinging jet flow CFD.

10. Algebraic Stress Models

Algebraic Stress Models (ASM) can provide a computationally inexpensive approach valid for some simple flows. The ASM models may be built with lower grid resolution in the wall region, which contributes to the computational efficiency. Rather than solve complete discretized differential transport equations this category of models solves algebraic equations which require fewer calculations. In cases where the turbulent velocity fluctuations change slowly compared to changes in the mean velocity, the Reynolds stresses can be approximated as algebraic functions of the dominant mean velocity derivatives in time and space. In a simple case, the ASM may use equations for calculating a length scale, which are particular to the problem geometry. This length scale is used to calculate turbulent viscosity, which is used with the Boussinesq approximation to determine the Reynolds stresses. Use of this approach requires enough advance knowledge of turbulent length and time scales for the problem of interest that the quantities may be calculated using algebraic equations, a potential source of large error. For simple geometries such as pipe flow or free jets, a set of equations for mixing length is available. Some ASMs simply drop the time and space derivatives of the Reynolds stresses from the equations, leaving only gradients of the mean flow velocity [74]. This approach assumes the turbulent convection and turbulent diffusion effects are either insignificant or are equal in magnitude. Unfortunately, for the impinging jet problem the boundary layer along the wall is not in equilibrium and this type of ASM is a crude approximation.

Comparative modeling by Funazaki and Hachiya showed that for an impingement problem their ASM overpredicted Nu by $\sim$30%, outperforming k–ε and RNG k–ε models which typically showed 50–55% error [93]. Numerical work by Souris et al. [82] found that the ASM had better free-jet modeling than the k–ε model, which generated better results in the wall region downstream. Both models overpredicted the centerline velocity decay, but the ASM overprediction was not as high. The error in jet width prediction of the ASM was as high as 35% close to the wall, better than the

59% error produced by the low Reynolds number version of the k–ε model. In contrast, this ASM model used the standard logarithmic Law of the Wall and generated poor predictions of velocity profile in the region closest to the wall (within the first quarter of the wall-jet thickness), with high jet thicknesses (up to 65% error at $r/D = 2.5$) and wall jet velocity magnitudes as much as 45% below experiment. These results do not mean the ASM correctly described the impinging flow, but rather the k–ε model resulted in gross errors, larger than the errors present when using the ASM. The ASM may be better than a number of poor k–ε models, but is not recommended as it does not yield accuracies that are commensurate with the required computational effort.

11. Complete RSM Modeling

The SMC Reynolds Stress Model (RSM), also known as the Reynolds Stress Transport Model (RSTM), tracks all six independent components of the Reynolds stress tensor, accounting for production, diffusive transport, dissipation, and turbulent transport. Common implementations require a number of constants to resolve terms such as a pressure–strain term and terms in the turbulence dissipation equation. Because the RSM model does not assume isotropic stresses, it can give much better predictions of fluid behavior in turning or rotating flows that those of the two-equations models. RSM modeling of impinging jets by Demuren [94] showed velocity predictions ranging from −40 to +40% of the experimentally measured velocities, and Reynolds stress errors of over 100%, which was attributed to a need for an extremely dense grid (denser than that utilized in the modeling). Craft et al. [75] presented computed centerline wall-normal r.m.s. turbulent velocity levels, which matched within 25% of experiment at $H/D = 2$, but had errors as large as 80–100% for $H/D = 6$. The RSM can predict the occurrence of a secondary peak in Nu, but not necessarily at the correct location [95]. This shows that although the various RSM implementations preserve all the Reynolds stress terms, they still use approximation equations based on a number of assumptions. That is, they eliminate the isotropy assumptions which yield the two-equation models but still rely upon other empirically generated equations to predict the stresses and do not give a "perfect" solution. Given the high computational cost compared to the eddy-viscosity models, these results are disappointing and the RSM is not recommended as an alternative.

12. The v^2–f Model

Durbin's v^2–f model, also known as the "normal velocity relaxation model," has shown some of the best predictions to date, with calculated Nu

values falling within the spread of experimental data [25,96]. The v^2–f model uses an eddy viscosity to increase stability (rather than using a full RSM) with two additional differential equations beyond those of the k–ε model, forming a four-equation model. It uses the turbulent stress normal to the streamlines (referred to as $\overline{v^2}$) to determine the turbulent eddy viscosity, rather than the scalar turbulence intensity used in the k–ε model. It incorporates upper and lower limits on the turbulent time and length scales. The "f" term in the model name refers to an included function to capture the effects of walls upon variations in $\overline{v^2}$. The additional equations are defined as

$$\frac{Dk}{Dt} = \frac{\partial}{\partial x_j}\left[(v+v')\frac{\partial k}{\partial x_j}\right] + 2v'S_{ij}S_{ij} - \varepsilon \tag{32}$$

$$v' = C_\mu \overline{v^2} T_{scale} \tag{33}$$

$$\frac{D\varepsilon}{Dt} = \frac{c'_{\varepsilon 1} 2v' S_{ij}S_{ij} - c_{\varepsilon 2}\varepsilon}{T_{scale}} + \frac{\partial}{\partial x_j}\left[\left(v+\frac{v'}{\sigma_\varepsilon}\right)\frac{\partial \varepsilon}{\partial x_j}\right] \tag{34}$$

$$\frac{D\overline{v^2}}{Dt} = kf_{wall} - \overline{v^2}\frac{\varepsilon}{k} + \frac{\partial}{\partial x_j}\left[(v+v')\frac{\partial \overline{v^2}}{\partial x_j}\right] \tag{35}$$

$$f_{wall} - L^2_{scale}\frac{\partial}{\partial x_j}\frac{\partial f}{\partial x_j} = \frac{(C_1 - 1)\left(\frac{2}{3} - \frac{\overline{v^2}}{k}\right)}{T_{scale}} + \frac{C_2 2v' S_{ij}S_{ij}}{k} \tag{36}$$

$$L_{scale} = C_L \max\left(\min\left(\frac{k^{\frac{3}{2}}}{\varepsilon}, \frac{1}{\sqrt{3}}\frac{k^{\frac{3}{2}}}{\overline{v^2}C_\mu\sqrt{2S_{ij}S_{ij}}}\right), C_\eta\left(\frac{v^3}{\varepsilon}\right)^{\frac{1}{4}}\right) \tag{37}$$

$$T_{scale} = \min\left(\max\left(\frac{k}{\varepsilon}, 6\sqrt{\frac{v}{\varepsilon}}\right), \frac{\alpha}{\sqrt{3}}\frac{k}{\overline{v^2}C_\mu\sqrt{2S_{ij}S_{ij}}}\right) \tag{38}$$

$$c'_{\varepsilon 1} = 1.44\left(1 + 0.045\sqrt{\frac{k}{\overline{v^2}}}\right) \tag{39}$$

where $C_\mu = 0.19$, $c_{\varepsilon 2} = 1.9$, $\sigma_\varepsilon = 1.3$, $C_1 = 1.4$, $C_2 = 0.3$, $C_\eta = 70.0$, $C_L = 0.3$, and $\alpha = 0.6$. Similar equations exist for predicting the transport of a scalar (e.g. thermal energy) with a Pr' as a function of v and v'.

As with the k–ω model, the v^2–f model requires a dense wall grid. In some cases the v^2–f model has been shown to predict realistic levels of turbulence in the decelerating jet core, but excessive turbulence levels in the shearing flow outside the core and in the wall jet [97]. Despite this difficulty and its moderately high computational cost, it is acknowledged as one of the best predictors of Nu distribution. It has an advantage over the standard k–ε series of models because it can predict the occurrence, position, and magnitude of the secondary Nu peak for low H/D. This model is highly recommended for the impinging jet problem, and its moderate computational cost is offset by its ability to closely match experimental results.

13. Hybrid Modeling

The impinging jet problem has at least three distinct flow regions with distinct flow physics. The computationally efficient two-equation models discussed previously are adjusted to perform best in one physical situation, with closure equations and coefficients based on a set of simple turbulent flows. Application to alternate geometries demonstrates the weakness of the model. No simple model has produced the ultimate answer, but by combining two or more models the CFD code can produce a compromise of sorts. For example, the model may calculate in which region the flow lies (free jet, stagnation, or wall jet) and use a model successfully tested for that particular region. The solution from the multiple models in multiple regions must then be combined at the boundaries in a smooth fashion to produce a Hybrid turbulence model. In doing so, the CFD program may utilize the strengths and minimize the weaknesses of each model.

Menter's Shear Stress Transport (SST) model is one of the most successful Hybrid models [98]. The SST model combines the k–ω model near the wall and the k–ε model farther from the wall to utilize the strengths of each. Smooth transition between the two is accomplished by use of a blending or weighting function based upon distance from the wall, formulated by Menter as

$$F_1 = \tanh\left[\left(\min\left\{\max\left[\frac{\sqrt{k}}{0.09\omega y}, \frac{500v}{y^2\omega}\right], \frac{4\rho\sigma_{\omega 2}k}{CD_{k\omega}y^2}\right\}\right)^4\right] \quad (40)$$

$$CD_{k\omega} = \max\left[2\rho\sigma_{\omega 2}\frac{1}{\omega}\frac{\partial k}{\partial x_j}\frac{\partial \omega}{\partial x_j}, 10^{-20}\right] \quad (41)$$

where $\sigma_{\omega 2} = 0.856$, an empirical constant associated with the transformed k–ε model. Menter's SST model uses a variant equation for determining turbulent viscosity based upon

$$v' = \frac{0.31k}{\max\left(0.34\omega, \Omega\left[\max\left(\frac{2\sqrt{k}}{0.09\omega y}, \frac{500v}{y^2\omega}\right)\right]^2\right)} \quad (42)$$

with the goal of improving predictions of turbulence in adverse pressure gradients. The SST model still requires a finely spaced mesh near the wall to produce accurate results. Validation comparisons by Esch *et al.* [77] showed Nu predictions within 20% of experimental results, and a Nu profile no farther than 5% above or below the profile predicted by the v^2–f model. The SST model also predicted mean velocities well, clearly better than the k–ε model and within the uncertainty of the experimental measurements. This indicates the SST model may provide predictions as good as those of the v^2–f model but at a lower computational cost, and it is recommended for this reason.

C. Conclusions Regarding Model Performance

Ultimately, all CFD results should be validated by comparison to reliable experimental results and to determine overall model error in predicting the real situation. The model must match the experimental conditions, including all of the geometry, fluid entry and exit conditions, and target surface properties. This matching must include not only the domain boundary average velocities, pressures, and temperatures, but also their turbulent components.

Experiments to measure turbulence levels and transfer coefficients continue. While many of these explore new geometries and provide more accurate measurements, some of these experiments reproduce the results of earlier work. Sufficient data exist for design of impinging jet devices with simple geometries, such as a single round or slot nozzle, or a simple jet array at large spacing. The most useful new experiments explore new geometries, in particular the less-than-ideal geometries required in functional hardware.

A large number of informative studies have been conducted using the k–ε model to attempt to predict the heat/mass transfer of impinging jets, with only limited success. Examination of RANS numerical modeling techniques showed that even with high-resolution grids, the various implementations of the k–ε, k–ω, RSM, and ASM models give large errors compared to experimental data sets. The v^2–f and SST models can produce better predictions of fluid properties in impinging jet flows and are recommended as the best compromise between solution speed and accuracy. Table III summarizes the relative performance of the various models, rated

TABLE III
Comparison of CFD Turbulence Models Used Impinging Jet Problems

Turbulence model	Computational cost (time required)	Impinging jet transfer coefficient prediction	Ability to predict secondary peak
k–ε	★ ★ ★ ★ Low cost	★ Poor: Nu errors of 15–60%	★ Poor
k–ω	★ ★ ★ ★ Low–moderate	★ ★ Poor–fair: anticipate Nu errors of at least 10–30%	★ ★ Fair: may have incorrect location or magnitude
Realizable k–ε and other k–ε variations	★ ★ ★ ★ Low	★ ★ Poor–fair: expect Nu errors of at 15–30%	★ ★ Poor–fair: may have incorrect location or magnitude
Algebraic stress model	★ ★ ★ ★ Low	★ ★ Poor–fair: anticipate Nu errors of at least 10–30%	★ Poor
Reynolds stress model (full SMC)	★ ★ Moderate–high	★ Poor: anticipate Nu errors of 25–100%	★ ★ Fair: may have incorrect location or magnitude
Shear stress transport (SST), hybrid method	★ ★ ★ Low–moderate	★ ★ ★ Good: typical Nu_0 errors of 20–40%	★ ★ Fair
$v^2 f$	★ ★ ★ Moderate	★ ★ ★ ★ Excellent: anticipate Nu errors of 2–30%	★ ★ ★ ★ Excellent
DNS/LES time-variant models	★ Extremely high (DNS available for low Re only)	★ ★ ★ ★ Good–excellent	★ ★ ★ ★ Good–excellent

[★ indicating undesirable model characteristics, to ★ ★ ★ ★ indicating excellent model characteristics.]

qualitatively on a scale from "★" indicating undesirable model characteristics, to "★ ★ ★ ★" indicating excellent model characteristics.

The actual computational cost will of course vary with model complexity and computing power. With the computing resources of a single desktop computer available at the time of writing, typically 3 GHz processors, for a high-resolution two-dimensional problem the steady time-averaged eddy viscosity models (k–ε, k-ω, SST, v^2–f) will have computation times of a few hours (3–12). In comparison, the more complex RSM could take 4–40 h depending on how smoothly the model converges. Based on recent work, unsteady LES models have computation times at least two orders of magnitude higher; a well-resolved three-dimensional LES impinging jet model could take weeks to provide a solution. As processing power continues to grow and parallel computation becomes more widely available, these calculation times will decrease accordingly.

Conclusions

The review of recent impinging jet research publications identified a series of engineering research tasks important to improving the design and resulting performance of impinging jets:

(1) Clearly resolve the physical mechanisms by which multiple peaks occur in the transfer coefficient profiles, and clarify which mechanism(s) dominate in various geometries and Reynolds number regimes.
(2) Develop a turbulence model, and associated wall treatment if necessary, that reliably and efficiently provides time-averaged transfer coefficients. Given the varied and inaccurate results of the alternatives, the SST and v^2–f models offer the best results for the least amount of computation time. Even so, they are imperfect. The improved turbulence model must correctly predict the jet spreading, turbulent flow effects in the stagnation region, and turbulent flow properties along the wall. Though inelegant, the solution by means of a hybrid model would serve this purpose if it included a turbulence model carefully adjusted to properly simulate the turning anisotropic flow field in the stagnation region.
(3) Develop alternate nozzle and installation geometries that provide higher efficiency, meaning improved Nu profiles at either a set flow or set blower power. Present work in swirling jets, pulsed jets, cross-shaped nozzles, tab nozzles, coaxial nozzles, and other geometries represent a small sample of the practical possibilities. In addition to simply raising the average Nu value, it is of practical use to design and test hardware to produce more uniform Nu patterns.

(4) Further explore the effects of jet interference in jet array geometries, both experimentally and numerically. This includes improved design of exit pathways for spent flow in array installations.

Appendix: Correlation Reference

A. Correlation List

The following correlations and equations for predicting Nu are provided as a reference for designers of impinging jet devices. In addition to providing design equations, they provide an example of typical Re ranges and configurations studied most frequently. The nomenclature is adjusted to maintain consistency with this review. Table A.1 summarizes the conditions.

Impingement Device/Nozzle Type: Slot Jet, Short Contoured Nozzle on Convex Surface

Source: Chan et al. [63]
Range of validity: Re from 5,600 to 13,200, H/B from 2 to 10, S/B from 0 to 13.6

Equation (s):

for H/B from 2 to 8: $Nu_0 = 0.514\ Re_B^{0.50}\ (H/B)^{0.124}$

for H/B from 8 to 10: $Nu_0 = 1.175\ Re_B^{0.54}\ (H/B)^{-0.401}$

for H/B from 2 to 8: $Nu_{avg} = 0.514\ Re_B^{0.50}(H/B)^{0.124}[1.068-(0.31/2)(S/B) + (0.079/3)(S/B)^2 - (0.01154/4)(S/B)^3 + (8.133 \times 10^{-4}/5)(S/B)^4 - (2.141 \times 10^{-5}/6)(S/B)^5]$

for H/B from 8 to 10: $Nu_{avg} = 1.175\ Re_B^{0.54}(H/B)^{-0.401}[1.016-(0.393/2)(S/B) + (0.1/3)(S/B)^2 - (0.01323/4)(S/B)^3 + (8.503 \times 10^{-4}/5)(S/B)^4 - (2.089 \times 10^{-5}/6)(S/B)^5]$

S = maximum lateral position on target surface (circumferential distance)
B = slot jet nozzle width

Impingement Device/Nozzle Type: Array of Round Jets, Orifice Nozzles

Source: Florschuetz et al. [33]
Range of validity:
Re from 2,500 to 70,000, $U_{crossflow}/U_{jet}$ from 0 to 0.8, H/D from 1 to 3
$p_x = p_{jet\ streamwise}/D$ from 5 to 15 for inline arrays, from 5 to 10 for staggered arrays,
$p_y = p_{jet\ spanwise}/D$ from 4 to 8, p_x/p_y from 0.625 to 3.75

TABLE A.1
Correlation list

Source	Nozzle type	Provides	Reynolds number, nozzle height range
Chan et al. [63]	Single slot nozzle (convex target), contoured	Nu_0, Nu_{avg}	$5{,}600 \leqslant Re \leqslant 13{,}200$ $2 \leqslant H/B \leqslant 10$
Florschuetz et al. [33]	Array of round nozzles, orifice	Nu_{avg}	$2{,}500 \leqslant Re \leqslant 70{,}000$ $1 \leqslant H/D \leqslant 3$
Goldstein and Seol [99]	Row of round nozzles (square orifice)	Nu_{avg}	$10{,}000 \leqslant Re \leqslant 40{,}000$ $0 \leqslant H/D \leqslant 6$
Goldstein and Behbahani [100]	Single round nozzle, orifice	Nu_{avg}	$34{,}000 \leqslant Re \leqslant 121{,}300$ $H/D = 6$ or 12
Goldstein et al. [10]	Single round nozzle, orifice	Nu_{avg}	$61{,}000 \leqslant Re \leqslant 124{,}000$ $6 \leqslant H/D \leqslant 12$
Gori and Bossi [48]	Single slot nozzle (on cylinder), pipe	Nu_{avg}	$4{,}000 \leqslant Re \leqslant 20{,}000$ $2 \leqslant H/B \leqslant 12$
Huang and El-Genk [101]	Single round nozzle, pipe	Nu_{avg}	$6{,}000 \leqslant Re \leqslant 60{,}000$ $1 \leqslant H/B \leqslant 12$
Huber and Viskanta [30]	Array of round nozzles, orifice	Nu_{avg}	$3{,}400 \leqslant Re \leqslant 20{,}500$ $0.25 \leqslant H/D \leqslant 6$
Lytle and Webb [102]	Single round nozzle, pipe	Nu_0 and Nu_{avg}	$3{,}600 \leqslant Re \leqslant 27{,}600$ $0.1 \leqslant H/D \leqslant 1$
Martin [2]	Single round nozzle, orifice or pipe	Nu_{avg}	$2{,}000 \leqslant Re \leqslant 400{,}000$ $2 \leqslant H/D \leqslant 12$
Martin [2]	Single slot nozzle, orifice or pipe	Nu_{avg}	$3{,}000 \leqslant Re \leqslant 90{,}000$ $2 \leqslant H/(2B) \leqslant 10$
Martin [2]	Array of round nozzles, orifice or pipe	Nu_{avg}	$2{,}000 \leqslant Re \leqslant 100{,}000$ $2 \leqslant H/D \leqslant 12$
Martin [2]	Array of slot nozzles, orifice or pipe	Nu_{avg}	$1{,}500 \leqslant Re \leqslant 40{,}000$ $1 \leqslant H/(2B) \leqslant 40$
Mohanty and Tawfek [103]	Single round nozzle, pipe and tapered nozzle	Nu_0	$4{,}860 \leqslant Re \leqslant 34{,}500$ $6 \leqslant H/D \leqslant 58$
San and Lai [27]	Array of round nozzles (staggered), orifice	Nu_0	$10{,}000 \leqslant Re \leqslant 30{,}000$ $2 \leqslant H/D \leqslant 5$
Tawfek [104]	Single round nozzle, pipe and tapered nozzle	Nu_{avg}	$3{,}400 \leqslant Re \leqslant 41{,}000$ $6 \leqslant H/D \leqslant 58$
Wen and Jang [105]	Single round nozzle, pipe	Nu_{avg}	$750 \leqslant Re \leqslant 27{,}000$ $3 \leqslant H/D \leqslant 16$

Equations:
$p_x = p_{jet\ streamwise}/D$
$p_y = p_{jet\ spanwise}/D$
$U_{crossflow}$ = magnitude of crossflow velocity in channel (density-weighted)
U_{jet} = magnitude of jet exit velocity (density-weighted)

$Nu = A\ Re^m \{1 - B[(H/D)(U_{crossflow}/U_{jet})]^n\}\ Pr^{1/3}$

$A_{inline} = 1.18(p_x^{-0.944})(p_y^{-0.642})([H/D]^{0.169})$

$m_{inline} = 0.612(p_x^{.059})(p_y^{.032})([H/D]^{-0.022})$

$B_{inline} = 0.437(p_x^{-0.095})(p_y^{-0.219})([H/D]^{0.275})$

$n_{inline} = 0.092(p_x^{-0.005})(p_y^{0.599})([H/D]^{1.04})$

$A_{staggered} = 1.87(p_x^{-0.771})(p_y^{-0.999})([H/D]^{-0.257})$

$m_{staggered} = 0.571(p_x^{.028})(p_y^{.092})([H/D]^{0.039})$

$B_{staggered} = 1.03(p_x^{-0.243})(p_y^{-0.307})([H/D]^{0.059})$

$n_{staggered} = 0.442(p_x^{.098})(p_y^{-0.003})([H/D]^{0.304})$

Impingement Device/Nozzle Type: Row of Round Jets, Square-Edged Orifice Nozzles

Source: Goldstein and Seol [99]
Range of validity: for H/D from 2 to 6, p_{jet}/D from 4 to 8, H/D from 0 to 6, Re from 10,000 to 40,000, orifice height $l = D$
Equation (s):

$$Nu_{avg} = \frac{2.9 \exp\left(-0.09(H/D)^{1.4}\right) Re^{0.7}}{22.8 + (p_{jet}/D)\sqrt{H/D}}$$

Impingement Device/Nozzle Type: Single Round Jet, Orifice Nozzle

Source: Goldstein and Behbahani [100]
Range of validity: r/D from 0.5 to 32, $Re = \sim 34{,}000$ to $121{,}300$, $L/D = 6$ or 12

Equation (s):
for $L/D = 6$, $Nu_{avg} = Re^{0.6}/[3.329 + 0.273\ (r_{max}/D)^{1.3}]$
for $L/D = 12$, $Nu_{avg} = Re^{0.6}/[4.577 + 0.4357\ (r_{max}/D)^{1.14}]$

Impingement Device/Nozzle Type: Single Round Jet, Orifice Nozzle

Source: Goldstein, Behbahani, and Heppelmann [10]
Range of validity: Re from 61,000 to 124,000, H/D from 6 to 12

Equation (s):

$$\text{target at constant temperature}: Nu_{avg} = \frac{24 - |(H/D) - 7.75|}{533 + 44(r/D)^{1.285}} Re^{0.76}$$

$$\text{target at constant heat flux}: Nu_{avg} = \frac{24 - |H/D - 7.75|}{533 + 44(r/D)^{1.394}} Re^{0.76}$$

Impingement Device/Nozzle Type: Single Slot Jet, Pipe Nozzle, Normal to a Cylindrical Target

Source: Gori and Bossi [48]
Range of validity: D/B from 1 to 4, Re from 4,000 to 20,000, H/B from 2 to 12

Equation (s):
for H/B from 2 to 8: $Nu_{avg} = 0.0516 \, (H/B)^{0.179} \, (D/B)^{0.214} \, Re^{0.753} \, Pr^{0.4}$
for H/B from 8 to 12: $Nu_{avg} = 0.0803 \, (H/B)^{-0.205} \, (D/B)^{0.162} \, Re^{0.800} \, Pr^{0.4}$

Impingement Device/Nozzle Type: Single Slot Jet, Pipe Nozzle

Source: Huang and El-Genk [102]
Range of validity: Re from 6000 to 60,000, H/B from 1 to 12

Equation (s):
$Nu_{avg} = Re^{0.76} \, Pr^{0.42} \, [a + b \, (H/D) + c \, (H/D)^2]$

$a = (1 \times 10^{-4}) \, (506 + 13.3 \, [r_{max}/D] - 19.6 \, [r_{max}/D]^2 + 2.41 \, [r_{max}/D]^3 - 0.0904 \, [r_{max}/D]^4)$

$b = (1 \times 10^{-4}) \, (32 - 24.3 \, [r_{max}/D] + 6.53 \, [r_{max}/D]^2 - 0.694 \, [r_{max}/D]^3 + 0.0257 \, [r_{max}/D]^4)$

$c = (-3.85 \times 10^{-4}) \, (1.147 + [r_{max}/D])^{-0.0904}$

Impingement Device/Nozzle Type: Array of Round Jets, Square Array, Orifice Nozzles, Confined Jets, Exit

Source: Huber and Viskanta [30]
Range of validity: H/D from 0.25 to 6, spacing p_{jet}/D from 4 to 8, Re from 3,400 to 20,500

Equation (s):

$$Nu_{avg} = 0.285 Re^{0.71} \ Pr^{0.33} (H/D)^{-0.123} (p_{jet}/D)^{-0.725}$$

Impingement Device/Nozzle Type: Single Round Jet, Pipe Nozzle

Source: Lytle and Webb [102]
Range of validity: H/D from 0.1 to 1, Nu_0 equations for Re from 3,700 to 30,000, Nu_{avg} equations for Re from 3,600 to 27,600

Equation (s):
for $H/D \leqslant 1.0$: $Nu_0 = 0.726 \ Re^{0.53} \ (H/D)^{-0.191}$

for $H/D \leqslant 0.5$: $Nu_0 = 0.663 \ Re^{0.53} \ (H/D)^{-0.248}$

for $H/D \leqslant 0.25$: $Nu_0 = 0.821 \ Re^{0.5} \ (H/D)^{-0.288}$

Radial peaks in Nu are located at: $r_{max}/D = 0.188 \ Re^{0.241} \ (H/D)^{0.224}$

Nu_{avg} out to $r/D = 1$: $Nu_{avg} = 0.424 \ Re^{0.57} \ (H/D)^{-0.33}$

Nu_{avg} out to $r/D = 2$: $Nu_{avg} = 0.150 \ Re^{0.67} \ (H/D)^{-0.36}$

Impingement Device/Nozzle Type: Single Round Nozzle

Source: Martin [2]
Range of validity: $2{,}000 < Re < 400{,}000$, $0.004 \leqslant f \leqslant 0.04$, $2.5 \leqslant r/D \leqslant 7.5$, and $2 \leqslant H/D \leqslant 12$.

Equation (s):

$$Nu_{avg} = Pr^{0.42} \frac{D}{r} \frac{1 - 1.1 D/r}{1 + 0.1(H/D - 6)D/r} F$$

for $2{,}000 < Re < 30{,}000$, $F = 1.36 \ Re^{0.574}$

for $30{,}000 < Re < 120{,}000$, $F = 0.54 \ Re^{0.667}$

for $120{,}000 < Re < 400{,}000$, $F = 0.151 \ Re^{0.775}$

Impingement Device/Nozzle Type: Single Slot Nozzle

Source: Martin [2]
Range of validity: $3000 < Re < 90{,}000$, $0.01 \leqslant f \leqslant 0.125$, $2 \leqslant x/S \leqslant 25$, and $2 \leqslant H/S \leqslant 10$.

Equation (s):

$$Nu_{avg} = Pr^{0.42} \frac{1.53}{(x/S)+(H/S)+1.39} Re^{\left(0.695 - \left[x/S + (H/S)^{1.33} + 3.06\right]^{-1}\right)},$$

S = twice slot width = nozzle hydraulic diameter

Impingement Device/Nozzle Type: Round Nozzle Array

Source: Martin [2]
Range of validity: $2{,}000 < Re < 100{,}000$, $0.004 \leqslant f \leqslant 0.04$, and $2 \leqslant H/D \leqslant 12$.

Equation (s):

$$Nu_{avg} = Pr^{0.42}\,(K)\,(G)\,(F)$$

$$K = \left(1 + \left(\frac{H/D}{0.6/\sqrt{f}}\right)^6\right)^{-0.05}$$

$$G = 2\sqrt{f}\,\frac{1 - 2.2\sqrt{f}}{1 + 0.2((H/D) - 6)\sqrt{f}}$$

$$F = 0.5\ Re^{2/3}$$

Equations are for a developed jet. For a sharp orifice jet, give the value of a contraction coefficient (discharge coefficient) and

- Multiply diameter D by square root of contraction coefficient;
- Multiply f by contraction coefficient;
- Divide U by contraction coefficient;
- Divide Re by square root of contraction coefficient;
- Divide $Pr^{0.42}$ by square root of contraction coefficient.

Impingement Device/Nozzle Type: Slot Nozzle Array

Source: Martin [2]
Range of validity: $1{,}500 < Re < 40{,}000$, $0.008 \leqslant f \leqslant 2.5\, f_0$, and $1 \leqslant H/S \leqslant 40$.

Equation (s):

$$Nu_{avg} = Pr^{0.42} \frac{2}{3} f_0^{3/4} \left(\frac{2Re}{f/f_0 + f_0/f} \right)^{2/3}$$

$$f_0 = \left[60 + 4(H/S - 2)^2 \right]^{-1/2}$$

S = twice slot width = nozzle hydraulic diameter

Equations are for a developed jet. For a sharp orifice jet, give the value of a contraction coefficient (discharge coefficient) and

- Multiply both S and f by contraction coefficient;
- Divide U by contraction coefficient;
- Divide $Pr^{0.42}$ by square root of contraction coefficient.

Impingement Device/Nozzle Type: Single Round Jet, Pipe and Tapered Nozzle

Source: Mohanty and Tawfek [103]
Range of validity: H/D from 6 to 41, Re from 4,860 to 34,500 (varies by case, see the following)

Equation (s):
for H/D from 10 to 16.7, Re from 4,860 to 15,300:
$Nu_0 = 0.15\, Re^{0.701}\, (H/D)^{-0.25}$

for H/D from 20 to 25, Re from 4,860 to 15,300:
$Nu_0 = 0.17\, Re^{0.701}\, (H/D)^{-0.182}$

for H/D from 6 to 58, Re from 6,900 to 24,900:
$Nu_0 = 0.388\, Re^{0.696}\, (H/D)^{-0.345}$

for H/D from 9 to 41.4, Re from 7,240 to 34,500:
$Nu_0 = 0.615\, Re^{0.67}\, (H/D)^{-0.38}$

Impingement Device/Nozzle Type: Array of Round Jets, Staggered, Orifice Nozzles

Source: San and Lai [27]
Range of validity: Re from 10,000 to 30,000, H/D from 2 to 5, p_{jet}/D from 4 to 16, orifice depth $l = D$

Equation (s):
For H/D from 2 to 3.5, p_{jet}/D from 6 to 16,
$Nu_0 = (p_{jet}/D) \exp(\alpha_1 + \alpha_2[p_{jet}/D]) \, Re^{0.6}$
$\alpha_1 = -0.504 - 1.662 \, (H/D) + 0.233 \, (H/D)^2$
$\alpha_2 = -0.281 + 0.116 \, (H/D) - 0.017 \, (H/D)^2$

For H/D from 3.5 to 6, p_{jet}/D from 4 to 8,
$Nu_0 = (p_{jet}/D) \exp(\alpha_1 + \alpha_2[p_{jet}/D]) \, Re^{0.4}$
$\alpha_1 = -2.627 + 0.546 \, (H/D) - 0.049 \, (H/D)^2$
$\alpha_2 = 0.132 - 0.093 \, (H/D) + 0.008 \, (H/D)^2$

For H/D from 3.5 to 6, p_{jet}/D from 8 to 16,
$Nu_0 = (p_{jet}/D) \exp(\alpha_1 + \alpha_2[p_{jet}/D]) \, Re^{0.5}$
$\alpha_1 = -4.752 + 1.007 \, (H/D) - 0.103 \, (H/D)^2$
$\alpha_2 = 0.229 - 0.132 \, (H/D) + 0.013 \, (H/D)^2$

Impingement Device/Nozzle Type: Single Round Jet, Pipe and Tapered Nozzle

Source: Tawfek [104]
Range of validity: r/D from 2 to 30, H/D from 6 to 58, Re from 3,400 to 41,000

Equation (s):
$$Nu_{avg} = 0.453 Pr^{1/3} \, Re^{0.691} \, (H/D)^{-0.22} (r/D)^{-0.38}$$

Impingement Device/Nozzle Type: Single Round Jet, Pipe Nozzle

Source: Wen and Jang [105]
Range of validity: H/D from 3 to 16, r/D from 0 to 7.14, Re from 750 to 27,000

Equation (s):
$$Nu_{avg} = 0.442 \, Re^{0.696} \, Pr^{1/3} \, (H/D)^{-0.20} \, (r/D)^{-0.41}$$

Table A.2 provides an additional set of references with correlations and test data based on experimental studies of specific designs. Though the correlations may not be applicable to a wide variety of design points or conditions, the references may prove useful for performance prediction of specialized jet impingement device designs.

B. Trends Within the Correlations

By mapping out the correlation predictions as a function of independent design variables we can rapidly see what range of Nusselt numbers may be produced by a jet impingement device as well as the influences of design changes. Figure A.1 shows correlation predictions of area-averaged Nu for a

TABLE A.2

Supplemental Correlation and References, for Special Cases not Considered in Table A.1

Source	Data/application	Parameter range
Bartoli et al. [106]	Wide slot jet on small cylindrical target	$1,500 \leqslant Re \leqslant 20,000$ $1.5 \leqslant H/B \leqslant 10$
Gau and Lee [53]	Interaction of impinging jet and ribbed target surface	$2,500 \leqslant Re \leqslant 11,000$ $2 \leqslant H/B \leqslant 16$
Goldstein et al. [107]	Recovery and entrainment effects	$61,000 \leqslant Re \leqslant 124,000$ $2 \leqslant H/D \leqslant 12$
Goldstein and Franchett [108]	Angled round jet on a flat target	$10,000 \leqslant Re \leqslant 35,000$ $4 \leqslant H/D \leqslant 10$
Hargrave et al. [109]	Calculation of Nu_0 as a function of Tu	$6837 \leqslant Re \leqslant 10742$ $0.014 \leqslant Tu \leqslant 0.256$
Hoogendorn [20]	Calculation of Nu_0 as a function of Tu	$2,000 \leqslant Re \leqslant 90,000$ $2 \leqslant H/D \leqslant 8$ $0.09 \leqslant Tu \leqslant 0.20$
Hrycak [110]	Array of round jets on concave cylinder	$25,000 \leqslant Re \leqslant 66,000$ $3 \leqslant H/D \leqslant 20$
Li and Garimella [111]	Effect of Pr on Nu_0 for various fluids	$4,000 \leqslant Re \leqslant 23,000$ $1 \leqslant H/D \leqslant 5$
Sparrow and Alhomoud [112]	Wide slot jet on small cylindrical target	$5,000 \leqslant Re \leqslant 60,000$ $0.25 \leqslant B/d \leqslant 0.5$
Sparrow and Lovell [41]	Angled round jet on a flat target	$2,500 \leqslant Re \leqslant 10,000$ $7 \leqslant H/D \leqslant 15$

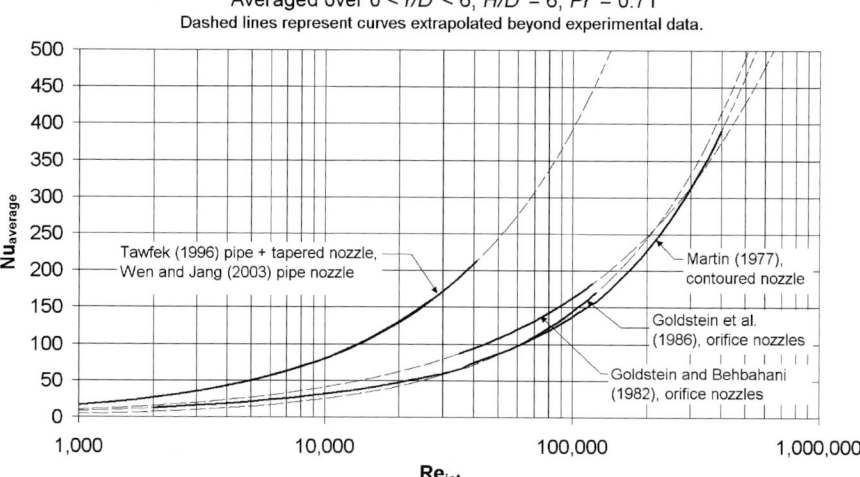

FIG. A.1. Sample comparison of correlations for single round jet impingement, $0 < r/D < 6$, $H/D = 6$.

particular round-jet configuration with $H/D = 6$, averaged over the span $0 < r/D < 6$. The large variation between correlations comes from the difference in nozzle types. The pipe nozzle correlations show a resulting Nu value two to three times higher than that of a contoured or orifice nozzle at the same Re, an effect resulting from higher initial turbulence in the fluid emerging from a pipe nozzle. In contrast, the orifice nozzles have little initial turbulence and rely more upon unsteady shear layer growth to promote turbulence an increase transfer rates. The apparent benefit of selecting a pipe nozzle design at this particular height and radial span must be weighed against the higher power required to drive fluid through a pipe nozzle.

By viewing the change in predicted Nu at a set value of Re we can assess the relative importance of changes in the target radial span ($r_{maximum}/D$) and nozzle height (H/D) upon the area-averaged value of Nu. Figure A.2 shows the dominant effect of r/D over H/D, with Nusselt number plotted relative to a selected reference case at $H/D = 6$, $0 < r/D < 5$. Though the opposing effects of jet deceleration and turbulence development are frequently studied and discussed, for a target of a set size at a fixed nozzle Reynolds number this effect is secondary to the choice of nozzle size expressed by the range of r/D.

Figure A.3 shows typical Nu values for heat transfer under an array of impinging jets with an inline jet pitch of $5D$ using three correlations suitable

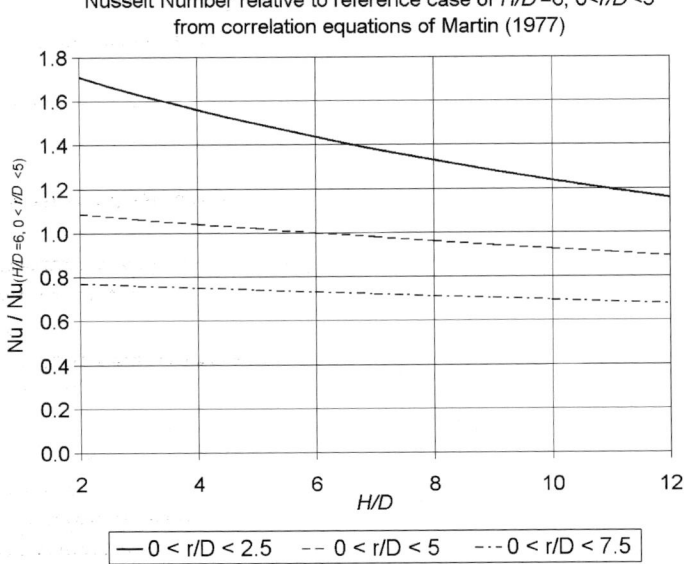

FIG. A.2. Influence of changes in jet height and target size at a set Reynolds number.

for three conditions. The correlation by Martin [2] includes the effect of jet interference and crossflow. The correlation by Florschuetz et al. [28] has been utilized based on a case with no crossflow penalty. The correlation by Huber and Viskanta [25] comes from an experiment using additional holes in the orifice plate to permit the exit of spent air. The predictions show a consistent trend, with the Huber and Viskanta data (based on a limited spent flow exit pathway) falling between the upper and lower brackets provided by the curves of Martin and the curves of Florschuetz et al. All three correlations show little degradation in Nu_{avg} with a 50% increase in H/D. For this case, the change in exit conditions between jet impingement without crossflow and jets in crossflow results in a degradation in Nu_{avg} of up to 25%.

Figure A.4 shows the effect of changing jet pitch for a sample case with an array positioned at $H/D = 3$. The influence of jet pitch on the average heat transfer rate is as important as that of Re. For this configuration it is seen that the heat transfer rate decreases as p/D increases from $p/D = 5$.

A similar effect is found for a design with a staggered jet array, as shown in Figure. A.5. As jet pitch p_{jet}/D increases beyond six, the transfer rate drops off, while for close jet spacing ($p_{jet}/D \leqslant 4$) the detrimental effects of jet

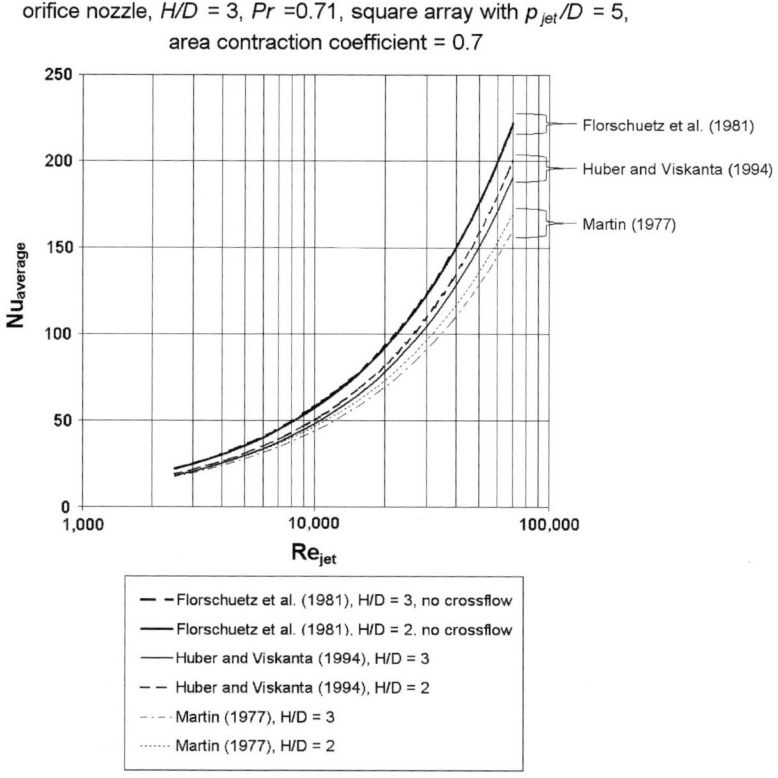

FIG. A.3. Sample comparison of correlations for inline array jet impingement, $p/D = 5$.

interference are seen. The two sources used experiments with somewhat different setups. Florschuetz *et al.* used a large array of orifice nozzles emerging into a channel, confined on one side to control the direction of the spent flow. San and Lai measured the cooling effects under the central jet of a small array (five orifices), with the spent flow allowed to travel in all radial directions. In addition, the aspect ratio of the two tests differed, so a direct comparison may not be meaningful, but it is noteworthy that the two configurations showed similar trends with changes in Re and p_{jet}/D.

Figure A.6 shows the effect of changing jet height at various radial spans on the Nu value for single round jet impingement, based on the equations of Martin [2]. The detrimental effect of raising nozzle height becomes less significant when the transfer rate is averaged over a larger target (larger $r_{maximum}/D$).

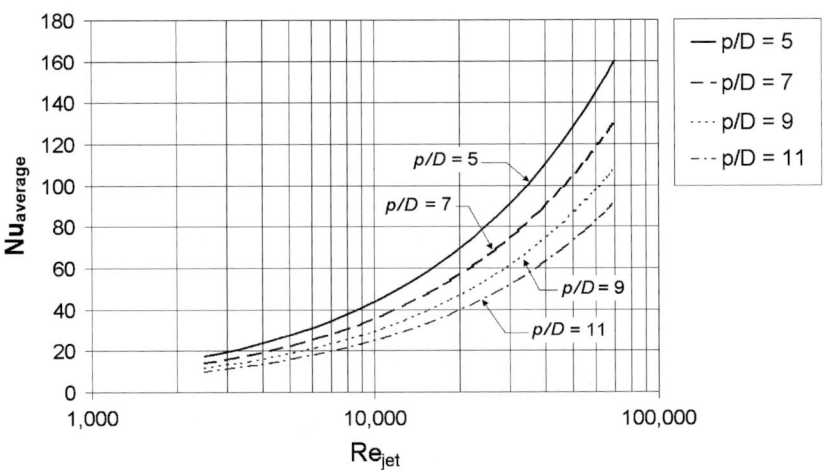

Fig. A.4. Effect of changing jet pitch on heat transfer rates for inline arrays of impinging jets.

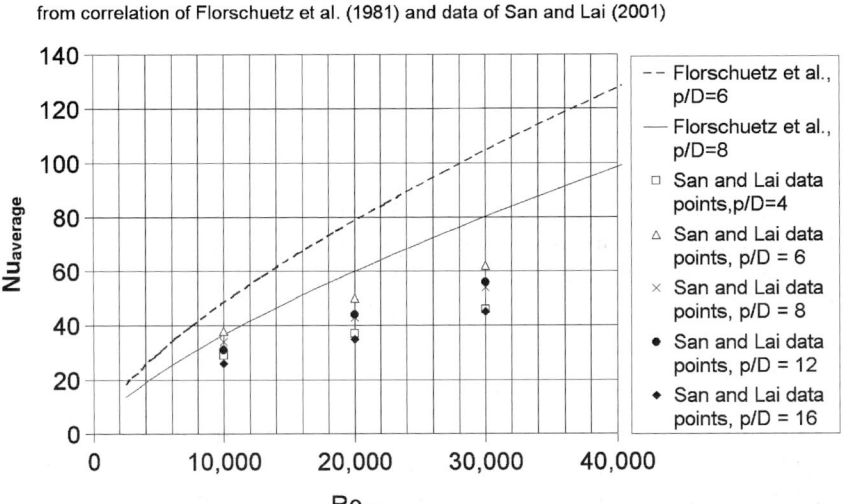

Fig. A.5. Effect of changing jet pitch for staggered arrays of impinging jets.

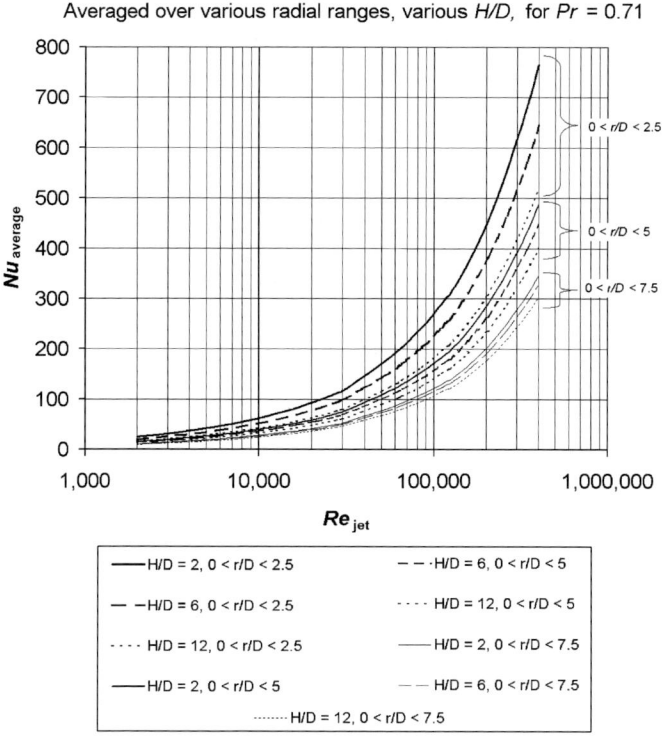

Fig. A.6. Influence of jet nozzle height on Nu for targets of various radial span.

Figures A.7 and A.8 show the relative importance of Re, nondimensional target curvature d/B, and nozzle height H/B for heat transfer on a convex curved cylinder. As with many other design variables, changes in nozzle size are coupled to all three of these non-dimensional variables. As B increases to span more of the target, H changes simultaneously, which also contributes to jet width. Heat transfer improves with submersion of more of the target surface in the stagnation zone, and less in the wall jet where separation may occur. Both H/B and d/B play an important role but are still not as influential as Re. A 100% increase in Re results in a change in Nu of typically 50%, while a 100% increase in H/B gives a change in Nu of typically 10%, and a 100% increase in d/B changes Nu by approximately 30%. The increase in Nu as H/B grows to six shows the effect of the turbulence development in the growing shear layers at the edges of the free jet emerging. The decrease in Nu as H/B grows to 10 illustrates the competing effect of velocity decay in the free jet.

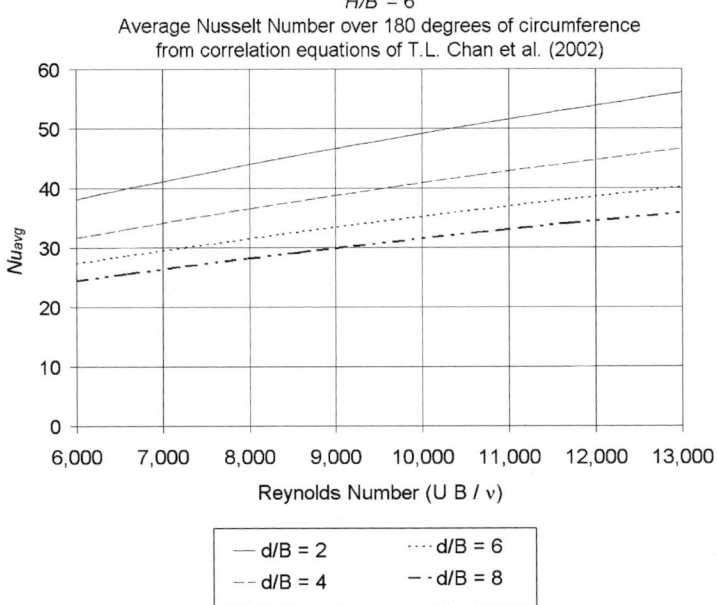

FIG. A.7. Slot jet impingement on a convex cylinder: influence of surface curvature.

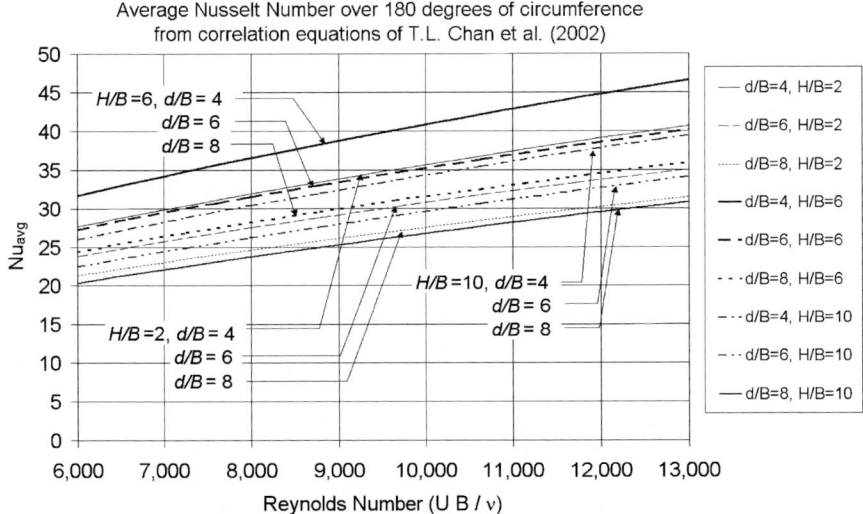

FIG. A.8. Slot jet impingement on a convex cylinder: Influence of curvature and jet height.

Nomenclature

A_f	target free area	Re	Reynolds number ($= U_0 D/\nu$ for a jet)
b	correlation exponent, used in $Nu \propto Re^b$	R_t	Turbulent Reynolds number (Eqn 26)
B	slot jet nozzle width	S_{ij}	strain rate tensor
C	mass concentration in fluid (with subscript), or equation coefficient	Sc	Schmidt number = fluid mass diffusivity/fluid viscosity = chemical diffusivity/fluid viscosity
$c'_{\varepsilon n}$	empirical constants for v^2–f model		
c_n	empirical constants for turbulence model (c_1, c_ε, c_μ, c_θ, c_w)	Sh	Sherwood number
c_p	specific heat of fluid	t	time
$C_{0\,jet}$	jet mass concentration	T	temperature
C_n	empirical constants for v^2–f model	T_{scale}	time scale
C_{wall}	wall surface concentration	$T_{0\,jet}$	jet adiabatic wall temperature, exiting nozzle
C_μ	turbulence model constant for calculating turbulent viscosity	T_f	far-field fluid temperature
$CD_{k\omega}$	cross-diffusion term	T_s	wall surface temperature
d	target diameter (twice target radius of curvature)	T_{wall}	wall surface temperature
		Tu	turbulence intensity (with subscript), or position along wall surface
D	nozzle diameter		
D_h	hydraulic diameter of nozzle		
D_i	mass diffusivity constant	U or u	fluid velocity component
f	relative nozzle area or $v^2 f$ model function	u_i	general velocity vector
		U_0	jet initial speed, average
f_ε	k–ε model function	v	fluid velocity
f_μ	k–ε model function	v^2	streamwise-normal velocity variance (from v^2–f model)
f_{wall}	$v^2 f$ model function		
F_1	wall blending function	x	coordinate direction (with subscript), or position along wall surface
G	jet mass flow per unit of target area		
h	convective heat transfer coefficient		
H	nozzle-to-target spacing (nozzle height)	y	distance from wall for turbulence model
		y^+	nondimensional distance from wall
k_c	fluid thermal conductivity		
k	turbulent kinetic energy	Yc	Yap correction (used with k–ε model)
k_i	mass transfer coefficient		
L_{scale}	length scale	z	axial position or height, measured off of target surface (distance from wall)
M	Mach number		
$\hat{n}$	wall-normal unit vector		
Nu	Nusselt number		
Nu_0	Nusselt number at stagnation point	**Greek Letters**	
		α	turbulence model constant
p	fluid pressure	β	turbulence model constant
p_{jet}	jet pitch (center-to-center distance)	δ_{ij}	identity tensor
Pr	Prandtl number = fluid thermal diffusivity/fluid viscosity	ε	turbulent kinetic energy dissipation rate
Q	volumetric flow rate	$\tilde{\varepsilon}$	isotropic turbulent kinetic energy dissipation rate
q''	heat flux		
r	radial position, measured from jet axis	θ	general scalar quantity such as temperature

μ	fluid viscosity	ω	fluctuation frequency (used in k–ω model)
ν	fluid kinematic viscosity		
ρ	fluid density		
σ	standard deviation function (normalized)	SUBSCRIPTS	
		Amb	ambient
σ_{ij}	steady stress tensor	Avg	average (area-weighted)
σ_n	empirical constants for k–ε model (σ_ε, σ_θ, σ_k)	Min	minimum
		max	maximum
τ_{ij}	turbulent stress tensor (Reynolds stress tensor)	t	turbulent (e.g. ν_t)
Ω	vorticity (scalar), or rotation rate tensor		

References

1. Ferrari, J., Lior, N., and Slycke, J. (2003). An evaluation of gas quenching of steel rings by multiple-jet impingement. *J. Mater. Process. Technol.* **136**, 190–201.
2. Martin, H. (1977). Heat and mass transfer between impinging gas jets and solid surfaces. *Adv. Heat Transfer* **13**, 1–60.
3. Jambunathan, K., Lai, E., Moss, M. A., and Button, B. L. (1992). A review of heat transfer data for single circular jet impingement. *Int. J. Heat Fluid Flow* **13**, 106–115.
4. Donaldson, C. D. and Snedeker, R. S. (1971). A study of free jet impingement. Part 1. Mean properties of free and impinging jets,. *J. Fluid Mech.* **45**, 281–319.
5. Viskanta, R. (1993). Heat transfer to impinging isothermal gas and flame jets. *Exp. Thermal Fluid Sci.* **6**, 111–134.
6. Zuckerman, N. and Lior, N. (2005). Impingement heat transfer: Correlations and numerical modeling. *J. Heat Transfer* **127**, 544–552.
7. Han, J.-C., Dutta, S., and Ekkad, S. (2000). "Gas Turbine Heat Transfer and Cooling Technology". Taylor & Francis, New York.
8. Taniguchi, H., Miyamae, S., Arai, N., and Lior, N. (2000). Power generation analysis for high temperature gas turbine in thermodynamic process. *AIAA J. Propul. Power* **16**, 557–561.
9. Maurel, S. and Solliec, C. (2001). A turbulent plane jet impinging nearby and far from a flat plate. *Exp. Fluids* **31**, 687–696.
10. Goldstein, R. J., Behbahani, A. I., and Heppelmann, K. K. (1986). Streamwise distribution of the recovery factor and the local heat transfer coefficient to an impinging circular air jet. *Int. J. Heat Mass Transfer* **29**, 1227–1235.
11. Kim, B. G., Yu, M. S., Cho, Y. I., and Cho, H. H. (2002). Distributions of recovery temperature on flat plate by underexpanded supersonic impinging jet. *J. Thermophys. Heat Transfer* **16**, 425–431.
12. Lee, J. and Lee, S. (2000). The effect of nozzle configuration on stagnation region heat transfer enhancement of axisymmetric jet impingement. *Int. J. Heat Mass Transfer* **43**, 3497–3509.
13. Kreith, F., Goswami, D. Y. (2005). "The CRC Handbook of Mechanical Engineering". Section 3, CRC Press, Boca Raton, FL.
14. Saleh, J. M. (2002). "Fluid Flow Handbook". McGraw-Hill, New York.
15. Ward-Smith, A. J. (1971). "Pressure Losses in Ducted Flows". Butterworths, London.
16. Gan, G. and Riffat, S. B. (1997). Pressure loss characteristics of orifice and perforated plates. *Exp. Thermal Fluid Sci.* **14**, 160–165.

17. Rahimi, M., Owen, I., and Mistry, J. (2003). Heat transfer between an under-expanded jet and a cylindrical surface. *Int. J. Heat Mass Transfer* **46**, 3135–3412.
18. Peyret, R. (1996). "Handbook of Computational Fluid Mechanics". Academic Press, San Diego, CA.
19. Gao, S. and Voke, P. R. (1995). Large-eddy simulation of turbulent heat transport in enclosed impinging jets. *Int. J. Heat Fluid Flow* **16**, 349–356.
20. Hoogendorn, C. J. (1977). The effect of turbulence on heat transfer at a stagnation point. *Int. J. Heat Mass Transfer* **20**, 1333–1338.
21. Abe, K. and Suga, K. (2001). Large eddy simulation of passive scalar in complex turbulence with flow impingement and flow separation. *Heat Transfer – Asian Res.* **30**, 402–418.
22. Cooper, D., Jackson, C., Launder, B. E., and Liao, G. X. (1993). Impinging jet studies for turbulence model assessment – I. Flow-field experiments,. *Int. J. Heat Mass Transfer* **36**, 2675–2684.
23. Gardon, R. and Akfirat, J. C. (1965). The role of turbulence in determining the heat-transfer characteristics of impinging jets. *Int. J. Heat Mass Transfer* **8**, 1261–1272.
24. Chung, Y. M. and Luo, K. H. (2002). Unsteady heat transfer analysis of an impinging jet. *ASME J. Heat Transfer* **124**, 1039–1048.
25. Behnia, M., Parneix, S., Shabany, Y., and Durbin, P. A. (1999). Numerical study of turbulent heat transfer in confined and unconfined impinging jets. *Int. J. Heat Fluid Flow* **20**, 1–9.
26. Narayanan, V., Seyed-Yagoobi, J., and Page, R. H. (2004). An experimental study of fluid mechanics and heat transfer in and impinging slot jet flow. *Int. J. Heat Mass Transfer* **47**, 1827–1845.
27. San, J. Y. and Lai, M. (2001). Optimum jet-to-jet spacing of heat transfer for staggered arrays of impinging air jets. *Int. J. Heat Mass Transfer* **44**, 3997–4007.
28. Slayzak, S. J., Viskanta, R., and Incropera, F. P. (1994). Effects of interaction between adjacent free surface planar jets on local heat transfer from the impingement surface. *Int. J. Heat Mass Transfer* **37**, 269–282.
29. Aldabbagh, L. B. Y. and Sezai, I. (2002). Numerical simulation of three-dimensional laminar multiple impinging square jets. *Int. J. Heat Fluid Flow* **23**, 509–518.
30. Huber, A. M. and Viskanta, R. (1994). Effect of jet–jet spacing on convective heat transfer to confined, impinging arrays of axisymmetric jets. *Int. J. Heat Mass Transfer* **37**, 2859–2869.
31. A. Chambers, D. R. H. Gillespie, P. T. Ireland, G. M. Dailey, (November 2003). Cooling performance of a narrow impingement channel including the introduction of cross flow upstream of the first hole. In: Proceedings of IMECE'03, 2003 ASME International Mechanical Engineering Congress, Washington, DC, November 15–12, 2003, IMECE2003-43975.
32. Florschuetz, L. W., Berry, R. A., and Metzger, D. E. (1980). Periodic streamwise variations of heat transfer coefficients for inline and staggered arrays of circular jets with crossflow of spent air. *J. Heat Transfer* **102**, 132–137.
33. Florschuetz, L. W., Truman, C. R., and Metzger, D. E. (1981). Streamwise flow and heat transfer distributions for jet array impingement with crossflow. *J. Heat Transfer* **103**, 337–342.
34. Florschuetz, L. W., Metzger, D. E., and Su, C. C. (1984). Heat transfer characteristics for jet array impingement with initial crossflow. *J. Heat Transfer* **106**, 34–41.
35. Chattopadhyay, H. and Saha, S. K. (2003). Turbulent flow and heat transfer from a slot jet impinging on a moving plate. *Int. J. Heat Fluid Flow* **24**, 685–697.
36. Bilen, K., Bakirci, K., Yapici, S., and Yavuz, T. (2002). Heat transfer from an impinging plate jet. *Int. J. Energy Res.* **26**, 305–320.

37. Lee, D. H., Won, S. E., Kim, Y. T., and Chung, Y. S. (2002). Turbulent heat transfer from a flat surface to a swirling round impinging jet. *Int. J. Heat Mass Transfer* **45**, 223–227.
38. Nozaki, A., Igarashi, Y., and Hishida, K. (2003). Heat transfer mechanism of a swirling impinging jet in a stagnation region. *Heat Transfer – Asian Res* **32**, 663–673.
39. Wen, M.-Y. (2005). Flow structures and heat transfer of swirling jet impinging on a flat surface with micro-vibrations. *Int. J. Heat Mass Transfer* **48**, 545–560.
40. Tawfek, A. A. (2002). Heat transfer studies of the oblique impingement of round jets upon a curved surface. *Heat Mass Transfer* **38**, 467–475.
41. Sparrow, E. M. and Lovell, B. J. (1980). Heat transfer characteristics of an obliquely impinging circular jet. *ASME J. Heat Transfer* **102**, 202–209.
42. Bart, G. C. J., van Ijzerloo, A. J., Geers, L. F. G., Hoek, L., and Hanjalic, K. (2002). Heat transfer of phase-locked modulated impinging-jet arrays. *Exp. Thermal Fluid Sci.* **25**, 299–304.
43. Göppert, S., Gürtler, T., Mocikat, H., and Herwig, H. (2004). Heat transfer under a precessing jet: Effects of unsteady jet impingement. *Int. J. Heat Mass Transfer* **47**, 2795–2806.
44. Hwang, S. D., Lee, C. H., and Cho, H. H. (2001). Heat transfer and flow structures in axisymmetric impinging jet controlled by vortex pairing. *Int. J. Heat Fluid Flow* **22**, 293–300.
45. Ichimiya, K. (2003). Heat transfer characteristics of an annular turbulent impinging jet with a confined wall measured by thermosensitive liquid crystal. *Heat Mass Transfer* **39**, 545–551.
46. Gau, C. and Chung, C. M. (1991). Surface curvature effect on slot-air-jet impingement cooling flow and heat transfer process. *J. Heat Transfer* **113**, 858–864.
47. Fleischer, A. S., Kramer, K., and Goldstein, R. J. (2001). Dynamics of the vortex structure of a jet impinging on a convex surface. *Exp. Thermal Fluid Sci.* **24**, 169–175.
48. Gori, F. and Bossi, L. (2003). Optimal slot height in the jet cooling of a circular cylinder. *Appl. Thermal Eng.* **23**, 859–870.
49. McDaniel, C. S. and Webb, B. W. (2000). Slot jet impingement heat transfer from circular cylinders. *Int. J. Heat Mass Transfer* **23**, 1975–1985.
50. Cornaro, C., Fleischer, A. S., and Goldstein, R. J. (1999). Flow visualization of a round jet impinging on cylindrical surfaces. *Exp. Thermal Fluid Sci.* **20**, 66–78.
51. Kayansayan, N. and Kucuka, S. (2001). Impingement cooling of a semi-cylindrical concave channel by confined slot-air-jet. *Exp. Thermal Fluid Sci.* **25**, 383–396.
52. Taslim, M. E., Setayeshgar, L., and Spring, S. D. (2001). An experimental evaluation of advanced leading edge impingement cooling concepts. *ASME J. Turbomachinery* **123**, 147–153.
53. Cau, C. and Lee, C. C. (1992). Impingement cooling flow structure and heat transfer along rib-roughened walls. *Int. J. Heat Mass Transfer* **35**, 3009–3020.
54. Rhee, D., Yoon, P., and Cho, H. H. (2003). Local heat/mass transfer and flow characteristics of array impinging jets with effusion holes ejecting spent air. *Int. J. Heat Mass Transfer* **46**, 1049–1061.
55. Li, X., Gaddis, J. L., and Wang, T. (2001). Modeling of heat transfer in a mist/steam impinging jet. *ASME J. Heat Transfer* **123**, 1086–1092.
56. Yokomine, T., Shimizu, A., Saitoh, A., and Higa, K. (2002). Heat transfer of multiple impinging jets with gas–solid suspensions. *Exp. Thermal Fluid Sci.* **26**, 617–626.
57. Yoshida, H., Suenaga, K., and Echigo, R. (1990). Turbulence structure and heat transfer of a two-dimensional impinging jet with gas–solid suspensions. *Int. J. Heat Mass Transfer* **33**, 859–867.

58. Malikov, G. K., Lobanov, D. L., Malikov, K. Y., Lisienko, V. G., Viskanta, R., and Fedorov, A. G. (2001). Direct flame impingement heating for rapid thermal materials processing. *Int. J. Heat Mass Transfer* **44**, 1751–1758.
59. Dong, L. L., Leung, C. W., and Cheung, C. S. (2004). Heat transfer and wall pressure characteristics of a twin premixed butane/air flame jets. *Int. J. Heat Mass Transfer* **47**, 489–500.
60. Baukal, C. E. and Gebhart, B. (1996). A review of empirical flame impingement heat transfer correlations. *Int. J. Heat Fluid Flow* **17**, 386–396.
61. Baukal, C. E. and Gebhart, B. (1996). A review of semi-analytical solutions for flame impingement heat transfer. *Int. J. Heat Mass Transfer* **39**, 2989–3002.
62. Ashforth-Frost, S., Jambunathan, K., and Whitney, C. F. (1997). Velocity and turbulence characteristics of a semiconfined orthogonally impinging slot jet. *Exp. Thermal Fluid Sci.* **14**, 60–67.
63. Chan, T. L., Leung, C. W., Jambunathan, K. J., Ashforth-Frost, S., Zhou, Y., and Liu, M. H. (2002). Heat transfer characteristics of a slot jet impinging on a semi-circular convex surface. *Int. J. Heat Mass Transfer* **45**, 993–1006.
64. Colucci, D. W. and Viskanta, R. (1996). Effect of nozzle geometry on local convective heat transfer to a confined impinging air jet. *Exp. Thermal Fluid Sci.* **13**, 71–80.
65. Wiberg, R. and Lior, N. (2004). Errors in thermochromic liquid crystal thermometry. *Rev. Sci. Instrum.* **75**(9), 2985–2994.
66. Wiberg, R., Muhammad-Klingmann, B., Ferrari, J., and Lior, N. (2000). Thermochromic coatings help characterize gas quenching. *ASM Int. Heat Treat. Prog.* **158**(4), 37–40.
67. Burmeister, L. C. (1993). "Convective Heat Transfer". Wiley, New York, NY.
68. Kang, S. H. and Greif, R. (1992). Flow and heat transfer to a circular cylinder with a hot impinging jet. *Int. J. Heat Mass Transfer* **35**, 2173–2183.
69. Polat, S., Huang, B., Mujumdar, A. S., and Douglas, W. J. M. (1989). Numerical flow and heat transfer under impinging jets: A review. *Ann. Rev. Fluid Mech. Heat Transfer* **2**, 157–197.
70. Beaubert, F. and Viazzo, S. (2002). Large eddy simulation of a plane impinging jet. *C. R. Mecanique* **330**, 803–810.
71. Hällqvist, T. (2003). "Numerical study of impinging jets with heat transfer, Licentiate Thesis". Royal Institute of Technology, Stockholm, Sweden, Universitetsservice US AB Stockholm. ISSN 0348-467X.
72. Olsson, M. and Fuchs, L. (1998). Large eddy simulations of a forced semiconfined circular impinging jet. *Phys. Fluids* **10**(2), 476–486.
73. Cziesla, T., Biswas, G., Chattopadhyay, H., and Mitra, N. K. (2001). Large-eddy simulation of flow and heat transfer in an impinging slot jet. *Int. J. Heat Fluid Flow* **22**, 500–508.
74. Chen, C.-J. and Jaw, S.-Y. (1998). "Fundamentals of Turbulence Modeling. Taylor and Francis". Washington, DC.
75. Craft, T. J., Graham, L. J. W., and Launder, B. E. (1993). Impinging jet studies for turbulence model assessment – II. An examination of the performance of four turbulence models. *Int. J. Heat Mass Transfer* **36**, 2685–2697.
76. Bouainouche, M., Bourabaa, N., and Desmet, B. (1997). Numerical study of the wall shear stress produced by the impingement of a plane turbulent jet on a plate. *Int. J. Numer. Methods Heat Fluid Flow* **7**, 548–564.
77. Esch, T., Menter, F., and Vieser, W. (2003). Heat transfer predictions based on two-equation turbulence models. *In* "The 6th ASME-JSME Thermal Engineering Joint Conference, March 16–20, 2003," TED-AJ03-542.

78. Heyerichs, K. and Pollard, A. (1996). Heat transfer in separated and impinging turbulent flows. *Int. J. Heat Mass Transfer* **39**, 2385–2400.
79. Craft, T. J., Launder, B. E., and Suga, K. (1996). Development and application of a cubic eddy-viscosity model of turbulence. *Int. J. Heat Fluid Flow* **17**, 108–115.
80. Turgeon, E. and Pelletier, D. (2001). Verification and validation of adaptive finite element method for impingement heat transfer. *J. Thermophys. Heat Transfer* **15**, 284–292.
81. Merci, B., Vierendeels, I., DeLange, C., and Dick, E. (2003). Numerical simulation of heat transfer of turbulent impinging jets with two-equation turbulence models. *Int. J. Numer. Methods Heat Fluid Flow* **13**, 110–132.
82. Souris, N., Liakos, H., Founti, M., Palyvos, J., and Markatos, N. (2002). Study of impinging turbulent jet flows using the isotropic low-reynolds number and the algebraic stress methods. *Comput. Mech.* **28**, 381–389.
83. Tzeng, P. Y., Soong, C. Y., and Hsieh, C. D. (1999). Numerical investigation of heat transfer under confined impinging turbulent slot jets. *Numerical Heat Transfer, Part A* **35**, 903–924.
84. Abe, K., Kondoh, T., and Nagano, Y. (1994). A new turbulence model for predicting fluid flow and heat transfer in separating and reattaching flows – I: Flow field calculations. *Int. J. Heat Mass Transfer* **37**, 139–151.
85. Hayase, T., Humphrey, J. A. C., and Greif, R. (1992). A consistently formulated quick scheme for fast and stable convergence using finite-volume iterative calculation procedures. *J. Comput. Phys.* **98**, 108–118.
86. Heck, U., Fritsching, K., and Bauckhage, K. (2001). Fluid flow and heat transfer in gas jet quenching of a cylinder. *Int. J. Numer. Methods Heat Fluid Flow* **11**, 36–49.
87. Wilcox, D. C. (2002). "Turbulence Modeling For CFD," 2nd edn. DCW Industries, La Cañada, CA.
88. Park, T. H., Choi, H. G., Yoo, J. Y., and Kim, S. J. (2003). Streamline upwind numerical simulation of two-dimensional confined impinging slot jets. *Int. J. Heat Mass Transfer* **46**, 251–262.
89. Chen, Q. and Modi, V. (1999). Mass transfer in turbulent impinging slot jets. *Int. J. Heat Mass Transfer* **42**, 873–887.
90. Shih, T.-H., Liou, W. W., Shabbir, A., Yang, Z., and Zhu, J. (1995). A new k–ε eddy-viscosity model for high reynolds number turbulent flows – model development and validation. *Comput. Fluids* **24**(3), 227–238.
91. Abdon, A. and Sunden, B. (2001). Numerical investigation of impingement heat transfer using linear and nonlinear two-equation turbulence models. *Numer. Heat Transfer, Pt. A* **40**, 563–578.
92. Park, T. S. and Sung, H. J. (2001). Development of a near-wall turbulence model and application to jet impingement heat transfer. *Int. J. Heat Fluid Flow* **22**, 10–18.
93. Funazaki, K. and Hachiya, K. (2003). Systematic numerical studies on heat transfer and aerodynamic characteristics of impingement cooling devices combined with pins, *In* "Proceedings of ASME Turbo Expo 2003, June 16–19, Atlanta, GA, USA".
94. Demuren, A. O. (1994). Calculations of 3D impinging jets in crossflow with Reynolds stress models. *In* "Heat Transfer in Turbomachinery" (R.J. Goldstein, D.E. Metzger and A.I. Leontiev, eds.), pp. 527–540. Begell House, Inc., New York.
95. Shi, Y., Ray, M. B., and Mujumdar, A. S. (2002). Computational study of impingement heat transfer under a turbulent slot jet. *Ind. Eng. Chem. Res.* **41**, 4643–4651.
96. Behnia, M., Parneix, S., and Durbin, P. A. (1998). Prediction of heat transfer in an axisymmetric turbulent jet impinging on a flat plate. *Int. J. Heat Mass Transfer* **41**, 1845–1855.

97. Thielen, L., Jonker, H. J. J., and Hanjalic, K. (2003). Symmetry breaking of flow and heat transfer in multiple impinging jets. *Int. J. Heat Fluid Flow* **24**, 444–453.
98. Menter, F. R. (1993). Zonal two equation k–ω turbulence models for aerodynamic flows. *In* "AIAA 24th Fluid Dynamics Conference, July 6–9, 1993, Orlando, FL". AIAA-93-2906.
99. Goldstein, R. J. and Seol, W. S. (1991). Heat transfer to a row of impinging circular air jets including the effect of entrainment. *Int. J. Heat Mass Transfer* **34**, 2133–2147.
100. Goldstein, R. J. and Behbahani, A. I. (1982). Impingement of a circular jet with and without crossflow. *Int. J. Heat Mass Transfer* **25**, 1377–1382.
101. Huang, L. and El-Genk, M. (1994). Heat transfer of an impinging jet on a flat surface. *Int. J. Heat Mass Transfer* **37**, 1915–1923.
102. Lytle, D. and Webb, B. W. (1994). Air jet impingement heat transfer at low nozzle-plate spacings. *Int. J. Heat Mass Transfer* **37**, 1687–1697.
103. Mohanty, A. K. and Tawfek, A. A. (1993). Heat transfer due to a round jet impinging normal to a flat surface. *Int. J. Heat Mass Transfer* **36**, 1639–1647.
104. Tawfek, A. A. (1996). Heat transfer and pressure distributions of an impinging jet on a flat surface. *Heat Mass Transfer* **32**, 49–54.
105. Wen, M.-Y. and Jang, K.-J. (2003). An impingement cooling on a flat surface by using circular jet with longitudinal swirling strips. *Int. J. Heat Mass Transfer* **46**, 4657–4667.
106. Bartoli, C., Di Marco, P., and Faggiani, S. (1993). Impingement heat transfer at a circular cylinder due to a submerged slot jet of water. *Exp. Thermal Fluid Sci.* **7**, 279–286.
107. Goldstein, R. J., Sobolik, K. A., and Seol, W. S. (1990). Effect of entrainment on the heat transfer to a heated circular air jet impinging on a flat surface. *ASME J. Heat Transfer* **112**, 608–611.
108. Goldstein, R. J. and Franchett, M. E. (1988). Heat transfer from a flat surface to an oblique impinging jet. *ASME J. Heat Transfer* **110**, 84–90.
109. Hargrave, G. K., Fairweather, M., and Kilham, J. K. (1986). Turbulence enhancement of stagnation point heat transfer on a circular cylinder. *Int. J. Heat Fluid Flow* **7**, 132–138.
110. Hrycak, P. (1981). Heat transfer from a row of impinging jet to concave cylindrical surfaces. *Int. J. Heat Mass Transfer* **24**, 407–419.
111. Li, C.-Y. and Garimella, S. V. (2001). Prandtl-number effects and generalized correlations for confined and submerged jet impingement. *Int. J. Heat Mass Transfer* **44**, 3471–3480.
112. Sparrow, E. M. and Alhomoud, A. (1984). Impingement heat transfer at a circular cylinder due to an offset or non-offset slot jet. *Int. J. Heat Mass Transfer* **27**, 2297–2306.

AUTHOR INDEX

Numerals in parentheses following the page numbers refer to reference numbers cited in the text

A

Abdel-Khalik, S. I., 509(94)
Abdon, A., 602(91)
Abe, K., 578(21), 600(84)
Abramowitz, M., 390(22)
Abramson, A. R., 238(125), 243(125)
Acrivos, A., 317(69; 70)
Advani, S. G., 351(104)
Agarwal, R., 369(131)
Ahuja, A. S., 356(117; 118), 357(117; 118), 358(117)
Ai, F., 271(35; 36), 285(35; 36), 288(36), 289(36), 290(36), 291(35), 292(35), 293(35), 294(35), 309(35)
Aichlmayr, H. T., 377(75), 385(18), 387(18), 388(18), 392(18), 393(18), 394(18), 400(18), 403(18)
Akers, W. W., 530(130), 546(130), 547(130), 548(130)
Akfirat, J. C., 579(23)
Aldabbagh, L. B. Y., 582(29)
Alfke, H., 257(133), 369(133)
Alhomoud, A., 618(112)
Allen, J. S., 476(45), 477(45)
Allen, M. P., 176(15)
Allen, P. B., 208(48; 49), 242(132)
Anderson, C. V. D. R., 242(133)
Anderson, H. C., 181(25)
Arai, N., 566(8)
Arcidiacono, S., 175(14)
Artioli, G., 194(40)
Asai, A., 500(83), 501(83)
Ashcroft, N. W., 170(3), 173(3), 182(3), 183(3), 184(3), 185(3), 190(3), 192(3)
Ashforth-Frost, S., 589(62; 63), 610(63), 611(63)

B

Badran, B., 470(28), 473(28)
Bae, K. D., 481(57)
Baek, S. S., 479(52), 481(57)
Baird, J. R., 531(132), 548(132)
Baird, R. M., 378(2)
Bakirci, K., 584(36)
Bala, P. R., 387(20)
Ball, P., 267(25)
Bandyopadhaya, A., 380(8), 393(8), 394(8), 403(8), 410(8), 411(8)
Barbu-Tudoran, L., 369(133)
Barone, P., 294(54)
Barrat, J.-L., 233(108), 237(108), 240(108), 243(108)
Bart, G. C. J., 585(42)
Bartoli, C., 618(106)
Batchelor, G. K., 318(71), 319(71), 341(93; 94), 342(95), 348(93), 349(94), 350(94; 95), 440(59), 442(59), 443(59), 444(59), 451(59)
Bauckhage, K., 600(86)
Bauer, R., 440(58)
Baukal, C. E., 588(60; 61)
Bea, J. E., 467(22)
Beasley, D. E., 413(50), 414(50), 415(50)
Beaubert, F., 593(70)
Beauchamp, P., 213(89), 243(89)
Bedeaux, D., 347(103)
Bedrov, R., 239(129)
Begg, E., 550(145), 551(145), 552(145)
Behbahani, A. I., 571(10), 611(10; 100), 612(100), 613(10)
Behnia, M., 579(25), 582(25), 605(25; 96), 620(25)
Bejan, A., 378(1)
Ben-Jacob, E., 267(20)
Benton, J. F., 494(73)
Berber, S., 213(79), 227(79), 243(79)
Berenson, P. J., 554(147)

Bernandes, N., 199(43)
Berre, M. L., 471(30)
Berry, R. A., 583(32)
Bhattacharya, A., 380(11)
Bhattacharya, P., 355(111)
Bilen, K., 584(36)
Biswas, G., 594(73)
Biswas, R., 213(69), 224(69), 225(69), 228(69), 229(69), 242(69), 245(69)
Bodapati, A., 213(81), 243(81)
Boegli, J. S., 383(16), 384(16), 392(16), 393(16), 405(16), 406(16), 415(16), 439(16), 440(16)
Bonnecaze, R. T., 326(79; 80), 327(79)
Borca-Tasciuc, T., 213(65), 223(65)
Bornea, D., 518(107)
Bossi, L., 586(48), 611(48), 613(48)
Bossis, G., 326(78), 339(89), 340(89), 341(89), 342(89), 343(89), 350(78)
Bottger, H., 237(114), 240(114), 244(114)
Bouainouche, M., 595(76)
Bourabaa, N., 595(76)
Brady, J. F., 326(78; 79; 80), 327(79), 350(78)
Brereton, G. J., 488(67), 489(67)
Briggs, M. E., 339(90; 91)
Brock, J. R., 352(106)
Brown, W. F., 344(101), 345(101), 346(101), 394(30), 396(30), 398(30), 402(30), 426(30)
Bruggeman Von, D. A. G., 313(65)
Brutin, D., 521(112), 523(112), 524(112)
Bucher, E., 219(92)
Buevich, Y. A., 355(109)
Buhrman, R. A., 264(11), 265(14)
Buonanno, G., 449(66; 68; 69), 450(69), 454(69)
Burmeister, L. C., 592(67)
Button, B. L., 566(3)

C

Cagin, T., 213(77; 78), 215(77), 216(77), 217(77; 78), 219(77), 220(77), 221(77), 227(77), 232(77), 241(77), 243(78), 245(77), 294(57)
Cahill, D. G., 169(1; 2), 185(32; 33), 203(32; 33), 218(32; 33), 294(54)
Calabrese, R. V., 339(90; 91)
Callaway, J., 200(44), 209(44)

Calmidi, V. V., 380(10; 11)
Cao, G., 213(80), 243(80)
Cao, Y., 467(22)
Car, R., 238(120), 240(120)
Carbonell, R. G., 385(19), 386(19), 392(19), 393(19), 416(19; 51), 425(51), 426(51), 427(19; 51), 428(51), 431(51), 434(19; 51), 435(51), 436(51), 451(51)
Carey, V. P., 487(64)
Carotenuto, A., 449(66; 68; 69), 450(69), 454(69)
Carslaw, H. S., 389(21), 392(21), 445(21), 448(21)
Castillo, J., 353(107), 354(107)
Cau, C., 587(53), 618(53)
Cavicchi, E., 267(22)
Challoner, A. R., 271(38)
Chan, C. K., 444(63), 445(63), 447(63), 448(63), 451(63)
Chan, C. T., 213(69), 224(69), 225(69), 228(69), 229(69), 242(69), 245(69)
Chan, T. L., 589(63), 610(63), 611(63)
Chantrenne, P., 233(108), 237(108), 240(108), 243(108)
Chattopadhyay, H., 584(35), 594(73)
Che, J., 213(77; 78), 215(77), 216(77), 217(77; 78), 219(77), 220(77), 221(77), 227(77), 232(77), 241(77), 243(78), 245(77), 294(57)
Chedester, R. C., 510(97), 511(97)
Chen, C.-J., 594(74), 603(74)
Chen, G., 213(70; 71; 89), 223(71), 228(71), 229(71), 231(71), 241(70), 243(89), 245(71), 333(84), 335(84), 336(84)
Chen, H. C., 473(34)
Chen, Q., 602(89)
Chen, X. L., 267(26)
Chen, Y., 213(67), 238(126), 241(67), 243(126)
Chen, Y. P., 533(122), 537(122)
Chen, Z. Q., 380(9)
Cheng, G. J., 450(73)
Cheng, P., 380(9), 410(46), 431(46), 432(46), 434(46; 57), 435(57), 436(57), 439(57), 451(46; 57), 461(148), 462(6; 8), 484(6; 8), 485(6; 8), 486(8), 488(66), 489(66), 505(6; 8), 508(6), 515(66), 519(108; 109; 110), 520(110), 521(109; 110), 522(110), 529(108; 109; 110), 533(122; 124), 537(122), 538(124),

540(124), 541(124), 542(124), 543(125), 544(125)
Cheng, S. C., 413(49)
Chernozatonskii, L. A., 237(112), 243(112)
Cheung, C. S., 588(59)
Cho, C. H., 481(57)
Cho, H. H., 572(11), 585(44), 587(54)
Cho, Y. I., 572(11)
Choi, B. Y., 479(52)
Choi, H. G., 601(88)
Choi, M., 294(55), 295(55), 305(55), 306(55), 307(55), 308(55)
Choi, S., 355(110), 359(119)
Choi, S. U. S., 213(88), 224(88), 271(29; 31; 32), 272(31), 273(31), 278(31), 279(31), 280(31), 281(31), 284(32), 287(32), 294(52), 304(52), 307(52), 356(115; 116)
Christen, D. K., 228(103)
Christensen, R. N., 531(120), 535(120), 537(120), 543(120), 547(120)
Chung, C. M., 586(46)
Chung, H. J., 479(51)
Chung, J. D., 203(47)
Chung, Y. M., 579(24)
Chung, Y. S., 584(37)
Ciccotti, G., 213(62), 224(62), 237(116)
Cole, R., 487(65)
Coleman, J. W., 529(117; 118), 530(117; 118), 534(118), 535(118), 543(117; 118), 549(141; 142), 550(141)
Collier, D. W., 404(39)
Colucci, D. W., 589(64)
Cooper, D., 578(22), 599(22)
Cornaro, C., 586(50)
Cornwell, K., 464(14), 465(14)
Craft, T. J., 594(75), 597(75), 599(75; 79), 604(75)
Crane, R. A., 412(47), 413(47), 414(47), 415(47)
Crilly, R. J., 488(67), 489(67)
Crosser, O. K., 277(47), 279(47), 280(47), 292(47), 314(47)
Cui, Y., 369(130)
Cummings, A., 237(113), 243(113)
Czarnetzki, W., 275(44)
Cziesla, T., 594(73)
Czitrovszky, A., 265(12)

D

Daly, B. C., 239(130; 131), 243(130; 131)
D'Amico, S., 366(128), 367(128)
Dapiaggi, M., 194(40)
Das, S. K., 275(45), 276(45), 291(45), 293(48), 295(45), 296(45), 297(45), 298(45), 299(48), 300(48), 362(124), 366(127), 367(127)
Datta, A., 356(113)
Davis, R. H., 322(76)
de Paiva, K. V., 472(33)
Deissler, R. G., 383(15; 16), 384(15; 16), 392(16), 393(16), 404(15; 16), 405(15; 16), 406(16), 415(15; 16), 439(16), 440(16)
DeLange, C., 599(81)
Demuren, A. O., 604(94)
Deng, P. G., 462(6; 8), 484(6; 8), 485(6; 8), 486(8), 505(6; 8), 508(6)
Deng, W., 213(77), 215(77), 216(77), 217(77), 219(77), 220(77), 221(77), 227(77), 232(77), 241(77), 245(77)
Depondt, B., 213(63)
Derewnicki, K. P., 498(81), 499(81)
Desmet, B., 595(76)
Di Marco, P., 618(106)
Dick, E., 599(81)
Dimitrijevic, N. M., 267(23; 24)
Ding, H.-Q., 178(18; 19)
Dixon, M., 226(96)
Domingues, G., 222(94)
Donaldson, C. D., 566(4)
Dong, J., 213(83), 221(83), 242(83), 245(83)
Dong, L. L., 588(59)
Donovan, B., 467(22)
Dorfman, J. R., 339(90; 91)
Douglas, W. J. M., 593(69)
Dove, M. T., 170(5), 192(5), 194(5; 41), 197(5), 204(5)
Dresselhaus, M. S., 294(58)
Drotboom, M., 265(13)
Du, X. Z., 530(127), 544(127), 545(127)
Duan, X., 369(130; 131)
Duckler, A. E., 518(107)
Duncan, A. B., 271(42), 274(42), 469(27)
Durbin, M. K., 356(113)
Durbin, P. A., 579(25), 582(25), 605(25; 96), 620(25)
Dutta, P., 356(113)
Dutta, S., 566(7), 586(7)

E

Eastman, J. A., 213(88), 224(88), 271(31; 32), 272(31), 273(31), 278(31), 279(31), 280(31), 281(31), 284(32), 287(32), 356(115; 116)
Echigo, R., 588(57)
Eggebrecht, J., 178(20)
Eian, C. S., 383(15), 384(15), 404(15), 405(15), 415(15)
Einstein, A., 185(31)
Ekkad, S., 566(7), 586(7)
Eklund, P. C., 294(58)
El-Genk, M., 611(101)
Ermis, K., 530(137), 548(137)
Evans, D. J., 226(95; 97; 98), 227(98)

F

Fabian, J., 242(132)
Faggiani, S., 618(106)
Faghri, A., 475(41), 550(145), 551(145), 552(145)
Fagri, A., 467(23), 468(23)
Fairweather, M., 618(109)
Fedorov, A. G., 588(58)
Feidt, M., 387(20)
Feldman, J. L., 208(48; 49), 242(132)
Ferrari, J., 565(1), 590(66), 592(1)
Feynman, R. P., 262(5; 6)
Fissan, H., 265(13)
Fleischer, A. S., 586(47; 50)
Fletcher, D. F., 531(132), 548(132)
Fletcher, L. S., 271(39; 40; 41; 42), 274(42)
Florschuetz, L. W., 583(32; 33; 34), 610(33), 611(33)
Ford, W. K., 169(2)
Founti, M., 599(82), 603(82)
Fourier, J., 257(1)
Franchett, M. E., 618(108)
Francis, M. K., 339(90; 91)
Frenkel, D., 176(16)
Fricke, H., 310(63), 312(63), 313(63)
Fritsching, K., 600(86)
Fritz, W., 463(12)
Fuchs, L., 593(72)
Fujii, Y., 199(42)

G

Gaddis, J. L., 587(55)
Gahler, F., 171(7)
Gale, J. D., 194(41)
Gallico, R., 237(116)
Gammon, R. W., 339(90; 91)
Gan, G., 573(16)
Gan, Y., 525(115), 527(115), 528(115)
Gao, M., 467(22)
Gao, S., 577(19), 593(19)
Gardon, R., 579(23)
Garimella, S., 529(117; 118), 530(117; 118; 119), 534(118; 119), 535(118), 536(119), 543(117; 118; 119), 549(141; 142), 550(141)
Garimella, S. V., 509(95), 618(111)
Garwin, L., 267(25)
Gaspard, P., 339(90; 91)
Gau, C., 586(46)
Gebhart, B., 588(60; 61)
Geers, L. F. G., 585(42)
Gemant, A., 405(40), 430(40)
Geng, X., 481(58), 482(58)
Germann, T. C., 171(8)
Gerner, F. M., 470(28), 473(28)
Gersten, B., 213(80), 243(80)
Ghiaasiaan, S. M., 509(94), 510(97), 511(97)
Giaever, I., 267(21)
Gillan, M. J., 212(57; 58), 213(57; 58), 221(57), 226(96)
Giovinco, G., 449(69), 450(69), 454(69)
Glatzmaier, G. C., 379(3), 390(3), 391(3; 25)
Gleiter, H., 264(10)
Glod, S., 498(82), 500(82), 504(89), 506(82)
Goddard, W. A. III, 213(77; 78), 215(77), 216(77), 217(77; 78), 219(77), 220(77), 221(77), 227(77), 232(77), 241(77), 243(78), 245(77), 294(57)
Goddard, W. A., 178(18; 19)
Goldstein, R. J., 571(10), 586(47; 50), 611(10; 99; 100), 612(99; 100), 613(10), 618(107; 108)
Goodson, K. E., 169(1; 2), 211(52), 511(99), 512(99)
Goodson, M., 339(92)
Gori, F., 392(26), 586(48), 611(48), 613(48)
Gorman, J., 369(132)
Göppert, S., 585(43)

Graham, L. J. W., 594(75), 597(75), 599(75), 604(75)
Granqvist, C. G., 264(11)
Grasselli, Y., 339(89), 340(89), 341(89), 342(89), 343(89)
Greegor, R. B., 189(36)
Green, M. S., 212(53)
Greif, R., 592(68), 600(85)
Groll, M., 467(21), 518(105), 519(105)
Gürtler, T., 585(43)
Grujicic, M., 213(80), 243(80)
Grulke, E. A., 294(52), 304(52), 307(52)
Guczi, L. D., 268(27)
Gupte, S. K., 351(104)

H

Hadley, G. R., 381(13), 382(13), 395(13), 402(13), 428(13), 431(13), 434(13), 451(13)
Hafskjold, B., 238(122)
Hallinan, K. P., 476(45), 477(45)
Halperin, W. P., 265(14)
Hamilton, R. L., 277(47), 279(47), 280(47), 292(47), 314(47)
Hammonds, K. D., 194(41)
Han, C. H., 479(51)
Han, J.-C., 566(7), 586(7)
Hanjalic, K., 585(42), 606(97)
Hansen, J. P., 227(99)
Harding, J. H., 213(59), 241(59)
Hargrave, G. K., 618(109)
Harrison, J., 385(17)
Hashin, Z., 318(72), 397(35)
Hasselman, D. P. H., 325(77)
Hatta, H., 322(75), 323(75), 325(75), 330(75)
Hattori, A., 362(123)
Haw, M. D., 339(87), 340(87)
Hayase, T., 600(85)
Haynes, B. S., 531(132), 548(132)
Heck, U., 600(86)
Heine, V., 194(41)
Heino, P., 238(121), 240(121)
Helfand, E., 212(56)
Heppelmann, K. K., 571(10), 611(10), 613(10)
Herwig, H., 585(43)
Hetsroni, G., 520(111)
Heyerichs, K., 598(78), 601(78)

Hibiki, T., 549(140)
Higa, K., 588(56)
Hirosaki, N., 213(76), 217(76), 226(76)
Hishida, K., 584(38)
Hiwatari, Y., 213(64), 227(64)
Hällqvist, T., 593(71)
Ho, C. M., 480(53; 54)
Ho, K. M., 213(69), 224(69), 225(69), 228(69), 229(69), 242(69), 245(69)
Ho, T. L., 267(19)
Hoek, L., 585(42)
Hoheisel, C., 213(62), 224(62)
Holland, M. G., 200(45), 209(45)
Honda, H., 461(148), 554(148), 555(148)
Hone, J., 294(51), 303(51)
Hoogendorn, C. J., 578(20), 618(20)
Hoover, W. G., 180(24), 200(46), 201(46), 213(46), 215(46), 216(46), 224(46)
Hopkins, R., 467(23), 468(23)
Hou, T. F., 477(48)
Hrycak, P., 618(110)
Hsieh, C. D., 599(83)
Hsu, C. C., 477(48)
Hsu, C. T., 380(9), 410(46), 431(46), 432(46), 434(46; 57), 435(57), 436(57), 439(57), 451(46; 57)
Hu, H. Y., 489(68), 490(68), 491(68), 515(68)
Huang, B., 593(69)
Huang, D., 470(29), 471(29)
Huang, L., 611(101)
Huang, X. Y., 484(62; 63)
Huang, Y., 369(130; 131)
Huber, A. M., 582(30), 611(30), 614(30)
Huisken, J., 353(108), 354(108)
Hulbert G., 213(66), 242(66)
Hummes, D., 265(13)
Humphrey, J. A. C., 600(85)
Huq, P., 351(104)
Hussein, M. I., 213(66), 242(66)
Huxtable, S., 294(54)
Hwang, S. D., 585(44)

I

Ichikawa, Y., 213(64), 227(64)
Ichimiya, K., 585(45)
Igarashi, Y., 584(38)
Iida, Y., 502(86; 87), 503(86), 504(86)
Ikeshoji, T., 238(122; 127)

Imamura, K., 237(115), 239(115; 130), 243(115; 130)
Incropera, F. P., 582(28), 620(28)
Inoue, R., 213(87), 224(87), 227(87)
Isaev, A. M., 355(109)
Ishii, H., 213(73; 84), 241(73; 84)
Ishii, M., 518(106)

J

Jackson, C., 578(22), 599(22)
Jaeger, J. C., 389(21), 392(21), 445(21), 448(21)
Jaguaribe, E. F., 413(50), 414(50), 415(50)
Jambunathan, K., 566(3), 589(62)
Jambunathan, K. J., 589(63), 610(63), 611(63)
Jang, K.-J., 611(105), 617(105)
Jang, S. P., 355(110)
Jang, W., 294(59), 295(59), 306(59)
Jani, P., 265(12)
Jaster, H., 547(131)
Jaw, S.-Y., 594(74), 603(74)
Jeffrey, D. J., 316(68), 332(68), 333(68), 334(68), 335(68), 395(33)
Jiang, L., 511(99), 512(99), 515(103), 517(103)
Jiang, L. N., 466(19)
Jin, C., 483(60)
Johnson, L. F., 325(77)
Johnson, W. C., 404(39)
Jonker, H. J. J., 606(97)
Jullien, R., 238(119), 240(119), 242(119)
Jun, T. K., 483(60)
Jund, P., 238(119), 240(119), 242(119)
Jung, J. Y., 505(91)

K

Kaburaki, H., 183(29), 192(37), 213(29), 216(29)
Kadau, K., 171(8)
Kaganer, M. G., 444(62), 445(62), 448(62), 451(62)
Kakimoto, K., 213(73; 84), 241(73; 84)
Kalinowski, M., 257(133), 369(133)
Kalman, H., 380(7), 382(7), 450(7)
Kamat, P. V., 267(23; 24)

Kandlikar, S. G., 462(9), 465(15), 509(9), 514(102)
Kang, C., 362(123)
Kang, S. H., 592(68)
Kang, S. W., 470(29), 471(29), 473(34; 35), 477(48)
Karasawa, N., 178(18; 19)
Kasagi, N., 510(98), 511(98)
Kato, K., 405(42), 439(42)
Kaviany, M., 169(133), 183(27; 28), 186(27; 34), 200(28), 201(28), 202(28), 203(47), 204(28), 207(28), 213(27; 28; 34; 66), 216(27), 217(27; 34), 219(34), 221(34), 222(34), 241(34), 242(34; 66), 247(28), 428(55)
Kawazoe, Y., 238(127)
Kayansayan, N., 587(51)
Keblinski, P., 178(20), 211(50), 213(61; 81; 88), 217(61), 224(88), 226(61), 233(61), 235(61), 236(61), 238(61), 239(61), 240(61), 242(61), 243(81), 244(61), 245(61), 294(56), 356(115)
Keller, J. B., 315(66)
Kenny, T. W., 511(99), 512(99)
Keon, K., 481(55)
Kew, P. A., 464(14), 465(14)
Khanafer, K., 363(125)
Khrustalev, D., 467(23), 468(23), 550(145), 551(145), 552(145)
Kikuchi, S., 433(74), 450(74)
Kilham, J. K., 618(109)
Killion, J. D., 549(141; 142), 550(141)
Kim, B. G., 572(11)
Kim, C. J., 480(53; 54), 483(60)
Kim, C. K., 479(51)
Kim, D., 294(59), 295(59), 306(59)
Kim, H. C., 481(55)
Kim, I. C., 329(82), 345(97; 98; 99; 100)
Kim, J. H., 364(126), 365(126), 494(73)
Kim, J. K., 479(51)
Kim, P., 294(59; 60), 295(59; 60), 306(59)
Kim, S. J., 601(88)
Kim, S. T., 328(81), 332(81)
Kim, Y. S., 481(57)
Kim, Y. T., 584(37)
Kissel, T., 257(133), 369(133)
Kitagawi, H., 213(76), 217(76), 226(76)
Kitscha, W. W., 448(65)
Kittel, C., 194(38)

Kladias, N., 380(8), 393(8), 394(8), 403(8), 410(8), 411(8)
Kloc, C., 219(92)
Kluge, M. D., 208(49)
Knight, W. D., 266(18)
Ko, M. H., 473(35)
Kocer, C., 213(76), 217(76), 226(76)
Konashi, K., 238(127)
Kondoh, T., 600(84)
Konstantinov, V. A., 219(93)
Koo, J. M., 511(99), 512(99)
Kosky, P. G., 547(131)
Kotake, S., 238(128)
Koyayashi, H., 462(5), 478(5)
Kraft, M., 339(92)
Kramer, K., 586(47)
Krupiczka, R., 393(27), 404(27), 405(27), 406(27), 410(27), 411(27), 415(27)
Kubo, R., 212(54)
Kucuka, S., 587(51)
Kuk, K., 481(56; 57)
Kulacki, F. A., 377(75)
Kulenovic, R., 518(105), 519(105)
Kumar, R., 366(128), 367(128)
Kunii, D., 381(12), 382(14), 383(14), 385(14), 392(12), 393(12), 394(12), 403(12), 407(44), 410(44), 411(44), 415(44), 416(12; 44)
Kurosaki, K., 213(85; 86)
Kwak, H. Y., 505(91)
Kwon, Y.-K., 213(79), 227(79), 243(79)

L

Lacroix, C., 387(20)
Ladd, A. J. C., 200(46), 201(46), 213(46), 215(46), 216(46), 224(46)
Lage, J. L., 379(5)
Lai, E., 566(3)
Lai, M., 581(27), 582(27), 611(27), 617(27)
Lallemand, M., 213(63), 467(21), 471(30), 472(33)
Landau, L. D., 442(60)
Launay, S., 471(30), 472(33)
Launder, B. E., 578(22), 594(75), 597(75), 599(22; 75; 79), 604(75)
Leal, L. G., 315(67)
Leblinski, P., 356(116)
Lee, C. C., 587(53), 618(53)

Lee, C. H., 585(44)
Lee, C. S., 479(51), 481(56; 57)
Lee, D. H., 584(37)
Lee, D. J., 494(74; 75), 495(74; 75)
Lee, H., 294(55), 295(55), 305(55), 306(55), 307(55), 308(55)
Lee, H. D., 479(51)
Lee, H. J., 479(51)
Lee, J., 573(12), 580(12)
Lee, J. D., 479(51)
Lee, J. Y., 505(91)
Lee, M., 515(104), 516(104), 517(104)
Lee, P. C., 512(100), 513(100), 514(100), 525(100)
Lee, S., 271(31), 272(31), 273(31), 278(31), 279(31), 280(31), 281(31), 573(12), 580(12)
Lee, S. H. K., 450(72)
Lee, S. P., 359(119)
Lee, S. W., 481(55)
Lee, Y. H., 213(69), 224(69), 225(69), 228(69), 229(69), 242(69), 245(69)
Lee, Y. K., 462(6; 8), 484(6; 8), 485(6; 8), 486(8), 505(6; 8), 508(6)
Leland, J., 469(25)
Lester, J. C., 326(78), 350(78)
Leung, C. W., 588(59), 589(63), 610(63), 611(63)
Li, C. H., 257(133)
Li, C.-Y., 618(111)
Li, D., 213(67), 238(126), 241(67), 243(126), 294(59), 295(59), 306(59)
Li, D. Y., 233(107), 238(107), 241(107), 243(107)
Li, H., 356(114)
Li, H. Y., 514(101), 515(101)
Li, J., 178(17), 183(29), 184(17), 187(17), 212(17), 213(17; 29; 60; 74), 214(17; 60), 216(17; 29), 217(17), 220(60), 224(17; 74), 225(74), 228(60), 229(60), 231(17; 60), 241(60), 488(66), 489(66), 515(66)
Li, J. M., 462(11), 463(11), 464(11), 465(11)
Li, Q., 271(33; 34), 281(33; 34), 286(34), 359(121), 360(121), 361(121)
Li, S., 271(31; 32), 272(31), 273(31), 278(31), 279(31), 280(31), 281(31), 284(32), 287(32)
Li, X., 587(55)
Liakos, H., 599(82), 603(82)

Liang, X.-G., 233(107), 238(107; 124), 241(107), 243(107; 124)
Liao, G. X., 578(22), 599(22)
Liao, Q., 552(146), 553(146), 554(146)
Lichteneker, K., 398(36), 402(36)
Lieber, C. M., 369(130; 131)
Lifshitz, E. M., 442(60)
Lightstone, M., 363(125)
Lim, H. T., 481(57)
Lin, H. C., 329(83), 330(83), 331(83), 333(83), 334(83), 335(83)
Lin, L., 501(84; 85)
Lin, L. C., 469(25)
Lin, L. W., 491(69), 492(69), 493(69), 509(69)
Lin, W. K., 494(76; 77), 496(76; 77)
Lindan, P. J. D., 212(57), 213(57; 59), 221(57), 241(59)
Lior, N., 565(1; 112), 566(6; 8), 589(65), 590(66), 592(1)
Liou, W. W., 602(90)
Lisienko, V. G., 588(58)
Liu, M. H., 589(63), 610(63), 611(63)
Liu, Y., 271(35; 36), 285(35; 36), 288(36), 289(36), 290(36), 291(35), 292(35), 293(35), 294(35), 309(35)
Lobanov, D. L., 588(58)
Lockwood, F. E., 294(52), 304(52), 307(52)
Lomdahl, P. S., 171(8)
Londry, F. A., 385(17)
Longtin, J. P., 470(28), 473(28)
Lord, S. M., 431(56)
Lorents, D., 174(13), 175(13)
Lovell, B. J., 584(41), 618(41)
Lu, S. Y., 328(81), 329(83), 330(83), 331(83), 332(81), 333(83), 334(83), 335(83)
Luchnikov, V. A., 218(91)
Lukes, J. R., 213(67), 233(107), 238(107; 123; 126), 241(67; 107), 243(107; 126)
Lund, K. O., 431(56)
Luo, K. H., 579(24)
Lurie, N. A., 199(42)
Lyle, F. W., 189(36)
Lytle, D., 611(102), 613(102), 614(102)

M

MacDonald, R. A., 237(109)
Mackowski, D. W., 353(107), 354(107)
Maeda, A., 227(100)
Mahajan, R. L., 380(10; 11)
Mahan, G. D., 169(2), 228(104), 229(104)
Maiti, A., 228(104), 229(104)
Majumdar, A., 169(1; 2), 228(102), 238(125; 126), 243(125; 126), 294(59; 60), 295(59; 60), 306(59)
Malhotra, R., 174(13), 175(13)
Malikov, G. K., 588(58)
Malikov, K. Y., 588(58)
Mantelli, B. H., 472(33)
Marek, R., 493(71), 494(71)
Mareschal, M., 213(63)
Marino, C., 392(26)
Maris, H. J., 169(2), 237(115), 239 (115; 130; 131), 243(115; 130; 131)
Markatos, N., 599(82), 603(82)
Martin, H., 566(2), 569(2), 570(2), 574(2), 592(2), 611(2), 614(2), 615(2), 616(2), 620(2), 621(2)
Maruyama, S., 175(14), 237(117; 118), 243(118), 294(61), 313(61)
Massarotti, N., 449(69), 450(69), 454(69)
Matsui, H., 238(127)
Maurel, S., 569(9)
Maxwell, J. C., 262(3), 310(3), 311(3), 394(32), 395(32)
Mazo, R. M., 339(88), 340(88), 344(88)
Mazur, P., 347(103)
McCarthy, J. J., 450(70; 71)
McDaniel, C. S., 586(49)
McDonald, I. R., 227(99)
McEuen, P. L., 294(60), 295(60)
McGaughey, A. J. H., 169(133), 173(11), 183(11; 27; 28), 186(27; 34), 197(11), 200(28), 201(28), 202(28), 203(11; 47), 204(28), 207(28), 211(11), 212(11), 213(27; 28; 34; 66), 214(11), 216(11; 27), 217(11; 27; 34), 219(34), 221(34), 222(34), 239(11), 241(34), 242(34; 66), 246(11), 247(28)
McQuarrie, D. A., 179(21), 180(21), 212(21)
Medvedev, N. N., 218(91)
Meier, F., 265(15)
Meincke, P. P. M., 265(16)
Menon, M., 237(112; 113), 243(112; 113)
Merci, B., 599(81)
Meredith, R. E., 276(46), 298(46), 313(46), 314(46)
Merlin, R., 169(2)

Mermin, N. D., 170(3), 173(3), 182(3), 183(3), 184(3), 185(3), 190(3), 192(3)
Mertz, R., 518(105), 519(105)
Messmer, J. H., 379(6), 380(6), 390(6)
Metzger, D. E., 583(32; 33; 34), 610(33), 611(33)
Michalski, J., 237(110), 240(110), 242(110)
Mikic, B. B., 449(67)
Mishima, K., 518(106), 549(140)
Mistry, J., 574(17)
Mitra, N. K., 594(73)
Miyamae, S., 566(8)
Mocikat, H., 585(43)
Modi, V., 602(89)
Mohanty, A. K., 611(103), 616(103)
Moran, B., 200(46), 201(46), 213(46), 215(46), 216(46), 224(46)
Morriss, G. P., 226(97; 98), 227(98)
Moser, K. W., 530(134), 532(134), 548(134)
Moss, M. A., 566(3)
Mosyak, A., 520(111)
Motoyama, S., 213(64), 227(64)
Mottola, E., 267(20)
Mountain, R. D., 237(109)
Mudawar, I., 522(113), 525(113), 526(113)
Muhammad-Klingmann, B., 590(66)
Mujumdar, A. S., 593(69), 604(95)
Muller-Plathe, F., 233(105), 234(105), 238(105), 239(105)
Munakata, T., 227(100)
Murakawa, A., 213(73; 84), 241(73; 84)
Murashov, V. V., 213(75), 227(75)
Myles, C. W., 213(83), 221(83), 242(83), 245(83)

N

Na, B., 530(134), 532(134), 548(134)
Naberukhin, Y. I., 218(91)
Nagano, Y., 600(84)
Nagasaka, Y., 270(28), 271(28), 272(28)
Nagashima, A., 270(28), 271(28), 272(28)
Nagy, A., 265(12)
Nakamura, Y., 213(76), 217(76), 226(76)
Nakanishi, K., 213(87), 224(87), 227(87)
Nalwa, H. S., 263(8), 267(8)
Narayanan, V., 579(26)
Neumann, S., 265(13)
Nguyen, H., 431(56)

Nield, D. A., 378(1), 398(37), 411(37), 429(37)
Nielsen, N. J., 462(4), 478(4)
Nishiguchi, N., 237(115), 239(115), 243(115)
Nose, S., 180(22; 23)
Novikov, V. N., 218(91)
Novotny, V., 265(16)
Noya, E. G., 237(112), 243(112)
Nozad, I., 385(19), 386(19), 392(19), 393(19), 416(19; 51), 425(51), 426(51), 427(19; 51), 428(51), 431(51), 434(19; 51), 435(51), 436(51), 451(51)
Nozaki, A., 584(38)

O

O'Brien, R. W., 321(74), 395(34), 440(59), 442(59), 443(59), 444(59), 451(59)
Ochoa-Tapia, J. A., 428(53)
Oe, A., 213(64), 227(64)
Ofuchi, K., 381(12), 392(12), 393(12), 394(12), 403(12), 416(12)
Ogata, S., 213(76), 217(76), 226(76)
Ogniewicz, Y., 444(64), 447(64)
Oguz, H. N., 481(58), 482(58)
Oh, Y. S., 479(52), 481(55; 56; 57)
Ohara, T., 187(35)
Okada, M., 360(122), 362(123)
Okamoto, T., 462(7)
Okuyama, K., 502(86; 87), 503(86), 504(86)
Oligschleger, C., 237(111), 240(111), 241(111), 242(111)
Olsson, M., 593(72)
Omini, M., 174(12)
Ooi, K. T., 474(40)
Ornstein, L. S., 346(102)
Osman, M. A., 227(101), 237(101; 113), 243(101; 113)
Owen, I., 574(17)
Oyama, K., 362(123)
Ozisik, L. R., 294(54)
Ozisik, R., 213(81), 243(81), 294(56)

P

Page, R. H., 579(26)
Palyvos, J., 599(82), 603(82)

Pan, C., 494(76; 77), 496(76; 77), 512(100), 513(100), 514(100; 101), 515(101), 525(100)
Pantelides, S. T., 228(104), 229(104)
Paolini, G. V., 213(59), 241(59)
Park, H. C., 505(91)
Park, S. O., 481(56)
Park, T. H., 601(88)
Park, T. S., 602(92)
Parker, S. C., 194(39)
Parneix, S., 579(25), 582(25), 605(25; 96), 620(25)
Parrinello, M., 181(26)
Patel, H. E., 293(48), 299(48), 300(48)
Pavel, M. C., 213(65), 223(65)
Pavlov, P. A., 498(80), 506(80)
Pelletier, D., 599(80)
Peng, X. F., 337(86), 338(86), 356(114), 489(68), 490(68), 491(68), 494(74; 75; 76; 77), 495(74; 75), 496(76; 77), 509(96), 515(68)
Perrin, B., 213(63; 72)
Peterson, G., 472(32)
Peterson, G. P., 257(133), 271(39; 40; 41; 42), 274(42), 469(27), 472(31)
Peyret, R., 576(18), 596(18)
Phelan, P. E., 355(111)
Phillips, R. J., 326(78), 350(78)
Phillpot, S. R., 169(2), 178(20), 211(50; 51; 52), 213(61; 88), 217(61), 224(88), 226(61), 233(61; 106), 235(61; 106), 236(61), 238(61; 106), 239(61), 240(61), 241(106), 242(61), 244(61), 245(61), 247(51), 356(115; 116)
Picu, R. C., 213(65), 223(65)
Pietrafesa, M., 392(26)
Pisano, A. P., 501(84)
Poetzsch, R. H. H., 237(114), 240(114), 244(114)
Pohl, R. O., 185(32; 33), 203(32; 33), 218(32; 33)
Polat, S., 593(69)
Pollack, G. L., 228(103)
Pollard, A., 598(78), 601(78)
Ponnappan, R., 469(25)
Porter, L., 213(60), 214(60), 220(60), 228(60), 229(60), 231(60), 241(60)
Porter, L. J., 192(37)
Poulikakos, D., 175(14), 498(82), 500(82), 504(89), 506(82)

Powell, R. W., 271(38)
Prasad, V., 380(8), 393(8), 394(8), 403(8), 410(8), 411(8)
Prasher, R., 355(111)
Preston, F. W., 393(28)
Prosperetti, A., 481(58), 482(58)
Pryde, A. K. A., 194(41)
Putra, N., 275(45), 276(45), 291(45), 295(45), 296(45), 297(45), 298(45), 362(124), 366(127), 367(127)
Pynn, R., 199(42)

Q

Qu, W., 522(113), 525(113), 526(113)
Quintard, M. S., 421(52)

R

Radcliff, T. D., 531(120), 535(120), 537(120), 543(120), 547(120)
Rahimi, M., 574(17)
Rahman, A., 181(26)
Ramirez, W. F., 379(3), 390(3), 391(3; 25)
Ray, M. B., 604(95)
Rayleigh, L., 312(64), 394(31)
Rhee, D., 587(54)
Richter, A. G., 356(113)
Riffat, S. B., 573(16)
Ristolainen, E., 238(121), 240(121)
Roblee, L. H. S., 378(2)
Rocha, A., 317(69; 70)
Roetzel, W., 275(44; 45), 276(45), 291(45), 295(45), 296(45), 297(45), 298(45), 359(120), 362(124), 366(127), 367(127)
Rose, J. W., 554(148), 555(148)
Rosner, D. E., 353(107), 354(107)
Rosson, H. F., 530(130), 546(130), 547(130), 548(130)
Roth, J., 171(7)

S

Saha, S. K., 584(35)
Sahraoui, M., 428(55)
Saitoh, A., 588(56)
Sakurain, K., 502(86), 503(86), 504(86)
Salamon, M. B., 174(13), 175(13)
Saleh, J. M., 573(14)

San, J. Y., 581(27), 582(27), 611(27), 617(27)
Sankey, O. F., 213(83), 221(83), 242(83), 245(83)
Santiago, J. G., 511(99), 512(99)
Sartre, V., 467(21), 471(30), 472(33)
Saulnier, J.-B., 213(63; 89), 222(94), 243(89)
Scandolo, S., 183(30), 213(30; 68), 217(30), 238(120), 240(120)
Schaper, A. K., 257(133), 369(133)
Schelling, P. K., 211(50; 51; 52), 213(61), 217(61), 226(61), 233(61; 106), 235(61; 106), 236(61), 238(61; 106), 239(61), 240(61), 241(106), 242(61), 244(61), 245(61), 247(51)
Schlünder, E. U., 409(45), 410(45), 411(45), 415(45), 416(45), 431(45), 434(45), 440(45; 58)
Schmidt, F., 265(13)
Schneider, M., 467(21)
Schon, G., 267(20)
Schon, J. C., 237(111), 240(111), 241(111), 242(111)
Schumann, T., 402(38), 404(38), 406(38), 415(38)
Schuster, J. R., 554(147)
Segal, Z., 520(111)
Sengers, J. V., 339(90; 91)
Seol, W. S., 611(99), 612(99), 618(107)
Setayeshgar, L., 587(52)
Seyed-Yagoobi, J., 579(26)
Sezai, I., 582(29)
Shabany, Y., 579(25), 582(25), 605(25), 620(25)
Shabbir, A., 602(90)
Shafii, M. B., 475(41)
Shah, M. M., 546(129)
Shenogin, S., 213(81), 243(81), 294(56)
Shi, B., 238(124), 243(124)
Shi, L., 294(59; 60), 295(59; 60), 306(59)
Shi, Y., 604(95)
Shih, J. W., 481(57)
Shih, T.-H., 602(90)
Shimizu, A., 588(56)
Shin, J. W., 481(56)
Shin, S. J., 481(56; 57)
Shirane, G., 199(42)
Shonnard, D. R., 428(54), 443(54)
Shtrikman, S., 318(72), 397(35)
Shuai, J., 518(105), 519(105)

Sievers, A. J., 265(17)
Silsbee, R. H., 267(22)
Sinha, S., 211(52)
Siu, W. W. M., 450(72)
Skripov, V. P., 497(79), 498(79; 80), 506(80)
Slavin, A. J., 385(17)
Slayzak, S. J., 582(28), 620(28)
Slycke, J., 565(1), 592(1)
Smit, B., 176(16)
Smith, G. D., 239(129)
Smith, J. M., 407(44), 410(44), 411(44), 415(44), 416(44)
Snedeker, R. S., 566(4)
Sobhan, C. B., 509(95)
Sobierska, E., 518(105), 519(105)
Sobolik, K. A., 618(107)
Soliman, M., 554(147)
Solliec, C., 569(9)
Song, H., 481(57)
Soong, C. Y., 599(83)
Soong, Y. J., 483(61)
Soukoulis, C. M., 213(69), 224(69), 225(69), 228(69), 229(69), 242(69), 245(69)
Souris, N., 599(82), 603(82)
Sparavigna, A., 174(12)
Sparrow, E. M., 584(41), 618(41; 112)
Spears, J. R., 488(67), 489(67)
Spring, S. D., 587(52)
Srivastava, D., 227(101), 237(101; 112; 113), 243(101; 112; 113)
Srivastava, G. P., 170(4), 192(4), 209(4), 210(4)
Srolovitz, D. J., 238(120), 240(120)
Stegun, I. A., 390(22)
Steinke, M. E., 514(102)
Stelzer, E. H. K., 353(108), 354(108)
Stenger, F. J., 476(43)
Straub, J., 493(70; 71), 494(71)
Stroeve, P., 428(53)
Su, C. C., 583(34)
Suenaga, K., 588(57)
Suga, K., 578(21), 599(79)
Sun, L., 213(82), 227(82), 243(82)
Sundararajan, T., 293(48), 299(48), 300(48)
Sunden, B., 602(91)
Sung, H. J., 602(92)
Suzuki, T., 360(122), 462(7)
Suzuki, Y., 510(98), 511(98)

Swift, D. L., 391(23), 392(23), 393(23), 394(23), 403(23), 405(23), 416(23), 444(23)

T

Tadrist, L., 521(112), 523(112), 524(112)
Taitel, Y., 518(107)
Take, K., 468(24)
Tamma, K. K., 242(133)
Tamura, S., 237(115), 239(115; 130; 131), 243(115; 130; 131)
Tanaka, H., 213(87), 224(87), 227(87)
Tanaka, Y., 237(115), 239(115; 131), 243(115; 131)
Tang, M., 192(37)
Taniguchi, H., 566(8)
Tanner, D. B., 265(17)
Taslim, M. E., 587(52)
Tassopoulos, M., 353(107), 354(107)
Tavman, I. H., 391(24), 431(24)
Tawfek, A. A., 584(40), 611(103; 104), 616(103), 617(104)
Taya, M., 322(75), 323(75), 325(75), 330(75)
Tea, N., 174(13), 175(13)
Tenenbaum, A., 237(116)
Teng, H., 543(125), 544(125)
Teubner, J., 219(92)
Thielen, L., 606(97)
Thiesen, P., 275(45), 276(45), 291(45), 295(45), 296(45), 297(45), 298(45)
Thome, J. R., 462(10), 509(10)
Thompson, C., 431(56)
Thompson, L. J., 271(32), 284(32), 287(32)
Tian, Q., 380(8), 393(8), 394(8), 403(8), 410(8), 411(8)
Tiano, W., 194(40)
Tien, C.-L., 233(107), 238(107; 123; 125), 241(107), 243(107; 125), 444(63), 445(63), 447(63), 448(63), 451(63)
Tierney, J. W., 378(2)
Tildesly, D. J., 176(15)
Tobias, C. W., 276(46), 298(46), 313(46), 314(46)
Tomanek, D., 213(79), 227(79), 243(79)
Tong, B. Y., 474(40)
Topin, F., 521(112), 523(112), 524(112)
Torquato, S., 329(82), 342(96), 343(96), 345(97; 98; 100), 453(75)
Torquatol, S., 345(99)

Touloukian, Y., 173(9; 10), 174(9; 10), 175(9), 214(9)
Trebin, H.-R., 171(7)
Tretiakov, K. V., 183(30), 213(30; 68), 217(30; 90)
Truman, C. R., 583(33), 610(33), 611(33)
Tsai, J. H., 491(69), 492(69), 493(69), 509(69)
Tsai, S. H., 473(34; 35)
Tsao, G. T., 413(48)
Tschaufeser, P., 194(39)
Tseng, F. G., 480(53; 54), 512(100), 513(100), 514(100; 101), 515(101), 525(100)
Turgeon, E., 599(80)
Tye, R. P., 260(2), 304(2), 306(2)
Tzeng, P. Y., 599(83)

U

Udell, K. S., 501(84)
Uhlenbeck, G. E., 346(102)
Uno, M., 213(85; 86)
Usrey., M., 294(54)

V

Vachon, R. I., 412(47), 413(47; 49), 414(47), 415(47)
Vafai, K., 363(125)
van Ijzerloo, A. J., 585(42)
Vargas, W. L., 450(70; 71)
Vassallo, P., 366(128), 367(128)
Viazzo, S., 593(70)
Vierendeels, I., 599(81)
Viskanta, R., 566(5), 569(5), 574(5), 582(28; 30), 588(5; 58), 589(64), 611(30), 614(30), 620(28)
Vogelsang, R., 213(62), 224(62)
Voke, P. R., 577(19), 593(19)
Volz, S. G., 213(63; 70; 71; 72; 89), 222(94), 223(71), 228(71), 229(71), 231(71), 241(70), 243(89), 245(71)
Vortmeyer, D., 405(43), 439(43)
Voss, V., 402(38), 404(38), 406(38), 415(38)

W

Waddams, A. L., 393(29)
Wakao, N., 405(42; 43), 439(42; 43)
Wakuri, S., 238(128)

Waldmann, L., 352(105)
Wang, B. X., 337(86), 338(86), 356(114), 462(11), 463(11), 464(11), 465(11), 489(68), 490(68), 491(68), 494(74; 75; 76; 77), 495(74; 75), 496(76; 77), 509(96), 515(68), 530(127), 544(127), 545(127)
Wang, C. Z., 213(69), 224(69), 225(69), 228(69), 229(69), 242(69), 245(69)
Wang, E. N., 511(99), 512(99)
Wang, F., 213(82), 227(82), 243(82)
Wang, H., 494(74; 75; 76; 77), 495(74; 75), 496(76; 77)
Wang, H. S., 554(148), 555(148)
Wang, J., 271(35; 36), 285(35; 36), 288(36), 289(36), 290(36), 291(35), 292(35), 293(35), 294(35), 309(35), 369(130)
Wang, T., 213(82), 227(82), 243(82), 587(55)
Wang, W. W. W., 531(120), 535(120), 537(120), 543(120), 547(120)
Wang, W. Y., 267(26)
Wang, X., 271(37), 273(37), 283(37), 284(37), 285(37), 298(37), 302(37)
Wang, Y. X., 472(31; 32)
Wang, Z., 213(82), 227(82), 243(82)
Ward-Smith, A. J., 573(15)
Warren, M. C., 194(41)
Watson, S. K., 185(33), 203(33), 218(33)
Webb, B. W., 586(49), 611(102), 613(102), 614(102)
Webb, K. L., 530(134), 532(134), 548(134)
Webb, R. L., 468(24), 530(137), 548(135; 137)
Weichold, M. H., 469(27)
Weiss, Y., 380(7), 382(7), 450(7)
Wen, C. S., 342(95), 350(95)
Wen, M.-Y., 584(39), 611(105), 617(105)
Westedt, U., 369(133)
Whitaker, 421(52)
Whitaker, S., 379(4), 380(4), 385(19), 386(19), 392(19), 393(19), 416(4; 19; 51), 417(4), 419(4), 421(4), 424(4), 425(51), 426(51), 427(19; 51), 428(4; 51; 53; 54), 431(51), 434(19; 51), 435(51), 436(51), 443(54), 451(51)
Whitney, C. F., 589(62)
Whitney, M., 294(51), 303(51)
Wiberg, R., 589(65), 590(66)
Widenfeld, G., 380(7), 382(7), 450(7)
Wilcox, D. C., 600(87)

Wilhelm, R. H., 404(39)
Wills, J. R., 320(73), 321(73), 324(73)
Wisniewski, D., 494(73)
Wolf, D., 178(20)
Wolfing, B., 219(92)
Won, S. E., 584(37)
Wong, K. W., 410(46), 431(46), 432(46), 434(46; 57), 435(57), 436(57), 439(57), 451(46; 57)
Wong, M., 466(19), 515(103; 104), 516(104), 517(103; 104)
Wong, T. N., 474(40), 484(62; 63)
Wong, Y. Y., 515(104), 516(104), 517(104)
Woodside, W., 379(6), 380(6), 390(6), 405(41)
Wooten, F., 208(49), 242(132)
Wu, H. Y., 461(148), 519(108; 109; 110), 520(110), 521(109; 110), 522(110), 529(108; 109; 110), 533(124), 538(124), 540(124), 541(124), 542(124)
Wu, Y., 213(67), 238(126), 241(67), 243(126)
Wyder, P., 265(15)
Wynkoop, R., 404(39)

X

Xi, T., 271(35; 36), 285(35; 36), 288(36), 289(36), 290(36), 291(35), 292(35), 293(35), 294(35), 309(35)
Xie, H., 271(35; 36), 285(35; 36), 288(36), 289(36), 290(36), 291(35), 292(35), 293(35), 294(35; 55), 295(55), 305(55), 306(55), 307(55), 308(55), 309(35)
Xu, J. L., 475(42), 484(62; 63), 525(115), 527(115), 528(115)
Xu, X., 271(37), 273(37), 283(37), 284(37), 285(37), 298(37), 302(37)
Xu, Y. P., 267(26)
Xuan, Y. M., 271(33; 34), 281(33; 34), 286(34), 359(121), 359(120), 360(121), 361(121)
Xue, L., 213(81), 243(81), 294(56), 356(116)
Xue, Q. Z., 336(85), 337(85)
Xue, S., 294(54)

Y

Yadigaroglu, G., 498(82), 500(82), 506(82)
Yagi., S., 382(14), 383(14), 385(14)

Yamada, K., 213(85; 86)
Yamaguchi, M., 192(37)
Yamamoto, N., 462(7)
Yamanaka, S., 213(85; 86)
Yang, J., 213(67), 238(126), 241(67), 243(126)
Yang, Z., 602(90)
Yano, K., 213(86)
Yao, Z., 294(59), 295(59), 306(59)
Yapici, S., 584(36)
Yavuz, T., 584(36)
Ybarra, P. G., 353(107), 354(107)
Yee, P., 266(18)
Yen, T. H., 510(98), 511(98)
Yip, S., 183(29), 192(37), 213(29; 60), 214(60), 216(29), 220(60), 228(60), 229(60), 231(60), 241(60)
Yip S., 213(74), 224(74), 225(74)
Yokomine, T., 588(56)
Yoo, J. Y., 601(88)
Yoon, J. B., 479(51)
Yoon, P., 587(54)
Yoon, Y.-G., 238(120), 240(120)
Yoshida, H., 588(57)
You, S. M., 364(126), 365(126)
Youn, W., 294(55), 295(55), 305(55), 306(55), 307(55), 308(55)
Yovanovich, M. M., 444(61; 64), 447(64), 448(65)
Yu, A. B., 450(73)
Yu, C., 294(59), 295(59), 306(59)
Yu, C. J., 356(113)
Yu, M. S., 572(11)
Yu, R. C., 174(13), 175(13)
Yu, W., 271(32), 284(32), 287(32), 294(52), 304(52), 307(52)
Yuan, H., 481(58), 482(58)

Z

Zaghdoudi, M. C., 467(21)
Zehner, P., 409(45), 410(45), 411(45), 415(45), 416(45), 431(45), 434(45), 440(45)
Zeller, H. R., 267(21)
Zettl, A., 294(51), 303(51)
Zhang, D. F., 267(26)
Zhang, L., 511(99), 512(99)
Zhang, M., 548(135)
Zhang, W., 213(82), 227(82), 243(82)
Zhang, X. M., 475(42)
Zhang, Y. W., 475(41)
Zhang, Z. G., 294(52), 304(52), 307(52)
Zhao, T. M., 483(61)
Zhao, T. S., 543(125), 544(125), 552(146), 553(146), 554(146)
Zhao, Z., 498(82), 500(82), 504(89), 506(82)
Zhou, D. W., 367(129)
Zhou, J., 525(115), 527(115), 528(115)
Zhou, L. P., 337(86), 338(86)
Zhou, Y., 589(63), 610(63), 611(63)
Zhou, Z. Q., 475(42)
Zhu, J., 602(90)
Zhu, Z., 213(82), 227(82), 243(82)
Ziman, J. M., 170(6), 174(6), 192(6)
Ziskind, G., 520(111)
Zohar, Y., 466(19), 515(103; 104), 516(104), 517(103; 104)
Zubarev, A. Y., 355(109)
Zuckerman, N., 565(112), 566(6)
Zulli, P., 450(73)
Zwanzig R., 212(55)

SUBJECT INDEX

Note: Page numbers in *italic* type indicate figures and tables

A

acetone, in sonofusion, 35
acoustic chamber
 bubbly liquid in, dynamics, 108–109
 design parameters for, *12*
 design considerations, 9–10
 results of simulations of, 12–17
 applied heat flux, 238–239
 applied temperature gradient, 237
ASM (algebraic stress models), in impinging jets, 603–604
ATILA code, 8–18, 91

B

Bjerknes force, 23–24, 151
Boltzmann transport equation, 334–335
bond number, 462–465
Bose–Einstein distribution, 228
Boussinesq approximation, 497
 in impinging jets model, 596, 602–603
Brownian motion
 Brownian motion simulation technique, 346
 coupled with thermal phoresis, effects, 338–359
Bruggemann equation, 298, 313
bubble cluster
 and liquid, standing acoustic wave field in, 109–114
 collapse, bubble concentration impact on, 125–130
 in an acoustic chamber, dynamics, 108–109
 linear dynamics, 107–108
 linear results, 111–114
 non-linear dynamics, final stage, 120–125
 non-linear dynamics, initial stage, 114–120
bubble dynamics
 about an ultra-thin heating wire, 494–497
 analysis, 53–54
 bubble cluster, linear dynamics, 107–108
 bubble implosions, transient phenomena during, 130–151
 bubble nucleation, in tensioned liquids, 151–161
 bubble-in-cell dynamics, 104–105
 bubbly liquids, dynamics, 106–107
 evaporation and condensation kinetics, 63–65
 high Mach number bubble dynamics, 65–79
 high Mach number period, equations of state for, 55–60
 homogeneous nucleation, role, 156
 linear and non-linear bubble cluster dynamics, 101–130
 liquid compressibility in, 105–106
 low Mach number bubble dynamics, 60–65
 low Mach number period of, equations of state for, 54–55
 mathematical model for, 103–106
 nucleation, activation threshold for, 154–155
 process description, 39–40
 vapor bubble in a liquid
  see separate entry
bubble growth and collapse about an ultra-thin heating wire, 497–499
bubble-in-cell dynamics, 104–105

C

Callaway–Holland BTE formulation, 201–203
CFD (computational fluid dynamics) models of impinging jets, 592–597, 602–603, 606–608
chemiluminescence model, 2
CHF (critical heat flux), 364–367

Clausius–Clapeyron equation, 490
'cold fusion', 1
compressor bleed flow, 566
conduction heat transfer and thermal conductivity of solids, 172–175
conduction models, 401–406
conductive thermal exchange in a plasma, 75–78
constriction resistance, 444–448
contact models, 440–450
CPL (capillary pumped loop), *476*
CVD (chemical vapor deposition), 479–480

D

D/D neutrons, production, 47–48
data-logging method, 270–273
DNS (direct numerical simulation) method, 593–594
Dulong–Petit theory, 173

E

effective conductivity mixture-rules, 429
effective thermal conductivity
 equations, development, 310–338
 measurements of, 276–298
 of saturated porous media, 377–453
 see also under saturated porous media
 physical bounds for, 399–400
electrical microdischarge, 2
Eller and Flynn model, 26
empirical contact parameter models, 428–440
 correlations, 428–431

F

'first passenger time' algorithm, 329
flexoelectrical effect, 2
flow condensation in, meso-/microchannels, 529–556
 annular/injection/slug-bubbly flow, 539–541
 condensation flow patterns, 529–544
 droplet/annular/injection/slug-bubbly flow, 539
 experimental investigations on, 530–533
 fully droplet flow, 539
 fully slug-bubbly flow, 541–544
 heat transfer coefficients, 544–548
 pressure drop characteristics, 548–550
fluid
 molecular and dissociated states, conservation equations for, 73
 plasma state, conservation equations for, 73
free submerged jet, flow field, 567, *569*
Fritz's correlation, 463

G

gas turbine compressor, 566
gaseous Argon, thermal conductivities, *260*
'giant response', 7
Gibbs energy analysis method, 356
GNC (genuine number of coincidences), 50
Grashof number, 303–304, 363
Greek–Kubo (GK) method, 171
Green's function/theorem, 323, 343
Green-Kubo method, 211–232, 237
 and direct methods, comparison, 244–245
'grid method', 346

H

Hamilton–Crosser model, 291
hard-sphere fluid model, 329
Hashin–Shtrikman method, 320
HCACF (heat current autocorrelation function), 212–245
heat and mass transfer in fluids with nanoparticle suspensions, 257–369
see also under nanoparticle suspensions
heat exchangers, miniaturization, 461–556
see also mesoscale and microscale phase-change heat transfer
Hertz–Knudsen–Langmuir formula, 63
heterogeneous nucleation in pool boiling, 487–491
homogeneous nucleation in pool boiling, 487–491
hot spot models, 2
'hot wire' method, 389–391
HWA (hot-wire anemometer) technique, 591
hybrid modeling, in impinging jets, 606–677
 'Hybrid Law of the Wall' model, 595

SUBJECT INDEX

I

impinging jets, in heat transfer, *see* jet impingement heat transfer
impinging jet device modeling, 565–582
   algebraic stress models (ASM), 603–604
   Boussinesq approximation, 596, 602–603
   CFD turbulence models, comparison, 592–597
   complete RSM modeling, 604
   empirical correlations, 592
   hybrid modeling, 606–607
   k–ε model, 596–600
   k–ε RNG model, 600
   k–ω model, 600–602
   laminar impingement, 592
   near-wall treatment, 594–596
   realizability limits, 602–603
   turbulent impingement models, 592–594
   v^2–f model, 604–606
'inert gas evaporation' method, 264
inkjet printheads, 478
intermediate solid–fluid conductive ratios, 400–415
   conduction models, 401–406
   lumped parameter models, 407–411
   resistor network models, 411–414

J

jet impingement heat transfer, 565–624
   correlations, 610–624
   flow regions, *568*
   impinging jet regions, 567–570
   jet geometry, 580
   jet pitch effect, 580–584
   see also jet–jet interaction
   nondimensional heat and mass transfer coefficients, 570–574
   nozzle structures, 585
   research methods, 588–609
   see also individual entry
   turbine cooling applications, 566
   turbulence generation and effects, 574–579
   vortex motion in, *579*
jet impingement heat transfer, research methods, 588–609
   experimental techniques, 588–591
   modeling jet device, 591–607

   see also impinging jet device modeling
jet–jet interaction, in jet impingement heat transfer, 580–584
   crossflow and target motion, 582–584
   jet height and spacing effects, *583*
   shear layer interference, 581–582
   small jet lengths and highly confined jets, 582
   spacing and interaction, 582

K

k–ε model, in impinging jets, 596–600
k–ε RNG model, in impinging jets, 600
k–ω model, in impinging jets, 600–602

L

LALPO (large-amplitude/large-period oscillation), 526
Langevin equation, 347
large solid–fluid conductivity ratios, 415–452
   closure and numerical solution, 422–426
   constriction resistance, 444–448
   contact models, 440–450
   see also separate entry
   empirical contact parameter models, 428–440
   see also separate entry
   method of volume averaging, 416–422
   numerical conduction models, 439–440
   numerical solution and experimental validation, 426–428
   solid mechanics, 448–450
   volume-averaged conduction equation, development, 416–422
layers-in-parallel mixture model, 429
layers-in-series mixture model, 429
LDA (laser Doppler anemometry) technique, 590
Legendre polynomials, 445
Lennard-Jones system, 182–184
   Lennard-Jones argon, 213–228
   phase comparisons, 187–192
LES (large Eddy simulation), in impinging jets model, 593–594
LHP (loop heat pipe), 476–477

linear and non-linear bubble cluster dynamics, 101–130
LTAF (liquid/two-phase alternating flow), 519
LTVAF (liquid/two-phase/vapor alternating flow), 519–522
lumped parameter models, 407–411, 431–439

M

Marangoni convection, 494, 497
Matheson rule, 239
Maxwell's equation, 298, 313, 329, 394–397, 427
MBSL (multibubble sonoluminescence), 1–20
mechanochemical model, 2
meso pulsating heat pipes, 474–476
meso-(mini-) heat pipes, 467–469
meso-(mini-)/microheat pipes, 467–477
meso-/microcapillary-pumped loop, 476–477
meso-/microchannels
 boiling flow patterns and instabilities, 515–529
 condensation in, theoretical studies, 550–556
 flow boiling in, 509–529
 flow condensation in, 529–556
 see also separate entry
 nucleation in, 509–515
meso-/microdevices, 462–487
 meso- and microsystems, classification, 462–465
 meso pulsating heat pipes, 474–476
 meso-(mini-) heat pipes, 467–469
 meso-(mini-)/microheat pipes, 467–477
 meso-/microcapillary-pumped loop, 476–477
 meso-/microchannel heat exchangers, 465–467
 microheat pipes, 469
 thermal bubble actuators, 477–487
 thermal bubble perturbators, 484–487
 thermal inkjet printheads, 478–481
 vapor bubble micropumps, 481–484
mesoscale and microscale phase-change heat transfer, 461–556

flow boiling in meso-/microchannels, 509–529
see also under meso-/microchannels
homogeneous and heterogeneous nucleation in pool boiling, 487–491
meso-/microdevices involving, 462–487
see also separate entry
phenomena description, 487–556
subcooled pool boiling under constant heat flux, 491–497
see also separate entry
subcooled pool boiling under pulse heating or transient heating, 497–509
see also separate entry
micro- and nano-scale molecular dynamics simulations, challenges at, 169–170
microbubble nucleation, 497
microheat pipes, 469
Mie–Grüneisen equation, 55–56, 82
Mishima–Hibiki correlation, 549
molecular dynamics simulations, phonon transport in, formulation and thermal conductivity prediction, 169–247
 conduction heat transfer and thermal conductivity of solids, 172–175
 expectations from, 245–246
 initialization, 181–182
 Lennard-Jones system, 182–184
 micro- and nano-scales, challenges at, 169–170
 motivation for using, 170–171
 nature of phonon transport, 208–211
 phonon gas and normal modes, 210–211
 phonon space analysis, 192–208
 see also separate entry
 quantum formulation and selection rules, 208–210
 real and phonon space analyses, 175–208
 see also separate entry
 real space analysis, 184–192
 see also separate entry
 scope, 171–172
 thermal conductivity prediction, 211–241
 see also separate entry
Monte Carlo method, 327, 340
MWCNs (multi-wall carbon nanotubes), 295–296

SUBJECT INDEX

N

nanoparticle suspensions, heat and mass transfer in fluids with, 257–369
  effective thermal conductivity, measurements of, 276–298
  experimental methods and test facilities, 270–276
  experimental studies and results, 264–310
  nanoparticle suspensions, preparation of the, 264–269
  theoretical investigations, 310–368
Navier–Stokes equations, 302, 352
neutron and SL signals, coincidence between, 48–52
neutron-induced bubble nucleation, 156–161
Nose–Hoover thermostat, 180
nuclear reactions, kinetics, 79–82
nucleation phenomena in sonofusion, 89–97
see also bubble dynamics
  bubble nucleation in, 95–97
  ion transport in, 92–94
  neutron interactions in, 94–95
  numerical conduction models, 439–440
    point contact, 440–444
Nusselt number, 360, 362–363

O

ONB (onset of nucleate boiling), 510, 525–526

P

Pade approximation, 343
Peclet number, 349–350, 358, 360
permeable-sphere particle model, 344
phase-change heat transfer processes, 461–556
see also mesoscale and microscale phase-change heat transfer
phonon dispersion, 204–208
phonon gas and normal modes, 210–211
phonon space analysis, 192–208
  harmonic approximation, 192–194
  lattice dynamics, 196–200
  normal modes, 194–196
  phonon dispersion, 204–208
  phonon relaxation time, 200–204

phonon transport in molecular dynamics simulations, 169–247
see also under molecular dynamics simulations
PHP (pulsating heat pipe), 474–476
PIV (particle image velocimetry) technique, 590–591
plasma
  conductive thermal exchange in, 75–78
  radiative thermal exchange in, 78–79
Poiseuille law, 356
PTV (particle tracking velocimetry) technique, 590–591

R

radial-grooved microheat pipe, 473
radiative thermal exchange in plasma, 78–79
RANS (Reynolds averaged Navier–stokes) models, 594–597, 607
Rayleigh–Plesset equation, 53, 63, 131
real and phonon space analyses, 175–208
  energy, temperature, and pressure, 178–179
  equations of motion, 179–181
  molecular dynamics simulation, 175–176
  simulation setup, 176–178
real space analysis, in MD simulations, 184–192
  Lennard-Jones phase comparisons, 187–192
  period of atomic oscillation and energy transfer, 185–187
  system parameters, prediction, 184–187
  unit cell size, 184–185
rectified diffusion, 24–31
resistor network models, 411–414
REV (representative elementary volume), 416–418, 425
RSM (Reynolds stress model), in impinging jets, 604
Runge–Kutta method, 123

S

SASPO (small-amplitude/short-period oscillation), 526–528
saturated porous media, effective thermal conductivity of, 377–453

experimental determination, 379
intermediate solid–fluid conductive ratios, 400–415
see also separate entry
large solid–fluid conductivity ratios, 415–452
see also separate entry
results and general trends, 392–394
small solid–fluid conductivity ratios, 394–400
see also separate entry
steady-state techniques, 379–385
transient techniques, 385–392
SBSL (single bubble sonoluminescence), 3–33
acoustic chamber design considerations, 9–10
analysis, 5
bubble stability considerations, 21–31
experimental results and considerations, 17–52
experimental techniques, 7
lessons learnt, 31–35
parameter optimization, 10–12
semi-empirical lumped parameter models, *434*
small solid–fluid conductivity ratios, 394–400
effective thermal conductivity problems, 395–396
Maxwell's formulae, 394–397
mixture rules, 397–399
physical bounds for the effective conductivity, 399–400
statistical approach, 396–397
solid mechanics, 448–450
sonofusion and sonoluminescence, 1–161
see also under bubble dynamics; sonoluminescence
1D HYDRO code analysis of, 82–89
applications, 97–101
choice of test liquid, 35
nucleation phenomena in, 89–97
see also separate entry
sonoluminescence and the search for sonofusion, 1–161
see also bubble dynamics; single bubble sonoluminescence; sonofusion
analytical modeling, 53–97
equations of state, 54–60
evidence of nuclear fusion in, 43
experimental system, 36–38

spherical liquid-filled flask
with dissipation, dynamics, 138–151
without dissipation, dynamics, 131–138
spherical particle packing arrangements, *442*, *447*
SST (shear stress transport) model, in impinging jets, 606–609
steady-state one-dimension method, *273*
steady-state techniques, for effective thermal conductivity, 379–385
Stokes–Einstein diffusion theory, 339
Stokesian dynamics method, 326
subcooled pool boiling under constant heat flux, 491–497
bubble dynamics about an ultra-thin heating wire, 494–497
subcooled pool boiling about a microfilm heater, 491–494
subcooled pool boiling under pulse heating or transient heating, 497–509
bubble growth and collapse about an ultra-thin heating wire, 497–499
bubble growth and collapse on a microfilm heater, 499–509
supersonic jet flow pattern, *572*
SUPG (streamline upwind Petrov Galerkin) stabilization, 68

T

temperature oscillation method, *275*, 291
thermal bubble actuators, 477–487
thermal bubble perturbators, 484–487
thermal conductivity prediction, in MD simulations, 211–241
applied heat flux, 238–239
applied temperature gradient, 237
convective and conductive contributions, 223–224
direct method, 232–241
formulation, 211–213
heat current autocorrelation function and thermal conductivity, 214–218
implementation, 235–239
multi-atom unit cell decomposition, 220–222
nonequilibrium Green-Kubo, 226–228
quantum corrections, 228–232
side effects, 239–241

SUBJECT INDEX

spectral methods, 224–226
thermal conductivity decomposition, 218–220
transient thermal diffusivity approach, 239
'thermal conductivity probe' method, 389
thermal inkjet printheads, 478–481
thermal phoresis, 338–359
thermistors, 382
TLC (thermodynamic liquid crystals), 589–590
touching-particle model, *434*
transient hot-wire method, 270–273
transient techniques, for effective thermal conductivity, 385–392
transient thermal diffusivity approach, 239
triboluminescence, 1
tritium production, 45–47
two-phase consolidation parameters, *431*

U

ultra-thin heating wire
  bubble dynamics about, 494–497
  bubble growth and collapse about, 497–499

V

v^2–f model, in impinging jets, 604–605
vacuum steady-state thermal conductivity test, *274*
vapor bubble in a liquid
  mechanical equilibrium, 152
  thermodynamical equilibrium, 152–154
vapor bubble micropumps, 481–484
Verlet leapfrog algorithm, 180, 183
VEROS (Vacuum Evaporation on Running Oil Substrate) technique, 268
volume-averaged conduction equation, development, 416–422

W

wire-bonded microheat pipe, *472*

Y

Young–Laplace equation, 462, 490, 516